MEASUREMENT SYSTEMS

MEASUREMENT SYSTEMS
Application and Design

Third Edition

Ernest O. Doebelin

Department of Mechanical Engineering
The Ohio State University

INTERNATIONAL EDITION

McGRAW-HILL BOOK COMPANY

Auckland Bogotá Guatemala Hamburg Johannesburg Lisbon
London Madrid Mexico New Delhi Panama Paris
San Juan São Paulo Singapore Sydney Tokyo

MEASUREMENT SYSTEMS

Application and Design

INTERNATIONAL EDITION

Copyright © 1983
Exclusive rights by McGraw-Hill Book Co. – Singapore for
manufacture and export. This book cannot be re-exported
from the country to which it is consigned by McGraw-Hill.

3rd printing 1986

This book was set in Times Roman by Santype-Byrd.
The editors were Rodger H. Klas and J. W. Maisel;
The production supervisor was Leroy A. Young.
New drawings were done by Danmark & Michaels, Inc.

Library of Congress Cataloging in Publication Data

Doebelin, Ernest O.
    Measurement systems.

    Includes bibliographical references and index.
     1. Measuring instruments. 2. Physical measurements.
I. Title.
QC100.5.D63   1983   681'.2   82-7163
ISBN 0-07-017337-0   AACR2

When ordering this title use ISBN 0-07-Y66263-0

Printed and Bound by KIN KEONG PRINTING CO. PTE. LTD. – Republic of Singapore.

CONTENTS

Part 1 General Concepts

Part 2   Measuring Devices

Part 3 Manipulation, Transmission, and Recording of Data

PREFACE TO THE THIRD EDITION

Although the needs of the second (1975) edition were met adequately by appending a final, updating chapter, I approached this third edition as if I were writing a new book. Since I and the many users of the earlier editions have found the approach and organization of the text basically sound, major surgery was not needed. However, every section was reviewed critically, and a decision was made to delete, revise, and/or augment. My judgments on these matters resulted in a somewhat lengthened text, but the organization should facilitate selection of topics and assignments for a particular course.

An obvious and simple improvement is the dispersal and integration into appropriate earlier chapters of the terminal, updating chapter and appendixes of the 1975 revision. All now-inappropriate references to vacuum-tube electronics have been replaced by their solid-state equivalents, and though the text is still not intended for electronics courses, some basic general material on operational amplifiers is included to augment the many op-amp applications present in the earlier editions. Part Two (sensors) has too many additions and changes in emphasis to enumerate here; however, they include new strain-gage fabrication technologies, many laser and other electro-optic sensors, synchro/digital conversion devices, pressure multiplexers, new mass flowmeters, and pyroelectric radiation detectors.

Perhaps more significant than the details just mentioned are several general themes which I tried to emphasize throughout the text. Two involve the computer technology that pervades all aspects of engineering today. This technology has two broad areas of application: the use of computers and special-purpose software as powerful, rapid analytical/design tools (computer-aided design) and the inclusion of dedicated computers of various scales (micro to mainframe) as components of operating machines and processes. Often the second area is called computer-aided manufacturing, though we would wish to include also nonmanufacturing applications, such as automation of testing and data processing in research and development laboratories. Several applications of powerful digital simulation languages to measurement system design/analysis are scattered throughout the text as illustrations of computer-aided design, while Chap. 13 on data systems emphasizes computers as components of operating processes. Also, many discussions of specific sensors in Part Two emphasize applications in manufacturing automation and productivity improvement, areas of vital national concern which often depend critically on proper integration of measurement system technology with computer methods. Of course, wide application of digital computers is also responsible for the increased emphasis in the text on digital sensors, analog-to-digital and digital-to-analog conversion methods, microprocessor-based systems, and digital data acquisition systems.

Some text users had suggested expansion of the general linear systems portion of Part One; however, this modification would have further lengthened an already long text or necessitated deletion of material which I felt was essential to a measurements book. While a measurements course is not an unreasonable place to teach general system dynamics, I can highly recommend instead a curricular plan which employs a separate, sophomore-level system dynamics course as a basis for later measurement, control, vibration, acoustics, electronics, and electromechanical systems courses. Although introduction of nontraditional courses such as system dynamics requires some curricular reshuffling, we have found the payoffs to be most worthwhile.

Ernest O. Doebelin

PREFACE TO THE REVISED EDITION

Since the first edition of this text emphasized basic principles, concepts, and methods of analysis which remain largely unchanged as time passes, this revision was planned as one of limited scope and low cost devoted mainly to the correction of errors, the addition of certain significant hardware developments, and recognition of the need for a gradual conversion from British to metric (SI) units. Since "new" hardware generally *augments* rather than *replaces* "old," the revision does not delete any material but adds a final chapter devoted to new hardware. Metric conversion is limited to inclusion of a conversion table and use of metric units in the new chapter. This approach recognizes the fact that practicing engineers will need to use *both* British and metric units for some years; thus it is unrealistic to train students only in metric. New references are added both at the end of each chapter (where available "white space" allows) and also after the new chapter.

Ernest O. Doebelin

PREFACE TO THE FIRST EDITION

The title of this book, "Measurement Systems: Application and Design," is sufficiently broad that one can read into it a wide range of possibilities with regard to content and approach. The author would like to use the preface to define these more sharply, in addition to presenting the way in which the material has been used and the purpose he feels it can serve in engineering education and practice.

Since measurement in one form or another is used regularly by all sorts of people in all sorts of jobs, one must first restrict the scope. This material has been used in connection with courses and laboratories in the mechanical engineering curriculum at The Ohio State University and is thus biased toward this audience. Sufficient material at a suitable level is included for two courses: an introductory treatment useful for a required undergraduate course and more advanced considerations suitable for an elective course at the advanced undergraduate or beginning graduate level. The inclusion of both types of material in a single text is in part a recognition of the variation in curricula from school to school. By including a wide scope of material it is hoped that individual instructors will be able to select topics and place emphasis so as to best utilize the previous preparation of their particular students.

The area of measurement is, of course, closely allied to that of laboratory teaching, and since this facet of engineering education is the subject of some controversy, there is certainly room for a wide variety of approaches. One can consider the general field to be composed of two parts: the hardware of measurement and the techniques of experimentation. It is difficult completely to separate the two, and some, by choice or because of pressure of time, will design courses to treat both types of material concurrently. At the author's school, three required courses are devoted to this general subject. The first is aimed mainly at developing an understanding of the operating principles of measurement hardware and the problems involved in the analysis, design, and application of such equipment. "Application" here (and in the book title) is not construed to encompass the detailed planning of comprehensive experiments but rather is limited mainly to consideration of the disturbing effect of the measuring instrument on the measured system and the influence of extraneous system variables on the instrument output. The method of presentation of this material is by 1-hr lectures each week and one 4-hr laboratory session each week. Lecturing is divided about equally between generalized theory (static characteristics, dynamic characteristics, etc.) and specific measurement hardware (vibration pickups, recorders, etc.) which may not be covered in the laboratory experiments. Laboratory experiments are preceded by about 1 hr of lecturing specific to the particular measurement area treated in that experiment. Experiments will not here be discussed in detail; however, the philosophy is that of having a relatively small number of experiments so as to allow sufficient time for adequate penetration and understanding. It is felt that it is impossible to cover all the significant hardware. Therefore experiments, while dealing with important specific areas, are designed mainly to serve as vehicles for the illustration of general concepts.

Following the course just described are two courses devoted to experimental analysis of engineering problems. These initially devote lecture time to development of systematic methods of planning, executing, and evaluating experiments. Then small groups of students undertake original projects which involve theoretical analysis, experiment planning, equipment design and construction, measurement-system design, experiment execution, evaluation, and report writing.

For the first course mentioned above, selected material from this book serves as text, while for the second and third courses it becomes a valuable reference. In addition to these three required courses, an additional elective in measurement systems, which essentially extends in breadth and depth from the first required course, is offered. This course has 3 hr of lecture and one 2-hr laboratory period per week. The objective is to develop increased competence in both the design and use of measurement equipment. This is implemented by consideration of both more advanced general concepts and also more sophisticated specific hardware. Experiments are designed mainly to provide familiarity with actual instrumentation equipment and problems involved in its use. Material from this book again serves as the text for this course.

Some explanation of the use of the word "system" in the title of this book may be in order. While the term "system engineering" has come to mean, at least to some, the planning and implementation of complex schemes on a grand scale, no such meaning is intended here. In fact, the author subscribes to the view that one person's component may be another person's system and that this varied use of the word is not objectionable. Thus to the designer of a large-scale data system, who essentially selects hardware from that available to achieve a compatible arrangement that meets the specified requirements, a tape recorder is legitimately considered a component. However, a designer of tape recorders would certainly insist that the machine is a most complex electromechanical system. Since this book is addressed to both these classes of application, the title word "system" is intended to be interpreted in either way, as appropriate.

The word "design" is used in the title to emphasize that many practicing engineers not only use instrumentation but also are engaged in designing it. While the design of an electronic amplifier has perhaps only limited mechanical aspects (packaging, shock mounting, cooling, manufacturing, etc.), much other equipment, particularly transducers, is as much mechanical as electrical. Although highly specialized electrical aspects of electromechanical system design are handled by electrical engineers, the electrical background of mechanical engineers is adequate to allow them to treat many electrical problems that are closely coupled to the mechanical aspects. Design is intended to consider not only the problems of individual components but also the assembly of available components (transducers, amplifiers, recorders, etc.) into a compatible system capable of meeting required specifications.

Some important features of the text include the following:

1. Consideration of measurement as applied to research and development operations and also to monitoring and control of industrial and military systems and processes.
2. A generalized treatment of error-compensating techniques.
3. Treatment of dynamic response for all types of inputs: periodic, transient, and random, on a uniform basis, utilizing frequency response.
4. Detailed consideration of problems involved in interconnecting components.
5. Discussion, including numerical values, of standards for all important quantities. These give the reader a feeling for the ultimate performance currently achievable.
6. Quotation of detailed numerical performance specifications of actual instruments.
7. Inclusion of significant material on important specific areas such as sound measurement, heat-flux sensors, gyroscopic instruments, hotwire anemometers, digital methods, random signals, mass flowmeters, amplifiers, and the use of feedback principles.

Since material for both introductory and advanced courses is included, it may be helpful to indicate the division. This is somewhat arbitrary since material considered advanced for a course at a given level in one curriculum might be thought elementary in another. For example, students who have had a course in dynamic systems analysis could cover the material on dynamic characteristics very rapidly. Thus the individual instructor must make the necessary selection. For those who would appreciate some assistance, the author offers the following suggestions.

Chapters 1 and 2 can be covered quite quickly and easily and should be included in an introductory course. In Chap. 3 the material on the proofs of the generalized loading equations, which starts at Fig. 3.29, can be omitted in a first course. The material on dynamic response utilizing Fourier transforms for transient response (Fig. 3.74) and on amplitude-modulated and -demodulated signals

(Figs. 3.81 to 3.90) can also be omitted, as can the treatment of random signals. In the remainder of this chapter the sections Requirements on the Instrument Transfer Function to Ensure Accurate Measurement and Experimental Determination of Measurement-System Parameters should be retained and the others omitted.

In Chap. 4 we begin to deal with specific devices, and the choice of topics to be omitted becomes less clear and somewhat a matter of personal preference. Sections 4.10 to 4.12 might reasonably be omitted and, in the remaining material, emphasis put on standards and calibration, potentiometers, strain gages, differential transformers, piezoelectric transducers, and accelerometers. Section 5.7 can be omitted, as can Secs. 6.6, 6.7, and 6.10. In Chap. 7 the material on hot-wire anemometers, electromagnetic flowmeters, and ultrasonic flowmeters can be left out. Radiation methods in Chap. 8 can be considerably cut, and heat-flux sensors omitted. Section 9.1 is the only essential part of Chap. 9. Section 10.1 is essential and would, in fact, usually be assigned much earlier, since it is needed in Chap. 4. Section 10.2 can be eliminated as can the hydraulic filter and statistical averaging of Sec. 10.3, and also Secs. 10.6 to 10.13. Chapter 11 may be omitted but most instructors will wish to cover all of Chap. 12, perhaps early in the course if an associated laboratory requires the use of recording equipment. Since Chap. 13 is short, entirely descriptive, and designed as a unifying conclusion, it should be included in an introductory course.

The author would like to acknowledge his appreciation to other workers in this field whose contributions are evidenced by the voluminous references and bibliography. Production of the manuscript was made as painless as possible by the faultless typing of Mrs. Maxine Fitzgerald. The contributions of students over the past 10 years to my understanding of methods of presenting material should not be minimized. Finally, the forbearance of a long-suffering wife and family is gratefully acknowledged.

Ernest O. Doebelin

PART
ONE
GENERAL CONCEPTS

CHAPTER
ONE

TYPES OF APPLICATIONS OF MEASUREMENT INSTRUMENTATION

1.1 INTRODUCTION

As background for our later detailed study of measuring instruments and their characteristics, it is useful first to discuss in a general way the uses to which such devices are put. Here we choose to classify these applications according to the following scheme:

1. Monitoring of processes and operations
2. Control of processes and operations
3. Experimental engineering analysis

Each class of application is now described in greater detail.

1.2 MONITORING OF PROCESSES AND OPERATIONS

Certain applications of measuring instruments may be characterized as having essentially a monitoring function. The thermometers, barometers, and anemometers used by the weather bureau serve in such a capacity. They simply indicate the condition of the environment, and their readings do not serve any control functions in the ordinary sense. Similarly, water, gas, and electric meters in the home keep track of the quantity of the commodity used so that the cost to the user may be computed. The film badges worn by workers in radioactive environments monitor the cumulative exposure of the wearer to radiations of various types.

1.3 CONTROL OF PROCESSES AND OPERATIONS

In another extremely important type of application for measuring instruments, the instrument serves as a component of an automatic control system. A functional block diagram illustrating the operation of such a system is shown in Fig. 1.1. Clearly, to control any variable by such a "feedback" scheme, it is first necessary to measure it; thus all such control systems must incorporate at least one measuring instrument.

Examples of this type of application are endless. A familiar one is the typical home-heating system employing some type of thermostatic control. A temperature-measuring instrument (often a bimetallic element) senses the room temperature, thus providing the information necessary for proper functioning of the control system. Much more sophisticated examples are found among the aircraft and missile control systems. A single control system may require information from many measuring instruments such as pitot-static tubes, angle-of-attack sensors, thermocouples, accelerometers, altimeters, and gyroscopes. Many industrial machine and process controllers also utilize multisensor measurement systems.

In attempting to classify applications within your own experiences according to the three categories of Sec. 1.1, you may find instances where the distinction among monitoring, control, and analysis functions is not clear-cut. Thus the category decided on may depend somewhat on your point of view. The data obtained by the weather bureau, for instance, serve mainly in a monitoring

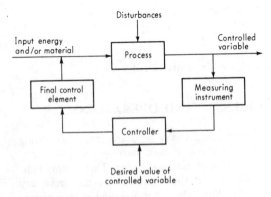

Figure 1.1 Feedback-control system.

function for the average person. For fruit growers, however, a report of cold weather may act in a control sense because it signals them to turn on smudge pots and apply other antifrost measures. Also, present weather data for large areas are correlated and analyzed to form the basis of short- and long-range weather predictions, so that you could say the instruments are supplying data for an engineering analysis. Once you recognize the possibility of a variety of interpretations, depending on the point of view, the apparent looseness of the classification should not cause any difficulty.

1.4 EXPERIMENTAL ENGINEERING ANALYSIS

In solving engineering problems, two general methods are available: theoretical and experimental. Many problems require the application of both methods. The relative amount of each depends on the nature of the problem. Problems on the frontiers of knowledge often require very extensive experimental studies since adequate theories are not available yet. Thus theory and experiment should be thought of as complementing each other, and the engineer who takes this attitude will, in general, be a more effective problem solver than one who neglects one or the other of these two approaches.

It may be helpful to summarize quickly the salient features of the theoretical and the experimental methods of attack. This is done in Figs. 1.2 and 1.3.

In considering the application of measuring instruments to problems of experimental engineering analysis, it may be helpful to have at hand a classification

---

1. Often give results that are of general use rather than for restricted application.

2. Invariably require the application of simplifying assumptions. Thus not the actual physical system but rather a simplified "mathematical model" of the system is studied. This means the theoretically predicted behavior is *always* different from the real behavior.

3. In some cases, may lead to complicated mathematical problems. This has blocked theoretical treatment of many problems in the past. Today, increasing availability of high-speed computing machines allows theoretical treatment of many problems that could not be so treated in the past.

4. Require only pencil, paper, computing machines, etc. Extensive laboratory facilites are not required. (Some computers are very complex and expensive, but they can be used for solving all kinds of problems. Much laboratory equipment, on the other hand, is special-purpose and suited only to a limited variety of tasks.)

5. No time delay engendered in building models, assembling and checking instrumentation, and gathering data.

Figure 1.2 Features of theoretical methods.

1. Often give results that apply only to the specific system being tested. However, techniques such as dimensional analysis may allow some generalization.

2. No simplifying assumptions necessary if tests are run on an actual system. The true behavior of the system is revealed.

3. *Accurate* measurements necessary to give a true picture. This may require expensive and complicated equipment. *The characteristics of all the measuring and recording equipment must be thoroughly understood.*

4. Actual system or a scale model required. If a scale model is used, similarity of all significant features must be preserved.

5. Considerable time required for design, construction, and debugging of apparatus.

Figure 1.3 Features of experimental methods.

of the types of problems encountered. This classification may be accomplished according to several different plans, but one which the author has found meaningful is given in Fig. 1.4.

1.5 CONCLUSION

Whatever the nature of the application, intelligent selection and use of measurement instrumentation depend on a broad knowledge of what is available and how the performance of the equipment may be best described in terms of the job

1. Testing the validity of theoretical predictions based on simplifying assumptions; improvement of theory, based on measured behavior.
   Example: frequency-response testing of mechanical linkage for resonant frequencies.

2. Formulation of generalized empirical relationships in situations where no adequate theory exists.
   Example: determination of friction factor for turbulent pipe flow.

3. Determination of material, component, and system parameters; variables; and performance indices.
   Examples: determination of yield point of a certain alloy steel, speed-torque curves for an electric motor, thermal efficiency of a steam turbine.

4. Study of phenomena with hopes of developing a theory.
   Example: electron microscopy of metal fatigue cracks.

5. Solution of mathematical equations by means of analogies.
   Example: solution of shaft torsion problems by measurements on soap bubbles.

Figure 1.4 Types of experimental-analysis problems.

to be done. New equipment is continuously being developed, but certain basic devices have proved their usefulness in broad areas and undoubtedly will be widely used for many years. A representative cross section of such devices is discussed in this text. These devices are of great interest in themselves; they also serve as the vehicle for the presentation and development of general techniques and principles needed in handling problems in measurement instrumentation. In addition, these general concepts are useful in treating any devices that may be developed in the future.

The treatment is also intended to be on a level that will be of service to not only the user, but also the designer of measurement instrumentation equipment. There are two main reasons for this emphasis. First, much experimental equipment (including measurement instruments) is often "homemade," especially in smaller companies where the high cost of specialized gear cannot always be justified. Second, the instrument industry is a large and growing one which utilizes many engineers in a design capacity. While the general techniques of mechanical and electrical design as applied to *machines* are also applicable to instruments, in many cases a rather different point of view is necessary in instrument design. This is due, in part, to the fact that the design of machines is mainly concerned with considerations of *power* and *efficiency*, whereas instrument design almost completely neglects these areas and concerns itself with the acquisition and manipulation of *information*. Since a considerable number of engineering graduates will work in the instrument industry, their education should include treatment of the most significant aspects of this area.

The third class of applications listed earlier, experimental engineering analysis, requires not only familiarity with measurement systems, but also some understanding of the planning, execution, and evaluation of experiments. While all these aspects of experimental studies might be treated in a single text or course, the author has chosen to concentrate on a thorough exposition of the measurement system aspect, referring the reader to the literature[1] for material on experiment design.

PROBLEMS

1.1 By consulting various technical journals in the library, find accounts of experimental studies carried out by engineers or scientists. Find three such articles, reference them completely, explain briefly what was accomplished, and attempt to classify them according to one or more categories of Fig. 1.4.

1.2 Give three specific examples of measuring-instrument applications in each of the following areas: (a) monitoring of processes and operations, (b) control of processes and operations, (c) experimental engineering analysis.

[1] Hilbert Schenck, Jr., "Theories of Engineering Experimentation," 3d ed., McGraw-Hill, New York, 1979; C. Lipson and J. Sheth, "Statistical Design and Analysis of Engineering Experiments," McGraw-Hill, New York, 1973; F. B. Wilson, Jr., "An Introduction to Scientific Research," McGraw-Hill, New York, 1952.

1.3 Compare and contrast the experimental and the theoretical approaches to the following problems:

(a) What is the tolerable vibration level to which astronauts may safely be exposed in launch vehicles?

(b) Find the relationship between applied force F and resulting friction torque T_f in the simple brake of Fig. P1.1.

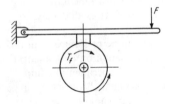

Figure P1.1

(c) Find the location of the center of mass of the rocket shown in Fig. P1.2 if the shapes, sizes, and materials of all the component parts are known.

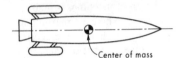

Center of mass   **Figure P1.2**

(d) At what angle with the horizontal should a projectile be launched to achieve the greatest horizontal range?

BIBLIOGRAPHY

Books

1. K. S. Lion: "Instrumentation in Scientific Research," McGraw-Hill, New York, 1959.
2. C. F. Hix and R. P. Alley: "Physical Laws and Effects," Wiley, New York, 1958.
3. P. K. Stein: "Measurement Engineering," Stein Engineering Services, Inc., Phoenix, Ariz., 1964.
4. C. S. Draper, Walter McKay, and Sidney Lees: "Instrument Engineering," vols. 1 to 3, McGraw-Hill, New York, 1955.
5. R. H. Cerni and L. E. Foster: "Instrumentation for Engineering Measurement," Wiley, New York, 1962.
6. J. P. Holman and W. J. Gajda: "Experimental Methods for Engineers," 3d ed., McGraw-Hill, New York, 1978.
7. T. G. Beckwith, W. L. Buck, and R. D. Marangoni: "Mechanical Measurements," 3d ed., Addison-Wesley, Reading, Mass., 1982.
8. N. H. Cook and E. Rabinowicz: "Physical Measurement and Analysis," Addison-Wesley, Reading, Mass., 1963.
9. D. P. Eckman: "Industrial Instrumentation," Wiley, New York, 1950.
10. G. L. Tuve: "Mechanical Engineering Experimentation," McGraw-Hill, New York, 1961.
11. D. M. Considine (ed.): "Process Instruments and Controls Handbook," McGraw-Hill, New York, 1957.
12. Transducer Compendium, Instrument Society of America, Pittsburgh, Pa.
13. H. K. P. Neubert: "Instrument Transducers," Oxford University Press, London, 1963.

Periodicals

1. *The Review of Scientific Instruments*
2. *Journal of Physics E: Scientific Instruments*
3. *Transactions of Instrument Society of America*
4. *Experimental Mechanics*
5. *Measurement Techniques* (USSR; English translation)
6. *Instruments and Experimental Techniques* (USSR; English translation)
7. *Industrial Laboratory* (USSR; English translation)
8. *Instruments and Control Systems*
9. *Control Engineering*
10. *Journal of Instrument Society of America*
11. *Archiv für Technisches Messen* (Germany)
12. *Journal of Research of the National Bureau of Standards*
13. *Transactions of the Society of Instrument Technology* (Great Britain)
14. *ASME Jour. of Dyn. Syst., Measurement and Control*
15. *Biomedical Engineering* (London)
16. *Medical Electronics and Data*
17. *Experimental Techniques*
18. *Optical Engineering*

TWO

GENERALIZED CONFIGURATIONS AND FUNCTIONAL DESCRIPTIONS OF MEASURING INSTRUMENTS

2.1 FUNCTIONAL ELEMENTS OF AN INSTRUMENT

It is possible and desirable to describe both the operation and the performance (degree of approach to perfection) of measuring instruments and associated equipment in a generalized way without recourse to specific physical hardware. The operation can be described in terms of the functional elements of instrument systems, and the performance is defined in terms of the static and dynamic performance characteristics. This section develops the concept of the functional elements of an instrument or instrument system.

If you examine diverse physical instruments with a view toward generalization, soon you recognize in the elements of the instruments a recurring pattern of similarity with regard to function. This leads to the concept of breaking down instruments into a limited number of types of elements according to the generalized function performed by the element. This breakdown can be made in a number of ways, and no standardized, universally accepted scheme is used at present. We present one such scheme which may help you to understand the operation of any new instrument with which you may come in contact and to plan the design of a new instrument.

Consider Fig. 2.1, which represents a possible arrangement of functional

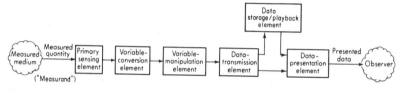

Figure 2.1 Functional elements of an instrument or a measurement system.

elements in an instrument and includes *all* the basic functions considered necessary for a description of any instrument. The *primary sensing element* is that which first receives energy from the measured medium and produces an output depending in some way on the measured quantity ("measurand"). It is important to note that an instrument *always* extracts some energy from the measured medium. Thus the measured quantity is *always* disturbed by the act of measurement, which makes a perfect measurement theoretically impossible. Good instruments are designed to minimize this effect, but it is always present to some degree.

The output signal of the primary sensing element is some physical variable, such as displacement or voltage. For the instrument to perform the desired function, it may be necessary to convert this variable to another more suitable variable while preserving the information content of the original signal. An element that performs such a function is called a *variable-conversion element*. It should be noted that not every instrument includes a variable-conversion element, but some require several. Also, the "elements" we speak of are *functional* elements, not physical elements. That is, Fig. 2.1 shows an instrument neatly separated into blocks, which may lead you to think of the physical apparatus as being precisely separable into subassemblies performing the specific functions shown. That is, in general, not the case; a specific piece of hardware may perform *several* of the basic functions, for instance.

In performing its intended task, an instrument may require that a signal represented by some physical variable be manipulated in some way. By manipulation, here we mean specifically a change in numerical value according to some definite rule but a preservation of the physical nature of the variable. Thus an electronic amplifier accepts a small voltage signal as input and produces an output signal that is also a voltage but is some constant times the input. An element that performs such a function is called a *variable-manipulation element*. Again, you should not be misled by Fig. 2.1. A variable-manipulation element does not necessarily *follow* a variable-conversion element, but may precede it, appear elsewhere in the chain, or not appear at all.

When the functional elements of an instrument are actually physically separated, it becomes necessary to transmit the data from one to another. An element performing this function is called a *data-transmission element*. It may be as simple as a shaft and bearing assembly or as complicated as a telemetry system for transmitting signals from satellites to ground equipment by radio.

If the information about the measured quantity is to be communicated to a human being for monitoring, control, or analysis purposes, it must be put into a form recognizable by one of the human senses. An element that performs this "translation" function is called a *data-presentation element*. This function includes the simple *indication* of a pointer moving over a scale and the *recording* of a pen moving over a chart. Indication and recording also may be performed in discrete increments (rather than smoothly), as exemplified by an optical flat used for measuring flatness of surfaces by light-interference principles and an electric type-writer for recording numerical data. While the majority of instruments communicate with people through the visual sense, the use of other senses such as hearing and touch is certainly conceivable.

Although data storage in the form of pen/ink recording is often employed, some applications require a distinct *data storage/playback* function which can easily re-create the stored data upon command. The magnetic tape recorder/reproducer is the classical example here. However, many recent instruments digitize the electric signals and store them in a computerlike digital memory.

Before we go on to some illustrative examples, let us emphasize again that Fig. 2.1 is intended as a vehicle for presenting the concept of functional elements,

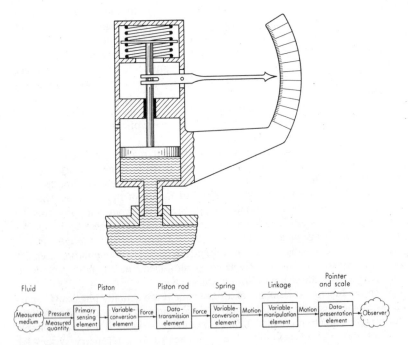

Figure 2.2 Pressure gage.

and not as a physical schematic of a generalized instrument. A given instrument may involve the basic functions in any number and combination; they need not appear in the order of Fig. 2.1. A given physical component may serve several of the basic functions.

As an example of the above concepts, consider the rudimentary pressure gage of Fig. 2.2. One of several possible valid interpretations is as follows: The primary sensing element is the piston, which also serves the function of variable conversion since it converts the fluid pressure (force per unit area) to a resultant force on the piston face. Force is transmitted by the piston rod to the spring, which converts force to a proportional displacement. This displacement of the piston rod is magnified (manipulated) by the linkage to give a larger pointer displacement. The pointer and scale indicate the pressure, thus serving as data-presentation elements. If it were necessary to locate the gage at some distance from the source of pressure, a small tube could serve as a data-transmission element.

Figure 2.3 depicts a pressure-type thermometer. The liquid-filled bulb acts as a primary sensor and variable-conversion element since a temperature change results in a pressure buildup within the bulb, because of the constrained thermal expansion of the filling fluid. This pressure is transmitted through the tube to a Bourdon-type pressure gage, which converts pressure to displacement. This displacement is manipulated by the linkage and gearing to give a larger pointer motion. A scale and pointer again serve for data presentation.

A remote-reading shaft-revolution counter is shown in Fig. 2.4. The micro-switch sensing arm and the camlike projection on the rotating shaft serve both a primary sensing and a variable-conversion function since rotary displacement is converted to linear displacement. The microswitch contacts also serve for variable conversion, changing a mechanical to an electrical oscillation (a sequence of

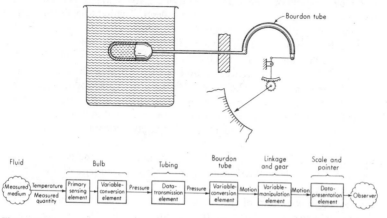

Figure 2.3 Pressure thermometer.

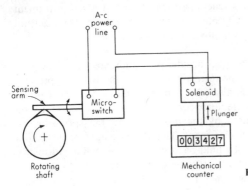

Figure 2.4 Digital revolution counter.

voltage pulses). These voltage pulses may be transmitted relatively long distances over wires to a solenoid. The solenoid reconverts the electrical pulses to mechanical reciprocation of the solenoid plunger, which serves as input to a mechanical counter. The counter itself involves variable conversion (reciprocating to rotary motion), variable manipulation (rotary motion to decimalized rotary motion), and data presentation.

As a final example, let us examine Fig. 2.5, which illustrates schematically a D'Arsonval galvanometer as used in oscillographs. A time-varying voltage to be recorded is applied to the ends of the two wires which transmit the voltage to a coil made up of a number of turns wound on a rigid frame. This coil is suspended in the field of a permanent magnet. The resistance of the coil converts the applied voltage to a proportional current (ideally). The interaction between the current and the magnetic field produces a torque on the coil, which gives another variable conversion. This torque is converted to an angular deflection by the torsion springs. A mirror rigidly attached to the coil frame converts the frame rotation to the rotation of a light beam which the mirror reflects. The light-beam rotation is twice the mirror rotation, which gives a motion magnification. The reflected beam intercepts a recording chart made of photosensitive material which is moved at a fixed and known rate, to give a time base. The combined horizontal motion of the light spot and vertical motion of the recording chart generates a graph of voltage versus time. The "optical lever arm" (the distance from the mirror to the recording chart) has a motion-magnifying effect, since the spot displacement per unit mirror rotation is directly proportional to it.

In this instrument, the coil and magnet assembly probably would be considered as the primary sensing element since the lead wires (which serve a transmission function) are not really part of the instrument, and the coil resistance (which acts in a variable-conversion function) is an intrinsic part of the coil. In any case, the assignment of precise names to specific components is not nearly as important as the recognition of the basic functions necessary to the successful operation of the instrument. By concentrating on these functions and the various physical devices available for accomplishing them, we develop our ability to

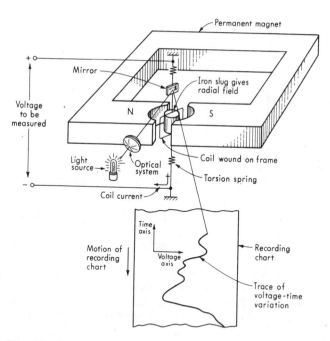

Figure 2.5 D'Arsonval galvanometer.

synthesize new combinations of elements leading to new and useful instruments. This ability is fundamental to all instrument design.

2.2 ACTIVE AND PASSIVE TRANSDUCERS

Once certain basic functions common to all instruments have been identified, then we see if it is possible to make some generalizations on *how* these functions may be performed. One such generalization is concerned with energy considerations. In performing any of the general functions indicated in Fig. 2.1, a physical component may act as an active transducer or a passive transducer.

A component whose output energy is supplied entirely or almost entirely by its input signal is commonly called a *passive transducer*. The output and input signals may involve energy of the same form (say, both mechanical), or there may be an energy conversion from one form to another (say, mechanical to electrical). (In much technical literature, the term "transducer" is restricted to devices involving energy *conversion*; but, conforming to the dictionary definition of the term, we do not make this restriction.)

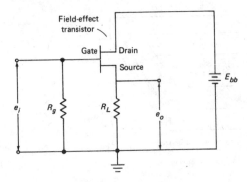

Figure 2.6 Electronic amplifier.

An *active transducer*, however, has an auxiliary source of power which supplies a major part of the output power while the input signal supplies only an insignificant portion. Again, there may or may not be a conversion of energy from one form to another.

In all the examples of Sec. 2.1, there is only one active transducer—the microswitch of Fig. 2.4; all other components are passive transducers. The power to drive the solenoid comes not from the rotating shaft, but from the ac power line, an auxiliary source of power. Some further examples of active transducers

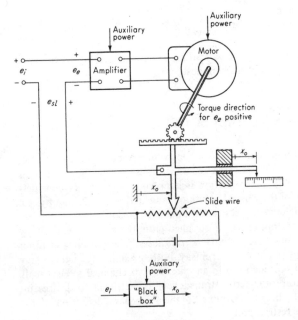

Figure 2.7 Instrument servomechanism.

may be in order. The electronic amplifier shown in Fig. 2.6 is a good one. The element supplying the input-signal voltage e_i need supply only a negligible amount of power since almost no current is drawn, owing to negligible gate current and a high R_g. However, the output element (the load resistance R_L) receives significant current and voltage and thus power. This power must be supplied by the battery E_{bb}, the auxiliary power source. Thus the input *controls* the output, but does not actually supply the output power.

Another active transducer of great practical importance, the *instrument servomechanism*, is shown in simplified form in Fig. 2.7. This is actually an instrument *system* made up of components, some of which are passive transducers and others active transducers. When it is considered as an entity, however, with input voltage e_i and output displacement x_o, it meets the definition of an active transducer and is profitably thought of as such. The purpose of this device is to cause the motion x_o to follow the variations of the voltage e_i in a proportional manner. Since the motor torque is proportional to the error voltage e_e, it is clear that the system can be at rest only if e_e is zero. This occurs only when $e_i = e_{sl}$; since e_{sl} is proportional to x_o, this means that x_o must be proportional to e_i in the static case. If e_i varies, x_o will tend to follow it, and by proper design accurate "tracking" of e_i by x_o should be possible. ·

2.3 ANALOG AND DIGITAL MODES OF OPERATION

It is possible further to classify how the basic functions may be performed by turning attention to the analog or digital nature of the signals that represent the information.

For analog signals, the precise value of the quantity (voltage, rotation angle, etc.) carrying the information is significant. However, digital signals are basically of a binary (on/off) nature, and variations in numerical value are associated with changes in the logical state ("true/false") of some combination of "switches." In a typical digital electronic system, *any* voltage in the range of $+2$ to $+5$ V produces the on state, while signals of 0 to $+0.8$ V correspond to off. Thus whether the voltage is 3 or 4 V is of *no* consequence. The same result is produced, and so the system is quite tolerant of spurious "noise" voltages which might contaminate the information signal. In a digitally represented value of, say, 5.763, the least significant digit (3) is carried by on/off signals of the same (large) size as for the most significant digit (5). Thus in an all-digital device such as a digital computer, there is no limit to the number of digits which can be accurately carried; we use whatever can be justified by the particular application. When *combined* analog/digital systems are used (often the case in measurement systems), the digital portions need not limit system accuracy. These limitations generally are associated with the analog portions and/or the analog/digital conversion devices.

The majority of primary sensing elements are of the analog type. The only digital device illustrated in this text up to this point is the revolution counter of

Fig. 2.4. This is clearly a digital device since it is impossible for this instrument to indicate, say, 0.79; it measures only in steps of 1. The importance of digital instruments is increasing, perhaps mainly because of the widespread use of digital computers in both data-reduction and automatic control systems. Since the digital computer works only with digital signals, any information supplied to it must be in digital form. The computer's output is also in digital form. Thus any communication with the computer at either the input or the output end must be in terms of digital signals. Since most measurement and control apparatus is of an analog nature, it is necessary to have both *analog-to-digital converters* (at the input to the computer) and *digital-to-analog converters* (at the output of the computer). These devices (which are discussed in greater detail in a later chapter) serve as "translators" that enable the computer to communicate with the outside world, which is largely of an analog nature.

2.4 NULL AND DEFLECTION METHODS

Another useful classification separates devices by their operation on a null or a deflection principle. In a *deflection-type* device, the measured quantity produces some physical effect that engenders a similar but opposing effect in some part of the instrument. The opposing effect is closely related to some variable (usually a mechanical displacement or deflection) that can be directly observed by some human sense. The opposing effect increases until a balance is achieved, at which point the "deflection" is measured and the value of the measured quantity inferred from this. The pressure gage of Fig. 2.2 exemplifies this type of device, since the pressure force engenders an opposing spring force as a result of an unbalance of forces on the piston rod (called the force-summing link), which causes a deflection of the spring. As the spring deflects, its force increases; thus a balance will be achieved at some deflection if the pressure is within the design range of the instrument.

In contrast to the deflection-type device, a *null-type* device attempts to maintain deflection at zero by suitable application of an effect opposing that generated by the measured quantity. Necessary to such an operation are a detector of unbalance and a means (manual or automatic) of restoring balance. Since deflection is kept at zero (ideally), determination of numerical values requires accurate knowledge of the magnitude of the opposing effect. A pressure gage operating on a null principle is depicted in simplified form in Fig. 2.8. By adding the proper standard weights to the platform of known weight, the pressure force on the face of the piston may be balanced by gravitational force. The condition of force balance is indicated by the platform remaining at rest between the upper and lower stops. Since the weights and the piston area are all known, the unknown pressure may be computed.

Upon comparing the null and deflection methods of measurement exemplified by the pressure gages described above, we note that, in the deflection instrument, accuracy depends on the calibration of the spring, whereas in the null instrument it depends on the accuracy of the standard weights. In this particular

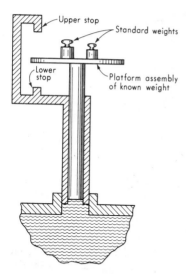

Figure 2.8 Deadweight pressure gage.

case (and for most measurements in general), the accuracy attainable by the null method is of a higher level than that by the deflection method. One reason is that the spring is not in itself a primary standard of force, but must be calibrated by standard weights, whereas in the null instrument a *direct* comparison of the unknown force with the standard is achieved. Another advantage of null methods is the fact that, since the measured quantity is balanced out, the detector of unbalance can be made very sensitive, because it need cover only a small range around zero. Also the detector need not be calibrated since it must detect only the presence and direction of unbalance, but not the amount. However, a deflection instrument must be larger, more rugged, and thus less sensitive if it is to measure large magnitudes.

The disadvantages of null methods appear mainly in dynamic measurements. Let us consider the pressure gages again. The difficulty in keeping the platform balanced for a fluctuating pressure should be apparent. The spring-type gage suffers not nearly so much in this respect. By use of automatic balancing devices (such as the instrument servomechanism of Fig. 2.7) the speed of null methods may be improved considerably, and instruments of this type are of great importance.

2.5 INPUT-OUTPUT CONFIGURATION OF MEASURING INSTRUMENTS AND INSTRUMENT SYSTEMS

Before we discuss instrument performance characteristics, it is desirable to develop a generalized configuration that brings out the significant input-output relationships present in all measuring apparatus. A scheme suggested by Draper,

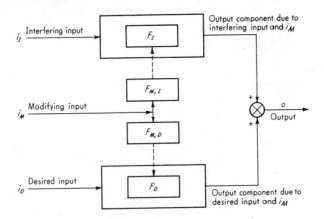

Figure 2.9 Generalized input-output configuration.

McKay, and Lees[1] is presented in somewhat modified form in Fig. 2.9. Input quantities are classified into three categories: desired inputs, interfering inputs, and modifying inputs. *Desired inputs* represent the quantities that the instrument is specifically intended to measure. *Interfering inputs* represent quantities to which the instrument is unintentionally sensitive. A desired input produces a component of output according to an input-output relation symbolized by F_D, where F_D denotes the mathematical operations necessary to obtain the output from the input. The symbol F_D may represent different concepts, depending on the particular input-output characteristic being described. Thus F_D might be a constant number K that gives the proportionality constant relating a constant static input to the corresponding static output for a linear instrument. For a nonlinear instrument, a simple constant is not adequate to relate static inputs and outputs; a mathematical *function* is required. To relate dynamic inputs and outputs, differential equations are necessary. If a description of the output "scatter," or dispersion, for repeated equal static inputs is desired, a statistical distribution function of some kind is needed. The symbol F_D encompasses all such concepts. The symbol F_I serves a similar function for an interfering input.

The third class of inputs might be thought of as being included among the interfering inputs, but a separate classification is actually more significant. This is the class of modifying inputs. *Modifying inputs* are the quantities that cause a change in the input-output relations for the desired and interfering inputs; that is, they cause a change in F_D and/or F_I. The symbols $F_{M,I}$ and $F_{M,D}$ represent (in the appropriate form) the specific manner in which i_M affects F_I and F_D, respectively. These symbols, $F_{M,I}$ and $F_{M,D}$, are interpreted in the same general way as F_I and F_D.

[1] C. S. Draper, Walter McKay, and Sidney Lees, "Instrument Engineering," vol. 3, p. 58, McGraw-Hill, New York, 1955.

The block diagram of Fig. 2.9 illustrates the above concepts. The circle with a cross in it is a conventional symbol for a *summing device*. The two plus signs as shown indicate that the output of the summing device is the instantaneous algebraic sum of its two inputs. Since an instrument system may have several inputs of each of the three types as well as several outputs, it may be necessary to draw more complex block diagrams than in Fig. 2.9. This extension is, however, straightforward.

The above concepts can be clarified by means of specific examples. Consider the mercury manometer used for differential-pressure measurement as shown in Fig. 2.10a. The desired inputs are the pressures p_1 and p_2 whose difference causes the output displacement x, which can be read off the calibrated scale. Figure 2.10b and c shows the action of two possible interfering inputs. In Fig. 2.10b the

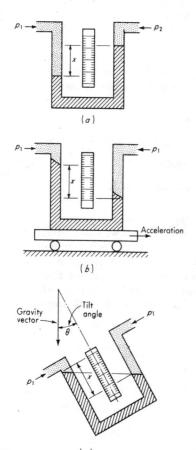

Figure 2.10 Spurious inputs for manometer.

manometer is mounted on some vehicle that is accelerating. A simple analysis shows that there will be an output x even though the differential pressure might be zero. Thus if you are trying to measure pressures under such circumstances, an error will be engendered because of the interfering acceleration input. Similarly, in Fig. 2.10c, if the manometer is not properly aligned with the gravity vector, it may give an output signal x even though no pressure difference exists. Thus the tilt angle θ is an interfering input. (It is also a modifying input.)

Modifying inputs for the manometer include ambient temperature and gravitational force. Ambient temperature manifests its influence in a number of ways. First, the calibrated scale changes length with temperature; thus the proportionality factor relating $p_1 - p_2$ to x is modified whenever temperature varies from its basic calibration value. Also, the density of mercury varies with temperature, which again leads to a change in the proportionality factor. A change in gravitational force resulting from changes in location of the manometer, such as moving it to another country or putting it aboard a spaceship, leads to a similar modification in the scale factor. Note that the effects of *both* the desired and the interfering inputs may be altered by the modifying inputs.

As another example, consider the electric-resistance strain-gage setup shown in Fig. 2.11. The gage consists of a fine-wire grid of resistance R_g firmly cemented to the specimen whose unit strain ϵ at a certain point is to be measured. When strained, the gage's resistance changes according to the relation

$$\Delta R_g = (GF)R_g \epsilon \tag{2.1}$$

where

$$\Delta R_g \triangleq \text{change in gage resistance, } \Omega^* \tag{2.2}$$

$$GF \triangleq \text{gage factor, dimensionless} \tag{2.3}$$

$$R_g \triangleq \text{gage resistance when unstrained, } \Omega \tag{2.4}$$

$$\epsilon \triangleq \text{unit strain, cm/cm} \tag{2.5}$$

The resistance change is proportional to the strain. Thus if we could measure the resistance, we could compute the strain. The resistance is measured by using the Wheatstone-bridge arrangement shown. When no load F is present, the bridge is balanced (e_o set to zero) by adjusting R_c. Application of load causes a strain, a ΔR_g, and thus unbalances the bridge, causing an output voltage e_o which is proportional to ϵ and can be measured on a meter or an oscilloscope. The voltage e_o is given by

$$e_o = -(GF)R_g \epsilon E_b \frac{R_a}{(R_g + R_a)^2} \tag{2.6}$$

The desired input here is clearly the strain ϵ which causes a proportional output voltage e_o. One interfering input which often causes trouble in such apparatus is the 60-Hz field caused by nearby power lines, electric motors, etc.

---

* The symbol $\triangleq$ means "equal by definition."

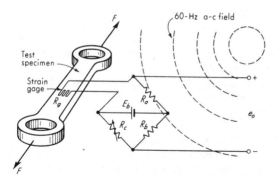

Figure 2.11 Interfering input for strain-gage circuit.

This field induces voltages in the strain-gage circuit, causing output voltages e_o even when the strain is zero. Another interfering input is the gage temperature. If this varies, it causes a change in gage resistance that will cause a voltage output even if there is no strain. Temperature has another interfering effect since it causes a differential expansion of the gage and the specimen, which gives rise to a strain ϵ and a voltage e_o even though no force F has been applied. Temperature also acts as a modifying input since the gage factor is sensitive to temperature. The battery voltage E_b is another modifying input. Both these are modifying inputs since they tend to change the proportionality factor between the desired input ϵ and the output e_o or between an interfering input (gage temperature) and output e_o.

Methods of Correction for Interfering and Modifying Inputs

In the design and/or use of measuring instruments, a number of methods for nullifying or reducing the effects of spurious inputs are available. We briefly describe some of the most widely used.

The *method of inherent insensitivity* proposes the obviously sound design philosophy that the elements of the instrument should *inherently* be sensitive to only the desired inputs. While usually this is not entirely possible, the simplicity of this approach encourages one to consider its application wherever feasible. In terms of the general configuration of Fig. 2.9, this approach requires that somehow F_I and/or $F_{M,D}$ be made as nearly equal to zero as possible. Thus, even though i_I and/or i_M may exist, they cannot affect the output. As an example of the application of this concept to the strain gage of Fig. 2.11, we might try to find some gage material that exhibits an extremely low temperature coefficient of resistance while retaining its sensitivity to strain. If such a material can be found, the problem of interfering temperature inputs is at least partially solved. Similarly, in mechanical apparatus that must maintain accurate dimensions in the face of ambient-temperature changes, the use of a material of very small temperature coefficient of expansion (such as the alloy Invar) may be helpful.

The *method of high-gain feedback* is exemplified by the system shown in Fig. 2.12*b*. Suppose we wish to measure a voltage e_i by applying it to a motor whose torque is applied to a spring, causing a displacement x_o, which may be measured on a calibrated scale. By proper design, the displacement x_o might be made proportional to the voltage e_i according to

$$x_o = (K_{Mo} K_{SP}) e_i \qquad (2.7)$$

where K_{Mo} and K_{SP} are appropriate constants. This arrangement, shown in Fig. 2.12*a*, is called an open-loop system. If modifying inputs i_{M1} and i_{M2} exist, they cause changes in K_{Mo} and K_{SP} that lead to errors in the relation between e_i and

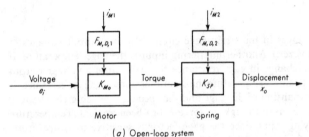

(*a*) Open-loop system

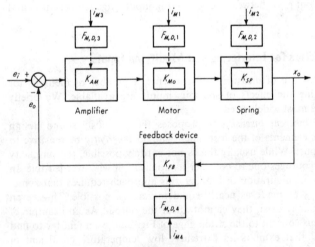

(*b*) Closed-loop or feedback system

Figure 2.12 Use of feedback to reduce effect of spurious inputs.

x_o. These errors are in *direct proportion* to the changes in K_{Mo} and K_{SP}. Suppose, instead, we construct a system as in Fig. 2.12b. Here the output x_o is measured by the feedback device, which produces a voltage e_o proportional to x_o. This voltage is subtracted from the input voltage e_i, and the difference is applied to an amplifier which drives the motor and thereby the spring to produce x_o. We may write

$$(e_i - e_o)K_{AM}K_{Mo}K_{SP} = (e_i - K_{FB}x_o)K_{AM}K_{Mo}K_{SP} = x_o \qquad (2.8)$$

$$e_i K_{AM}K_{Mo}K_{SP} = (1 + K_{AM}K_{Mo}K_{SP}K_{FB})x_o \qquad (2.9)$$

$$x_o = \frac{K_{AM}K_{Mo}K_{SP}}{1 + K_{AM}K_{Mo}K_{SP}K_{FB}} e_i \qquad (2.10)$$

Suppose, now, that we design K_{AM} to be very large (a "high-gain" system), so that $K_{AM}K_{Mo}K_{SP}K_{FB} \gg 1$. Then

$$x_o \approx \frac{1}{K_{FB}} e_i \qquad (2.11)$$

The significance of Eq. (2.11) is that the effect of variations in K_{Mo}, K_{SP}, and K_{AM} (as a result of modifying inputs i_{M1}, i_{M2}, and i_{M3}) on the relation between input e_i and output x_o has been made negligible. *We now require only that K_{FB} stay constant (unaffected by i_{M4}) in order to maintain constant input-output calibration as shown by Eq. (2.11).*

You may question whether much really has been gained by this somewhat elaborate scheme, since we merely transferred the requirements for stability from K_{Mo} and K_{SP} to K_{FB}. In practice, however, this method often leads to great improvements in accuracy. One reason is that, since the amplifier supplies most of the power needed, the feedback device can be designed with low power-handling capacity. In general, this leads to greater accuracy and linearity in the feedback-device characteristics. Also, the input signal e_i need carry only negligible power; thus the feedback system extracts less energy from the measured medium than the corresponding open-loop system. This, of course, results in less distortion of the measured quantity because of the presence of the measuring instrument. Finally, if the open-loop chain consists of several (perhaps many) devices, each susceptible to its own spurious inputs, then *all* these bad effects can be negated by the use of high amplification and a stable, accurate feedback device.

Before we pass on to other methods, we should mention that application of the feedback principle is not without its own peculiar problems. The main one is dynamic instability, wherein excessively high amplification leads to destructive oscillations. The study of the design of feedback systems is a whole field in itself, and many texts treating this subject are available.[1]

The *method of calculated output corrections* requires one to measure or estimate the magnitudes of the interfering and/or modifying inputs and to know quantitatively how they affect the output. With this information, it is possible to calculate corrections which may be added to or subtracted from the indicated

[1] E. O. Doebelin, "Dynamic Analysis and Feedback Control," McGraw-Hill, New York, 1962.

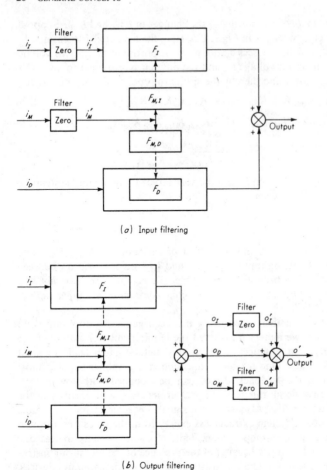

(a) Input filtering

(b) Output filtering

Figure 2.13 General principle of filtering.

output so as to leave (ideally) only that component associated with the desired input. Thus, in the manometer of Fig. 2.10, the effects of temperature on both the calibrated scale's length and the density of mercury may be quite accurately computed if the temperature is known. The local gravitational acceleration is also known for a given elevation and latitude, so that this effect may be corrected by calculation. Since many measurement systems today can afford to include a microcomputer to carry out various functions, if we also provide sensors for the spurious inputs, the microcomputer can implement the method of calculated output corrections on an automatic basis.

The *method of signal filtering* is based on the possibility of introducing certain elements ("filters") into the instrument which in some fashion block the spurious signals, so that their effects on the output are removed or reduced. The filter may be applied to any suitable signal in the instrument, be it input, output, or intermediate signal. The concept of signal filtering is shown schematically in Fig. 2.13 for the cases of input and output filtering. The application to intermediate signals should be obvious. In Fig. 2.13a the inputs i_I and i_M are caused to pass through filters whose input-output relation is (ideally) zero. Thus i_I' and i_M' are zero even if i_I and i_M are not zero. The concept of output filtering is illustrated in Fig. 2.13b. Here the output o, though really one signal, is thought of as a superposition of o_I (output due to interfering input), o_D (output due to desired input), and o_M (output due to modifying input). If it is possible to construct filters that selectively block o_I and o_M but allow o_D to pass through, this may be symbolized as in Fig. 2.13b and results in o' consisting entirely of o_D.

The filters necessary in the application of this method may take several forms; they are best illustrated by examples. If put directly in the path of a spurious input, a filter can be designed (ideally) to block completely the passage of the signal. If, however, it is inserted at a point where the signal contains both desired and spurious components, the filter must be designed to be selective. That is, it must pass the desired components essentially unaltered while effectively suppressing all others.

Often it is necessary to attach delicate instruments to structures that vibrate. Electromechanical devices for navigation and control of aircraft or missiles are outstanding examples. Figure 2.14a shows how the interfering vibration input may be filtered out by use of suitable spring mounts. The mass-spring system is actually a mechanical filter which passes on to the instrument only a negligible fraction of the motion of the vibrating structure.

The interfering tilt-angle input to the manometer of Fig. 2.10c may be effectively filtered out by means of the gimbal-mounting scheme of Fig. 2.14b. If the gimbal bearings are essentially frictionless, the rotations θ_1 and θ_2 cannot be communicated to the manometer; thus it always hangs vertical.

In Fig. 2.14c the thermocouple reference junction is shielded from ambient-temperature fluctuations by means of thermal insulation. Such an arrangement acts as a filter for temperature or heat-flow inputs.

The strain-gage circuit of Fig. 2.14d is shielded from the interfering 60-Hz field by enclosing it in a metal box of some sort. This solution corresponds to filtering the interfering *input*. Another possible solution, which corresponds to selective filtering of the *output*, is shown in Fig. 2.14e. For this approach to be effective, it is essential that the frequencies in the desired signal occupy a frequency range considerably separated from those in the undesired component of the signal. In the present example, suppose the strains to be measured are mainly steady and never vary more rapidly than 2 Hz. Then it is possible to insert a simple *RC* filter, as shown, that will pass the desired signals but almost completely block the 60-Hz interference.

Figure 2.14f shows the pressure gage of Fig. 2.2 modified by the insertion of a flow restriction between the source of pressure and the piston chamber. Such an

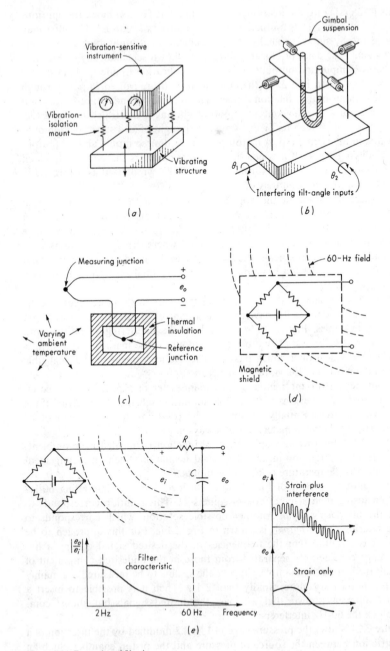

Figure 2.14 Examples of filtering.

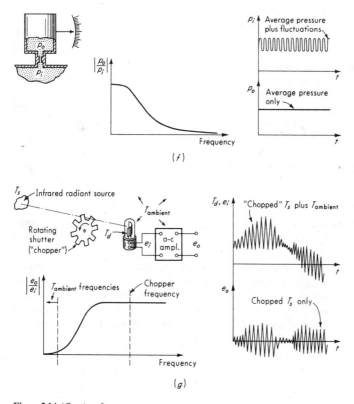

Figure 2.14 (*Continued*)

arrangement is useful, for example, if you wish to measure only the average pressure in a large air tank that is being supplied by a reciprocating compressor. The pulsations in the air pressure may be smoothed by the pneumatic filtering effect of the flow restriction and associated volume. The variation of the output-input amplitude ratio $|p_o/p_i|$ with frequency is similar to that for the electrical RC filter of Fig. 2.14e. Thus steady or slowly varying input pressures are accurately measured while rapid variations are strongly attenuated. The flow restriction may be in the form of a needle valve, which allows easy adjustment of the filtering effect.

A "chopped" radiometer is shown in simplified form in Fig. 2.14g. This device senses the temperature T_s of some body in terms of the infrared radiant energy emitted. The emitted energy is focused on a detector of some sort and causes the temperature T_d of the detector, and thus its output voltage e_i, to vary. The difficulty with such devices is that the ambient temperature, as well as T_s, affects T_d. This effect is serious since the radiant energy to be measured causes very small changes in T_d; thus small ambient drifts can completely mask the desired input. An ingenious solution to this problem interposes a rotating shutter between the radiant source and the detector, so that the desired input is "chopped," or modulated, at a known frequency. This frequency is chosen to be much higher than the frequencies at which ambient drifts may occur. The output signal e_i of the detector thus is a superposition of slow ambient fluctuations and a high-frequency wave whose amplitude varies in proportion to variations in T_s. Since the desired and interfering components are thus widely separated in frequency, they may be selectively filtered. In this case, we desire a filter that rejects constant and slowly varying signals, but faithfully reproduces rapid variations. Such a characteristic is typical of an ordinary ac amplifier, and since amplification is necessary in such instruments in any case, the use of an ac amplifier as shown solves two problems at once.

In summing up the method of signal filtering, it may be said that, in general, it is usually possible to design filters of mechanical, electrical, thermal, pneumatic, etc., nature which separate signals according to their frequency content in some specific manner. Figure 2.15 summarizes the most common useful forms of such devices.

The *method of opposing inputs* consists of intentionally introducing into the instrument interfering and/or modifying inputs that tend to cancel the bad effects of the unavoidable spurious inputs. Figure 2.16 shows schematically the concept for interfering inputs. The extension to modifying inputs should be obvious. The intentionally introduced input is designed so that the signals o_{I1} and o_{I2} are

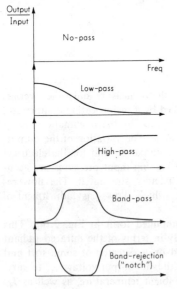

Figure 2.15 Basic filter types.

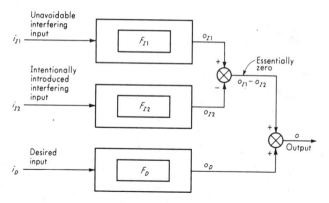

Figure 2.16 Method of opposing inputs.

essentially equal, but act in opposite sense; thus the net contribution $o_{I1} - o_{I2}$ to the output is essentially zero. This method actually might be considered as a variation on the method of calculated output corrections. However, the "calculation" and application of the correction are achieved automatically owing to the structure of the system, rather than by numerical calculation by a human operator. Thus the two methods are similar; however, the distinction between them is a worthwhile one since it helps to organize your thinking in inventing new applications of these generalized correction concepts.

Some examples of the method of opposing inputs are shown in Fig. 2.17. A millivoltmeter, shown in Fig. 2.17a, is basically a *current*-sensitive device. However, as long as the total circuit resistance is constant, its scale can be calibrated in voltage, since voltage and current are proportional. A modifying input here is the ambient temperature, since it causes the coil resistance R_{coil} to change, thereby altering the proportionality factor between current and voltage. To correct for this error, the compensating resistance R_{comp} is introduced into the circuit, and its material is carefully chosen to have a temperature coefficient of resistance *opposite* to that of R_{coil}. Thus when the temperature changes, the total resistance of the circuit is unaffected and the calibration of the meter remains accurate.

Figure 2.17b shows a static-pressure-probe design due to L. Prandtl. As the fluid flows over the surface of the probe, the velocity of the fluid must increase since these streamlines are longer than those in the undistorted flow. This velocity increase causes a drop in static pressure, so that a tap in the surface of the probe gives an incorrect reading. This underpressure error varies with the distance d_1 of the tap from the probe tip. Prandtl recognized that the probe support will have a stagnation point (line) along its front edge and that this overpressure will be felt upstream, the effect decreasing as the distance d_2 increases. By properly choosing distances d_1 and d_2 (by experimental test), these

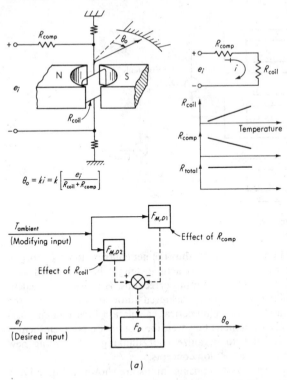

$$\theta_o = ki = k\left[\frac{e_i}{R_{coil} + R_{comp}}\right]$$

(a)

Figure 2.17 Examples of method of opposing inputs. (c) *Courtesy of National Instrument Laboratories, Inc., Washington, D. C.*

two effects can be made exactly to cancel, giving a true static-pressure value at the tap.

A device for the measurement of the mass flow rate of gases is shown in Fig. 2.17c. The mass flow rate of gas through an orifice may be found by measuring the pressure drop across the orifice, perhaps by means of a U-tube manometer. Unfortunately, the mass flow rate also depends on the density of the gas, which varies with pressure and temperature. Thus the pressure-drop measuring device usually cannot be calibrated to give the mass flow rate, since variations in gas temperature and pressure yield different mass flow rates for the same orifice pressure drop. The instrument of Fig. 2.17c overcomes this problem in an ingenious fashion. The flow rate through the orifice also depends on its flow area. Thus if the flow area could be varied in just the right way, this variation could compensate for pressure and temperature changes so that a given orifice pressure drop could *always* correspond to the same mass flow rate. This is accomplished by attaching the specially shaped metering pin to a gas-filled bellows as shown.

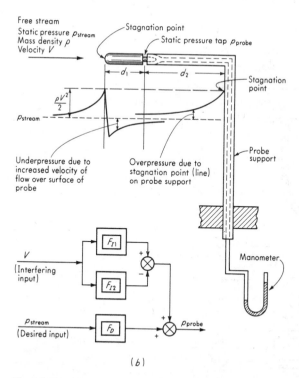

Figure 2.17 (*Continued*)

When the temperature drops (causing an increase in density and therefore in mass flow rate), the gas in the bellows contracts, which moves the metering pin into the orifice and thereby reduces the flow area. This returns the mass flow rate to its proper value. Similarly, should the pressure of the flowing gas increase, causing an increase in density and mass flow rate, the gas-filled bellows would be compressed again, reducing the flow area and correcting the mass flow rate. The proper shape for the metering pin is revealed by a detailed analysis of the system.

A final example of the method of opposing inputs is the rate gyroscope of Fig. 2.17d. Such devices are widely used in aerospace vehicles for the generation of stabilization signals in the control system. The action of the device is that a vehicle rotation at angular velocity $\dot{\theta}_i$ causes a proportional displacement θ_o of the gimbal relative to the case. This rotation θ_o is measured by some motion pickup (not shown in Fig. 2.17d). Thus a signal proportional to vehicle angular velocity is available, and this is useful in stabilizing the vehicle. When the vehicle undergoes rapid motion changes, however, the angle θ_o tends to oscillate, giving an incorrect angular-velocity signal. To control these oscillations, the gimbal rotation θ_o is damped by the shearing action of a viscous silicone fluid in a

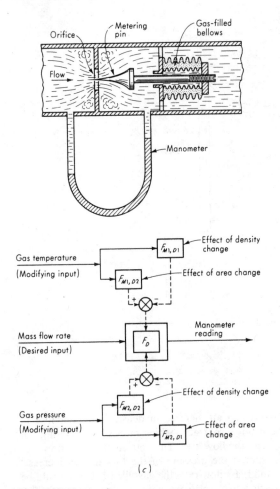

Figure 2.17 (Continued)

narrow damping gap. The damping effect varies with the viscosity of the fluid and the thickness of the damping gap. Although the viscosity of the silicone fluid is fairly constant, it does vary with ambient temperature, causing an undesirable change in damping characteristics. To compensate for this, a nylon cylinder is used in the gyro of Fig. 2.17d. When the temperature increases, viscosity drops, causing a loss of damping. Simultaneously, however, the nylon cylinder expands, narrowing the damping gap and thus restoring the damping to its proper value. By proper choice of materials and geometry, the two effects can be made to very nearly cancel over the operating temperature range of the equipment.

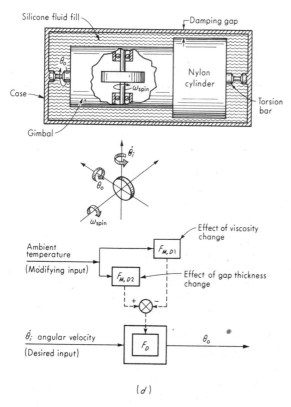

(d)

Figure 2.17 *(Continued)*

2.6 CONCLUSION

In this chapter we developed useful generalizations with regard to the functional elements and the input-output configurations of measuring instruments and systems. In the analysis of a given instrument or in the design of a new one, the starting point is the separation of the overall operation into its functional elements. Here you must take a broad view of *what* must be done, but not be concerned with *how* it is actually accomplished. Once the general functional concepts have been clarified, the details of operation may be considered fruitfully. The ideas of active and passive transducers, analog and digital modes of operation, and null versus deflection methods give a systematic approach for either analysis or design.

Finally, compensation of spurious inputs and detailed evaluation of performance are facilitated by application of input-output block diagrams. These configuration diagrams show clearly which physical analyses must be made to evaluate performance with respect to accurate measurement of the desired inputs and rejection of spurious inputs. The evaluation of the relative quality of different instruments (or the same instrument with different numerical parameter values) requires the definition of performance criteria against which competitive designs may be compared. This is the subject of Chap. 3.

PROBLEMS

2.1 Make block diagrams such as Fig. 2.1, showing the functional elements of the instruments depicted in the following:

    (a) Fig. 2.7.

    (b) Fig. 2.8.

    (c) Fig. 2.10a.

    (d) Fig. 2.11. Take F as input and e_o as output.

    (e) Fig. 2.14g. Take T_s as input and e_o as output.

    (f) Fig. 2.17b. Take V as input and manometer Δh as output.

    (g) Fig. 2.17d. Take θ_i as input and θ_o as output.

2.2 Identify the active transducers, if any, in the instruments of (a) Fig. 2.8, (b) Fig. 2.10a, (c) Fig. 2.11, (d) Fig. 2.17b, (e) Fig. 2.17c.

2.3 Consider a man, driving a car along a road, who sees the opportunity to pass and decides to accelerate.

    (a) If the light waves entering his eyes are considered input and accelerator-pedal travel is taken as output, is the man functioning as an active or a passive transducer?

    (b) If accelerator-pedal travel is considered input and car velocity as output, is the automobile engine an active or a passive transducer?

2.4 Give an example of a null method of force measurement.

2.5 Give an example of a null method of voltage measurement.

2.6 Sketch and explain two possible modifications of the system of Fig. 2.4 that will allow measurement to $\frac{1}{10}$ revolution.

2.7 Identify desired, interfering, and modifying inputs for the systems of (a) Fig. 2.2, (b) Fig. 2.3, (c) Fig. 2.4, (d) Fig. 2.5.

2.8 Why is tilt angle in Fig. 2.10c a modifying input?

2.9 Suppose in Eq. (2.7) that $K_{MO} = K_{SP} = e_i = 1.0$. Now let K_{MO} change by 10 percent to 1.1. What is the change in x_o? In Eq. (2.10), let $K_{MO} = K_{SP} = K_{FB} = e_i = 1.0$, and $K_{AM} = 100$. Now let K_{MO} change by 10 percent to 1.1. What is the change in x_o? Investigate the effect of similar changes in K_{AM}, K_{SP}, and K_{FB}.

2.10 The natural frequency of oscillation of the balance wheel in a watch depends on the moment of inertia of the wheel and the spring constant of the (torsional) hairspring. A temperature rise results in a reduced spring constant, which lowers the oscillation frequency. Propose a compensating means for this effect. Non-temperature-sensitive hairspring material is not an acceptable solution.

THREE

GENERALIZED PERFORMANCE CHARACTERISTICS OF INSTRUMENTS

3.1 INTRODUCTION

If you are trying to choose, from commercially available instruments, the one most suitable for a proposed measurement, or, alternatively, if you are engaged in the design of instruments for specific measuring tasks, then the subject of performance criteria assumes major proportions. That is, to make intelligent decisions, there must be some quantitative bases for comparing one instrument (or proposed design) with the possible alternatives. Chapter 2 has served as a useful preliminary to these considerations since there we developed systematic methods for breaking down the overall problem into its component parts. Now we propose to study in considerable detail the performance of measuring instruments and systems with regard to how well they measure the desired inputs and how thoroughly they reject the spurious inputs.

The treatment of instrument performance characteristics generally has been broken down into the subareas of *static characteristics* and *dynamic characteristics*, and this plan is followed here. The reasons for such a classification are several. First, some applications involve the measurement of quantities that are constant or vary only quite slowly. Under these conditions, it is possible to define a set of performance criteria that give a meaningful description of the quality of measurement without becoming concerned with dynamic descriptions involving differential equations. These criteria are called the *static* characteristics. Many other measurement problems involve rapidly varying quantities. Here the dynamic relations between the instrument input and output must be examined, generally by the use of differential equations. Performance criteria based on these dynamic relations constitute the *dynamic* characteristics.

Actually, static characteristics also influence the quality of measurement under dynamic conditions, but the static characteristics generally show up as nonlinear or statistical effects in the otherwise linear differential equations giving the dynamic characteristics. These effects would make the differential equations unmanageable, and so the conventional approach is to treat the two aspects of the problem separately. Thus the differential equations of dynamic performance generally neglect the effects of dry friction, backlash, hysteresis, statistical scatter, etc., even though these effects affect the dynamic behavior. These phenomena are more conveniently studied as static characteristics, and the overall performance of an instrument is then judged by a semiquantitative superposition of the static and dynamic characteristics. This approach is, of course, approximate but a necessary expedient.

3.2 STATIC CHARACTERISTICS

We begin our study of static performance characteristics by considering the meaning of the term "static calibration."

Meaning of Static Calibration

All the static performance characteristics are obtained by one form or another of a process called static calibration. So it is appropriate at this point to develop a clear concept of what is meant by this term.

In general, *static calibration* refers to a situation in which all inputs (desired, interfering, modifying) except one are kept at some constant values. Then the one input under study is varied over some range of constant values, which causes the output(s) to vary over some range of constant values. The input-output relations developed in this way comprise a static calibration *valid under the stated constant conditions of all the other inputs*. This procedure may be repeated, by varying in turn each input considered to be of interest and thus developing a family of static input-output relations. Then we might hope to describe the overall instrument static behavior by some suitable form of superposition of these individual effects. In some cases, if overall rather than individual effects were desired, the calibration procedure would specify the variation of several inputs simultaneously. Also if you examine any practical instrument critically, you will find many modifying and/or interfering inputs, each of which might have quite small effects and which would be impractical to control. Thus the statement "*all* other inputs are held constant" refers to an ideal situation which can be only approached, but never reached, in practice. *Measurement method* describes the ideal situation while *measurement process* describes the (imperfect) physical realization of the measurement method.

The statement that one input is varied and all others are held constant implies that all these inputs are determined (measured) independently of the instrument being calibrated. For interfering or modifying inputs (whose effects on the output should be relatively small in a good instrument), the measurement of these inputs usually need not be at an extremely high accuracy level. For example, suppose a pressure gage has temperature as an interfering input to the extent that a temperature change of 100°C causes a pressure error of 0.100 percent. Now, if we had measured the 100°C interfering input with a thermometer which itself had an error of 2.0 percent, the pressure error actually would have been 0.102 percent. It should be clear that the difference between an error of 0.100 and 0.102 percent is entirely negligible in most engineering situations. However, when calibrating the response of the instrument to its *desired* inputs, you must exercise considerable care in choosing the means of determining the numerical values of these inputs. That is, if a pressure gage is inherently capable of an accuracy of 0.1 percent, you must certainly be able to determine its input pressure during calibration with an accuracy somewhat greater than this. In other words, it is impossible to calibrate an instrument to an accuracy greater than that of the standard with which it is compared. A rule often followed is that the calibration standard should be at least about 10 times as accurate as the instrument being calibrated.

While we do not discuss standards in detail at this point, it is of utmost importance that the person performing the calibration be able to answer the question: How do I know that this standard is capable of its stated accuracy? The ability to trace the accuracy of a standard back to its ultimate source in the fundamental standards of the National Bureau of Standards is termed *traceability*.

In performing a calibration, the following steps are necessary:

1. Examine the construction of the instrument, and identify and list all the possible inputs.
2. Decide, as best you can, which of the inputs will be significant in the application for which the instrument is to be calibrated.
3. Procure apparatus that will allow you to vary all significant inputs over the ranges considered necessary.
4. By holding some inputs constant, varying others, and recording the output(s), develop the desired static input-output relations.

Now we are ready for a more detailed discussion of specific static characteristics. These characteristics may be classified as either general or special. General static characteristics are of interest in *every* instrument. Special static characteristics are of interest in only a particular instrument. We concentrate mainly on general characteristics, leaving the treatment of special characteristics to later sections of the text in which specific instruments are discussed.

Accuracy, Precision, and Bias

When we measure some physical quantity with an instrument and obtain a numerical value, usually we are concerned with how close this value may be to the "true" value. It is first necessary to understand that this so-called true value is, in general, unknown and unknowable, since perfectly exact definitions of the physical quantities to be measured are impossible. This can be illustrated by specific example, for instance, the length of a cylindrical rod. When we ask ourselves what we *really* mean by the length of this rod, we must consider such questions as these:

1. Are the two ends of the rod planes?
2. If they are planes, are they parallel?
3. If they are not planes, what sort of surfaces are they?
4. What about surface roughness?

We see that complex problems are introduced when we deal with a real object rather than an abstract, geometric solid. The term "true value," then, refers to a value that would be obtained if the quantity under consideration were measured by an *exemplar method*,[1] that is, a method agreed on by experts as being sufficiently accurate for the purposes to which the data ultimately will be put.

We must also be concerned about whether we are describing the characteristics of a single reading of an instrument or of a measurement process. If we speak of a single measurement, the *error* is the difference between the measurement and the corresponding true value, which is taken to be positive if the measurement is greater than the true value. When using an instrument, however, we are concerned with the characteristics of the measurement process associated with that instrument. That is, we may take a single reading, but this is a sample from a statistical population generated by the measurement process. If we know the characteristics of the process, we can *put bounds on* the error of the single measurement, although we cannot tell what the error itself is, since this would imply that we knew the true value. Thus we are interested in being able to make statements about the accuracy (lack of error) of our readings. This can be done in terms of the concepts of *precision* and *bias* of the measurement process.

The measurement process consists of actually carrying out, as well as possible, the instructions for performing the measurement, which are the measurement method. (Since calibration is essentially a refined form of measurement, these remarks apply equally to the process of calibration.) If this process is repeated over and over under *assumed* identical conditions, we get a large number of readings from the instrument. Usually these readings will not all be the same, and so we note immediately that we may *try* to ensure identical conditions for each trial, but it is never exactly possible. The data generated in this fashion may be used to describe the measurement process so that, if it is used in the future, we may be able to attach some numerical estimates of error to its outputs.

If the output data are to give a meaningful description of the measurement process, the data must form what is called a *random sequence*. Another way of

---

[1] Churchill Eisenhart, Realistic Evaluation of the Precision and Accuracy of Instrument Calibration Systems, *J. Res. Natl. Bur. Std., C*, vol. 67C, no. 2, April–June 1963.

saying this is that the process must be in a state of *statistical control*.[1] The concept of the state of statistical control is not a particularly simple one, but we try to explain its essence briefly. First we note that it is meaningless to speak of the accuracy of an instrument as an isolated device. We must always consider the instrument plus its environment and method of use, that is, the instrument plus its inputs. This aggregate constitutes the measurement process. Every instrument has an infinite number of inputs; that is, the causes that can conceivably affect the output, if only very slightly, are limitless. Such effects as atmospheric pressure, temperature, and humidity are among the more obvious. But if we are willing to "split hairs," we can uncover a multitude of other physical causes that could affect the instrument with varying degrees of severity. In defining a calibration procedure for a specific instrument, we specify that certain inputs must be held "constant" within certain limits. These inputs, it is hoped, are the ones that contribute the largest components to the overall error of the instrument. The remaining infinite number of inputs is left uncontrolled, and it is hoped that each of these individually contributes only a very small effect and that in the aggregate their effect on the instrument output will be of a random nature. If this is indeed the case, the process is said to be in statistical control. Experimental proof that a process is in statistical control is not easy to come by; in fact, *strict* statistical control is unlikely of practical achievement. Thus we can only approximate this situation.

Lack of control is sometimes obvious, however, if we repeat a measurement and plot the result (output) versus the trial number. Figure 3.1a shows such a graph for the calibration of a particular instrument. In this instance, it was ascertained after some study that the instrument actually was much more sensitive to temperature than had been thought. The original calibration was carried out in a room without temperature control. Thus the room temperature varied from a low in the morning to a peak in the early afternoon and then dropped again in the late afternoon. Since the 10 trials covered a period of about one day, the trend of the curve is understandable. By performing the calibration in a temperature-controlled room, the graph of Fig. 3.1b was obtained. For the detection of more subtle deviations from statistical control, the methods of statistical quality-control charts are useful.[2]

If the measurement process is in reasonably good statistical control and if we repeat a given measurement (or calibration point) over and over, we will generate a set of data exhibiting random scatter. As an example, consider the pressure gage of Fig. 3.2. Suppose we wish to determine the relationship between the desired input (pressure) and the output (scale reading). Other inputs which could be significant and which might have to be controlled during the pressure calibration include temperature, acceleration, and vibration. Temperature can cause expansion and contraction of instrument parts in such a way that the scale

---

[1] Ibid.

[2] E. B. Wilson, Jr., "An Introduction to Scientific Research," chap. 9, McGraw-Hill, New York, 1952.

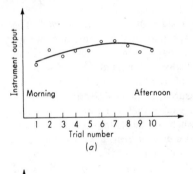

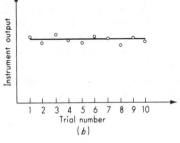

Figure 3.1 Effect of uncontrolled input on calibration.

reading will change even though the pressure has remained constant. An instrument acceleration along the axis of the piston rod will cause a scale reading even though pressure again has remained unchanged. This input is significant if the pressure gage is to be used aboard a vehicle of some kind. A small amount of vibration actually may be helpful to the operation of an instrument, since vibration may reduce the effects of static friction. Thus if the pressure gage is to be attached to a reciprocating air compressor (which always has some vibration), it may be more accurate under these conditions than it would be under calibration conditions where no vibration was provided. These examples illustrate the general importance of carefully considering the relationship between the calibration conditions and the actual application conditions.

Suppose, now, that we have procured a sufficiently accurate pressure standard and have arranged to maintain the other inputs reasonably close to the actual application conditions. Repeated calibration at a given pressure (say, 10 kPa) might give the data of Fig. 3.3. Suppose we now order the readings from the lowest (9.81) to the highest (10.42) and see how many readings fall in each interval of, say, 0.05 kPa, starting at 9.80. The result can be represented graphically as in Fig. 3.4a. Suppose we now define the quantity Z by

$$Z \triangleq \frac{\text{(number of readings in an interval)/(total number of readings)}}{\text{width of interval}} \quad (3.1)$$

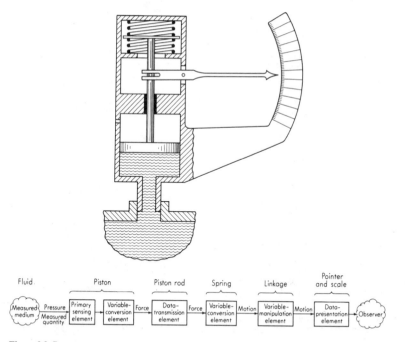

Figure 3.2 Pressure gage.

and we plot a "bar graph" with height Z for each interval. Such a "histogram" is shown in Fig. 3.4b. It should be clear from Eq. (3.1) that the area of a particular "bar" is numerically equal to the probability that a specific reading will fall in the associated interval. The area of the entire histogram must then be 1.0 (100 percent = 1.0), since there is 100 percent probability that the reading will fall somewhere between the lowest and highest values, at least based on the data available. If it were now possible to take an infinite number of readings, each with an infinite number of significant digits, we could make the chosen intervals as small as we pleased and still have each interval contain a finite number of readings. Thus the steps in the graph of Fig. 3.4b would become smaller and smaller, with the graph approaching a smooth curve in the limit. If we take this limiting abstract case as a mathematical model for the real physical situation, the function $Z = f(x)$ is called the *probability density function* for the mathematical model of the real physical process (see Fig. 3.5a). From the basic definition of Z, clearly

$$\text{Probability of reading lying between } a \text{ and } b \triangleq P(a < x < b) = \int_a^b f(x)\, dx$$

$$(3.2)$$

True pressure = $10.000 \pm .001$ kPa
Acceleration = 0
Vibration level = 0
Ambient temperature = $20 \pm 1°C$

Trial number	Scale reading, kPa
1	10.02
2	10.20
3	10.26
4	10.20
5	10.22
6	10.13
7	9.97
8	10.12
9	10.09
10	9.90
11	10.05
12	10.17
13	10.42
14	10.21
15	10.23
16	10.11
17	9.98
18	10.10
19	10.04
20	9.81

Figure 3.3 Pressure-gage calibration data.

The probability information sometimes is given in terms of the *cumulative distribution function* $F(x)$, which is defined by

$$F(x) \triangleq \text{probability that reading is less than any chosen value of } x$$

$$F(x) = \int_{-\infty}^{x} f(x) \, dx \qquad (3.3)$$

and is shown in Fig. 3.5b.

From the infinite number of forms possible for probability density functions, a relatively small number are useful mathematical models for practical applications; in fact, *one* particular form is quite dominant. The most useful density function or distribution is the normal or *Gaussian* function, which is given by

$$f(x) = \frac{1}{\sqrt{2\pi}\,\sigma} e^{-(x-\mu)^2/(2\sigma^2)} \qquad -\infty < x < +\infty \qquad (3.4)$$

Equation (3.4) defines a whole family of curves depending on the particular numerical values of μ (the mean value) and σ (the standard deviation). The shape of the curve is determined entirely by σ, with μ serving only to locate the position of the curve along the x axis. The cumulative distribution function $F(x)$ cannot be

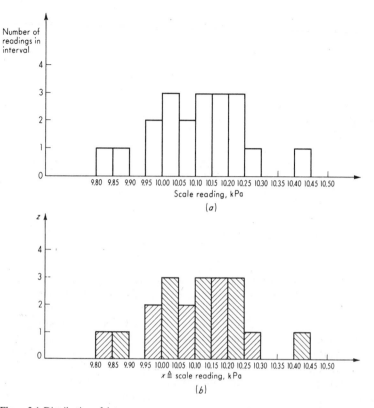

Figure 3.4 Distribution of data.

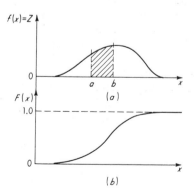

Figure 3.5 Probability distribution function.

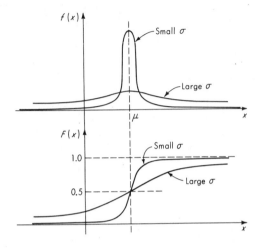

Figure 3.6 Gaussian distribution.

written explicitly in this case because the integral of Eq. (3.3) cannot be carried out; however, the function has been tabulated by performing the integration by numerical means. Figure 3.6 shows that a small value of σ indicates a high probability that a "reading" will be found close to μ. Equation (3.4) also shows that there is a small probability that very large $(\rightarrow \pm \infty)$ readings will occur. This is one of the reasons why a true Gaussian distribution can never occur in the real world; physical variables are always limited to finite values. There is *zero* probability, for example, that the pointer on a pressure gage will read 100 kPa when the range of the gage is only 20 kPa. Real distributions must thus, in general, have their "tails" cut off, as in Fig. 3.7.

Although actual data may not conform *exactly* to the Gaussian distribution, very often they are sufficiently close to allow use of the Gaussian model in engineering work. It would be desirable to have available tests that would indicate whether the data were "reasonably" close to Gaussian, and two such procedures are explained briefly. We must admit however, that in much practical work the time and effort necessary for such tests cannot be justified, and the Gaussian model is simply *assumed* until troubles arise which justify a closer study of the particular situation.

The first method of testing for an approximate Gaussian distribution involves the use of probability graph paper. If we take the cumulative distribution

Figure 3.7 Non-Gaussian distribution.

function for a Gaussian distribution and suitably distort the vertical scale of the graph, the curve can be made to plot as a straight line, as shown in Fig. 3.8. (This, of course, can be done with any curvilinear relation, not just probability curves.) Such graph paper is commercially available and may be used to give a rough, qualitative test for conformity to the Gaussian distribution. For example, consider the data of Fig. 3.3. These data may be plotted on Gaussian probability graph paper as follows: First lay out on the uniformly graduated horizontal axis a numerical scale that includes all the pressure readings. Now the probability graph paper represents the cumulative distribution, so that the ordinate of any point represents the probability that a reading will be less than the abscissa of that particular point. This probability, in terms of the sample of data available, is simply the percentage (in decimal form) of points that fell at or below that particular value. Figure 3.9a shows the resulting plot. Note that the highest point (10.42) cannot be plotted since 100 percent cannot appear on the ordinate scale. Also shown in Fig. 3.9a is the "perfect Gaussian line," the straight line that would be perfectly followed by data from an infinitely large sample of Gaussian data which had the same μ and σ values as our actual data sample. To plot this line, we must estimate μ from the *sample mean value* $\bar{X}$, using

$$\bar{X} \triangleq \frac{\sum_{i=1}^{N} X_i}{N} \tag{3.5}$$

where
$$X_i \triangleq \text{individual reading} \tag{3.6}$$

$$N \triangleq \text{total reading of readings} \tag{3.7}$$

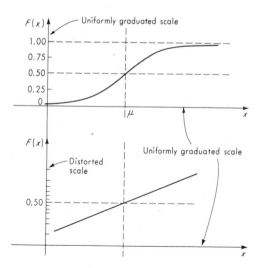

Figure 3.8 Rectification of Gaussian curve.

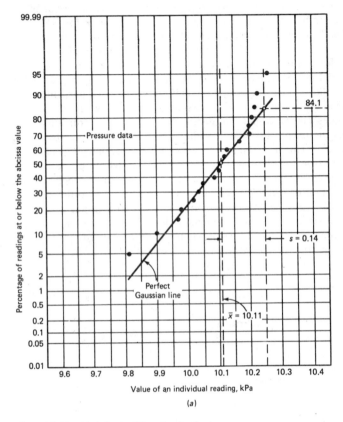

Figure 3.9 Graphical check of Gaussian distribution.

and σ from the *sample standard deviation s*, using

$$s \triangleq \sqrt{\frac{\sum_{i=1}^{N} (X_i - \bar{X})^2}{N - 1}} \tag{3.8}$$

The data of Fig. 3.3 give $\bar{X} = 10.11$ and $s = 0.14$ kPa. Two points that may be used to plot any perfect Gaussian line are $(\bar{X}, 50\%)$ and $(\bar{X} + s, 84.1\%)$, which yield the line of Fig. 3.9a for our data. Superimposing this line on our actual data, we may judge visually and qualitatively whether our data are "close to" or "far from" Gaussian. Note that there is *no hope* of ever *proving* real-world data to be Gaussian. There are always physical constraints which require real data to be at least somewhat non-Gaussian.

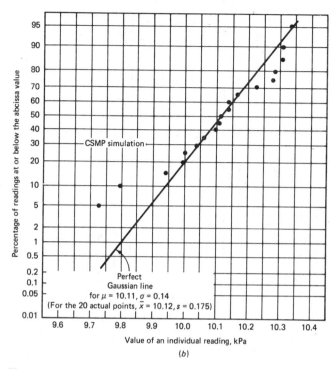

Figure 3.9 (*Continued*)

Since our graphical test is clearly subjective, it may be useful to provide some additional data to help reach the required decision. Figure 3.9*b* and *c* shows "data" generated by a digital-computer Gaussian random-number generator which is part of the IBM digital simulation program called CSMP (CSMP is discussed in greater detail later in this chapter). Such random-number algorithms are excellent simulations of perfect Gaussian distributions. Figure 3.9*b* graphs data generated by CSMP when the program was asked to produce a sample of 20 readings with $\mu = 10.11$ and $\sigma = 0.14$. Note that even though the sample was drawn from a perfect Gaussian distribution, the points do not fall on the perfect Gaussian line. This does not mean that the computer algorithm is faulty. It simply shows that a sample of size 20 is too small to give a statistically reliable prediction. In Fig. 3.9*c*, the sample size was increased to 100, which gives a clear improvement but still not perfection. The results of Fig. 3.9*a* would be considered by most people to indicate a reasonable approximation to a Gaussian distribution. However, this test is clearly only qualitative, and its main usefulness perhaps is in revealing *gross* departures from the theoretical distribution. Such

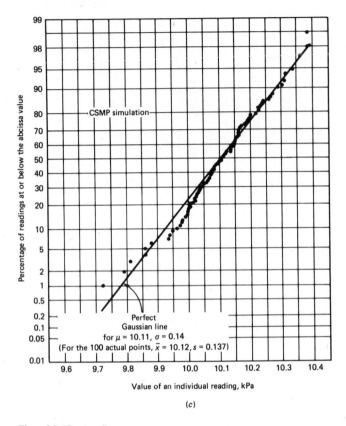

Figure 3.9 (*Continued*)

deviations would lead us to examine the instrument and measurement process more closely before attempting to make any statistical statements about accuracy. For a perfect Gaussian distribution, it can be shown that

$$68\% \text{ of the readings lie within } \pm 1\sigma \text{ of } \mu$$

$$95\% \text{ of the readings lie within } \pm 2\sigma \text{ of } \mu \qquad (3.9)$$

$$99.7\% \text{ of the readings lie within } \pm 3\sigma \text{ of } \mu$$

Thus if we assume that our real distribution is nearly Gaussian, we might predict, for instance, that if more readings were taken, 99.7 percent would fall within ± 0.42 kPa of 10.11. The estimates $\bar{X}$ and s of μ and σ are themselves

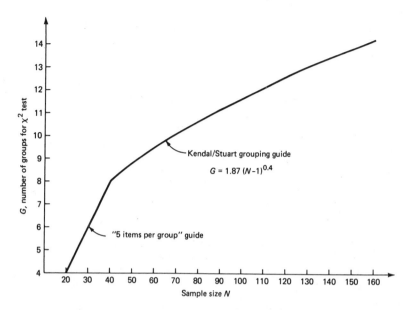

Figure 3.10 Grouping guide for chi-square test.

random variables and can be improved by taking more readings. For example, the standard deviation $s_{\bar{X}}$ of $\bar{X}$ may be found from[1]

$$s_{\bar{X}} = \frac{s}{\sqrt{N-1}} \qquad (3.10)$$

which clearly shows a reduction in uncertainty for $\bar{X}$ as sample size N increases.

A somewhat more quantitative method for deciding whether data are close to Gaussian involves the chi-square (χ^2) goodness-of-fit test. Here we first order the data from lowest to highest and gather them into groups. Statistics texts usually are not very specific about how this grouping is done. We offer Fig. 3.10 as a guide; also sample sizes of 20 or more are recommended. For $20 \leq N \leq 40$, group the data to get at least 5 points per group; for $N > 40$, try for equal-size groups, using the number[2] of groups given by Fig. 3.10. For the test to be carried out at all, there must be at least four groups, and so we see that a sample of at least about 20 readings should be taken. The larger the sample, the more signifi-

[1] A. M. Mood, "Introduction to the Theory of Statistics," pp. 133, 159, McGraw-Hill, New York, 1950.

[2] M. C. Kendal and A. Stuart, "*The Advanced Theory of Statistics*," vol. 2, Griffin, London, 1961.

Group number	Range of x	Range of w	n_0	n_e	$\dfrac{(n_0 - n_e)^3}{n_e}$
1	$-\infty$ to 10.03	$-\infty$ to -0.572	5	5.66	0.077
2	10.03 to 10.115	-0.572 to 0.0357	5	4.62	0.031
3	10.115 to 10.215	0.0357 to 0.75	6	5.18	0.130
4	10.215 to ∞	0.75 to ∞	4	4.532	0.062
					$\chi^2 = 0.300$

Figure 3.11 Tabulation for chi-square test.

cant the test will be. The quantity χ^2 is defined as follows:

$$\chi^2 \triangleq \sum_{i=1}^{n} \frac{(n_0 - n_e)^2}{n_e} \tag{3.11}$$

where $n_0 \triangleq$ number of readings actually observed in given range (group)

$n_e \triangleq$ number of readings that would be observed in same range if distribution were Gaussian with $\mu = \bar{x}$ and $\sigma = s$

$n \triangleq$ number of groups

It is necessary to explain how the number n_e is calculated. We mentioned earlier that tables of the cumulative Gaussian distribution are available.[1] These tables allow us to calculate the number n_e as follows: Entries in the table are values of $F(w)$, where

$$F(w) \triangleq \text{probability that reading falls in range from } -\infty \text{ to } w \tag{3.12}$$

$$w \triangleq \frac{x - \mu}{\sigma} \tag{3.13}$$

Definition (3.13) puts the tables on a nondimensional basis, so that one table serves for all possible values of μ and σ. As an example, suppose we wish to calculate n_e for the first range, $-\infty$ to 10.03. This range of x corresponds to a range on w of $-\infty$ to -0.572, since we use Eq. (3.13) with $\mu = \bar{x} = 10.11$ and $\sigma = s = 0.14$ to calculate w. Now, the probability that a reading falls in the range $-\infty$ to -0.572 is the same as the probability that it falls in the range $+0.572$ to $+\infty$, since the Gaussian curve is perfectly symmetric about $w = 0$. Looking at the table for $w = 0.572$, we find $P(-\infty < w < 0.572)$ is 0.717. Thus, $P(0.572 < w < \infty) = 1.0000 - 0.717 = 0.283$, and in a sample of 20 trials we would expect that $(20)(0.283) = 5.66$ readings would fall in the range $-\infty < x < 10.03$. Actually, we found that exactly five readings fell in this range. All other entries in Fig. 3.11 are found in the same manner.

---

[1] R. S. Burlington, "Handbook of Mathematical Tables and Formulas," 2d ed., p. 257, McGraw-Hill, New York, 1940.

To make the final interpretation of this test, we must know the number of "degrees of freedom." This is numerically equal to the number of groups minus 3, and so in the present example we have 1 degree of freedom. The significance of the numerical value of χ^2 is given in Fig. 3.12, which is interpreted as follows: If we had a perfectly Gaussian distribution with $\mu = 10.11$ and $\sigma = 0.14$ from which we drew a sample of 20 readings, we would *not*, in general, get a χ^2 of zero. That is, because of the random nature of the readings, any finite sample will not exhibit the same properties as its "parent" population. And the smaller the sample, the more likely it is that, just by chance, there is taken a sample that *appears* to be non-Gaussian. Thus, even though we might be drawing a sample from a *perfect*

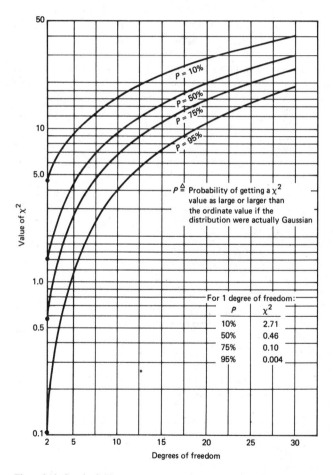

Figure 3.12 Graph of chi-square test.

Gaussian distribution, for, say, 10 degrees of freedom (data gathered into 13 groups), Fig. 3.12 shows that 95 percent of the time we would calculate a χ^2 value as large as or larger than 3.94. If our *actual* data gave a χ^2 value of, say, 2.71, we would have strong evidence that the distribution was close to Gaussian. If, however, we calculated a χ^2 of, say, 20.2, there is less than a 10 percent chance that we would have a distribution close to Gaussian. The data of Fig. 3.11 give $\chi^2 = 0.30$ for 1 degree of freedom, so the probability of a close-to-Gaussian distribution is between 50 and 75 percent. This is not very conclusive, however, since the sample size is so small (the χ^2 test not very sensitive); and since Fig. 3.9a looked somewhat encouraging, we might be willing to proceed under the Gaussian assumption.

Having considered the problem of determining the normality of scattered data, we return to the main business of this section, that is, definition of the terms "accuracy," "precision," and "bias." Up to now, we have been examining the situation in which a single true value is applied repeatedly and the resulting measured values are recorded and analyzed. In an actual instrument calibration, the true value is varied, in increments, over some range, causing the measured value also to vary over a range. Very often there is no multiple repetition of a given true value. The procedure is merely to cover the desired range in both the increasing and the decreasing directions. Thus a given true value is applied, at

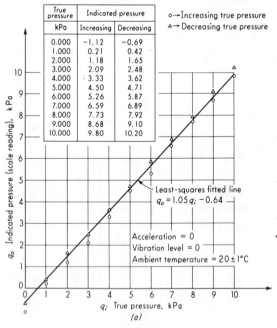

True pressure	Indicated pressure	
kPa	Increasing	Decreasing
0.000	−1.12	−0.69
1.000	0.21	0.42
2.000	1.18	1.65
3.000	2.09	2.48
4.000	3.33	3.62
5.000	4.50	4.71
6.000	5.26	5.87
7.000	6.59	6.89
8.000	7.73	7.92
9.000	8.68	9.10
10.000	9.80	10.20

o → Increasing true pressure
Δ → Decreasing true pressure

Least-squares fitted line
$q_o = 1.05 q_i - 0.64$

Acceleration = 0
Vibration level = 0
Ambient temperature = $20 \pm 1°C$

q_o Indicated pressure (scale reading), kPa

q_i True pressure, kPa

(a)

Figure 3.13 Pressure-gage calibration.

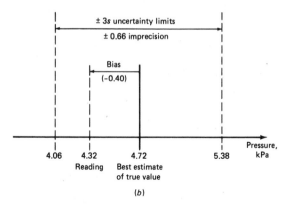

Figure 3.13 (*Continued*)

most, twice if we choose to use the same set of true values for both increasing and decreasing readings.

As an example, suppose we wish to calibrate the pressure gage of Fig. 3.2 for the relation between the desired input (pressure) and the output (scale reading). Figure 3.13a gives the data for such a calibration over the range 0 to 10 kPa. In this instrument (as in most but not all), the input-output relation is ideally a straight line. The *average calibration curve* for such an instrument generally is taken as a straight line which fits the scattered data points best as defined by some chosen criterion. The most common is the least-squares criterion, which minimizes the sum of the squares of the vertical deviations of the data points from the fitted line. (The least-squares procedure also can be used to fit curves other than straight lines to scattered data.) The equation for the straight line is taken as

$$q_o = mq_i + b \qquad (3.14)$$

where
$$q_o \triangleq \text{output quantity (dependent variable)} \qquad (3.15)$$
$$q_i \triangleq \text{input quantity (independent variable)} \qquad (3.16)$$
$$m \triangleq \text{slope of line} \qquad (3.17)$$
$$b \triangleq \text{intercept of line on vertical axis} \qquad (3.18)$$

The equations for calculating m and b may be found in several references[1]:

$$m = \frac{N\Sigma q_i q_o - (\Sigma q_i)(\Sigma q_o)}{N\Sigma q_i^2 - (\Sigma q_i)^2} \qquad (3.19)$$

$$b = \frac{(\Sigma q_o)(\Sigma q_i^2) - (\Sigma q_i q_o)(\Sigma q_i)}{N\Sigma q_i^2 - (\Sigma q_i)^2} \qquad (3.20)$$

[1] H. D. Young, "Statistical Treatment of Experimental Data," p. 121, McGraw-Hill, New York, 1962.

where $\qquad\qquad N \triangleq$ total number of data points $\qquad\qquad$ (3.21)

In this example, calculation gives $m = 1.05$ and $b = -0.64$ kPa. Since these values are derived from scattered data, it would be useful to have some idea of their possible variation. The standard deviations of m and b may be found from

$$s_m^2 = \frac{N s_{q_o}^2}{N\Sigma q_i^2 - (\Sigma q_i)^2} \qquad\qquad (3.22)$$

$$s_b^2 = \frac{s_{q_o}^2 \Sigma q_i^2}{N\Sigma q_i^2 - (\Sigma q_i)^2} \qquad\qquad (3.23)$$

where $\qquad\qquad s_{q_o}^2 = \frac{1}{N} \Sigma (m q_i + b - q_o)^2 \qquad\qquad$ (3.24)

The symbol s_{q_o} represents the standard deviation of q_o. That is, if q_i were fixed and then repeated over and over, q_o would give scattered values, with the amount of scatter being indicated by s_{q_o}. If we assume that this s_{q_o} would be the same for *any* value of q_i, we can calculate s_{q_o}, using *all* the data points of Fig. 3.13*a* and *without* having to repeat any one q_i many times. For this example, calculation gives $s_{q_o} = 0.23$ kPa. Then $s_m = 0.0154$ and $s_b = 0.091$ kPa. Assuming a Gaussian distribution and the 99.7 percent limits ($\pm 3s$), we could give m as 1.05 ± 0.05 and b as -0.64 ± 0.27 kPa.

In *using* the calibration results, the situation is such that q_o (the indicated pressure) is known and we wish to make a statement about q_i (the true pressure). The least-squares line gives

$$q_i = \frac{q_o + 0.64}{1.05} \qquad\qquad (3.25)$$

However, the q_i value computed in this way must have some plus-or-minus error limits put on it. These can be obtained since s_{q_i} can be computed from

$$s_{q_i}^2 = \frac{1}{N} \Sigma \left(\frac{q_o - b}{m} - q_i\right)^2 = \frac{s_{q_o}^2}{m^2} \qquad\qquad (3.26)$$

which in this example gives $s_{q_i} = 0.22$ kPa. Thus if we were using this gage to measure an unknown pressure and got a reading of 4.32 kPa, our estimate of the true pressure would be 4.72 ± 0.66 kPa if we wished to use the $\pm 3s$ limits.

Another common method of giving bounds on the error employs the *probable error* e_p. This is defined by

$$e_p \triangleq 0.674s \qquad\qquad (3.27)$$

A range of $\pm e_p$ includes the true value 50 percent of the time. In this case, the above value would be quoted as 4.72 ± 0.15 kPa. It should be clear that when we use statements of this kind, it is extremely important to state whether probable errors or $\pm 3s$ limits are used.

We should note that in computing s_{q_o} either of two approaches could be used. We might use data such as in Fig. 3.13a and apply Eq. (3.24) or, alternatively, repeat a given q_i many times and compute s_{q_o} from Eq. (3.8). If s_{q_o} is actually the same for all values of q_i (as assumed above), these two methods should give the same answer for large samples. In computing s_{q_i}, however, the second method is not feasible because we cannot, in general, fix q_o in a calibration and then repeat that point over and over to get scattered values of q_i. This is because q_i is truly an independent variable (subject to choice), whereas q_o is dependent (not subject to choice). Thus, in computing s_{q_i}, an approach such as Eq. (3.26) is necessary.

A calibration such as that of Fig. 3.13a allows decomposition of the total error of a measurement process into two parts, the *bias* and the *imprecision* (Fig. 3.13b). That is, if we get a reading of 4.32 kPa, the true value is given as 4.72 ± 0.66 kPa (3s limits), the bias would be −0.40 kPa, and the imprecision ±0.66 kPa (3s limits). Of course, once the instrument has been calibrated, the bias can be removed, and the only remaining error is that due to imprecision. The bias is also called the *systematic error* (since it is the same for each reading and thus can be removed by calibration). The error due to imprecision is called the *random error*, or *nonrepeatability*, since it is, in general, different for every reading and we can only put bounds on it, but cannot remove it. Thus calibration is the process of removing bias and defining imprecision numerically. The *total inaccuracy* of the process is defined by the combination of bias and imprecision. If the bias is known, the total inaccuracy is entirely due to imprecision and can be specified by a single number such as s_{q_i}.

A more refined method[1] of specifying uncertainty, which recognizes that s_{q_i} values based on small samples ($N < 30$) are less reliable than those based on large samples, is available. By using the statistical t distribution, computed s_{q_i} values are adjusted to reflect the effect of sample size. This reference gives a very comprehensive treatment of measurement uncertainty and is recommended for those wishing further details.

In actual engineering practice, the accuracy of an instrument usually is given by a single numerical value; very often it is not made clear just what the precise meaning of this number is meant to be. Often, even though a calibration, as in Fig. 3.13, has been carried out, s_{q_i} is not calculated. The error is taken as the largest horizontal deviation of any data point from the fitted line. In Fig. 3.13 this occurs at $q_i = 0$ and amounts to 0.48 kPa. The inaccuracy in this case thus might be quoted as ±4.8 percent of full scale. Note that this corresponds to about ±$2s_{q_i}$ in this case. This practice is no doubt due to the practical viewpoint that when a measurement is taken, all we really want is to say that it cannot be incorrect by more than some specific value; thus the "easy way out" is simply to give a single number. This would be legitimate if the bias were known to be zero (removed by calibration) and if the plus-or-minus limit given were specified as ±s, ±2s, ±3s, or ±e_p, since all these terms recognize the random nature of the

[1] R. B. Abernethy, "Measurement Uncertainty Handbook," Instrument Society of America, Research Triangle Park, N.C., 1980.

error. However, if the bias is unknown (and not zero), the quotation of a single number for the total inaccuracy is somewhat unsatisfactory, although it may be a necessary expedient.

One reason for this is that if we are trying to estimate the overall inaccuracy of a measurement system made up of a number of components, each of which has a known inaccuracy, the method of combining the individual inaccuracies is different for systematic errors (biases) than for random errors (imprecisions). Thus, if the number given for the total inaccuracy of a given component contains both bias and imprecision in unknown proportions, the calculation of overall system inaccuracy is confused. However, in many cases there is no alternative, and by calculation from theory, past experience, and/or judgment the experimenter must arrive at the best available estimate of the total inaccuracy, or *uncertainty* (as it is sometimes called), to be attached to the reading. In such cases, a useful viewpoint is that we are willing to bet with certain odds (say 19 to 1) that the error falls within the given limits. Then such limits may be combined as if they were imprecisions in calculations of overall system error.[1]

Irrespective of the precise *meaning* to be attached to accuracy figures provided, say, by instrument manufacturers, the *form* of such specifications is fairly uniform. More often than not, accuracy is quoted as a percentage figure based on the full-scale reading of the instrument. Thus if a pressure gage has a range from 0 to 10 kPa and a quoted inaccuracy of ±1.0 percent of full scale, this is to be interpreted as meaning that no error greater than ±0.1 kPa can be expected for any reading that might be taken on this gage, provided it is "properly" used. The manufacturer may or may not be explicit about the conditions required for "proper use." Note that for an actual reading of 1 kPa, a 0.1-kPa error is 10 percent *of the reading*.

Another method sometimes utilized gives the error as a percentage of the particular reading with a qualifying statement to apply to the low end of the scale. For example, a spring scale might be described as having an inaccuracy of ±0.5 percent of reading or ±0.1 N, whichever is greater. Thus for readings less than 20 N, the error is constant at ±0.1 N, while for larger readings the error is proportional to the reading.

Combination of Component Errors in Overall System-Accuracy Calculations

A measurement system is often made up of a chain of components, each of which is subject to individual inaccuracy. If the individual inaccuracies are known, how is the overall inaccuracy computed? A similar problem occurs in experiments that use the results (measurements) from several different instruments to compute some quantity. If the inaccuracy of each instrument is known, how is the inaccur-

---

[1] S. J. Kline and F. A. McClintock, Describing Uncertainties in Single Sample Experiments, *Mech. Eng.*, vol. 75, p. 3, January 1953; L. W. Thrasher and R. C. Binder, A Practical Application of Uncertainty Calculations to Measured Data, *Trans. ASME*, p. 373, February 1957.

acy of the computed result estimated? Or, inversely, if there must be a certain accuracy in a computed result, what errors are allowable in the individual instruments?

To answer the above questions, consider the problem of computing a quantity N, where N is a known function of the n *independent* variables $u_1, u_2, u_3, \ldots, u_n$. That is,

$$N = f(u_1, u_2, u_3, \ldots, u_n) \tag{3.28}$$

The u's are the measured quantities (instrument or component outputs) and are in error by $\pm \Delta u_1, \pm \Delta u_2, \pm \Delta u_3, \ldots, \pm \Delta u_n$, respectively. These errors will cause an error ΔN in the computed result N. The Δu's may be considered as absolute limits on the errors, as statistical bounds such as e_p's or $3s$ limits, or as uncertainties on which we are willing to give certain odds as including the actual error. However, the method of computing ΔN and the interpretation of its meaning are different for the first case as compared with the second and third. If the Δu's are considered as absolute limits on the individual errors and we wish to calculate similar absolute limits on the error in N, we could calculate

$$N \pm \Delta N = f(u_1 \pm \Delta u_1, u_2 \pm \Delta u_2, u_3 \pm \Delta u_3, \ldots, u_n \pm \Delta u_n) \tag{3.29}$$

By subtracting N in Eq. (3.28) from $N \pm \Delta N$ in Eq. (3.29) we finally obtain $\pm \Delta N$. This procedure is needlessly time-consuming, however, and an approximate solution valid for engineering purposes may be obtained by application of the Taylor series. Expanding the function f in a Taylor series, we get

$$f(u_1 \pm \Delta u_1, u_2 \pm \Delta u_2, \ldots, u_n \pm \Delta u_n) = f(u_1, u_2, \ldots, u_n)$$

$$+ \Delta u_1 \frac{\partial f}{\partial u_1} + \Delta u_2 \frac{\partial f}{\partial u_2} + \cdots + \Delta u_n \frac{\partial f}{\partial u_n}$$

$$+ \frac{1}{2} \left[(\Delta u_1)^2 \frac{\partial^2 f}{\partial u_1^2} + \cdots \right] + \cdots \tag{3.30}$$

where all the partial derivatives are to be evaluated at the known values of u_1, $u_2, \ldots, u_n$. That is, if the measurements have been made, the u's are all known as numbers and may be substituted into the expressions for the partial derivatives to give other numbers. In actual practice, the Δu's will all be small quantities, and thus terms such as $(\Delta u)^2$ will be negligible. Then Eq. (3.30) may be given approximately as

$$f(u_1 + \Delta u_1, u_2 + \Delta u_2, \ldots, u_n + \Delta u_n) = f(u_1, u_2, \ldots, u_n)$$

$$+ \Delta u_1 \frac{\partial f}{\partial u_1} + \Delta u_2 \frac{\partial f}{\partial u_2} + \cdots + \Delta u_n \frac{\partial f}{\partial u_n} \tag{3.31}$$

So the absolute error E_a is given by

$$E_a = \Delta N = \left| \Delta u_1 \frac{\partial f}{\partial u_1} \right| + \left| \Delta u_2 \frac{\partial f}{\partial u_2} \right| + \cdots + \left| \Delta u_n \frac{\partial f}{\partial u_n} \right| \tag{3.32}$$

The absolute-value signs are used because some of the partial derivatives might be negative, and for a positive Δu such a term would *reduce* the total error. Since an error Δu is, in general, just as likely to be positive as negative, to estimate the maximum possible error, the absolute-value signs must be used as in Eq. (3.32). The form of Eq. (3.32) is very useful since it shows which variables (u's) exert the strongest influence on the accuracy of the overall result. That is, if, say, $\partial f / \partial u_3$ is a large number compared with the other partial derivatives, then a small Δu_3 can have a large effect on the total E_a. If the relative or percentage error E_r is desired, clearly it is given by

$$E_r = \frac{\Delta N}{N} \times 100 = \frac{100 E_a}{N} \tag{3.33}$$

So the computed result may be expressed as either $N \pm E_a$ or $N \pm E_r$ percent, and the interpretation is that we are *certain* this error will not be exceeded since this is how the Δu's were defined.

In carrying out the above computations, questions of significant figures and rounding will occur. While hand calculators allow us to easily carry many digits (without the need to *think* about how many are really meaningful), even here, rounding may not be entirely foolish. The tradeoff involved is between the time it takes to properly round and the time it takes (plus the greater probability of misentering a digit) to enter a long string of digits. Each individual will have to personally resolve this tradeoff. Be sure to note, however, that irrespective of what is done at intermediate steps, the *final result* must *always* be rounded to a number of digits consistent with the accuracy of the basic data. Here we briefly review these matters for those not familiar with such procedures. A significant figure is any one of the digits 1, 2, 3, 4, 5, 6, 7, 8, 9; zero is a significant figure except when it is used to fix the decimal point or to fill the places of unknown or discarded digits. Thus in the number 0.000532, the significant figures are 5, 3, and 2, while in the number 2,076 *all* the digits, including the zero, are significant. For a number such as 2,300, the zeros may or may not be significant. To convey which figures are significant, we write this as 2.3×10^3 if two significant figures are intended, 2.30×10^3 if three, 2.300×10^3 if four, and so forth.

In computations often we deal with quantities having unequal numbers of significant figures. For example, it may be necessary to perform 4.62×0.317856. The first number is assumed good to three significant figures, while the second is good to six. It can be shown that the product will be good to only three significant figures. Therefore, to save work, the six-figure number should be rounded before multiplication. A number of rules have been proposed for this rounding procedure, and we now state one that is widely used:

> To round a number to n significant figures, discard all digits to the right of the nth place. If the discarded number is less than one-half a unit in the nth place, leave the nth digit unchanged. If the discarded number is greater than one-half a unit in the nth place, increase the nth digit by 1. If the discarded number is exactly one-half a unit in the nth place, leave the nth digit unchanged if it is an even number and add 1 to it if it is odd.

To determine the extent to which numbers should be rounded, the following rules may be applied.

Addition

For addition, retain one more decimal digit in the more accurate numbers than is contained in the least accurate number. (The more accurate numbers are those with the greatest number of significant figures.) Then round the result to the same decimal place as the least accurate number.

Example

2.635		2.64
0.9		0.9
1.52	$\longrightarrow$	1.52
0.7345		0.73
		5.79 $\longrightarrow$ 5.8

Subtraction

For subtraction, round the more accurate number to the same number of decimal places as the less accurate one before subtracting. Give the result to the same number of decimal places as the less accurate figure.

Example

$$
\begin{array}{r} 7.6345 \\ -0.031 \end{array} \longrightarrow \begin{array}{r} 7.634 \\ -0.031 \\ \hline 7.603 \end{array} \longrightarrow 7.603
$$

Multiplication and Division

For multiplication and division, round the more accurate numbers to one more significant figure than the least accurate before computing. Round the result to the same number of significant figures as the least accurate number.

Example

$$
\frac{(1.2)(6.335)(0.0072)}{3.14159} \to \frac{(1.2)(6.34)(0.0072)}{3.14} \to 0.0174 \to 0.017
$$

Now we have a step-by-step procedure for computing the overall error:

1. Tabulate all data points, each with its plus-or-minus error attached. All errors should be expressed to two significant figures. (Actually, one significant figure often is adequate.) The reason is that the errors themselves are not generally known very accurately, and so it is foolish to carry many significant figures.
2. If the quantity to be computed is N, where $N = f(u_1, u_2, \ldots, u_n)$, compute the partial derivatives $\partial f/\partial u_1, \partial f/\partial u_2, \ldots, \partial f/\partial u_n$ and evaluate each to three significant figures by substituting in the basic data $u_1, u_2, \ldots, u_n$.
3. Using Eq. (3.32), compute E_a and round to two significant figures.
4. Compute N from Eq. (3.28) to one more decimal place than the rounded E_a of step 3. Thus if $E_a = \pm 0.062$, N should be computed as, say, 7.0516. Then this value is rounded to the same number of decimal places as E_a, in this case to 7.052. In computing N, treat u_1, u_2, etc., as exact numbers; that is, they each have an infinite number of significant figures. This viewpoint is necessary because we must be able to compute N to as many significant figures as required by E_a according to the above rule.
5. The result may be quoted as

$$7.052 \pm 0.062 \qquad \text{absolute terms}$$

or

$$7.052 \pm 0.88 \text{ percent} \qquad \text{relative terms}$$

When the problem is one in which a certain overall accuracy is known to be required and we wish to know what component accuracies are needed, the following method may be employed. It should be apparent that this problem is mathematically indeterminate since an infinite number of combinations of individual accuracies could result in the same overall accuracy. The means of resolving this difficulty are to be found in the *method of equal effects*. This principle merely assumes that each source of error contributes an equal amount to the total error. Mathematically, if

$$\Delta N = \left| \frac{\partial f}{\partial u_1} \Delta u_1 \right| + \left| \frac{\partial f}{\partial u_2} \Delta u_2 \right| + \cdots + \left| \frac{\partial f}{\partial u_n} \Delta u_n \right|$$

then, if each term is assumed to be equal, we may write

$$\left| \frac{\partial f}{\partial u_1} \Delta u_1 \right| = \left| \frac{\partial f}{\partial u_2} \Delta u_2 \right| = \cdots = \left| \frac{\partial f}{\partial u_n} \Delta u_n \right| = \frac{\Delta N}{n} \qquad (3.34)$$

Now the allowable overall error ΔN is known, and so are n and $u_1, u_2, \ldots, u_n$.

Thus

$$\frac{\partial f}{\partial u_i} \Delta u_i = \frac{\Delta N}{n}$$

$$\Delta u_i = \frac{\Delta N}{n(\partial f/\partial u_i)} \qquad i = 1, 2, 3, \dots, n \qquad (3.35)$$

and the allowable error Δu_i in each measurement may be calculated. (The partial derivatives are evaluated at the known values of $u_1, u_2, \dots, u_n$.) If a particular Δu_i turns out to be smaller than what can be achieved by the instruments available, it may be possible to relax this requirement if some *other* Δu_i can be made smaller than the value given by Eq. (3.35). That is, some instruments may give better accuracy than required by Eq. (3.35) while others may be unable to meet the requirements of Eq. (3.35). In such cases, it may still be possible to meet the overall accuracy requirement; this may be checked by the formulas given.

When the Δu's are considered not as absolute limits of error, but rather as statistical bounds such as $\pm 3s$ limits, probable errors, or uncertainties, the formulas for computing overall errors must be modified. It can be shown[1] that the proper method of combining such errors is according to the root-sum square (rss) formula

$$E_{a_{rss}} = \sqrt{\left(\Delta u_1 \frac{\partial f}{\partial u_1}\right)^2 + \left(\Delta u_2 \frac{\partial f}{\partial u_2}\right)^2 + \cdots + \left(\Delta u_n \frac{\partial f}{\partial u_n}\right)^2} \qquad (3.36)$$

The overall error $E_{a_{rss}}$ then has the same meaning as the individual errors. That is, if Δu_i represents a $\pm 3s$ limit on u_i, then $E_{a_{rss}}$ represents a $\pm 3s$ limit on N, and 99.7 percent of the values of N can be expected to fall within these limits. Equation (3.36) always gives a smaller value of error than does Eq. (3.32). Equation (3.35) also must be modified when this viewpoint is taken:

$$\Delta u_i = \frac{\Delta N}{\sqrt{n} \, (\partial f/\partial u_i)} \qquad (3.37)$$

As an example of the above procedures, consider an experiment for measuring, by means of a dynamometer, the average power transmitted by a rotating shaft. The formula for power can be written as

$$\text{Watts} = \frac{2\pi RFL}{t} \quad \text{or} \quad \text{hp} = \frac{2\pi RFL}{550t} \qquad (3.38)$$

---

[1] J. B. Scarborough, "Numerical Mathematical Analysis," 3d ed., p. 429, Johns Hopkins, Baltimore, Md., 1955.

where $R \triangleq$ revolutions of shaft during time t

   $F \triangleq$ force at end of torque arm, lbf (N)

   $L \triangleq$ length of torque arm, ft (m)   (3.39)

   $t \triangleq$ time length of run, s

A sketch of the experimental setup is shown in Fig. 3.14. The revolution counter is of the type shown in Fig. 2.4 and can be turned on and off with an electric switch. The instants of turning on and off are recorded by a stopwatch. If it is assumed the counter does not miss any counts, the maximum error in R is ± 1, because of the digital nature of the device (see Fig. 3.15).

There is a related error, however, in determining the time t since perfect synchronization of the starting and stopping of watch and counter is not possible. The stopwatch might be known to be a quite accurate time-measuring instrument, but this does not guarantee that it will always measure the time interval intended. In assigning an error to t, then, we are not helped much by the watch manufacturer's guarantee of 0.10 percent inaccuracy if our synchronization error is much larger than this. This synchronization error is certainly not precisely known since it involves human factors. An experiment to determine its statistical characteristics would be a more expensive and involved undertaking than the power measurement of which it is a part. So we are in the rather common position of having to rely on experience and judgment in arriving at an estimate of the proper numerical value, and we begin to appreciate that some of the statistical niceties and fine points of theory considered earlier may appear somewhat academic in such a situation. They are always useful in terms of the understanding of basic concepts that they develop; however, they cannot be relied on to give clear-cut answers in situations where the basic data are ill-defined. In the present case, suppose it is decided that a total starting and stopping error is taken

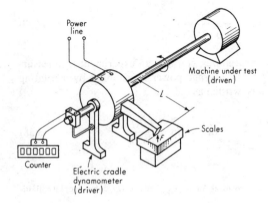

Figure 3.14 Dynamometer test setup.

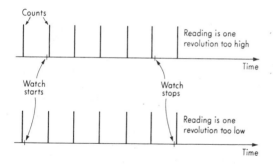

Figure 3.15 Revolution-counting error.

as ± 0.50 s. Whether this is to be considered as an absolute limit or as a $\pm 3s$ limit is somewhat meaningless when the basic number is arrived at in such an arbitrary fashion.

The scales used to measure the force F can be statistically calibrated with vagaries, depending on the care taken in this particular measurement. Suppose we use a fairly rough procedure and decide an on error of ± 0.05 in.

The scales used to measure the force F can be statistically calibrated with deadweights, which yields a set of data analogous to that of Fig. 3.13. Suppose this is done and an s_{q_i} of 0.0133 lbf is obtained. The $\pm 3s$ limits would then be ± 0.040 lbf. Again, however, the situation is not this simple. When actually used, the scales will be subject to vibration, which may reduce frictional effects and increase precision. At the same time, the pointer on the scale will not stand perfectly still when the dynamometer is running; thus in reading the scale we must perform a mental averaging process, which may introduce more error. Such effects are clearly difficult to quantify, and again we must make a decision based partly on experience and judgment. Suppose we assume the two mentioned effects cancel and thus take ± 0.040 lbf as the force measurement error.

For a specific run, if the data are

$$R = 1,202 \pm 1.0 \text{ r}$$

$$F = 10.12 \pm 0.040 \text{ lbf}$$

$$L = 15.63 \pm 0.050 \text{ in}$$

$$t = 60.0 \pm 0.50 \text{ s}$$

(3.40)

then the calculation proceeds as follows: In terms of inch units, we have

$$\text{hp} = \frac{2\pi}{(550)(12)} \frac{FLR}{t} = K \frac{FLR}{t}$$

(3.41)

Then, computing the various partial derivatives to three significant figures gives

$$\frac{\partial(hp)}{\partial F} = \frac{KLR}{t} = \frac{(0.000952)(15.63)(1,202)}{60} = 0.298 \text{ hp/lbf} \qquad (3.42)$$

$$\frac{\partial(hp)}{\partial R} = \frac{KFL}{t} = \frac{(0.000952)(10.12)(15.63)}{60} = 0.00251 \text{ hp/r} \qquad (3.43)$$

$$\frac{\partial(hp)}{\partial L} = \frac{KFR}{t} = \frac{(0.000952)(10.12)(1,202)}{60} = 0.193 \text{ hp/in} \qquad (3.44)$$

$$\frac{\partial(hp)}{\partial t} = -\frac{KFLR}{t^2} = -\frac{(0.000952)(10.12)(15.63)(1,202)}{3,600}$$

$$= -0.0500 \text{ hp/s} \qquad (3.45)$$

If we now choose to consider the component errors as absolute limits and wish to compute the absolute limits on the overall error, we use Eq. (3.32) and get

$$E_a = (0.298)(0.040) + (0.00251)(1.0) + (0.193)(0.050) + (0.05)(0.50) \qquad (3.46)$$

$$E_a = 0.0119 + 0.00251 + 0.00965 + 0.025 = 0.049 \text{ hp} \qquad (3.47)$$

We now compute a rough value of horsepower as

$$hp = \frac{(0.000952)(10.12)(15.63)(1,202)}{60} = 3.02 \qquad (3.48)$$

If we follow the rule saying that horsepower should be computed to one more decimal place than E_a, we must calculate hp to four decimal places, which means five significant figures in this case. Thus

$$hp = \frac{(2.0000)(3.1416)(10.120)(15.630)(1,202.0)}{(550.00)(12.000)(60.000)} = 3.0167 \qquad (3.49)$$

which we round off to 3.017. Then the result may be quoted as hp = 3.017 ± 0.049 hp or hp = 3.017 ± 1.6 percent. If the component errors were considered as having only one significant figure (which might well be the case here, where they are in considerable doubt), the above computations would be simplified since fewer significant figures would need to be carried. The final result thus might be given more realistically as hp = 3.02 ± 0.05.

If the individual errors are thought of as $\pm 3s$ limits, then Eq. (3.36) should be used to compute $\pm 3s$ limits on hp. Let us carry this out to see the numerical significance.

$$E_{a_{rss}} = \sqrt{0.0119^2 + 0.00251^2 + 0.00965^2 + 0.025^2} = 0.029 \text{ hp} \qquad (3.50)$$

We see that $E_{a_{rss}}$ is significantly smaller than E_a. We might say that the error is *possibly* as large as 0.049 but *probably* not larger than 0.029 hp. If the individual errors had been accurately known to be $\pm 3s$ limits, the word "probably" would have precise statistical meaning; otherwise, not.

Finally, suppose we wish to measure hp to 0.5 percent accuracy in the previous example. What accuracies are needed in the individual measurements? We can use either Eq. (3.35) (if we wish to be conservative) or Eq. (3.37) (if we wish to give ourselves every chance of showing the measurement to be possible). Using Eq. (3.37), we get

$$\Delta F = \frac{(3.02)(0.005)}{\sqrt{4}\,(0.298)} = 0.025 \text{ lbf}$$

$$\Delta R = \frac{(3.02)(0.005)}{\sqrt{4}\,(0.0025)} = 3.0 \text{ r}$$

$$\Delta L = \frac{(3.02)(0.005)}{\sqrt{4}\,(0.193)} = 0.039 \text{ in}$$

$$\Delta t = \frac{(3.02)(0.005)}{\sqrt{4}\,(0.05)} = 0.15 \text{ s}$$

(3.51)

[If we use Eq. (3.35), all these allowable errors are cut in half.] If it is found that the best instrument and technique available for measuring, say, F are good to only 0.04 lbf rather than the 0.025 lbf called for by Eq. (3.51), this does not necessarily mean that horsepower cannot be measured to 0.5 percent. However, it does mean that one or more of the other quantities R, L, and t *must* be measured *more* accurately than required by Eq. (3.51). Making one or more of these measurements more accurately may offset the excessive error in the F measurement. The given formulas allow calculation of whether this will be true.

Static Sensitivity

When an input-output calibration such as that of Fig. 3.13 has been performed, the *static sensitivity* of the instrument can be defined as the slope of the calibration curve. If the curve is not nominally a straight line, the sensitivity will vary with the input value, as shown in Fig. 3.16b. To get a meaningful definition of sensitivity, the output quantity must be taken as the actual physical output, not the meaning attached to the scale numbers. That is, in Fig. 3.13 the output quantity is plotted as kilopascals; however, the actual physical output is an angular rotation of the pointer. Thus to define sensitivity properly, we must know the angular spacing of the kilopascal marks on the scale of the pressure gage. Suppose this is 5 angular degrees/kPa. Since we already calculated the slope in kilopascals per kilopascal as 1.05 in Fig. 3.13, we get the instrument static sensitivity as $(5)(1.05) = 5.25$ angular degrees/kPa. In this form the sensitivity allows comparison of this pressure gage with others as regards its ability to detect pressure changes.

While the instrument's sensitivity to its desired input is of primary concern, its sensitivity to interfering and/or modifying inputs also may be of interest. As an example, consider temperature as an input to the pressure gage mentioned above.

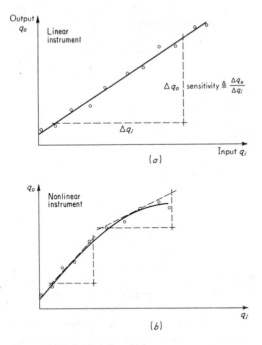

Figure 3.16 Definition of sensitivity.

Temperature can cause a relative expansion and contraction that will result in a change in output reading even though the pressure has not changed. In this sense, it is an interfering input. Also, temperature can alter the modulus of elasticity of the pressure-gage spring, thereby affecting the pressure sensitivity. In this sense, it is a modifying input. The first effect is often called a *zero drift* while the second is a *sensitivity drift* or *scale-factor drift*. These effects can be evaluated numerically by running suitable calibration tests. To evaluate zero drift, the pressure is held at zero while the temperature is varied over a range and the output reading re-corded. For reasonably small temperature ranges, the effect is often nearly linear; then we can quote the zero drift as, say, 0.01 angular degree/C°. Sensitivity drift may be found by fixing the temperature and running a pressure calibration to determine pressure sensitivity. Repeating this for various temperatures should show the effect of temperature on pressure sensitivity. Again, if this is nearly linear, we can specify sensitivity drift as, say, 0.0005 (angular degree/kPa)/C°.

Figure 3.17 shows how the superposition of these two effects determines the total error due to temperature. If the instrument is used for measurement only and the temperature is known, numerical knowledge of zero drift and sensitivity drift allows correction of the readings. If such corrections are not feasible, then knowledge of the drifts is used mainly to estimate overall system errors due to temperature.

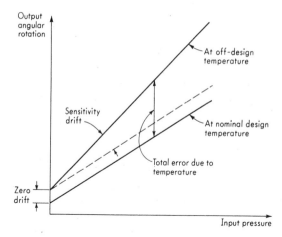

Figure 3.17 Zero and sensitivity drift.

The generalized input-output configuration of Fig. 2.9 can, for this example, be made specific, as in Fig. 3.18. New symbology introduced here includes the *transfer function* and the *variable multiplier*. The output of the variable-multiplier symbol is taken as the product of the two inputs. The output of the transfer-function symbol is taken as the input times the function inside the box. In this case, all the "functions" are merely constants. However, later we generalize this concept to include dynamic relations derived from differential equations relating input and output.

Linearity

If an instrument's calibration curve for desired input is not a straight line, the instrument may still be highly accurate. In many applications, however, linear behavior is most desirable. The conversion from a scale reading to the corresponding measured value of input quantity is most convenient if we merely have to multiply by a fixed constant rather than consult a nonlinear calibration curve or compute from a nonlinear calibration equation. Also, when the instrument is part of a larger data or control system, linear behavior of the parts often simplifies design and analysis of the whole. Thus specifications relating to the degree of conformity to straight-line behavior are common.

Several definitions[1] of linearity are possible. However, *independent linearity* seems to be preferable in many cases. Here the reference straight line is the least-squares fit, as in Fig. 3.13. Thus the linearity is simply a measure of the maximum deviation of any calibration points from this straight line. This may be

---

[1] L. P. Entin, Instrument Uncertainties: I and II, *Control Eng.*, December 1959, February 1960.

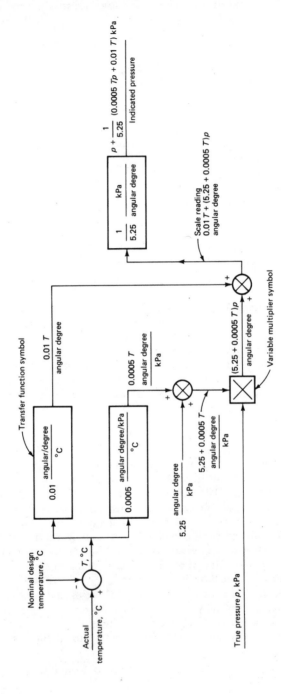

Figure 3.18 Block diagram of pressure gage.

(The −0.64 kPa zero bias of Fig. 3.13 is assumed to have been removed by gage adjustment.)

expressed as a percentage of the actual reading, a percentage of full-scale reading, or a combination of the two. The last method is probably the most realistic and leads to the following type of specification:

Independent nonlinearity $= \pm A$ percent of reading or
$\pm B$ percent of full scale, whichever is greater   (3.52)

The first part ($\pm A$ percent of reading) of the specification recognizes the desirability of a constant-percentage nonlinearity, while the second ($\pm B$ percent of full scale) recognizes the impossibility of testing for extremely small deviations near zero. That is, if a fixed percentage of reading is specified, the absolute deviations approach zero as the readings approach zero. Since the test equipment should be about 10 times as accurate as the instrument under test, this leads to impossible requirements on the test equipment. Figure 3.19 shows the type of tolerance band allowed by specifications of the form (3.52).

Note that in instruments considered essentially linear, the specification of nonlinearity is equivalent to a specification of overall inaccuracy when the common (nonstatistical) definition of inaccuracy is used. Thus in many commercial linear instruments, only a linearity specification (and not an accuracy specification) may be given. The reverse (an accuracy specification but not a linearity specification) may be true if nominally linear behavior is implied by the quotation of a fixed sensitivity figure.

In addition to overall accuracy requirements, linearity specifications often are useful in dividing the total error into its component parts. Such a division is sometimes advantageous in choosing and/or applying measuring systems for a particular application in which, perhaps, one type of error is more important

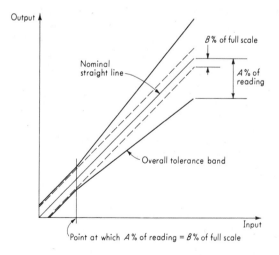

Figure 3.19 Linearity specification.

than another. In such cases, different definitions of linearity may be especially suitable for certain types of systems. The Scientific Apparatus Makers Association standard load-cell (force-measuring device) terminology,[1] for instance, defines linearity as follows: "The maximum deviation of the calibration curve from a straight line drawn between no-load and full-scale load outputs, expressed as a percentage of the full-scale output and measured on increasing load only." The breakdown of total inaccuracy into its component parts is carried further in the next few sections, where hysteresis, resolution, etc., are considered.

Threshold, Resolution, Hysteresis, and Dead Space

Consider a situation in which the pressure gage of Fig. 3.2 has the input pressure slowly and smoothly varied from zero to full scale and then back to zero. If there were no friction due to sliding of moving parts, the input-output graph might appear as in Fig. 3.20a. The noncoincidence of loading and unloading curves is due to the internal friction or hysteretic damping of the stressed parts (mainly the spring). That is, not all the energy put into the stressed parts upon loading is recoverable upon unloading, because of the second law of thermodynamics, which rules out perfectly reversible processes in the real world. Certain materials exhibit a minimum of internal friction, and they should be given consideration in designing highly stressed instrument parts, provided that their other properties are suitable for the specific application. For instruments with a usable range on both sides of zero, the behavior is as shown in Fig. 3.20b.

If it were possible to reduce internal friction to zero but external sliding friction were still present, the results might be as in Fig. 3.20c and d, where a constant coulomb (dry) friction force is assumed. If there is any free play or looseness in the mechanism of an instrument, a curve of similar shape will result.

Hysteresis effects also show up in electrical phenomena. One example is found in the relation between output voltage and input field current in a dc generator, which is similar in shape to Fig. 3.20b. This effect is due to the magnetic hysteresis of the iron in the field coils.

In a given instrument, a number of causes such as those just mentioned may combine to give an overall hysteresis effect which might result in an input-output relation as in Fig. 3.20e. The numerical value of hysteresis can be specified in terms of either input or output and usually is given as a percentage of full scale. When the total hysteresis has a large component of internal friction, time effects during hysteresis testing may confuse matters, since sometimes significant relaxation and recovery effects are present. Thus in going from one point to another in Fig. 3.20e, we may get a different output reading immediately after changing the input than if some time elapses before the reading is taken. If this is the case, the time sequence of the test must be clearly specified if reproducible results are to be obtained.

---

[1] Standard Load Cell Terminology and Definitions: II, Scientific Apparatus Makers Association, Chicago, Jan. 11, 1962.

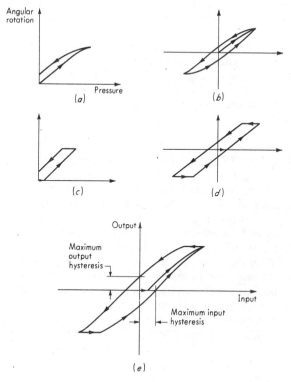

Figure 3.20 Hysteresis effects.

If the instrument input is increased very gradually from zero, there will be some minimum value below which no output change can be detected. This minimum value defines the *threshold* of the instrument. In specifying threshold, the first detectable output change often is described as being any "noticeable" or "measurable" change. Since these terms are somewhat vague, to improve reproducibility of threshold data it may be preferable to state a definite numerical value for output change for which the corresponding input is to be called the threshold.

If the input is increased slowly from some arbitrary (nonzero) input value, again the output does not change at all until a certain input increment is exceeded. This increment is called the *resolution*; again, to reduce ambiguity, it is defined as the input increment that gives some small but definite numerical change in the output. Thus resolution defines the smallest measurable input *change* while threshold defines the smallest measurable *input*. Both threshold and resolution may be given either in absolute terms or as a percentage of full-scale reading. An instrument with large hysteresis does not necessarily have poor resolution. Internal friction in a spring can give a large hysteresis, but even small changes in input (force) cause corresponding changes in deflection, giving high resolution.

The terms "dead space," "dead band," and "dead zone" sometimes are used interchangeably with the term "hysteresis." However, they may be defined as the total range of input values possible for a given output and thus may be numerically twice the hysteresis as defined in Fig. 3.20e. Since none of these terms is completely standardized, you should always be sure which definition is meant.

Scale Readability

Since the majority of instruments that have analog (rather than digital) output are read by a human observer noting the position of a "pointer" on a calibrated scale, usually it is desirable for data takers to state their opinions as to how closely they believe they can read this scale. This characteristic, *which depends on both the instrument and the observer*, is called the *scale readability*. While this characteristic logically should be *implied* by the number of significant figures recorded in the data, it is probably good practice for the observer to stop and think about this before taking data and to then *record* the scale readability. It may also be appropriate at this point to suggest that all data, including scale readabilities, be given in decimal rather than fractional form. Since some instrument scales are calibrated in $\frac{1}{4}$'s, $\frac{1}{2}$'s, etc., this requires data takers to convert to decimal form before recording data. This procedure is considered preferable to recording a piece of data as, say, $21\frac{1}{4}$ and then *later* trying to decide whether 21.250 or 21.3 was meant.

Span

The range of variable that an instrument is designed to measure is sometimes called the *span*. Equivalent terminology also in use states the "low operating limit" and "high operating limit." For essentially linear instruments, the term "linear operating range" is also common. A related term, which, however, implies dynamic fidelity also, is the *dynamic range*. This is the ratio of the largest to the smallest dynamic input that the instrument will faithfully measure. The number representing the dynamic range often is given in decibels, where the decibel (dB) value of a number N is defined as dB $\triangleq 20 \log N$. Thus a dynamic range of 60 dB indicates the instrument can handle a range of input sizes of 1,000 to 1.

Generalized Static Stiffness and Input Impedance

We mentioned that the introduction of any measuring instrument into a measured medium always results in the extraction of some energy from the medium, thereby changing the value of the measured quantity from its undisturbed state and thus making perfect measurements theoretically impossible. Since the instrument designer wishes to approach perfection as nearly as practicable, some numerical means of characterizing this "loading" effect of the instrument on the measured medium would be helpful in comparing competitive instrument designs. The concepts of *stiffness* and *input impedance*[1] serve such a function. While both terms are useful for both static and dynamic conditions, here we consider their static aspects only.

In Fig. 2.1 and subsequent schematic and block diagrams, the connection of functional elements by single lines perhaps gives the impression that the transfer of information and energy is described by a single variable only. Closer examination reveals that energy transfers require the specification of two variable quantities for their description. The definitions of stiffness and generalized input impedance are in terms of two such variables. At the input of each component in a measuring system, there exists a variable q_{i1} with which we are primarily involved, insofar as information transmission is concerned. At the same point, however, there is associated with q_{i1} another variable q_{i2} such that the product $q_{i1}q_{i2}$ has the dimensions of power and represents an instantaneous rate of energy withdrawal from the preceding element. When these two signals are identified, we can define the generalized input impedance Z_{gi} by

$$Z_{gi} \triangleq \frac{q_{i1}}{q_{i2}} \tag{3.53}$$

if q_{i1} is an "effort variable." ("Effort variable" is defined shortly.) [At this point we consider only systems where (3.53) is an ordinary algebraic equation. However, the concept of impedance can be extended easily to dynamic situations, and then (3.53) must be given a more general interpretation.] Using (3.53), we see that the power drain is $P = q_{i1}^2/Z_{gi}$ and that *a large input impedance is needed to keep the power drain small*. The concept of generalized impedance (and, of course, the terminology itself) is a generalization of electric impedance, and we first give some examples from this perhaps somewhat more familiar field.

Consider a voltmeter of the common type shown in Fig. 2.17. Suppose this meter is to be applied to a circuit in order to measure an unknown voltage E, as in Fig. 3.21*a*. As soon as the meter is attached to terminals a and b, the circuit is changed and the value of E is no longer the same. For the meter alone, the input variable of direct interest (q_{i1}) is the terminal voltage E_m. If we look for an associated variable (q_{i2}) which, when multiplied by q_{i1}, gives the power withdrawal, we find the meter current i_m meets these requirements. In this example, then, $Z_{gi} = E_m/i_m = R_m$, the meter resistance.

To further illustrate the significance of input impedance, let us determine just how much error is caused when the meter is connected to the circuit. To facilitate this, first we cite without proof a very useful network theorem called Thévenin's theorem.[2] Consider any network made up of linear, bilateral impedances and power sources. A linear impedance is one whose elements (R, L, C) do not change

[1] R. G. Boiten, The Mechanics of Instrumentation, *Proc. Inst. Mech. Engrs.* (*London*), vol. 177, no. 10, 1963.

[2] K. Y. Tang, "Alternating Current Circuits," 2d ed., p. 202, International Textbook Company, Scranton, Pa., 1952.

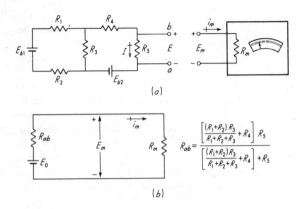

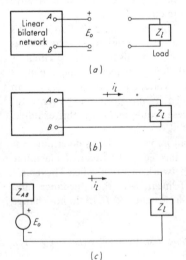

Figure 3.21 Voltmeter loading effect.

value with the magnitude of the current or voltage. Most resistances, capacitances, and air-core inductances are linear; iron-core inductances are nonlinear. A bilateral impedance transmits energy equally well in either direction. Resistances, capacitances, and inductances are essentially bilateral. Field-effect transistors (FETs) are unilateral since they effectively transmit energy in only one direction (from gate to source, *not* the reverse).

A linear bilateral network is shown in Fig. 3.22*a* as a "black box" with terminals *A* and *B*. A load of impedance Z_l may be connected across the terminals *A* and *B*. When the load Z_l is *not* connected, a voltage will, in general, exist at terminals *A* and *B*. This is called E_o, the open-circuit output voltage of the

Figure 3.22 Thévenin's theorem.

network. Also with Z_l *not* connected, it is possible to determine the impedance Z_{AB} between terminals A and B. When this is done, any power sources in the network are to be replaced by their internal impedance. If their internal impedance is assumed to be zero, they are replaced by a short circuit (just a wire with no resistance). Thévenin's theorem then states: If the load Z_l is connected as shown in Fig. 3.22b, a current i_l will flow. This current will be the *same* as the current that flows in the fictitious equivalent circuit of Fig. 3.22c. Thus the network, no matter how complex it is, may be replaced by a single impedance (the output impedance) Z_{AB} in series with a single voltage source E_o.

Applying Thévenin's theorem to Fig. 3.21a, we get Fig. 3.21b. We see here that the value E_m indicated by the meter is *not* the true value E, but rather

$$E_m = \frac{R_m}{R_{ab} + R_m} E \tag{3.54}$$

and that if E_m is to approach E, we must have $R_m \gg R_{ab}$. Thus our earlier statement about the desirability of high input impedance can now be made more specific. *The input impedance must be high relative to the output impedance of the system to which the load is connected.* Assuming that it is possible to define generalized input and output impedances Z_{gi} and Z_{go} in nonelectric as well as electric systems, we may generalize Eq. (3.54) to

$$q_{i1m} = \frac{Z_{gi}}{Z_{go} + Z_{gi}} q_{i1u} = \frac{1}{Z_{go}/Z_{gi} + 1} q_{i1u} \tag{3.55}$$

where $q_{i1m} \triangleq$ measured value of effort variable

$q_{i1u} \triangleq$ undisturbed value of effort variable

Of course, if we knew both Z_{gi} and Z_{go}, we could correct q_{i1m} by means of Eq. (3.55). However, this would be inconvenient; also Z_{go} is not always known, especially in nonelectric systems, in which definition of *both* Z_{gi} and Z_{go} is not always straightforward. Thus a high value of Z_{gi} is desirable since then corrections are unnecessary and the actual values of either Z_{gi} or Z_{go} need not be known.

To achieve a high value of input impedance for any instrument, not just voltmeters, a number of paths are open to the designer. Now we describe three, using the voltmeter as a specific example. The most obvious approach is to leave the configuration of the instrument unchanged, but to alter the numerical values of physical parameters so that the input impedance is increased. In the voltmeter of Fig. 3.21, this is accomplished simply by winding the coil in such a way (higher resistance material and/or more turns) that R_m is increased. While this accomplishes the desired result, certain undesirable effects also appear. Since this type of voltmeter is basically a *current*-sensitive rather than a voltage-sensitive device, an increase in R_m will *reduce* the magnetic torque available from a given impressed voltage. Thus if the spring constant of the restraining springs is not changed, the angular deflection for a given voltage (the sensitivity) is reduced. To bring the sensitivity back to its former value, we must reduce the spring constant. Also, because of lower torque levels, pivot bearings with less friction must be employed. These design changes generally result in a less rugged and less reliable instrument, so that this method of increasing input impedance is limited in the degree of improvement possible before other performance features are compromised. This situation occurs in most instruments, not just in this specific example.

If input impedance is to be increased without compromising other characteristics, different approaches are needed. One of general utility employs a change in configuration of the instrument so as to include an auxiliary power source. The concept is that a rugged instrument requires a fair amount of power to actuate its output elements, but that this power need *not* necessarily be taken from the measured medium. Rather, the low power signal from the primary sensing element may *control* the output of the auxiliary power source so as to realize a power-amplifying effect.

To continue our voltmeter example, this approach is exemplified by a transistor version of the classic vacuum-tube voltmeter (VTVM) shown in rudimentary form in Fig. 3.23a. When the input voltage is zero (short circuit across the E_m terminals), adjustment of R_a allows setting of the meter voltage E_m' to zero. If an input E_m is applied, the gate bias is no longer the same, the currents i_1 and i_2 are

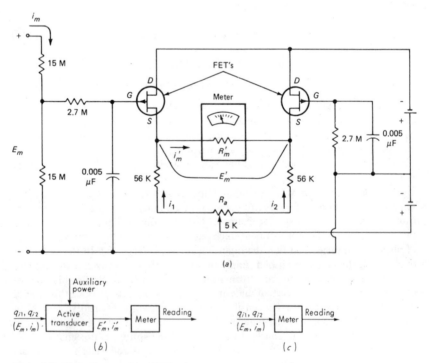

Figure 3.23 Field-effect transistor (FET) voltmeter.

no longer equal and a meter voltage E'_m exists. While the meter current i'_m may still be as large as in a conventional meter, the current that determines the input impedance is i_m, which will be very small. Thus a very high input impedance can be realized while a rugged meter element is used. A block diagram for the FET voltmeter is seen in Fig. 3.23b, which may be compared with that for an ordinary meter in Fig. 3.23c.

Still another approach to the problem of increasing input impedance uses the principle of feedback or null balance. For the specific area of voltage measurements, this technique is exemplified by the potentiometer. The simplest form of this instrument is shown in Fig. 3.24. Clearly, in Fig. 3.24a each position of the sliding contact corresponds to a definite voltage between terminals a and b. Thus the scale can be calibrated, and any voltage between zero and the battery voltage can be obtained by properly positioning the slider. If we now connect an unknown voltage E_m and a galvanometer (current detector), as in Fig. 3.24b, then if E_m is less than the battery voltage, there will be some point on the slider scale at which the voltage picked off the slide wire just equals the unknown E_m. This point of null balance can be detected by a zero deflection of the galvanometer since the net loop voltage a, c, d, b, a will be zero. Then the unknown voltage can be read from the calibrated scale.

We should note that under conditions of perfect balance the current drawn from the unknown voltage source is exactly zero, which yields an infinite input

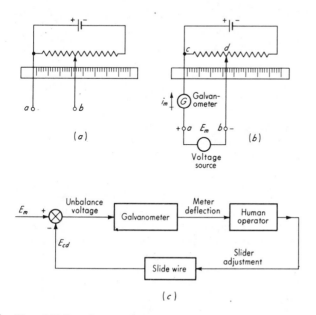

Figure 3.24 Potentiometer voltage measurement.

impedance. In actual practice, there must always remain some unknown unbalance current since a galvanometer always has a threshold below which currents cannot be detected. The interpretation of potentiometric voltage measurement as a feedback scheme is shown in Fig. 3.24c. While the manual balancing described above is adequate when the unknown voltage is relatively constant, the procedure may be made automatic by use of the instrument servomechanism (in this case called a self-balancing potentiometer), as shown in Fig. 2.7. Then, by providing a pen and recording chart, varying voltages may be accurately measured.

In generalizing the concepts of input impedance, a reasonable starting point might be a listing of q_{i1} and corresponding q_{i2} variables for some common measurement situations. In general, the quantity q_{i1}, which is of primary concern, may be either a *flow variable* or an *effort variable*.[1] Briefly, energy transfer across the boundaries of a system may be defined in terms of two variables, the product of which gives the instantaneous power. One of these variables, the flow variable, is an *extensive* variable, in the sense that its magnitude depends on the extent of the system taking part in the energy exchange. The other variable, the effort variable, is an *intensive* variable, whose magnitude is independent of the amount of material being considered. In the literature, flow variables are also called "through" variables, and effort variables are called "across" variables. When q_{i1} is an effort variable, Eq. (3.53) and subsequent developments apply.

However, if q_{i1} is a flow variable, the situation is somewhat different. Then it is appropriate to define a *generalized input admittance* Y_{gi} as

$$Y_{gi} \triangleq \frac{\text{flow variable}}{\text{effort variable}} = \frac{q_{i1}}{q_{i2}} \tag{3.56}$$

rather than a generalized input impedance Z_{gi},

$$Z_{gi} = \frac{\text{effort variable}}{\text{flow variable}} \tag{3.57}$$

Now we can write the power drain of the instrument from the measured medium in terms of the measured variable q_{i1} as

$$P = q_{i1} q_{i2} = \frac{q_{i1}^2}{Y_{gi}} \tag{3.58}$$

and we note that now a large value of input admittance Y_{gi} is required to minimize the power drain. A familiar electrical example of this situation is the ammeter. In Fig. 3.25a, we are interested in measuring the current by means of an ammeter inserted into the circuit as shown. Applying Thévenin's theorem, we can reduce Fig. 3.25a to Fig. 3.25b. We see that the measured value of the current is given by

$$I_m = \frac{E_{ab}}{R_{ab} + R_m} = \frac{E_{ab}}{1/Y_{ab} + 1/Y_m} \tag{3.59}$$

---

[1] E. O. Doebelin, "System Modeling and Response," chap. 7, Wiley, New York, 1980.

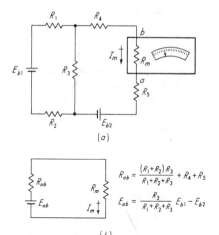

$$R_{ob} = \frac{(R_1 + R_2) R_3}{R_1 + R_2 + R_3} + R_4 + R_5$$

$$E_{ob} = \frac{R_3}{R_1 + R_2 + R_3} E_{b1} - E_{b2}$$

(b)

Figure 3.25 Ammeter loading effect.

whereas the true (undisturbed) value of the current would be

$$I_u = \frac{E_{ab}}{R_{ab}} = \frac{E_{ab}}{1/Y_{ab}} \tag{3.60}$$

So it is clear that if I_m is to approach I_u, we must use an ammeter with $Y_m \gg Y_{ab}$; that is, the meter resistance must be sufficiently *low*, just the *opposite* of that desired in a voltmeter. This result can be generalized to apply to all effort variables (such as voltage) and all flow variables (such as current). Equation (3.55) applies to those cases in which the measured variable q_{i1} is an effort variable, and Eq. (3.61) gives the corresponding relationship when q_{i1} is a flow variable:

$$q_{i1m} = \frac{1}{Y_{go}/Y_{gi} + 1} q_{i1u} \tag{3.61}$$

where $Y_{go} \triangleq$ generalized output admittance of preceding element

$Y_{gi} \triangleq$ generalized input admittance of instrument

$q_{i1m} \triangleq$ measured value of flow variable

$q_{i1u} \triangleq$ undisturbed value of flow variable

For some instruments, in the case of a static input, the *power* drain from the preceding element is zero in the steady state, although some total *energy* is removed in going from one steady state to another. In such an instance, the concepts of impedance and admittance are not as directly useful as one would like, and it is appropriate to consider the concepts of *static stiffness* and *static compliance*. These make it possible to characterize the *energy* drain (in the same way that impedance and admittance define the *power* drain) in those situations in which impedance or admittance becomes infinite and thus not directly meaningful. The terms "stiffness" and "compliance" come from the terminology of mechanical systems, which afford some of the best examples of the application of these concepts. However, we generalize their definitions to include all types of physical systems, just as we did with impedance and admittance.

Consider the system of Fig. 3.26a as an idealized model of some elastic structure under applied load f_{appl}. This load will cause forces in the various structural members. Suppose we wish to measure the force in the member represented by the spring k_2. A common method of force measurement employs a calibrated elastic link whose deflection is proportional to force; thus a deflection measurement allows a force measurement. Such a device, with spring constant k_m, is shown in Fig. 3.26b. To measure the force in link k_2, we insert the

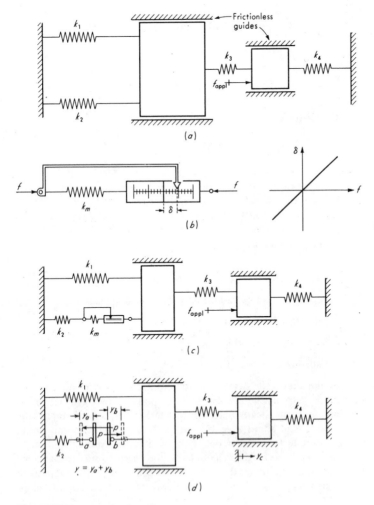

Figure 3.26 Force-gage loading effect.

force-measuring device "in series" with the link k_2, as shown in Fig. 3.26c. The usual difficulty is encountered here in that the insertion of the measuring instrument alters the condition of the measured system and thus changes the measured variable from its undisturbed value. We wish to assess the nature and amount of this error, and we use this example to introduce the concept of static stiffness.

The measured variable, force, is an effort variable. Thus if we try to use the impedance concept, we must find an associated flow variable whose product with force will give power. Since mechanical power has dimensions newton-meters per second, we have

$$\text{Flow variable} = \frac{\text{power}}{\text{effort variable}} = \frac{N \cdot m/s}{N} = \frac{(m)}{s} = \text{velocity} \qquad (3.62)$$

Mechanical impedance is thus given by

$$\text{Mechanical impedance} = \frac{\text{effort variable}}{\text{flow variable}} = \frac{\text{force}}{\text{velocity}} \qquad (3.63)$$

If we calculate the static mechanical impedance of an elastic system by applying a constant force and noting the resulting velocity, we get

$$\text{Static mechanical impedance} = \frac{\text{force}}{0} = \infty \qquad (3.64)$$

This difficulty may be overcome by using energy rather than power in the definition of the variable associated with the measured variable. If this is done, a new term for the ratio of the two variables must be introduced, since the use of mechanical impedance as the ratio of force to velocity is well established. We thus define

$$\text{Mechanical static stiffness} \triangleq \frac{\text{force}}{\text{displacement}} = \frac{\text{force}}{\int (\text{velocity}) \, dt} \qquad (3.65)$$

since

$$\text{Energy} = (\text{force})(\text{displacement}) \qquad (3.66)$$

Thus, in general, whenever the measured variable is an effort variable and the static impedance is infinite, instead of using impedance, we use a generalized static stiffness S_g defined by

$$S_g \triangleq \frac{\text{effort variable}}{\int (\text{flow variable}) \, dt} \qquad (3.67)$$

If this is done, it can be shown that the same formulas can be used for calculating the error due to inserting the measuring instrument as were utilized for impedance, except S is employed instead of Z. Thus, Eq. (3.55) becomes

$$q_{i1m} = \frac{S_{gi}}{S_{go} + S_{gi}} q_{i1u} = \frac{1}{S_{go}/S_{gi} + 1} q_{i1u} \qquad (3.68)$$

where $q_{i1m} \triangleq$ measured value of effort variable
$\quad q_{i1u} \triangleq$ undisturbed value of effort variable
$\quad S_{gi} \triangleq$ generalized static input stiffness of measuring instrument
$\quad S_{go} \triangleq$ generalized static output stiffness of measured system

Let us apply these general concepts to the specific case at hand. The output stiffness of the system of Fig. 3.26a at the point of insertion of the measuring device is simply the ratio of force p to deflection y at terminals a and b in Fig. 3.26d. This stiffness can be found theoretically by applying a fictitious load p and calculating the resulting y; or if the structure (or a scale model) has been constructed, we can obtain the stiffness experimentally by applying known loads and measuring the resulting deflections. A theoretical analysis might proceed as below:

$$\Sigma \text{ forces} = 0$$

$$p - y_b k_1 + k_3(y_c - y_b) = 0 \tag{3.69}$$

$$f_{\text{appl}} - k_3(y_c - y_b) - k_4 y_c = 0 \tag{3.70}$$

$$(-k_1 - k_3)y_b + k_3 y_c = -p \tag{3.71}$$

$$k_3 y_b + (-k_3 - k_4)y_c = -f_{\text{appl}} \tag{3.72}$$

Using determinants yields

$$y_b = \frac{\begin{vmatrix} -p & k_3 \\ -f_{\text{appl}} & -(k_3 + k_4) \end{vmatrix}}{\begin{vmatrix} -(k_1 + k_3) & k_3 \\ k_3 & -(k_3 + k_4) \end{vmatrix}} = \frac{p(k_3 + k_4) + f_{\text{appl}}(k_3)}{(k_3 + k_4)(k_1 + k_3) - k_3^2} \tag{3.73}$$

The output stiffness is now obtained from Eq. (3.73) by letting f_{appl} be zero:

$$S_{go} = \frac{p}{y} = \frac{p}{y_a + y_b} = \frac{p}{p/k_2 + p(k_3 + k_4)/[(k_3 + k_4)(k_1 + k_3) - k_3^2]} \tag{3.74}$$

$$S_{go} = \frac{1}{1/k_2 + (k_3 + k_4)/[(k_3 + k_4)(k_1 + k_3) - k_3^2]} \tag{3.75}$$

The input stiffness of the measuring instrument is given by

$$S_{gi} = \frac{\text{force}}{\text{displacement}} = k_m \tag{3.76}$$

We may now apply Eq. (3.68) to get

$$\frac{\text{Measured value of force}}{\text{True value of force}} = \frac{k_m}{1/[1/k_2 + (k_3 + k_4)/(k_1 k_3 + k_1 k_4 + k_3 k_4)] + k_m} \tag{3.77}$$

From Eq. (3.68) it is apparent that in general we should like to have $S_{gi} \gg S_{go}$ in order to have the measured value close to the true value. In this example, this requirement corresponds to

$$k_m \gg \frac{1}{1/k_2 + (k_3 + k_4)/(k_1 k_3 + k_1 k_4 + k_3 k_4)} \tag{3.78}$$

Thus the measuring device must have a sufficiently stiff spring.

We saw earlier that when the measured variable is not an effort variable, admittance (rather than impedance) is a more convenient tool. Again, however, under static conditions it may happen that admittance is infinite; thus a concept analogous to stiffness is needed to facilitate the treatment of such situations. For such cases, the generalized compliance C_g is defined by

$$C_g \triangleq \frac{\text{flow variable}}{\int (\text{effort variable}) \, dt} \tag{3.79}$$

As a mechanical example, suppose we wish to measure the displacement x in Fig. 3.27a by means of a dial indicator. Such indicators generally have a spring load to ensure positive contact with the body whose motion is being measured. This spring load adds a force to the measured system, thereby causing error in the motion measurement. It is clear that the indicator spring load should be as light as possible, but we wish to make more quantitative statements about measurement accuracy. Again, it is possible to show the applicability of the admittance relations [Eq. (3.61)] if we replace admittance by compliance:

$$q_{i1m} = \frac{1}{C_{go}/C_{gi} + 1} \, q_{i1u} \tag{3.80}$$

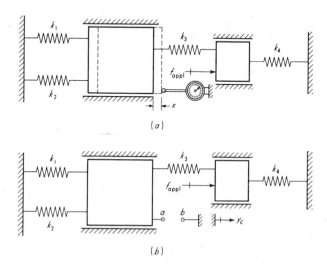

Figure 3.27 Displacement-gage loading effect.

where $q_{i1m} \triangleq$ measured value of flow variable

$\quad q_{i1u} \triangleq$ undisturbed value of flow variable

$\quad C_{gi} \triangleq$ generalized static input compliance of measuring instrument

$\quad C_{go} \triangleq$ generalized static output compliance of measured system

We note that for accurate measurement we require $C_{gi} \gg C_{go}$.

In our example the measured variable is displacement, which is a flow variable. If we try to use admittance concepts, we can find the associated effort variable in the usual way:

$$\text{Power} = (\text{effort variable})(\text{flow variable}) \tag{3.81}$$

$$\frac{\text{N} \cdot \text{m}}{\text{s}} = (\text{effort variable})(\text{displacement, m}) \tag{3.82}$$

$$\text{Effort variable} = \frac{\text{N}}{\text{s}} = \text{rate of change of force} \tag{3.83}$$

The admittance is then

$$Y = \frac{\text{flow variable}}{\text{effort variable}} = \frac{\text{displacement}}{\text{rate of change of force}} \tag{3.84}$$

If we apply this definition to the case of a static load on a spring, we get $Y = \infty$. In this case, however, the compliance would be

$$C_g = \frac{\text{flow variable}}{\int (\text{effort variable}) \, dt} = \frac{\text{displacement}}{\text{force}} = \frac{1}{\text{spring constant}} \tag{3.85}$$

In our example, the output compliance of the measured system is the ratio of displacement to force for terminals a and b of Fig. 3.27b. If we apply a fictitious force p between terminals a and b, a displacement y will occur; it may be computed as follows:

$$\Sigma \text{ forces} = 0$$

$$p - y(k_1 + k_2) + k_3(y_c - y) = 0 \tag{3.86}$$

$$f_{\text{appl}} - k_3(y_c - y) - k_4 y_c = 0 \tag{3.87}$$

$$(-k_1 - k_2 - k_3)y + k_3 y_c = -p \tag{3.88}$$

$$k_3 y + (-k_3 - k_4)y_c = -f_{\text{appl}} \tag{3.89}$$

$$y = \frac{\begin{vmatrix} -p & k_3 \\ -f_{\text{appl}} & -k_3 - k_4 \end{vmatrix}}{\begin{vmatrix} -k_1 - k_2 - k_3 & k_3 \\ k_3 & -k_3 - k_4 \end{vmatrix}}$$

$$= \frac{p(k_3 + k_4) + f_{\text{appl}}(k_3)}{(k_3 + k_4)(k_1 + k_2 + k_3) - k_3^2} = \frac{p(k_3 + k_4) + f_{\text{appl}}(k_3)}{(k_3 + k_4)(k_1 + k_2) + k_3 k_4} \tag{3.90}$$

We can now get C_{go} from Eq. (3.90) by letting $f_{appl} = 0$:

$$C_{go} = \frac{y}{p} = \frac{k_3 + k_4}{(k_3 + k_4)(k_1 + k_2) + k_3 k_4}$$

$$= \frac{1}{k_1 + k_2 + k_3 k_4/(k_3 + k_4)} \tag{3.91}$$

If the spring constant of the dial indicator is k_m, the input compliance of the measuring instrument is given by

$$C_{gi} = \frac{1}{k_m} \tag{3.92}$$

We then have

$$\frac{\text{Measured value of deflection}}{\text{True value of deflection}} = \frac{1}{k_m/[k_1 + k_2 + k_3 k_4/(k_3 + k_4)] + 1} \tag{3.93}$$

and k_m must be sufficiently small to get accurate displacement measurement.

To illustrate the general applicability of the concepts of impedance, admittance, stiffness, and compliance to measurement problems, Fig. 3.28 has been compiled. In the first column are listed some of the physical quantities commonly measured, each one identified as a flow variable or an effort variable. Then the appropriate associated variables that give either power or energy, when multiplied with the measured variable, are listed. The last four columns indicate the dimensions of the appropriate loading criteria for that particular measurement. For effort variables, both impedance and stiffness are given; which one to use depends on the nature of the specific instrument. The fact that admittance and compliance are *not* given for effort variables means not that they could not be defined, but merely that there is no need to consider them when the methods explained earlier in this section are used. Similar statements apply to those measured variables that are flow variables.

The basic formulas (3.55), (3.61), (3.68), and (3.80) were given without a detailed proof. At this point we wish to show the justification for these results and to clarify the physical meaning of impedance, admittance, stiffness, and compliance for physical systems in general. This discussion also indicates more clearly how we calculate theoretically or measure experimentally these important system characteristics.

Consider first Fig. 3.29, showing two separate elements which we subsequently wish to interconnect in the order shown. Our objective is to determine the characteristics of the *individual* devices that must be known in order to predict their overall operation when connected. First we assume that each element may be characterized as a "two-port."[1] This means essentially that the only significant energy exchanges that take place between the device and others which might be connected to it occur at two places ("ports"), which we denote as the input and

[1] Ibid.

Measured variable	Associated variable		Impedance	Admittance	Stiffness	Compliance
	Power based	Energy based				
Voltage (effort)	Current	Charge	$\dfrac{V}{A}$	$\dfrac{A}{V}$	$\dfrac{V}{C}$	$\dfrac{A}{V \cdot s}$
Current (flow)	Voltage	$\int (\text{voltage})\, dt$				
Force (effort)	Translational velocity	Translational displacement	$\dfrac{N}{m/s}$			
Translational displacement (flow)	$\dfrac{d}{dt}(\text{force})$	Force		$\dfrac{m}{N/s}$	$\dfrac{N}{m}$	$\dfrac{m}{N}$
Torque (effort)	Rotational velocity	Rotational displacement	$\dfrac{N \cdot m}{rad/s}$		$\dfrac{N \cdot m}{rad}$	
Rotational displacement (flow)	$\dfrac{d}{dt}(\text{torque})$	Torque	$\dfrac{rad}{N \cdot m/s}$		$\dfrac{rad}{N \cdot m}$	
Translational velocity (flow)	Force	$\int (\text{force})\, dt$		$\dfrac{m/s}{N}$		$\dfrac{m/s}{N \cdot s}$
Rotational velocity (flow)	Torque	$\int (\text{torque})\, dt$		$\dfrac{rad/s}{N \cdot m}$		$\dfrac{rad/s}{N \cdot m/s}$

Variable	Expression 1	Units 1	Expression 2	Units 2
Translational acceleration (flow)	$\int (\text{force})\, dt$	$\dfrac{m/s^2}{N \cdot s}$	$\int \left[\int (\text{force})\, dt\right] dt$	$\dfrac{m/s^2}{N \cdot s^2}$
Rotational acceleration (flow)	$\int (\text{torque})\, dt$	$\dfrac{rad/s^2}{N \cdot m/s}$	$\int \left[\int (\text{torque})\, dt\right] dt$	$\dfrac{rad/s^2}{N \cdot m \cdot s^2}$
Fluid pressure (effort)	Volume flow rate	$\dfrac{Pa}{m^3/s}$	Volume flow	$\dfrac{Pa}{m^3}$
Volume flow rate (flow)	Fluid pressure	$\dfrac{m^3/s}{Pa}$	$\int (\text{fluid pressure})\, dt$	$\dfrac{m^3/s}{Pa \cdot s}$
Temperature (effort)	Heat-transfer rate per unit temperature difference	$\dfrac{C^\circ}{J/(s \cdot C^\circ)}$	Total heat transfer per unit temperature difference	$\dfrac{C^\circ}{J/C^\circ}$

Figure 3.28 Loading parameters for common variables.

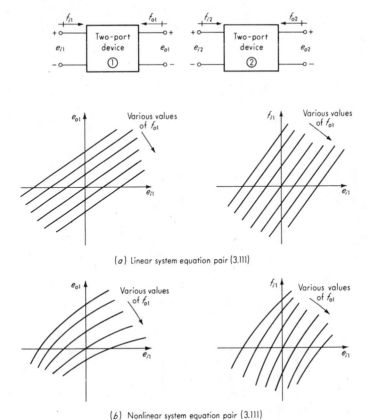

(a) Linear system equation pair (3.111)

(b) Nonlinear system equation pair (3.111)

Figure 3.29 Generalized loading configuration.

the output. Sometimes devices which have three or more ports may legitimately be treated as two ports by restricting the generality of the system model. For example, an electronic amplifier has an input voltage (and current) and an output voltage (and current), but it also has a third "port," the connection to its power supply. However, if we insist that our amplifier always use the *same* (rather than a variety of) power supplies, it can be included within the boundaries of the amplifier itself, which allows characterization of such amplifiers as two-port.

In Fig. 3.29, the e's represent effort variables whereas f's denote flow variables. Any one of the four variables associated with each device may be considered as being determined by the values of the other variables. That is, we may consider any one variable as a *dependent* variable whose value is determined by the values of several independent variables. It is important to note that the quantities in question (the e's and f's) are all available at the "terminals" of the device and may be measured experimentally *without any knowledge of the internal details of the device. Thus complex devices for which no adequate theory exists may be studied experimentally by such methods.*

Suppose we choose to consider first the quantity e_{o1} as a dependent variable. Then, at first, we might think that the independent variables would be e_{i1}, f_{i1}, and f_{o1}, that is,

$$e_{o1} = e_{o1}(e_{i1}, f_{i1}, f_{o1}) \tag{3.94}$$

This, however, is not the case, since e_{i1}, f_{i1}, and f_{o1} are not all independent. If any two of these three are assigned values, the value of the third is determined by the system and is not open to independent choice. The truth of this statement may be demonstrated as follows: Similarly to Eq. (3.94), we could write

$$f_{o1} = f_{o1}(e_{i1}, f_{i1}, e_{o1}) \tag{3.95}$$

$$e_{i1} = e_{i1}(f_{i1}, f_{o1}, e_{o1}) \tag{3.96}$$

$$f_{i1} = f_{i1}(e_{i1}, e_{o1}, f_{o1}) \tag{3.97}$$

We note now, however, from (3.95) that f_{o1} depends on e_{i1} and f_{i1}. Thus it would not be correct to have e_{i1}, f_{i1}, and f_{o1} as independent variables in (3.94). Similar inconsistencies can be found in all four formulas. So it is clear that three independent variables are too many. If we try two instead, we can write

$$e_{o1} = e_{o1}(e_{i1}, f_{i1}) \tag{3.98}$$

or

$$e_{o1} = e_{o1}(e_{i1}, f_{o1}) \tag{3.99}$$

or

$$e_{o1} = e_{o1}(f_{i1}, f_{o1}) \tag{3.100}$$

Also

$$f_{o1} = f_{o1}(e_{i1}, e_{o1}) \tag{3.101}$$

or

$$f_{o1} = f_{o1}(e_{i1}, f_{i1}) \tag{3.102}$$

or

$$f_{o1} = f_{o1}(e_{o1}, f_{i1}) \tag{3.103}$$

Also

$$e_{i1} = e_{i1}(e_{o1}, f_{i1}) \tag{3.104}$$

or

$$e_{i1} = e_{i1}(e_{o1}, f_{o1}) \tag{3.105}$$

or

$$e_{i1} = e_{i1}(f_{i1}, f_{o1}) \tag{3.106}$$

Also

$$f_{i1} = f_{i1}(e_{i1}, e_{o1}) \tag{3.107}$$

or

$$f_{i1} = f_{i1}(e_{i1}, f_{o1}) \tag{3.108}$$

or

$$f_{i1} = f_{i1}(e_{o1}, f_{o1}) \tag{3.109}$$

We can choose any one of the above equations and immediately find its companion (another equation with the *same* independent variables) and thus define the system in terms of two "input" quantities (the independent variables) and two "output" quantities (the dependent variables). Enumerating all possibilities, we get

$$e_{o1} = e_{o1}(e_{i1}, f_{i1})$$
$$f_{o1} = f_{o1}(e_{i1}, f_{i1})$$

(3.110)

$$e_{o1} = e_{o1}(e_{i1}, f_{o1})$$
$$f_{i1} = f_{i1}(e_{i1}, f_{o1})$$

(3.111)

$$e_{o1} = e_{o1}(f_{i1}, f_{o1})$$
$$e_{i1} = e_{i1}(f_{i1}, f_{o1})$$

(3.112)

$$f_{o1} = f_{o1}(e_{i1}, e_{o1})$$
$$f_{i1} = f_{i1}(e_{i1}, e_{o1})$$

(3.113)

$$f_{o1} = f_{o1}(e_{o1}, f_{i1})$$
$$e_{i1} = e_{i1}(e_{o1}, f_{i1})$$

(3.114)

$$e_{i1} = e_{i1}(e_{o1}, f_{o1})$$
$$f_{i1} = f_{i1}(e_{o1}, f_{o1})$$

(3.115)

Thus, in general, we can choose values for any two of the four quantities, and then the values of the other two are determined. If the physical system is strictly linear, its static input-output characteristics can be displayed graphically for any one of the equation pairs (3.110) to (3.115) in a fashion similar to Fig. 3.29a. If it is nonlinear in a continuous (smooth) fashion, the curves might be as in Fig. 3.29b.

We are now in a position to derive Eq. (3.55). Taking the first equation of pair (3.111), we may write

$$de_{o1} = \left.\frac{\partial e_{o1}}{\partial e_{i1}}\right|_{f_{o1} = \text{const}} de_{i1} + \left.\frac{\partial e_{o1}}{\partial f_{o1}}\right|_{e_{i1} = \text{const}} df_{o1}$$

(3.116)

We define

$$\left.\frac{\partial e_{o1}}{\partial e_{i1}}\right|_{f_{o1} = \text{const}} \triangleq \text{no-load static transfer function} \triangleq K$$

(3.117)

$$\left.\frac{\partial e_{o1}}{\partial f_{o1}}\right|_{e_{i1} = \text{const}} \triangleq \text{generalized output impedance } Z_{go}$$

(3.118)

The physical interpretation here is that we consider the system originally in equilibrium with e_{i1}, f_{i1}, e_{o1}, and f_{o1} all at some constant values. Now, if e_{i1} changes by de_{i1} and if f_{o1} should stay constant [the simplest case is where f_{o1} is constant *at zero* because device 1 is "open circuit" (unloaded) at its output], the change in output is given by

$$de_{o1} = K\, de_{i1}$$

(3.119)

If, however, the output of device 1 *is* connected to the input of device 2, then f_{o1} will not be constant and de_{o1} will be different from $K\, de_{i1}$. The amount of "loading error" depends on df_{o1}; to find it, we must know the input impedance of device 2. To define this, we use the second equation of pair (3.112) and apply it to device 2:

$$de_{i2} = \frac{\partial e_{i2}}{\partial f_{i2}}\bigg|_{f_{o2}=\text{const}} df_{i2} + \frac{\partial e_{i2}}{\partial f_{o2}}\bigg|_{f_{i2}=\text{const}} df_{o2} \qquad (3.120)$$

If device 2 has no third device connected to its output, it would have either de_{o2} or $df_{o2} = 0$. Suppose, for example, that $df_{o2} = 0$. Then

$$de_{i2} = \frac{\partial e_{i2}}{\partial f_{i2}}\bigg|_{f_{o2}=\text{const}} df_{i2} \qquad (3.121)$$

We now define

$$\frac{\partial e_{i2}}{\partial f_{i2}}\bigg|_{f_{o2}=\text{const}} \triangleq \text{generalized input impedance } Z_{gi} \qquad (3.122)$$

From Fig. 3.29, we note that when devices 1 and 2 are connected, $e_{o1} = e_{i2}$ and $f_{o1} = -f_{i2}$. We thus get

$$de_{o1} = K\, de_{i1} + Z_{go}\, df_{o1} = K\, de_{i1} - \frac{Z_{go}}{Z_{gi}}\, de_{o1} \qquad (3.123)$$

$$\left(1 + \frac{Z_{go}}{Z_{gi}}\right) de_{o1} = K\, de_{i1} \qquad (3.124)$$

Now, $K\, de_{i1}$ is the value de_{o1} would have if device 2 were not connected to device 1. This corresponds to q_{i1u} of Eq. (3.55). So we can write

$$de_{o1} = \frac{1}{Z_{go}/Z_{gi} + 1}\, K\, de_{i1} \qquad (3.125)$$

thereby proving Eq. (3.55). Equations (3.61), (3.68), and (3.80) can all be established in similar fashion.

When the physical devices are strictly linear, the small changes de_{o1}, de_{i1}, etc., can be replaced by the actual quantities e_{o1}, e_{i1}, etc., and the partial derivatives are constant for all values of the other independent variable. That is, in Eq. (3.116), for example, the term

$$\frac{\partial e_{o1}}{\partial e_{i1}}\bigg|_{f_{o1}=\text{const}}$$

would be numerically the same for any and all values of f_{o1} that we might choose. We generally choose the simplest value of f_{o1} with which to work theoretically or experimentally. This is $f_{o1} \equiv 0$. The other partial derivatives are handled similarly. Thus, in linear systems, on one hand, the various impedances, admittances, transfer functions, etc., are all constant and *independent of the size* of the various signals e_{i1}, e_{o1}, etc., in the devices.

In nonlinear devices, on the other hand, the system terminal characteristics vary with the size of the signal and give accurate results only if the signal variations are limited to *small* excursions from some operating point. That is, we assume, in Fig. 3.29b, for example, that the independent variables f_{o1} and e_{i1} are set at some fixed values $f_{o1,0}$ and $e_{i1,0}$, respectively, as shown in Fig. 3.30a. Then the changes in the output quantities caused by small changes in the input quantities can be predicted by Eq. (3.116), by using the curve slopes of Fig. 3.30a and b as the numerical values of the partial derivatives. A similar interpretation of Eq. (3.120) for the second device leads to Fig. 3.30c and d. The "operating points" cannot be chosen arbitrarily. They must correspond to a stable equilibrium condition when the two devices are interconnected, since then $e_{o1,0}$ must equal $e_{i2,0}$ and $f_{o1,0}$ must equal $-f_{i2,0}$.

As an example of the above procedures, consider a situation in which we

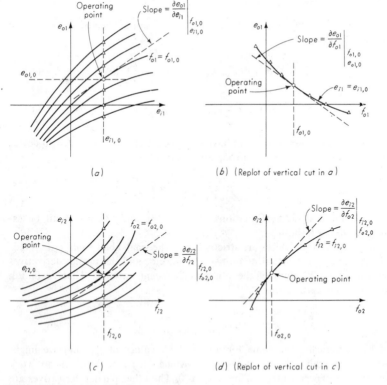

(a)

(b) (Replot of vertical cut in a)

(c)

(d) (Replot of vertical cut in c)

Figure 3.30 Definition of loading parameters.

wish to measure the volume flow rate of a motor-driven pump by means of a flowmeter connected at its discharge line. With no flowmeter attached, the pump is assumed to discharge to atmospheric pressure (0 psig) at a certain flow rate. When the flowmeter is attached, it will cause a pressure drop across itself; thus, if the flowmeter discharges to atmosphere, the pump discharge is now above atmosphere. Depending on the type of pump, this increase in discharge pressure will cause a greater or lesser change in flow rate; thus the flowmeter will not be measuring the no-load flow accurately. A generalization of this problem might be stated as follows: Suppose the pump (with flowmeter attached) is driven at a speed ω_0 with a shaft torque T_0. This will result in pump discharge pressure $p_{p,0}$ and a volume flow rate Q_0. If the pump speed is changed by an amount $d\omega$, the flow rate will change by an amount dQ. How much different will this dQ be from the dQ that would occur for the same $d\omega$ if the flowmeter were not present?

Figure 3.31 illustrates this situation and its analysis by means of admittance techniques. Since the measured variable Q is a flow variable, we search the equation list (3.110) to (3.115) for equations that will allow us to define the output admittance Y_{go} of the pump and the input admittance Y_{gi} of the flowmeter. Also, since we are concerned with a change in speed ω (rather than in torque T), we desire an equation containing the flow variable (rather than the effort variable) as one of the independent variables. For the pump, the first equation of pair (3.114) gives

$$dQ = \left.\frac{\partial Q}{\partial \omega}\right|_{p_p = p_{p,0}} d\omega + \left.\frac{\partial Q}{\partial p_p}\right|_{\omega = \omega_0} dp_p \qquad (3.126)$$

We define

$$\left.\frac{\partial Q}{\partial \omega}\right|_{p_p = p_{p,0}} \triangleq \text{no-load static transfer function} \triangleq K_{Q,\omega} \qquad (3.127)$$

$$\left.\frac{\partial Q}{\partial p_p}\right|_{\omega = \omega_0} \triangleq \text{output admittance} \triangleq Y_{go} \qquad (3.128)$$

The numerical values of these parameters may be established by experiments, which give the graphical results of Fig. 3.31c. Turning now to the flowmeter, we use the second equation of pair (3.113) to get

$$-dQ = \left.\frac{\partial Q}{\partial p_p}\right|_{p_d = p_{d,0}} dp_p + \left.\frac{\partial Q}{\partial p_d}\right|_{p_p = p_{p,0}} dp_d \qquad (3.129)$$

We define

$$\left.\frac{\partial Q}{\partial p_p}\right|_{p_d = p_{d,0}} \triangleq \text{input admittance} \triangleq Y_{gi} \qquad (3.130)$$

$$\left.\frac{\partial Q}{\partial p_d}\right|_{p_p = p_{p,0}} \triangleq \text{flowmeter output admittance} \triangleq Y_{go,f} \qquad (3.131)$$

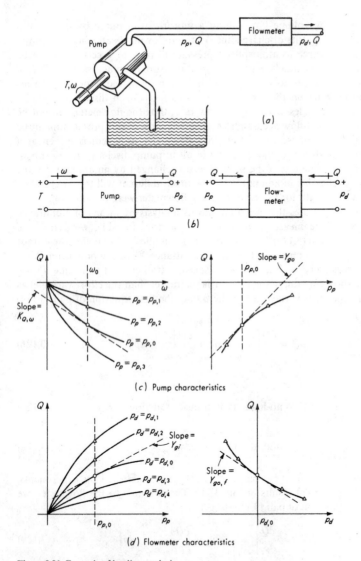

(a)

(b)

(c) Pump characteristics

(d) Flowmeter characteristics

Figure 3.31 Example of loading analysis.

Figure 3.31d illustrates the method of obtaining these parameters from experimental tests on the flowmeter.

If we assume that the flowmeter output is not attached to the input of some other device but discharges directly to atmosphere, then $p_d = $ constant and

$dp_d = 0$. Combining Eqs. (3.126) and (3.129), we get

$$dQ = K_{Q,\omega}\, d\omega + Y_{go}\, dp_p = K_{Q,\omega}\, d\omega - \frac{Y_{go}}{Y_{gi}}\, dQ \qquad (3.132)$$

$$\left(1 + \frac{Y_{go}}{Y_{gi}}\right) dQ = K_{Q,\omega}\, d\omega \qquad (3.133)$$

$$dQ = \frac{1}{1 + Y_{go}/Y_{gi}}\, K_{Q,\omega}\, d\omega \qquad (3.134)$$

If the flowmeter were not present, dp_p in Eq. (3.126) would be zero, which gives

$$dQ = K_{Q,\omega}\, d\omega \qquad (3.135)$$

Thus the flowmeter causes an "error" (flow change) that depends on the numerical value of Y_{go}/Y_{gi}. If Y_{go}/Y_{gi} is very small compared with 1, the error becomes negligible.

In concluding this section on loading effects, the following comments are appropriate. In every measurement, the instrument input causes a load on the measured-medium output. If the instrument system consists of several interconnected stages (the "elements" of Fig. 2.1), furthermore, there may be significant loading effects between stages. This is often the case when the "elements" are general-purpose devices such as amplifiers and recorders which are connected in various ways at different times to create a measurement system suited to a particular problem. In using such a "building block" approach, loading problems must be carefully considered, and methods such as have been outlined above must be used if satisfactory and predictable results are to be obtained. If, however, the elements are merely functional components (permanently connected) of a specific instrument, then loading between elements may be a necessary consideration for the instrument *designer* but not for the user. The user need be concerned only with the loading situation at the interface between measured medium and primary sensing element.

While the impedance-type concepts discussed in this section are extremely useful in studying the disturbing effects of measuring instruments on measured media, some such effects do not lend themselves to this type of approach. For example, if a pitot tube is inserted into a flow field to measure flow velocity, the presence of the tube distorts the velocity field without necessarily extracting any energy or power. Thus, our impedance-type concepts are not directly applicable to such situations.

A related example, in which impedance-type concepts are useful (though not in the way described in this section), is given by Boiten.[1] An instrument for measuring subsoil pressures due to surface loading is useful in soil mechanics studies related to the construction of roads, airfields, etc. Figure 3.32 shows the essential features of such a device. While it is usually desirable that a force-

---

[1] Boiten, The Mechanics of Instrumentation, *Proc. Inst. Mech. Engrs.*

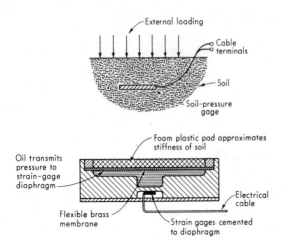

Figure 3.32 Soil-pressure gage.

measuring device have the highest possible stiffness, this criterion is *not* correct for this particular application. The reason is that the presence of the pressure gage in the soil causes a distortion of the pressure field if the stiffness of the gage is either higher or lower than the stiffness of the surrounding soil. That is, a very stiff gage will read too high when the external loading is applied, since the soil around it will compress more than the gage; thus the gage carries a disproportionate share of the total load. Similarly, an excessively compliant gage will read too low a pressure. The best gage is one whose stiffness just *matches* that of the soil in which it is to be used and thus does not distort the pressure field in the soil.

The feature distinguishing this example from those treated earlier in this section is that it deals with a distributed-parameter ("field") type of model for the physical system, whereas our earlier discussions were all in terms of lumped-parameter ("network") models. The pitot-tube example also falls in the distributed-parameter category. While impedance concepts have been applied to distributed-parameter models, we do not pursue this subject further in this text.

Concluding Remarks on Static Characteristics

In this section we presented the most significant static characteristics commonly used in describing instrument performance. It is not possible to list *all* the specific static characteristics that might be pertinent in a particular instrument; rather, those of general interest were considered. It should also be emphasized that the terminology of the measurement field has not been thoroughly standardized. Therefore, you should be careful to determine the precise definition intended when an apparently familiar term is encountered.

Also, errors that are not intrinsically associated with the instrument itself, but are due to human factors involved in taking readings or incurred by incorrect installation of the instrument, were not considered in detail. However, such errors certainly could be included in our general framework merely by extending the concept of inputs to include human factors and installation effects.

3.3 DYNAMIC CHARACTERISTICS

The study of measurement system dynamics is just one of many practical applications of the more general area, often called *system dynamics*, which is treated in depth in other texts[1] and is in some curricula a prerequisite for courses in measurement, control, vibration, etc. Since not all readers have had a system dynamics course, we include those fundamentals necessary for our purposes.

Generalized Mathematical Model of Measurement System

As in so many other areas of engineering application (vibration theory, circuit theory, automatic-control theory, aircraft stability and control theory, etc.), the most widely useful mathematical model for the study of measurement-system dynamic response is the ordinary linear differential equation with constant coefficients. We assume that the relation between any particular input (desired, interfering, or modifying) and the output can, by application of suitable simplifying assumptions, be put in the form

$$a_n \frac{d^n q_o}{dt^n} + a_{n-1} \frac{d^{n-1} q_o}{dt^{n-1}} + \cdots + a_1 \frac{dq_o}{dt} + a_0 q_o = b_m \frac{d^m q_i}{dt^m}$$

$$+ b_{m-1} \frac{d^{m-1} q_i}{dt^{m-1}} + \cdots + b_1 \frac{dq_i}{dt} + b_0 q_i \quad (3.136)$$

where $q_o \triangleq$ output quantity
$q_i \triangleq$ input quantity
$t \triangleq$ time
a's, b's $\triangleq$ combinations of system physical parameters, assumed constant

If we define the differential operator $D \triangleq d/dt$, then Eq. (3.136) can be written as

$$(a_n D^n + a_{n-1} D^{n-1} + \cdots + a_1 D + a_0) q_o$$

$$= (b_m D^m + b_{m-1} D^{m-1} + \cdots + b_1 D + b_0) q_i \quad (3.137)$$

The solution of equations of this type has been put on a systematic basis by using either the "classical" method of D operators or the Laplace-transform[2] method. With the D-operator method, the complete solution q_o is obtained in two

---

[1] E. O. Doebelin, "System Dynamics," Merrill, Columbus, Ohio, 1972; "System Modeling and Response," Wiley, New York, 1980.

[2] Doebelin, "System Modeling and Response."

separate parts as

$$q_o = q_{ocf} + q_{opi} \tag{3.138}$$

where $q_{ocf} \triangleq$ complementary-function part of solution
$\quad q_{opi} \triangleq$ particular-integral part of solution

The solution q_{ocf} has n arbitrary constants; q_{opi} has none. These n arbitrary constants may be evaluated numerically by imposing n initial conditions on Eq. (3.138). The solution q_{ocf} is obtained by calculating the n roots of the algebraic *characteristic equation*

$$a_n D^n + a_{n-1} D^{n-1} + \cdots + a_1 D + a_0 = 0 \tag{3.139}$$

where the operator D is treated as if it were an algebraic unknown. Once these roots $s_1, s_2, \ldots, s_n$ have been found, the complementary-function solution is immediately written by following the rules stated below:

1. *Real roots, unrepeated.* For each real unrepeated root s one term of the solution is written as Ce^{st}, where C is an arbitrary constant. Thus, for example, roots -1.7, $+3.2$, and 0 give a solution $C_1 e^{-1.7t} + C_2 e^{3.2t} + C_3$.
2. *Real roots, repeated.* For each root s which appears p times, the solution is written as $(C_0 + C_1 t + C_2 t^2 + \cdots + C_{p-1} t^{p-1})e^{st}$. Thus, if there are roots -1, -1, $+2$, $+2$, $+2$, 0, 0, the solution is written as $(C_0 + C_1 t)e^{-t} + (C_2 + C_3 t + C_4 t^2)e^{2t} + C_5 + C_6 t$.
3. *Complex roots, unrepeated.* A complex root has the general form $a + ib$. It can be shown that if the a's of Eq. (3.139) are themselves real numbers (which they generally will be, since they are physical quantities such as mass, spring rate, etc.), then when any complex roots occur, they will always occur in pairs of the form $a \pm ib$. For each such root pair, the corresponding solution is

$$Ce^{at} \sin(bt + \phi)$$

where C and ϕ are the two arbitrary constants. Thus roots $-3 \pm i4$, $2 \pm i5$, and $0 \pm i7$ give a solution $C_0 e^{-3t} \sin(4t + \phi_0) + C_1 e^{2t} \sin(5t + \phi_1) + C_2 \sin(7t + \phi_2)$.
4. *Complex roots, repeated.* For each pair of complex roots $a \pm ib$ which appears p times the solution is $C_0 e^{at} \sin(bt + \phi_0) + C_1 t e^{at} \sin(bt + \phi_1) + \cdots + C_{p-1} t^{p-1} e^{at} \sin(bt + \phi_{p-1})$. Roots $-3 \pm i2$, $-3 \pm i2$, and $-3 \pm i2$ thus give a solution $C_0 e^{-3t} \sin(2t + \phi_0) + C_1 t e^{-3t} \sin(2t + \phi_1) + C_2 t^2 e^{-3t} \sin(2t + \phi_2)$.

The complete complementary-function solution is simply the algebraic sum of the individual parts found from the four rules. Whereas the above method for finding q_{ocf} *always* works, no universal method for finding the particular solution q_{opi} exists. This is because q_{opi} depends on the form of q_i, and we can always define a sufficiently "pathological" form of q_i to prevent solution for q_{opi}. However, if q_i is restricted to functions of prime engineering interest, a relatively simple method for finding q_{opi} is available. This is the *method of undetermined coefficients*, which is briefly reviewed. Since the method does not work for all q_i, the first question to be answered is whether it will work for the q_i of interest. For a given q_i, the right side of Eq. (3.136) is some known function of time $f(t)$. To test whether this method can be applied, we repeatedly differentiate $f(t)$ and then examine the functions created by these differentiations. There are three possibilities:

1. After a certain-order derivative, all higher derivatives are zero.
2. After a certain-order derivative, all higher derivatives have the same functional form as some lower-order derivative.
3. Upon repeated differentiation, new functional forms continue to arise.

If case 1 or 2 occurs, the method will work. If case 3 occurs, this method will not work, and others must be tried. If the method is applicable, the solution q_{opi} is immediately written as

$$q_{opi} = Af(t) + Bf'(t) + Cf''(t) + \cdots \qquad (3.140)$$

where the right side includes one term for each functionally different form found by examining $f(t)$ and all its derivatives. The constants A, B, C, etc., can be found *immediately* (they do *not* depend on the initial conditions) by substituting q_{opi} [as given in Eq. (3.140)] into Eq. (3.136) and requiring (3.136) to be an *identity*. This procedure always generates as many simultaneous algebraic equations in the unknowns A, B, C, etc., as there are unknowns; thus the equations can be solved for A, B, C, etc.

Digital Simulation Methods for Dynamic Response Analysis

In studying the dynamic performance of measurement systems, the classical operator and Laplace-transform methods outlined or referenced earlier are capable of providing analytical solutions even for high-order differential equations. However, digital simulation becomes an increasingly attractive alternative as system complexity increases, and becomes a necessity if nonlinear and/or time-variant effects are to be examined. Since a detailed treatment of digital simulation application is not within the scope of this text, we use it only as a working tool whenever appropriate. Fortunately, the basic methods are rather easy to comprehend, so the reader with little or no previous experience should have no difficulty following the examples. We use the IBM CSMP language, which is well documented[1] for the reader who desires greater detail.

Operational Transfer Function

In the analysis, design, and application of measurement systems, the concept of the operational transfer function is very useful. The operational transfer function

[1] Doebelin, "System Dynamics" and "System Modeling and Response"; F. H. Speckhart and W. L. Green, "A Guide to Using CSMP," Prentice-Hall, Englewood Cliffs, N.J., 1976.

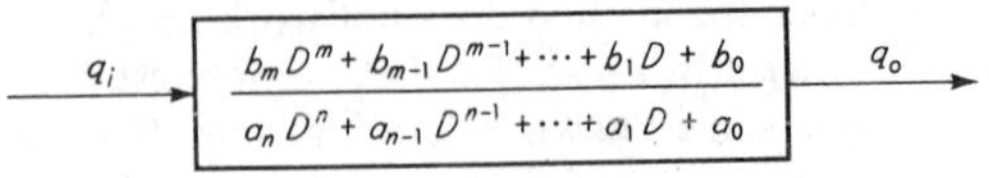

$$\frac{b_m D^m + b_{m-1} D^{m-1} + \cdots + b_1 D + b_0}{a_n D^n + a_{n-1} D^{n-1} + \cdots + a_1 D + a_0}$$

with input q_i and output q_o.

Figure 3.33 General operational transfer function.

relating output q_o to input q_i is defined by treating Eq. (3.137) as if it were an algebraic relation and forming the ratio of output to input:

$$\text{Operational transfer function} \triangleq \frac{q_o}{q_i} D$$

$$\triangleq \frac{b_m D^m + b_{m-1} D^{m-1} + \cdots + b_1 D + b_0}{a_n D^n + a_{n-1} D^{n-1} + \cdots + a_1 D + a_0}$$

$$(3.141)$$

In writing transfer functions, we always write $(q_o/q_i)(D)$, not just q_o/q_i, to emphasize that the transfer function is a *general* relation between q_o and q_i and very definitely *not* the instantaneous ratio of the time-varying quantities q_o and q_i.

One of the several useful features of transfer functions is their utility for graphic symbolic depiction of system dynamic characteristics by means of block diagrams. That is, if we wish to depict graphically a device with transfer function (3.141), we can draw a block diagram as in Fig. 3.33. Furthermore, the transfer function is helpful in determining the overall characteristics of a system made up of components whose individual transfer functions are known. This combination is most simply achieved when there is negligible loading[1] (the input impedance of the second device is much higher than the output impedance of the first, etc.) between the connected devices. For this case, the overall transfer function is simply the product of the individual ones, since the output of the preceding device becomes the input of the following one. Figure 3.34 illustrates this procedure. When significant loading *is* present, we may apply the impedance concepts of Sec. 3.2 (extended to the dynamic case) or simply analyze the complete system "from scratch" without using the individual transfer functions.

In the technical literature, the Laplace-transform method is in common use for the study of linear systems. When such methods are employed, the *Laplace transfer function* is defined as the ratio of the Laplace transform of the output quantity to the Laplace transform of the input quantity when all initial conditions are zero. Thus, analogous to Eq. (3.141), the Laplace transfer function would be written as

$$\frac{q_o(s)}{q_i(s)} \triangleq \frac{q_o}{q_i}(s) \triangleq \frac{b_m s^m + b_{m-1} s^{m-1} + \cdots + b_1 s + b_0}{a_n s^n + a_{n-1} s^{n-1} + \cdots + a_1 s + a_0} \qquad (3.142)$$

where $s \triangleq \sigma + i\omega$ is the complex variable of the Laplace transform. We note that, as far as the *form* of the transfer function is concerned, we can shift from the

---

[1] Doebelin, "System Dynamics," p. 353.

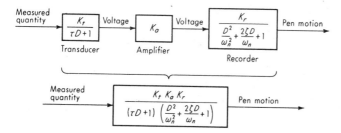

Figure 3.34 Combination of individual transfer functions.

Laplace form to the D-operator form (or vice versa) simply by interchanging s and D. Thus, if we encounter a block diagram using the Laplace notation, we can *always* convert to the D notation by a simple substitution. Then all the methods we subsequently develop may be applied to the operational transfer function.

Sinusoidal Transfer Function

In studying the quality of measurement under dynamic conditions, we analyze the response of measurement systems to certain "standard" inputs. One of the most important of such responses is the steady-state response to a sinusoidal input. Here the input q_i is of the form $A_i \sin \omega t$. If we wait for all transient effects to die out (the complementary-function solution of a stable linear system always dies out eventually), we see that the output quantity q_o will be a sine wave of exactly the same frequency (ω) as the input. However, the amplitude of the output may differ from that of the input, and a phase shift may be present. These results are easily shown by obtaining the particular (steady-state) solution by the method of undetermined coefficients. Since the frequency is the same, the relation between the input and output sine waves is completely specified by giving their amplitude ratio and phase shift. Both quantities, in general, change when the driving frequency ω changes. Thus the *frequency response* of a system consists of curves of amplitude ratio and phase shift as a function of frequency. Figure 3.35 illustrates these concepts.

While the frequency response of any linear system may be obtained by getting the particular solution of its differential equation with

$$q_i = A_i \sin \omega t$$

much quicker and easier methods are available. These methods depend on the concept of the sinusoidal transfer function. The sinusoidal transfer function of a system is obtained by substituting $i\omega$ for D in the operational transfer function:

$$\text{Sinusoidal transfer function} \triangleq \frac{q_o}{q_i} (i\omega)$$

$$\triangleq \frac{b_m(i\omega)^m + b_{m-1}(i\omega)^{m-1} + \cdots + b_1 i\omega + b_0}{a_n(i\omega)^n + a_{n-1}(i\omega)^{n-1} + \cdots + a_1 i\omega + a_0} \quad (3.143)$$

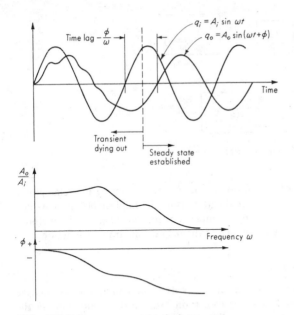

Figure 3.35 Frequency-response terminology.

where $i \triangleq \sqrt{-1}$ and $\omega \triangleq$ frequency in radians per unit time. For any given frequency ω, Eq. (3.143) shows that $(q_o/q_i)(i\omega)$ is a complex number, which can always be put in the polar form $M \angle \phi$. We prove that *the magnitude M of the complex number is the amplitude ratio A_o/A_i while the angle ϕ is the phase angle by which the output q_o leads the input q_i.* (If the output *lags* the input, ϕ is negative.)

The proof of the above statement is most readily demonstrated by the rotating-vector or phasor method of representing sinusoidal quantities. By a well-known trigonometric identity, we may write, in general,

$$Ae^{i\theta} = A(\cos\theta + i\sin\theta) = A\cos\theta + iA\sin\theta \qquad (3.144)$$

The complex number represented by the right side can be exhibited graphically as in Fig. 3.36a. If now we apply this general result to the specific problem of representing q_o and q_i, we get the following:

For input q_i, $\qquad A = A_i \qquad$ and $\qquad \theta = \omega t$

$$A_i e^{i\omega t} = A_i\cos\omega t + iA_i\sin\omega t \qquad (3.145)$$

For output q_o, $\qquad A = A_o \qquad$ and $\qquad \theta = \omega t + \phi$

$$A_o e^{i(\omega t + \phi)} = A_o\cos(\omega t + \phi) + iA_o\sin(\omega t + \phi) \qquad (3.146)$$

Note that the frequency ω of sinusoidal oscillation is also the angular velocity of rotation of the phasors of Fig. 3.36b. The phasors both rotate at the same angular velocity ω, but maintain a fixed angle ϕ between them.

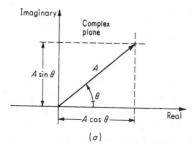

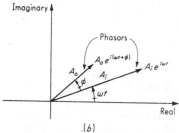

Figure 3.36 Phasor representation of sine waves.

In carrying out our proof, we need to be able to differentiate phasor quantities. Since the amplitude A and the quantity i are constants, we have, in general,

$$\frac{d}{dt}(Ae^{i\theta}) = \left(i\,\frac{d\theta}{dt}\right)Ae^{i\theta} \tag{3.147}$$

and, in particular, if $A = A_i$ and $\theta = \omega t$,

$$\frac{d}{dt}(A_i e^{i\omega t}) = i\omega A_i e^{i\omega t} \tag{3.148}$$

or, if $A = A_o$ and $\theta = \omega t + \phi$,

$$\frac{d}{dt}(A_o e^{i(\omega t + \phi)}) = i\omega A_o e^{i(\omega t + \phi)} \tag{3.149}$$

Clearly, for any higher derivative, we would get

$$\frac{d^n}{dt^n}(A_i e^{i\omega t}) = (i\omega)^n A_i e^{i\omega t} \tag{3.150}$$

and

$$\frac{d^n}{dt^n}(A_o e^{i(\omega t + \phi)}) = (i\omega)^n A_o e^{i(\omega t + \phi)} \tag{3.151}$$

Thus, differentiating a phasor quantity n times with respect to time t may be achieved simply by multiplying it by $(i\omega)^n$.

Suppose we consider Eq. (3.137) for the sinusoidal steady-state case. Then every term on each side of the equation will be a sinusoidally varying quantity, since repeated differentiation of sine waves gives only more sine waves (or cosines, which can be replaced by sines with a phase angle). We convert the differential equation (3.137) to a complex algebraic equation by replacing each sinusoidal term by its phasor representation. This is *not* a matter of simple substitution, since the sinusoidal terms are not *equal* to the phasor quantities; rather, they are *represented* by the phasor quantities. So we must be careful to show that, when the new phasor (complex-number) equation is satisfied, we are guaranteed that the original system differential equation is also satisfied. Then we can perform any desired manipulations on the complex-number equation with assurance that correct results will be obtained. This is done by first replacing the sinusoidal terms by their phasor representations:

$$a_n(i\omega)^n A_o e^{i(\omega t + \phi)} + a_{n-1}(i\omega)^{n-1} A_o e^{i(\omega t + \phi)} + \cdots$$

$$+ a_1(i\omega)A_o e^{i(\omega t + \phi)} + a_0 A_o e^{i(\omega t + \phi)} = b_m(i\omega)^m A_i e^{i\omega t}$$

$$+ b_{m-1}(i\omega)^{m-1} A_i e^{i\omega t} + \cdots + b_1(i\omega)A_i e^{i\omega t} + b_0 A_i e^{i\omega t} \qquad (3.152)$$

This complex-number equation can be satisfied only if the real parts on the left equal the real parts on the right, and similarly for the imaginary parts. Thus, if Eq. (3.152) is enforced, we are guaranteed that the equation given by the imaginary parts also will be satisfied. If we obtain the first few terms in this equation, the pattern should be obvious. We have

$$\text{Im } (a_0 A_o e^{i(\omega t + \phi)}) = a_0 A_o \sin(\omega t + \phi) = a_0 q_o \quad \text{lowest-order terms}$$

$$\text{Im } (b_0 A_i e^{i\omega t}) = b_0 A_i \sin \omega t = b_0 q_i \quad \begin{array}{l}\text{in the original}\\ \text{differential equations}\end{array}$$

$$\text{Im } [a_1(i\omega)A_o e^{i(\omega t + \phi)}] = \text{Im } \{a_1(i\omega)A_o [\cos(\omega t + \phi) + i \sin(\omega t + \phi)]\}$$

$$= a_1 \omega A_o \cos(\omega t + \phi) = a_1 D q_o \quad \begin{array}{l}\text{next terms}\\ \text{in original}\\ \text{differential}\end{array}$$

$$\text{Im } [b_1(i\omega)A_i e^{i\omega t} = \text{Im } [b_1(i\omega)A_i(\cos \omega t + i \sin \omega t)] \quad \text{equation}$$

$$= b_1 \omega A_i \cos \omega t = b_1 D q_i$$

It should be clear that requiring Eq. (3.152) to hold is *equivalent* to requiring (3.137) to hold, even though they are *not* the same equation.

We now manipulate Eq. (3.152) as follows to prove our final result:

$$[a_n(i\omega)^n + a_{n-1}(i\omega)^{n-1} + \cdots + a_1(i\omega) + a_0]A_o e^{i(\omega t + \phi)}$$

$$= [b_m(i\omega)^m + b_{m-1}(i\omega)^{m-1} + \cdots + b_1(i\omega) + b_0]A_i e^{i\omega t} \qquad (3.153)$$

$$\frac{A_o e^{i(\omega t + \phi)}}{A_i e^{i\omega t}} = \frac{b_m(i\omega)^m + b_{m-1}(i\omega)^{m-1} + \cdots + b_1(i\omega) + b_0}{a_n(i\omega)^n + a_{n-1}(i\omega)^{n-1} + \cdots + a_1(i\omega) + a_0} \triangleq \frac{q_o}{q_i}(i\omega) \qquad (3.154)$$

Now,

$$\frac{A_o e^{i(\omega t + \phi)}}{A_i e^{i\omega t}} = \frac{A_o}{A_i} e^{i\phi} = \frac{A_o}{A_i}(\cos \phi + i \sin \phi) \qquad (3.155)$$

$$\cos \phi + i \sin \phi = \sqrt{\cos^2 \phi + \sin^2 \phi} \ \angle \phi = 1 \angle \phi$$

and thus

$$\frac{q_o}{q_i}(i\omega) = \frac{A_o}{A_i} \angle \phi = M \angle \phi \qquad (3.156)$$

Equation (3.156) states that at any chosen frequency ω, the magnitude of the complex number $(q_o/q_i)(i\omega)$ is numerically the amplitude ratio A_o/A_i while the angle of the complex number is the angle by which the output leads the input. Therefore our desired result is proved.

Zero-Order Instrument

While the general mathematical model of Eq. (3.136) is adequate for handling any linear measurement system, certain special cases occur so frequently in practice that they warrant separate consideration. Furthermore, more complicated systems can be studied profitably as combinations of these simple special cases.

The simplest possible special case of Eq. (3.136) occurs when all the a's and b's other than a_0 and b_0 are assumed to be zero. The differential equation then degenerates into the simple algebraic equation

$$a_0 q_o = b_0 q_i \tag{3.157}$$

Any instrument or system that closely obeys Eq. (3.157) over its intended range of operating conditions is defined to be a *zero-order instrument*. Actually, two constants a_0 and b_0 are not necessary, and so we define the static sensitivity (or steady-state "gain") as follows:

$$q_o = \frac{b_0}{a_0} q_i = Kq_i \tag{3.158}$$

$$K \triangleq \frac{b_0}{a_0} \triangleq \text{static sensitivity} \tag{3.159}$$

Since the equation $q_o = Kq_i$ is algebraic, it is clear that, no matter how q_i might vary with time, the instrument output (reading) follows it *perfectly* with no distortion or time lag of any sort. *Thus, the zero-order instrument represents ideal or perfect dynamic performance.*

A practical example of a zero-order instrument is the displacement-measuring potentiometer. Here (see Fig. 3.37) a strip of resistance material is excited with a voltage and provided with a sliding contact. If the resistance is distributed linearly along length L, we may write

$$e_o = \frac{x_i}{L} E_b = Kx_i \tag{3.160}$$

where $K \triangleq E_b/L$ volts per inch.

If you examine this measuring device more critically, you will find that it is not *exactly* a zero-order instrument. This is simply a manifestation of the *universal* rule that no mathematical model can *exactly* represent *any* physical system. In our present example, we would find that, if we wish to *use* a potentiometer for motion measurements, we must attach to the output terminals some voltage-measuring device (such as an oscilloscope). Such a device will always draw some current (however small) from the potentiometer. Thus, when x_i changes, the potentiometer winding current will also change. This in itself would

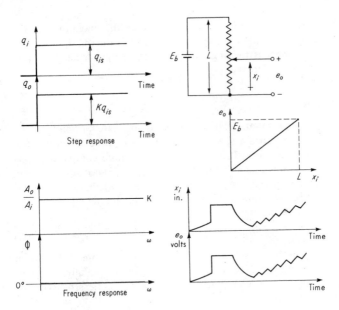

Figure 3.37 Zero-order instrument.

cause no dynamic distortion or lag *if the potentiometer were a pure resistance.* However, the idea of a pure resistance is a *mathematical model,* not a real system; thus the potentiometer will have some (however small) inductance and capacitance. If x_i is varied relatively slowly, these parasitic inductance and capacitance effects will not be apparent. However, for sufficiently fast variation of x_i, these effects are no longer negligible and cause dynamic errors between x_i and e_o. The reasons why a potentiometer is normally called a zero-order instrument are as follows:

1. The parasitic inductance and capacitance can be made very small by design.
2. The speeds ("frequencies") of motion to be measured are not high enough to make the inductive or capacitive effects noticeable.

Another aspect of nonideal behavior in a real potentiometer comes to light when we realize that the sliding contact must be attached to the body whose motion is to be measured. Thus, there is a mechanical loading effect, due to the inertia of the sliding contact and its friction, which will cause the measured motion x_i to be different from that which would occur if the potentiometer were not present. Thus this effect is different *in kind* from the inductive and capacitive phenomena mentioned earlier, since they affected the relation [Eq. (3.160)] between e_o and x_i whereas the mechanical loading has no effect on this relation but, rather, makes x_i different from the undisturbed case.

First-Order Instrument

If in Eq. (3.136) all a's and b's other than a_1, a_0, and b_0 are taken as zero, we get

$$a_1 \frac{dq_o}{dt} + a_o q_o = b_0 q_i \qquad (3.161)$$

Any instrument that follows this equation is, by definition, a first-order instrument. There may be some conflict here between mathematical terminology and common engineering usage. In mathematics, a first-order *equation* has the general form

$$a_1 \frac{dq_o}{dt} + a_0 q_o = (b_m D^m + b_{m-1} D^{m-1} + \cdots + b_1 D + b_0) q_i \qquad (3.162)$$

where m could have any numerical value. However, through long usage, in engineering we commonly understand a first-order *instrument* to be defined by Eq. (3.161). Since in technical presentations both words *and equations* generally are employed, confusion on this point is rarely a problem.

While Eq. (3.161) has three parameters a_1, a_0, and b_0, only two are really essential since the whole equation could always be divided through by a_1, a_0, or b_0, thus making the coefficient of one of the terms numerically equal to 1. The most useful procedure is to divide through by a_0, which gives

$$\frac{a_1}{a_0} \frac{dq_o}{dt} + q_o = \frac{b_0}{a_0} q_i \qquad (3.163)$$

which becomes $$(\tau D + 1) q_o = K q_i \qquad (3.164)$$

when we define $$K \triangleq \frac{b_0}{a_0} \triangleq \text{static sensitivity} \qquad (3.165)$$

$$\tau \triangleq \frac{a_1}{a_0} \triangleq \text{time constant} \qquad (3.166)$$

The time constant τ always has the dimensions of time, while the static sensitivity K has the dimensions of output divided by input. For *any*-order instrument, K always is defined as b_0/a_0 and always has the same physical meaning, that is, the amount of output per unit input when the input is static (constant), because under such conditions all the derivative terms in the differential equation are zero. The operational transfer function of any first-order instrument is

$$\frac{q_o}{q_i}(D) = \frac{K}{\tau D + 1} \qquad (3.167)$$

As an example of a first-order instrument, let us consider the liquid-in-glass thermometer of Fig. 3.38. The input (measured) quantity here is the temperature $T_i(t)$ of the fluid surrounding the bulb of the thermometer, and the output is the displacement x_o of the thermometer fluid in the capillary tube. We assume the

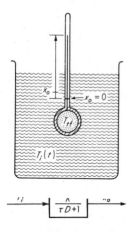

$$\xrightarrow{\,x_i\,} \boxed{\dfrac{\wedge}{\tau D+1}} \xrightarrow{\,x_o\,}$$

Figure 3.38 First-order instrument.

temperature $T_i(t)$ is uniform throughout the fluid at any given time, but may vary with time in an arbitrary fashion. The principle of operation of such a thermometer is the thermal expansion of the filling fluid which drives the liquid column up or down in response to temperature changes. Since this liquid column has inertia, mechanical lags will be involved in moving the liquid from one level to another. However, we assume that this lag is negligible compared with the thermal lag involved in transferring heat from the surrounding fluid through the bulb wall and into the thermometer fluid. This assumption rests (as all such assumptions necessarily must) on experience, judgment, order-of-magnitude calculations, and, ultimately, experimental verification (or refutation) of the results predicted by the analysis. Assumption of negligible mechanical lag allows us to relate the temperature of the fluid in the bulb to the reading x_o by the instantaneous (algebraic) equation

$$x_o = \frac{K_{ex} V_b}{A_c} T_{tf} \tag{3.168}$$

where $x_o \triangleq$ displacement from reference mark, m

$T_{tf} \triangleq$ temperature of fluid in bulb (assumed uniform throughout bulb volume), $T_{tf} = 0$ when $x_o = 0$, °C

$K_{ex} \triangleq$ differential expansion coefficient of thermometer fluid and bulb glass, $m^3/(m^3 \cdot C°)$

$V_b \triangleq$ volume of bulb, m^3

$A_c \triangleq$ cross-sectional area of capillary tube, m^2

To get a differential equation relating input and output in this thermometer, we consider conservation of energy over an infinitesimal time dt for the thermometer bulb:

Heat in $-$ heat out $=$ energy stored

$$U A_b (T_i - T_{tf}) \, dt - 0(\text{assume no heat loss}) = V_b \rho C \, dT_{tf} \tag{3.169}$$

where $U \triangleq$ overall heat-transfer coefficient across bulb wall, W/(m$^2 \cdot$ °C)
   $A_b \triangleq$ heat-transfer area of bulb wall, m^2
   $\rho \triangleq$ mass density of thermometer fluid, kg/m^3
   $C \triangleq$ specific heat of thermometer fluid, J/(kg $\cdot$ °C)
Equation (3.169) involves many assumptions:

1. The bulb wall and fluid films on each side are pure resistance to heat transfer with no heat-storage capacity. This will be a good assumption if the heat-storage capacity (mass) (specific heat) of the bulb wall and fluid films is small *compared* with $C\rho V_b$ for the bulb.
2. The overall coefficient U is constant. Actually, film coefficients and bulb-wall conductivity all change with temperature, but these changes are quite small as long as the temperature does not vary over wide ranges.
3. The heat-transfer area A_b is constant. Actually, expansion and contraction would cause this to vary, but this effect should be quite small.
4. No heat is lost from the thermometer bulb by conduction up the stem. Heat loss will be small if the stem is of small diameter, made of a poor conductor, and immersed in the fluid over a great length and if the exposed end is subjected to an air temperature not much different from T_i and T_{tf}.
5. The mass of fluid in the bulb is constant. Actually mass must enter or leave the bulb whenever the level in the capillary tube changes. For a fine capillary and a large bulb, this effect should be small.
6. The specific heat C is constant. Again, this fluid property varies with temperature, but the variation is slight except for large temperature changes.

The above list of assumptions is not complete, but should give some appreciation of the discrepancies between a mathematical model and the real system it represents. Many of these assumptions could be relaxed to get a more accurate model, but we would pay a heavy price in increased mathematical complexity. The choice of assumptions that are *just good enough* for the needs of the job at hand is one of the most difficult and important tasks of the engineer.

Returning to Eq. (3.169), we may write it as

$$V_b \rho C \frac{dT_{tf}}{dt} + U A_b T_{tf} = U A_b T_i \tag{3.170}$$

Using Eq. (3.168), we get

$$\frac{\rho C A_c}{K_{ex}} \frac{dx_o}{dt} + \frac{U A_b A_c}{K_{ex} V_b} x_o = U A_b T_i \tag{3.171}$$

which we recognize to be the form of Eq. (3.163), and so we immediately define

$$K \triangleq \frac{K_{ex} V_b}{A_c} \qquad \text{m/C°} \tag{3.172}$$

$$\tau \triangleq \frac{\rho C V_b}{U A_b} \qquad \text{s} \tag{3.173}$$

Having shown a concrete example of a first-order instrument, let us return to the problem of examining the dynamic response of first-order instruments in general. Once you have obtained the differential equation relating the input and output of an instrument, you can study its dynamic performance by taking the input (quantity to be measured) to be some known function of time and then solving the differential equation for the output as a function of time. If the output is closely proportional to the input at all times, the dynamic accuracy is good. The fundamental difficulty in this approach lies in the fact that, in actual practice, the quantities to be measured usually do not follow some simple mathematical function, but rather are of a random nature. Fortunately, however, much can be learned about instrument performance by examining the response to certain, rather simple "standard" input functions. That is, just as you are able to analyze not the real *system*, but rather an idealized model of it, so also you can work not with the real *inputs* to a system, but rather with simplified representations of them. This simplification of inputs (just as that of systems) can be carried out at several different levels, which leads to either simple, rather inaccurate input functions that are readily handled mathematically or complex, more accurate representations that lead to mathematical difficulties.

We commence our study by considering several quite simple standard inputs that are in wide use. Although these inputs are, in general, only crude approximations to the actual inputs, they are extremely useful for studying the effects of parameter changes in a given instrument or for comparing the *relative* performance of two competitive measurement systems.

Step Response of First-Order Instruments

To apply a step input to a system, we assume that initially it is in equilibrium, with $q_i = q_o = 0$, when at time $t = 0$ the input quantity increases instantly by an amount q_{is} (see Fig. 3.39). For $t > 0$, Eq. (3.164) becomes

$$(\tau D + 1)q_o = Kq_{is} \tag{3.174}$$

It can be shown generally (by mathematical reasoning) or in any specific physical problem, such as the thermometer (by physical reasoning), that the initial condition for this situation is $q_o = 0$ for $t = 0^+$ ($t = 0^+$ means an infinitesimal time after $t = 0$). The complementary-function solution is

$$q_{ocf} = Ce^{-t/\tau} \tag{3.175}$$

while the particular solution is

$$q_{opi} = Kq_{is} \tag{3.176}$$

giving the complete solution as

$$q_o = Ce^{-t/\tau} + Kq_{is} \tag{3.177}$$

Applying the initial condition, we get

$$0 = C + Kq_{is}$$

$$C = -Kq_{is}$$

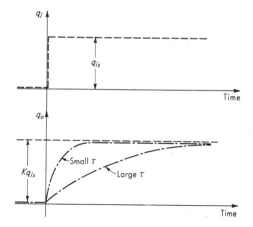

Figure 3.39 Step-function response of first-order instrument.

which gives finally

$$q_o = Kq_{is}(1 - e^{-t/\tau}) \qquad (3.178)$$

Examination of Eq. (3.178) shows that the speed of response depends on *only* the value of τ and is faster if τ is smaller. Thus in first-order instruments we strive to minimize τ for faithful dynamic measurements.

These results may be nondimensionalized by writing

$$\frac{q_o}{Kq_{is}} = 1 - e^{-t/\tau} \qquad (3.179)$$

and then plotting $q_o/(Kq_{is})$ versus t/τ, as in Fig. 3.40a. This curve is then universal for any value of K, q_{is}, or τ that might be encountered. We could also define the measurement error e_m as

$$e_m \triangleq q_i - \frac{q_o}{K} \qquad (3.180)$$

$$e_m = q_{is} - q_{is}(1 - e^{-t/\tau})$$

and nondimensionalize for plotting in Fig. 3.40b as

$$\frac{e_m}{q_{is}} = e^{-t/\tau} \qquad (3.181)$$

A dynamic characteristic useful in characterizing the speed of response of any instrument is the *settling time*. This is the time (after application of a step input) for the instrument to reach and stay within a stated plus-and-minus tolerance band around its final value. A small settling time thus indicates fast response. It is obvious that the numerical value of a settling time depends on the percentage tolerance band used; you must always state this. Thus you speak of, say, a 5

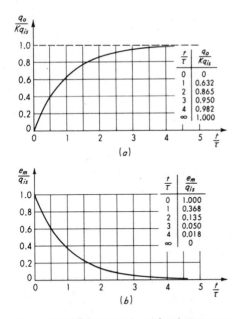

Figure 3.40 Nondimensional step-function response of first-order instrument.

percent settling time. For a first-order instrument a 5 percent settling time is equal to three time constants (see Fig. 3.41). Other percentages may be and are used in actual practice.

Knowing now that fast response requires a small value of τ, we can examine any specific first-order instrument to see what physical changes would be needed to reduce τ. If we use our thermometer example, Eq. (3.173) shows that τ may be reduced by

1. Reducing ρ, C, and V_b
2. Increasing U and A_b

Since ρ and C are properties of the fluid filling the thermometer, they cannot be varied independently of each other, and so for small τ we search for fluids with a small ρC product. The bulb volume V_b may be reduced, but this will also reduce A_b unless some extended-surface heat-transfer augmentation (such as fins on the bulb) is introduced. Even more significant is the effect of reduced V_b on the static sensitivity K, as given by Eq. (3.172). We see that attempts to reduce τ by decreasing V_b will result in reductions in K. Thus increased speed of response is traded off for lower sensitivity. This tradeoff is not unusual and will be observed in many other instruments.

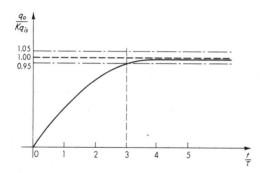

Figure 3.41 Settling-time definition.

The fact that τ depends on U means that we cannot state that a certain *thermometer* has a certain time constant, but only that a specific thermometer *used in a certain fluid under certain heat-transfer conditions* (say, free or forced convection) has a certain time constant. This is because U depends partly on the value of the film coefficient of heat transfer at the outside of the bulb, which varies greatly with changes in fluid (liquid or gas), flow velocity, etc. For example, a thermometer in stirred oil might have a time constant of 5 s while the same thermometer in stagnant air would have a τ of perhaps 100 s. Thus you must always be careful in giving (or using) performance data to be sure that the conditions of use correspond to those in force during calibration or that proper corrections are applied.

To illustrate the nature of nonlinear instrument responses, linearization techniques available for approximate analysis, and the utility of the digital simulation methods mentioned above, consider the vacuum furnace of Fig. 3.42a. The wall temperature T_i is steady at $400°C$ when a thermometer at $300°C$ is suddenly inserted, subjecting it to a $100°C$ step change. Because of the vacuum environment, heat transfer from the furnace to the thermometer is assumed to be strictly by radiation, and Eq. (3.169) assumes the form

$$MC\, dT_o = E(T_i^4 - T_o^4)\, dt \qquad \frac{dT_o}{dt} = \frac{E}{MC}(T_i^4 - T_o^4) = 10^{-8}(T_i^4 - T_o^4) \quad (3.182)$$

where some typical numerical values have been inserted for M, C, and E. While nonlinear Eq. (3.182) presents difficulties for analytical solution, digital simulation obtains a near-perfect numerical solution with very little effort, and approximate analytical linearizations are also available. The most common linearizing approximation is the Taylor-series method.[1] If Eq. (3.182) is considered as an isolated model relating T_o to T_i, then T_i plays the role of a given input (rather than an unknown) and the term T_i^4 need not be linearized to allow analytical solution. The Taylor-series approach then gives

$$T_o^4 \approx T_{oo}^4 + 4T_{oo}^3(T_o - T_{oo}) = -4.50 \times 10^{10} + (1.715 \times 10^8)T_o \quad (3.183)$$

[1] Doebelin, "System Modeling and Response," p. 24.

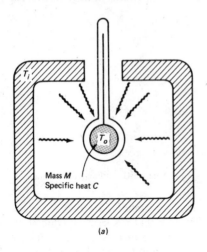

(a)

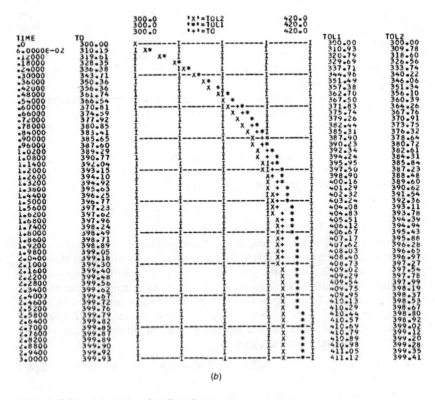

(b)

Figure 3.42 Digital simulation of nonlinear instrument.

where the operating point T_{oo} has been chosen as 350°C, midway between 300 and 400°C. The linearized approximate version of Eq. (3.182) is then

$$0.583 \frac{dT_o}{dt} + T_o = 411.8 \qquad (3.184)$$

This can easily be solved analytically and clearly has a time constant $\tau = 0.583$ s.

If the system under study included not only the thermometer response model but also, say, a temperature control system for the furnace, then T_i would now play the role of an *unknown* and T_i^4 would also need to be linearized:

$$\frac{dT_o}{dt} = 10^{-8}[T_{io}^4 + 4T_{io}^3(T_i - T_{io}) - T_{oo}^4 - 4T_{oo}^3(T_o - T_{oo})] \qquad (3.185)$$

$$0.583 \frac{dT_o}{dt} + T_o = T_i = 400 \qquad (3.186)$$

A final linearizing scheme, which does not employ Taylor series, expands

$$T_i^4 - T_o^4 = (T_i^2 + T_o^2)(T_i + T_o)(T_i - T_o) \qquad (3.187)$$

and then takes $(T_i^2 + T_o^2)(T_i + T_o)$ constant at the operating point (350°C) values to give

$$0.583 \frac{dT_o}{dt} + T_o = T_i = 400$$

which is the same result as Eq. (3.186).

Even though the linearized approximate equations could easily be solved analytically, for convenience we use the digital simulation language CSMP to solve and graph (3.182), (3.184), and (3.186):

PARAM $K = 1.E - 08$,TI $= 400.$,TOI $= 300.$ gives parameter values

TODOT $= K*(TI**4 - TO**4)$ computes highest derivative of unknown

TO $=$ INTGRL(TOI,TODOT) numerically integrates TODOT to get TO, starting from initial value TOI

TODTL1 $= 706.2 - 1.715*TOL1$ same as above for the
TOL1 $=$ INTGRL(TOI,TODTL1) TO linearized equation

TODTL2 $= 686. - 1.715*TOL2$ same as above for the
  TI and TO linearized
TOL2 $=$ INTGRL(TOI,TODTL2) equation

TIMER FINTIM $= 3.0$,DELT $= .001$,OUTDEL $= .06$ gives finish time, computing increment and plotting increment

OUTPUT   TO,TOL1,TOL2   requests graphs of TO, TOL1, TOL2
                           versus time

PAGE   GROUP,WIDTH = 50   requests all three curves to same
                             scale, 50 characters wide

END

Because CSMP (and other similar languages) have many "built-in" convenience features, we are able to solve the above linear and nonlinear equations with very little time and effort. Most of the statements are explained above. In the TIMER statement, we chose FINTIM = 3.0 s since our linearized models have $\tau = 0.583$ s and a first-order step response is nearly complete in 5τ. Computing increment DELT usually is taken near FINTIM/2000, but its value is not critical since CSMP self-adjusts to optimize accuracy and speed. Plotting increment OUTDEL generally is taken as FINTIM/50 to just fill the printer page unless this obscures details, in which case the curves are spread over several pages by using a smaller OUTDEL.

Figure 3.42b shows that the two approximate linearizations which follow Eq. (3.186) give very good results, while that of Eq. (3.184) suffers from an incorrect final value. It appears, then, that a linearized first-order model with $\tau = 0.583$ s would be acceptable for many purposes under the given conditions. Changing conditions (larger step inputs, input forms other than steps, etc.) may decrease the accuracy of the linearized model; however, all such situations can be easily studied by using CSMP and appropriate decisions can be made.

Ramp Response of First-Order Instruments

To apply a ramp input to a system, we assume that initially the system is in equilibrium, with $q_i = q_o = 0$, when at $t = 0$ the input q_i suddenly starts to change at a constant rate $\dot{q}_{is}$. We thus have

$$q_i = \begin{cases} q_o = 0 & t \leq 0 \\ \dot{q}_{is} t & t \geq 0 \end{cases} \tag{3.188}$$

and therefore $(\tau D + 1)q_o = K\dot{q}_{is}t$

The necessary initial condition again can be shown to be $q_o = 0$ for $t = 0^+$. Solution of Eq. (3.188) gives

$$q_{ocf} = Ce^{-t/\tau}$$

$$q_{opi} = K\dot{q}_{is}(t - \tau)$$

$$q_o = Ce^{-t/\tau} + K\dot{q}_{is}(t - \tau)$$

and applying the initial condition gives

$$q_o = K\dot{q}_{is}(\tau e^{-t/\tau} + t - \tau) \tag{3.189}$$

We again define measurement error e_m by

$$e_m \triangleq q_i - \frac{q_o}{K} = \dot{q}_{is}t - \dot{q}_{is}\tau e^{-t/\tau} - \dot{q}_{is}t + \dot{q}_{is}\tau \qquad (3.190)$$

$$e_m = \underbrace{-\dot{q}_{is}\tau e^{-t/\tau}}_{\substack{\text{transient error} \\ e_{m,t}}} + \underbrace{\dot{q}_{is}\tau}_{\substack{\text{steady-} \\ \text{state} \\ \text{error} \\ e_{m,ss}}} \qquad (3.191)$$

We note that the first term of e_m gradually will disappear as time goes by, and so it is called the *transient error*. The second term, however, persists forever and is thus called the *steady-state error*. The transient error disappears more quickly if τ is small. The steady-state error is directly proportional to τ; thus small τ is desirable here also. Steady-state error also increases directly with $\dot{q}_{is}$, the rate of change of the measured quantity. In steady state, the horizontal (time) displacement between input and output curves is seen to be τ, and so we may make the interpretation that the instrument is reading what the input *was* τ seconds ago. The above results, together with a nondimensionalized representation, are given graphically in Fig. 3.43.

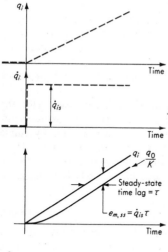

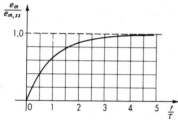

Figure 3.43 Ramp response of first-order instrument.

Frequency Response of First-Order Instruments

Equation (3.143) may be applied directly to the problem of finding the response of first-order systems to sinusoidal inputs. We have

$$\frac{q_o}{q_i}(i\omega) = \frac{K}{i\omega\tau + 1} = \frac{K}{\sqrt{\omega^2\tau^2 + 1}} \angle (\tan^{-1} - \omega\tau) \qquad (3.192)$$

Thus the amplitude ratio is

$$\frac{A_o}{A_i} = \left| \frac{q_o}{q_i}(i\omega) \right| = \frac{K}{\sqrt{\omega^2\tau^2 + 1}} \qquad (3.193)$$

and the phase angle

$$\phi = \angle \frac{q_o}{q_i}(i\omega) = \tan^{-1} - \omega\tau \qquad (3.194)$$

The ideal frequency response (zero-order instrument) would have

$$\frac{q_o}{q_i}(i\omega) = K \angle 0° \qquad (3.195)$$

Thus a first-order instrument approaches perfection if Eq. (3.192) approaches Eq. (3.195). We see this occur if the product $\omega\tau$ is sufficiently small. Thus for *any* τ there will be some frequency of input ω below which measurement is accurate; or, alternatively, if a q_i of high frequency ω must be measured, the instrument used must have a sufficiently small τ. Again, we see that accurate dynamic measurement requires a small time constant.

If we were concerned with the measurement of *pure* sine waves only, the above considerations would not be very pertinent since if we knew the frequency and τ, we could easily correct for amplitude attenuation and phase shift by simple calculations. In actual practice, however, q_i is often a combination of several sine waves of different frequencies. An example will show the importance of adequate frequency response under such conditions. Suppose we must measure a q_i given by

$$q_i = 1 \sin 2t + 0.3 \sin 20t \qquad (3.196)$$

(where t is in seconds) with a first-order instrument whose τ is 0.2 s. Since this is a linear system, we may use the superposition principle to find q_o. We first evaluate the sinusoidal transfer function at the two frequencies of interest:

$$\left. \frac{q_o}{q_i}(i\omega) \right|_{\omega = 2} = \frac{K}{\sqrt{0.16 + 1}} \angle -21.8° = 0.93K \angle -21.8° \qquad (3.197)$$

$$\left. \frac{q_o}{q_i}(i\omega) \right|_{\omega = 20} = \frac{K}{\sqrt{16 + 1}} \angle -76° = 0.24K \angle -76° \qquad (3.198)$$

We can then write q_o as

$$q_o = (1)(0.93K) \sin (2t - 21.8°) + (0.3)(0.24K) \sin (20t - 76°) \qquad (3.199)$$

$$\frac{q_o}{K} = 0.93 \sin (2t - 21.8°) + 0.072 \sin (20t - 76°) \qquad (3.200)$$

Since ideally $q_o/K = q_i$, comparison of Eq. (3.200) with (3.196) shows the presence of considerable measurement error. A graph of these two equations in Fig. 3.44*b* shows that the instrument gives a severely distorted measurement of the input. Furthermore, the high-frequency (20 rad/s) component present in the instrument output is now so small relative to the low-frequency component that any attempts at correction are not only inconvenient, but also inaccurate.

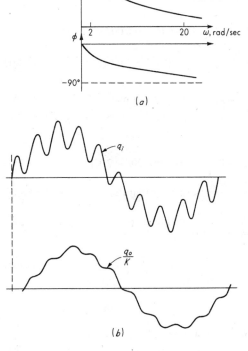

Figure 3.44 Example of inadequate frequency response.

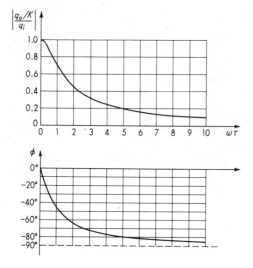

Figure 3.45 Frequency response of first-order instrument.

Suppose we consider use of an instrument with $\tau = 0.002$ s. Then we have

$$\frac{q_o}{q_i}(i\omega)\bigg|_{\omega=2} = \frac{K}{\sqrt{1.6 \times 10^{-5} + 1}} \angle -0.23° = 1.00K \angle -0.23° \qquad (3.201)$$

$$\frac{q_o}{q_i}(i\omega)\bigg|_{\omega=20} = \frac{K}{\sqrt{1.6 \times 10^{-3} + 1}} \angle -2.3° = 1.00K \angle -2.3° \qquad (3.202)$$

which yields

$$\frac{q_o}{K} = 1.00 \sin(2t - 0.23°) + 0.3 \sin(20t - 2.3°) \qquad (3.203)$$

Comparison of Eq. (3.196) and (3.203) shows clearly that this instrument faithfully measures the given q_i.

A nondimensional representation of the frequency response of any first-order system may be obtained by writing Eq. (3.192) as

$$\frac{q_o/K}{q_i}(i\omega) = \frac{1}{\sqrt{(\omega\tau)^2 + 1}} \angle (\tan^{-1} - \omega\tau) \qquad (3.204)$$

and plotting as in Fig. 3.45.

Impulse Response of First-Order Instruments

The final standard input we consider is the *impulse function*. Consider the pulse function $p(t)$ defined graphically in Fig. 3.46a. The impulse function of "strength"

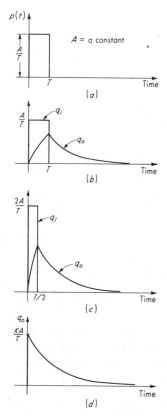

Figure 3.46 Impulse response of first-order instrument.

(area) A is defined by the limiting process

$$\text{Impulse function of strength } A \triangleq \lim_{T \to 0} p(t) \qquad (3.205)$$

We see that this "function" has rather peculiar properties. Its time duration is infinitesimal, its peak is infinitely high, and its area is A. If A is taken as 1, it is called the *unit* impulse function $\delta(t)$. Thus an impulse function of any strength A may be written as $A\delta(t)$. This rather peculiar function plays an important role in system dynamic analysis, as we see in greater detail later.

We now find the response of a first-order instrument to an impulse input. We do this by finding the response to the pulse $p(t)$ and then applying the limiting process to the result. For $0 < t < T$ we have

$$(\tau D + 1)q_o = Kq_i = \frac{KA}{T} \qquad (3.206)$$

Since, up until time T, this is no different from a *step* input of size A/T, our initial condition is $q_o = 0$ at $t = 0^+$, and the complete solution is

$$q_o = \frac{KA}{T}(1 - e^{-t/\tau}) \quad . \tag{3.207}$$

However, this solution is valid only up to time T. At this time we have

$$q_o \Big|_{t=T} = \frac{KA}{T}(1 - e^{-T/\tau}) \tag{3.208}$$

Now for $t > T$, our differential equation is

$$(\tau D + 1)q_o = Kq_i = 0 \tag{3.209}$$

which gives

$$q_o = Ce^{-t/\tau} \tag{3.210}$$

The constant C is found by imposing initial condition (3.208),

$$\frac{KA}{T}(1 - e^{-T/\tau}) = Ce^{-T/\tau} \tag{3.211}$$

$$C = \frac{KA(1 - e^{-T/\tau})}{Te^{-T/\tau}} \tag{3.212}$$

giving finally

$$q_o = \frac{KA(1 - e^{-T/\tau})e^{-t/\tau}}{Te^{-T/\tau}} \tag{3.213}$$

Figure 3.46b shows a typical response, and Fig. 3.46c shows the effect of cutting T in half. As T is made shorter and shorter, the first part ($t < T$) of the response becomes of negligible consequence, so that we can get an expression for q_o by taking the limit of Eq. (3.213) as $T \to 0$.

$$\lim_{T \to 0} \left\{ \frac{KA(1 - e^{-T/\tau})}{Te^{-T/\tau}} \right\} e^{-t/\tau} = KAe^{-t/\tau} \lim_{T \to 0} \frac{1 - e^{-T/\tau}}{Te^{-T/\tau}} \tag{3.214}$$

$$\lim_{T \to 0} \frac{1 - e^{-T/\tau}}{T} = \frac{0}{0} \quad \text{an indeterminate form}$$

Applying L'Hospital's rule yields

$$\lim_{T \to 0} \frac{1 - e^{-T/\tau}}{T} = \lim_{T \to 0} \frac{(1/\tau)e^{-T/\tau}}{1} = \frac{1}{\tau} \tag{3.215}$$

Thus we have finally for the impulse response of a first-order instrument

$$q_0 = \frac{KA}{\tau} e^{-t/\tau} \tag{3.216}$$

which is plotted in Fig. 3.46d.

We note that the output q_o is also "peculiar" in that it has an infinite (vertical) slope at $t = 0$ and thus goes from zero to a finite value in infinitesimal time. Such behavior is clearly impossible for a physical system since it requires energy transfer at an infinite rate. In our thermometer example, for instance, to cause the temperature of the fluid in the bulb *suddenly* to rise a finite amount requires an infinite rate of heat transfer. Mathematically, this infinite rate of heat transfer is provided by having the input $T_i(t)$ be infinite, i.e., an impulse function. In actuality, of course, T_i cannot go to infinity; however, if it is large enough and of sufficiently short duration (relative to the response speed of the system), the system may respond very nearly as it would for a perfect impulse.

To illustrate this, suppose in Fig. 3.46a we take $A = 1$ and $T = 0.01\tau$. The response to this approximate unit impulse is

$$q_o = \begin{cases} \dfrac{100K}{\tau}(1 - e^{-t/\tau}) & 0 \leq t \leq T \quad\quad (3.217) \\[4mm] \dfrac{100K(1 - e^{-0.01})e^{-t/\tau}}{\tau e^{-0.01}} & T \leq t \leq \infty \quad\quad (3.218) \end{cases}$$

Figure 3.47 gives a tabular and graphical comparison of the exact and approximate response, showing excellent agreement. The agreement is quite acceptable in most cases if T/τ is even as large as 0.1. It can also be shown that *the shape of the pulse is immaterial*; as long as its duration is sufficiently short, only its *area* matters. The plausibility of this statement may be shown by integrating the terms in the differential equation as follows:

$$\tau \frac{dq_o}{dt} + q_o = Kq_i \quad\quad (3.219)$$

$$\int_0^{0^+} \tau \, dq_o + \int_0^{0^+} q_o \, dt = \int_0^{0^+} Kq_i \, dt \quad\quad (3.220)$$

$$\tau \left(q_o \Big|_{0^+} - q_o \Big|_0 \right) + 0 = K \text{ (area under } q_i \text{ curve from } t = 0 \text{ to } t = 0^+ \text{)} \quad (3.221)$$

$$q_o \Big|_{0^+} = \frac{K}{\tau} \text{ (area of impulse)} \quad\quad (3.222)$$

This analysis holds strictly for an exact impulse and is a good approximation for a pulse of arbitrary shape if its duration is sufficiently short. It should be noted that, since the right side of the differential equation (3.219) is zero for $t > 0^+$, an impulse (or a short pulse) is equivalent to a zero forcing function and a nonzero initial $(t = 0^+)$ condition. That is, the solution of

$$(\tau D + 1)q_o = 0$$

$$q_o = \frac{K}{\tau} \quad \text{at } t = 0^+ \quad\quad (3.223)$$

is exactly the same as the impulse response.

Another interesting aspect of the impulse function is its relation to the step function. Since a perfect step function is also physically unrealizable because it

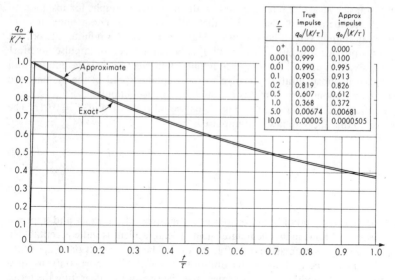

$\dfrac{t}{\tau}$	True impulse $q_0/(K/\tau)$	Approx impulse $q_0/(K/\tau)$
0^+	1.000	0.000
0.001	0.999	0.100
0.01	0.990	0.995
0.1	0.905	0.913
0.2	0.819	0.826
0.5	0.607	0.612
1.0	0.368	0.372
5.0	0.00674	0.00681
10.0	0.00005	0.0000505

Figure 3.47 Exact and approximate impulse response.

changes from one level to another in infinitesimal time, consider an approximation such as in Fig. 3.48. If this approximate step function is fed into a differentiating device, the output will be a pulse-type function. As the approximate step function is made to approach the mathematical ideal more and more

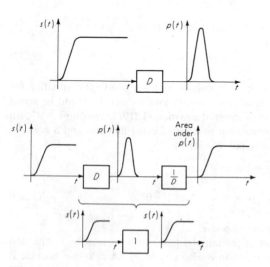

Figure 3.48 Approximate step and impulse functions.

closely, the output of the differentiating device will approach a perfect impulse function. *In this sense, the impulse function may be thought of as the derivative of the step function*, even though the discontinuities in the step function preclude the rigorous application of the basic definition of the derivative. In Fig. 3.48, the truth of these assertions is demonstrated by passing the output of the differentiating device through an integrating device $(1/D)$.

Second-Order Instrument

A *second-order instrument* is one that follows the equation

$$a_2 \frac{d^2 q_o}{dt^2} + a_1 \frac{dq_o}{dt} + a_0 q_o = b_0 q_i \tag{3.224}$$

Again, a second-order *equation* could have more terms on the right-hand side, but in common engineering usage, Eq. (3.224) is generally accepted as defining a second-order *instrument*.

The essential parameters in Eq. (3.224) can be reduced to three:

$$K \triangleq \frac{b_0}{a_0} \triangleq \text{static sensitivity} \tag{3.225}$$

$$\omega_n \triangleq \sqrt{\frac{a_0}{a_2}} \triangleq \text{undamped natural frequency, rad/time} \tag{2.226}$$

$$\zeta \triangleq \frac{a_1}{2\sqrt{a_0 a_2}} \triangleq \text{damping ratio, dimensionless} \tag{3.227}$$

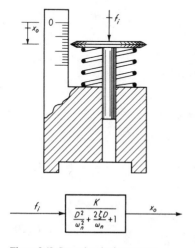

Figure 3.49 Second-order instrument.

which gives

$$\left(\frac{D^2}{\omega_n^2} + \frac{2\zeta D}{\omega_n} + 1\right) q_0 = K q_i \tag{3.228}$$

The operational transfer function is thus

$$\frac{q_o}{q_i}(D) = \frac{K}{D^2/\omega_n^2 + 2\zeta D/\omega_n + 1} \tag{3.229}$$

A good example of a second-order instrument is the force-measuring spring scale of Fig. 3.49. We assume the applied force f_i has frequency components only well below the natural frequency of the spring itself. Then the main dynamic effect of the spring may be taken into account by adding one-third of the spring's mass to the main moving mass. This total mass we call M. The spring is assumed linear with spring constant K_s newtons per meter. Although in a real scale there might be considerable dry friction, we assume perfect film lubrication and therefore a viscous damping effect with constant B (in newtons per meter per second).

The scale can be adjusted so that $x_o = 0$ when $f_i = 0$ (gravity force will then drop out of the equation), which yields

$$\Sigma \text{ forces} = (\text{mass})(\text{acceleration})$$

$$f_i - B\frac{dx_o}{dt} - K_s x_o = M\frac{d^2 x_o}{dt^2} \tag{3.230}$$

$$(MD^2 + BD + K_s)x_o = f_i \tag{3.231}$$

Noting this to fit the second-order model, we immediately define

$$K \triangleq \frac{1}{K_s} \qquad \text{m/N} \tag{3.232}$$

$$\omega_n \triangleq \sqrt{\frac{K_s}{M}} \qquad \text{rad/s} \tag{3.233}$$

$$\zeta \triangleq \frac{B}{2\sqrt{K_s M}} \tag{3.234}$$

Step Response of Second-Order Instruments

For a step input of size q_{is} we get

$$\left(\frac{D^2}{\omega_n^2} + \frac{2\zeta D}{\omega_n} + 1\right) q_o = K q_{is} \tag{3.235}$$

with initial conditions

$$q_o = 0 \qquad \text{at } t = 0^+$$

$$\frac{dq_o}{dt} = 0 \qquad \text{at } t = 0^+ \tag{3.236}$$

The particular solution of Eq. (3.235) is clearly $q_{opi} = Kq_{is}$. The complementary-function solution takes on one of three possible forms, depending on whether the roots of the characteristic equation are real and unrepeated (overdamped case), real and repeated (critically damped case), or complex (underdamped case). The complete solutions of Eq. (3.235) with initial conditions (3.236) are, in nondimensional form,

$$\frac{q_o}{Kq_{is}} = -\frac{\zeta + \sqrt{\zeta^2 - 1}}{2\sqrt{\zeta^2 - 1}} e^{(-\zeta + \sqrt{\zeta^2 - 1})\,\omega_n t}$$

$$+ \frac{\zeta - \sqrt{\zeta^2 - 1}}{2\sqrt{\zeta^2 - 1}} e^{(-\zeta - \sqrt{\zeta^2 - 1})\,\omega_n t} + 1 \qquad \text{overdamped} \qquad (3.237)$$

$$\frac{q_o}{Kq_{is}} = -(1 + \omega_n t)e^{-\omega_n t} + 1 \qquad \text{critically damped} \qquad (3.238)$$

$$\frac{q_o}{Kq_{is}} = -\frac{e^{-\zeta\omega_n t}}{\sqrt{1 - \zeta^2}} \sin\left(\sqrt{1 - \zeta^2}\,\omega_n t + \phi\right) + 1 \qquad \text{underdamped} \qquad (3.239)$$

$$\phi \triangleq \sin^{-1}\sqrt{1 - \zeta^2}$$

Since t and ω_n always appear as the product $\omega_n t$, the curves of $q_o/(Kq_{is})$ may be plotted against $\omega_n t$, which makes them universal for any ω_n, as in Fig. 3.50. This fact also shows that ω_n is *a direct indication of speed of response.* For a given ζ, doubling ω_n will halve the response time since $\omega_n t$ [and thus $q_o/(Kq_{is})$] achieves the same value at one-half the time. The effect of ζ is not clearly perceived from the equations, but is evident from the graphs. An increase in ζ reduces oscillation,

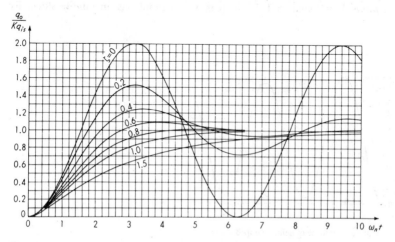

Figure 3.50 Nondimensional step-function response of second-order instrument.

but also slows the response in the sense that the first crossing of the final value is retarded. A settling time actually may be a better indication of response speed; however, then the optimum value of ζ will vary with the chosen tolerance band. For example, if we choose a 10 percent settling time, the curve for $\zeta = 0.6$ gives a settling time of about $2.4/\omega_n$, and this is optimum, since either larger or smaller ζ gives a longer settling time. However, if we had chosen a 5 percent settling time, a ζ between 0.7 and 0.8 would give the shortest value. In choosing a proper ζ value for a practical application, the situation is further complicated by the fact that the real inputs will not be step functions and their *actual* form influences what will be the best ζ value. If the actual inputs are quite variable in form, some compromise must be struck. Many commercial instruments use $\zeta = 0.6$ to 0.7. We show shortly that this range of ζ gives good frequency response over the widest frequency range.

Terminated-Ramp Response of Second-Order Instruments

Under certain circumstances, the response of second-order instruments to perfect step inputs is misleading. The best example is perhaps found in piezoelectric pressure pickups, accelerometers, etc. While these devices are discussed in detail later, at present it is sufficient to state that usually they have an extremely high natural frequency and very little damping ($\zeta < 0.01$, often). Based on a perfect step input, such an instrument appears highly undesirable because of its large overshoot and strong oscillation (Fig. 3.51). Actually, these instruments may give excellent response. The explanation of this apparent inconsistency lies in the fact that *perfect* step inputs do not occur in nature, since a macroscopic quantity cannot change a finite amount in an infinitesimal time. Thus a more realistic input than the step is the *terminated-ramp input*, defined in Fig. 3.52. This input has a *finite* slope equal to $1/T$, whereas a step input has an infinite slope. By

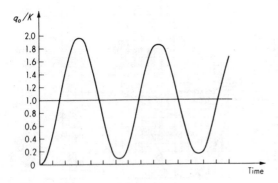

Figure 3.51 Step response of lightly damped system.

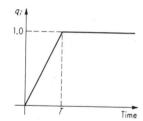

Figure 3.52 Terminated-ramp input.

letting T get smaller and smaller, we can approach the perfect step input. For a second-order system, we would have

$$\left(\frac{D^2}{\omega_n^2} + \frac{2\zeta D}{\omega_n} + 1\right) q_o = K q_i \tag{3.240}$$

$$q_i = \begin{cases} \dfrac{t}{T} & 0 \le t \le T \\ 1.0 & T \le t < \infty \end{cases} \tag{3.241}$$

$$q_o = \frac{dq_o}{dt} = 0 \qquad \text{at } t = 0^+ \tag{3.242}$$

Since we are concerned here with lightly damped systems, we obtain the solution for only the underdamped case:

$$\frac{q_o}{K} = \frac{t}{T} - \frac{2\zeta}{\omega_n T} + \frac{1}{\omega_n T \sqrt{1 - \zeta^2}} \, e^{-\zeta\omega_n t} \sin\left(\sqrt{1 - \zeta^2} \, \omega_n t + \phi\right) \quad 0 \le t \le T \tag{3.243}$$

$$\frac{q_o}{K} = \frac{t}{T} - \frac{2\zeta}{\omega_n T} + \frac{1}{\omega_n T \sqrt{1 - \zeta^2}} \, e^{-\zeta\omega_n t} \sin\left(\sqrt{1 - \zeta^2} \, \omega_n t + \phi\right)$$

$$- \left\{ \frac{t}{T} - 1 - \frac{2\zeta}{\omega_n T} + \frac{1}{\omega_n T \sqrt{1 - \zeta^2}} \, e^{-\zeta\omega_n(t - T)} \sin\left[\sqrt{1 - \zeta^2} \, \omega_n(t - T) + \phi\right] \right\}$$

$$T \le t < \infty \tag{3.244}$$

$$\phi \triangleq 2 \tan^{-1} \frac{\sqrt{1 - \zeta^2}}{\zeta} \tag{3.245}$$

From Eq. (3.243) we note immediately that, for $0 \le t \le T$, the following is true:

1. There is a steady-state error of size $2\zeta/(\omega_n T)$.
2. The transient error can be no larger than $1/(\omega_n T \sqrt{1 - \zeta^2})$.

Thus if $\zeta = 0$ (no damping), the steady-state error is zero and the "transient" error is a sustained sine wave of amplitude $1/(\omega_n T)$. *Therefore, if ω_n is sufficiently*

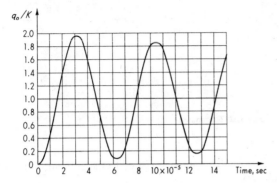

Figure 3.53 Step response of lightly damped system.

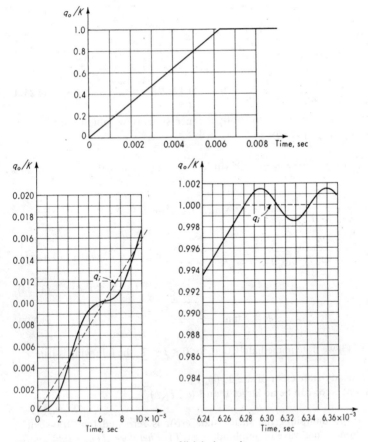

Figure 3.54 Terminated-ramp response of lightly damped system.

132

large relative to $1/T$, *the transient error can be made very small even if the damping is practically nonexistent.* This result is based on Eq. (3.243), but similar results are obtained from (3.244) for $T \leq t \leq \infty$, since the transient induced at $t = 0$ by the increasing ramp is essentially the same as that induced at $t = T$ by a decreasing ramp. That is, the q_i of Fig. 3.52 is really a superposition of an increasing ramp starting at $t = 0$ and a decreasing ramp starting at $t = T$.

As a numerical example, suppose a pressure pickup with $\zeta = 0.01$ and $\omega_n = 100,000$ rad/s is subjected to terminated-ramp-type inputs with $T = 0.00628$ s. The step response of such an instrument is shown in Fig. 3.53 and indicates the severe overshooting and oscillation which could lead us to reject the instrument. Figure 3.54, however, shows the terminated-ramp response corresponding to the *actual* input. It is clear that the response is almost perfect. In fact, if we allow a transient error $1/(\omega_n T)$ of 1 percent, T can be as short as 0.001.

Ramp Response of Second-Order Instruments

The differential equation here is

$$\left(\frac{D^2}{\omega_n^2} + \frac{2\zeta D}{\omega_n} + 1 \right) q_o = K \dot{q}_{is} t \tag{3.246}$$

$$q_o = \frac{dq_o}{dt} = 0 \qquad \text{at } t = 0^+$$

The solutions are found to be

$$\frac{q_o}{K} = \dot{q}_{is} t - \frac{2\zeta \dot{q}_{is}}{\omega_n} \left(1 + \frac{2\zeta^2 - 1 - 2\zeta \sqrt{\zeta^2 - 1}}{4\zeta \sqrt{\zeta^2 - 1}} e^{(-\zeta + \sqrt{\zeta^2 - 1})\omega_n t} \right.$$

$$\left. + \frac{-2\zeta^2 + 1 - 2\zeta \sqrt{\zeta^2 - 1}}{4\zeta \sqrt{\zeta^2 - 1}} e^{(-\zeta + \sqrt{\zeta^2 - 1})\omega_n t} \right) \qquad \text{overdamped} \tag{3.247}$$

$$\frac{q_o}{K} = \dot{q}_{is} t - \frac{2\dot{q}_{is}}{\omega_n} \left[1 - e^{-\omega_n t} \left(1 + \frac{\omega_n t}{2} \right) \right] \qquad \text{critically damped} \tag{3.248}$$

$$\frac{q_o}{K} = \dot{q}_{is} t - \frac{2\zeta \dot{q}_{is}}{\omega_n} \left[1 - \frac{e^{-\zeta \omega_n t}}{2\zeta \sqrt{1 - \zeta^2}} \sin \left(\sqrt{1 - \zeta^2} \, \omega_n t + \phi \right) \right] \tag{3.249}$$

$$\tan \phi = \frac{2\zeta \sqrt{1 - \zeta^2}}{2\zeta^2 - 1} \qquad \text{underdamped}$$

Figure 3.55 shows the general character of the response. There is a steady-state error $2\zeta \dot{q}_{is}/\omega_n$. Since the value of $\dot{q}_{is}$ is set by the measured quantity, the steady-state error can be reduced only by reducing ζ and increasing ω_n. For a given ω_n, reduction in ζ results in larger oscillations. There is also a steady-state time lag $2\zeta/\omega_n$. Figure 3.56 gives a set of nondimensionalized curves that summarize system behavior.

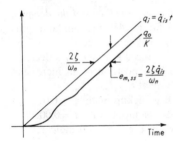

Figure 3.55 Ramp response of second-order instrument.

Frequency Response of Second-Order Instruments

The sinusoidal transfer function is

$$\frac{q_o}{q_i}(i\omega) = \frac{K}{(i\omega/\omega_n)^2 + 2\zeta i\omega/\omega_n + 1} \qquad (3.250)$$

which can be put in the form

$$\frac{q_o/K}{q_i}(i\omega) = \frac{1}{\sqrt{[1 - (\omega/\omega_n)^2]^2 + 4\zeta^2\omega^2/\omega_n^2}} \angle \phi \qquad (3.251)$$

$$\phi \triangleq \tan^{-1}\frac{2\zeta}{\omega/\omega_n - \omega_n/\omega} \qquad (3.252)$$

Figure 3.57 gives the nondimensionalized frequency-response curves. Clearly, increasing ω_n will increase the range of frequencies for which the amplitude-ratio curve is relatively flat; thus a high ω_n is needed to measure accurately high-frequency q_i's. An optimum range of values for ζ is indicated by both amplitude-ratio and phase-angle curves. The widest flat amplitude ratio exists for ζ of about 0.6 to 0.7. While zero phase angle would be ideal, it is rarely possible to realize this even approximately. Actually, if the main interest is in q_o reproducing the

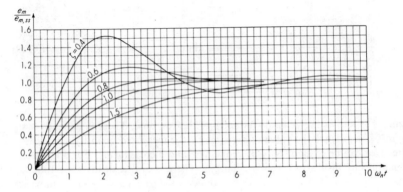

Figure 3.56 Nondimensional ramp response.

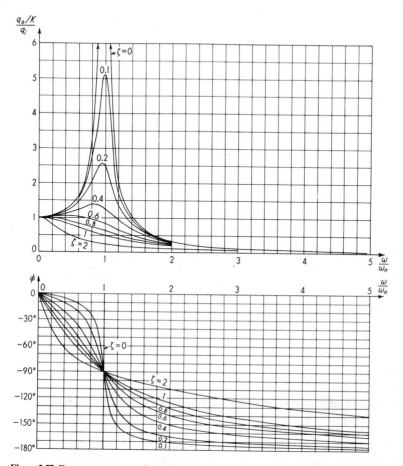

Figure 3.57 Frequency response of second-order instrument.

correct *shape* of q_i and if a time delay is acceptable, we show shortly that ϕ need not be zero; rather, it should vary *linearly* with frequency ω. Examining the phase curves of Fig. 3.57, we note that the curves for $\zeta = 0.6$ to 0.7 are nearly straight for the widest frequency range. These considerations lead to the widely accepted choice of $\zeta = 0.6$ to 0.7 as the optimum value of damping for second-order instruments. There are exceptions, however, as noted in the section on terminated-ramp response.

Impulse Response of Second-Order Instruments

In the section on first-order instruments we showed that the impulse response is equivalent to the free (unforced) response if the initial ($t = 0^+$) conditions produced by the impulse are taken into account. To find the initial conditions

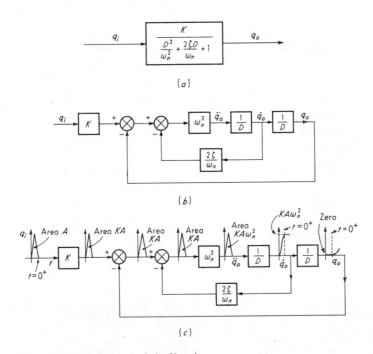

Figure 3.58 Block-diagram analysis of impulse response.

produced by applying an impulse of area A to a second-order instrument, redraw the block diagram of Fig. 3.58a as in Fig. 3.58b. (The equivalence of the two diagrams is easily demonstrated by tracing through the signals in Fig. 3.58b to get the differential equation relating q_o to q_i.) In Fig. 3.58c the impulse is applied at q_i, and the "propagation" of this input signal is traced through the rest of the diagram. This analysis shows that at $t = 0^+$ we have $q_o = 0$ and $\dot{q}_o = KA\omega_n^2$. The differential equation to be solved is then

$$\left(\frac{D^2}{\omega_n^2} + \frac{2\zeta D}{\omega_n} + 1\right) q_o = 0$$

$$q_o = 0 \qquad \frac{dq_o}{dt} = KA\omega_n^2 \qquad \text{at } t = 0^+ \qquad (3.253)$$

The solutions are found to be

$$\frac{q_o}{KA\omega_n} = \frac{1}{2\sqrt{\zeta^2 - 1}} \left(e^{(-\zeta + \sqrt{\zeta^2 - 1})\omega_n t} - e^{(-\zeta - \sqrt{\zeta^2 - 1})\omega_n t}\right) \qquad \text{overdamped}$$

$$(3.254)$$

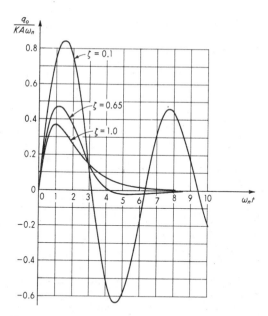

Figure 3.59 Nondimensional impulse response of second-order instrument.

$$\frac{q_o}{KA\omega_n} = \omega_n t e^{-\omega_n t} \qquad \text{critically damped} \tag{3.255}$$

$$\frac{q_o}{KA\omega_n} = \frac{1}{\sqrt{1-\zeta^2}} e^{-\zeta\omega_n t} \sin\left(\sqrt{1-\zeta^2}\,\omega_n t\right) \qquad \text{underdamped} \tag{3.256}$$

Figure 3.59 displays these results graphically.

Dead-Time Elements

Some components of measuring systems are adequately represented as dead-time elements. A *dead-time element* is defined as a system in which the output is exactly the same form as the input, but occurs τ_{dt} seconds (the dead time) later. Mathematically,

$$q_o(t) = K q_i(t - \tau_{dt}) \qquad t \geq \tau_{dt} \tag{3.257}$$

This type of element is also called a *pure delay* or *transport lag*. An example of such an effect is found in pneumatic signal-transmission systems. A pressure signal at one end of a length of pneumatic tubing will cause no response at all at the other end until the pressure wave has had time to propagate the distance between them. Because this speed of propagation is the same as the speed of

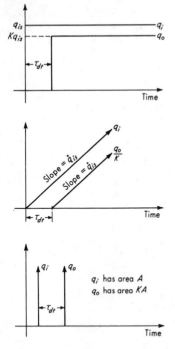

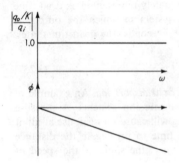

Figure 3.60 Dead-time responses.

sound, a 350-m length of tubing will have a dead time of about 1 s, since the speed of sound in standard air is 345 m/s.

The response of dead-time elements to the standard inputs is easily found. For steps, ramps, and impulses the results are given in Fig. 3.60. For sinusoidal input we have

$$q_i = A_i \sin \omega t$$

Figure 3.61 Dead-time frequency response.

and from Eq. (3.257)

$$q_o = KA_i \sin \omega(t - \tau_{dt}) \tag{3.258}$$

$$q_o = KA_i \sin (\omega t - \omega\tau_{dt}) = KA_i \sin (\omega t + \phi) \tag{3.259}$$

Thus

$$\frac{q_o/K}{q_i} (i\omega) = 1 \angle \phi = e^{-i\omega \tau_{dt}} \tag{3.260}$$

The frequency-response curves for a dead-time element are shown in Fig. 3.61.

Logarithmic Plotting of Frequency-Response Curves

We will find the frequency response of measurement systems extremely useful, so that rapid methods for getting the amplitude-ratio and phase-angle curves would be helpful. Certain logarithmic methods are in wide use and are explained now.

The sinusoidal transfer function of a measurement system generally can be put in the form

$$\frac{q_o}{q_i}(i\omega) = \frac{\begin{array}{c} K(i\omega)^n(i\omega\tau_1 + 1) \cdots (i\omega\tau_m + 1)[(i\omega/\omega_{n1})^2 + 2\zeta_1 i\omega/\omega_{n1} + 1] \\ \cdots [(i\omega/\omega_{nr})^2 + 2\zeta_r i\omega/\omega_{nr} + 1](e^{-i\omega\tau_{dt1}}) \cdots (e^{-i\omega\tau_{dtp}}) \end{array}}{\begin{array}{c} (i\omega\tau_I + 1) \cdots (i\omega\tau_M + 1)[(i\omega/\omega_{nI})^2 + 2\zeta_I i\omega/\omega_{nI} + 1] \\ \cdots [(i\omega/\omega_{nR})^2 + 2\zeta_R i\omega/\omega_{nR} + 1] \end{array}} \tag{3.261}$$

This follows from the fact that the polynomials in the numerator and denominator of Eq. (3.141) can, in general, be *factored* into terms of the form D^n, $\tau D + 1$, and $D^2/\omega_n^2 + 2\zeta D/\omega_n + 1$. Replacing D by $i\omega$ then gives Eq. (3.261) when dead-time elements are also included.

Since Eq. (3.261) is in the form of a *product* of complex numbers, the use of logarithms suggests itself as a means of replacing multiplication by addition. That is,

$$\frac{q_o}{q_i}(i\omega) = G_1(i\omega)G_2(i\omega) \cdots G_u(i\omega) \tag{3.262}$$

where the $G(i\omega)$ functions represent the various terms of Eq. (3.261). The amplitude ratio would be given by

$$\left|\frac{q_o}{q_i}(i\omega)\right| = |G_1(i\omega)||G_2(i\omega)| \cdots |G_u(i\omega)| \tag{3.263}$$

A widely used logarithmic method uses the *decibel notation* to express amplitude ratios. An amplitude ratio A is given in decibels (dB) by

$$\text{Decibel value} \triangleq \text{dB} \triangleq 20 \log A \tag{3.264}$$

Then

$$20 \log \left| \frac{q_o}{q_i} (i\omega) \right| = 20 \log \left[|G_1(i\omega)| |G_2(i\omega)| \cdots |G_u(i\omega)| \right] \qquad (3.265)$$

$$20 \log \left| \frac{q_o}{q_i} (i\omega) \right| = 20 \log |G_1(i\omega)| + 20 \log |G_2(i\omega)|$$

$$+ \cdots + 20 \log |G_u(i\omega)| \qquad (3.266)$$

Thus, if we get the amplitude-ratio curves for the individual terms in Eq. (3.261) in the decibel form, we can obtain the overall decibel curve by simple graphical addition. The phase-angle curves also are obtained by simple addition since we add phase angles when multiplying complex numbers.

We now show how to obtain, with a minimum of effort, the amplitude-ratio (dB) and phase-angle curves for each of the types of terms in Eq. (3.261). To put the curves in their simplest form, we plot against the logarithm of frequency rather than against frequency itself. The simplest term is the sensitivity K, which is a real number $K \angle 0°$; thus its decibel curve is a straight horizontal line through the decibel value of K, while its phase-angle curve is a straight horizontal line through zero degrees (see Fig. 3.62).

The next type of term is $(i\omega)^n$, where $n = \pm 1, \pm 2, \ldots$. The phase angle of such terms is constant with frequency and is given by $90n°$. The amplitude ratio is ω^n, so that the decibel value is

$$20 \log \omega^n = 20n \log \omega$$

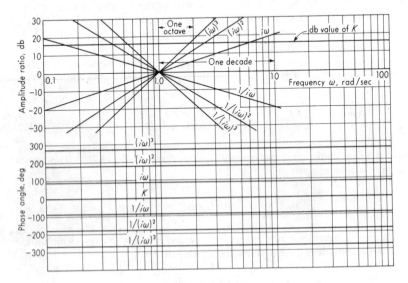

Figure 3.62 Integrator and differentiator frequency response.

Since we plot against log ω, the decibel curves become straight lines of slope $20n$ decibels per decade (See Fig. 3.62). A *decade* is defined as any 10-to-1 frequency range; an *octave* is any 2-to-1 range.

Terms of the form $i\omega\tau + 1$ and $1/(\iota\omega\tau + 1)$ give, respectively,

$$dB = 20 \log \sqrt{(\omega\tau)^2 + 1} \tag{3.267}$$

and

$$dB = -20 \log \sqrt{(\omega\tau)^2 + 1} \tag{3.268}$$

When $\omega\tau \gg 1$, these become

$$dB \approx 20 \log \omega\tau = 20 \log \tau + 20 \log \omega \tag{3.269}$$

and

$$dB \approx -20 \log \omega\tau = -20 \log \tau - 20 \log \omega \tag{3.270}$$

We see that both represent straight lines of slope ± 20 dB/decade, and these straight lines will be the high-frequency asymptotes of the actual amplitude-ratio curves. Similarly, for $\omega\tau \ll 1$,

$$dB \approx 20 \log 1 = 0 \tag{3.271}$$

and

$$dB \approx -20 \log 1 = 0 \tag{3.272}$$

so that the low-frequency asymptote is simply the 0-dB line. The two straight-line asymptotes will meet at $\omega\tau = 1$ because this is where (3.269) and (3.270) are zero. The point $\omega = 1/\tau$ is called the *breakpoint*, or *corner frequency*. In plotting curves for such terms, we first locate the breakpoint and then draw the two asymptotes. The true curve is obtained by correcting the straight-line asymptotes at several points, using the data of Fig. 3.63. The phase-angle curves may be quickly plotted by using the data of Fig. 3.64. A numerical example illustrating these methods is given in Fig. 3.65.

Terms of the form $[(i\omega/\omega_n)^2 + 2\zeta i\omega/\omega_n + 1]^{\pm 1}$ have low-frequency asymptotes of 0 dB and high-frequency asymptotes of slope ± 40 dB/decade. They intersect at $\omega = \omega_n$. The exact curves for a given ζ are obtained by applying the

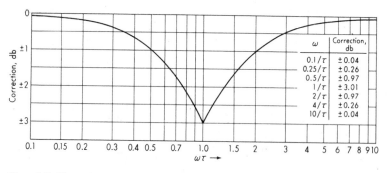

Figure 3.63 First-order-system amplitude-ratio corrections.

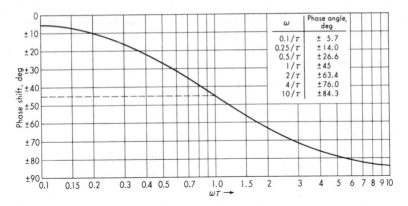

Figure 3.64 First-order-system phase angle.

corrections of Fig. 3.66. The phase-angle curves are obtained from Fig. 3.67. Figure 3.68 gives numerical examples.

The final type of term considered is the dead-time term $e^{-i\omega\tau_{dt}}$. Since the amplitude ratio is 1.0 for all frequencies, the decibel curve is simply the 0-dB line. The phase-angle curve is easily plotted from $\phi = -\omega\tau_{dt}$ for any given dead time.

To illustrate the procedure for combining the individual terms to obtain the overall frequency-response curves, we consider the following example:

$$\frac{q_o}{q_i}(i\omega) = \frac{4.4(i\omega)}{(i\omega + 1)(0.2i\omega + 1)} \tag{3.273}$$

Figure 3.69 shows the procedure and results.

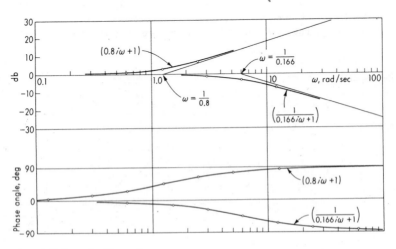

Figure 3.65 Example of first-order terms.

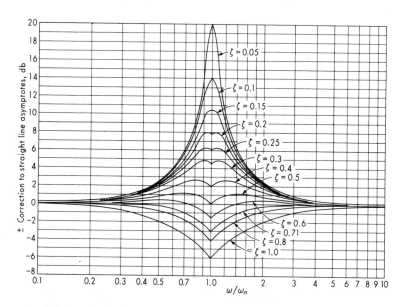

Figure 3.66 Second-order-system amplitude-ratio corrections.

While knowledge of these graphical methods will continue to be an important tool for the engineer, computer-aided design and analysis techniques should be employed when appropriate and if available. Most digital-computer libraries have "canned" programs already available to calculate system frequency response by using one or both Eqs. (3.143) and (3.261).

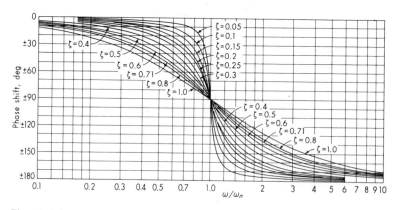

Figure 3.67 Second-order-system phase angle.

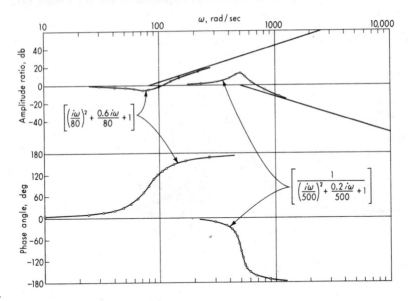

Figure 3.68 Example of second-order terms.

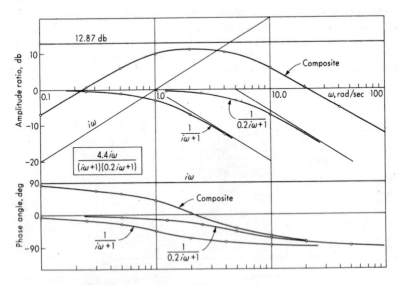

Figure 3.69 Example of frequency-response plot.

Response of a General Form of Instrument to a Periodic Input

Our approach to the dynamic response of measurement systems has, to this point, been limited in two ways. First, we considered only rather simple types of instruments (zero-order, first-order, and second-order). Second, we subjected these instruments to rather simple inputs (steps, ramps, sine waves, and impulses). At this point, by applying more advanced mathematical tools, we begin to remove both limitations. We see that the concept of frequency response plays a central role in these developments. The first step involves the study of the response of a general (linear, time-invariant) instrument to periodic inputs.

By a periodic function we mean one that repeats itself cyclically over and over, as in Fig. 3.70. If this function meets the *Dirichlet conditions*—(it must be single-valued, be finite, and have a finite number of discontinuities and maxima and minima in one cycle—it may be represented by a *Fourier series*.[1] That is,

$$q_i(t) = q_{i,\,av} + \frac{1}{L}\left(\sum_{n=1}^{\infty} a_n \cos \frac{n\pi t}{L} + \sum_{n=1}^{\infty} b_n \sin \frac{n\pi t}{L} \right) \tag{3.274}$$

where
$$q_{i,\,av} \triangleq \text{average value of } q_i = \frac{1}{2L} \int_{-L}^{L} q_i(t)\, dt \tag{3.275}$$

$$a_n \triangleq \int_{-L}^{L} q_i(t) \cos \frac{n\pi t}{L}\, dt \tag{3.276}$$

$$b_n \triangleq \int_{-L}^{L} q_i(t) \sin \frac{n\pi t}{L}\, dt \tag{3.277}$$

(The origin of the t coordinate may be chosen wherever most convenient.) Then we see that any periodic function satisfying the Dirichlet conditions can be replaced by a sum of terms consisting of a constant and sine and cosine waves of various frequencies. If we wish, any *pair* of sine and cosine terms of the *same* frequency can be replaced by a *single* sine wave of the same frequency at some phase angle, since

$$A \cos \omega t + B \sin \omega t \equiv C \sin (\omega t + \alpha) \tag{3.278}$$

where
$$C \triangleq \sqrt{A^2 + B^2} \tag{3.279}$$

$$\alpha \triangleq \tan^{-1} \frac{A}{B} \tag{3.280}$$

The Fourier series usually will be an *infinite* series, and to get a *perfect* reconstruction of $q_i(t)$ from the series, an infinite number of terms would have to be added. Fortunately, perfect reproduction is not required in engineering applications; thus generally $q_i(t)$ is *approximated* by a truncated (cut off after a certain number of terms) series. Just how many terms to use depends on both the form of

---

[1] B. J. Ley, S. G. Lutz, and C. F. Rehberg, "Linear Circuit Analysis," chap. 6, McGraw-Hill, New York, 1959.

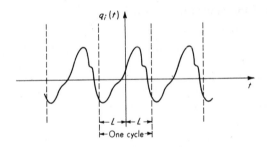

Figure 3.70 General periodic function.

$q_i(t)$ (if it has very sharp changes more terms are required) and the use to which the information is to be put. Often, less than 10 "harmonics" (the first 10 different frequencies) is adequate.

The method of obtaining the desired terms in the Fourier series depends on the nature of $q_i(t)$. If $q_i(t)$ is given as a known mathematical formula, Eqs. (3.274) to (3.277) may be employed. If the required integrations cannot be performed analytically because of the complexity of $q_i(t)$ or because it is given by a graph or table rather than a formula, then various approximate numerical methods are available.[1] Most computer libraries nowadays include one or several easy-to-use Fourier-series programs, so you rarely need to write one. Some hand calculators even provide modest Fourier-series capability.

Once the Fourier series for a particular $q_i(t)$ has been found, the *steady-state* response of any instrument to this input may be determined by use of frequency-response techniques and the principle of superposition. That is, the response for each individual sinusoidal term is found, and then they are added algebraically to get the total response. By use of Eqs. (3.278) to (3.280) all terms in the Fourier series can be put in the form

$$A_{ik} \sin(\omega_k t + \alpha_k)$$

We now define the complex number $Q_i(i\omega_k)$ by

$$Q_i(i\omega_k) \triangleq A_{ik} \angle \alpha_k \qquad (3.281)$$

For example, the constant term -7.2 becomes $7.2 \angle 180°$, and the term $9.3 \sin(20t + 37°)$ becomes $9.3 \angle 37°$. When the Fourier series representing $q_i(t)$ is expressed in this form, it is called $Q_i(i\omega)$, the *input-frequency spectrum*. Thus, if

$$q_i(t) = A_{i0} + A_{i1} \sin(\omega_1 t + \alpha_1) + A_{i2} \sin(\omega_2 t + \alpha_2) + \cdots \qquad (3.282)$$

then $\qquad Q_i(i\omega) = |A_{i0}| \angle (0° \text{ or } 180°) + A_{i1} \angle \alpha_1 + A_{i2} \angle \alpha_2 + \cdots \qquad (3.283)$

[1] J. B. Scarborough, "Numerical Mathematical Analysis," 3d ed., chap. 17, Johns Hopkins, Baltimore, 1958; R. K. Otnes and L. Enochson, "Applied Time Series Analysis," vol. 1, p. 230, Wiley, New York, 1978.

Such a spectrum, which exists only at isolated frequencies, is called a *discrete spectrum*.

Now, we recall that the sinusoidal transfer function of the system, $(q_o/q_i)(i\omega)$, is also a complex number for any given frequency. An alternative name for the sinusoidal transfer function is the *system frequency response*. If we pick any frequency ω_k and multiply the complex numbers $Q_i(i\omega_k)$ and $(q_o/q_i)(i\omega_k)$, we get another complex number that we define as $Q_o(i\omega_k)$. If we do this for *all* frequencies and add all the $Q_o(i\omega_k)$, the sum is called $Q_o(i\omega)$, the *output frequency spectrum*, and it has the form

$$Q_o(i\omega) = A_{o0} \angle (0° \text{ or } 180°) + A_{o1} \angle \beta_1 + A_{o2} \angle \beta_2 + \cdots \qquad (3.284)$$

This is now interpreted as in Eqs. (3.282) and (3.283) to give

$$q_o(t) = \pm A_{o0} + A_{o1} \sin(\omega_1 t + \beta_1) + A_{o2} \sin(\omega_2 t + \beta_2) + \cdots \qquad (3.285)$$

Since $q_i(t)$ is periodic, the frequencies ω_2, ω_3, etc., are all integer multiples of ω_1, and thus $q_o(t)$ also will be a periodic function. For accurate measurement, $q_i(t)$

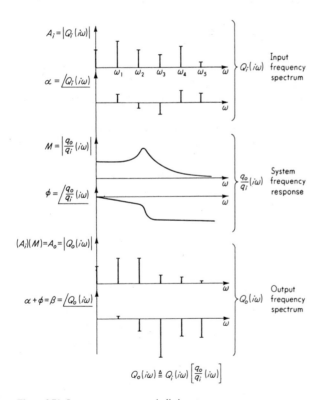

$$Q_o(i\omega) \triangleq Q_i(i\omega) \left[\frac{q_o}{q_i}(i\omega) \right]$$

Figure 3.71 System response to periodic input.

and $q_o(t)$ must have nearly identical waveforms, of course. The validity of the statement

$$Q_o(i\omega) = Q_i(i\omega)\left[\frac{q_o}{q_i}(i\omega)\right] \tag{3.286}$$

used in the above manipulations follows easily from the basic definition of sinusoidal transfer function, the superposition theorem, and the rules for multiplying complex numbers. Figure 3.71 illustrates the method graphically.

The above method is extremely useful for understanding the basic relationships governing the choice of a suitable measuring instrument for a periodic signal. However, other approaches may be more useful for actual numerical calculation of specific cases. Digital simulation, such as CSMP can yield very complete information with little effort, as our next example shows. Figure 3.72a represents the periodic volume flow rate of a simple reciprocating pump running at constant speed:

$$\text{Flow rate} = Q_p|\sin \omega_p t| \quad \text{m}^3/\text{s} \tag{3.287}$$

Application of Eqs. (3.274) to (3.280) gives the frequency spectrum of this signal for $\omega_k = 0, 2\omega_p, 4\omega_p, \ldots, 2k\omega_p$ ($k = 1, 2, 3, \ldots$):

$$Q_i(0) = \frac{2Q_p}{\pi} \qquad Q_i(i\omega_k) = \left|\frac{4Q_p}{\pi(1 - 4k^2)}\right| \angle -90° \tag{3.288}$$

Figure 3.72b shows this spectrum for $Q_p = 1.0$ and $\omega_p = 100$ rad/s. We now use CSMP to compare the exact Q_i with a six-term truncated Fourier series, examine four candidate second-order flowmeter instruments (with various values of ω_n and ζ) for measurement quality, and observe both "starting transients" and

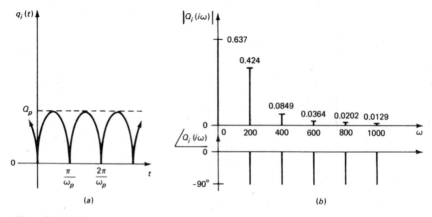

Figure 3.72 Reciprocating-pump flow rate and Fourier series.

"perodic steady-state" behavior of the instruments. (The calculation method of Fig. 3.71 gives *no* clue as to transient behavior.)

QI = ABS(SIN(100.*TIME))   Fortran statement for $Q_i(t)$

Q1 = CMPXPL(0.0,0.0,.65,1000.,QI)   CSMP statement for second-order system with input QI, output Q1, initial value Q1 = 0.0, initial value $\dot{Q}1$ = 0.0, ζ = 0.65, ω_n = 1000

Q01 = Q1*1.E06   multiplies Q1 by ω_n^2 since CSMP second-order model *always* takes $K = 1/\omega_n^2$ and we want instrument to have $K = 1.0$

Q2 = CMPXPL(0.0,0.0,.65,100.,QI) ⎫ same as above but for a different
Q02 = Q2*1.E04   ⎬ instrument, with ω_n = 100

Q3 = CMPXPL(0.0,0.0,.05,5000.,OI) ⎫ still another instrument
Q03 = Q3*25.E06   ⎭

Q4 = CMPXPL(0.0, 0.0,.1,200.,QI) ⎫ fourth and last instrument
Q04 = Q4*40000.   ⎭

QIFS = .63662 − .42442*COS(200.*TIME) − .084881*COS(400.*TIME)
− ˑ... .036378*COS(600.*TIME) − .020210*COS(800.*TIME) − ...
.012860*COS(1000.*TIME)   Fortran statement for six-term Fourier-series approximation to QI. The symbol "..." is the CSMP convention for continuation cards.

Q1FS = CMPXPL(0.0,0.0,.65,1000.,QIFS) ⎫ gets response of one of earlier instruments to QI ap-
                                      ⎬ proximation
Q01FS = Q1FS*1.E06   ⎭

TIMER FINTIM = .10,DELT = .0001,OUTDEL = .0005   FINTIM chosen to observe about 3 cycles of QI

OUTPUT TIME,QI,Q01,Q02,Q03,Q04   sets up graph to compare various instruments' responses to QI

PAGE GROUP,XYPLOT,HEIGHT = 5.5,WIDTH = 6.5   specifies a 5.5 in by 6.5 in electrostatic plotter graph with all curves plotted to same scale

OUTPUT TIME,QI,QIFS,Q01,Q01FS

another graph to compare QI with QIFS and instrument responses to each

PAGE GROUP,XYPLOT,HEIGHT = 5.5,WIDTH = 6.5

END

Note that CSMP uses Fortran statements freely intermixed with CSMP, but unlike with Fortran, the statement *sequence* is irrelevant for most CSMP programs since CSMP has an internal *sorting algorithm*. That is, all the above cards (except for END) could be put in *any* sequence and the CSMP program would run identically and correctly. At first glance, it appears that numerical integration (statement INTGRL), the heart of any digital simulation language, has not been used above, even though differential equations (second-order instruments) *are* being solved. Of course, INTGRL *is* being used, but this is not apparent since it is embedded in CMPXPL. This is another convenience of CSMP: common dynamic systems (first-order, second-order, etc.) can be modeled by *single statements* (such as CMPXPL) rather than by writing out the complete differential equation.

For graphical output in this example, rather than using printer plots, which are not accurate enough for our present needs, we use XYPLOT to call for graphing on a high-resolution electrostatic plotter. When this option is used, the

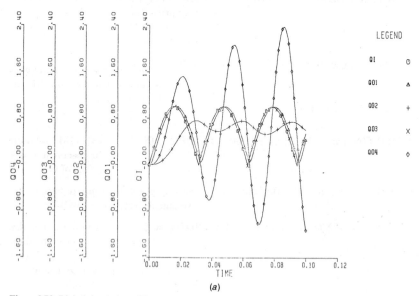

Figure 3.73 Digital simulation of flowmeter response.

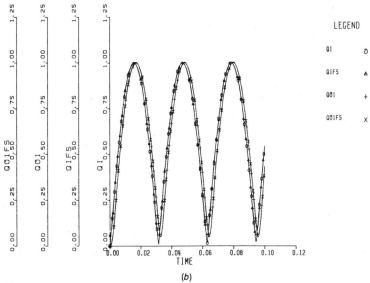

Figure 3.73 (*Continued*)

OUTPUT statement must have the x-axis variable (in our case TIME) as the first item in the list; all other listed quantities are graphed against it. This option also allows cross plotting (say, Q01 versus QI), which is not available with the printer plot. In Fig. 3.73*a* we see that an instrument with $\zeta = 0.65$, $\omega_n = 1{,}000$ (since its frequency response is quite flat for all the harmonics shown in Fig. 3.72*b*) gives a quite accurate measurement of QI's shape and size, but seems to be delayed in time. Changing to $\zeta = 0.05$, $\omega_n = 5{,}000$ (the "poor" damping does not cause trouble because of the high ω_n) gives almost perfect response with reduced time delay. Clearly, $\zeta = 0.65$, $\omega_n = 100$ is inadequate since its amplitude ratio has departed from flatness even at $\omega = 200$. Particularly bad is $\zeta = 0.1$, $\omega_n = 200$, which "resonates" with the QI fundamental frequency, building up to its periodic steady state in about 3 cycles. Figure 3.73*b* shows that a six-term Fourier series gives a close approximation to QI itself; thus calculations for Q01FS should be very close to Q01, as seen on the graphs.

Response of a General Form of Instrument to a Transient Input

By a transient input, we mean a $q_i(t)$ that is identically zero for all values of time greater than some finite value t_o, that is, an input that eventually dies out. For transient inputs of specific mathematical form, usually we can solve the differential equation and get $q_o(t)$ directly. For q_i's given by experimental data or, more important, if we wish to bring out certain important results of a *general* (not restricted to a specific type of q_i) nature, the methods of Fourier transforms or Laplace transforms are useful.[1] We now present the methods of applying these techniques, without proof of their validity.

The *direct Fourier transform* $Q_i(i\omega)$ (or the Laplace transform with $s = i\omega$) of the transient input $q_i(t)$ which is zero for $t < 0$ is given by

$$Q_i(\omega) \triangleq \int_0^\infty q_i(t) \cos \omega t \, dt - i \int_0^\infty q_i(t) \sin \omega t \, dt \qquad (3.289)$$

where ω can take all values from $-\infty$ to $+\infty$. Equation (3.289) is said to transform the input function from the time domain $[q_i(t)]$ to the frequency domain $[Q_i(i\omega)]$. The function $Q_i(i\omega)$ is also called the *frequency spectrum* of the input and plays the same role for transient inputs as Eq. (3.283) does for periodic inputs. However, whereas $Q_i(i\omega)$ is a *discrete* spectrum for $q_i(t)$ periodic, it is a *continuous* spectrum for $q_i(t)$ transient. That is, if you carry out Eq. (3.289) for a given $q_i(t)$, you will find $Q_i(i\omega)$ to be a complex number that varies with (is a function of) frequency ω and exists for *all* ω, not just at isolated points. As an example, consider the transient input of Fig. 3.74a. Applying Eq. (3.289), we get

$$Q_i(i\omega) = \int_0^T A \cos \omega t \, dt - i \int_0^T A \sin \omega t \, dt \qquad (3.290)$$

$$Q_i(i\omega) = \underbrace{\frac{A \sin \omega T}{\omega}}_{\text{real part}} + i \underbrace{\frac{A}{\omega}(-1 + \cos \omega T)}_{\text{imaginary part}} \qquad (3.291)$$

or, alternatively,

$$Q_i(i\omega) = \underbrace{\frac{\sqrt{2}A}{\omega}\sqrt{1 - \cos \omega T}}_{\text{magnitude}} \underbrace{\angle \alpha}_{\text{angle}} \qquad (3.292)$$

where

$$\tan \alpha = \frac{\cos \omega T - 1}{\sin \omega T} \qquad (3.293)$$

Plots of this frequency spectrum are given in Fig. 3.74b and c. [While $Q_i(i\omega)$ exists for both positive and negative values of ω, because of symmetry, often we can get the desired results by considering only the range $0 \le \omega < +\infty$. The stated symmetry consists of the following:

1. Re $Q_i(-i\omega) =$ Re $Q_i(+i\omega)$
2. Im $Q_i(-i\omega) = -$ Im $Q_i(+i\omega)$
3. Magnitude $Q_i(-i\omega) =$ magnitude $Q_i(+i\omega)$
4. Angle $Q_i(-i\omega) = -$ angle $Q_i(+i\omega)$

In most of our graphs and calculation methods, we employ the range $0 \le \omega < +\infty$, but $Q_i(-i\omega)$ always exists and can be found from a given $Q_i(+i\omega)$ by

---

[1] E. O. Doebelin, "System Modeling and Response," Wiley, New York, 1980; M. F. Gardner and J. L. Barnes, "Transients in Linear Systems," Wiley, New York, 1942; A. Papoulis, "The Fourier Integral and Its Applications," McGraw-Hill, New York, 1962.

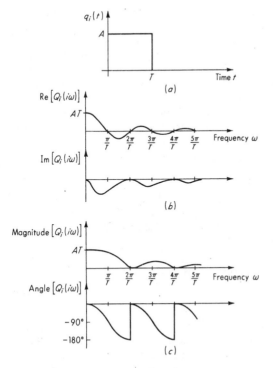

Figure 3.74 Frequency spectrum of transient.

application of the above symmetry rules.] These graphs indicate the "frequency content" of the transient input, just as the Fourier series indicates the frequency content of a periodic input. Thus we see that if T is small, large values of $Q_i(i\omega)$ persist out to higher frequencies than if T is large. Therefore a short-duration pulse is said to have greater high-frequency content than a long one. It is important to point out that the concept of frequency content for transients is not as clear-cut as for periodic functions; $q_i(t)$ for a transient *cannot* be built up by simply adding distinct sine waves, because $Q_i(i\omega)$ is now a *continuous function and no distinct frequencies exist.*

A further illustration of this distinction may be found by examining the dimensions of $Q_i(i\omega)$ in both cases. As an example, consider a $q_i(t)$ which is a pressure in kPa. If this pressure is periodic, Eqs. (3.279), etc., show that $Q_i(i\omega)$ has the same dimensions as $q_i(t)$, that is, kPa. Now, however, if the pressure is a transient, Eq. (3.289) gives

$$Q_i(i\omega) = \int_0^\infty (\text{kPa}) \underbrace{\cos \omega t}_{\text{dimensionless}} (\text{s}) - i \, (\text{same dimensions}) \qquad (3.294)$$

We see that $Q_i(i\omega)$ now has dimensions of $N \cdot s/m^2$, or, reinterpreting this, kPa/(rad/s). That is, $Q_i(i\omega)$ is thought of as the amount of signal *per unit frequency increment*, rather than as the actual amount of signal at a discrete frequency. This is analogous to the concept of distributed (rather than concentrated) loads in strength of materials. When a beam has sand (or water, etc.) piled on it, the applied load at any particular *point* is zero, but over an *area* the load is the force density times the area. Similarly, a transient signal has no discrete frequencies, but does contain a certain amount of signal within any frequency *band*. Thus, for a transient, $Q_i(i\omega)$ may be thought of as the *density* of signal per frequency bandwidth rather than as the signal itself.

The main purpose of using Eq. (3.289) is to convert functions from the time domain to the frequency domain, perform certain desired operations (which are *easier* or more *revealing* in the frequency domain than in the time domain), and then convert the information back to the time domain, since this is the more familiar and directly applicable (in an engineering sense) form. The conversion from frequency domain to time domain is accomplished by the *inverse-Fourier-* (or Laplace-) *transform formula* given by

$$q_i(t) = \frac{2}{\pi} \int_0^\infty \text{Re} \left[Q_i(i\omega) \right] \cos \omega t \, d\omega \qquad t > 0$$

$$q_i(t) \equiv 0 \qquad t < 0$$

(3.295)

Since these transformations are unique, if a $Q_i(i\omega)$ for a given $q_i(t)$ is found from Eq. (3.289), it should be possible to reconstruct the original $q_i(t)$ from the $Q_i(i\omega)$ by Eq. (3.295). Carrying this out for our example, we get

$$q_i(t) = \frac{2}{\pi} \int_0^\infty \frac{A \sin \omega T}{\omega} \cos \omega t \, d\omega \qquad t > 0$$

$$q_i(t) \equiv 0 \qquad t < 0$$

(3.296)

After some transformation, this can be put in a standard form found in integral tables and gives

$$q_i(t) = \begin{cases} 0 & t < 0 \\ A & 0 < t < T \\ A/2 & t = T \\ 0 & T < t \end{cases}$$

(3.297)

This function is shown in Fig. 3.75, and we see that it is practically identical to Fig. 3.74a. Actually, in Fig. 3.74a, we were not mathematically precise in defining $q_i(t)$ at $t = T$ since the graph shows $q_i(t)$ taking on *all* values between 0 and A. The usual practice for such discontinuities is to define the function as single-valued and equal to the midpoint. The Fourier transform is set up on this basis and thus always gives results similar to (3.297). The Fourier *series* for a periodic function with step discontinuities also behaves in this fashion; that is, it converges

Figure 3.75 Single-valued definition of transient.

to the midpoint. Thus in using numerical schemes, if an ordinate falls right on a discontinuity, the midpoint should be used as the numerical entry in the computation schedule.

To get a better feeling for the above methods and to show how they are graphically or numerically applied to functions (data) for which mathematical formulas are not available, let us consider the following development. Figure 3.76a shows a typical transient $q_i(t)$ as might be experimentally recorded in, say, a shock test. The first thing we note is that, for all practical purposes, the transient is ended in a finite time t_0. Thus, since $q_i(t)$ is a multiplying factor in the integrand and becomes zero for $t > t_0$, we may write Eq. (3.289) as

$$Q_i(i\omega) = \int_0^{t_0} q_i(t) \cos \omega t \, dt - i \int_0^{t_0} q_i(t) \sin \omega t \, dt \qquad (3.298)$$

We obtain $Q_i(i\omega)$ one point (frequency) at a time as follows

1. Choose a numerical value of ω, say ω_1.
2. Now, $\cos \omega_1 t$ is a perfectly definite curve and may be plotted against t.
3. Multiply $q_i(t)$ and $\cos \omega_i(t)$ point by point to get the curve $q_i(t) \cos \omega_1 t$.
4. Integrate, by any suitable numerical, graphical, or mechanical means, the curve $q_i(t) \cos \omega_1 t$ from $t = 0$ to $t = t_0$. Call the integral (area under curve) a_1.
5. Repeat the above procedure for $q_i(t) \sin \omega_1 t$ and call the integral b_1.
6. Then $Q_i(i\omega_1)$ is $a_1 + ib_1$.
7. Repeat for as many ω's as desired to generate the curves for $Q_i(i\omega)$ versus ω.

Figure 3.76 illustrates these procedures.

To appreciate the difference in "frequency content" between a "slow" transient and a "fast" one, we consider Fig. 3.77. Here $Q_i(i\omega)$ is found (for a high value of ω) for both a slow transient (Fig. 3.77a) and a fast one (Fig. 3.77b). It is clear that $Q_i(i\omega)$ will be nearly zero for ω's at or above the chosen ω_1 for the slow transient; thus its frequency content is limited to lower frequencies. The fast transient, however, has a nonzero value for $Q_i(i\omega_1)$ and thus "contains" frequencies at and somewhat above this value. For any real-world transient, we can always find *some* ω_1 high enough to make $Q_i(i\omega_1) \approx 0$; that is, all *real* transients are limited in frequency content at the high end. An "unreal" (only mathematically possible) transient is the impulse function, which we can easily show to contain *all* frequencies from 0 to ∞ and all in equal "strength." For an impulse

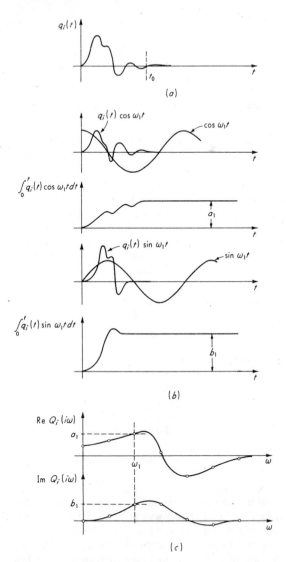

Figure 3.76 Graphical interpretation of direct Fourier transform.

of area A

$$Q_i(i\omega) = \int_0^\infty A\delta(t) \cos \omega t \ dt - i \int_0^\infty A\delta(t) \sin \omega t \ dt \qquad (3.299)$$

$$Q_i(i\omega) = A - i0 = A \qquad \text{for any finite } \omega \qquad (3.300)$$

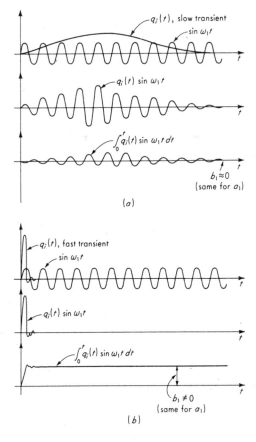

Figure 3.77 Frequency content of fast and slow transients.

Thus Fig. 3.78 shows the frequency content of an impulse. This property of an impulse makes it most useful as a "test signal" for investigating unknown systems, since all frequencies will be excited equally and the true nature of the system will be revealed by its response. We develop this important concept in greater detail later.

The process of *inverse* transformation also can be interpreted graphically. The defining equation (3.295) may, in actual practice, be written as

$$q_i(t) = \frac{2}{\pi} \int_0^{\omega_0} \text{Re} \left[Q_i(i\omega) \right] \cos \omega t \, d\omega \qquad t > 0$$

$$q_i(t) \equiv 0 \qquad t < 0$$

$$(3.301)$$

since all $Q_i(i\omega)$ representing physical quantities become approximately equal to zero for ω greater than some finite value ω_0. This follows directly from the fact

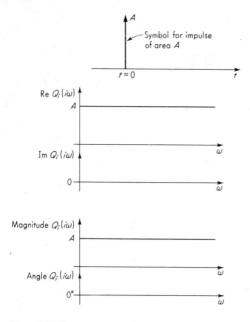

Figure 3.78 Frequency spectrum of impulse.

that the $Q_i(i\omega)$ come from the $q_i(t)$ and they cannot contain infinitely high frequencies. A step-by-step procedure for finding $q_i(t)$ one point at a time from a given $Q_i(i\omega)$ is as follows:

1. Choose a numerical value of t, say t_1.
2. Now $\cos \omega t_1$ is a perfectly definite curve and may be plotted against ω.
3. Multiply $\text{Re}[Q_i(i\omega)]$ and $\cos \omega t_1$ point by point to get the curve $\text{Re}[Q_i(i\omega)] \cos \omega t_1$.
4. Integrate, by any suitable numerical, graphical, or mechanical means, the curve $\text{Re}[Q_i(i\omega)] \cos \omega t_1$ from $\omega = 0$ to $\omega = \omega_0$. The integral (area under curve) is $(\pi/2)q_i(t_1)$, from Eq. (3.301). Plot $q_i(t_1)$ versus t.
5. Repeat for as many t's as desired to generate the curve $q_i(t)$ versus t.

Figure 3.79 illustrates this procedure.

The main usefulness of the above transform methods is based on the important result[1] relating the Fourier transform of the input signal $Q_i(i\omega)$, the system frequency response $(q_0/q_i)(i\omega)$, and the Fourier transform of the output signal

---

[1] J. A. Aseltine, "Transform Method in Linear System Analysis," McGraw-Hill, New York, 1958.

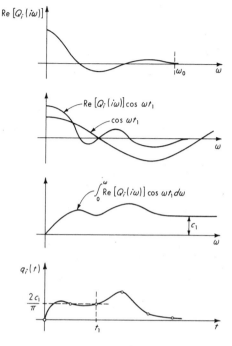

Figure 3.79 Graphical interpretation of inverse Fourier transform.

$Q_0(i\omega)$:

$$Q_0(i\omega) = Q_i(i\omega) \left[\frac{q_0}{q_i} (i\omega) \right] \qquad (3.302)$$

The restriction on this result is that the initial conditions must be zero; the system starts from rest. Many important practical problems meet this requirement. (Also, for nonzero initial conditions, the method may still be applied if the input is suitably modified.[1])

The meaning of Eq. (3.302) is that when the frequency-response curves for a system are known and a transient input $q_i(t)$ is applied, we can transform $q_i(t)$ to $Q_i(i\omega)$, multiply $Q_i(\omega)$ and $(q_0/q_i)(i\omega)$ point by point to get $Q_0(i\omega)$, and then inverse-transform $Q_0(i\omega)$ to $q_0(t)$ to get the output or response of the system. Figure 3.80 illustrates the procedure.

From a measurement-system point of view, Eq. (3.302) has the following important interpretation. For accurate measurement, $q_0(t) \approx K q_i(t)$, and since the Fourier transforms are unique [only one possible $F(i\omega)$ for each $f(t)$ and vice versa], this requires $Q_0(i\omega) \approx K Q_i(i\omega)$. Since $Q_0(i\omega)$ is obtained by multiplying $Q_i(i\omega)$ and $(q_0/q_i)(i\omega)$, this means $(q_0/q_i)(i\omega)$ must be $K \angle 0°$ over the entire range

[1] Ibid.

160 GENERAL CONCEPTS

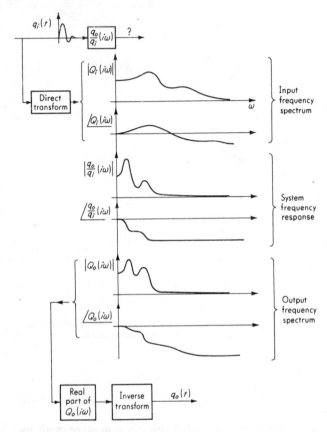

Figure 3.80 System response to transient input.

of frequencies for which $Q_i(i\omega)$ is not practically zero *but can be anything elsewhere.* The requirement that $(q_o/q_i)(i\omega)$ be $K \angle 0°$ for *all* frequencies for perfect measurement is obvious without the use of transform methods. The condition that this need be so only for a definite, finite *range* of frequencies (corresponding to the frequency content of the input) is the contribution of the transform methods and is of great practical significance since it puts much more realistic demands on the measurement system. A further relaxation of these requirements (which allows phase shift) is developed later in this chapter.

While Eq. (3.289) was useful for defining and calculating $Q_i(i\omega)$ for the purposes of this chapter, an alternative approach that uses a Laplace transform as an intermediate step may expedite numerical calculations for transients expressible as mathematical formulas. In this method, $q_i(t)$ is analytically Laplace-transformed to $Q_i(s)$ by using available transform theorems and tables. Then, by simply substituting $s = i\omega$, $Q_i(i\omega)$ can be calculated numerically for a desired range of frequency by using a digital-computer program. Such programs (including graphical output at a CRT terminal) are particularly quick and easy to write if you have available a high-level engineering language such as SPEAKEASY.[1]

Frequency Spectra of Amplitude-Modulated Signals

Interest in amplitude-modulated signals stems mainly from two considerations:

1. Physical data that are to be measured and interpreted sometimes are amplitude-modulated.
2. Certain types of measurement systems intentionally introduce amplitude modulation for one or more benefits.

While, in general, the signal that modulates the amplitude of a carrier wave may be of any form (single sine wave, general periodic function, random wave, transient, etc.) and the carrier may be given different forms (sine wave, square wave, etc.), perhaps the process is most easily understood for a single sine wave modulating a sinusoidal carrier. The modulation process is basically one of multiplying the signal carrying the information by a carrier wave of constant frequency and amplitude (see Fig. 3.81). For our simple example, we have

$$\text{Output} = (A_s \sin \omega_s t)(A_c \sin \omega_c t) \tag{3.303}$$

where $A_s \triangleq$ amplitude of signal
$\omega_s \triangleq$ frequency of signal
$A_c \triangleq$ amplitude of carrier
$\omega_c \triangleq$ frequency of carrier

The frequency ω_c is greater (usually considerably greater) than ω_s. For such a situation, the output has the shape shown in Fig. 3.82a. The frequency spectrum of such a signal is obtained easily from the following trigonometric identity:

$$\sin \alpha \sin \beta \equiv \tfrac{1}{2} \cos (\alpha - \beta) - \tfrac{1}{2} \cos (\alpha + \beta) \tag{3.304}$$

Applying this to Eq. (3.303), we get

$$\text{Output} = \frac{A_s A_c}{2} [\cos(\omega_c - \omega_s)t - \cos(\omega_c + \omega_s)t] \tag{3.305}$$

$$\text{Output} = \frac{A_s A_c}{2} \sin[(\omega_c - \omega_s)t + 90°] + \frac{A_s A_c}{2} \sin[(\omega_c + \omega_s)t - 90°]$$

$$\tag{3.306}$$

We see that the frequency spectrum of this signal is a discrete spectrum existing only at the frequencies $\omega_c - \omega_s$ and $\omega_c + \omega_s$, the so-called *side frequencies*. If

[1] Doebelin, "System Modeling and Response," pp. 102–111.

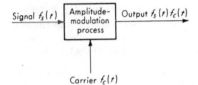

Figure 3.81 Amplitude modulation.

such a signal is the input $q_i(t)$ to a measurement system, we can find the steady-state output easily by the methods of Fig. 3.71.

Some applications of these concepts may be appropriate at this point. In a first, rough consideration of the vibration and noise of shafts with gears, perhaps we would expect the important frequencies to correspond to the rotational speeds

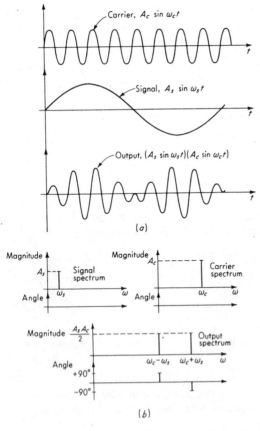

Figure 3.82 Frequency spectrum of amplitude-modulated signals.

of the shafts and to the tooth-meshing frequencies. For example, a shaft with a 20-tooth gear running at 200 r/s would be expected to generate noise at 200 and 4,000 Hz. However, actual noise measurements in such situations may show the peak noise to occur at frequencies different from those expected from the above crude analysis.[1] These discrepancies often may be resolved by the application of amplitude-modulation concepts as follows:

For a pair of absolutely true-running gears, the tooth forces (which cause vibration and thus noise) would have a fundamental frequency equal to the tooth-meshing frequency. These forces would not be pure sine waves, but *would be periodic*. Thus we could get a Fourier series for them. For simplicity, let us assume these forces to be pure sine waves of fixed amplitude. In an actual set of gears, there is always some eccentricity or "runout"; that is, the gears are closer together at some points in their rotation than at others. It is postulated that this runout leads to a force amplitude that *varies* as the gear rotates; that is, the tooth-force amplitude is *modulated* as a function of rotational position. If this is so, the frequencies of generated noise (corresponding to tooth-force frequencies) would be expected to be the side frequencies generated by modulating the 4,000-Hz tooth-meshing frequency with the 200-Hz (once per rotation) runout frequency. These frequencies (3,800 and 4,200 Hz in our example) actually have been measured, which confirms the original conjecture. Without the amplitude-modulation concepts, the engineer would have been hard pressed to explain these frequencies in the measured data.

Another interesting example is found in the carrier amplifier. To measure easily and record very small voltages coming from transducers (such as strain gages) requires a very-high-gain amplifier. Because of drift problems, a high-gain amplifier is easier to build as an ac, rather than a dc unit. An ac amplifier, however, does not amplify constant or slowly varying voltages and so would appear to be unsuitable for measuring static strains. This problem is overcome by exciting the strain-gage bridge with alternating voltage (say 5 V at 3,000 Hz) rather than direct. Thus when the bridge is unbalanced by strain-induced resistance changes, the output voltage will be a 3,000 Hz ac voltage whose amplitude will be modulated by the strain changes. Thus if we are measuring strains which vary from, say, 0 Hz (static) to 10 Hz, the amplifier will have an input-frequency spectrum bounded by 2,990 and 3,010 Hz. This range of frequencies is easily handled by an ac amplifier. Figure 3.83 illustrates these concepts. This "shifting" of the information frequencies from one part of the frequency range to another is the basis of many useful applications of amplitude modulation.

As a final example, suppose that the wires leading from the bridge to the amplifier in Fig. 3.83 are subjected to a stray 60-Hz field from surrounding ac machinery and a 60-Hz noise, or "hum," is superimposed (additively) on the desired signals. This 60-Hz noise easily could be larger than the desired strain

[1] P. K. Stein, "Measurement Engineering," vol. 1, sec. 17, Stein Engineering Services, Phoenix, 1962; A. L. Gu and R. H. Badgley, Prediction of Vibration Sidebands in Gear Meshes, *ASME Paper* 74-DET-95, 1974.

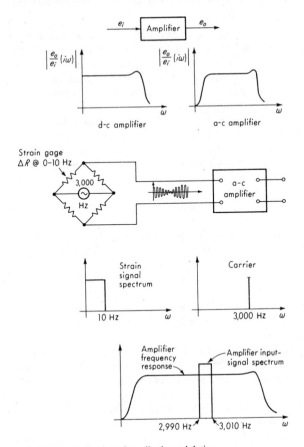

Figure 3.83 Application of amplitude modulation.

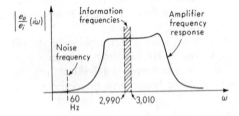

Figure 3.84 Noise rejection through amplitude modulation.

signals. With a carrier system, however, this noise may be eliminated merely by designing the ac amplifier so that it does not respond to 60 Hz. Since the desired band of frequencies is 2,990 to 3,010 Hz, making the low-frequency cutoff of the amplifier greater than 60 Hz is not difficult. Figure 3.84 illustrates this situation.

We now extend the amplitude-modulation concept to signals other than just a single sine wave. If the modulating signal is a periodic function $f_i(t)$, it may be expanded in a Fourier series to get the output of the modulator as [see Eq. (3.274)]

$$\text{Output} = \left[f_{i,\,\text{av}} + \frac{1}{L} \left(\sum_{n=1}^{\infty} a_n \cos \frac{n\pi t}{L} + \sum_{n=1}^{\infty} b_n \sin \frac{n\pi t}{L} \right) \right] A_c \sin \omega_c t \quad (3.307)$$

which can be written as

$$\begin{aligned}
\text{Output} = A_0 A_c \sin \omega_c t &+ (A_1 A_c \cos \omega_1 t \sin \omega_c t \\
&+ A_2 A_c \cos \omega_2 t \sin \omega_c t + \cdots) + (B_1 A_c \sin \omega_1 t \sin \omega_c t \\
&+ B_2 A_c \sin \omega_2 t \sin \omega_c t + \cdots)
\end{aligned} \quad (3.308)$$

Now,
$$\sin \alpha \sin \beta \equiv \tfrac{1}{2} \cos (\alpha - \beta) - \tfrac{1}{2} \cos (\alpha + \beta)$$

and
$$\sin \alpha \cos \beta \equiv \tfrac{1}{2} \sin (\alpha + \beta) + \tfrac{1}{2} \sin (\alpha + \beta)$$

and so

$$\begin{aligned}
\text{Output} = A_0 A_c \sin \omega_c t &+ C_1 \{ \sin [(\omega_c + \omega_1)t - \alpha_1] \\
&+ \sin [(\omega_c - \omega_1)t + \alpha_1] \} + \cdots
\end{aligned} \quad (3.309)$$

where
$$C_1 \triangleq \frac{A_c}{2} \sqrt{A_1^2 + B_1^2}$$

$$\alpha_1 \triangleq \tan^{-1} \frac{B_1}{A_1} \quad (3.310)$$

We see that the spectrum of the output signal is a discrete spectrum containing the frequencies ω_c, $\omega_c \pm \omega_1$, $\omega_c \pm \omega_2$, $\omega_c \pm \omega_3$, etc. That is, each frequency component of the modulating signal produces one pair of side frequencies (see Fig. 3.85). If the output of the modulator is applied to the input of a system with known frequency response, the methods of Fig. 3.71 can be used again to find the steady-state output.

If the modulating signal is a transient, the spectrum of the modulator output may be determined with the help of the modulation theorem[1] for Fourier (or Laplace) transforms. If the modulating signal is a transient $f_i(t)$, it will have a Fourier transform $F_i(i\omega)$, which can be obtained in the usual ways. The modulation theorem leads to the following result if $f_i(t)$ is multiplied by the carrier

[1] G. A. Korn and T. M. Korn, "Mathematical Handbook for Scientists and Engineers," p. 219, McGraw-Hill, New York, 1961; Papoulis, "Fourier Integral and Its Applications," p. 15.

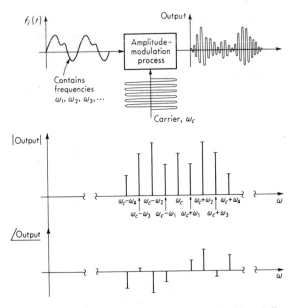

Figure 3.85 Frequency spectrum when modulating signal is periodic.

$A_c \sin \omega_c t$ to produce the modulated output:

$$\text{Fourier transform of modulated output} \triangleq |Q_i(i\omega)| \angle Q_i(i\omega) \qquad (3.311)$$

where
$$|Q_i(i\omega)| = \frac{A_c}{2} \text{magnitude}\{F_i[i(\omega - \omega_c)]\} \qquad (3.312)$$

$$\angle Q_i(i\omega) = \text{angle}\{F_i[i(\omega - \omega_c)]\} - 90° \qquad (3.313)$$

and $0 \leq \omega < \infty$. Note that the argument (independent variable) of the F_i function is now $i(\omega - \omega_c)$, so that if we want to find $Q_i(i4,000)$ and $\omega_c = 3,000$, we must evaluate $F_i(i1,000)$. Also note that, to get $Q_i(i\omega)$ for $0 < \omega < \infty$, we must know $F_i(i\omega)$ for *negative* ω's, since any $\omega < \omega_c$ gives $F_i[i(\omega - \omega_c)]$ a negative argument. While generally we have worked with positive ω's, the transform for negative ω's always exists and is easily found from the previously given symmetry rules. The spectrum given by Eq. (3.311) will be a continuous one, and if the modulated output is applied as an input to some system with known frequency response, then the corresponding output can be obtained by the methods of Fig. 3.80.

For measurement systems in which amplitude modulation is intentionally introduced to allow the use of carrier-amplifier techniques, the carrier frequency must be considerably greater (usually 5 to 10 times) than any significant frequencies present in the modulating signal. For such a situation, the pertinent

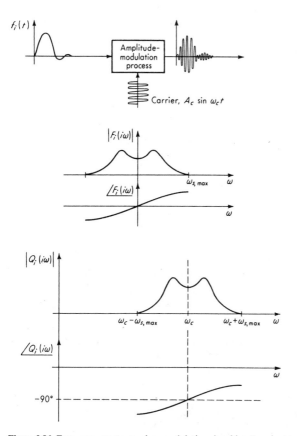

Figure 3.86 Frequency spectrum when modulating signal is a transient.

frequency spectra are as shown in Fig. 3.86. We see again that the amplitude-modulation process shifts the frequency spectrum by the amount ω_c.

When amplitude modulation is intentionally introduced to facilitate data handling in one way or another, it generally plays the role of an intermediate step, and the amplitude-modulated signal usually is not considered a suitable final readout. Rather, the original form of the modulating signal (the basic measured data from, say, a transducer) should be recovered. The process for accomplishing this involves *demodulation* (or *detection*, as it is sometimes called) and filtering. Demodulation may be full-wave, half-wave, phase-sensitive, or non-phase-sensitive (Fig. 3.87). Here we treat the form giving the best reproduction of the original data, full-wave phase-sensitive demodulation, and consider only the process, not the hardware for accomplishing it. Again, it is necessary to consider whether the form of the original signal was a single sine wave, periodic function, or transient. For a single sine wave $A_s \sin \omega_s t$ which is modulating a carrier

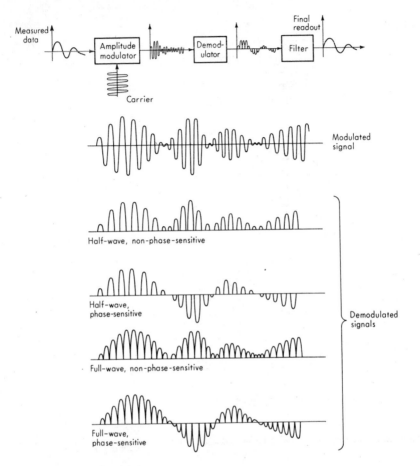

Figure 3.87 Types of modulation.

$A_c \sin \omega_c t$, the expression for the full-wave phase-sensitive demodulated signal is

$$\text{Demodulator output} = (A_s \sin \omega_s t)|A_c \sin \omega_c t| \qquad (3.314)$$

as seen from Fig. 3.88a. Now $|A_c \sin \omega_c t|$ is a periodic function and may be expanded in a Fourier series by application of Eq. (3.274). The results are

$$|A_c \sin \omega_c t| = \frac{2A_c}{\pi}\left(1 - \frac{2}{3}\cos 2\omega_c t - \frac{2}{15}\cos 4\omega_c t + \cdots\right.$$

$$\left. + \frac{2}{1-4n^2}\cos 2n\omega_c t + \cdots\right) \qquad n = 3, 4, 5. \ldots \quad (3.315)$$

Equation (3.314) can then be written as

$$\text{Demodulator output} = (A_s \sin \omega_s t) \left[\frac{2A_c}{\pi} \left(1 - \frac{2}{3} \cos 2\omega_c t \right. \right.$$

$$\left. \left. - \frac{2}{15} \cos 4\omega_c t + \cdots \right) \right] \tag{3.316}$$

which, when multiplied, gives

$$\text{Demodulator output} = \frac{2A_c A_s}{\pi} \sin \omega_s t - \frac{4A_c A_s}{3\pi} \sin \omega_s t \cos 2\omega_c t$$

$$- \frac{4A_c A_s}{15\pi} \sin \omega_s t \cos 4\omega_c t + \cdots \tag{3.317}$$

Now, by a trigonometric identity, terms of the form $(\sin \omega_s t)(\cos 2n\omega_c t)$ can be written as $[\sin (2n\omega_c + \omega_s)t - \sin (2n\omega_c - \omega_s)t]/2$. Thus we can write Eq. (3.317) as

$$\text{Demodulator output} = \frac{2A_c A_s}{\pi} \sin \omega_s t - \frac{2A_c A_s}{3\pi} [\sin (2\omega_c + \omega_s)t$$

$$- \sin (2\omega_c - \omega_s)t] - \frac{2A_c A_s}{15\pi} [\sin (4\omega_c + \omega_s)t - \sin (4\omega_c - \omega_s)t] + \cdots$$

$$\tag{3.318}$$

From this we see that the frequency spectrum of the demodulator output signal is a discrete spectrum with frequency content at ω_s, $2\omega_c \pm \omega_s$, $4\omega_c \pm \omega_s$, etc., as shown in Fig. 3.88b. If this signal were an input to a system of known frequency response (such as the filter of Fig. 3.87), the output of this system would be found by the methods of Fig. 3.71. If the output of the filter is to look like the original data, the filter must be designed to *reject* frequencies $2\omega_c \pm \omega_s$, $4\omega_c \pm \omega_s$, etc., while *passing* with a minimum of distortion the signal frequency ω_s. The design of such a low-pass filter is made simpler if the passband and the rejection band are more widely separated. This is the basis of our earlier statement that carrier frequencies usually are chosen to be 5 to 10 times the highest expected signal frequency.

When the modulating signal is a periodic wave rather than a single sine wave, a procedure similar to that just used is employed, except now the modulating signal is *also* expressed as a Fourier series of the form

$$\text{Modulating signal} = A_{s0} + A_{s1} \sin (\omega_s t + \alpha_1) + A_{s2} \sin (2\omega_s t + \alpha_2) + \cdots$$

$$\tag{3.319}$$

When this is multiplied by $|A_c \sin \omega_c t|$ as given by Eq. (3.315), we find exactly the same situation as for a single sine wave, but it must be applied for *each* signal frequency $(0, \omega_s, 2\omega_s, 3\omega_s, \text{etc.})$. The frequency spectrum of the demodulated

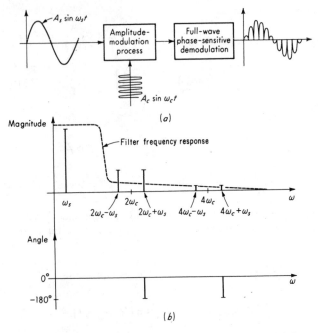

(a)

(b)

Figure 3.88 Frequency spectrum of full-wave phase-sensitive demodulation.

signal thus will be a discrete spectrum with frequency content at $\omega = 0, (\omega_s, 2\omega_c,$ $2\omega_c \pm \omega_s, 4\omega_c, 4\omega_c \pm \omega_s, 6\omega_c, 6\omega_c \pm \omega_s,$ etc.$), (2\omega_s, 2\omega_c \pm 2\omega_s, 4\omega_c \pm 2\omega_s, 6\omega_c$ $\pm 2\omega_s,$ etc.$),$ etc. Figure 3.89 illustrates these concepts. Again, if such a signal is applied to a system of known frequency response, the methods of Fig. 3.71 allow calculation of the output.

When the modulating signal is a transient $f_i(t)$, the demodulated signal will be $f_i(t)| A_c \sin \omega_c t|$, which can be written as

$$\text{Demodulated signal} = f_i(t)\left[\frac{2A_c}{\pi}\left(1 - \frac{2}{3}\cos 2\omega_c t - \frac{2}{15}\cos 4\omega_c t + \cdots\right)\right]$$

(3.320)

or, after multiplying,

$$\text{Demodulated signal} = \frac{2A_c}{\pi}f_i(t) - \frac{4A_c}{3\pi}f_i(t)\cos 2\omega_c t$$

$$- \frac{4A_c}{15\pi}f_i(t)\cos 4\omega_c t + \cdots$$

(3.321)

Application of the modulation theorem to each of the modulated terms of Eq.

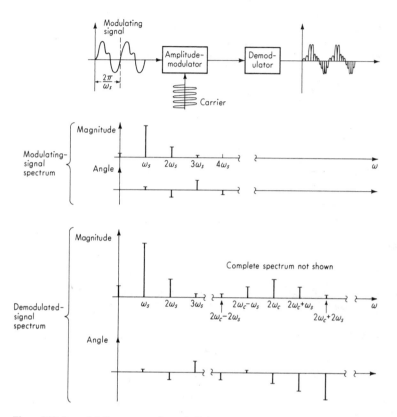

Figure 3.89 Demodulation spectrum for periodic input.

(3.321) leads to the result

Fourier transform of demodulated signal $\triangleq |Q_i(i\omega)| \angle Q_i(i\omega)$

where

$$|Q_i(i\omega)| = \frac{2A_c}{\pi}|F_i(i\omega)| + \frac{2A_c}{3\pi}|F_i[i(\omega - 2\omega_c)]|$$

$$+ \frac{2A_c}{15\pi}|F_i[i(\omega - 4\omega_c)]| + \cdots \tag{3.322}$$

and $0 \le \omega \le \infty$. The expression for $\angle Q_i(i\omega)$ is not easily given since in Fourier-transforming Eq. (3.321) (each term may be treated separately since superposition is allowed), we find $Q_i(i\omega)$ as a *sum* of complex numbers rather than a product. To find the overall angle of a sum of complex numbers, we must express *each* number as $a_i + ib_i$ and then add the a's and b's to get the overall $a + ib$. The angle is then $\tan^{-1}(b/a)$. The angle due to transforming $f_i(t)$ is $\angle F_i(i\omega)$, the angle

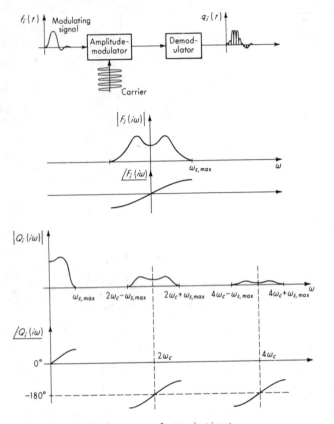

Figure 3.90 Demodulation spectrum for transient input.

due to transforming $-f_i(t) \cos 2\omega_c t$ is $-180° + \angle F_i[i(\omega - 2\omega_c)]$, the angle due to transforming $-f_i(t) \cos 4\omega_c t$ is $-180° + \angle F_i[i(\omega - 4\omega_c)]$, etc. These results allow calculation of $\angle Q_i(i\omega)$. The frequency spectrum of such a signal is shown in Fig. 3.90. Since it is a continuous spectrum, the response of a system to such an input may be found by the methods of Fig. 3.80. In Fig. 3.90 it is assumed that the $F_i(i\omega)$ is practically zero for $\omega > \omega_{s,\,max}$ and that $\omega_c > \omega_{s,\,max}$. For such a situation, the phase-angle calculation is greatly simplified, since the individual terms of Eq. (3.322) do not coexist over any frequency range. That is, for $0 < \omega < \omega_{s,\,max}$, only the first term is nonzero; for $(2\omega_c - \omega_{s,\,max}) < \omega < (2\omega_c + \omega_{s,\,max})$, only the second term is nonzero; etc. Thus the overall phase angle is determined by one term only within each of the specified frequency bands.

This concludes our treatment of amplitude-modulated and -demodulated signals. The methods developed can be readily applied to the other common variations such as half-wave demodulation, non-phase-sensitive demodulation, non-sinusoidal-carrier waveforms (square wave, for instance), etc. Non-phase-sensitive demodulation cannot detect a sign change in the modulating signal. Half-wave systems shift the side frequencies of demodulated signals to $\omega_c \pm \omega_s$ rather than $2\omega_c \pm \omega_s$, thus making the filtering problem more difficult. They also have less amplitude than full-wave systems. Nonsinusoidal carriers may be useful in reducing heating effects in resistive transducers such as strain gages. The carrier waveform is such as to give a high ratio of peak value to rms (effective heating) value. The output signal is related to peak value while the power dissipated in the strain gage is related to rms value; thus a high peak/rms ratio increases output for a given allowable heating level. A carrier in the form of a train of high, narrow pulses satisfies this sort of criterion.

Characteristics of Random Signals

The final class of signals we consider is the so-called random or stochastic type. Such signals are of increasing importance since they serve as more realistic mathematical models of many physical processes than do deterministic signals. By a random signal we mean one that can be described only statistically before it actually occurs; that is, it cannot be described by a specific function of time prior to its occurrence. Of course, *all* signals in the real world have some degree of randomness, so that we should make it clear that we are *always* dealing with random signals, even though we may take a specific time function (periodic, transient, etc.) as a *model* of what is really going on to simplify analysis in some types of problems.

Figure 3.91*a* shows a time record of a typical random signal such as might be measured by a vibration pickup mounted on a booster-rocket structure subjected to acoustic-pressure forces generated by the rocket exhaust. These pressures are strongly random in that no specific frequencies are apparent in a time record. The stresses caused by such pressures can lead to fatigue failure of the structure. Thus

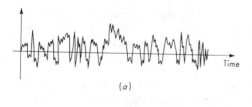

(*a*)

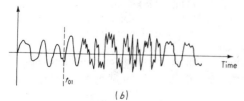

(*b*)

Figure 3.91 Random signals.

engineers are concerned with means to analyze the effects of such random forcing functions on structures and machinery. Since experimental methods are needed in much of this work, accurate measurement of random signals is important. We will see that frequency-response techniques are most useful for this type of problem.

First, we should note that, as far as deciding on the requirements for a measuring system is concerned, *after* a random process has occurred and there is a time record of it, we can define the function as zero for some time $t > t_0$, treat it as a transient, and calculate its Fourier transform to determine its frequency content and thus the required frequency response of the instrumentation. The only problem here is the selection of the cutoff time t_0 so as to have an adequate statistical "sample" of the random function. In Fig. 3.91*b*, for example, if we based our calculation on the record for $0 < t < t_{01}$, we would miss completely the high-frequency character apparent in the record for $t > t_{01}$. The existence of some minimum valid cutoff time t_0 implies that the random process is *stationary*, that is, its statistical properties (such as average value, mean-square value, etc.) do not change with time. When this is true, there will exist some t_0 corresponding to a chosen level of confidence in the results.

We should make it clear that, in dealing with random processes, theoretically an *infinite* record length is needed to give precise results, and results based on finite-length records must always be qualified by statistical statements referring to the *probability* of the result being correct within a certain percentage. This situation leads to an engineering tradeoff since long records are desirable from accuracy considerations, but are undesirable in terms of the cost involved in obtaining and analyzing them. Also, in many situations the maximum available length of record is limited by the lifetime of the device under study, which may be quite short in the case of a missile or rocket. If a long enough record is available, the minimum allowable t_0 can be found by choosing a small t_0 and calculating the Fourier transform. Then a larger t_0 is chosen and the Fourier transform again evaluated. If the second transform differs significantly from the first, the first t_0 was not long enough. By choosing successively longer t_0's, you will find some range of t_0 beyond which further increases in t_0 cause no significant change in the transform. Any t_0 beyond this range thus would be considered acceptably long. (In actual practice, more refined statistical procedures may be required, and are available,[1] to resolve questions of this sort.)

While the above concepts are adequate for understanding the requirements put on measurement systems for random variables, they do not cover the means available for statistically *describing* the signals. That is, while the exact form of a random function cannot be predicted ahead of time, certain of its statistical characteristics *can* be predicted; these may be useful in predicting (in a statistical way) the output of some physical system that has the random variable as an input. Now we develop some of the more common methods of statistically describing random signals. These methods fall mainly into two groups: those con-

---

[1] Doebelin," System Modeling and Response,",pp. 111, 267.

cerned with describing the magnitude of the variable and those concerned with describing the rapidity of change (frequency content) of the variable.

If we call a random variable $q_i'(t)$, the *average* or *mean value* $\overline{q_i'(t)}$ is defined by

$$\overline{q_i'(t)} \triangleq \lim_{T \to \infty} \left(\frac{1}{T}\right) \int_0^T q_i'(t) \, dt \tag{3.323}$$

Since this can be thought of as a constant component of the total signal, it is not random and usually is subtracted from the total signal, to give a signal with zero mean value. Also, many real random processes inherently have zero mean value. For these reasons, from here on we consider only signals $q_i(t)$ with zero mean value. An indication of the magnitude of the random variable is the *mean-squared value* $\overline{q_i^2(t)}$ given by

$$\overline{q_i^2(t)} \triangleq \lim_{T \to \infty} \left(\frac{1}{T}\right) \int_0^T q_i^2(t) \, dt \tag{3.324}$$

The mean-squared value has dimensions of $[q_i(t)]^2$. Thus to get a measure of the size of $q_i(t)$ itself, the *root-mean-square* (rms) value $q_i(t)_{rms}$ is defined by

$$q_i(t)_{rms} \triangleq \sqrt{\overline{q_i^2(t)}} \tag{3.325}$$

The above definitions are illustrated in Fig. 3.92. The quantities $\overline{q_i^2(t)}$ and $q_i(t)_{rms}$ give an indication of the overall size of $q_i(t)$, but no clue as to the distribution of "amplitude," that is, the probability of occurrence of large or small values of $q_i(t)$. The specification of this important information is provided by the *amplitude-distribution function* (probability density function) $W_1(q_i)$. To define this function, we consider Fig. 3.93. We define the probability P that $q_i(t)$ will be found between some specific value q_{i1} and $q_{i1} + \Delta q_i$ by

$$\text{Probability} \mid q_{i1} < q_i < (q_{i1} + \Delta q_i)] \triangleq P[q_{i1}, (q_{i1} + \Delta q_i)] = \lim_{T \to \infty} \frac{\sum \Delta t_i}{T} \tag{3.326}$$

where $\sum \Delta t_i$ represents the total time spent by $q_i(t)$ within the band Δq_i during time interval T. We now define $W_1(q_i)$ by

$$\text{Amplitude-distribution function} \triangleq W_1(q_i) \triangleq \lim_{\Delta q_i \to 0} \frac{P[q_i, (q_i + \Delta q_i)]}{\Delta q_i} \tag{3.327}$$

From this definition it should be clear that

$$W_1(q_i) \, dq_i = \text{probability that } q_i \text{ lies in } dq_i \tag{3.328}$$

and thus

$$\int_{q_{i1}}^{q_{i2}} W_1(q_i) \, dq_i = \text{probability } q_i \text{ lies between } q_{i1} \text{ and } q_{i2} \tag{3.329}$$

The function $W_1(q_i)$ theoretically can take on an infinite number of different

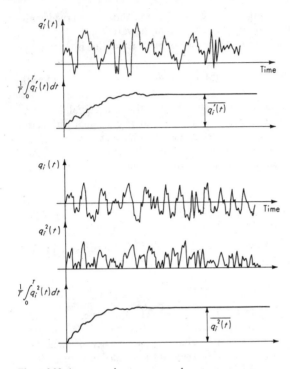

Figure 3.92 Average and mean-square values.

forms; however, certain forms have been found to be adequate mathematical models for real physical processes. The most common is the Gaussian or normal distribution given by

$$W_1(q_i) = \frac{1}{\sqrt{2\pi}\,\sigma}\, e^{-q_i^2/(2\sigma^2)} \tag{3.330}$$

where $\sigma \triangleq$ standard deviation. Whether a given random process closely approximates this form usually must be found experimentally by means of instruments based on Eqs. (3.326) and (3.327). Since the limiting processes in these equations can never be exactly realized in a physical instrument, again we must be satisfied with statements regarding the *probability* that a process is Gaussian rather than the *certainty* that it is.

While the quantities $q_i(t)_{\text{rms}}$ and $W_1(q_i)$ usually are sufficient to describe the magnitude of a random variable, they give no indication as to the *rapidity* of variation in time. That is, two random processes could both be Gaussian with the same numerical value of σ, but one could vary much more rapidly than the other.

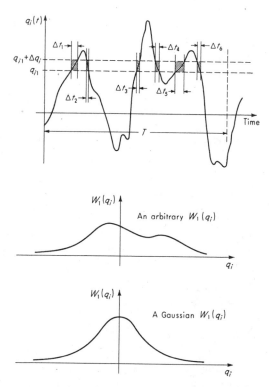

Figure 3.93 Definition of probability density function.

To describe the time aspect of random variables, the concepts of *autocorrelation function* and *mean-square spectral density* (power spectral density) are employed.[1] The autocorrelation function $R(\tau)$ of a random variable $q_i(t)$ is given by

$$R(\tau) \triangleq \lim_{T \to \infty} \left(\frac{1}{T}\right) \int_0^T q_i(t)q_i(t + \tau)\, dt \qquad (3.331)$$

The function $q_i(t + \tau)$ is simply $q_i(t)$ shifted in time by τ seconds. Thus, to find $R(\tau)$, we select a value of τ, say 2 s, plot against t the functions $q_i(t)$ and $q_i(t + 2)$, multiply them together point by point, integrate the product curve from 0 to T, and divide by T. Then this procedure is repeated for other values of τ to generate the curve $R(\tau)$ versus τ. The shifting of $q_i(t)$ by τ sec could be accomplished by writing $q_i(t)$ on magnetic tape at one point and reading it off at another. The time

---

[1] S. H. Crandall, "Random Vibration," Wiley, New York, 1958; Doeblin, op. cit., p. 120.

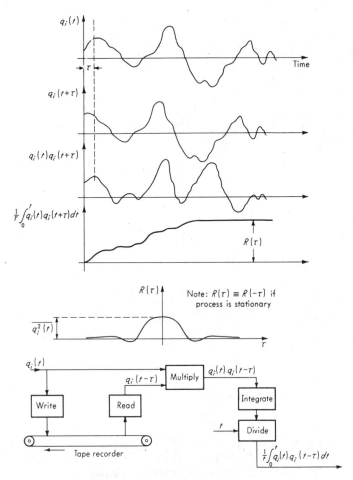

Figure 3.94 Definition of autocorrelation function.

delay τ then is simply

$$\tau = \frac{\text{distance between read and write heads}}{\text{tape velocity}}$$

The multiplication of $q_i(t)$ and $q_i(t + \tau)$, the integration, and the division by T can all be accomplished by standard electronic analog-computer components. However, today the entire $R(\tau)$ calculation would usually be done digitally. Figure 3.94 illustrates these concepts. For $\tau = 0$, Eq. (3.331) gives $R(0) = \overline{q_i^2(t)}$; that is, the autocorrelation function is numerically equal to the mean-square value for $\tau = 0$.

To appreciate the relation between $R(\tau)$ and the rapidity of variation of $q_i(t)$, consider Fig. 3.95, where both a slowly varying and a rapidly varying $q_i(t)$ are shown. For *any* $q_i(t)$, fast or slow, when $\tau = 0$, $q_i(t)q_i(t + \tau)$ is *positive* for all t. Thus integration gives the largest possible value, $\overline{q_i^2(t)}$. Any shift $(\tau \neq 0)$ will "misalign" the positive and negative parts of $q_i(t)$ and $q_i(t + \tau)$, causing the product curve to be sometimes positive, sometimes negative. Thus integration of this curve gives a smaller value than for $\tau = 0$. If $q_i(t)$ is rapidly varying, only a small shift (small τ) is needed to cause this misalignment, whereas a slowly varying $q_i(t)$ requires a larger shift before $R(\tau)$ drops off significantly. Thus a sharp peak in $R(\tau)$ at $\tau = 0$ indicates the presence of rapid variation (strong high-frequency content) in $q_i(t)$.

The mean-square spectral density (power spectral density) is another method of determining the frequency content of a random signal. The mean-square spectral density is proportional to the Fourier transform of $R(\tau)$ and conveys in the frequency domain exactly the same information as $R(\tau)$ conveys in the time domain. While they are mathematically related, in actual practice where they must be determined experimentally, one or the other may be preferable. It appears that in random-vibration work the mean-square spectral-density approach is largely preferred. To develop this concept, let us consider first a periodic function expanded into a Fourier series to give

$$q_i(t) = A_{i1} \sin (\omega_1 t + \alpha_1) + A_{i2} \sin (2\omega_1 t + \alpha_2) + \cdots \quad (3.332)$$

It is easy to show that the total mean-square value of $q_i(t)$ is equal to the sum of the individual mean-square values for each of the harmonic terms:

$$\overline{q_i^2(t)} = \overline{q_{i1}^2(t)} + \overline{q_{i2}^2(t)} + \cdots \quad (3.333)$$

Thus the contribution of each frequency to the overall mean-square value is easily found.

We now develop, for a *random* function, a related technique that will show how the total mean-square value is "distributed" over the frequency range. First

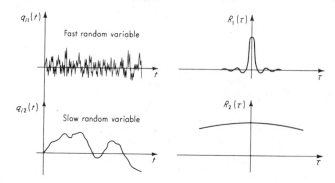

Figure 3.95 Frequency significance of autocorrelation function.

we note that for a random function, no isolated, discrete frequencies exist; thus the frequency spectrum is a continuous one. The concept of mean-square spectral density perhaps is most clearly visualized in terms of the analog instrumentation which could be used to measure it experimentally. Figure 3.96a shows the arrangement necessary to measure the *overall* mean-square value of a signal $q_i(t)$. To find out how much each part of the frequency range contributes to this overall value, we simply filter out (with a narrow-bandpass filter of bandwidth $\Delta\omega$) all frequencies other than the narrow band of interest and then perform the squaring and averaging operations on what remains, as in Fig. 3.96b. The results of this operation are thus the mean-square value of that part of the signal $q_i(t)$ lying within the chosen frequency band. We call this value $\overline{q_{i,\,\omega}^2}$. The filter is adjustable in the sense that we can shift its passband anywhere along the frequency axis, and so we can obtain $\overline{q_{i,\,\omega}^2}$ for any chosen center frequency ω. A narrow passband is desirable for resolving closely spaced peaks in the spectrum, but is undesirable in terms of the increased time required to cover a given frequency range in small steps rather than large. Thus a compromise is needed. The mean-square spectral density $\phi(\omega)$ is defined by

$$\phi(\omega) \triangleq \frac{\overline{q_{i,\,\omega}^2}}{\Delta\omega} \tag{3.334}$$

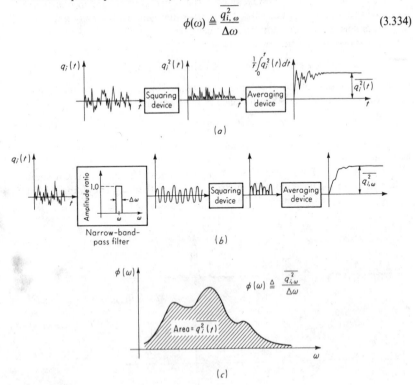

Figure 3.96 Definition of mean-square spectral density.

and represents the "density" (amount per unit frequency bandwidth) of the mean-square value, since $\phi(\omega)\Delta\omega = q_{i,\omega}^2$. If we evaluate $\phi(\omega)$ for a whole range of frequencies, we can plot it as a curve versus ω, as in Fig. 3.96c. Note that the dimensions of $\phi(\omega)$ are those of $q_i^2(t)/(\text{rad/s})$. The total area under the $\phi(\omega)$-versus-ω curve will be the total mean-square value $\overline{q_i^2(t)}$.

You will find that in the literature the term "power spectral density" is used almost exclusively for the quantity we call mean-square spectral density, except that power spectral density is π times $\phi(\omega)$. This is a carry-over from communications engineering where the concept was originally developed and where the signal $q_i(t)$ is a voltage applied to a 1-ohm resistor. Under these conditions, $\phi(\omega)$ would have dimensions of power/(rad/s). Actually, however, the concept of $\phi(\omega)$ is a *mathematical* one related to the mean-square value, and *not* a physical one related to electrical engineering. In physical applications the dimensions of $\phi(\omega)$ would be $(°C)^2/(\text{rad/s})$ if $q_i(t)$ were temperature, $g^2/(\text{rad/s})$ if $q_i(t)$ were acceleration, etc., which are in general *not* power/(rad/s). However, the term "power spectral density" seems to be firmly entrenched and probably will continue to prevail. Here we merely suggest that mean-square spectral density might be more appropriate terminology. Thus when dealing with, say, random pressures, you would refer to the mean-square spectral density of pressure.

Perhaps the main interest in the mean-square spectral density lies in the fact that if a $q_i(t)$ with a known $\phi_i(\omega)$ is applied as input to a linear system of known frequency response, the mean-square spectral density $\phi_o(\omega)$ of the output is easily computed from the relation[1]

$$\phi_o(\omega) = \phi_i(\omega)\left|\frac{q_o}{q_i}(i\omega)\right|^2 \tag{3.335}$$

We can thus compute the $\phi_o(\omega)$ curve point by point (see Fig. 3.97). The area under this curve will be the total mean-square value of $q_o(t)$. Furthermore, if the input is Gaussian, the output will also be Gaussian. It is then possible to make statements such as

$$|q_o(t)| > \sqrt{\overline{q_o^2(t)}} \qquad 31.7\% \text{ of the time}$$

$$|q_o(t)| > 2\sqrt{\overline{q_o^2(t)}} \qquad 4.6\% \text{ of the time} \tag{3.336}$$

$$|q_o(t)| > 3\sqrt{\overline{q_o^2(t)}} \qquad 0.3\% \text{ of the time}$$

Other useful results of this nature can be found in the literature.[2]

A particular form of $\phi(\omega)$ is of great utility in practice. This is the *white noise*. For a mathematically perfect white noise, $\phi(\omega)$ is equal to a constant for all frequencies; that is, it contains all frequencies in equal amounts (see Fig. 3.98).

---

[1] J. S. Bendat, "Principles and Application of Random Noise Theory," p. 199, Wiley, New York, 1958.

[2] J. S. Bendat, L. D. Enochson, and A. G. Piersol, Analytical Study of Vibration Data Reduction Methods, *NASA* N64-15529, 1963.

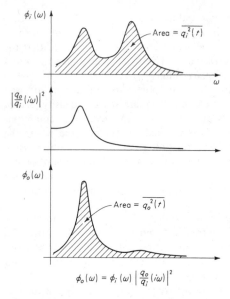

Figure 3.97 System response to random input.

This makes it useful as a test signal, just as the impulse is useful as a transient test signal because of its uniform frequency content. From Eq. (3.335), if $\phi_i(\omega) = C$, then

$$\phi_o(\omega) = C \left| \frac{q_o}{q_i} (i\omega) \right|^2 \tag{3.337}$$

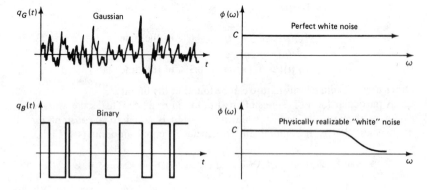

Figure 3.98 White noise.

and thus

$$\left| \frac{q_o}{q_i}(i\omega) \right| = \sqrt{\frac{\phi_o(\omega)}{C}} \qquad (3.338)$$

While it is not possible to build a generator of perfect white noise, the flat range of $\phi(\omega)$ of a practical generator can be quite large, and as long as its flat range extends beyond the frequency response of the system being tested, the "nonwhiteness" will not present any difficulty. If a white-noise generator is available we can "construct" almost any $\phi(\omega)$ that we wish simply by passing the white noise through a suitably designed filter according to Eq. (3.337). That is, $|q_o/q_i(i\omega)|^2$ for the filter must have the shape desired in $\phi(\omega)$. This technique is widely used in computer simulation studies of aircraft gust response, control-system evaluation, etc., where the frequency characteristics of some random-input quantity are known and it is desired to study their effect on some system.

In addition to the Gaussian (continuous) white noise we emphasize here, a flat mean-square spectral density also can be achieved with *binary random noise*, a signal that switches between two fixed levels at time intervals which vary randomly. A further variation is pseudorandom binary noise; details on the application of these types of signals are available in the literature.[1]

Most of the previous material has referred to statistical properties of a *single* random variable. Useful practical results may be derived by consideration of two random variables. The *joint amplitude-distribution function* (joint probability density function) of two random variables $q_1(t)$ and $q_2(t)$ is given by

$$W_1(q_1, q_2) = \lim_{T \to \infty} \lim_{\substack{\Delta q_1 \to 0 \\ \Delta q_2 \to 0}} \frac{1}{T \, \Delta q_1 \Delta q_2} \, \Sigma \, \Delta t_i \qquad (3.339)$$

where $\Sigma \, \Delta t_i$ represents the total time (during time T) that $q_1(t)$ and $q_2(t)$ spent *simultaneously* in the bands $q_1 + \Delta q_1$ and $q_2 + \Delta q_2$. Figure 3.99 illustrates these concepts. From the basic definition it should be clear that

$$\text{Probability } (q_{1a} < q_1 < q_{1b}, q_{2a} < q_2 < q_{2b}) = \int_{q_{2a}}^{q_{2b}} \int_{q_{1a}}^{q_{1b}} W_1(q_1, q_2) \, dq_1 \, dq_2$$

$$(3.340)$$

Again, $W_1(q_1, q_2)$ can take an infinite variety of forms. The most useful is probably the bivariate Gaussian (normal) distribution.[2] The main purpose in experimentally measuring $W_1(q_1, q_2)$ is, just as for $W_1(q_i)$, to determine whether the physical data follow approximately some simple mathematical form such as the Gaussian. If this can be proved, many useful theoretical results can be applied. Also, certain calculations can be made directly from $W_1(q_1, q_2)$. For example, if q_1 and q_2 represent the random vibratory motions of two adjacent machine parts, knowledge of $W_1(q_1, q_2)$ allows calculation of the probability that the two parts

[1] Doebelin, "System Modeling and Response," p. 273.
[2] A. M. Mood, "Introduction to the Theory of Statistics," p. 165, McGraw-Hill, New York, 1950.

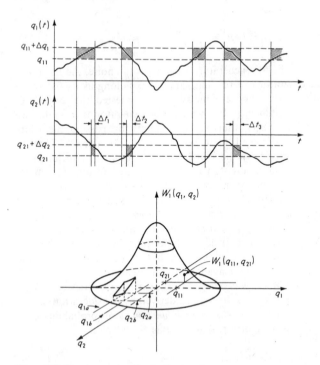

Figure 3.99 Bivariate probability density function.

will strike each other. In actual practice, engineering applications of $W_1(q_1, q_2)$ are somewhat limited because of the difficulty of measuring this function.

The *cross-correlation function* $R_{q_1q_2}(\tau)$ for two random variables $q_1(t)$ and $q_2(t)$ is defined by

$$R_{q_1q_2}(\tau) = \lim_{T \to \infty} \left(\frac{1}{T}\right) \int_0^T q_1(t)q_2(t + \tau) \, dt \qquad (3.341)$$

An example[1] of its application might be as follows: Suppose a source of vibratory motion $q_1(t)$ exists at one point in a structure and causes a vibratory response motion $q_2(t)$ at another point in the structure. Suppose also that the transmission of vibration from the first point to the second could occur by either (or both) of two mechanisms. The first mechanism is by acoustic (air-pressure) wave propagation through the air separating the two points; the second is by elastic wave propagation through the metallic structure connecting the two points. Since the propagation velocities of waves in air and metal are greatly different, we would expect that an input at q_1 would cause an output at q_2 that would be delayed by

[1] Bendat, Enochson, and Piersol, op. cit.

different times, depending on whether the transmission was mainly through the air or mainly through the metal. If we know the transmission path length and the wave velocity, these delays can be calculated. Suppose the air-path delay is 0.01 s and the structure-path delay is 0.002 s. If we now experimentally measure $R_{q_1q_2}(\tau)$ and find a large peak at $\tau = 0.01$ s and a smaller one at $\tau = 0.002$ s, we conclude that the acoustic transmission is responsible for most of the vibration at q_2. Thus, since $R_{q_1q_2}$ is a measure of the correlation ("relatedness") between two signals delayed by various amounts, it can be used as a diagnostic tool for investigating the presence and/or nature of the relation, as in the above example. Developments in LSI (large-scale integration) electronics and microprocessors are making available low-cost correlators which can serve as the basis for various useful instrument systems.[1]

Information equivalent to that contained in the cross-correlation function but in a different (and often more practically useful) form is found in the *cross-spectral density* (cross-power spectral density) $\phi_{q_1q_2}(\omega)$ given by

$$\phi_{q_1q_2}(\omega) = C_{q_1q_2}(\omega) - iQ_{q_1q_2}(\omega) \tag{3.342}$$

where $\quad C_{q_1q_2}(\omega) \triangleq \text{cospectrum}$

$$\triangleq \lim_{T \to \infty} \lim_{\Delta\omega \to 0} \frac{1}{T \, \Delta\omega} \int_0^T (q_{1\Delta\omega})(q_{2\Delta\omega}) \, dt \tag{3.343}$$

$Q_{q_1q_2}(\omega) \triangleq \text{quad spectrum}$

$$\triangleq \lim_{T \to \infty} \lim_{\Delta\omega \to 0} \frac{1}{T \, \Delta\omega} \int_0^T (q_{1\Delta\omega})_{90°}(q_{2\Delta\omega}) \, dt \tag{3.344}$$

$$q_{1\Delta\omega} \triangleq \text{output of narrow- } (\Delta\omega) \text{ band filter whose input is } q_1(t) \tag{3.345}$$

$$q_{2\Delta\omega} \triangleq \text{output of narrow- } (\Delta\omega) \text{ band filter whose input is } q_2(t)$$

$$(q_{1\Delta\omega})_{90°} \triangleq \text{signal } q_{1\Delta\omega} \text{ with phase shift of } 90° \tag{3.346}$$

A block diagram illustrating the definition of $\phi_{q_1q_2}(\omega)$ is given in Fig. 3.100. Note that $\phi_{q_1q_2}(\omega)$ is a *complex* quantity whereas $\phi(\omega)$ is real. Perhaps the main application of the cross-spectral density is in the experimental determination of the sinusoidal transfer function $(q_o/q_i)(i\omega)$ of a linear system. From Eq. (3.335) we see that by using the ordinary mean-square spectral density, we can find only the magnitude (not the phase angle) of the transfer function if $\phi_i(\omega)$ is known and $\phi_o(\omega)$ is measured. The cross-spectral density determines both magnitude and phase according to the following equation:

$$\frac{q_o}{q_i}(i\omega) = \frac{\phi_{q_iq_o}(\omega)}{\phi_{q_i}(\omega)} \tag{3.347}$$

[1] M. S. Beck, Correlation in Instruments: Cross Correlation Flowmeters, *J. Phys. Eng.: Sci. Instrum.*, vol. 14, pp. 7–19, 1981.

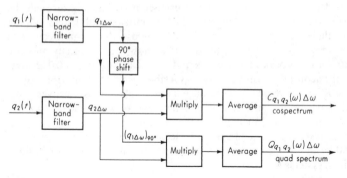

Figure 3.100 Cross-spectral density.

where $\phi_{q_iq_o}(\omega) \triangleq$ cross-spectral density of q_i and q_o

$\phi_{q_i}(\omega) \triangleq$ mean-square spectral density of q_i

We see that it is necessary to measure one (ordinary) mean-square spectral density and one cross-spectral density.

This concludes our treatment of random signals. The most important, commonly useful results have been presented, and the main terminology has been developed. Further theoretical and practical details may be found in the literature. Today frequency spectrum calculations for all types of dynamic signals and statistical calculations for random signals most often are done digitally by using fast Fourier transform (FFT) methods. These are implemented either on general-purpose digital computers using ready-made programs or on special-purpose computers marketed as signal and/or system analyzers. We discuss these latter instruments in greater detail in a later chapter.

Requirements on Instrument Transfer Function to Ensure Accurate Measurement

Having expressed up to this point many different forms of signals in terms of the common denominator of frequency content, we can now give a general statement of the requirements that a measurement system must meet in order to measure accurately a given form of input.

1. For perfect shape reproduction with no time delay between q_i and q_o and with:
   a. Periodic inputs. $(q_o/q_i)(i\omega)$ must equal $K \angle 0°$ for all frequencies contained in q_i with significant amplitude.
   b. Transient inputs. $(q_o/q_i)(i\omega)$ must equal $K \angle 0°$ for the entire frequency range in which the Fourier transform of $q_i(t)$ has significant magnitude.
   c. Amplitude-modulated signals. Same criteria as in parts a and b, depending on whether the modulating signal is periodic or transient.

d. Demodulated signals. $(q_o/q_i)(i\omega)$ for everything following the demodulator should be $K \angle 0°$ for all significant frequency bands of the modulating signal and should be zero for all carrier and side frequency bands produced by the modulation process.

e. Random signals. $(q_o/q_i)(i\omega)$ must equal $K \angle 0°$ over the entire frequency band where $\phi_{q_i}(\omega)$ is significantly larger than zero.

While proper choice of parameters allows many measurement systems to meet the requirement of flat amplitude ratio, the *simultaneous* achievement of near-zero phase angle over the same frequency range is rarely possible (second-order instruments with small ζ and large ω_n, such as piezoelectric devices, are an exception). A relaxed criterion which *can* be met by most practical systems allows the phase angle to be nonzero, but requires that it vary *linearly* with frequency over the range of flat amplitude ratio. This requirement results in q_o being a "perfect" reproduction of q_i; however, q_o will be delayed in time by a specific amount, as if the system contained a dead-time effect τ_{dt}. For many applications, the fact that q_o appears on our oscilloscope screen or recorder chart τ_{dt} seconds "late" is of no importance whatever. Thus requiring a system frequency response of $K \angle (-\omega\tau_{dt})$ is widely acceptable. There are, however, two situations in which such time (phase) shift might cause difficulty. In *multichannel* systems, unless each channel has the *same* τ_{dt} (not likely), the following happens: if you pick, say, a gas-temperature value from channel 1 and a gas-pressure value from channel 4, by using the same chart paper time value, any computed gas density value would be *wrong* because the temperature and pressure values, while aligned on the chart, were not simultaneous in actual time occurrence. Of course, if you know the τ_{dt} values for each channel, suitable corrections can be applied, perhaps including those in a computer program used for data reduction. In fact, by using microprocessor technology, these corrections could be made part of the multi-channel recorder itself, with each channel being properly shifted in time *before* it is written on the chart.

The second type of application where $K \angle (-\omega\tau_{dt})$ suffers relative to $K \angle 0°$ occurs when the measuring system is embedded in a feedback control loop. Here phase lag detracts from system stability and should be minimized.

Revising our earlier accuracy statement, we now say:

2. For perfect shape reproduction with time delay τ_{dt} between q_i and q_o and with:
   a. Periodic inputs.
   b. Transient inputs.
   c. Amplitude-modulated signals.
   d. Demodulated signals.
   e. Random signals.

   Same as 1 except that wherever $(q_o/q_i)(i\omega) = K \angle 0°$ is required, now $(q_o/q_i)(i\omega) = K \angle (-\omega\tau_{dt})$ is required. *That is, amplitude ratio is constant but phase lag increases linearly with frequency ω.*

The validity of the above statements is readily perceived by consideration of Fig. 3.101. To get the output in the frequency domain, the frequency-domain

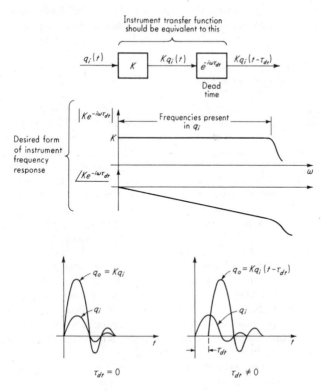

Figure 3.101 Requirements for accurate measurement.

input is always multiplied by the sinusoidal transfer function. If this transfer function is $K \angle (-\omega \tau_{dt})$ the output will have a magnitude equal to K times the input magnitude and an angle equal to·the angle of the input minus $\omega \tau_{dt}$. This is exactly what we would get if we passed $q_i(t)$ through a pure gain K followed by a dead time τ_{dt}. Thus when we inverse-transform to get from $Q_o(i\omega)$ to $q_o(t)$, we are *bound* to get $Kq_i(t)$ delayed by τ_{dt} seconds. Thus, while the *actual* instrument transfer function will be of the form of Eq. (3.261), over the pertinent range of frequencies it must effectively amount to $K \angle (-\omega \tau_{dt})$ if accurate waveform reproduction is to be expected. This is the basis for the selection of $\zeta \approx 0.6$ to 0.7 in second-order instruments since this range of ζ values makes the amplitude ratio most nearly constant and the phase-angle-versus-frequency curve most nearly linear over the widest possible frequency range for a given ω_n.

It should be noted that the above accuracy criteria do not actually state any *numerical* results. That is, no statement of the form "If the amplitude ratio is flat within $\pm x$ percent and the phase angle linear within $\pm y$ percent over the range of frequencies in which $|Q_i(i\omega)|$(Fourier-series term coefficient or Fourier-transform magnitude) is greater than z percent of its maximum value, then $q_o(t)$ will be within $\pm w$ percent of $Kq_i(t - \tau_{dt})$ at all times" is or *can be* made. While this sort of statement would be exceedingly useful, unfortunately it cannot be made in any *general* sense. For specific forms of $q_i(t)$ and $(q_o/q_i)(i\omega)$, we could investigate mathematically the effect of variations in numerical values of system parameters on dynamic accuracy by using analytical or computer simulation methods.

We can use CSMP to give a convincing demonstration that requiring essentially a flat amplitude ratio and a straight-line phase angle yields a measurement system that behaves like a dead time. Let the signal to be measured be a Gaussian "white" noise with flat power spectral density of 1.0 from 0 to 20 rad/s with a smooth rolloff to essentially zero at 100 rad/s and mean value zero. This type of random signal is easily obtained[1] on CSMP. If we require the measurement-system amplitude ratio flat within ± 5 percent to $\omega = 100$, a first-order instrument with $\tau \approx 0.005$ s or a second-order one with $\zeta = 0.65$, $\omega_n \approx 100$ should be adequate. For a first-order system.

$$\phi = -\tan^{-1} \omega\tau \approx -\omega\tau \qquad \text{for } \omega \to 0$$

$$\tau_{dt} \triangleq \tau = 0.005 \text{ s} \tag{3.348}$$

For a second-order system,

$$\phi = \tan^{-1} \frac{2\zeta}{\omega/\omega_n - \omega_n/\omega} \approx -2\zeta \frac{\omega}{\omega_n} \qquad \text{for } \omega \to 0$$

$$\tau_{dt} = \frac{2\zeta}{\omega_n} = 0.013 \text{ s} \tag{3.349}$$

A CSMP program might go as follows:

METHOD   RKSFX   fixed step-size numerical integration algorithm

```
YP = IMPULS(0.0,.0275)
YG = GAUSS(1,0.0,3.1E08)          generate random sig-
YH = ZHOLD(YP,YG)                 nal YRAND
YF1 = CMPXPL(0.0,0.0,.67,63.6,YH)
```

```
YRAND = CMPXPL(0.0,0.0,.67,63.6,YF1)
Q01 = REALPL(0.0,.005,YRAND)      first-order system with in-
                                  put YRAND, output Q01,
                                  initial value 0.0, $\tau = 0.005$,
                                  $K = 1.0$
```

[1] Doebelin, op. cit., p. 210.

Q2 = CMPXPL(0.0,0.0,.65,100.,YRAND)   second-order system with input YRAND, output Q02, initial conditions both zero, $\zeta = 0.65$, $\omega_n = 100.$, $K = 1.0$

Q02 = Q2*10000.

QDT1 = DELAY(50,.005,YRAND)   perfect dead time with input YRAND, output QDT1, $\tau_{dt} = 0.005$

QDT2 = DELAY(130,.013,YRAND)   same as above except $\tau_{dt} = 0.013$

TIMER   FINTIM = .275,DELT = .0001,OUTDEL = .001

OUTPUT   YRAND,Q01,Q02,QDT1,QDT2

PAGE GROUP

END

TIMER   FINTIM = 1.0,DELT = .0001,OUTDEL = .005   rerun problem with changes in timing and graphing

OUTPUT   YRAND,Q02

PAGE GROUP

END

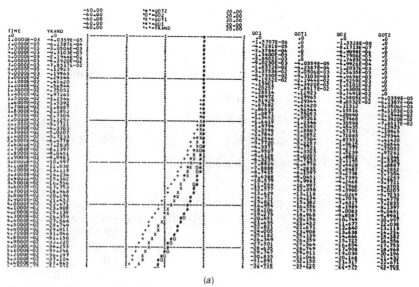

(a)

Figure 3.102 Digital simulation of frequency-response specification.

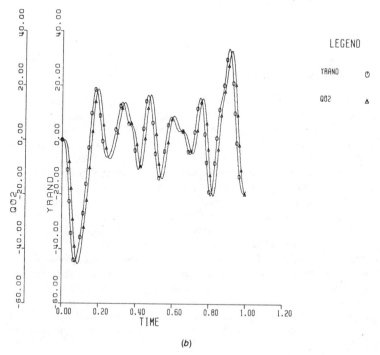

(b)

Figure 3.102 (*Continued*)

Figure 3.102*a* shows that once the starting transient (which extends from $t = 0$ to about $t = 0.03$ s) is over, Q01 ≈ QDT1, Q02 ≈ QDT2, and Q01 and Q02 are each nearly equal to YRAND, but delayed, respectively, by 0.005 and 0.013 s. Since the spread-out time scale of Fig. 3.102*a* (chosen to emphasize the time-delay effects) does not clearly show how YRAND varies with time, we rerun the problem for FINTIM = 1.0, giving several oscillations which begin to show the random nature of YRAND and, again, that Q02 is a nearly perfect measurement of YRAND except for the 0.013-s time shift (Fig. 3.102*b*).

Numerical Correction of Dynamic Data

Theoretically, if $q_o(t)$ (the actual measured data) is known and if $(q_o/q_i)(i\omega)$ for the measurement system is known, we can always reconstruct a *perfect* record of $q_i(t)$ by the following process:

1. Transform $q_o(t)$ to $Q_o(i\omega)$.
2. Apply the formula

$$Q_o(i\omega) = Q_i(i\omega) \frac{q_o}{q_i}(i\omega)$$

in the inverse sense as

$$Q_i(i\omega) = \frac{Q_o(i\omega)}{(q_o/q_i)(i\omega)}$$

to find $Q_i(i\omega)$.

3. Inverse-transform $Q_i(i\omega)$ to $q_i(t)$.

This procedure theoretically will give the exact $q_i(t)$ whether the measurement system meets the $K \angle (-\omega\tau_{dt})$ requirements or not. In actual practice, of course, while the measurement system does not have to meet $K \angle -\omega\tau_{dt}$, it *does* have to respond fairly strongly to all frequencies present in q_i; otherwise, some parts of the q_o frequency spectrum will be so small as to be submerged in the unavoidable "noise" present in all systems and thus be unrecoverable by the above mathematical process. As general-purpose digital computers are used more in data processing and as computing power is built into more "instruments," such dynamic correction becomes increasingly practical and is a usable alternative in those situations where measurement systems meeting $K \angle (-\omega\tau_{dt})$ cannot be constructed with the present state of the art.

An important variation of the above process has been successfully applied in cases where the primary sensor is inadequate but can be cascaded with frequency-sensitive analog elements whose transfer functions make up the deficiencies in the primary sensor. The above computations are then, in a sense, automatically and continuously carried out by the compensating equipment to reconstruct $q_i(t)$. This subject is discussed in detail later under the topic Dynamic Compensation.

Experimental Determination of Measurement-System Parameters

While theoretical analysis of instruments is vital to reveal the basic relationships involved in the operation of a device, it is rarely accurate enough to provide usable numerical values for critical parameters such as sensitivity, time constant, natural frequency, etc. Thus calibration of instrument systems is a necessity. We discussed static calibration; here we concentrate on dynamic characteristics.

For zero-order instruments, the response is instantaneous and so no dynamic characteristics exist. The only parameter to be determined is the static sensitivity K, which is found by static calibration.

For first-order instruments, the static sensitivity K also is found by static calibration. There is only one parameter pertinent to dynamic response, the time constant τ, and this may be found by a variety of methods. One common method applies a step input and measures τ as the time to achieve 63.2 percent of the final value. This method is influenced by inaccuracies in the determination of the $t = 0$ point and also gives no check as to whether the instrument is really first-order. A preferred method uses the data from a step-function test replotted semilogarithmically to get a better estimate of τ and to check conformity to true first-order response. This method goes as follows: From Eq. (3.178) we can write

$$\frac{q_o - Kq_{is}}{Kq_{is}} = -e^{-t/\tau} \tag{3.350}$$

$$1 - \frac{q_o}{Kq_{is}} = e^{-t/\tau} \tag{3.351}$$

Now we define

$$Z \triangleq \log_e \left(1 - \frac{q_o}{Kq_{is}}\right) \tag{3.352}$$

and then

$$Z = \frac{-t}{\tau} \qquad \frac{dZ}{dt} = \frac{-1}{\tau} \tag{3.353}$$

Thus if we plot Z versus t, we get a straight line whose slope is numerically $-1/\tau$. Figure 3.103 illustrates the procedure. This gives a more accurate value of τ since the best line through *all* the data points is used rather than just two points, as in the 63.2 percent method. Furthermore, if the data points fall nearly on a straight line, we are assured that the instrument is behaving as a first-order type. If the data deviate considerably from a straight line, we know the instrument is not truly first-order and a τ value obtained by the 63.2 percent method would be quite misleading.

An even stronger verification (or refutation) of first-order dynamic characteristics is available from frequency-response testing, although at considerable cost of time and money if the system is not completely electrical, since nonelectrical sine-wave generators are neither common nor necessarily cheap. If the equipment is available, the system is subjected to sinusoidal inputs over a wide frequency

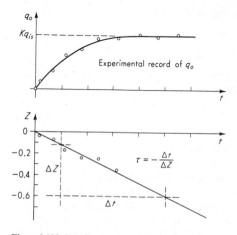

Figure 3.103 Step-function test of first-order system.

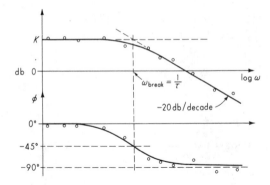

Figure 3.104 Frequency-response test of first-order system.

range, and the input and output are recorded. Amplitude ratio and phase angle are plotted on the logarithmic scales. If the system is truly first-order, the amplitude ratio follows the typical low- and high-frequency asymptotes (slope 0 and -20 dB/decade) and the phase angle approaches $-90°$ asymptotically. If these characteristics are present, the numerical value of τ is found by determining ω at the breakpoint and using $\tau = 1/\omega_{break}$ (see Fig. 3.104). Deviations from the above amplitude and/or phase characteristics indicate non-first-order behavior.

For second-order systems, K is found from static calibration, and ζ and ω_n can be obtained in a number of ways from step or frequency-response tests. Figure 3.105a shows a typical step-function response for an underdamped second-order system. The values of ζ and ω_n may be found from the relations

$$\zeta = \sqrt{\frac{1}{(\pi/\log_e (a/A))^2 + 1}} \tag{3.354}$$

$$\omega_n = \frac{2\pi}{T \sqrt{1 - \zeta^2}} \tag{3.355}$$

When a system is lightly damped, any fast transient input will produce a response similar to Fig. 3.105b. Then ζ can be closely approximated by

$$\zeta \approx \frac{\log_e (x_1/x_n)}{2\pi n} \tag{3.356}$$

This approximation assumes $\sqrt{1 - \zeta^2} \approx 1.0$, which is quite accurate when $\zeta < 0.1$, and again ω_n can be found from Eq. (3.355). In applying Eq. (3.355), if several cycles of oscillation appear in the record, it is more accurate to determine the period T as the average of as many distinct cycles as are available rather than from a single cycle. If a system is strictly linear and second-order, the value of n in Eq. (3.356) is immaterial; the same value of ζ will be found for any number of cycles. Thus if ζ is calculated for, say, $n = 1, 2, 4,$ and 6 and *different* numerical

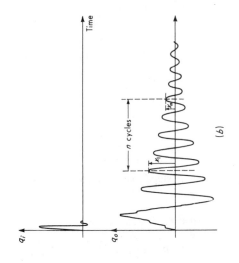

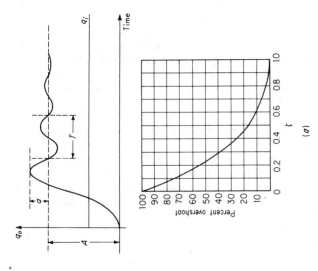

Figure 3.105 Step and pulse tests for second-order system.

values of ζ are obtained, we know the system is not following the postulated mathematical model. For overdamped systems ($\zeta > 1.0$), no oscillations exist, and the determination of ζ and ω_n becomes more difficult. Usually it is easier to express the system response in terms of two time constants τ_1 and τ_2, rather than ζ and ω_n. From Eq. (3.237) we can write

$$\frac{q_o}{Kq_{is}} = \frac{\tau_1}{\tau_2 - \tau_1} e^{-t/\tau_1} - \frac{\tau_2}{\tau_2 - \tau_1} e^{-t/\tau_2} + 1 \tag{3.357}$$

where

$$\tau_1 \triangleq \frac{1}{(\zeta - \sqrt{\zeta^2 - 1})\omega_n} \tag{3.358}$$

$$\tau_2 \triangleq \frac{1}{(\zeta + \sqrt{\zeta^2 - 1})\omega_n} \tag{3.359}$$

To find τ_1 and τ_2 from a step-function response curve, we may proceed as follows[1]:

1. Define the *percent incomplete response* R_{pi} as

$$R_{pi} \triangleq \left(1 - \frac{q_o}{Kq_{is}}\right) 100$$

2. Plot R_{pi} on a logarithmic scale versus time t on a linear scale. This curve will approach a straight line for large t if the system is second-order. Extend this line back to $t = 0$, and note the value P_1 where this line intersects the R_{pi} scale. Now, τ_1 is the time at which the straight-line asymptote has the value $0.368P_1$.
3. Now plot on the same graph a new curve which is the difference between the straight-line asymptote and R_{pi}. If this new curve is not a straight line, the system is not second-order. If it is a straight line, the time at which this line has the value $0.368(P_1 - 100)$ is numerically equal to τ_2.

Figure 3.106 illustrates this procedure. Once τ_1 and τ_2 are found, ζ and ω_n can be determined from Eqs. (3.358) and (3.359) if desired. Other methods[2] for finding τ_1 and τ_2 are available in the literature. Frequency-response methods also may be used to find ζ and ω_n or τ_1 and τ_2. Figure 3.107 shows the application of these techniques. The methods shown use the amplitude-ratio curve only. If phase-angle curves are available, they constitute a valuable check on conformance to the postulated model.

For measurement systems of arbitrary form (as contrasted to first- and second-order types), description of the dynamic behavior in terms of frequency response usually is desired. This information may be obtained by sinusoidal,

[1] N. A. Anderson, Step-Analysis Method of Finding Time Constant, *Instr. Contr. Syst.*, p. 130, November 1963.

[2] G. M. Hoerner, Second-Order System Characteristics from Initial Step Response, *Constr. Eng.*, p. 93, December 1962.

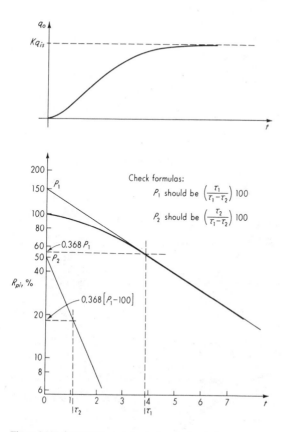

Check formulas:

P_1 should be $\left(\dfrac{\tau_1}{\tau_1 - \tau_2}\right) 100$

P_2 should be $\left(\dfrac{\tau_2}{\tau_1 - \tau_2}\right) 100$

Figure 3.106 Step test for overdamped second-order systems.

pulse, or random signal testing, following the general methods[1] used to experimentally determine mathematical models for physical systems. When the physical system being studied is a *measurement* system, the output signal q_o is itself generally useful and no separate output sensor is required. However, we do usually need to measure the input signal q_i with a separate sensor, which serves as the calibration standard and whose accuracy is known, and, if possible, is about 10 times better than that of the system being calibrated. If we can obtain $(q_o/q_i)(i\omega)$ thus for the measurement system, this defines the frequency range over which corrections are negligible and provides the data needed to make dynamic corrections (using the transform methods of the previous section) if we wish to use the instrument in its nonflat range of frequency response.

[1] Doebelin, op. cit., chap. 6.

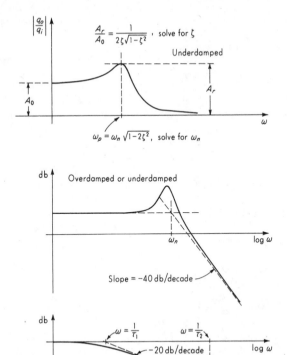

Figure 3.107 Frequency-response test of second-order system.

Loading Effects Under Dynamic Conditions

The treatment of loading effects by means of impedance, admittance, etc., was discussed in Sec. 3.2 for static conditions. All these results can be immediately transferred to the case of dynamic operation by generalizing the definitions in terms of transfer functions. The basic equations relating the undisturbed value q_{i1u} and the actual measured value q_{i1m} at the input of a device are

$$q_{i1m} = \frac{1}{Z_{go}/Z_{gi} + 1} \cdot q_{i1u} \tag{3.55}$$

$$q_{i1m} = \frac{1}{Y_{go}/Y_{gi} + 1} \, q_{i1u} \tag{3.61}$$

$$q_{i1m} = \frac{1}{S_{go}/S_{gi} + 1} q_{i1u} \qquad (3.68)$$

$$q_{i1m} = \frac{1}{C_{go}/C_{gi} + 1} q_{i1u} \qquad (3.80)$$

The quantities Z, Y, S, and C previously were considered to be the ratios of small changes in two related system variables under stated conditions. To generalize these concepts, we now define the quantities Z, Y, S, and C as *transfer functions* relating the same two variables under the same conditions except that now dynamic operation is considered. That is, we must get (theoretically or experimentally) $Z(D)$, $Y(D)$, $S(D)$, and $C(D)$ if we wish to use operational transfer functions and $Z(i\omega)$, $Y(i\omega)$, $S(i\omega)$, and $C(i\omega)$ if we wish to use frequency-response methods.

Usually the frequency-response form is most useful if these quantities must be found experimentally. This means, then, that in finding, say $Z(i\omega)$, one of the two variables involved in the definition of Z plays the role of an "input" quantity which we vary sinusoidally at different frequencies. This causes a sinusoidal change in the other ("output") variable, and thus we can speak of an amplitude ratio and phase angle between these two quantities, making $Z(i\omega)$ now a complex number that varies with frequency. (If the system is somewhat nonlinear, the effective approximate Z becomes a function also of input amplitude. This situation was adequately described under static conditions in Sec. 3.2.) In Eq. (3.55), for example, both Z_{go} and Z_{gi} would now be complex numbers; if these were known, we could calculate the amplitude and phase of q_{i1m} if the amplitude, phase, and frequency of a sinusoidal q_{i1u} were given. The quantity q_{i1m} then would be the *actual* input (q_i) to the measuring device, and we could calculate q_o if the transfer function $(q_o/q_i)(i\omega)$ were known. That is,

$$Q_o(i\omega) = \frac{1}{Z_{go}(i\omega)/Z_{gi}(i\omega) + 1} \left[\frac{q_o}{q_i}(i\omega) \right] Q_{i1u}(i\omega) \qquad (3.360)$$

Thus we could define a *loaded transfer function* $(q_o/q_{i1u})(i\omega)$ as

$$\frac{q_o}{q_{i1u}}(i\omega) \triangleq \frac{1}{Z_{go}(i\omega)/Z_{gi}(i\omega) + 1} \frac{q_o}{q_i}(i\omega) \qquad (3.361)$$

where $q_o \triangleq$ actual output of measuring device that has no load at *its* output
$q_{i1u} \triangleq$ measured variable value that would exist if measuring device caused *no* loading on measured medium

Equations (3.61), (3.68), and (3.80) may be modified in similar fashion. Also, if differential equations relating $q_o(t)$ are desired, we may write

$$\frac{q_o}{q_{i1u}}(D) = \frac{1}{Z_{go}(D)/Z_{gi}(D) + 1} \frac{q_o}{q_i}(D) \qquad (3.362)$$

and then obtain the differential equation in the usual way by "cross-multiplying":

$$[Z_{go}(D) + Z_{gi}(D)](a_n D^n + a_{n-1}D^{n-1} + \cdots + a_1 D + a_0)q_o$$
$$= [Z_{gi}(D)](b_m D^m + b_{m-1}D^{m-1} + \cdots + b_1 D + b_0)q_{i1u} \qquad (3.363)$$

An example of the above methods will be helpful. Consider a device for measuring translational velocity, as in Fig. 3.108a. The unloaded transfer function relating the output displacement x_o and the input (measured) velocity v_i is obtained as follows:

$$B_i(\dot{x}_i - \dot{x}_o) - K_{is}x_o = M_i \ddot{x}_o \qquad (3.364)$$

$$\frac{x_o}{v_i}(D) = \frac{K_i}{D^2/\omega_{ni}^2 + 2\zeta_i D/\omega_{ni} + 1} \qquad (3.365)$$

where $K_i \triangleq$ instrument static sensitivity $\triangleq \dfrac{B_i}{K_{is}}$ m/(m/s) $\qquad (3.366)$

$\zeta_i \triangleq$ instrument damping ratio $\triangleq \dfrac{B_i}{2\sqrt{K_{is}M_i}} \qquad (3.367)$

$\omega_{ni} \triangleq$ instrument undamped natural frequency $\triangleq \sqrt{\dfrac{K_{is}}{M_i}}$ rad/s $\qquad (3.368)$

We see that the instrument is second-order and thus will measure v_i accurately for frequencies sufficiently low relative to ω_{ni}. Suppose we now attach this instrument to a vibrating system whose velocity we wish to measure, as in Fig. 3.108b. The presence of the measuring instrument will distort the velocity we are trying to measure. The character of this distortion may be assessed by application of Eq. (3.61), since the measured quantity is velocity (a flow variable; see Fig. 3.28) and thus admittance is the appropriate quantity to use. We determine the input admittance $Y_{gi}(D) = (v/f)(D)$ from Fig. 3.108c as follows:

$$f - K_{is}x_o = M_i \ddot{x}_o \qquad (3.369)$$

Also
$$f = B_i(v - \dot{x}_o) \qquad (3.370)$$

and, eliminating x_o, we get

$$Y_{gi}(D) = \frac{v}{f}(D) = \frac{(1/B_i)(D^2/\omega_{ni}^2 + 2\zeta_i D/\omega_{ni} + 1)}{D^2/\omega_{ni}^2 + 1} \qquad (3.371)$$

Figure 3.108c also shows the frequency characteristics of this input admittance. The output admittance $Y_{go}(D) = (v/f)(D)$ of the measured system is obtained from Fig. 3.108d:

$$f - B\dot{x} - K_s x = M\ddot{x} \qquad (3.372)$$

$$Y_{go}(D) = \frac{v}{f}(D) = \frac{(1/K_s)D}{D^2/\omega_n^2 + 2\zeta D/\omega_n + 1} \qquad (3.373)$$

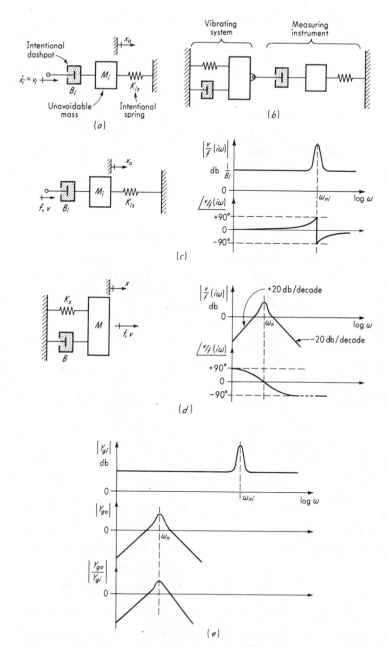

Figure 3.108 Example of dynamic loading analysis.

The frequency characteristic of this output admittance is shown in Fig. 3.108d. We may now write

$$\frac{x_o}{v_{i1u}}(D) = \frac{1}{Y_{go}(D)/Y_{gi}(D) + 1} \frac{x_o}{v_i}(D) \tag{3.374}$$

$$\frac{x_o}{v_{i1u}}(D) = \left[\cfrac{1}{\underbrace{\cfrac{(1/K_s)D}{D^2/\omega_n^2 + 2\zeta D/\omega_n + 1} \cfrac{(D^2/\omega_{ni}^2) + 1}{(1/B_i)(D^2/\omega_{ni}^2 + 2\zeta_i D/\omega_{ni} + 1)}}_{\text{loading effect}} + 1} \right]$$

$$\times \left[\frac{K_i}{D^2/\omega_{ni}^2 + 2\zeta_i D/\omega_{ni} + 1} \right] \tag{3.375}$$

where $x_o \triangleq$ actual output of measuring device
$v_{i1u} \triangleq$ velocity that would exist if measuring device caused no loading

Figure 3.108e shows that in this example the loading effect is most serious for frequencies near the natural frequency of the measured system, but approaches zero for both very low and very high frequencies. Since the loading effects can be expressed in frequency terms, they can be handled for all kinds of inputs by using appropriate Fourier series, transform, or mean-square spectral density.

PROBLEMS

3.1 Consider the system of Fig. 2.3.

(*a*) Explain how you would carry out a static calibration to determine the relation between the desired input and the output.

(*b*) The temperature of the air surrounding the capillary tube is an interfering input. Explain how you would calibrate the relation between this input and the output.

(*c*) The elevation difference between the Bourdon tube and the bulb is another interfering input. Discuss means for its calibration.

3.2 Does the system of Fig. 2.4 require calibration? Explain.

3.3 What fundamental difficulties arise in trying to define the true temperature of a physical body?

3.4. Slide a coin along a smooth surface, trying to make it come to rest at a drawn line. Measure the distance of the coin from the line. Repeat 100 times and check the resulting data for conformance to a Gaussian distribution, using probability graph paper.

3.5 Using the data generated in Prob. 3.4, apply the chi-squared test for conformance to a Gaussian distribution.

3.6 In Eq. (2.6), solve for the strain ϵ in terms of the other parameters; $\epsilon = f(GF, R_g, E_b, R_a, e_o)$. Then take the natural log of both sides; $\ln \epsilon = \ln f$. Now take the differential of both sides so that terms such as $d\epsilon/\epsilon, de_o/e_o, dRa/R_a$, etc., are formed. This will give the percentage error $d\epsilon/\epsilon$ in ϵ as a function of the percentage errors in the other parameters. If GF, R_g, E_b, R_a, and e_o are all measured to ± 1 percent error, what is the possible error in the computed value of ϵ?

3.7 Is the logarithmic differentiation method of Prob. 3.6 applicable to all forms of functional relations? Explain. *Hint*: Apply it to the relation $w = \sin x + 5y^3 - 6e^z$.

3.8 The discharge coefficient C_q of an orifice can be found by collecting the water that flows through during a timed interval when it is under a constant head h. The formula is

$$C_q = \frac{W}{t\rho A \sqrt{2gh}}$$

Find C_q and its possible error if:

$W = 865 \pm 0.5$ lbm	$A = \pi d^2/4$	$d = 0.500 \pm 0.001$ in
$t = 600.0 \pm 2$ s	$g = 32.17 \pm 0.1\%$ ft/s^2	
$\rho = 62.36 \pm 0.1\%$ lbm/ft^3	$h = 12.02 \pm 0.01$ ft	

considering both the following:

    (a) The errors are the absolute limits.
    (b) The errors are $\pm 3s$ limits.

3.9 In Prob. 3.8 if C_q must be measured within ± 0.5 percent for the numerical mean values given, what errors are allowable in the measured data? Use the method of equal effects.

3.10 Static calibration of an instrument gives the data of Fig. P3.1. Calculate (a) the best-fit straight line, (b) s_m and s_b, (c) s_{qi}, (d) q_i and its error limits if the instrument is used after calibration and reads $q_o = 5.72$.

q_i	q_o Increasing values	q_o Decreasing values
0	−0.07	+0.01
5	1.08	1.16
10	2.05	2.10
15	3.27	3.29
20	4.28	4.36
25	5.41	5.45
30	6.43	6.53
35	7.57	7.61
40	8.66	8.75

Figure P3.1

3.11 In Fig. 3.21, what percentage error may be expected in measuring the voltage across R_5 if $R_1 = R_2 = R_3 = R_4 = R_5 = 100 \ \Omega$ and $R_m = 1,000 \ \Omega$? If $R_m = 10,000$ ohms?

3.12 Repeat Prob. 3.11, except now the voltage across R_3 is to be measured.

3.13 In Fig. 3.25, what percentage error may be expected in measuring the current through R_5 if $R_1 = R_2 = R_3 = R_4 = R_5 = 100$ and $R_m = 10 \ \Omega$? If $R_m = 1 \ \Omega$?

3.14 Repeat Prob. 3.13, except now the current through R_3 is to be measured.

3.15 In Fig. 3.26, what percentage error may be expected in measuring the force in k_2 if $k_1 = k_2 = k_3 = k_4 = 100$ and $k_m = 1,000$ N/cm? If $k_m = 10,000$ N/cm?

3.16 Repeat Prob. 3.15, except now the force in k_3 is to be measured.

3.17 In Fig. 3.27, what percentage error may be expected in measuring the deflection x if $k_1 = k_2 = k_3 = k_4 = 1$ and $k_m = 0.1$ N/cm? If $k_m = 0.01$ N/cm?

3.18 Repeat Prob. 3.17, except now the motion of the right-hand block is to be measured.

3.19 Using methods similar to those used in proving Eq. (3.55), prove (a) Eq. (3.61), (b) Eq. (3.68), (c) Eq. (3.80).

3.20 A mercury thermometer has a capillary tube of 0.010-in diameter. If the bulb is made of a zero-expansion material, what volume must it have if a sensitivity of 0.10 in/F° is desired? Assume operation near 70°F. If the bulb is spherical and is immersed in stationary air, estimate the time constant.

3.21 A balloon carrying a first-order thermometer with a 15-s time constant rises through the atmosphere at 6 m/s. Assume temperature varies with altitude at 0.15 °C/30 m. The balloon radios temperature and altitude readings back to the ground. At 3000 m the balloon says the temperature is 0°C. What is the true altitude at which 0°C occurs?

3.22 A first-order instrument must measure signals with frequency content up to 100 Hz with an amplitude inaccuracy of 5 percent. What is the maximum allowable time constant? What will be the phase shift at 50 and 100 Hz?

3.23 For the spring scale of Fig. 3.48, discuss the tradeoff between sensitivity and speed of response resulting from changes in K_s.

3.24 Derive Eqs. (3.243) to (3.245).

3.25 Find the transfer function of a spring scale (Fig. 3.49) whose mass is negligible. Show that the steady-state time lag for a ramp input is the same whether mass is zero or not.

3.26 We wish to design (choose $\tau_1, \tau_2, \zeta, \omega_n$) the measurement system of Fig. P3.2a so as to achieve an amplitude ratio which is flat ± 5 percent for the frequency range 0 to 100 Hz. (Note that such a problem does not have a single unique answer.) Strive for the *largest* τ's and *smallest* ω_n which will meet the specifications. (Why?) Note that the rising amplitude ratio associated with a small ζ can compensate for the dropoff due to the τ's and that $\zeta = 0.65$ is not necessarily optimum in this case. When you find a set of values which is satisfactory:

(a) Check the linearity of phase angle with frequency, and estimate the effective time delay between q_o and q_i.

(b). If $q_i(t)$ is a transient of the form shown in Fig. P3.2b, use CSMP (or other available digital simulation) to find the smallest T for which the instantaneous error between q_o and q_i does not exceed 5 percent of the full-scale q_i value.

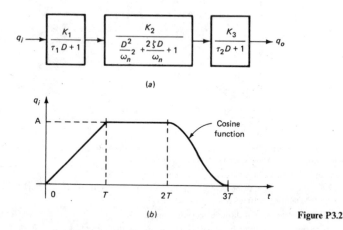

(a)

(b)

Figure P3.2

3.27 Find $Q_i(i\omega)$ for the $q_i(t)$ of Fig. P3.3 by the exact analytical method.

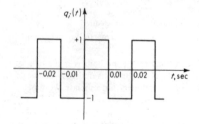

Figure P3.3

3.28 Repeat Prob. 3.27 for Fig. P3.4.

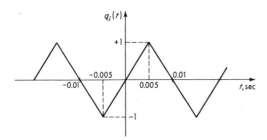

Figure P3.4

3.29 If the $q_i(t)$ of Prob. 3.27 is the input to a first-order system with a gain of 1 and a time constant of 0.001 s find $Q_o(i\omega)$ and $q_o(t)$ for the periodic steady state.

3.30 Repeat Prob. 3.29, except use $q_i(t)$ from Prob. 3.28.

3.31 In Fig. 3.82 let the carrier be a square wave as in Fig. P3.5. Find the frequency spectrum of the output signal of the modulator.

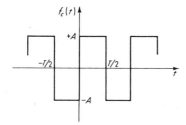

Figure P3.5

3.32 Repeat Prob. 3.31 if the carrier is a square wave as in Fig. P3.6.

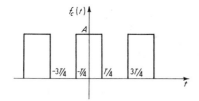

Figure P3.6

3.33 In an analog-computer study, it is desired to simulate a random atmospheric turbulence whose mean-square spectral density $\phi_t(\omega)$ is adequately represented as $10/(1 + 0.0001\omega^2)$, where ω is in radians per second. A white-noise generator having $\phi_{wn}(\omega) = 10$ is available. Select a suitable filter configuration and numerical values to follow the generator and produce the desired $\phi_t(\omega)$. The output of the noise generator should "see" a filter input resistance of 10,000 Ω.

3.34 Tests on a gyroscope show that it can withstand any random vibration along a given axis if the frequency content is between 0 and 1,000 rad/s and the rms acceleration is less than 80 in/s². This gyro is to be mounted in a rocket where it will be subjected to acoustic-pressure-induced vibration.

The transfer function between pressure and acceleration and the mean-square spectral density of pressure are as given in Fig. P3.7. Will this gyro withstand the vibration?

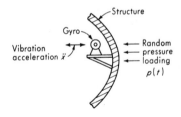

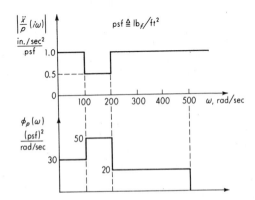

Figure P3.7

3.35 Derive Eq. (3.354).

3.36 Derive Eq. (3.356).

3.37 Explain how the sinusoidal transfer function of a system may be obtained from measured records of $q_i(t)$ and $q_o(t)$ if q_i is a transient of any shape whatever.

3.38 Reanalyze the force-measuring problem of Fig. 3.26 for *dynamic* operation, assuming the two blocks have masses M_1 and M_2. That is, get the operational transfer function analogous to Eq. (3.77).

3.39 Reanalyze the displacement-measuring problem of Fig. 3.27 for *dynamic* operation, assuming the two blocks have masses. M_1 and M_2. That is, get the operational transfer function analogous to Eq. (3.93).

3.40 Reanalyze the voltage-measuring problem of Fig. 3.21 for dynamic operation; i.e., replace the batteries with sources of time-varying voltage. Also, let the voltage-measuring device be an oscilloscope with R_m shunted by a capacitor C_m.

3.41 Reanalyze the current-measuring problem of Fig. 3.25 for dynamic operation; i.e., replace the batteries with sources of time-varying voltage. Also, let the current-measuring device be a galvanometer which has an inductance L_m in series with R_m.

BIBLIOGRAPHY

1. H. E. Koenig and W. A. Blackwell: "Electromechanical System Theory," McGraw-Hill, New York, 1961.

2. C. L. Cuccia: "Harmonics, Sidebands, and Transients in Communication Engineering," McGraw-Hill, New York, 1952.
3. T. N. Whitehead: "The Design and Use of Instruments and Accurate Mechanisms," Dover, New York, 1954.
4. V. L. Lebedev: Random Processes in Electrical and Mechanical Systems, *NASA, Tech. Transl.*, F–61, 1961.
5. C. C. Perry: The Least Squares Method, *Mach. Des.*, p. 210, May 12, 1960.
6. V. R. Boulton: Economics of Instrumentation Precision, *Aerospace Eng.*, p. 30, March 1961.
7. C. T. Morrow: Averaging Time and Data-Reduction Time for Random Vibration Spectra, *J. Acoust. Soc. Am.*, vol. 30, no. 6, p. 572, June 1958.
8. N. R. Goodman et al.: Frequency Response from Stationary Noise: Two Case Histories, *Technometrics*, p. 245, May 1961.
9. R. L. Hammon: An Application of Random Process Theory to Gyro Drift Analysis, *IRE Trans. PGANE*, vol. ANE-7, no. 3, September 1960.
10. D. E. Cartwright et al: Digital Techniques for the Study of Sea Waves, Ship Motion and Allied Processes, *Trans. Soc. Instr. Tech. (London)*, p. 1, March 1962.
11. J. T. Broch: Automatic Recording of Amplitude Density Curves, *B & K Tech. Rev.*, B & K Instruments, Cleveland, Ohio, no. 4, 1959.
12. J. T. Broch: Recording of Narrow Band Noise, *B & K Tech. Rev.*, B & K Instruments, Cleveland, Ohio, no. 4, 1960.
13. K. R. Thorson and Q. R. Bohne: Application of Power Spectral Methods in Airplane and Missile Design, *J. Aero/Scope Sci.*, p. 107, February 1960.
14. J. C. Laurence: Intensity, Scale and Spectra of Turbulence in Mixing Region of Free Subsonic Jet, *NACA, Tech. Notes* 3561, 1955.
15. J. R. Rice et al.: On the Prediction of Some Random Loading Characteristics Relevant to Fatigue, *NASA*, CR-56152, 1964.
16. W. A. Wildhack et al.: Accuracy in Measurements and Calibrations, *NBS, Tech. Notes* 262, 1965.
17. H. W. Maynard: An Evaluation of Ten Fast Fourier Transform (FFT) Programs, US Army Elect. Command, Ft. Monmouth, N. J., 1973.
18. H. McNeill: Digital Data Reduction Methods for Aircraft Engine Noise Analysis, *Sound Vib.* April 1972.
19. S. G. Cline: New Capabilities and Digital Low-Frequency Spectrum Analysis, *Hewlett-Packard J.* June 1972.
20. Special Issue on the Fast Fourier Transform. *IEEE Trans. on Audio and Electroacoustics*, vol. AV-17, June 1969.
21. Real Time Signal Processing in the Frequency Domain, Monograph No. 3, Federal Scientific Corp., N. Y., 1972.
22. K. N. Fieldhouse: Techniques for Identifying Sources of Noise and Vibration, *Sound & Vib.*, December 1970.
23. J. S. Bendat and A. G. Piersol: "Measurement and Analysis of Random Data," Wiley, New York, 1966.
24. G. A. Korn: "Random Process Simulation and Measurement," McGraw-Hill, New York, 1966.
25. E. O. Doebelin: "System Dynamics," Merrill, Columbus, Ohio, 1972.
26. E. O. Doebelin: "System Modeling and Response," New York, 1980.
27. J. S. Bendat and A. G. Piersol: "Engineering Applications of Correlation and Spectral Analysis," Wiley, New York, 1980.

PART
TWO

MEASURING DEVICES

FOUR

MOTION MEASUREMENT

4.1 INTRODUCTION

We commence our study of specific measuring devices with motion measurement since it is based on two of the fundamental quantities in nature (length and time) and because so many other quantities (such as force, pressure, temperature, etc.) are often measured by transducing them to motion and then measuring this resulting motion. As indicated in the chapter title, our main interest is in motion (a *changing* displacement). However, many of the displacement sensors described are used in manufacturing processes where gaging of part dimensions (*fixed* lengths) is required.

We are also mainly (though not exclusively) concerned with electromechanical transducers which convert motion quantities into electrical quantities. The intent is not to present a catalog listing of the myriad physical effects which have been, or might be, used as the basis of a motion transducer, but rather to provide sufficient detail for practical application of the relatively small number of transducer types which form the basis of the majority of practical measurements. The above-mentioned catalog-listing type of information is extremely useful to one who has a measurement problem not solvable by standard techniques and who must therefore invent and/or develop a new instrument. Material of this type is available in several references.[1]

---

[1] K. S. Lion, "Instrumentation in Scientific Research," McGraw-Hill, New York, 1959; C. F. Hix, Jr., and R. P. Alley, "Physical Laws and Effects", Wiley, New York, 1958.

4.2 FUNDAMENTAL STANDARDS

The four fundamental quantities of the International Measuring System, for which independent standards have been defined, are length, time, mass, and temperature. Units and standards for all other quantities are *derived* from these. In motion measurement, the fundamental quantities are length and time. Prior to 1960 the standard of length was the carefully preserved platinum-iridium International Meter Bar at Sèvres, France. In 1960 the meter was redefined in terms of the wavelength of a krypton-86 lamp as the length equal to 1,650,763.73 wavelengths in vacuum corresponding to the transition between the energy levels $2p_{10}$ and $5d_5$ of the atom krypton 86. This standard is believed to be reproducible to about 2 parts in 10^8 and can be applied at this precision level to measurements of length in the range of about 10^{-8} to 40 in.[1]

The above National Prototype Standard is not available for routine calibration work. Rather, to protect such top-level standards from deterioration, the National Bureau of Standards has set up National Reference Standards and, below these, Working Standards. Further down the line in accuracy are the so-called Interlaboratory Standards, which are standards sent in to the National Bureau of Standards for calibration and certification by factories and laboratories all over the country. These last-mentioned standards are usually readily available to the working engineer for calibration of motion transducers.

The fundamental unit of time is the second, which was redefined for scientific use as 1/31,556,925.9747 of the tropical year at 12^h ephemeris time, 0 January 1900,[2] by the International Committee on Weights and Measures in 1956. A serious fault in this definition is that no one can measure an interval of time by direct comparison with the interval of time defining the second. Rather, lengthy astronomical measurements over several years are necessary to relate the current value of the mean solar second to the basic standard. These measurements and calculations result in an estimated probable error of about 1 part in 10^9, which is quite poor compared with the precision implied in the basic definition of the second. To remedy this difficulty, metrologists in 1964 again redefined the second in terms of the frequencies of atomic resonators.[3] Now the second is defined as the interval of time corresponding to 9,192,631,770 cycles of the atomic resonant frequency of cesium 133. Figure 4.1a, b, and c (taken from *NBS Tech. Note* 262, Accuracy in Measurements and Calibrations, 1965) gives data on the accuracy of the length and time standards, while Fig. 4.1d (*NBS Spec. Publ.* 445-1., The National Measurement System for Time and Frequency, A.S. Risley, 1976) gives an interesting historical perspective on timekeeping accuracy.

The above short discussion is concerned with the *fundamental* standards of

[1] J. S. Beers, A Gage Block Measurement Process Using Single Wavelength Interferometry, *NBS Monograph* 152, December 1975.

[2] A. G. McNish, Fundamentals of Measurement, *Electro-Technol.* (*New York*), p. 113, May 1963.

[3] H. Hellwig, Frequency Standards and Clocks: A Tutorial Introduction, *NBS Tech. Note* 616, 1972.

length and time, rather than the practical working standards with which most engineers will be concerned. These practical standards and associated calibration procedures are discussed in each specific section, such as relative displacement, acceleration, etc.

4.3 RELATIVE DISPLACEMENT, TRANSLATIONAL AND ROTATIONAL

We consider here devices for measuring the translation along a line of one point relative to another and the plane rotation about a single axis of one line relative to another. Such displacement measurements are of great interest as such and because they form the basis of many transducers for measuring pressure, force, acceleration, temperature, etc., as shown in Fig. 4.2.

Calibration

Static calibration of translational devices often can be satisfactorily accomplished by using ordinary dial indicators or micrometers as the standard. When

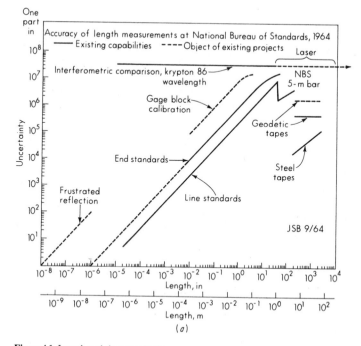

Figure 4.1 Length and time standards.

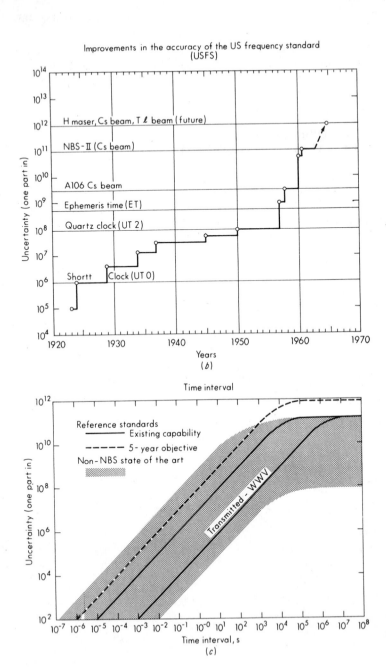

Figure 4.1 (*Continued*)

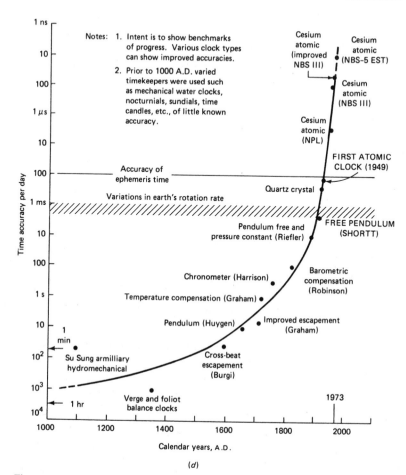

Figure 4.1 (*Continued*)

used directly to measure the displacement of the transducer, these devices usually are suitable to read to the nearest 0.0001 in or 0.01 mm. If smaller increments are necessary, lever arrangements (about a 10:1 ratio is fairly easy to achieve) or wedge-type mechanisms (about 100:1) can be employed for motion reduction.[1] The Mikrokator,[2] a unique mechanical gage of high sensitivity, may also be useful in measuring small motions down to a few millionths of an inch.

---

[1] H. C. Roberts, "Mechanical Measurements by Electrical Methods," chap. 13, Instruments Publishing, Pittsburgh, 1951; M. Barrangon, Calibration to One Millionth, *Mech. Eng.*, p. 38, October 1965.

[2] C. E. Johansson Gage Co., Dearborn, Mich.

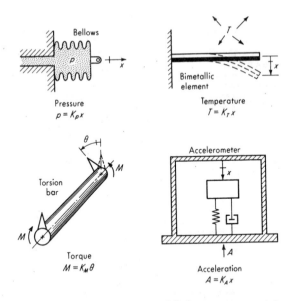

Figure 4.2 Transducer applications of displacement measurement.

If accuracy to 0.0001 in or better is required, such equipment should itself be calibrated against gage blocks, or (for maximum accuracy) gage blocks should be used *directly* to calibrate the transducer. *Gage blocks* are small blocks of hard, dimensionally stable steel or other material, made up in sets which can be stacked up to provide accurate dimensions over a wide range and in small steps. They are the basic working length standards of industry. As purchased from the manufacturer, their dimensions are accurate to ± 8 μin for working grade blocks, $\pm 4\mu$in for reference grade, and ± 2 μin for all blocks up to 1 in ($\pm 2\mu$in/in for blocks longer than 1 in) for master blocks. If these tolerances are too large, the blocks can be sent to the National Bureau of Standards and calibrated[1] against light wavelengths to the nearest 10^{-7} in. Some precision-manufacturing operations currently require and use the latter calibration service. When transducers are calibrated to very high accuracies, it is extremely important to control all interfering and/or modifying inputs such as ambient temperature,[2] electrical excitation to the transducer, etc.

Rotational or angular displacement is not itself a fundamental quantity since it is based on length, and so a fundamental standard is not necessary. However, reference and working standards for angles (and thus angular displacement) are desirable and available. The basic standards (against which other standards or

[1] P. E. Pontius, Measurement Assurance Program, Long Gage Blocks, *NBS Monograph* 149, November 1975.

[2] J. B. Bryan et al., Thermal Effects in Dimensional Metrology, *ASME Paper* 65-Prod − 13, 1965.

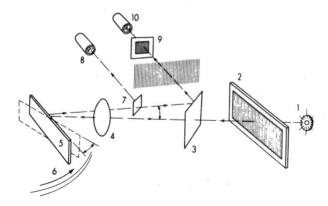

The Midarm system is comprised of an optical unit and an electronic unit. The simplified diagram shows how Midarm operates. Light from a monochromatic light point source (1) passes through a grid (2), a beam splitter (3), a collimating lens (4), and strikes a mirror (5) mounted on the rotating specimen to be tested (6). The image is reflected back into the system where it is directed by a beam splitter (7) to the reference photosensor (8). The image is also reflected by the beam splitter (3) through a second grid (9) to the control photosensor (10). As the test specimen (mirror) rotates, the image of the first grid passes across the second grid which allows minimum and maximum amounts of light to reach the control photosensor. The output voltage of the photosensor has a period of 12.8 arc-sec, the angle subtended by the grid spaces. The Midarm has digital output pulses at 12.8 arc-sec and an analog voltage output of 30 volts/arc-sec.

Figure 4.3 Midarm system.

instruments may be calibrated) are called *angle blocks*.[1] These are carefully made steel blocks about $\frac{5}{8}$ in wide and 3 in long, with a specified angle between the two contact surfaces. Just as for length gage blocks, these angle blocks can be stacked to "build up" any desired angle accurately and in small increments. The blocks can be calibrated to an accuracy of 0.1 second of arc by the National Bureau of Standards.[2]

Developments in inertial guidance systems have required angle and angular rate measurements on rotating components to an accuracy approaching or exceeding the capability of National Bureau of Standards calibration.[3] Combinations of optical and electronic principles have led to the development of instruments such as the Midarm[4] system to meet these requirements. This instrument

[1] C. E. Haven and A. G. Strong, Assembled Polygon for the Calibration of Angle Blocks, *Natl. Bur. Std.* (*U. S.*), *Handbook 77*, vol. 3, p. 318, 1961.

[2] Independent Standards Laboratory, *Instr. Contr. Syst.*, p. 478, March 1961.

[3] R. L. Hall, Analysis Technique for Calibration Angular Measuring Devices, paper X76-752/201, *AIAA Guid. & Cont. Conf.*, San Diego, August 16–18, 1976.

[4] Razdow Laboratories, Inc., Newark, N. J.

will, with relative convenience, measure angular displacement with an accuracy of 0.05 second of arc and a repeatability of 0.02 second of arc (see Fig. 4.3).

Rotational transducers rarely require such accuracy for calibration, nor can the laborious and expensive techniques necessary to realize these limits be economically justified. Thus most static calibration of angular-displacement transducers can be adequately carried out by using more convenient and readily available equipment. Examples[5] of such equipment which should be available in a precision machine shop are the circular division tester (range 360°, microscope reads to 0.1 minute of arc, precision of scale disk ±20 seconds of arc), the optical dividing head (range 360°, scale reads to 1.0 minute of arc, working accuracy ±20 seconds of arc), and the division tester with telescope and collimator (accuracy ±2 seconds of arc). In some applications, even cruder devices such as ordinary machine-tool index heads, calibrated dials, etc., may be perfectly adequate.

Resistive Potentiometers

Basically, a resistive potentiometer consists of a resistance element provided with a movable contact. (See Fig. 4.4.) The contact motion can be translation, rotation,

----

[5] Carl Zeiss, Inc., New York.

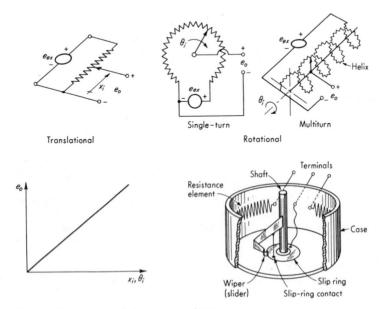

Figure 4.4 Potentiometer displacement transducer.

or a combination of the two (helical motion in a multiturn rotational device), thus allowing measurement of rotary and translatory displacements. Translatory devices have strokes from about 0.1 to 20 in and rotational ones range from about 10° to as much as 60 full turns. The resistance element is excited with either dc or ac voltage, and the output voltage is (ideally) a linear function of the input displacement. Resistance elements in common use may be classified as wire-wound, conductive plastic, hybrid, or cermet.

If the distribution of resistance with respect to translational or angular travel of the wiper (moving contact) is linear, the output voltage e_o will faithfully duplicate the input motion x_i or θ_i if the terminals at e_o are open-circuit (no current drawn at the output). (For ac excitation, x_i or θ_i amplitude-modulate e_{ex}, and e_o does not look like the input motion.) The usual situation, however, is one in which the potentiometer output voltage is the input to a meter or recorder that draws some current from the potentiometer. Thus a more realistic circuit is as shown in Fig. 4.5a. Analysis of this circuit gives

$$\frac{e_o}{e_{ex}} = \frac{1}{1/(x_i/x_t) + (R_p/R_m)(1 - x_i/x_t)} \tag{4.1}$$

which becomes for ideal ($R_p/R_m = 0$ for an open circuit) conditions

$$\frac{e_o}{e_{ex}} = \frac{x_i}{x_t} \tag{4.2}$$

Thus for no "loading" the input-output curve is a straight line. In actual practice, $R_m \neq \infty$ and Eq. (4.1) shows a nonlinear relation between e_o and x_i. This devi-

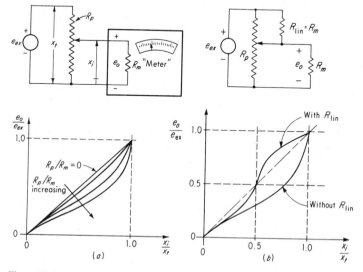

Figure 4.5 Potentiometer loading effect.

ation from linearity is shown in Fig. 4.5a. The maximum error is about 12 percent of full scale if $R_p/R_m = 1.0$ and drops to about 1.5 percent when $R_p/R_m = 0.1$. For values of $R_p/R_m < 0.1$, the position of maximum error occurs in the neighborhood of $x_i/x_t = 0.67$, and the maximum error is approximately 15 R_p/R_m percent of full scale.

We see that to achieve good linearity, for a "meter" of a given resistance R_m, we should choose a potentiometer of sufficiently *low* resistance relative to R_m. This requirement conflicts with the desire for high sensitivity. Since e_o is directly proportional to e_{ex}, it would seem possible to get any sensitivity desired simply by increasing e_{ex}. This is not actually the case, however, since potentiometers have definite power ratings related to their heat-dissipating capacity. Thus a manufacturer may design a series of potentiometers, say single-turn 2-in-diameter, with a wide range (perhaps 100 to 100,000 Ω) of total resistance R_p, but all these will be essentially the same size and mechanical configuration, giving the same heat-transfer capability and thus the same power rating, say about 5 W at 20°C ambient. If the heat dissipation is limited to P watts, the maximim allowable excitation voltage is given by

$$\max e_{ex} = \sqrt{PR_p} \qquad (4.3)$$

Thus a low value of R_p allows only a small e_{ex} and therefore a small sensitivity. Choice of R_p thus must be influenced by a tradeoff between loading and sensitivity considerations. The maximum available sensitivity of potentiometers varies considerably from type to type and also with size in a given type. It can be calculated from the manufacturer's data on maximum allowable voltage, current, or power and the maximum stroke. The shorter-stroke devices generally have higher sensitivity. *Extreme* values are of the order of 15 V/deg for short-stroke rotational types ("sector" potentiometers) and 300 V/in for short stroke (about $\frac{1}{4}$ in) translational pots. It must be emphasized that these are maximum values and that the usual application involves a much smaller (10 to 100 times smaller) sensitivity. Figure 4.5b shows a method for improving linearity without increasing R_m.

The resolution of potentiometers is strongly influenced by the construction of the resistance element. An obvious approach is to use a single slide-wire as the resistance which gives an essentially continuous stepless resistance variation as the wiper travels over it. Such potentiometers are available, but are limited to rather small resistance values since the length of wire is limited by the desired stroke in a translational device and by space restrictions (diameter) in a rotational one. Resistance of a given length of wire can be increased by decreasing the diameter, but this is limited by strength and wear considerations.

To get sufficiently high resistance values in small space, the wirewound resistance element is widely used. The resistance wire is wound on a mandrel or card which is then formed into a circle or helix if a rotational device is desired (see Fig. 4.6). With such a construction the variation of resistance is not a linear continuous change, but actually proceeds in small steps as the wiper moves from one turn of wire to the next (see Fig. 4.7). This phenomenon results in a fundamental limitation on the resolution in terms of resistance-wire size. For instance, if a

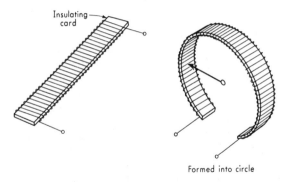

Figure 4.6 Construction of wirewound resistance elements.

translational device has 500 turns of resistance wire on a card 1 in long, motion changes smaller than 0.002 in cannot be detected. (This is slightly conservative since the wiper, in going from one turn to the next, goes through an intermediate position in which it is touching *both* turns at once.[1] The resolution thus actually varies from one position to another, with the *worst* value being that given by a simple counting of turns per inch.) The actual limit for wire spacing according to current practice is between 500 and 1,000 turns per inch.[2] For translational devices, resolution is thus limited to 0.001 to 0.002 in, while single-turn rotational devices can trade off increased diameter D for increased angular resolution according to the relation

$$\text{Best angular resolution} = \frac{0.12 \text{ to } 0.24}{D} \qquad \text{degrees} \qquad D \text{ in inches} \qquad (4.4)$$

[1] H. Gray, How to Specify Resolution for Potentiometer Servos, *Contr. Eng.*, p. 129, November 1959; C. A. Mounteer, The Effective Resolution of Wire-Wound Potentiometers, *Giannini Tech. Notes*, G. M. Giannini Co., Pasadena, Calif., January–February 1959.

[2] S. A. Davis and B. K. Ledgerwood, "Electromechanical Components for Servomechanisms," p. 53, McGraw-Hill, New York, 1961.

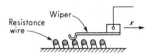

Figure 4.7 Resolution of wirewound potentiometers.

It should be noted that resolution is intimately related to total resistance since the fine wire required to get close wire spacing will naturally have a high resistance. Thus one cannot choose total resistance and resolution independently.

Nonwirewound resistance elements provide improved resolution and life; however, they are more temperature-sensitive, have a high (and variable) wiper contact resistance, and can tolerate only moderate wiper currents. Elements of both cermet (combination of ceramic and metallic materials) and conductive plastic (mixture of plastic resin with proprietary conductive powders) are in the form of flat strips or films and thus present a smooth surface to the wiper. They are conventionally described as having "infinitesimal resolution." However, resolution cannot be actually measured and numerically specified (as it can in wire-wound units) because the output-voltage deviations from ideal straight-line behavior are somewhat random rather than the largely reproducible discrete steps of the wirewound pot. *Output smoothness* (really *roughness*) is a specification that attempts to quantify this effect in nonwirewound pots and is given as the ratio of the peak amplitude of the random variations to e_{ex}, with typical values being 0.1 percent. While resolution, usually quoted as a theoretical value calculated from wire spacing, is numerically available for wirewound pots, a measurable noise specification (analogous to output smoothness for nonwirewound) called *equivalent noise resistance* R_{en} is often quoted. The test involves measuring the pot's output-voltage fluctuations while the wiper is moved over the winding with a constant wiper current applied, and thus the test includes effects of both resolution and various random-noise effects. The numerical value derived from the test is quoted as a fictitious equivalent contact resistance variation R_{en} corresponding to the largest fluctuations. For a given pot application (e_{ex}, R_p, and R_m known), one can use this R_{en} value to calculate the worst spurious voltage at e_o as the product of wiper current and R_{en}. Typical R_{en} values are 5 to 100 Ω.

A useful feature of conductive plastic elements is that they may be contoured along one edge (see Fig. 4.8) by milling or laser processes to adjust the resistance distribution from the as-molded state. These manufacturing processes are servocontrolled, so that the procedure is cost-effectively applied *individually* to each complete potentiometer to tailor the resistance element to compensate for all other sources of nonlinearity in that particular unit. This manufacturing process also is used to improve the conformity of the resistance element to *desired* nonlinear functions for pots used as computing elements (see Chap. 10).

Application of a narrow track of conductive plastic onto a conventional wirewound element produces the *hybrid potentiometer*. This combines most of the best features of each technology, but at a somewhat increased cost.

Another approach to increased resolution involves the use of multiturn potentiometers. The resistance element is in the form of a helix, and the wiper travels along a "lead screw." The number of wires per inch of element is still limited, as mentioned above, but an increase in resolution can be obtained by introducing gearing between the shaft whose motion is to be measured and the potentiometer shaft. For example, one rotation of the measured shaft could cause 10 rotations of the potentiometer shaft; thus the resolution of measured-shaft

Figure 4.8 Conductive-plastic potentiometer. (*Courtesy Waters Mfg., Wayland,
Mass.*)

motion is increased by a factor of 10. Multiturn potentiometers are available up
to about 60 turns in wirewound, nonwirewound, and hybrid types. For translational devices, various motion-amplifying mechanisms could be used in similar
fashion.

Most potentiometers used for motion measurement are intended to give a
linear input-output relation and are used as purchased, without calibration. Thus
a specification of linearity is essentially equivalent to one of accuracy. Potentiometers are available in a wide range of linearities and corresponding prices.
Linearity depends greatly on the uniformity of the resistance winding, but errors
in this can be corrected by adding fixed resistances in series and/or parallel at
proper locations on the winding. This procedure (called *tapping*) also can correct
loading errors so as to give a linear relation for a heavily loaded potentiometer.[1]
The best nonlinearities commercially available range from 1 percent of full scale
for $\frac{1}{2}$-in-diameter single-turn pots through 0.02 percent for a 2-in-diameter multiturn to 0.002 percent for a 10-in-diameter multiturn. The best nonlinearities of
translational pots are about 0.05 to 0.10 percent of full scale. It should be noted
that accuracy can be no better (and is generally worse) than one-half the resolution; thus resolution places a limit on accuracy.

Noise in potentiometers refers to spurious output-voltage fluctuations occurring during motion of the slider and includes the effects of resolution. In
addition, various mechanical and electrical defects produce noise. In a wirewound
pot, motion of the slider over the resistance wires may cause bouncing of the

---

[1] Ibid., p. 59.

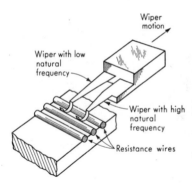

Figure 4.9 Antivibration wiper construction.

contact at certain speeds, thus resulting in intermittent contact. This phenomenon becomes particularly significant if the speed and wire spacing are such as to produce forces of frequency near the resonant frequency of the spring-loaded contact. Contacts sometimes are made in the configuration of Fig. 4.9 to overcome this problem. Here the resonant frequency of each section of the wiper is different. Thus if one section is resonating at a certain speed, the other will be off resonance and making continuous contact. Another possibility lies in filling the interior of the potentiometer with a damping fluid or coating the wipers with a layer of elastomeric damping material to limit resonant amplitudes. This also generally increases the shock and vibration tolerance of the unit. Another source of noise is found in dirt and wear products which come between the contact surface and the winding. Even if no dirt or wear products are present, the contact resistance of a moving contact varies during motion, and if any load current is flowing through this contact, a spurious iR voltage appears in the output. This effect also occurs at the slip-ring contact. Numerical values of noise voltage quoted in specifications generally include all sources of noise and correspond to a definite speed and current.[1]

The dynamic characteristic of potentiometers (if we consider displacement as input and voltage as output) is essentially that of a zero-order instrument since the impedance of the winding is almost purely resistive at the motion frequencies for which the device is usable. However, the mechanical loading imposed on the measured motion by the inertia and friction of the potentiometer's moving parts should be carefully considered. The friction is usually mostly dry friction, and the manufacturer generally supplies numerical values of the starting and running friction force or torque. These values vary over a wide range, depending on the construction of the potentiometer. Special low-friction rotary pots have starting torques as small as $0.003 \text{ oz} \cdot \text{in}$. More conventional instruments may have 0.1 to $0.5 \text{ oz} \cdot \text{in}$ or more. Translational pots have friction values from less than 1 oz to

---

[1] PPMA Conference Report, *Electromech. Des.*, p. 8, April 1964.

over 1 lb. Inertia values for both rotary and translatory pots vary widely with size. A typical $\frac{7}{8}$-in-diameter single-turn pot has a moment of inertia of 0.12 $g \cdot cm^2$, while a 2-in diameter 10-turn pot has about $18 \, g \cdot cm^2$. Moving masses of translatory pots have weights ranging from fractions of an ounce to several ounces.

Finally, selection of potentiometers should take into account various environmental factors such as high or low temperatures, shock and vibration, humidity, and altitude. These may act as modifying and/or interfering inputs so as seriously to degrade instrument performance. While the life of potentiometers varies greatly with type and environment, representative values might be 2 million rotation cycles for wirewound, 10 million for hybrid, and 50 million for conductive plastic. Recent automotive engine designs often employ many sensing devices, including potentiometric displacement sensors. Design of low-cost sensors for the severe under-the-hood environment has been particularly challenging.[1] Here a term called "dither life" is significant. Dither refers to conditions where case vibration caused by the engine superimposes high-frequency and low-amplitude wiper oscillations on the intended measured motion, a situation conducive to rapid localized wear. Dither life refers to cycles of the low-amplitude vibratory motion, which accumulate much more rapidly than strokes of gross motion. Conductive plastic pots have proved successful in many of these applications, providing dither lives in excess of 100 million cycles.

An interesting recent variation on the potentiometer principle is shown in Fig. 4.10.[2] Certain rotary-motion applications require a single-turn device which provides output *all the way to 360°* before repeating, a capability denied to "ordinary" pots by the need for at least a small gap between the two ends of the winding (see Fig. 4.6). The device of Fig. 4.10 has two conductive plastic elements combined with a built-in electronic switching network, which combines the outputs of the two windings so that output voltage proportional to angle is available over the full 0 to 360° range with an accuracy better than 0.36°. Life exceeds 10^8 at 1500-r/min maximum speed.

Resistance Strain Gages[3]

Consider a conductor of uniform cross-sectional area A and length L, made of a material with resistivity ρ. The resistance R of such a conductor is given by

$$R = \frac{\rho L}{A} \qquad (4.5)$$

If this conductor is now stretched or compressed, its resistance will change be-

---

[1] W. Wheeler, Precision Position Sensors in Automotive Applications, *SAE Paper* 780209, 1978.

[2] U. S. Patent 4,203,074, May 13, 1980.

[3] J. W. Dally and W. F. Riley, "Experimental Stress Analysis," 2d ed., McGraw-Hill, New York, 1978.

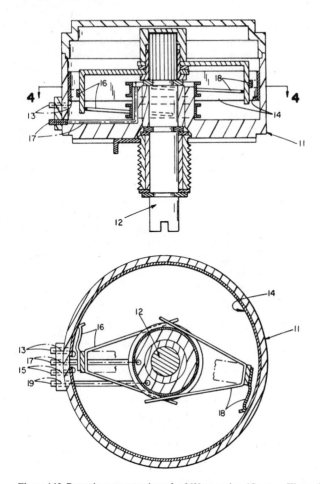

Figure 4.10 Potentiometer transducer for 360° operation. (*Courtesy Waters Mfg., Wayland, Mass.*)

cause of dimensional changes (length and cross-sectional area) and because of a fundamental property of materials called *piezoresistance*[1] (pronounced pī-ēzō-resistance), which indicates a dependence of resistivity ρ on the mechanical strain. To find how a change dR in R depends on the basic parameters, we differentiate Eq. (4.5) to get

$$dR = \frac{A(\rho \, dL + L \, dp) - \rho L \, dA}{A^2} \tag{4.6}$$

---

[1] C. M. Harris and C. E. Crede (eds.), "Shock and Vibration Handbook," vol. 1, pp. 16–35, McGraw-Hill, New York, 1961.

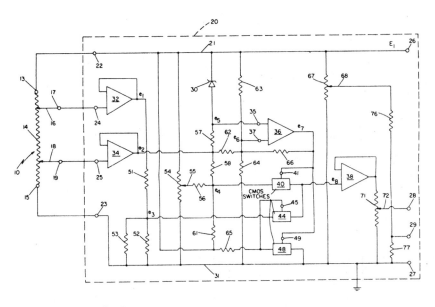

Figure 4.10 (*Continued*)

Since volume $V = AL$, $dV = A\,dL + L\,dA$. Also

$$dV = L(1 + \epsilon)A(1 - \epsilon v)^2 - AL \tag{4.7}$$

where $\epsilon \triangleq$ unit strain and $v \triangleq$ Poisson's ratio. Since ϵ is small, $(1 - v\epsilon)^2 \approx 1 - 2v\epsilon$ and Eq. (4.7) becomes

$$dV = AL\,\epsilon\,(1 - 2v) = A\,dL + L\,dA \tag{4.8}$$

and since $\epsilon \triangleq dL/L$,

$$A\,dL(1 - 2v) = A\,dL + L\,dA \tag{4.9}$$

$$- 2vA\,dL = L\,dA \tag{4.10}$$

Substituting in Eq. (4.6) yields

$$dR = \frac{\rho A\,dL + LA\,d\rho + 2v\rho A\,dL}{A^2} \tag{4.11}$$

and thus

$$dR = \frac{\rho\,dL(1 + 2v)}{A} + \frac{L\,d\rho}{A} \tag{4.12}$$

Dividing by Eq. (4.5) gives

$$\frac{dR}{R} = \frac{dL}{L}(1 + 2v) + \frac{d\rho}{\rho} \tag{4.13}$$

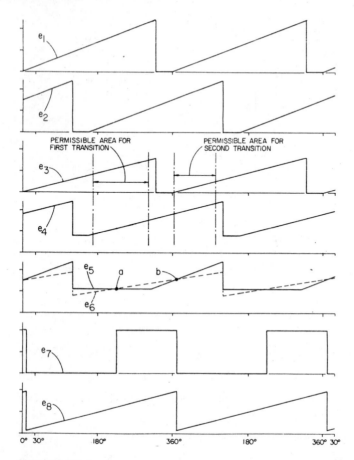

Figure 4.10 (*Continued*)

and finally

$$\text{Gage factor} \triangleq \frac{dR/R}{dL/L} = \underbrace{1}_{\substack{\text{resistance} \\ \text{change due} \\ \text{to length} \\ \text{change}}} + \underbrace{2v}_{\substack{\text{resistance} \\ \text{change due to} \\ \text{area change}}} + \underbrace{\frac{d\rho/\rho}{dL/L}}_{\substack{\text{resistance change due} \\ \text{to piezoresistance} \\ \text{effect}}} \tag{4.14}$$

Thus if the gage factor is known, measurement of dR/R allows measurement of the strain $dL/L = \epsilon$. This is the principle of the resistance strain gage. The term

$(d\rho/\rho)/(dL/L)$ can also be expressed as $\pi_1 E$, where

$$\pi_1 \triangleq \text{longitudinal piezoresistance coefficient}$$

$$E \triangleq \text{modulus of elasticity}$$

The material property π_1 can be either positive or negative. Poisson's ratio is always between 0 and 0.5 for all materials.

The basic principle of the resistance strain gage is implemented in several different ways:

1. Unbonded metal-wire gage
2. Bonded metal-wire gage
3. Bonded metal-foil gage
4. Vacuum-deposited thin-metal-film gage
5. Sputter-deposited thin-metal-film gage
6. Bonded semiconductor gage
7. Diffused semiconductor gage

Strain gages, in general, are applied in two types of tasks: in experimental stress analysis of machines and structures and in construction of force, torque, pressure, flow, and acceleration transducers. The unbonded metal-wire gage,[1] used almost exclusively for transducer applications, employs a set of preloaded resistance wires connected in a Wheatstone bridge, as in Fig. 4.11. At initial preload, the strains and resistances of the four wires are nominally equal, which gives a balanced bridge and $e_o = 0$ (see Chap. 10 for bridge-circuit behavior). Application of a small (full scale ≈ 0.04 mm) input motion increases tension in two wires and decreases it in two others (wires *never* go slack), causing corresponding resistance changes, bridge unbalance, and output voltage proportional to input motion. The wires may be made of various copper-nickel, chrome-nickel or nickel-iron alloys, are about 0.03 mm in diameter, can sustain a maximum force of only about 0.002 N, and have a gage factor of 2 to 4. Electric resistance of each bridge arm is 120 to 1,000 Ω, maximum excitation voltage is 5 to 10 V, and full-scale output typically is 20 to 50 mV.

The bonded metal-wire gage (today largely superseded by the bonded metal-foil construction) has been applied to both stress analysis and transducers. A grid of fine wire is cemented to the specimen surface, where strain is to be measured. Embedded in a matrix of cement, the wires cannot buckle and thus faithfully follow both the tension and the compression strains of the specimen. Since materials and wire sizes are similar to those of the unbonded gage, gage factors and resistances are comparable.

Bonded metal-foil gages using identical or similar materials to wire gages are used today for most general-purpose stress-analysis tasks and many transducers.

---

[1] "Universal Transducing Cell," Gould/Statham, Oxnard, Calif.; L. Statham, Bonded vs. Unbonded Strain Gage Transducers, *Inst. & Cont. Syst.*, p. 123, September 1962.

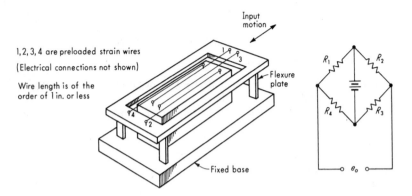

Figure 4.11 Unbonded strain gage.

The sensing elements are formed from sheets less than 0.0002 in thick by photo-etching processes, which allows great flexibility with regard to shape. In Fig. 4.12, for example, the three linear grid gages are designed with "fat" end turns. This local increase in area reduces the transverse sensitivity, a spurious input since the gage is intended to measure the strain component along the length of the grid elements. In a wire gage, these end turns would necessarily have the same cross section as the longitudinal elements, which increases the spurious transverse sensitivity. The manufacturing process also easily provides convenient soldering tabs (integral to the sensing grid) on all four gages of Fig. 4.12.

Evaporation-deposited thin-metal-film gages are used mostly for transducers, as are the sputter-deposited variety. Both processes begin with a suitable elastic metal element to transduce the measured quantity to local strain, just as if bonded foil gages were to be used. To use a pressure transducer as an example, this would be a thin, circular, metal diaphragm. Both the evaporation and the sputtering processes form all the gage elements directly on the strain surface; they

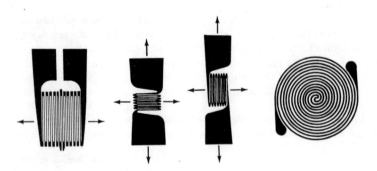

Figure 4.12 Foil strain gages.

are not separately attached, as with bonded gages. In the evaporation process,[1] the diaphragm is placed in a vacuum chamber with some insulating material. Heat is applied until the insulating material vaporizes and then condenses, forming a thin dielectric film on the diaphragm. Then suitably shaped templates are placed over the diaphragm, and the evaporation/condensation process is repeated with the metallic gage material, forming the desired gage pattern on top of the insulating substrate.

In the sputtering process, a thin dielectric layer is again deposited in vacuum over the entire diaphragm surface; however, the detailed mechanism of deposition is different from the evaporation method. Then a complete layer (no templates) of metallic gage material is sputtered on top of the dielectric substrate. Now the diaphragms are removed from the vacuum chamber, and microimaging techniques using photosensitive masking materials are used to define the gage patterns. Returning the diaphragms to the vacuum chamber, we now use sputter-etching to remove all the unmasked metal layer, leaving behind the completed gage pattern. Resistances and gage factors of film gages usually are similar to those of bonded foil. Since no organic cements as used with bonded foil gages are employed, thin-film gages exhibit improved time and temperature stability. Recent developments[2] in the sputtering technique have provided useful strain, temperature, and corrosion sensors for difficult jet-engine turbine blade measurements.

Bonded semiconductor gages are used mainly in transducers; however, they may find occasional application in stress analysis if the available strain is very small. They are made by slicing small sections from specially processed silicon crystals and are available in both *N*- and *P*-type. The *P*-type gages increase resistance with applied tensile strain while the *N*-type decrease resistance. Their main feature is a very high gage factor—as much as 150. Equation (4.14) shows that most of this gage factor must come from piezoresistance effects, whereas dimensional change explains most of the gage factor of metallic gages. Transducers based on semiconductor gages are often called *piezoresistive* transducers. Unfortunately, the high gage factor is accompanied by high temperature sensitivity, nonlinearity, and mounting difficulties. Solutions for these problems exist, and they can be accepted in the transducer manufacturing environment, but they are difficult to live with in routine stress analysis.

In diffused semiconductor gages (used exclusively in transducers), the diffusion process employed in integrated-circuit manufacture is utilized. In a pressure transducer, for example, the diaphragm would be silicon rather than metal and the strain gage effect would be realized by depositing impurities in the diaphragm

---

[1] R. L. Cheney, Color It Strained, *Ind. Res.*, p. 63, April 1976; P. R. Perino, Thin-Film Strain Gage Transducers, *Inst. & Cont. Syst.* p. 119, December 1965; A. Bray and A. Calcatelli, Vacuum Deposited Films at High Temperature, *Inst. & Cont. Syst.* p. 121, November 1966.

[2] R. R. Dils, Microsensors Can Give Reliable Data on Turbine Blade Behavior, *Mach. Des.*, p. 198, March 6, 1980; R. R. Dils and P. S. Follansbee, Superalloy Sensors, *Basic Combustion Tech. Rep.* 76–04, Pratt and Whitney Aircraft, E. Hartford, Conn., 1976.

to form intrinsic strain gages at the desired locations. This type of construction may allow lower manufacturing costs in some designs, since a large number of diaphragms can be made on a single silicon wafer.

The remainder of this discussion focuses on mainly bonded metal-foil gages, since they are the type most likely to be employed by the individual engineer both in stress analysis and for construction of homemade transducers. These gages come mounted on a flexible insulating carrier film (polyimide, glass-reinforced phenolic, etc.) about 0.001 in thick; thus the 0.0002-in-thick metal grid will be slightly more than 0.001 in above the specimen surface, as a result of adhesive film thickness. This displacement is significant only when we are measuring bending strain in very thin specimens, since then the gage could be feeling a strain considerably different from that of the specimen. In stress analysis, the goal is measurement of stress at a geometric point; this is impossible in a strain gage because the grid covers a finite area and thus the gage reads an average stress over this area. If the strain gradient is linear, then this average value also can be associated with the midpoint of the gage length; if not, then the point at which the gage's reading applies is somewhat uncertain. However, this uncertainty diminishes with gage size, which makes small gages desirable where strain gradients are sharp (stress concentrations, etc.). Minimum practical gage size is constrained by manufacturing limitations and handling and attachment problems; the smallest gages are about 0.015 in (0.38 mm) long. Gages can be applied to curved surfaces; the minimum safe bend radius can be as small as 0.06 in in some gages.

Typical gage resistances are 120, 350, and 1,000 Ω, with the allowable gage current[1] determined by heat-transfer conditions but typically 5 to 40 mA; gage factors are 2 to 4. Resistance of an individual gage is easily measured, but measurement of the gage factor requires cementing the gage to a specimen for which strain can be accurately calculated *from theory*. Since gages cannot be removed and reused, the gage-factor number supplied with a purchased gage has *not* been measured for that individual gage, but rather is an average value obtained by sample-testing the production of that type of gage. Thus we rely on the statistical quality control of the manufacturer in maintaining the accuracy of the gage factor. A ± 1 percent accuracy is typical, and this is a fundamental limit on accuracy in stress-analysis applications. Note that this does *not* limit accuracy of strain-gage *transducers*, since such transducers are calibrated "end to end" (pressure in to voltage out, say, in a pressure transducer) and one need not even know the gage factor. The maximum measurable strain varies from 0.5 to 4 percent; however, special "postyield" gage devices allow measurement up to 0.1 in/in. Fatigue life of gages varies with conditions; however, 10 million cycles at $\pm 1,500$ microstrain (45,000 lb/in^2 stress in steel, a common full-scale design value for foil-gage transducers) is typical. Bonded semiconductor gages[2] can work at

---

[1] Optimizing Strain Gage Excitation Levels, TN-127-3, Micro-Measurements Inc., Romulus, Mich.

[2] M. Lebow and R. Caris, Semiconductor Gages or Foil Gages for Transducers, Lebow Assoc., Oak Park, Mich., May 1965.

much lower full-scale strain values (typically 20 $\mu\epsilon$), which allows design of rugged and fast-responding transducers for low-range inputs where foil gages would require a very compliant elastic element.

Many different adhesives have been developed for fastening strain gages to specimens. Gages and fastening methods are available to cover temperature ranges from $-452°F$ ($-269°C$) to $1500°F$ ($816°C$). Extreme high temperatures may require welding or flame-spraying techniques rather than adhesive joining. Some adhesives cure at room temperature; others require baking. Cure times vary from a few minutes to several days. The quality of the adhesive joint obviously is critical to the proper operation of the strain gage, since we rely completely on it to transmit the specimen strain to the gage grid. These questions are particularly critical in extreme temperature and humidity conditions and for long-time installations. Protective and/or waterproofing coatings often are applied to increase reliability.

In addition to single-element gages, gage combinations called *rosettes*[1] (Fig. 4.13) are available in many configurations for specific stress analysis or transducer applications. While individual gages conceivably could be cemented down in the same patterns, precise relative orientation of the several gages is critical in most of these applications, and this is much more easily obtained in rosette manufacture than by the user with single gages. One rosette commonly used in stress analysis solves the problem of a surface stress whose magnitude and direction are totally unknown. Theory shows that measurements with a 3-gage rosette allow calculation of all the desired information. Since such measurements are intended to define stress at a *point*, ideally the three gages should be superimposed on that point. This "sandwich" construction (called *stacked rosette*) is feasible; however, it places the top gage farther from the specimen surface and increases its self-heating, since it is better insulated from the underlying metal specimen which acts as a heat sink. If these disadvantages outweigh the advantage of "point" measurement, we can use the planar rosette design, which usually is available (see Fig. 4.13).

Temperature is an important interfering input for strain gages since resistance changes with *both* strain and temperature.[2] Since strain-induced resistance changes are quite small, the temperature effect can assume major proportions. Another aspect of temperature sensitivity is found in the possible differential thermal expansion of the gage and the underlying material. This can cause a strain and resistance change in the gage even though the material is not subjected to an external load. These temperature effects can be compensated in various ways. In Fig. 4.14, a "dummy" gage (identical to the active gage) is cemented to a piece of the same material as is the active gage and placed so as to assume the same temperature. The dummy and active gages are placed in adjacent legs of a Wheatstone bridge; thus resistance changes due to the temperature coefficient of resistance and differential thermal expansion will have no effect on the bridge

[1] C. C. Perry, Care and Feeding of Strain Gage Rosettes, *VRE Tech. Ed. News*, Vishay Inc., Malvern, Pa., no. 23, September 1978.

[2] Strain Gage Temperature Effects, TN-128-2, Micro-Measurements Inc., Romulus, Mich., 1976.

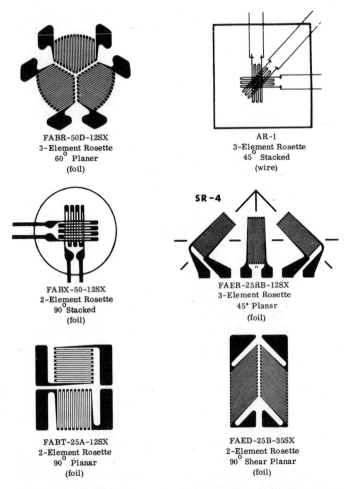

FABR-50D-12SX
3-Element Rosette
60° Planer
(foil)

AR-1
3-Element Rosette
45° Stacked
(wire)

FABX-50-12SX
2-Element Rosette
90° Stacked
(foil)

SR-4

FAER-25RB-12SX
3-Element Rosette
45° Planar
(foil)

FABT-25A-12SX
2-Element Rosette
90° Planar
(foil)

FAED-25B-35SX
2-Element Rosette
90° Shear Planar
(foil)

Figure 4.13 Strain-gage rosettes. (*Courtesy BLH Electronics, Waltham, Mass.*)

output voltage, whereas resistance changes due to an applied load will unbalance the bridge in the usual way (see Chap. 10 on bridge circuits). Another approach to this problem involves special, inherently temperature-compensated gages. These gages are designed to be used on a specific material and have expansion and resistance properties such that the two effects very nearly cancel and no dummy gage is required.

Temperature also can act as a modifying input in that it may change the gage factor. With metallic gages this effect usually is quite small, except at extremely low or high temperatures. Semiconductor gages are more seriously affected in this

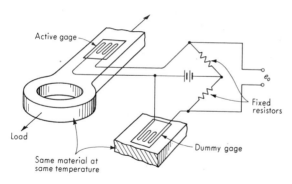

Figure 4.14 Strain-gage temperature compensation.

way; however, compensation is possible. Although the above temperature problems must be considered carefully in each application, strain gages have been successfully employed from liquid-helium temperature (7°R) to the order of 2000°F. However, these extreme (especially the high temperature) applications require special techniques and yield results of lower accuracy than are obtained in routine room-temperature situations.

Used directly, the bonded strain gage is useful for measuring only very small displacements (strains). However, larger displacements may be measured by bonding the gage to a flexible element, such as a thin cantilever beam, and applying the unknown displacement to the end of the beam, as in Fig. 4.15. For such an application, the gage factor need not be accurately known since the overall system can be calibrated by applying known displacements to the end of the beam and measuring the resulting bridge output voltage. The configuration shown is temperature-compensated without the need for dummy gages and has 4 times the sensitivity of a single gage because of judicious application of bridge-circuit properties. Such transducers may be accurate to 0.1 percent of full scale.

The dynamic response of bonded strain gages with respect to faithfully reproducing as a resistance variation the strain variation of the underlying surface is very good.[1] The dynamic effects of wave propagation in the cement and strain wires seem to be negligible for frequencies up to at least 50,000 Hz, and so a zero-order dynamic model is generally adequate. The loading effect of the cement and strain wires on the underlying structure is generally negligible except for very thin members, in which the stiffening effect of the strain gage may reduce the measured strain to a value considerably lower than that present without the gage. Compliance techniques can be applied to study this effect.

The voltage output from metallic strain-gage circuits is quite small (a few microvolts to a few millivolts), and so amplification is generally needed. As an

[1] P. K. Stein, "Measurement Engineering," vol. 2, Stein Engineering Services, Inc., Phoenix, 1962; M. G. Pottinger, Effect of Gage Length in Dynamic Strain Measurement, ARL 69-0014, Wright-Patterson AFB, Ohio, January, 1969.

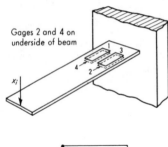

Gages 2 and 4 on
underside of beam

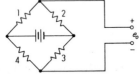

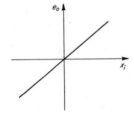

Figure 4.15 Beam displacement transducer.

example, consider the measurement of a stress level of 1,000 lb/in^2 in steel with a single active gage of 120-Ω resistance and a gage factor of 2.0. If a bridge circuit of all equal arms is used, the maximum allowable bridge voltage for 30-mA gage current is

$$e_{ex} = (240)(0.030) = 7.2 \text{ V} \tag{4.15}$$

The strain ϵ is $1,000/30 \times 10^6 = 3.33 \times 10^{-5}$ in/in, so that

$$\Delta R = (\text{gage factor})(\epsilon)(R) = (2)(3.33 \times 10^{-5})(120) = 7.99 \times 10^{-3} \ \Omega \tag{4.16}$$

For the given bridge arrangement,

$$e_o = e_{ex} \frac{1}{4R} \Delta R = \frac{(7.2)(7.99 \times 10^{-3})}{480} = 0.12 \text{ mV} \tag{4.17}$$

Based on limitations of the gage alone, the smallest detectable strain depends on the thermal or Johnson-noise[1] voltage generated in every resistance because of the random motion of its electrons. This random voltage is essentially a white

[1] E. B. Wilson, Jr., "An Introduction to Scientific Research," p. 116, McGraw-Hill, New York, 1952.

noise of spectral density $4kTR$ volts squared per hertz, where

$$k \triangleq \text{Boltzmann's constant} = 1.38 \times 10^{-23} \text{ J/K} \qquad (4.18)$$

$$T \triangleq \text{absolute temperature of resistor, K} \qquad (4.19)$$

$$R \triangleq \text{resistance, } \Omega \qquad (4.20)$$

Thus if this voltage were measured by a hypothetical noise-free oscilloscope with a bandwidth of Δf Hz, the measured rms voltage would be

$$E_{\text{noise, rms}} = \sqrt{4kTR\,\Delta f} \qquad \text{volts} \qquad (4.21)$$

As an example, a strain gage of $R = 120 \ \Omega$ at 300 K over a bandwidth of 100,000 Hz would put out an rms noise voltage of 0.45 μV. Comparing this with our earlier calculation of the signal due to 1,000 lb/in^2 stress, we see that the signal to noise ratio would be $120 : 0.45 = 267 : 1$. Suppose, however, that we wish to measure 1 lb/in^2 stress rather than 1,000. The signal is then 0.12 μV, which is less than the noise; therefore the signal would be lost in the noise. Amplification under these conditions is of no use since the signal and noise are both amplified. This simple example does not cover other methods that have been developed to reduce this limitation, but it should be understood that the limitation is a fundamental one and can be reduced but not overcome. Similar random fluctuations limit the measurable threshold of all physical variables.[1] In practical strain-gage measurement systems, Johnson noise of resistances other than the strain gage and other sources of noise in transistors, actually limit the system resolution.

To meet the needs of severe environments, difficult installation conditions, and/or high temperatures as typified by pipelines, offshore drilling platforms, tunnels, dams, nuclear containment vessels, steam lines, gas turbines, etc., strain gages enclosed permanently in protective metal housings which are attached by spot welding to the specimen have been developed.[2] These gages incorporate a number of unique features (see Fig. 4.16). The strain filament and heavier (0.007-in diameter) leadout wire are of unitized construction with a carefully controlled taper; this minimizes joint fatigue problems and erratic electric connections. Active and dummy filaments are in close proximity; the dummy gage experiences no strain since the helix angle of its winding matches Poisson's ratio. Both filaments are encased in a strain tube made by welding a tubular shell to a flat flange. The filament is mechanically coupled to the strain tube, but electrically isolated from it by highly compacted magnesium oxide powder, by using a high-speed centrifuge and swaging operation. When the flange is spot-welded to the specimen, the specimen strain is faithfully transmitted to the strain tube and in turn by the powder to the gage filament for both tensile and compressive stresses.

Our next device is not a strain gage, but is included here because of its use in experimental stress and failure analysis and the increasing importance of fracture-mechanics methods. In such studies, the measurement in standard test specimens

---

[1] Ibid.

[2] Ailtech Co., City of Industry, Calif.

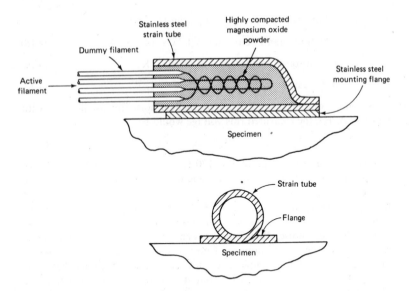

Figure 4.16 Hermetically sealed, weldable strain gage.

of crack length and growth with time is of great interest. The KRAK-GAGE[1] uses a bonded thin metal foil as in ordinary strain gages, but of a geometry suited to the particular fracture-mechanics specimen under study. A constant current is supplied to the gage terminals. As the crack grows, the gage resistance increases, producing a voltage which increases linearly with crack length. Gages with full-scale lengths of 5 to 100 mm are available. Another non-strain-gage device that seems appropriate to mention here is the fatigue-life gage.[2] This gage appears similar to a bonded-foil strain gage and is manufactured by similar processes; however, its electrical behavior is quite different. For cyclic straining of a fixed amplitude, gage resistance permanently and irreversibly (but not linearly) increases with each cycle until fatigue failure occurs. The starting resistance is usually 100 Ω, while the end-of-life resistance is about 108 Ω. If the strain amplitude is increased, the resistance change is greater per cycle. Applications of this device to fatigue-failure studies are discussed in Dorsey.

Differential Transformers

Figure 4.17 shows schematic and circuit diagrams for translational and rotational linear variable-differential-transformer (LVDT) displacement pickups. The excitation of such devices is normally a sinusoidal voltage of 3 to 15 V rms amplitude

[1] Hartrun Corp., Chaska, Minn.
[2] J. Dorsey, Engineering Concepts in Fatigue Life Gage Use, AN-127, Micro-Measurements Inc., Romulus, Mich., May 1971.

and frequency of 60 to 20,000 Hz. The two identical secondary coils have induced in them sinusoidal voltages of the same frequency as the excitation; however, the amplitude varies with the position of the iron core. When the secondaries are connected in series opposition, a null position exists ($x_i \triangleq 0$) at which the net output e_o is essentially zero. Motion of the core from null then causes a larger mutual inductance (coupling) for one coil and a smaller mutual inductance for the other, and the amplitude of e_o becomes a nearly linear function of core position for a considerable range either side of null. The voltage e_o undergoes a 180° phase shift in going through null. The output e_o is generally out of phase with the excitation e_{ex}; however, this varies with the frequency of e_{ex}, and for each differential transformer there exists a particular frequency (numerical value supplied by the manufacturer) at which this phase shift is zero. If the differential transformer is used with some readout system that requires a small phase shift between e_o and e_{ex} (some carrier-amplifier systems require this), excitation at the correct frequency can solve this problem. If the output voltage is applied directly to an ac meter or an oscilloscope, this phase shift is not a problem.

The origin of this phase shift can be seen from analysis of Fig. 4.18. Applying Kirchhoff's voltage-loop law, we get

$$i_p R_p + L_p \frac{di_p}{dt} - e_{ex} = 0 \tag{4.22}$$

Now the voltage induced in the secondary coils is given by

$$e_{s1} = M_1 \frac{di_p}{dt} \tag{4.23}$$

$$e_{s2} = M_2 \frac{di_p}{dt} \tag{4.24}$$

where M_1 and M_2 are the respective mutual inductances. The net secondary voltage e_s is then given by

$$e_s = e_{s1} - e_{s2} = (M_1 - M_2) \frac{di_p}{dt} \tag{4.25}$$

The net mutual inductance $M_1 - M_2$ is the quantity that varies linearly with core motion. If the output is open circuit (no voltage-measuring device attached), we have for a fixed core position

$$e_o = e_s = (M_1 - M_2) \frac{D}{L_p D + R_p} e_{ex} \tag{4.26}$$

and thus
$$\frac{e_o}{e_{ex}}(D) = \frac{[(M_1 - M_2)/R_p]D}{\tau_p D + 1} \qquad \tau_p \triangleq \frac{L_p}{R_p} \tag{4.27}$$

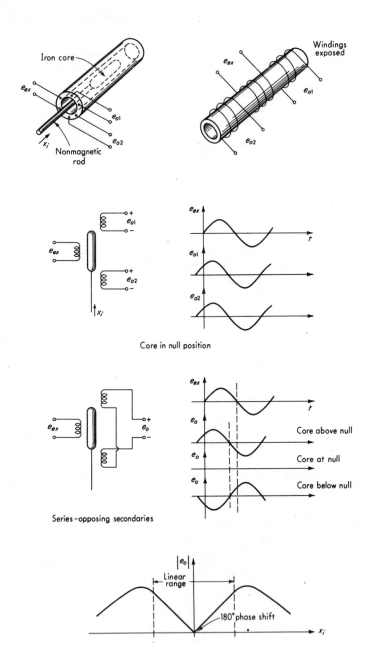

Figure 4.17 Differential transformer.

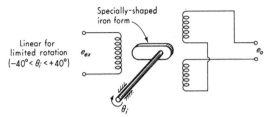

Rotational differential transformer

Figure 4.17 (*Continued*)

In terms of frequency response,

$$\frac{e_o}{e_{ex}}(i\omega) = \frac{\omega(M_1 - M_2)/R_p}{\sqrt{(\omega\tau_p)^2 + 1}} \angle \phi \qquad \phi = 90° - \tan^{-1} \omega\tau_p \qquad (4.28)$$

which demonstrates the phase shift between e_o and e_{ex}. If a voltage-measuring device of input resistance R_m is attached to the output terminals, a current i_s will flow, and we can write

$$i_p R_p + L_p D i_p - (M_1 - M_2)D i_s - e_{ex} = 0 \qquad (4.29)$$

$$(M_1 - M_2)D i_p + (R_s + R_m)i_s + L_s D i_s = 0 \qquad (4.30)$$

which lead to

$$\frac{e_o}{e_{ex}}(D) = \frac{R_m(M_2 - M_1)D}{[(M_1 - M_2)^2 + L_p L_s]D^2 + [L_p(R_s + R_m) + L_s R_p]D + (R_s + R_m)R_p}$$

$$(4.31)$$

Since the frequency response of $(e_o/e_{ex})(i\omega)$ has a phase angle of $+90°$ at low frequencies and $-90°$ at high, somewhere in between it will be zero, as mentioned

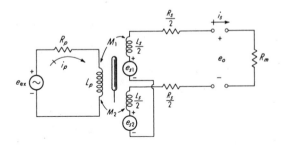

Figure 4.18 Circuit analysis.

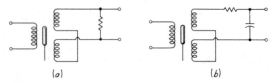

(a)   (b)

Two possible methods for retarding a leading phase angle

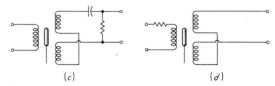

(c)   (d)

Two possible methods for advancing a lagging phase angle

Figure 4.19 Phase-angle-adjustment circuits.

earlier. If, for some reason, the excitation frequency cannot be adjusted to this value, the same effect may be achieved for a given frequency by one of the methods[1] shown in Fig. 4.19.

While the output voltage at the null position is ideally zero, harmonics in the excitation voltage and stray capacitance coupling between the primary and secondary usually result in a small but nonzero null voltage. Under usual conditions this is less than 1 percent of the full-scale output voltage and may be quite acceptable. Methods of reducing this null when it is objectionable are available. First, the preferred connection shown in Fig. 4.20a should be used if a balanced (center-tapped) excitation-voltage source is available. The grounding shown tends to reduce capacitance-coupling effects. If a center-tapped voltage source is not available, the arrangement of Fig. 4.20b can be used. With the core at the null position and the output-measuring device connected, the potentiometer is adjusted until the minimum null reading is obtained. The values of R and R_p are not critical, but should be as low as possible without loading (drawing excessive current from) the excitation source.

The output of a differential transformer is a sine wave whose amplitude is proportional to the core motion. If this output is applied to an ac voltmeter, the meter reading can be directly calibrated in motion units. This arrangement is perfectly satisfactory for measurement of static or very slowly varying displacements, except that the meter will give exactly the same reading for displacements of equal amount on *either* side of the null since the meter is not sensitive to the 180° phase change at null. Thus we cannot tell to which side of null the reading applies without some independent check. Furthermore, if rapid core motions are

---

[1] A. Miller, Differential Transformers, *The Right Angle*, The Sanborn Co., Waltham, Mass., August 1956, November 1956; Schaevitz Engineering, Pennsauken, N. J., *Bull.* AA-1A.

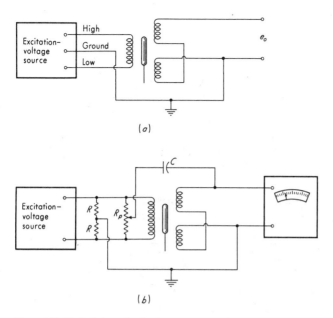

Figure 4.20 Methods for null reduction.

to be measured, the meter cannot follow or record them, and an oscillograph or oscilloscope must be used as a readout device. These instruments record the actual waveform of the output as an amplitude-modulated sine wave, which is usually undesirable. What is desired is an output-voltage record that looks like the mechanical motion being measured. To achieve the desired results, demodulation and filtering must be performed; if it is necessary to detect unambiguously the motions on both sides of null, the demodulation must be phase-sensitive. Many different circuits are available for performing these operations. We show here only one arrangement, which is quite simple. To use this approach, all four output leads of the LVDT must be accessible (some have the series-opposition connection *internal* to the case and would thus not be applicable to the following discussion).

Figure 4.21*a* shows the circuit arrangement for phase-sensitive demodulation using semiconductor diodes. Ideally, these pass current only in one direction; thus when *f* is positive and *e* is negative, the current path is *efgcdhe*, while when *f* is negative and *e* positive, the path is *ehcdgfe*. The current through *R* is therefore always from *c* to *d*. A similar situation exists in the lower diode bridge. For static or very slowly varying core displacements, the voltage e_o may be applied directly to a dc voltmeter. The meter will act as an electromechanical low-pass filter, with the needle assuming a position corresponding to the average value of the rectified sine wave e_o. If motions both sides of null are to be measured, a meter with zero

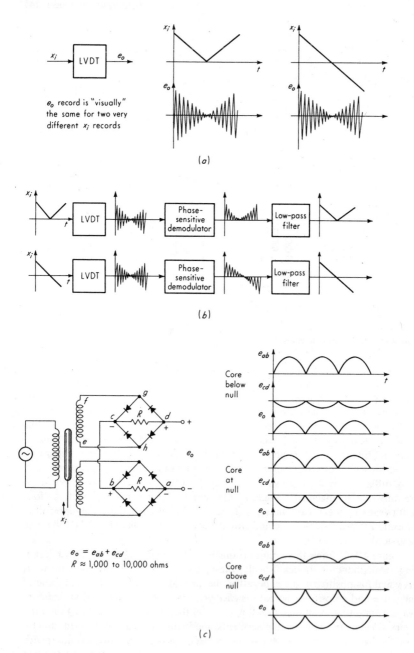

Figure 4.21 Demodulation and filtering.

in the center of the scale will eliminate the need for switching lead wires when e_o goes negative. When rapid core motions are to be measured, this dc meter arrangement is useless since the meter movement cannot follow variations more rapid than about 1 Hz. It is then necessary to connect e_o of Fig. 4.21c to the input of a low-pass filter which will pass the frequencies present in x_i but reject all those (higher) frequencies produced by the modulation process. The design of such a filter is eased by making the LVDT excitation frequency much higher than the x_i frequencies.

If a frequency ratio of 10 : 1 or more is feasible, a simple RC filter as in Fig. 4.22a may be adequate. The output of this filter then becomes the input to an oscillograph or oscilloscope. For example, suppose we wish to measure a transient x_i whose Fourier transform has dropped to insignificant magnitude for all frequencies higher than 1,000 Hz. Suppose also an LVDT system with an excitation frequency of 10,000 Hz is available. The frequencies produced by the modulation process will thus lie in the band 19,000 to 21,000 Hz. Suppose that we desire the "ripple" due to frequencies at 19,000 Hz and higher to be no more than 5 percent of its unfiltered value. The filter time constant $\tau_f = R_f C_f$ can be calculated then as

$$0.05 = \frac{1}{\sqrt{[(19,000)(6.28)\tau_f]^2 + 1}} \tag{4.32}$$

$$\tau_f = 0.00017 \text{ s} \tag{4.33}$$

At the highest motion frequency (1,000 Hz) this filter has an amplitude ratio of 0.68 and a phase shift of $-47°$; thus it will distort the high-frequency portion of the x_i transient considerably. A more selective (sharper cutoff) filter would help this situation. Consider the double RC filter of Fig. 4.22b. The value of τ_f for a 5 percent ripple is now obtained from

$$0.05 = \frac{1}{[(19,000)(6.28)\tau_f]^2 + 1} \tag{4.34}$$

$$\tau_f = 0.000037 \text{ s} \tag{4.35}$$

Now, at 1,000 Hz the amplitude ratio is 0.94 and the phase angle is $-26°$. Since the phase angle of this filter from $\omega = 0$ to $\omega = 6,280$ rad/s is nearly linear and the amplitude ratio is nearly flat (1.0 to 0.94), the waveform of the transient will be faithfully reproduced, but with a delay (dead time) of about $26/[(57.3)(6,280)] = 72\mu s$. The above calculations give the desired value of τ_f but not R_f and C_f directly. In going from the demodulator circuit to the filter and then to the oscilloscope, transfer functions of the individual elements can be multiplied only if no significant loading of the successive stages is present. As a rule of thumb, the impedance level should go up about 10 to 1 for each successive stage. A typical oscilloscope input resistance is about 10^6 Ω. This suggests that in Fig. 4.22b, $10R_f$ could be about 10^5 Ω and R_f thus 10^4 Ω. In Fig. 4.21c the demodulator R can be of the order of 10^3 Ω, and so the overall chain should not

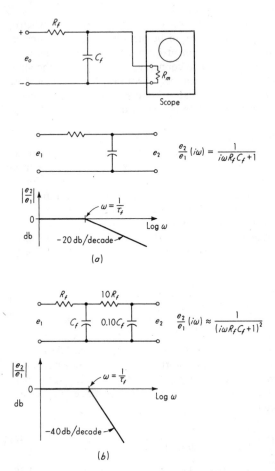

$$\frac{e_2}{e_1}(i\omega) = \frac{1}{i\omega R_f C_f + 1}$$

(a)

$$\frac{e_2}{e_1}(i\omega) \approx \frac{1}{(i\omega R_f C_f + 1)^2}$$

(b)

Figure 4.22 Filter frequency response.

show much loading effect and our above calculations should be fairly accurate. In any case, experimental checks of the system should be performed to verify the final design. If R_f of 10^4 Ω is satisfactory, then C_f will be $(37 \times 10^{-6})/10^4 = 0.0037$ μF.

The full-range stroke of commercially available translational LVDTs ranges from about ± 0.005 to about ± 3 in, with other sizes available as specials. The nonlinearity of standard units is of the order of 0.5 percent full scale, with 0.1 percent possible by selection. Sensitivity with normal excitation voltage of 3 to 6 V is of the order of 0.6 to 30 mV per 0.001 in, depending on frequency of excitation (higher frequency gives more sensitivity) and stroke (smaller strokes usually have higher sensitivity). Some special units have sensitivity as high as 1 to 1.5 V per 0.001 in. Since the coupling variation due to core motion is a continuous phenomenon, the resolution of LVDTs is infinitesimal. Amplification of the output voltage allows detection of motions down to a few microinches. There is no physical contact between the core and the coil form; thus there is no friction or wear. There are, however, small radial and longitudinal magnetic forces on the core if it is not centered radially and at the null position. These are in the nature of magnetic "spring" forces in that they increase with motion from the equilibrium point. They are rarely more than 0.1 to 0.3 g and thus are often negligible. Rotary LVDTs have a nonlinearity of about ± 1 percent of full scale for travel of $\pm 40°$ and ± 3 percent for $\pm 60°$. The sensitivity is of the order of 10 to 20 mV/deg. The moving mass (core) of LVDTs is quite small, ranging from less than 0.1 in small units to 5 g or more in larger ones. There is a small radial clearance (air gap) between the core and the hole in which it moves. Motion in the radial direction produces a small output signal, but this undesirable transverse sensitivity is usually less than 1 percent of the longitudinal sensitivity.

The dynamic response of LVDTs is limited mainly by the excitation frequency, since it must be much higher than the core-motion frequencies so as to be able to distinguish between them in the amplitude-modulated output signal. For adequate demodulation and filtering, a frequency ratio much less than 10 : 1 presents problems. Since few differential transformers are designed to be excited by more than 20,000 Hz, the useful range of motion frequencies is limited to about 2,000 Hz. This is adequate for many applications. Although the transformer principle of the LVDT clearly requires ac excitation, so-called DCDTs are available from manufacturers. The "DC" part of the terminology refers to the

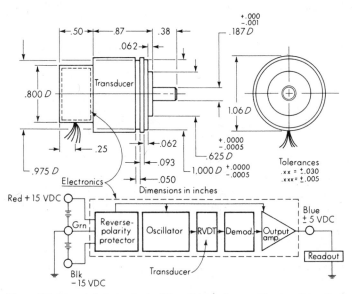

Figure 4.23 Integrated electronics in differential transformer. (*Courtesy Schaevitz Eng., Pennsauken, N. J.*)

fact that the user applies only a dc excitation to the device. The apparent discrepancy is resolved when we discover that microelectronic technology has allowed us to build into the transducer case a complete electronic system including oscillator (produces ac excitation from dc power), demodulator, amplifier, and low-pass filter (see Fig. 4.23). LVDTs also have been combined with pneumatic servomechanisms to achieve a noncontacting displacement probe (see Fig. 4.55).

Synchros and Induction Potentiometers

The term "synchro" is applied to a family of ac electromechanical devices which, in various forms, perform the functions of angle measurement, voltage and/or angle addition and subtraction, remote angle transmission, and computation of rectangular components of vectors. In this section we are concerned with only the angle-measuring function; equipment for performing the other functions is covered in later appropriate chapters.

Synchros for angle measurement are most utilized as components of servomechanisms (automatic motion-control feedback systems) where they are used to measure and compare the actual rotational position of a load with its commanded position, as in Fig. 4.24. To perform this function, two different types of synchros, the control transmitter and the control transformer, are used. The error voltage signal e_e is an ac voltage of the same frequency as the excitation and of amplitude proportional (for small error angles) to the error angle $\theta_R - \theta_B$. Its phase changes by 180° at the null point; thus the direction of the error is detected. When $\theta_R = \theta_B$, the error voltage (and thus the amplifier output and motor

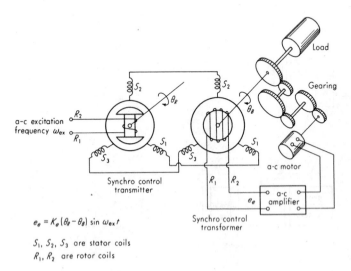

$$e_e = K_e(\theta_R - \theta_B)\sin \omega_{ex} t$$

S_1, S_2, S_3 are stator coils

R_1, R_2 are rotor coils

Figure 4.24 AC servomechanism.

input) is zero and the system stays at rest. If a command rotation θ_R is now put in, $e_e \neq 0$ and the motor will rotate so as to return θ_B to correspondence with θ_R.

The physical constructions of the control transmitter and control transformer are identical except that the transmitter has a salient-pole ("dumbbell") rotor while the transformer has a cylindrical rotor. The construction is similar to that of a wound-rotor induction motor. Figure 4.25a shows the coil arrangement of the transmitter alone. Basically, rotation of the rotor changes the mutual inductance (coupling) between the rotor coil and the stator coils. For a given stator coil, the open-circuit output voltage is sinusoidal in time and varies in amplitude with rotor position, also sinusoidally, as shown in Fig. 4.25b. The three voltage signals from the stator coils uniquely define the angular position of the rotor. When these three voltages are applied to the stator coils of a control transformer, they produce a resultant magnetomotive force aligned in the same direction as that of the transmitter rotor. The rotor of the transformer acts as a "search coil" in detecting the direction of its stator field. If the axis of this coil is aligned with the field, the maximum voltage is induced into the transformer rotor coil. If the axis is perpendicular to the field, zero voltage is induced, giving the null position mentioned above. The output-voltage amplitude actually varies sinusoidally with

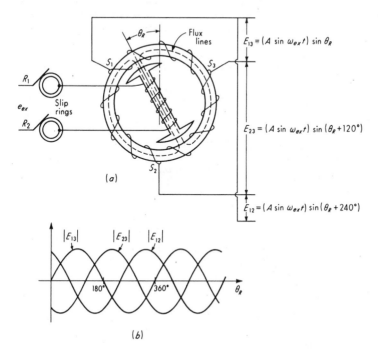

$$E_{13} = (A \sin \omega_{ex} t) \sin \theta_R$$

$$E_{23} = (A \sin \omega_{ex} t) \sin (\theta_R + 120°)$$

$$E_{12} = (A \sin \omega_{ex} t) \sin (\theta_R + 240°)$$

Figure 4.25 Synchro.

the misalignment angle, but for small angles the sine and the angle are nearly equal, giving a linear output.

In an induction potentiometer, there is one winding on the rotor and one on the stator. (Additional dummy windings are sometimes used to improve accuracy, however.) Both these windings are concentrated; thus for simplicity we show them as single-turn coils in Fig. 4.26. The primary winding (rotor) is excited with alternating current. This induces a voltage into the secondary (stator). The amplitude of this output voltage varies with the mutual inductance (coupling) between the two coils, and this varies with the angle of rotation. For single-turn (concentrated) coils, the variation with angle would be sinusoidal and only a small linear range around null would be obtained. By carefully *distributing* the rotor and stator windings, a linear relation for up to $\pm 90°$ rotation may be obtained.

While synchros and induction pots could be designed to work at a variety of excitation frequencies, standard commercial units generally are available only for 60 or 400 Hz. The physical size ranges from about $\frac{1}{2}$- to 3-in diameter. Sensitivities of both synchros and induction pots are of the order of 1 V/deg rotation while the residual voltage at null is of the order of 10 to 100 mV. For a standard synchro transmitter-transformer pair, the misalignment of the two shafts when rotated from an originally established electrical null to any other null position within a complete rotation is of the order of 10 angular minutes. This type of error puts a basic limit on the positioning accuracy of servosystems using synchros. Synchro pairs and induction pots are capable of continuous rotation,

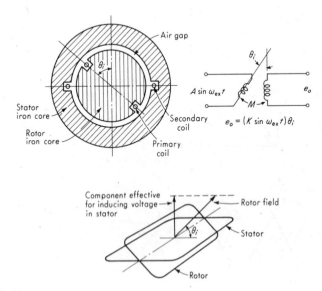

Figure 4.26 Induction potentiometer.

although the linear range of induction pots is limited to about ± 60 to $\pm 90°$. Within this range the nonlinearity is of the order of 0.25 percent.

Just as in LVDTs, synchros and induction pots require some sort of phase-sensitive demodulation to obtain a signal of the same form as the mechanical-motion input. When used in ac servomechanisms, the conventional two-phase ac servomotor itself accomplishes this function without any additional equipment. In a strictly measurement (as opposed to control) application, some sort of phase-sensitive demodulator is needed if an electric output signal of the same form as the mechanical input is required. The dynamic response is limited by the excitation frequency and demodulator filtering requirements, just as in LVDTs. The mechanical loading of these rotary components on the measured system is mainly the inertia of the rotor.

While synchros were originally developed for use with analog systems, many applications (machine tools, antennas, etc.) today use digital approaches. Even though a digital shaft-angle encoder might seem the logical sensor for an otherwise digital system, synchros (and the closely related resolver) have a number of advantages[1] which make them desirable. To interface them with the digital part of the system, we require a synchro-to-digital converter,[2] a piece of solid-state electronic equipment (no moving parts) which we show in block-diagram form in Fig. 4.27. The ac signals from the synchro are first applied to a special transformer, a Scott T, which produces ac voltages with amplitudes proportional to the sine and cosine of the synchro shaft angle θ:

$$V_1 = KE_o \sin \omega t \sin \theta \qquad V_2 = KE_o \sin \omega t \cos \theta \qquad (4.36)$$

[Note that if a resolver (rather than a synchro) were used, the Scott T transformer would be unnecessary since the resolver produces voltages of form (4.36) directly.] The synchro-to-digital converter is an all-electronic feedback system which compares the synchro angle θ with its digital version ϕ. If ϕ is not equal to θ, the converter provides a correction signal to the counter containing ϕ so as to drive its contents toward the correct value. For those familiar with feedback systems, the control system is a Type II (two integrations), so there will be zero steady-state error between command θ and response ϕ for both step and ramp inputs of θ.

The voltages V_1 and V_2 are digitally multiplied by $\sin \phi$ and $\cos \phi$ to give

$$V_3 = KE_o \sin \omega t \sin \theta \cos \phi \qquad (4.37)$$

$$V_4 = KE_o \sin \omega t \cos \theta \sin \phi \qquad (4.38)$$

[1] W. M. Cullum, Angle-Sensing Transducers for Shaft-to-Digital Conversion, *Tech. Bull.* 121, North Atlantic, Hauppauge, N. Y., 1979.

[2] Ibid.; Data Acquisition Components and Subsystems Catalog, pp. 12–27, Analog Devices, Norwood, Mass., 1980; G. Boyes, Synchro and Resolver Conversion, Analog Devices, Norwood Mass., 1980.

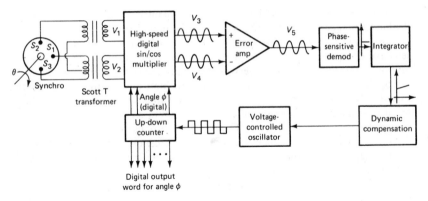

Figure 4.27 Synchro-to-digital converter.

and then these two voltages are subtracted in the error amplifier to give

$$V_5 = KE_o \sin \omega t \, (\sin \theta \cos \phi - \cos \theta \sin \phi)$$

$$= KE_o \sin \omega t \sin (\theta - \phi) \tag{4.39}$$

The amplitude of the ac voltage V_5 is proportional to $\sin (\theta - \phi) \approx \theta - \phi$, the error between θ and ϕ. This ac error signal is phase-sensitive demodulated to obtain a dc error signal, which is then applied to an op-amp type of integrator. (The next block, dynamic compensation, is not fundamental to the operating principle, but is necessary in most feedback systems to optimize performance.) Since a constant voltage applied to a voltage-controlled oscillator produces a proportional and constant output frequency which will ramp the counter at a fixed rate, this provides the second integration in the control loop. For the Analog Devices SDC1704 operated at 400 Hz and referenced above, the manufacturer gives this transfer function:

$$\frac{\phi}{\theta} (D) = \frac{0.0082D + 1}{3.39 \times 10^{-8} D^3 + 2.73 \times 10^{-5} D^2 + 0.00661D + 1} \tag{4.40}$$

This clearly has zero steady-state error for a step input, while solution of the differential equation (particular solution only) for $\theta = At$ shows zero error for a ramp also. Actually, zero ramp error is maintained only for $A < 12$ r/s. The resolution of this 14-bit device is 1.3 minutes of arc, and the accuracy is ± 2.9 minutes of arc ± 1 LSB (LSB = least significant bit = 1.3 minutes of arc). This device also provides a dc voltage proportional to $\dot{\theta}$, useful in stabilizing a servosystem which uses the synchro for angle information.

Variable-Inductance and Variable-Reluctance Pickups

Closely related to LVDTs and synchros, but in practice distinguished from them by name, is a family of motion pickups variously called variable-inductance, variable-reluctance, or variable-permeance (permeance is the reciprocal of reluctance) pickups or transducers. The terminology used for these pickups is not uniform nor necessarily descriptive of their basic principles of operation. We are concerned here mainly with describing some common examples rather than trying to develop a systematic nomenclature.

Figure 4.28a shows the arrangement of a typical translational variable-inductance transducer. Outwardly the physical size and shape are very similar to those of an LVDT. Again, there is a movable iron core which provides the mechanical input. However, only two inductance coils are present; they generally form two legs of a bridge which is excited with alternating current of 5 to 30 V at 60 to 5,000 Hz. With the core at the null position, the inductance of the two coils is equal, the bridge is balanced, and e_o is zero. A core motion from null causes a change in the reluctance of the magnetic paths for each of the coils, increasing one and decreasing the other. This reluctance change causes a proportional change in inductance for each coil, a bridge unbalance, and thus an output voltage e_o. By careful construction, e_o can be made a nearly linear function of x_i over the rated displacement range.

Two alternative methods of forming the bridge circuit are shown in Fig. 4.28b. The total transducer impedance (Z_1 plus Z_2) at the excitation frequency is of the order of 100 to 1,000 Ω. The resistors R are usually about the same value as Z_1 and Z_2, and the input impedance of the voltage-measuring device at e_o should be at least $10R$. If the bridge output must be worked into a low-

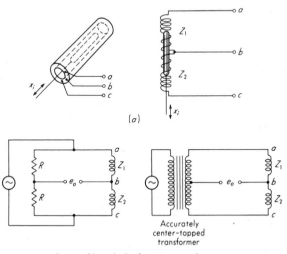

(a)

Two possible methods of exciting transducer
and forming bridge circuit

Accurately
center-tapped
transformer

(b)

Figure 4.28 Variable-inductance pickup.

impedance load, R must be quite small. To get high sensitivity, high excitation voltage is needed; this causes a high power loss (heating) in the resistors R. To solve this problem, a center-tapped transformer circuit may be used. Here the bridge is mainly inductive, and less power is consumed with corresponding less heating.

Such variable-inductance transducers are available in strokes of about 0.1 to as much as 200 in. The resolution is infinitesimal, and the nonlinearity ranges from about 1 percent of full scale for standard units to 0.02 percent for special units of rather long stroke. Sensitivity is of the order of 5 to 40 V/in. Rotary versions using specially shaped rotating cores have a nonlinearity of the order of 0.5 to 1 percent of full scale over a $\pm 45°$ range. The sensitivity is about 0.1 V/deg.

Figure 4.29 shows another common version of the variable-reluctance principle. This particular application is an accelerometer for measurement of accelerations in the range $\pm 4g$. Since the force required to accelerate a mass is proportional to the acceleration, the springs supporting the mass in Fig. 4.29 deflect

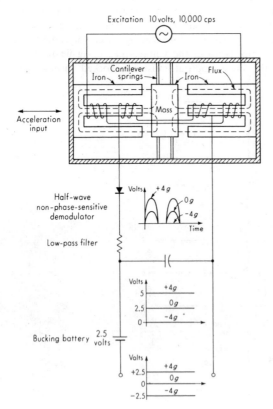

Figure 4.29 Variable-reluctance accelerometer.

in proportion to the acceleration; thus a displacement measurement allows an acceleration measurement. The mass is of iron and thus serves as both an inertial element for transducing acceleration to force and a magnetic circuit element for transducing motion to reluctance.

We consider the complete instrument here since it has several features of general interest with regard to displacement measurement. Ordinarily, such an instrument would be constructed so that the iron core would be halfway between the two E frames when the acceleration was zero, thus giving zero output voltage for zero acceleration. However, to detect motion on both sizes of zero (corresponding to plus or minus accelerations), a fairly involved phase-sensitive demodulator would be required. It was desired to save the cost, weight, and space of this demodulator, and so another solution (which can also be used with LVDTs and similar devices) was proposed. With zero-acceleration input, the iron core and springs were adjusted so that the core was offset to one side by an amount equal to the spring deflection corresponding to $4g$ acceleration. Thus, with no acceleration applied, the output voltage was not zero but some specific value (2.5 V in this particular case). Then, when $+4g$ of acceleration was applied, the output went to 5.0 V; and when $-4g$ was applied, the output went to zero. In this way a relatively simple demodulator and filter circuit can be used to provide direction-sensitive motion measurement. The main drawback of this scheme (which argues against its use except when necessary) is the loss of linearity. This is because the greatest linearity is found around the null position. Thus for a given total stroke it is better to put one-half of it on each side of null rather than all on one side, as in the above scheme.

Let us return to the basic motion-measuring principle of Fig. 4.29. The primary coils set up a flux dependent on the reluctance of the magnetic path. The main reluctance is the air gap. When the core is in the neutral position, the flux is the same for both halves of the secondary coil; and since they are connected in series opposition, the net output voltage is zero. A motion of the core increases the reluctance (air gap) on one side and decreases it on the other, causing more voltage to be induced into one half of the secondary coil than the other and thus a net output voltage. Motion in the other direction causes the reverse action, with a 180° phase shift occurring at null. The output voltage is half-wave, non-phase-sensitive rectified (demodulated) and filtered to produce an output of the same form as the acceleration input. If the 2.5-V output for zero-acceleration input is objectionable, it can be bucked out with a 2.5-V battery of opposite polarity connected externally to the accelerometer. The actual full-scale motion of the mass in this particular instrument is just a few thousandths of an inch, which gives a displacement sensitivity for the variable-reluctance element of almost 1,000 V/in.

The final variable-reluctance element we consider is the Microsyn,[1] a rotary component shown in Fig. 4.30 and widely used in sensitive gyroscopic instruments. The sketch shows the instrument in the null position where the voltages

[1] P. H. Savet (ed.), "Gyroscopes: Theory and Design," p. 332, McGraw-Hill, New York, 1961.

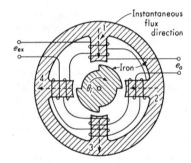

Figure 4.30 Microsyn.

induced in coils 1 and 3 (which aid each other) are just balanced by those of coils 2 and 4 (which also aid each other but oppose 1 and 3). Motion of the input shaft from the null (say clockwise) increases the reluctance (decreases the induced voltage) of coils 1 and 3 and decreases the reluctance (increases the voltage) of coils 2 and 4, thus giving a net output voltage e_o. Motion in the opposite direction causes a similar effect, except the output voltage has a 180° phase shift. If a direction-sensitive dc output is required, a phase-sensitive demodulator is necessary.

The excitation voltage is 5 to 50 V at 60 to 5,000 Hz. Sensitivity is of the order of 0.2 to 5 V/degree rotation. Nonlinearity is about 0.5 percent of full scale for $\pm 7°$ rotation and 1.0 percent for $\pm 10°$. The null voltage is extremely small, less than the output signal generated by 0.01° of rotation; thus very small motions can be detected. The magnetic-reaction torque is also extremely small. Since there are no coils on the rotor, no slip rings (with their attendant friction) are needed.

Eddy-Current Noncontacting Transducers

In this type of transducer (Fig. 4.31), the probe usually contains two coils, one (active) which is influenced by the presence of a conducting target and a second (balance) which serves to complete a bridge circuit and provide temperature compensation. Bridge excitation is high-frequency (about 1 MHz) ac. Magnetic flux lines from the active coil pass into the conductive target surface, producing in the target eddy currents whose density is greatest at the surface and which become negligibly small about three "skin depths" below the surface. Figure 4.32[1] gives formulas for computing skin depth δ and graphs of these formulas for the common excitation frequency of 1 MHz. While thinner targets can be successfully employed, a minimum of three skin depths is recommended to reduce temperature effects. As the target comes closer to the probe, the eddy currents become stronger, which changes the impedance of the active coil and causes a

---

[1] Kaman Sciences Corp., Colorado Springs, Colo., *Appl. Note* 108, 1979.

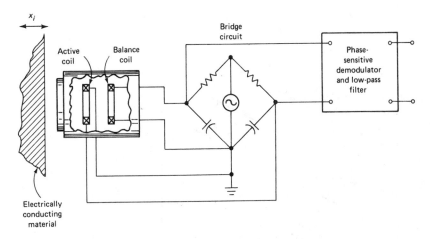

Figure 4.31 Eddy-current noncontacting transducer.

bridge unbalance related to target position. This unbalance voltage is demodulated, low-pass filtered (and sometimes linearized) to produce a dc output proportional to target displacement. The high excitation frequency not only allows the use of thin targets, but also provides good system frequency response (up to 100 kHz).

Probes are commercially available[1] with full-scale ranges from about 0.25 (probe about 2-mm diameter, 20 mm long) to 30 mm (76 mm-diameter, 40 mm long) nonlinearity of 0.5 percent, and maximum resolution 0.0001 mm. Targets are not supplied with the probes since the majority of applications involve noncontact measurement of existing machine parts (thus the part itself serves as target). Since target material, shape, etc., influence output, generally it is necessary to statically calibrate the system with the specific target to be used. For nonconductive targets, you must fasten a piece of conductive material of sufficient thickness to the surface. Commercially available adhesive-backed aluminum-foil tape[2] is convenient for this purpose. The recommended measuring range of a given probe begins at a "standoff" distance equal to about 20 percent of the probe's stated range. That is, a probe rated at 0 to 1 mm range should be used at target-probe distances of 0.2 to 1.2 mm.

Flat targets should be about the same diameter as the probe or larger, if possible. Targets larger than the probe have little effect on the output; however, output drops to about 50 percent for target diameter one-half of probe diameter.[3] Curved-surface targets such as the periphery of a circular shaft behave similarly

[1] Kaman Sciences Corp., Colorado Springs, Colo.; Bentley Nevada, Minden, Nev.
[2] Mystik Tape #7453, Northfield, Ill.
[3] Kaman Sciences Corp., Colorado Springs, Colo., *Appl. Note* 104.

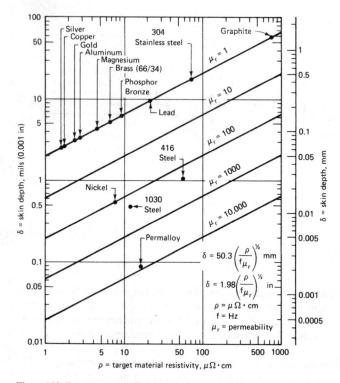

Figure 4.32 Target-material effect on eddy-current transducer.

to flat surfaces if the shaft diameter exceeds four transducer diameters.[1] Special four-probe systems[2] for measuring orbital motions of rotating shafts and various centering and alignment operations are available. While rotating-shaft measurements are routinely accomplished, special care may be necessary to deal with "electrical runout"[3] in shafts made of ferromagnetic material (such as steel). This refers to a variation in magnetic permeability around the periphery of the shaft (resulting from inhomogeneities of heat treatment, hardness, etc.) which causes an electrical output even for a perfectly true-running shaft, thus giving false motion readings. If electrical runout is excessive, filtering, differential measurements using two diametrically opposed probes, nickel plating of the shaft, or shaft-surface grinding to remove hardness variations may help.

[1] Ibid.
[2] KD-2700 Alignment System, Kaman Sciences Corp., Colorado Springs.
[3] Kaman Sciences Corp., Colorado Springs. *Appl. Note* 109, 1979.

Capacitance Pickups

A rotational or translatory motion may be used in many ways to change the capacitance of a variable capacitor.[1] The resulting capacitance change can be converted to a usable electrical signal by means of a variety of circuitry. We consider here a few typical applications. Capacitance-type pickups often require electronics somewhat more complex than for the more common types of transducers. However, their mechanical simplicity, very small mechanical loading effects, and high sensitivity make them attractive in a number of applications, and some commercial general-purpose devices now available are quite convenient to use.

The most common form of variable capacitor used in motion transducers is the parallel-plate capacitor with a variable air gap. Theory gives the capacitance of such an arrangement as

$$C = \frac{0.225A}{x} \qquad (4.41)$$

where $C \triangleq$ capacitance, pF

$A \triangleq$ plate area, in^2

$x \triangleq$ plate separation, in

For example, the capacitance of an air capacitor with 1-in^2 plates separated by 0.01 in is 22.5 pF. The impedance $1/(i\omega C)$ of this capacitor at a frequency of, say, 10,000 Hz has a magnitude of 708,000 Ω. This very high impedance level of capacitance gages is responsible for some of their main problems with respect to spurious noise voltages, sensitivity to length and position of connecting cables, and requirement for high-input-impedance electronics. From Eq. (4.41) we also note that the variation of capacitance with plate separation x is nonlinear (hyperbolic); thus the percentage change in x from a chosen "neutral" position must be small if good linearity is to be achieved. The sensitivity of capacitance to changes in plate separation may be computed from Eq. (4.41):

$$\frac{dC}{dx} = -\frac{0.225\,A}{x^2} \qquad (4.42)$$

We note that the sensitivity increases as x decreases. However, the *percentage* change in C is equal to the *percentage* change in x for small changes about *any*

---

[1] Harris and Crede, "Shock and Vibration Handbook," vol. 1, p. 14–1; H. C. Roberts, "Mechanical Measurements by Electrical Methods," chaps. 3 and 9, Instruments Publishing, Pittsburgh, 1951; R. R. Batcher and W. Moulic, "The Electronic Control Handbook," chaps. 2 and 3, Instruments Publishing, Pittsburgh, 1946.

neutral position, as shown by

$$\frac{dC}{dx} = -\frac{C}{x} \tag{4.43}$$

$$\frac{dC}{C} = -\frac{dx}{x} \tag{4.44}$$

Perhaps the simplest useful circuit is that employed with capacitor microphones (see Sec. 6.10 on microphones for a detailed analysis). Figure 4.33 shows the arrangement. When the capacitor plates are stationary with a separation x_0, no current flows and $e_o = E_b$. If there is then a relative displacement x_i from the x_0 position, a voltage e_o is produced and is related to x_i by

$$\frac{e_o}{x_i}(D) = \frac{K\tau D}{\tau D + 1} \tag{4.45}$$

where

$$K \triangleq \frac{E_b}{x_0} \quad \text{V/in} \tag{4.46}$$

$$\tau \triangleq 0.225 \times 10^{-12} \frac{AR}{x_0} \quad \text{s} \tag{4.47}$$

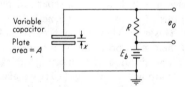

$$x_i \triangleq x - x_0$$

Small motions, $\dfrac{x_i}{x_0} < 0.10$

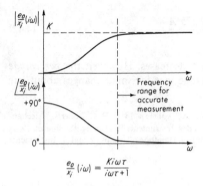

$$\frac{e_o}{x_i}(i\omega) = \frac{Ki\omega\tau}{i\omega\tau + 1}$$

Figure 4.33 Capacitive transducer.

Equation (4.45) shows that this arrangement does not allow measurement of static displacements since e_o is zero in steady state for any value of x_i. For sufficiently rapid variations in x_i, however, the signal e_o will faithfully measure the motion. This is most easily seen from the frequency response

$$\frac{e_o}{x_i}(i\omega) = \frac{K\tau i\omega}{i\omega\tau + 1} \qquad (4.48)$$

which becomes, for $\omega\tau \gg 1$,

$$\frac{e_o}{x_i}(i\omega) \approx K \qquad (4.49)$$

Thus e_o follows x_i accurately under these conditions. A microphone usually need not measure sound pressures slower than about 20 Hz, and so the above arrangement is perfectly satisfactory. To make $\omega\tau \gg 1$ for low frequencies requires a large τ. For a given capacitor and x_0, the value of τ can be increased only by increasing R. Typically, R will be 10^6 Ω or more. Thus to prevent loading of the capacitance transducer circuit, the readout device connected to the e_o terminals must have a high (10^7 Ω or more) input impedance, such as provided by FET electronics.

The use of a variable differential (three-terminal) capacitor with a bridge circuit is shown in the Equibar[1] differential pressure transducer of Fig. 4.34. Spherical depressions of a depth of about 0.001 in are ground into the glass disks; then these depressions are gold-coated to form the fixed plates of a differential capacitor. A thin, stainless-steel diaphragm is clamped between the disks and serves as the movable plate. With equal pressures applied to both ports, the diaphragm is in a neutral position, the bridge is balanced, and e_o is zero. If one pressure is greater than the other, the diaphragm deflects in proportion, giving an output at e_o in proportion to the differential pressure. For the opposite pressure difference, e_o exhibits a 180° phase change. A direction-sensitive dc output can be obtained by conventional phase-sensitive demodulation and filtering. Balance resistors necessary for initially nulling the bridge are not shown in Fig. 4.34. This method (as opposed to that of Fig. 4.33) allows measurement of static deflections. Such differential-capacitor arrangements also exhibit considerably greater linearity than do single-capacitor types.[2]

An ingenious method[3] of circumventing the nonlinear relationship [Eq. 4.41] between x and C is shown in Fig. 4.35. This technique employs an operational amplifier, perhaps the most common building block in linear electronics. The assumptions necessary to an analysis of this circuit depend on the well-known

[1] Trans-Sonics, Inc., Burlington, Mass.

[2] N. H. Cook and E. Rabinowicz, "Physical Measurement and Analysis," p. 142, Addison-Wesley, Reading, Mass., 1963.

[3] Wayne Kerr Corp., Philadelphia.

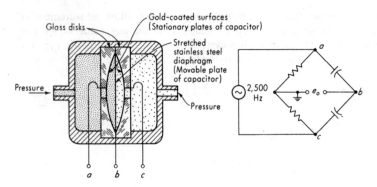

Figure 4.34 Differential-capacitor pressure pickup.

characteristics of op-amps:

1. The input impedance is so high that the amplifier input current may be taken as zero relative to other currents.
2. The gain is so high that if the output voltage of the amplifier is not saturated, the input voltage is extremely small and may be taken as zero relaltive to other voltages. For example, a typical amplifier has linear output for the range ± 10 V and a gain of 10^7 V/V. Thus the maximum input for linear operation is 10^{-6} V.

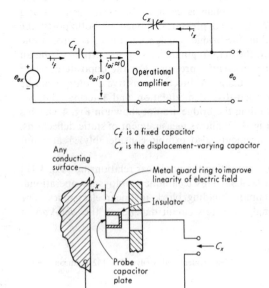

Figure 4.35 Feedback-type capacitive pickup.

Using these assumptions in Fig. 4.35, we can write

$$\frac{1}{C_f} \int i_f \, dt = e_{ex} - e_{ai} = e_{ex} \tag{4.50}$$

$$\frac{1}{C_x} \int i_x \, dt = e_o - e_{ai} = e_o \tag{4.51}$$

$$i_f + i_x - i_{ai} = 0 = i_f + i_x \tag{4.52}$$

Manipulation then gives

$$e_o = \frac{1}{C_x} \int i_x \, dt = -\frac{1}{C_x} \int i_f \, dt = -\frac{C_f}{C_x} e_{ex} \tag{4.53}$$

$$e_o = -\frac{C_f \, x e_{ex}}{0.225 A} = K x \tag{4.54}$$

Equation (4.54) shows that the output voltage is now *directly* proportional to the plate separation x; thus linearity is achieved for both large and small motions. In the commercial instrument described above, e_{ex} is a 50-kHz sine wave of fixed amplitude. The output e_o is also a 50-kHz sine wave which is rectified and applied to a dc voltmeter calibrated directly in distance units.

For vibratory displacements, e_o will be an amplitude-modulated wave as in Fig. 4.36. The average value of this wave after rectification is still the mean separation of the plates and still can be read by the same meter as used for static displacements. The vibration amplitude around this mean position is extracted by applying the e_o signal also to a demodulator and a low-pass filter with cutoff at 10 kHz. The output of this filter is applied to a peak-to-peak voltmeter directly calibrated in vibration amplitude and also is available at a jack for connection to an oscilloscope for viewing of the vibration waveform. The instrument is provided with six different probes ranging from 0.0447 to 1.0 in in diameter of the capacitance plate and covering the full-scale displacement ranges of 0.001, 0.005, 0.01, and 0.5 in, respectively. The overall system accuracy is of the order of 2 percent of full scale. Any flat conductive surface may serve as the second plate of the variable capacitor. Thus in vibrating machine parts, the parts themselves

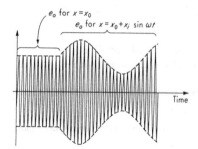

e_o for $x = x_0$

e_o for $x = x_0 + x_i \sin \omega t$

Time

Figure 4.36 Waveform for sinusoidal displacement.

often perform this function. The resolution of 0.5 percent of full scale indicates that (with the 0.001-in full-scale probe) it is possible to detect motion as small as 5 μin.

Another linearization technique[1] which is the basis of a line[2] of capacitive pressure transducers and accelerometers is the pulse-width-modulation scheme shown in Fig. 4.37. The sensing capacitors are arranged in the transducer in a differential manner as shown. Voltages e_1 and e_2 switch back and forth at about 400 kHz between the excitation voltage (say 6 V dc) and ground; the output voltage is the average (low-pass filtered) difference between e_1 and e_2. At the transducer null position, $e_1 - e_2$ is a symmetrical square wave of zero average value, which produces no output. Motion changes the relative width (time duration) of the $+6$- and -6-V portions of $e_1 - e_2$, which gives a positive or negative average value depending on whether x_i was positive or negative. Circuit behavior is as follows. The four solid-state switches are switched simultaneously by the output of two op-amp comparators whenever e_3 or e_4 passes through the fixed reference voltage of 3 V. Note in the diagram that we are observing the circuit at an instant when e_3 is connected to ground ($e_3 = 0$) and capacitor C_1 is discharged, while C_2 is being charged by the 6-V supply through R, with a time constant RC_2. As C_2 charges, e_4 rises exponentially (nearly linearly for early times); and when e_4 reaches 3 V (the reference voltage level), the two inputs of comparator 4 are now equal, causing its output to throw all four switches to their opposite positions. This "instantaneously" discharges C_2 and begins a similar charging process for C_1. Thus an oscillation is established whose period $t_1 + t_2$ (about 2.5 μs in the referenced device) is determined by $R, C_1 + C_2$, and e_{ref}, but whose duty cycle varies as C_1 and C_2 change. In steady state, the output voltage e_o is the average of $e_1 - e_2$. Thus

$$e_o = \frac{e_{ex}t_1}{t_1 + t_2} - \frac{e_{ex}t_2}{t_1 + t_2} = e_{ex}\frac{C_1 - C_2}{C_1 + C_2} \tag{4.55}$$

In the differential capacitor,

$$C_1 = \frac{C_o x_o}{x_o - x_i} \qquad C_2 = \frac{C_o x_o}{x_o + x_i}$$

which finally gives the desired linear relation between e_o and x_i:

$$e_o = e_{ex}\frac{x_i}{x_o}$$

Linearization of capacitance transducers also can be accomplished by using a feedback system[1] which adjusts capacitor-current amplitude so that it stays con-

[1] S. Y. Lee, Variable Capacitance Signal Transduction and the Comparison with Other Transduction Schemes, Setra Systems Inc., Natick, Mass.

[2] Setra Systems Inc., Natick, Mass.

[3] L. Michelson, Greater Precision for Noncontact Sensors, *Mach. Des.*, p. 117, Dec. 6, 1979; Lion Precision Corp., Newton, Mass.

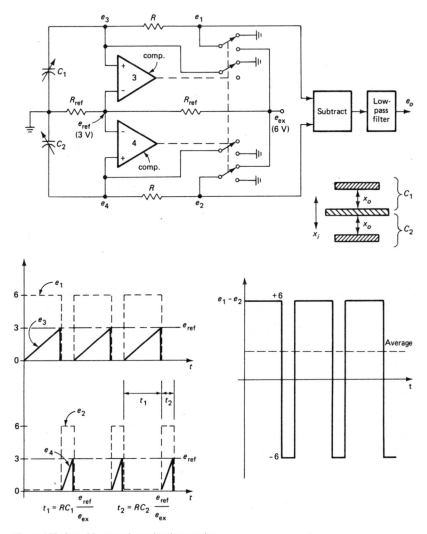

Figure 4.37 Capacitive transducer signal processing.

stant at a reference value for all displacements x_i. This is accomplished by obtaining a dc signal proportional to capacitor current (from a demodulator), comparing this current with the reference current, and adjusting the voltage amplitude of the system excitation oscillator (typically 100 kHz fixed frequency) until the two currents agree. If the capacitor-current amplitude is kept constant irrespective of capacitor motion, then the capacitor-voltage amplitude is linear

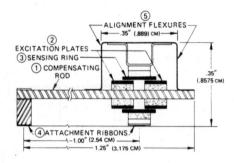

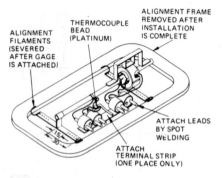

Figure 4.38 Capacitive high-temperature strain gage. Top, capacitive strain gage—cross-section view. Bottom, capacitive strain gage in alignment frame. (*Courtesy Hi Tec Corp., Westford, Mass.*)

with x_i:

$$\frac{i_c}{e_c}(i\omega) = i\omega C \qquad |e_c| = \frac{|i_c|}{\omega C} = \frac{|i_c|\, x_i}{\omega C_o\, x_o} = K x_i \tag{4.56}$$

Figure 4.38 shows a capacitive displacement transducer marketed as a high-temperature strain gage. A differential capacitor in the form of three thin rings is used, with the two inner rings supported by a temperature-compensating rod made of the same material as the test specimen. Since the measuring axis is displaced rather far above the surface where strain is to be measured, the outer ring is carried in specially designed flexures, so that the effect of any bending in the specimen is compensated. Maximum strain is $\pm 20,000$ $\mu\epsilon$, corresponding to a total displacement of ± 0.02 in over the 1-in gage length (a model with $\frac{1}{4}$-in gage length and $\pm 80,000$ $\mu\epsilon$ is also available). Temperature range is cryogenic to 1500°F; spurious apparent strain due to temperature is typically ± 300 $\mu\epsilon$ at 1500°F. Electronics using ac excitation at 3.39 kHz provides nonlinearity of ± 0.5 percent at 3,000 $\mu\epsilon$ and 70°F; linearization details are not given. The gage exerts a load of 20 g on the specimen at 5,000 $\mu\epsilon$.

Piezoelectric Transducers

When certain solid materials are deformed, they generate within them an electric charge. This effect is reversible in that if a charge is applied, the material will mechanically deform in response. These actions are given the name *piezoelectric* (pī-ēzō-electric) *effect.*[1] This electromechanical energy-conversion principle is applied usefully in both directions. The mechanical-input/electrical-output direction is the basis of many instruments used for measuring acceleration, force, and pressure. It also can be utilized as a means of generating high-voltage low-current electric power, such as is used in spark-ignition engines and electrostatic dust filters. The electrical-input/mechanical-output direction is applied in small vibration shakers, sonar systems for acoustic ranging and direction detection, industrial ultrasonic nondestructive test equipment, ultrasonic flowmeters, and micromotion actuators.[2]

The materials that exhibit a significant and useful piezoelectric effect fall into two main groups: natural (quartz, rochelle salt) and synthetic (lithium sulfate, ammonium dihydrogen phosphate) crystals and polarized ferroelectric ceramics (barium titanate, etc.). Because of their natural asymmetrical structure, the crystal materials exhibit the effect without further processing. The ferroelectric ceramics must be artificially polarized by applying a strong electric field to the material (while it is heated to a temperature above the Curie point of that material) and then slowly cooling with the field still applied. (The *Curie temperature* is the temperature above which a material loses its ferroelectric properties; thus it limits the highest temperature at which such materials may be used.) When the external field is removed from the cooled material, a remanent polarization is retained and the material exhibits the piezoelectric effect.

The piezoelectric effect can be made to respond to (or cause) mechanical deformations of the material in many different modes, such as thickness expansion, transverse expansion, thickness shear, and face shear. The mode of motion effected depends on the shape and orientation of the body relative to the crystal axes and the location of the electrodes. Metal electrodes are plated onto selected faces of the piezoelectric material so that lead wires can be attached for bringing in or leading out the electric charge. Since the piezoelectric materials are insulators, the electrodes also become the plates of a capacitor. A piezoelectric element used for converting mechanical motion to electric signals thus may be thought of as a charge generator and a capacitor. Mechanical deformation generates a charge; this charge then results in a definite voltage appearing between the electrodes according to the usual law for capacitors, $E = Q/C$. The piezoelectric effect is direction-sensitive in that tension produces a definite voltage polarity while compression produces the opposite.

We illustrate the main characteristics of piezoelectric motion-to-voltage transducers by considering only one common mode of deformation, thickness expansion. For this mode the physical arrangement is as in Fig. 4.39b. Various double-subscripted physical constants numerically describe the phenomena occurring. The convention is that the first subscript refers to the direction of the electrical effect and the second to that of the mechanical effect, by using the axis-numbering system of Fig. 4.39a.

[1] Harris and Crede, op. cit., chap. 16, pt. 2.
[2] Inchworm Translator, Burleigh Instruments, Fisher, N. Y.

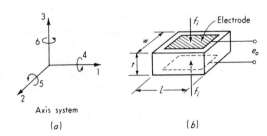

Axis system

(a)

(b)

Figure 4.39 Piezoelectric transducer.

Two main families of constants, the g constants and the d constants, are considered. For a barium titanate thickness-expansion device, the pertinent g constant is g_{33}, which is defined as

$$g_{33} \triangleq \frac{\text{field produced in direction 3}}{\text{stress applied in direction 3}} = \frac{e_o/t}{f_i/(wl)} \qquad (4.57)$$

Thus if we know g for a given material and the dimension t, we can calculate the output voltage per unit applied stress. Typical g values are 12×10^{-3} (V/m)/(N/m^2) for barium titanate and 50×10^{-3} for quartz. Thus, for example, a quartz crystal 0.1 in thick would have a sensitivity of 0.88 V/(lb/in^2), illustrating the large voltage output for small stress typical of piezoelectric devices.

To relate applied force to generated charge, the d constants can be defined as

$$d_{33} \triangleq \frac{\text{charge generated in direction 3}}{\text{force applied in direction 3}} = \frac{Q}{f_i} \qquad (4.58)$$

Actually, d_{33} can be calculated from g_{33} if the dielectric constant ϵ of the material is known, since

$$C = \frac{\epsilon wl}{t} \qquad (4.59)$$

$$g_{33} \triangleq \frac{\text{field}}{\text{stress}} = \frac{e_o wl}{t f_i} = \frac{e_o C}{\epsilon f_i} = \frac{Q}{\epsilon f_i} = \frac{d_{33}}{\epsilon} \qquad (4.60)$$

$$d_{33} = \epsilon g_{33} \qquad (4.61)$$

The dielectric constant of quartz is about 4.06×10^{-11} F/m while for barium titanate it is $1,250 \times 10^{-11}$. For quartz, then,

$$d_{11} = \epsilon g_{11} = (4.06 \times 10^{-11})(50 \times 10^{-3}) = 2.03 \text{ pC/N} \qquad (4.62)$$

(The subscripts 11 are used because in quartz the thickness-expansion mode is along the crystallographic axis conventionally called axis 1.) Sometimes it is desired to express the output charge or voltage in terms of deflection (rather than stress or force) of the crystal, since it is really the *deformation* that causes the charge generation. To do this, we must know the modulus of elasticity, which is 8.6×10^{10} N/m^2 for quartz and 12×10^{10} for barium titanate.

With the above brief introduction as background, we proceed to consider piezoelectric elements as displacement transducers. The ultimate purpose is generally force, pressure, or acceleration measurement, but here we consider only the conversion from displacement to voltage. For analysis purposes it is necessary to consider the transducer, connecting cable, and associated amplifier as a unit. The transducer impedance is generally very high; thus the amplifier is usually a high-impedance type used for buffering purposes rather than voltage gain. Charge amplifiers (see Chap. 10) are commonly used also. The cable capacitance can be significant, especially for long cables. For the transducer alone, if a static deflection x_i is applied and maintained, a transducer terminal voltage will be developed but the charge will slowly leak off through the leakage resistance of the transducer. Since R_{leak} is generally very large (the order of 10^{11} Ω), this decay would be very slow, perhaps allowing at least a quasi-static response. However, when an external voltage-measuring device of low input impedance is connected to the transducer, the charge leaks off very rapidly, preventing the measurement of static displacements. Even relatively high-impedance amplifiers generally do not allow static measurements. Some commercially available[1] systems using quartz transducers (very high leakage resistance) and electrometer input amplifiers (very high input impedance) achieve an effective total resistance of 10^{14} Ω, which gives a sufficiently slow leakage to allow static measurements.

To put the above discussion on a quantitative basis, we consider Fig. 4.40. The charge generated by the crystal can be expressed as

$$q = K_q x_i \tag{4.63}$$

where

$$K_q \triangleq C/cm \tag{4.64}$$

$$x_i \triangleq \text{deflection, cm} \tag{4.65}$$

The resistances and capacitances of Fig. 4.40b can be combined as in 4.40c. We also convert the charge generator to a more familiar current generator according to

$$i_{cr} = \frac{dq}{dt} = K_q \left(\frac{dx_i}{dt}\right) \tag{4.66}$$

We may then write

$$i_{cr} = i_C + i_R \tag{4.67}$$

$$e_o = e_C = \frac{\int i_C \, dt}{C} = \frac{\int (i_{cr} - i_R) \, dt}{C} \tag{4.68}$$

$$C\left(\frac{de_o}{dt}\right) = i_{cr} - i_R = K_q\left(\frac{dx_i}{dt}\right) - \frac{e_o}{R} \tag{4.69}$$

$$\frac{e_o}{x_i}(D) = \frac{K\tau D}{\tau D + 1} \tag{4.70}$$

[1] Kistler Instrument Corp., Amherst, N. Y.

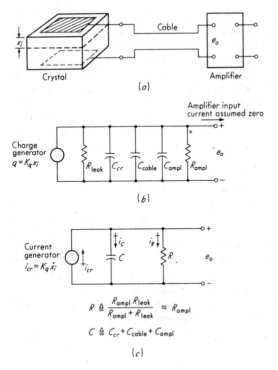

Figure 4.40 Equivalent circuit for piezoelectric transducer.

$$\text{where} \qquad K \triangleq \text{sensitivity} \triangleq \frac{K_q}{C}, \text{V/cm} \qquad (4.71)$$

$$\tau \triangleq \text{time constant} \triangleq RC, \text{s} \qquad (4.72)$$

We see that, just as in the capacitance pickup of Fig. 4.33, the steady-state response to a constant x_i is zero; thus we cannot measure static displacements. For a flat amplitude response within, say, 5 percent, the frequency must exceed ω_1, where

$$0.95^2 = \frac{(\omega_1\tau)^2}{(\omega_1\tau)^2 + 1} \qquad (4.73)$$

$$\omega_1 = \frac{3.04}{\tau} \qquad (4.74)$$

Thus a large τ gives an accurate response at lower frequencies.

The response of these transducers is further illuminated by considering the displacement input of Fig. 4.41. The differential equation is

$$(\tau D + 1)e_o = (K\tau D)x_i \qquad (4.75)$$

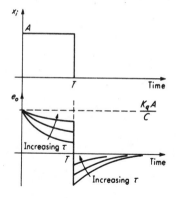

Figure 4.41 Pulse response of piezoelectric transducer.

Since $x_i = A$ for $0 < t < T$, this becomes

$$(\tau D + 1)e_o = 0 \qquad (4.76)$$

Now at $t = 0^+$ the displacement x_i is A, and so the charge *suddenly* increases to $K_q A/C$. Thus our initial condition is

$$e_o = \frac{K_q A}{C} \qquad \text{at } t = 0^+ \qquad (4.77)$$

Solving Eq. (4.76) with initial condition (4.77) gives

$$e_o = \frac{K_q A}{C} e^{-t/\tau} \qquad 0 < t < T \qquad (4.78)$$

Equation (4.78) holds until $t = T$. At this instant we must stop using it because of the change in x_i. For $T < t < \infty$ the differential equation is

$$(\tau D + 1)e_o = 0 \qquad (4.79)$$

At $t = T^-$, Eq. (4.78) is still valid and

$$e_o = \frac{K_q A}{C} e^{-T/\tau} \qquad (4.80)$$

Now, at $t = T$, suddenly x_i drops an amount A, causing a sudden decrease in charge of $K_q A$ and a sudden decrease in e_o of $K_q A/C$ from its value at $t = T^-$. Thus at $t = T^+$, e_o is given by

$$e_o = \frac{K_q A}{C} (e^{-T/\tau} - 1) \qquad (4.81)$$

which becomes the initial condition for Eq. (4.79). The solution then becomes

$$e_o = \frac{K_q A}{C} (e^{-T/\tau} - 1)e^{-(t-T)/\tau} \qquad T < t < \infty \qquad (4.82)$$

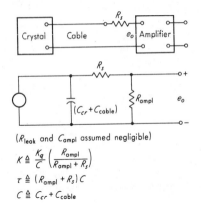

$(R_{leak}$ and C_{ampl} assumed negligible)

$$K \triangleq \frac{K_q}{C} \left(\frac{R_{ampl}}{R_{ampl} + R_s} \right)$$

$$\tau \triangleq (R_{ampl} + R_s) C$$

$$C \triangleq C_{cr} + C_{cable}$$

Figure 4.42 Use of series resistor to increase time constant.

Figure 4.41 shows the complete process for three different values of τ. It is clear that a large τ is desirable for faithful reproduction of x_i. If the decay and "undershoot" at $t = T$ are to be kept within, say, 5 percent of the true value, τ must be at least $20T$. If an increase of τ is required in a specific application, it may be achieved by increasing either or both R and C. An increase in C is easily obtained by connecting an external shunt capacitor across the transducer terminals, since shunt capacitors add directly. The price paid for this increase in τ is a loss of sensitivity according to $K = K_q/C$. Often this may be tolerated because of the initial high sensitivity of piezoelectric devices. An increase in R generally requires an amplifier of greater input resistance. If sensitivity can be sacrificed, a series resistor connected external to the amplifier, as shown in Fig. 4.42, will increase τ without the need of obtaining a different amplifier.

Though perhaps not quite as versatile with respect to computing/compensating operations as the strain-gage/bridge-circuit combination, piezoelectric elements do provide some similar functions, as exemplified by the three-component force transducer of Fig. 4.43.[1] Two types of thin quartz sensing disks are used here. One type is cut to be sensitive to tension/compression, and the other to shear. One pair of tension/compression disks is used to measure the z component of force, while two pairs of shear disks, each properly oriented, measure the x and y components. Use of pairs of disks rather than single disks simplifies electrode attachment, eliminates the need for insulating layers which reduce rigidity, and increases sensitivity. Note that the disk pairs are electrically in parallel so that the charges produced by each element are summed. This signal-summing action can be extended to more than two elements and is not limited to elements within a single transducer. Thus, say, the x component of force from one transducer could be electrically summed with the y component

[1] G. H. Gautschi, Piezoelectric Multicomponent Force Transducers and Measuring Systems, Kistler Instrument Corp., Amherst N. Y., 1978.

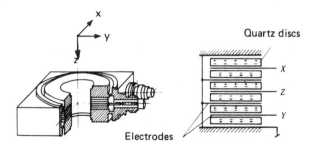

Figure 4.43 Three-component piezoelectric force transducer.

from another. While piezoelectric elements respond to both positive and negative strains (with charge of opposite polarity), it is mechanically difficult to couple tension stresses into the elements, since adhesive connections usually are not considered sufficiently reliable. The usual procedure is thus to install the elements (such as the disks above) under heavy compressive preload, sufficient to preclude relaxation of the preload under any expected external tension and large enough to provide reliable frictional transfer of the x and y shear components without slippage. Here the leakage of charge for steady loads is an advantage, since the preload persists mechanically but produces zero output voltage. Applied positive and negative z loads then produce corresponding plus and minus voltages since the output voltage corresponds to the *change* in strain away from the preload condition.

Detailed data on static and dynamic performance characteristics of piezoelectric transducers are deferred to the respective sections on force, pressure, and acceleration measurement, where these data will be more meaningful.

Electro-Optical Devices

We describe several displacement-measuring devices that combine optical and electronic principles. Each is designed for a specific class of applications, but all share the general advantages of optical measurement,[1] noncontacting operation with negligible force exerted on the moving object since the radiation pressure of light is miniscule.

Our first example, the Fotonic sensor,[2] uses fiber optics to measure displacements in the range of microinches to tenths of inches with a relatively simple optical/electronic arrangement. The probe (Fig. 4.44), 0.02 to 0.3 in in diameter and 3 in long, consists of a bundle of several hundred optical fibers, each a few thousandths of an inch in diameter. The fiber bundle extends for 1 to 3 ft to the electronics chassis, where it is divided into two equal groups of fibers. One group

[1] New Dimensions in Optical Gaging, *Inst. Tech.*, p. 9, June 1980.

[2] C. Menadier, C. Kissinger, and H. Adkins, The Fotonic Sensor, *Inst. & Cont. Syst.*, p. 114, June 1967; Mechanical Technology Inc., Latham, N. Y.

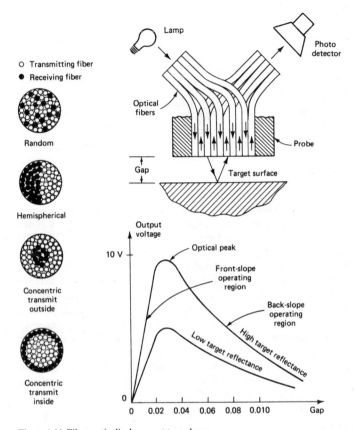

Figure 4.44 Fiber-optic displacement transducer.

(transmitting fibers) is exposed to a light source and thus carries light to the probe tip, where it is emitted and reflected by the target surface. While targets of high reflectivity give greater output, even rather dull surfaces can be measured. The reflected light is picked up by the other (receiving) group of fibers, transmitted to the electronics package, and focused on a suitable photodetector whose electronics then produces a dc output related to probe-target gap. To minimize cross-talk between adjacent fibers, the fibers are of a core/cladding construction; two glasses of different refractive index are used to obtain total internal reflection.

Probes of this general type display two useful measuring ranges: front slope and back slope. At zero gap, no light can escape from the transmitting fibers and output is zero. As the gap opens up, more of the target surface is illuminated and reflection increases, giving a very sensitive and nearly linear range of measurement (front slope). As gap increases, finally the entire target is illuminated, giving peak output. Motion beyond this point causes a *reduction* in response since both target illumination and the fraction of reflected light gathered by the sensor decrease roughly according to an inverse square law. This "back slope" region is also useful for measurement, but is less linear and sensitive. Theory shows that maximum sensitivity in the front-slope region is achieved by arranging the fibers in a precise geometric pattern in which each receiving fiber is surrounded by four transmitting fibers. This is not practical from a manufacturing viewpoint, so a random distribution, which has been found to give nearly identical results, is usually used. Three other arrangements (Fig. 4.44) which allow tradeoffs among various features are also available, however. Typical characteristics for a standard (random distribution) probe of 0.109-in diameter for the front-slope region are as follows: sensitivity 1.6 mV/μin, linear (± 1 percent) range 0.001 to 0.004 in, static resolution 1 μin, frequency response flat ± 3 dB from dc to 50 kHz, full-scale output 0 to 10 V dc, output-signal ripple 50 mV p-p. Since the target surface is usually an existing machine part, static calibration with that specific surface is generally required; however, automatic compensation for slowly changing surface reflectivity is sometimes possible.[1]

Lasers are the basis of many measurement systems used both in industrial manufacturing and research laboratories. Several schemes[2] have been devised using lasers for motion and dimension measurement. One such[3] is shown in Fig. 4.45. Here a single narrow helium-neon laser beam is scanned over a workspace by a five-sided prism rotating at 1800 r/min, giving one measurement scan in $\frac{1}{150}$ s. A special collimating lens produces parallel rays which sweep through the workspace at a linear rate proportional to the prism's rotational speed. Thus the position within the workspace of the cylindrical target (whose diameter is to be measured) is not critical. (If the target moves while it is being scanned, an error will result; however, the rapid measurement cycle reduces this effect. If target motion is oscillatory, averaging can be used to minimize error.)

As a result of the shadow cast by the target, the photodetector output voltage exhibits a "notch" whose width in time is proportional to the target width in space. Measurement of this time interval (by gating an electronic counter) is made more precise by electrically double-differentiating the photodetector signal to produce two narrow spikes. Differentiation also allows ac signal processing and makes the system less sensitive to gradual illumination changes resulting from drift in laser power, smoke in the workspace, etc. A single system oscillator (typically 18 MHz) is used both as the counter clock and the ac frequency reference for the hysteresis-synchronous motor used to rotate the prism. Thus any drifts in oscillator frequency cause motor-speed changes and counter time-base changes which nullify each other. Additional circuitry is provided to reject

[1] G. J. Philips and F. Hirschfeld, Rotating Machinery Bearing Analysis, *Mech. Eng.*, p. 28, July 1980.

[2] G. B. Foster, Lasers as Dimension Transducers, *Instrum. Technol.*, p. 47, September 1975; T. R. Pryor and J. C. Cruz, Laser-Based Gauging/Inspection, *Electro-optical Syst. Des.*, p. 26. May 1975.

[3] R. Moore, Laser-Based Noncontact Gauging System, *Electro-optical Syst. Des.*, p. 46, March 1978; Zygo Corp., Middlefield, Conn.

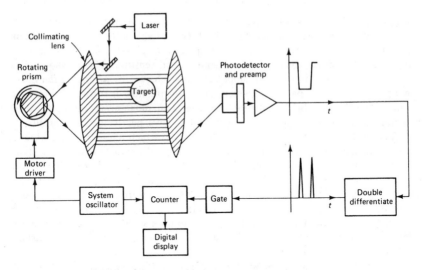

Figure 4.45 Laser dimensional gage.

certain types of "bad" measurements, such as no target present, signal too low for accuracy because of excess smoke in workspace, etc. The system holds the last "good" measurement until such situations clear up. While each application exhibits its own characteristics, accuracy on the order of 0.0001 in has been achieved in plastics extrusion, bar-mill diameter control, centerless grinding machine control, and remote diameter measurements of nuclear fuel rods.[1]

We now briefly treat the *laser interferometer*, a very precise motion measuring system. The use of light-interference principles as a measurement tool certainly can be considered a part of classical physics. Michelson, in the 1890s, used the scheme of Fig. 4.46, which is the basis of all subsequent developments. Using the wave model of light, we would expect the observer to see cycles of light and darkness as the motion of the movable mirror shifted the phase of beam 2 with respect to fixed beam 1, causing alternate reinforcement and interference of the two beams. If we know the light wavelength to be, say, 0.5×10^{-6} m, then each 0.25×10^{-6} m of mirror movement corresponds to one complete cycle (light to dark to light) of illumination. By counting the number of illumination cycles, we can calculate the distance between any two positions of the movable mirror.

This seemingly simple measurement principle is fraught with difficulties that have kept it a tedious standards-laboratory procedure of limited application (rather than a practical "machine shop" tool) until the recent appearance of several laser-based developments. The laser itself provides a much higher-quality monochromatic (single frequency) light source than did earlier "lamps," since its

[1] Moore, ibid.

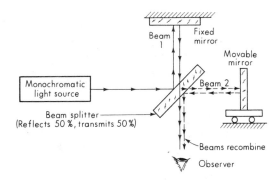

Figure 4.46 Original Michelson interferometer.

light is "coherent" (stays in phase with itself) over much greater distances and its frequency is very stable and precisely known (to about one part in 10^7). When light travels through air, its frequency is unaffected, but its wavelength changes whenever air pressure, temperature, or humidity causes alterations in the air's refractive index and thus in the speed of light. The pressure effect is about 4 parts in 10^7 per millimeter of mercury, while the temperature effect is nearly 11 parts in 10^7 per degree centigrade. Many applications do not warrant corrections for these effects; but if they are deemed necessary, some commercial systems[1] provide temperature, pressure, and humidity sensors that automatically scale the instrument readout to provide continuous correction. Since the lengths of objects being measured are *themselves* temperature-dependent, automatic sensing of this temperature and correction of readings to some standard (often 20°C) are of practical interest and are provided on some systems.

Clever optical, mechanical, and electronic design have built on the above simple principles to provide measuring systems of much improved portability, precision, range of applicability, and ease of use. Figure 4.47 shows one such system which we briefly discuss (Refs. 2 and 3 give more details). The helium-neon laser used is unusual in that it produces light at two distinct optical frequencies f_1 and f_2, both in the neighborhood of 5×10^{14} Hz, but separated in frequency by about 2 MHz and of opposite polarizations. As the beam leaves the laser, it is split in half. One half goes directly to a polarizer and photodetector to create an electric reference signal; the other half proceeds out to the external optics. The reference-beam polarizer gives the two frequencies *equal* polarizations so that they may exhibit ordinary constructive and destructive interference. This interference can be thought of in terms of the intensity-versus-time graph of Fig.

[1] Hewlett-Packard Model 5510A.
[2] A Two-Hundred-Foot Yardstick with Graduations Every Microinch, *Hewlett-Packard J.*, August 1970.
[3] Remote Laser Interferometry, *Hewlett-Packard J.*, December 1971.

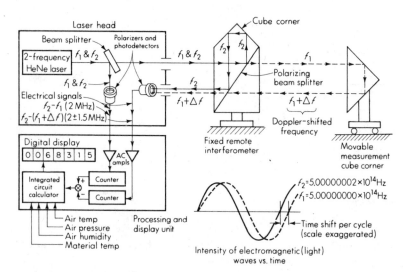

Figure 4.47 Modern laser interferometer.

4.47. There, two waves initially "in phase" get slightly "out of phase" in one cycle, since the frequencies are slightly different. (Note that the time shift per cycle is a *very* small fraction of one cycle; thus, the waves shown are nearly identical and interfere *constructively*, giving a high illumination to the reference-beam photodetector and a large electric output.) *Destructive* interference (giving low illumination and small electric output) will occur 1.25×10^8 cycles (0.25×10^{-6} s) later when the time shift per cycle has accumulated a total phase shift of one-half cycle. The "light" on the reference photodetector thus "flickers" at a 2-MHz rate, producing a 2-MHz electric signal.

Turning our attention to the measuring beam, we see that it proceeds out from the laser and encounters first the fixed remote interferometer's polarizing beam splitter, which efficiently reflects the f_2 component around a cube corner (prism) and transmits the f_1 component to the measurement cube corner. If this cube corner is stationary, no frequency change occurs in f_1 between the incident and reflected beam. However, if the cube corner moves, the reflected beam exhibits a Doppler shift of frequency proportional to the velocity of motion [about 3.3 MHz/(m/s)]. When this $f_1 + \Delta f$ beam is recombined with the f_2 beam in the remote interferometer and impinged on the photodector, again we get interference effects that produce an electric signal of frequency 0.5 to 3.5 MHz, depending on the velocity. The reference and measurement signals are ac-amplified and applied to a subtracting counter that reads zero if there is no motion, but which accumulates counts in proportion to the distance traveled from a chosen reference position. For example, a motion at 1 cm/s for 1 (total motion 1 cm) gives 33,000 counts and thus a resolution of about 3×10^{-7} m.

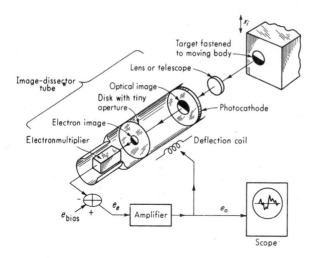

Figure 4.48 Optron displacement pickup.

The "ac" interferometer described above exhibits a number of advantages relative to older "fringe-counting" instruments. The Doppler technique allows use of simpler and more reliable ac (rather than dc) amplifiers, gives unambiguous indication of the *direction* of motion, and allows *velocity* measurement in addition to displacement. Locating the interferometer remote from the laser head reduces thermal deformation problems due to the laser's heat. By varying the external optical hardware, the system[1] can be adapted to measure length, flatness, straightness, and squareness to high precision (a few parts per million) over long distances (up to 200 m). In multiaxis machine tools, a single laser beam may be suitably reflected to serve the measurement needs of up to eight axes of motion.[1, 2]

To measure uniaxial or biaxial motion in a plane perpendicular to the line of sight, optical trackers of various types are available. Perhaps the most common technique utilizes an image-dissector tube and electron multiplier with a servo lock-on technique. Figure 4.48 shows the arrangement for a uniaxial measurement using this approach.[3] The moving object must itself exhibit a "target" (such as a machined edge) with an optical-contrast ratio of at least 3 : 1, or else we fasten to it a black-white target with the dividing line perpendicular to the direction of motion. Illumination of average room level or more is required and must be nonpulsating. In microscope or telescope optics (depending on the stand-

---

[1] A. F. Rudé and M. J. Ward, Laser Transducer Systems for High-Accuracy Machine Positioning, *Hewlett-Packard J.*, p. 2, February 1976.

[2] W. E. Olson and R. B. Smith, Electronics for the Laser Transducer, *Hewlett-Packard J.*, p. 7, February 1976.

[3] Optron Corp., Woodbridge, Conn.

off distance and range of motion), an image of the target is focused on the photocathode of the image-dissector tube. The rear surface of the photocathode emits electrons in relation to the light intensity striking the front surface, thus forming an "electron image" of the target. An applied electric field accelerates the electron image down the tube and focuses it on an aperture plate with a tiny central hole. Thus only a small portion of the electron image is seen by the electron multiplier, which produces an output current proportional to the intensity of this image. If the electron multiplier sees "all white," it produces a large output; if "all black," a small output. If the target black-white dividing line lies within the aperture, an intermediate level of output is produced. If we choose, say, a 50 : 50 black-white ratio, this would produce a definite output signal, say e_{bias}. We now subtract the electron-multiplier output from the fixed value e_{bias} to generate an error signal e_e, which will go positive or negative depending on whether more black or more white is present. Since the image-dissector tube has a deflection coil (just like an oscilloscope CRT) which can position the electron image, we connect our amplified error signal to this coil with polarity such that if there is "too much black," the coil drives the image in the white direction, and vice-versa. Thus a feedback system is created which keeps the black-white dividing line nearly centered on the aperture at all times. The output signal, proportional to target displacement x_i, is obtained from the amplifier output e_o. As in all servosystems, accuracy and speed require a high loop gain. However, this is possible here without instability since all dynamic elements in the control loop have small lags because of the absence of mechanical moving parts.

The system of Fig. 4.48 can be modified for simultaneous biaxial (vertical and horizontal) motion by using an image-dissector tube with two perpendicular (x_v, x_h) deflection coils, a target with both horizontal and vertical black-white discontinuities (Fig. 4.49a), and a timesharing scheme. This extension to biaxial motion is not quite as obvious as it might seem at first, since the electron multiplier has only one output signal, not a horizontal and a vertical, and this output signal really is related to the ratio of black to white *area*, not displacement in any particular direction. To appreciate the nature of the problem, consider Fig. 4.49b, where the servosystem would be satisfied (since we have equal black and white areas) even though the target is *not* centered in x_h, or Fig. 4.49c, where the vertical position is incorrect, but would not be corrected because again the 50 : 50 area ratio produces no corrective x_v deflection-coil signal.

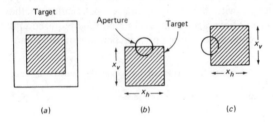

Figure 4.49 Target geometry for two-dimensional tracking.

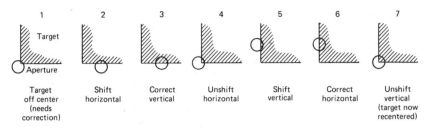

Figure 4.50 Operation cycle for two-dimensional tracking.

Figure 4.50 explains the solution to these difficulties used in the commercial instrument.[1] An electronic switching system rapidly (50 kHz) switches the output of the phototube between the horizontal and vertical channels of the image deflection system. Hold amplifiers in each channel preserve the most recent data, while the other channel is being updated, until fresh data are available for that channel. A single cycle of servosystem action goes as follows: The cycle starts with condition 1 in Fig. 4.50, where the target has moved off center and its image position needs correction. In step 2 the electronic system applies a fixed horizontal shift, sufficient to always position the image so that phototube output is related to vertical error only, as shown. Using this vertical-error signal, the vertical channel-deflection system corrects the vertical error as in step 3. In step 4 the fixed horizontal shift is removed, and a similar procedure (steps 5, 6, 7) is applied to the horizontal error, which completes the cycle. The electronic switching and holding system is designed so that the horizontal- and vertical-motion output signals do not show the "shift voltages" since they do not, of course, correspond to actual target motion.

If target "lock-on" is momentarily lost, the large "all-white" phototube signal detects this and switches the instrument to a search mode of operation; then the instrument scans the entire field until signal reduction (indicative of the presence of some "dark") tells it that the target has been found, whereupon it returns to the servomode. Instruments of this type can have full-scale displacement ranges from 0.002 inch to many feet (depending on the target distance and optics used), nonlinearity better than 1 percent and frequency response (uniaxial) beyond 100 kHz. At close range, resolution of a few microinches may be achieved. An extensometer version is available which measures the *differential* motion between two targets separated by, say, a 1-in gage length and mounted on a material test specimen. An "orthogonal sweep" version applies a linear ramp to the horizontal deflection coils while the vertical axis is servocontrolled as usual. Using a horizontal black-white target extending along the side of a vertically vibrating beam, we can now measure and display on an oscilloscope the dynamic mode shape of the beam vibration.

The development of self-scanning photodiode arrays[2] has made possible

[1] Optron Corp., Woodbridge, Conn.
[2] EG&G Reticon, Sunnyvale, Calif.

solid-state cameras[1] for applications in pattern recognition, size and position measurement, etc. Figure 4.51 a shows a linear array of specific dimensions. However, the manufacturing process allows considerable flexibility in size and geometry; area arrays containing 10,000 elements in a 100×100 matrix with $60\text{-}\mu$m center-to-center spacing, for example, allow imaging of two-dimensional objects. On a single silicon chip, such arrays include a row (or rows) of individual light-sensitive photodiodes, each with its own charge-storage capacitor and solid-state multiplex switch. Also contained on the chip is a shift register for serial readout of the individual element signals. The diodes transduce incident light into charge, which is stored on the capacitor until readout. The charge is proportional to the product of light intensity and exposure time; however, a saturation level does exist. For an array as in Fig. 4.51a, saturation typically occurs at about $6\mu\text{W} \cdot \text{s/cm}^2$. Scan frequency (the frequency at which the individual elements are interrogated and read out) can be varied over a wide range (up to about 40 MHz in some models), which allows a tradeoff between sensitivity and speed, since slower scan rates allow the same illumination to produce more charge each scan cycle.

Figure 4.51b shows how a linear array can be used to construct a line-scan camera for dimensional measurement. The object to be measured is back-lighted (front lighting is also feasible) so as to produce a light-dark pattern with transitions at the object's edges. Conventional optics focus an image on the photodiode array. Dimensional resolution at the array is, of course, limited by the diode spacing, say 0.001 in. However, if the optics is such that a 5-in object produces a 1-in image, then resolution of the object's dimension is only 0.005 in. If the object is moving (as is often the case in industrial inspection), then additional errors are possible; however, higher scan frequency may alleviate such problems. Figure 4.51b shows two output signals, "video" and "data." The video signal is a "boxcar" (stepwise changing) function that shows the time-integrated illumination of each individual picture element (called *pixel*) over one scan cycle. Note that the light (and dark) regions do not give an identical response for each pixel. No illumination field is perfectly uniform; and even if it were, the individual pixel sensitivities cannot be made perfectly equal. To get an unambiguous dimension measurement, the video signal is compared with a judiciously chosen threshold level to produce distinct switching in the "data" signal.

Many useful variations of the above apparatus are possible. Area scanning can be done with an area array or by using an oscillating mirror to deflect the image past a linear array. Wide objects can be measured by using two cameras separated by a known distance. Color applications are possible by using three cameras with proper-color filters. Cameras can be interfaced with microprocessors which have digital representations of objects stored in ROM (read-only memory). The measured objects' "signature" can be compared with a "standard"

[1] The Selection of a Line Scan Camera for Industrial Measurement Applications, Reticon, Sunnyvale, Calif., 1974; Line-Scan Camera Subsystem Model CCD1300, Brochure 253-11-0007-076/12M, Fairchild Camera and Inst. Corp., Mountain View, Calif., 1976.

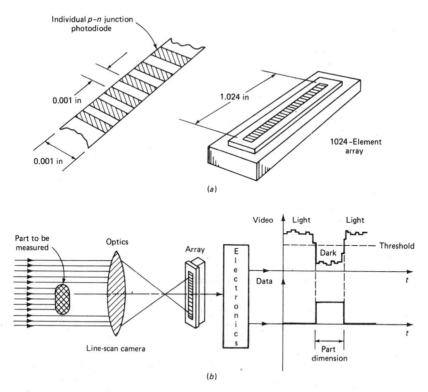

Figure 4.51 Self-scanning diode arrays and cameras. (*a*) 1024 element array; (*b*) line-scan camera.

stored in ROM to check whether they match, thereby identifying the object or perhaps detecting defects in it.

Photographic, Videotape, and Holographic Techniques

The application of still and motion-picture photography often allows qualitative and quantitative analysis of complex motions that would be difficult by other methods.[1] Here we touch briefly on some of the more common applications.

Perhaps the simplest application of still photography is the single-flash "stop-action" technique.[2] The objective is to "freeze" a motion at a particular phase of its occurrence to allow detailed visual study of some physical phenomenon. The equipment usually employed consists of a still camera, a stroboscopic light source, and some means of triggering a single flash of the strobe light

[1] W. G. Hyzer, "Engineering and Scientific High-Speed Photography," Macmillan, New York, 1962.

[2] "Handbook of High-Speed Photography," General Radio Co., West Concord, Mass.

at the desired instant. If the experiment can be performed in a darkened room, the procedure consists of manually opening the camera shutter, allowing the phenomenon to occur, triggering the light at the desired instant, and then manually closing the shutter. If triggering, focus, and exposure are correct, a photo of the phenomenon, "frozen" at the instant of the light flash, is obtained. Such photos can be most helpful in understanding complex physical processes in fluid motion or moving machine parts. Of course, the effective freezing of the motion depends on the flash duration being sufficiently short compared with the velocity of the motion. Flash durations of the order of 1 to 3 μs are readily available. Thus, for example, a velocity of 300 m/s will cause a "blurring" of 0.03 to 0.09 cm at the object. The actual blurring on the film will be this value times the image/object ratio of the camera setup. If a 1-cm object shows up on the film as 0.1 cm, for example, the above blurring would amount to 0.003 to 0.009 cm on the film, which generally would be considered acceptable. If the experiment cannot be carried out in a darkened room, the opening and closing of the shutter must be synchronized with the flash and the open time of the shutter must be short enough so as not to overexpose the film from the room light.

If a displacement-time record is desired, a multiple-flash still-camera technique may be employed. The setup is essentially the same as above except the strobe light flashes repetitively at a known rate. The result is a multiple-exposure photo showing the moving object in successive positions which are separated by known increments of time. By including a calibrated length scale in the photo (preferably in the plane of the motion) numerical values of displacement at specific time intervals may be measured.

High-speed motion-picture photography is used to study motions that occur too rapidly for the eye to analyze properly. This is accomplished by taking the pictures at a high camera picture frequency (frames per second) and then projecting the film at a low projector picture frequency. The lowest usable projector frequency is about 16 frames per second since lower frequencies result in flickering because the human eye's persistence of vision is about 0.06 s. The highest usable camera picture frequency depends on the construction of the camera. Relatively small, portable cameras are available with picture frequencies up to about 20,000 frames per second, and these are in relatively wide use in industry. Larger, more complex, and expensive cameras[1] are available where the application dictates their higher speed. Their picture frequencies are up to several million per second. The time magnification is defined as the ratio of the camera frequency to the projector frequency; thus if 16 is the projector frequency, magnifications of several hundred thousand to one are achievable while 1,500 : 1 is not unusual in common industrial practice. Aside from picture frequency, the shutter speed of the camera must be sufficiently high to prevent blurring of the individual frames, just as in still photography. For 16-mm film a blur of 0.005 cm is considered acceptable. The shutter-speed requirement can be greatly relaxed by the

[1] G. R. Van Horn, Rotating Drum Cameras for High Speed Photography, *Electro-optical Syst. Des.*, p. 26, December, 1977.

use of a synchronized short-duration electronic flash as the light source, since the flash duration then controls blur no matter what the shutter speed. Selection of a proper camera picture rate may be judged roughly by the rule that the projection of the complete motion to be visually analyzed should take about 2 to 10 s. Thus a motion occurring in 0.001 s requires a time magnification of 2,000 to 10,000. For vibratory motions the camera picture rate must be several (preferably 5 to 10) times the highest vibration frequency.

A convenient alternative to high-speed cinematography for applications allowing use of modest (60 or 120 per second) frame rates is found in the special videotape systems[1] designed for industrial measurement. These are identical in principle to units used for "instant replay" in commercial TV. However, their detail design provides a 2 to 4 times higher frame rate, as well as elimination of rolling noise bars, jitter, and picture distortion not tolerable in instrumentation applications. Also, to "freeze" rapid motion, illumination is stroboscopic with about a 20-μs flash duration, once per frame. Note that time resolution of motion depends on the frame rate, *not* on the strobe duration. Thus these systems compete with photographic methods for only the slower motions; however, many industrial problems lie in this range. For these cases, they provide considerably greater convenience of use and quicker availability of results. One system,[2] combining new developments in video sensors, magnetic tape recording and digital processing, provides frame rates to 2,000 per second (full size) and 12,000 per second ($\frac{1}{6}$ size).

Optical holography refers to a wide variety[3] of laser-based optical techniques applicable to vibration measurement, flow visualization, particle diagnostics, etc. Here we briefly describe *holographic interferometry*[4] for vibration measurement, one of the most widely used techniques. Classical (prelaser) interferometry was limited to measurement of small motions of specularly reflecting (polished) flat surfaces, whereas current laser techniques can deal with three-dimensional objects with arbitrarily curved and diffusely reflecting surfaces, in other words, almost any real object.

Figure 4.52 shows the essential features of a simple holographic apparatus. A single laser beam is divided by a beam splitter into two separate beams, which retain a fixed phase relationship with each other over some distance related to the *coherence length* of that particular laser. (Coherence length is the distance over which a laser stays in phase with itself. A long coherence length allows us to

---

[1] E. J. Aleks, Instant Replay in the Product Development Lab, *Mach. Des.*, p. 120, Oct. 21, 1976; Video Logic Corp., Sunnyvale, Calif.;Unilux, Inc., Hackensack, N. J.

[2] Spin Physics, Inc, San Diego.

[3] R. K. Erf (ed.), "Holographic Nondestructive Testing," Academic, New York, 1974; Holographic Instrumentation Applications, *NASA SP*-248, 1970.

[4] D. J. Monnier, Practical Applications for Holography in Industry, SESA Fall Meeting, 1973; W. A. Penn, Holographic Interferometry, Plus Publ. Code No. 119, General Electric Electronics Lab., Syracuse, N. Y., 1977; D. A. Cain, C. D. Johnson, G. M. Moyer, Applications of Optical Holographic Interferometry, Naval Underwater Syst. Cent., Newport, R.I., TM # TD12-134-73, 1973; H. J. Caulfield (ed.), "Handbook of Optical Holography," sec. 10.4, Academic, New York, 1979.

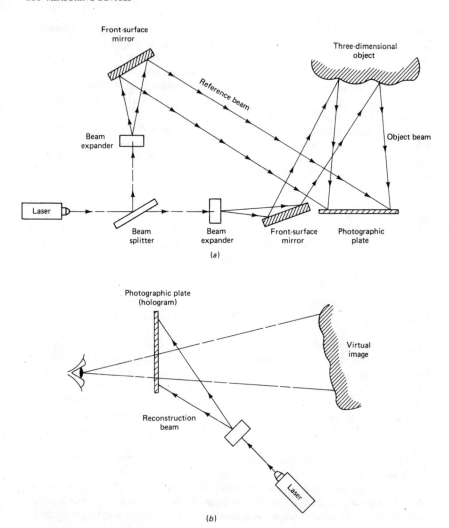

Figure 4.52 Basic holographic apparatus. (*a*) Hologram construction; (*b*) hologram reconstruction.

measure larger objects. Typically a 20-mW helium-neon laser might have a co-herence length of 0.25 m and be capable of measuring objects up to about 0.5 m in all dimensions.) After the laser beam is split, the two beams are each diverged in beam spreaders so that an area (rather than a "point") can be illuminated. (Note that the undiverged beams present *safety hazards* and must be treated with respect.) One diverged beam, the reference beam, is directed to illuminate a high-resolution photographic plate, while the other (object) beam illuminates the measured object, which reflects the light to the same photographic plate. If the measured object is stationary, when the object and reference beams recombine at the photographic plate, because of phase shifts resulting from the different path lengths of rays illuminating the object, an intensity pattern called a *hologram* is recorded on the plate, which is then chemically developed as in ordinary photography. The developed plate can be observed with the unaided eye, but bears no resemblance to the object. The intensity pattern spacings ($\approx 5 \times 10^{-4}$ mm) are too small to see; however, the recorded pattern contains complete information about the measured object. To observe the object, it is now necessary to *reconstruct* the three-dimensional image by illuminating the hologram with the same laser light used to record it, whereupon a virtual image (indistinguishable from the actual object) appears to the unaided eye.

At this point we have a method (holography) for producing realistic (three-dimensional) images of objects—quite an accomplishment in its own right, but not yet a motion-measuring scheme. In holographic interferometry we record *two* holograms on the same plate. If the object has moved and/or deformed between the two exposures, these two holograms will be different. When we reconstruct this double hologram, the two patterns will interfere, causing light and dark fringes (superimposed on the object's image) which now *are* visible to the unaided eye and can be photographed. The fringes represent contours of equal displacement along the viewing axis, with each successive fringe corresponding to $\frac{1}{4}$ to $\frac{1}{2}$ (depending on the geometry of the optical layout) wavelength of light, 6 to 12 μin for a helium-neon laser. To study static and dynamic motions and deformations,

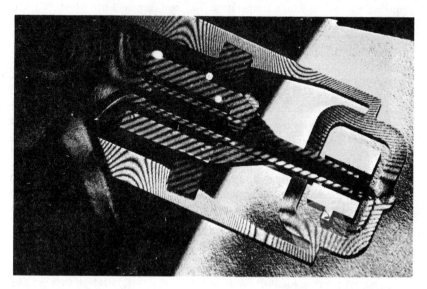

Figure 4.53 Holographic vibration analysis of disk-memory component. (*Courtesy Spectron Development Labs, Costa Mesa, Calif.*)

various specific techniques[1] (static double exposure, dynamic time average, real time, pulse[2]) have been developed. Figure 4.53 shows typical results obtained in vibration analysis of a machine component.[3]

Photoelastic, Brittle-Coating, and Moiré Fringe Stress-Analysis Techniques

Since these methods are really strain- rather than stress-sensitive and since strain is a small displacement, we include a brief treatment in this section on displacement measurement.

Photoelastic methods[4] depend on the property of birefringence under load exhibited by certain natural or synthetic transparent materials. Birefringence (double refraction) under load refers to the phenomenon in which light travels at different speeds in a transparent material, depending on both the direction of travel relative to the directions of the principle stresses and the magnitude of the difference between the principal stresses, for a two-dimensional stress field. By constructing models (from suitable transparent materials) of the same shape as the part to be stress-analyzed and shining suitably polarized light through them while they are subjected to loads proportional to those expected in actual service, a pattern of light and dark fringes appears which shows the stress distribution throughout the piece and allows numerical calculation of stresses at any chosen point.

By use of the "frozen stress" technique, the method can be extended to three-dimensional problems. A three-dimensional plastic model is subjected to simultaneous load and high temperature. The load is maintained while the specimen is slowly cooled to room temperature whereupon the load is released. It is found that a residual stress pattern identical to that produced by the load is "frozen" into the specimen. Furthermore, now the model may be carefully sliced in various directions to produce flat (two-dimensional) slabs, which can be photoelastically analyzed to determine three-dimensional stresses. By combining photoelastic and high-speed photographic techniques, the method can be extended to dynamic studies such as the propagation of shock waves through solid bodies.

In *reflective photoelasticity* there is no need to construct a plastic model; the part itself is given a thin "coating" of photoelastic material. Unfortunately, sprayed or brushed coatings, the most convenient application technique, exhibit too much thickness variation and are used only in rough qualitative studies. For accurate work, a thin (0.01 to 0.125 in) sheet of photoelastic plastic is bonded to the part with a reflective adhesive and thus participates in the surface strain when the part is loaded. Polarized light is directed onto the part, reflects from the shiny

[1] Erf, op. cit.

[2] R. Levin, Industrial Holographic Applications, *Electro-optical Syst. Des.*, p. 31, April 1976.

[3] J. D. Trolinger and R. S. Reynolds, Stresses during Small Motions, *Ind. Res./Dev.*, pp. 133–136, May 1979.

[4] M. M. Frocht, "Photoelasticity," vols. 1 and 2, Wiley, New York, 1941, 1948; Dally and Riley, "Experimental Stress Analysis," Photoelastic Stress Analysis Techniques and Products, *Bull.* SFC-200, Photoelastic Inc., Malvern, Pa.

undersurface, and produces fringe patterns that are analyzed by a reflection polariscope. Flat surfaces are easily fitted with available sheet material. Complex surfaces, however, require casting of a thin flat sheet, careful "form fitting" to the part while the sheet is still "limp," allowing the formed sheet to harden, and then cementing to the part with reflective adhesive.

Photoelastic methods, as compared with bonded resistance strain gages, give an overall picture of the stress distribution in a part. This is very helpful in locating and numerically evaluating stress concentrations and in redesigning the part for optimum material use. Also the method does not disturb the local stress field as a strain gage might.

In the brittle-coating stress-analysis technique,[1] a special lacquerlike material is sprayed on the actual part to be analyzed and the coating allowed to dry. Application of load causes visible cracking of the brittle coating. The direction of the cracking shows the direction of maximum stress, and the spacing of the cracks indicates magnitude. Under favorable conditions, numerical values of stress can be calculated from measured crack spacing to about 10 percent accuracy. The main features of the method are its simplicity, low cost, and speed in giving an overall picture of stress distribution. Often it is used in conjunction with electric-resistance strain gages, with the brittle coating locating the points of maximum stress and its direction so that strain gages can be applied at the proper places and in the proper orientation for accurate strain measurement.

Moiré fringe methods[2] utilize a fine (200 to 2000 lines/in) square grid of equidistant lines bonded to the surface to be analyzed. When a master (undeformed) grid is superimposed (directly or by projection) on the grid as deformed by loading the part, a fringe pattern appears which can be analyzed to obtain surface displacements or strains.

Displacement-to-Pressure (Nozzle-Flapper) Transducer

The nozzle-flapper-transducer principle is widely used in precision gaging equipment and as a basic component of pneumatic and hydraulic measurement and control apparatus. Figure 4.54 shows the general arrangement. Fluid at a regulated pressure is supplied to a fixed flow restriction and a variable flow restriction connected in series. The variable flow restriction is varied by moving the "flapper" to change the distance x_i. This causes a change in output pressure p_o which, for a limited range of motion, is nearly proportional to x_i and extremely sensitive to it. Thus a pressure-measuring device connected to p_o can be calibrated to read x_i. Ideally (pressure-containing chambers rigid; fluid incompressible) a sudden change in x_i would cause an instantaneous change in p_o. Actually, the dynamics are approximately those of a linear, first-order system for small changes in x_i. The time constant is determined for gases by the compressibility of

[1] C. C. Perry and H. R. Lissner, "The Strain Gage Primer," chap. 13, McGraw-Hill, New York, 1955; Dally and Riley, op. cit.

[2] *Tech. Bull.* TDG-2, Photolastic, Inc., Malvern Pa.

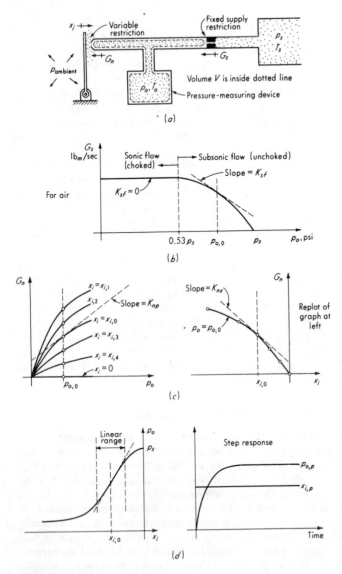

Figure 4.54 Nozzle-flapper transducer.

the gas; for liquids the elastic deformation of the pressure-sensing device often controls.

We analyze the system of Fig. 4.54a for the case of a gaseous medium since a majority of practical applications utilize low-pressure ($p_s \approx 20$ to 30 lb/in^2 gage) air as the working fluid. The principle of conservation of mass is applied to the volume V by stating that during a time interval dt the difference between entering mass and leaving mass must show up as an additional mass storage in V.

It is necessary to obtain expressions for the mass flow rates G_s and G_n. We assume that supply pressure p_s and temperature T_s are constant. Then G_s depends on p_o only; however, the dependence is nonlinear, and so we employ a linearized (perturbation) analysis which will be valid for small changes from an operating point. We can write

$$G_s = G_s(p_o) \approx G_{s,\,0} + \frac{dG_s}{dp_o}\bigg|_{p_o = p_{o,\,0}} (p_o - p_{o,\,0}) = G_{s,\,0} + K_{sf}\,p_{o,\,p} \qquad (4.83)$$

where
$$G_{s,\,0} \triangleq \text{value of } G_s \text{ at equilibrium operating point} \qquad (4.84)$$

$$p_{o,\,0} \triangleq \text{value of } p_o \text{ at equilibrium operating point} \qquad (4.85)$$

$$p_{o,\,p} \triangleq \text{small change (perturbation) in } p_o \text{ from } p_{o,\,0} \qquad (4.86)$$

$$K_{sf} \triangleq \text{value of } dG_s/dp_o \text{ at } p_{o,\,0} \text{ (a constant)} \qquad (4.87)$$

The function $G_s(p_o)$ can be found theoretically from fluid mechanics and thermodynamics (with the help of an experimental orifice-discharge coefficient) or entirely by experiment for a given orifice. Its general shape is given in Fig. 4.54b.

In finding the nozzle mass flow rate G_n, we assume that the process from p_s, T_s to p_o, T_o is a perfect-gas, work-free, adiabatic process. Also, the velocity of the gas at pressure p_o in volume V is assumed zero. Thus the gas at p_o is essentially in a stagnation state, and since the stagnation enthalpy for a perfect-gas, work-free, adiabatic process is constant, the temperature T_o is nearly the same as T_s and remains nearly constant. If T_o may be assumed constant, the nozzle mass flow rate depends on only p_o and x_i. The relationship among G_n, p_o, and x_i can be found from theory (with experimental corrections) or from experiment alone for a specific device. The relationship is again nonlinear, and so a perturbation analysis is in order:

$$G_n(p_o, x_i) \approx G_{n,\,0} + \frac{\partial G_n}{\partial p_o}\bigg|_{\substack{x_{i,\,0} \\ p_{o,\,0}}} (p_o - p_{o,\,0}) + \frac{\partial G_n}{\partial x_i}\bigg|_{\substack{x_{i,\,0} \\ p_{o,\,0}}} (x_i - x_{i,\,0}) \qquad (4.88)$$

$$G_n \approx G_{n,\,0} + K_{np}\,p_{o,\,p} + K_{nx}\,x_{i,\,p} \qquad (4.89)$$

Figure 4.54c shows how K_{np} and K_{nx} could be found from experimental data.

The mass storage in volume V can be treated by using the perfect-gas law $p_o V = MRT_0$. We assume V, R, and T_o to be constant. Then

$$p_o = \frac{RT_o}{V}\,M \qquad (4.90)$$

$$p_{o,\,0} + p_{o,\,p} = \frac{RT_o}{V}\,(M_0 + M_p) \qquad (4.91)$$

$$\frac{dp_{o,\,p}}{dt} = \frac{RT_o}{V}\,\frac{dM_p}{dt} \qquad (4.92)$$

By conservation of mass during a time interval dt,

$$\text{Mass in} - \text{mass out} = \text{additional mass stored}$$

$$(G_{s,0} + K_{sf}\, p_{o,p})\, dt - (G_{n,0} + K_{np}\, p_{o,p} + K_{nx}\, x_{i,p})\, dt = dM_p = \frac{V}{RT_o}\, dp_{o,p} \quad (4.93)$$

If the operating point $p_{o,0}$, $x_{i,0}$ is an equilibrium condition, then $G_{s,0} = G_{n,0}$. So we have

$$\frac{V}{RT_o}\frac{dp_{o,p}}{dt} + (K_{np} - K_{sf})p_{o,p} = (-K_{nx})x_{i,p} \quad (4.94)$$

This is clearly a first-order system, and so we define

$$K \triangleq \frac{-K_{nx}}{K_{np} - K_{sf}} \quad \text{(lbf/in}^2)/\text{in} \quad (4.95)$$

$$\tau \triangleq \frac{V}{RT_o(K_{np} - K_{sf})} \quad \text{s} \quad (4.96)$$

to give
$$(\tau D + 1)p_{o,p} = K x_{i,p} \quad (4.97)$$

and
$$\frac{p_{o,p}}{x_{i,p}}(D) = \frac{K}{\tau D + 1} \quad (4.98)$$

To improve speed of response (decrease τ), the volume V should be minimized. Since T_o is usually the ambient temperature, R and T_o are not available for adjustment. An increase in $K_{np} - K_{sf}$ will decrease τ but at the expense of sensitivity, as shown by Eq. (4.95).

A relatively crude device of this type made up for student laboratory use had a nozzle diameter of $\frac{1}{32}$ in and a volume V of the order of 1 in³. For a supply-orifice diameter of $\frac{1}{32}$ in and a supply pressure of 25 lbf/in² gage, this device had a K (at the most sensitive part of its range) of about 2,000 (lbf/in²)/in and a τ of about 0.12 s. It is quite linear over a range of about ± 0.002 in around $x_{i,0} = 0.004$ in. By changing only the supply-orifice diameter to $\frac{1}{64}$ in, the sensitivity was raised to about 8,000 (lbf/in²)/in, while τ increased to 0.24 s. The linear range was now about ± 0.0005 in around $x_{i,0} = 0.0015$ in. Since 1 lbf/in² = 27.7 in of water, a water manometer used to read p_o gives a (easily readable) 0.1-in change for an x_i change of only 0.45×10^{-6} in when the sensitivity is 8,000 (lbf/in²)/in. This illustrates the great sensitivity of this transducer.

A useful approximate expression for the static sensitivity may be easily obtained by assuming incompressible flow. The results are accurate for liquids and a good estimate for gases if pressure changes are not large. The mass flow through the supply orifice is now

$$G_s = \frac{C_d \pi d_s^2}{4} \sqrt{2\rho(p_s - p_o)} \quad (4.99)$$

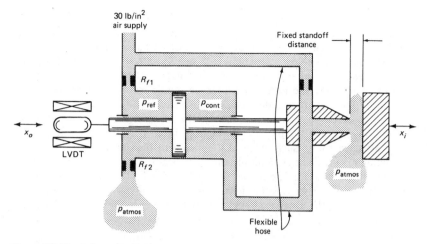

Figure 4.55 Nozzle-flapper/LVDT noncontact gage.

where $C_d \triangleq$ discharge coefficient

$d_s \triangleq$ supply-orifice diameter

$\rho \triangleq$ fluid mass density

and we neglect the velocity of approach since the orifice area is very small compared with the upstream passage. The flow area for the nozzle is taken as the peripheral area of a cylinder of height x_i and diameter d_n, the nozzle diameter. This is true only for small values of x_i. The discharge coefficient for this configuration may be different from that for the supply orifice and may vary somewhat with x_i, but here we take it to be the same as C_d in Eq. (4.99). We have then

$$G_n = C_d \pi d_n x_i \sqrt{2\rho(p_o - p_{\text{ambient}})} \qquad (4.100)$$

For steady state, $G_n = G_s$ so that (taking $p_{\text{ambient}} = 0$ lbf/in.² gage) we get

$$p_o = \frac{p_s}{1 + 16(d_n^2 x_i^2/d_s^4)} \qquad (4.101)$$

The sensitivity dp_o/dx_i varies with x_i and is found to have its maximum value at $x_i = 0.14 d_s^2/d_n$. This maximum value is

$$K_{\text{max}} = \frac{2.6 d_n p_s}{(d_s)^2} \qquad \text{(lbf/in}^2\text{)/in} \qquad (4.102)$$

Thus we see that large d_n, large p_s, and small d_s lead to high sensitivity.

An interesting application of the nozzle flapper in noncontact displacement measurement is commercially available[1] and shown in Fig. 4.55. Here the nozzle

[1] B. C. Ames Co., Waltham, Mass.; Schaevitz Engineering, Pennsauken, N. J.

flapper, whose linear range is severely limited, is used not as the final readout device, but rather as a highly sensitive null detector in a pneumatic servosystem which maintains the measuring probe at a preselected standoff distance from the moving object. Fixed-flow restrictions R_{f1} and R_{f2} are selected so that the constant reference pressure p_{ref} is about 15 lbf/in^2 gage. If the opposing pressure p_{cont} is not equal to this, the piston and attached nozzle will move until the nozzle-flapper output pressure p_{cont} does balance p_{ref}. This will always occur at the same standoff distance, so whenever x_i moves, the servosystem moves x_o an equal amount, maintaining the preselected standoff distance. Any displacement transducer attached to x_o (such as the LVDT shown) thus measures x_i in a noncontacting manner. The standoff distance is typically 0.004 to 0.015, the air force exerted on the measured object is 1 to 10 g, and overall repeatability is about 0.0001 in. For time-varying motions, a dynamic analysis[1] of the servosystem allows design for the usual accuracy/stability tradeoff typical of feedback control systems.

Digital Displacement Transducers (Translational and Rotary Encoders)

Since transducers often communicate with digital computers, transducers with digital output would be particularly convenient. However, very few such devices exist, so usually we use an analog transducer to produce a voltage signal and an electronic analog-to-digital converter to realize the desired digital data. Digital transducers called *encoders*[2] do exist, however, for translational and rotary displacement and are in wide use. Three major classes—tachometer, incremental, and absolute—are available in various detail forms; Fig. 4.56 shows their characteristic output signals. A tachometer encoder has only a single output signal which consists of a pulse for each increment of displacement. If motion were always in one direction, a digital counter could accumulate these pulses to determine displacement from a known starting point. However, any reversed motion would produce identical pulses, causing errors. Thus this type of encoder usually is used for speed, rather than displacement, measurement in situations where rotation never reverses.

The incremental encoder[3] solves the reverse-motion problem by employing at least two (and sometimes a third) signal-generating elements. By mechanically displacing the two tracks, one of the electric signals is shifted $\frac{1}{4}$ cycle relative to the other, allowing detection of motion direction by noting which signal rises first. Thus the up-down pulse counter can be signaled to *subtract* pulses whenever motion reverses. A third output, which produces a single pulse per revolution at a distinct point, is sometimes provided as a zero reference. An incremental encoder has the advantage of being able to rotate through as many revolutions as the

[1] E. O. Doebelin, "Dynamic Analysis and Feedback Control," McGraw-Hill, New York, 1962.
[2] C. Hudson, A Guide to Optical Shaft Encoders, *Inst. & Cont. Syst.*, May 1978.
[3] Techniques for Digitizing Rotary and Linear Motion, Dynamics Research Corp., Wilmington, Mass., 1976.

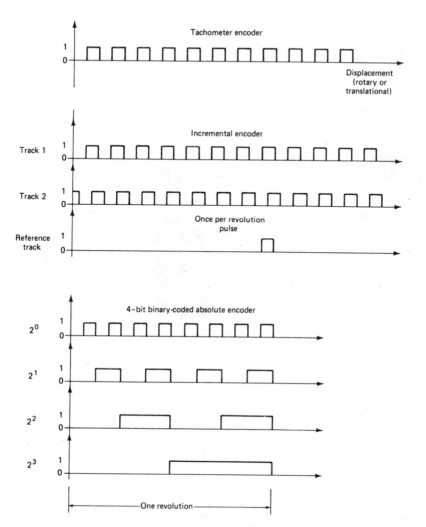

Figure 4.56 Three major classes of encoders.

application requires. However, any false pulses resulting from electric noise will cause errors that persist even when the noise disappears, and loss of system power also causes total loss of position data with no recovery when power is reapplied.

Absolute encoders generally are limited to a single revolution and utilize multiple tracks and outputs, which are read out in parallel to produce a binary representation of the angular position of the shaft. Since there is a one-to-one

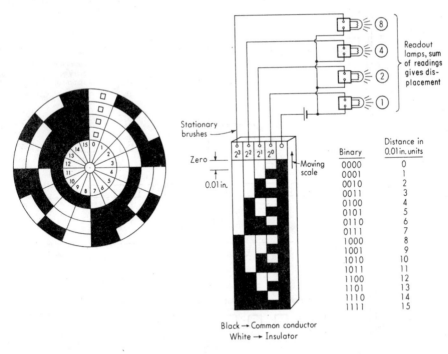

Figure 4.57 Translational and rotary encoders.

correspondence between shaft position and binary output, position data are recovered when power is restored after an outage, and transient electric noise
causes only transient measurement error.

Encoders of all three types can be constructed as contact devices (stationary
brushes rub on rotating code disk, Fig. 4.57) or as noncontacting devices using
either magnetic or optical principles. In Fig. 4.57 the readout lamps are shown
for explanatory purposes only; the voltages on the four lamp lines could be sent
directly to a computer. If a visual readout *were* desired, these four voltages would
be applied to a binary-to-decimal conversion module and then read out deci-
mally on a display, to avoid the need to "mentally sum" lamp readings. For the
finest resolution, optical encoders are generally required. Commercial rotary instruments that resolve a fraction of a second of arc are available (see Fig. 4.58);
translational encoders may achieve 1-μm resolution. Any transducer with resolution above about 18 bits ($1/2^{18}$) will require extreme care[1] in installation and
use. Tachometer and incremental encoders often employ a grating principle in

---

[1] D. H. Breslow, Installation and Maintenance of High Resolution Optical Shaft Encoders, Itek
Corp., Newton, Mass., 1978.

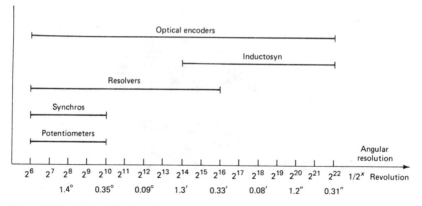

Figure 4.58 Resolution of displacement sensors.

which two glass disks (one fixed, the other rotating), with identical opaque/clear patterns photographically deposited, are mounted side by side with about 0.01-in clearance between. Parallel light is projected through the two disks toward photosensors on the far side. When opaque segments are aligned, a minimum (logical 0) signal is produced while alignment of clear segments gives a maximum (logical 1) signal. Absolute encoders may use the demagnification method in which the light is sharply focused, rather than parallel, and only one disk is employed, with the narrow light beam and photosensor acting in the same fashion as the brushes in a contacting encoder.

While the code pattern of Fig. 4.57 is most convenient for explaining how motion is represented in the familiar natural binary system, many commercial encoders use different code patterns (such as the Gray code of Fig. 4.59) to avoid errors resulting from small misalignments possible in any real device. For example, at the midpoint of the natural binary scale, if the shaded area of the 2^2 bit were displaced slightly left, instead of going from 0111 (7) to 1000 (8), the count would go from 0111 (7) to 0011 (3). The Gray code shown in the same figure does not suffer from this type of problem since only one bit changes at each transition. Since the Gray-code output may not be compatible with the readout device, conversion from Gray to natural binary (or vice versa) may be necessary and is easily accomplished by using standard logic gates as shown in Fig. 4.59.

Figure 4.58 includes data on the Inductosyn, a high-resolution incremental encoder based on the electromagnetic coupling between a fixed scale provided with an ac excited serpentine conductor (produced by printed-circuit techniques) and a similar but smaller sensing winding which travels over the scale (Fig. 4.60). When alignment is as in Fig. 4.60b, output is at a positive maximum. A displacement of $s/2$ results in minimum output, s gives negative maximum, $3s/2$ gives minimum again, and $2s$ returns the output to positive maximum. The output variation over the $2s$ cycle length is essentially cosinusoidal. A "coarse" digital

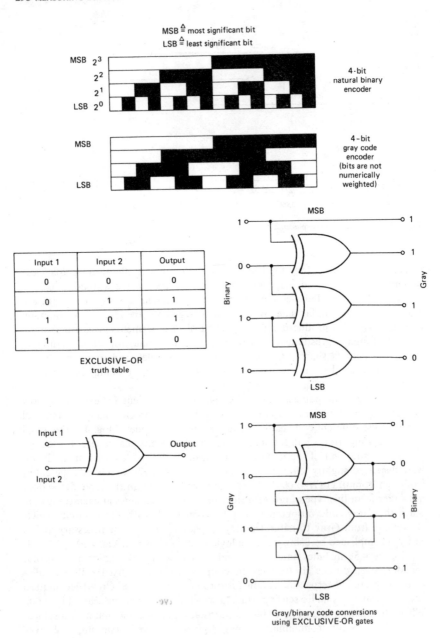

Figure 4.59 Natural and Gray binary codes.

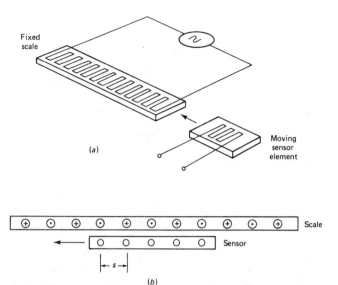

Figure 4.60 Inductosyn transducer. (*Courtesy Farrand Controls, Valhalla, N. Y.*)

output is obtained by counting the cycles of spacing 2s, while fine resolution is obtained by electronically digitizing the analog voltage variation within each cycle. As in other incremental encoders, to detect *direction* of motion, the sensor element includes a second winding displaced $s/2$ from the first, providing a sinusoidal signal. With both a sine and cosine output available, the device behaves essentially as a resolver and can use resolver-to-digital type of electronics[1] similar to those discussed earlier under synchros. Inductosyns are available in both translational (best accuracy $\pm$ 0.001 mm) and rotary (best accuracy $\pm$ 0.5 second of arc) forms. The standard spacings s available are 0.1 in, 0.05 in, and 2 mm.

Ultrasonic Transducers

We show three examples of displacement measurement utilizing wave-propagation principles and producing digital output. Such sensors often are called *ultrasonic* since the signals employed are generally outside the frequency range of human hearing. Our first example, intended mainly for full-scale ranges of 1 to 10 ft or more, utilizes a permanent magnet which moves relative to a magnetostrictive wire enclosed in a nonferrous protective tube (see Fig. 4.61). Electronic circuitry drives a current pulse through the wire. At the magnet location, magnetostrictive action generates in the wire a stress pulse which propa-

---

[1] Inductosyn to Digital Converter, Analog Devices, Norwood, Mass.

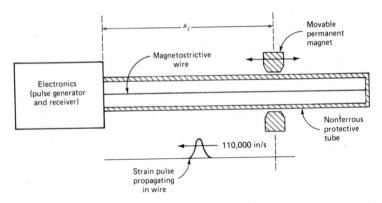

Figure 4.61 Ultrasonic displacement transducer. (*Courtesy Temposonics, Inc., Plainview, N. Y.*)

gates to the receiver location at a fixed speed of about 110,000 in/s. At the receiver location, a pickup coil senses the arrival of the pulse. The time interval between the initiating current pulse and the arrival of the sensed stress pulse is clearly proportional to the displacement x_i, with a sensitivity of about 9 μs/in. By using the two pulses to gate on and off a 10-MHz counter, a digital output reading with resolution of 0.011 in can be obtained. To get a continuous analog signal, the pulse is applied repetitively and the transit time interval is used to modulate the width of a rectangular pulse train. Low-pass filtering of this pulse train produces an analog voltage whose amplitude is proportional to pulse width and thus to x_i. The repetition rate is sufficiently high that the low-pass filter can be designed to be flat (-3 dB) to about 200 Hz for a 2-ft-stroke unit. Nonlinearity better than 0.05 percent of full scale is obtained.

Our next example is based on the automatic focusing system of a popular camera. The motor drive which focuses the lens gets its information from an ultrasonic rangefinder (distance measuring) system. The manufacturer has made available an experimenter's kit and manual[1] which allow study of ultrasonic distance measurement; our description is based on this manual. The heart of the system is an electrostatic transducer (similar in principle to the capacitor microphone of Chap. 6) which is, however, alternately used as a loudspeaker and a microphone. The principle of all such systems, of course, is that of classical sonar where an acoustic signal is sent out and its reflected return timed to allow calculation of the target's distance by use of the known propagation velocity (1100 ft/s for standard air). Whereas simpler ultrasonic ranging systems (such as those used for tank liquid-level measurement) might use a single-frequency signal, the camera rangefinder must deal with targets of variable and unpredictable

---

[1] Polaroid Ultrasonic Ranging Experimenter's Kit, Polaroid Corp., Cambridge, Mass., 1980.

shape and size. This led to use of a multifrequency signal to ensure that a reflected echo would occur reliably. This signal consists of 8 cycles of 60 kHz, 8 cycles of 57 kHz, 16 cycles of 53 kHz, and 24 cycles of 50 kHz for a total of 56 cycles and about 1 ms. By using the transducer as a loudspeaker, this electric signal is applied, causing a similar acoustic signal to propagate toward the target. Before the echo returns, the transducer connections are electronically switched, so that it acts as a microphone to transduce the reflected pressure wave into an electric signal whose time of arrival can be detected and the target distance thereby calculated. The system is designed to measure distances over the range 0.9 to 35 ft with a resolution of 0.1 ft. Since the variation in echo signal strength for targets from 3 to 35 ft is almost a million to one, the microphone's tuned amplifier self-adjusts its gain (16 settings) and bandwidth (8 settings) to maintain signal levels more nearly constant in subsequent circuits. In the kit mentioned above, distance measurements are made and read out digitally on an LED display 5 times per second; however, circuit adjustments can speed this up to about 20 measurements per second.

Our final example is a digitizer[1] used to pick dimensions off of solid objects and input the digitized values to various computer-aided design programs. Then the computer can display an accurate "three-dimensional" picture of the object on a graphics terminal; and since the object's geometry is now numerically specified, the computer can perform any engineering calculations (surface areas, volumes, stress calculations, etc.) that are in the computer's memory. A pen whose tip contains a tiny electric spark source is used to trace the surface of the object. The spark discharge generates a sharp acoustic pulse which propagates in all directions at known speed. Microphones at the four corners of a 75 × 150 cm plane surface (the pen can be as much as 180 cm away from the plane in the perpendicular direction) receive the pulses. Timing circuitry allows calculation of the four slant ranges, from which x, y, z coordinates can be calculated by a built-in microprocessor or the user's own general-purpose computer. Resolution is about 0.1 mm and up to 140 points per second can be digitized. If the spark source is attached at a particular point on a moving object, the translational displacement, velocity, and acceleration of this point could be calculated since the time intervals between digitized displacement values are known.

4.4 RELATIVE VELOCITY, TRANSLATIONAL AND ROTATIONAL

We consider here devices for measuring the velocity of translation, along a line, of one point relative to another and the plane rotational velocity about a single axis of one line relative to another.

---

[1] Science Accessories Corp., Southport, Conn.

Calibration

The measurement of rotational (angular) velocity is probably more common than that of translational velocity. Since translation generally can be obtained from rotation by suitable gearing or mechanisms, we consider mainly the calibration of rotational devices.

Perhaps the most difficult area of angular rate measurements is the extremely slow rotations associated with inertial guidance equipment. Angular velocities of 1/day (earth's rate) and less are of interest. Electro-optical devices such as the Midarm system (discussed under displacement calibration) enable measurement (and thus calibration of other less accurate transducers) of these low angular velocities with an accuracy of the order of 0.0002 degree/h, using a measuring time of 1 min.

For higher angular and linear velocities, perhaps the most convenient calibration scheme uses a combination of a toothed wheel, a simple magnetic proximity pickup and an electronic EPUT (events per unit time) meter (see Fig. 4.62). The angular rotation is provided by some adjustable-speed drive of adequate stability. The toothed iron wheel passing under the proximity pickup produces an electric pulse each time one tooth passes. These pulses are fed to the EPUT meter, which counts them over an accurate period (say 1.00000 s), displays the result visually for a few seconds to enable reading, and then repeats the process. The stability of the rotational drive is easily checked by observing the variation of the EPUT meter readings from one sample to another. The inaccuracy of pulse counting is ± 1 pulse plus the error in the counter time base, which is of the order of 1 ppm. The overall accuracy achieved depends on the stability of the motion source, the angular velocity being measured, and the number of teeth on the wheel. If the motion source were *absolutely* stable (no change in velocity whatever), very accurate measurement could be achieved simply by counting pulses

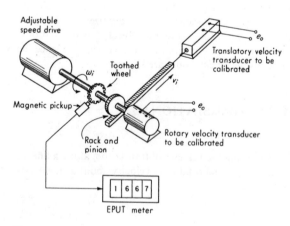

Figure 4.62 Velocity-calibration setup.

over a long time, since then the average velocity and the instantaneous velocity would be identical. If the motion source has some drift, however, the time sample must be fairly short. For example, a shaft rotating at 1,000 r/min with a 100-tooth wheel produces 1,667 pulses in a 1-s sample period. The inaccuracy here would be 1 part in 1,667 (the 1-ppm time-base error is totally negligible), or 0.06 percent. If the shaft rotated at 10 r/min, the error would be 6 percent. Slow rotations can be measured accurately by such means if the toothed wheel is placed on a shaft which is sufficiently geared up from the shaft driving the transducer being calibrated.

The above procedure uses relatively simple equipment and generally provides entirely adequate accuracy. Other simpler and less accurate procedures can be employed if they are adequate for their intended purpose. These usually consist of simply comparing the reading of a velocity transducer known to be accurate with the reading of the transducer to be calibrated when both are experiencing the same velocity input.

Velocity by Electrical Differentiation of Displacement Voltage Signals

The output of any displacement transducer may be applied to the input of a suitable differentiating circuit to obtain a voltage proportional to velocity (see Chap. 10 for differentiating circuits). The main problem is that differentiation accentuates any low-amplitude, high-frequency noise present in the displacement signal. Thus a carbon-film potentiometer would be preferable to the wirewound type, and demodulated and filtered signals from ac transducers may cause trouble because of the remaining ripple at carrier frequency. Workable systems using electrical differentiation are possible, however, with adequate attention to details.

Average Velocity from Measured Δx and Δt

Often a value of average velocity over a short distance or time interval is adequate, and a continuous velocity/time record is not required. A useful basic method is somehow (optically, magnetically, etc.) to generate a pulse when the moving object passes two locations whose spacing is accurately known. If the velocity were constant, any spacing could be used, with large spacing, of course, leading to greater accuracy. If the velocity is varying, the spacing Δx should be small enough that the average velocity over Δx is not very different from the velocity at either end of Δx. The same technique is applicable to rotational motion.

Figure 4.63a shows the application of a variable-reluctance proximity pickup such as was used in Fig. 4.62. When magnetic material passes close in front of the face of the pickup, the reluctance of the magnetic path changes with time, generating a voltage in the coil. These pickups are simple and cheap and give a large output voltage (often several volts) under typical operating conditions. The output voltage increases with velocity and closeness of the external moving iron to the pickup. Display of the two pulses on a single sweep of an oscilloscope with

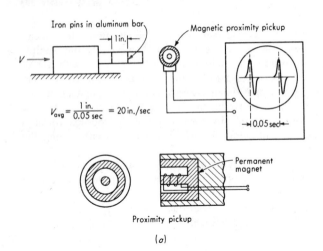

$$V_{avg} = \frac{1\ in.}{0.05\ sec} = 20\ in./sec$$

Iron pins in aluminum bar

1 in.

V

Magnetic proximity pickup

0.05 sec

Permanent magnet

Proximity pickup

(a)

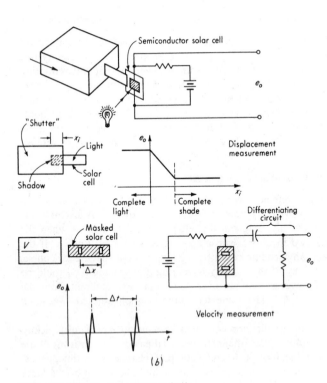

Semiconductor solar cell

e_o

"Shutter"

x_i

Light

Solar cell

Shadow

e_o

Displacement measurement

Complete light

Complete shade

x_i

Masked solar cell

V

Δx

Differentiating circuit

e_o

e_o

Δt

t

Velocity measurement

(b)

Figure 4.63 Velocity measurement as $\Delta x / \Delta t$.

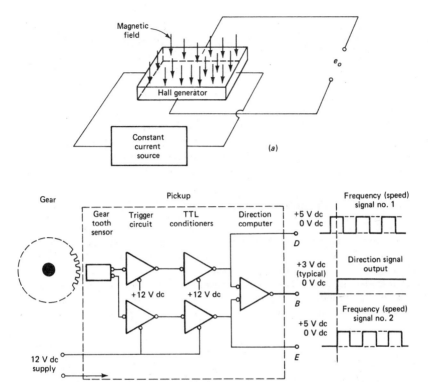

Figure 4.64 Hall-effect proximity pickup.

a calibrated time base allows measurement of the average velocity. Greater accuracy may be achieved by applying the voltage pulses to an electronic time-interval meter. In some velocity-measurement or proximity-detecting applications, pickups of this type are unsatisfactory because signal size decreases with velocity and is zero for a stationary object. We might then use proximity pickups of the eddy-current or Hall-effect type. Eddy-current proximity pickups are similar in principle to the eddy-current displacement transducers discussed earlier, except that a switching (rather than a linearly proportional) output is desired. When a conductive metal object approaches, the pickup's oscillator experiences a reduction in amplitude sufficient to trigger a switching signal.[1] Hall-effect devices utilize the Hall principle of Fig. 4.64a, where a suitable material (originally gold, but today a semiconductor) supplied with constant current produces an output

[1] Mini-Prox II, Electro Corp., Sarasota, Fla.

voltage e_o when a transverse magnetic field is applied. Proximity pickups based on this principle may utilize a permanent magnet [1] attached to the moving object or else require a ferrous target whose approach changes the reluctance of an internal magnetic cirucit whose flux the Hall sensor feels.[2] In either case, the Hall element output voltage triggers a standard TTL logic signal as output. Figure 4.64b shows a version using two mechanically offset Hall sensors which provide also a direction-of-rotation signal.

Electro-optical techniques also can be employed to generate the pulses necessary in such measurement schemes; Fig. 4.63b shows one version that uses solar cell transducers. Photography of the motion, by using a stroboscopic lamp flashing at a known rate, also provides velocity data of this type.

Mechanical Flyball Angular-Velocity Sensor

A classical rotary speed-measuring device still in wide use today, especially as a measuring element of industrial speed-control systems for engines, turbines, etc., is the flyball. Figure 4.65 shows the general arrangement schematically. Since the centrifugal force varies as the square of input velocity ω_i, the output x_o will not vary linearly with speed if an ordinary linear spring is used. For *small* changes in ω_i, a linearized model may be used to show that the transfer function between ω_i and x_o is essentially of the form

$$\frac{x_o}{\omega_i}(D) = \frac{K}{D^2/\omega_n^2 + 2\zeta D/\omega_n + 1} \tag{4.103}$$

The nonlinear static relation between ω_i and x_o for large speed changes may be acceptable in some systems. Where it is not, a nonlinear spring with $F_{spring} = K_s x_o^2$ can be used to get a linear overall characteristic since, at balance,

$$\text{Centrifugal force} = \text{spring force}$$

$$F_c = K_c \omega_i^2 = F_{spring} = K_s X_o^2 \tag{4.104}$$

and thus $x_o = \sqrt{K_c/K_s}\, \omega_i$, a linear relationship.

A variation[3] on this principle uses a pneumatic force-balance system to replace the spring and produces a standard 3 to 15 lbf/in^2 gage air-pressure signal proportional to ω_i^2. Since this is the standard pressure range of industrial-process-control systems, this speed transducer can be directly incorporated into such systems.

---

[1] "Handbook for Applying Solid State Hall Effect Sensors," Honeywell Micro Switch Division, Freeport, Ill., 1976.

[2] Zero Velocity Digital Pickup, Airpax Corp., Ft. Lauderdale, Fla.

[3] G. C. Carroll, "Industrial Process Measuring Instruments," p. 187, McGraw-Hill, New York, 1962.

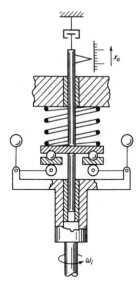

Figure 4.65 Flyball velocity pickup.

Mechanical Revolution Counters and Timers

When continuous reading and an electric output signal are not required, a variety of mechanical revolution counters (with or without built-in timers) are available. They are generally supplied with a variety of rubber-tipped wheels which transmit by friction the motion to be measured to the counter input shaft.

Magnetic and Photoelectric Pulse-Counting Methods

The arrangement of Fig. 4.62, using magnetic pickups (or photocells and light sources with slotted wheels or black-and-white targets) and discussed under Calibration, is often used for measurement since the equipment needed is quite widely available in industry today. Tachometer encoders (Fig. 4.56) are another possibility. If an analog signal (varying ac voltage) proportional to speed is desired, electronic frequency-to-voltage converters (see Chap. 10) can be connected to the pickup output terminals.

Stroboscopic Methods[1]

Rotational velocity may be conveniently measured by using electronic stroboscopic lamps which flash at a known and adjustable rate. The light is directed onto the rotating member which itself usually has spokes, gear teeth, or some

---

[1] "Handbook of Stroboscopy," GenRad, Concord, Mass.

other feature enabling "lock on." If not, a simple black-and-white paper target can be attached. The frequency of lamp flashing is adjusted until the "target" appears motionless. At this setting, the lamp frequency and motion frequency are identical, and the numerical value can be read from the lamp's calibrated dial to an inaccuracy of about ± 1 percent of the reading (0.01 percent in some units with crystal-controlled time base). The range of lamp frequency of a typical unit[1] is 110 to 25,000 flashes per minute. Speeds greater than 25,000 r/min can be measured by the following technique. Synchronism can be achieved at any flashing rate r that is an integral submultiple of the speed to be measured, n. The flashing rate is adjusted until synchronism is achieved at the largest possible flashing rate, say r_1. Then the flashing rate is slowly decreased until synchronism is again achieved at a rate r_2. The unknown speed n is then given by

$$n = \frac{r_1 r_2}{r_1 - r_2} \tag{4.105}$$

For very high speeds, r_1 and r_2 are close together, giving poor accuracy. Accuracy can be improved by reducing the flashing rate below r_2 until synchronism is again achieved. This procedure can be continued until synchronism is obtained N times $(r_1, r_2, r_3, \ldots, r_N)$. The speed n is given by

$$n = \frac{r_1 r_N (N - 1)}{r_1 - r_N} \tag{4.106}$$

This procedure can extend the upper range to about 250,000 r/min.

Translational-Velocity Transducers (Moving-Coil and Moving-Magnet Pickups)

The moving-coil pickup of Fig. 4.66 is based on the law of induced voltage.

$$e_o = (Blv_i)10^{-8} \tag{4.107}$$

where e_o = terminal voltage, V
B = flux density, G
l = length of coil, cm
v_i = relative velocity of coil and magnet, cm/s

Since B and l are constant, the output voltage follows the input velocity linearly and reverses polarity when the velocity changes sign. Such pickups are widely used for the measurement of vibratory velocities. Since the flux density available from permanent magnets is limited to the order of 10,000 G, an increase in sensitivity can be achieved only by an increase in the length of wire in the coil. To keep the coil small, this requires fine wire and thus high resistance. High-resistance coils require a high-resistance voltage-measuring device at e_o to prevent loading. A typical pickup of about 500-Ω resistance has a sensitivity of

---

[1] Strobotac, GenRad, Concord, Mass.

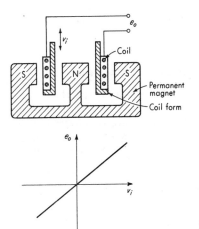

Figure 4.66 Moving-coil velocity pickup.

0.15 V/(in/s) and a full-scale displacement of 0.15 in with a nonlinearity of ±1 percent. A more sensitive coil used in a seismometer (instrument to measure earth shocks) has 500,000-Ω resistance and a sensitivity of 115 V/(in/s).

The transducer shown in Fig. 4.67 uses a permanent-magnet core moving inside a form wound with two coils connected as shown. Units are available in

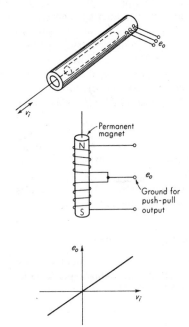

Figure 4.67 Velocity pickup. (*Courtesy Schaevitz Engineering, Pennsauken, N. J.*)

full-range strokes from about 0.05 to 20 in. Sensitivity varies from about 0.5 to 0.05 V/(in/s), and nonlinearity is about 1 percent.

DC Tachometer Generators for Rotary-Velocity Measurement

An ordinary dc generator (using either a permanent magnet or separately excited field) produces an output voltage roughly proportional to speed. By emphasizing certain aspects of design, such a device can be made an accurate instrument for measuring speed rather than a machine for producing power. The basic principle is again Eq. (4.107), which when applied to the rotational configuration of a dc generator becomes

$$e_o = \frac{n_p n_c \phi N}{60 n_{pp}} 10^{-8} \qquad (4.108)$$

where $e_o \triangleq$ average output voltage, V
$n_p \triangleq$ number of poles
$n_c \triangleq$ number of conductors in armature
$\phi \triangleq$ flux per pole, lines
$N \triangleq$ speed, r/min
$n_{pp} \triangleq$ number of parallel paths between positive and negative brushes

The voltage e_o is a dc voltage proportional to speed which reverses polarity when the angular velocity reverses. A small superimposed ripple voltage is present because of the finite number of conductors. While low-pass filtering is effective in reducing ripple at high speeds, this is not usually practical at low speeds and approaches such as gearing should be tried. A typical high-accuracy unit[1] (permanent magnet) has a sensitivity of 7 V per/1,000 r/min, a rated speed of 5,000 r/min, nonlinearity of 0.07 percent over a range 0 to 3,600 r/min, ripple voltage 2 percent of average voltage for speeds above 100 r/min, friction torque of 0.2 in · oz, rotor inertia of 7 g · cm^2, output impedance of 2,800 Ω, and a total weight of 3 oz.

A special dc tachometer[2] of unique design for use where a limited ($\pm 15°$) angular travel is acceptable exhibits a very high sensitivity. A 1-in-diameter model gives 500 V per 1,000 r/min while a 3-in-diameter gives 30,000 V per 1,000 r/min. The nonlinearity is ± 9 percent for $\pm 15°$ travel, and the operating torque is 500 g · cm. In this generator the permanent magnet rotates while the coil is stationary, and no commutator is needed because of the limited travel.

AC Tachometer Generators for Rotary-Velocity Measurement

An ac two-phase squirrel-cage induction motor can be used as a tachometer by exciting one phase with its usual ac voltage and taking the voltage appearing at the second phase as output. With the rotor stationary, the output voltage is

[1] General Precision Inc., Kearfott Div., Little Falls, N. J.
[2] Armstrong Whitworth Equipment, Hucclecote, Gloucester, England.

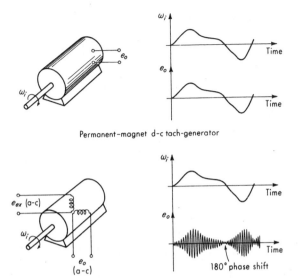

Permanent-magnet d-c tach-generator

A-c tach-generator

Figure 4.68 Tachometer generators.

essentially zero. Rotation in one direction causes at the output an ac voltage of the same frequency as the excitation and of an amplitude proportional to the instantaneous speed. This output voltage is in phase with the excitation. Reversal of rotation causes the same action, except the phase of the output shifts 180° (see Fig. 4.68). While squirrel-cage rotors sometimes are used, the most accurate units employ a drag-cup rotor. This does not change the basic operating characteristics.

A typical high-accuracy unit[1] is excited by 115 V/400-Hz voltage, has a sensitivity of 2.8 V per 1,000 r/min, nonlinearity of 0.05 percent from 0 to 3,600 r/min, negligible rotor friction, rotor inertia of 7 g · cm^2, and a total weight of 6.7 oz. Most commercial ac tachometers are designed to be used in ac servomechanisms which conventionally operate on either 60 or 400 Hz, and so they are generally designed for operation at these frequencies. For general-purpose motion measurement, the frequency response of such units is limited (as are all ac or "carrier"-type devices) by the carrier frequency to about one-tenth to one-fifth of the carrier frequency. It is usually possible, however, to excite a tachometer designed for 400 Hz at considerably higher frequencies if necessary, although some of or all the performance characteristics may change value.

Eddy-Current Drag-Cup Tachometer

Figure 4.69 shows schematically an eddy-current tachometer. Rotation of the magnet induces voltages into the cup which thereby produce circulating eddy

[1] General Precision Inc., Kearfott Div., Little Falls, N.J.

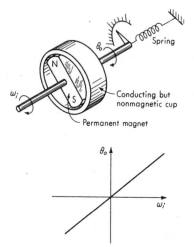

Figure 4.69 Drag-cup velocity pickup.

currents in the cup material. These eddy currents interact with the magnet field to produce a torque on the cup in proportion to the relative velocity of magnet and cup. This causes the cup to turn through an angle θ_o until the linear spring torque just balances the magnetic torque. Thus in steady state the angle θ_o is directly proportional to ω_i, the input velocity. If an electric signal is desired, any low-torque displacement transducer can be used to measure θ_o. Dynamic operation is governed by the rotary inertia of parts moving with θ_o, spring stiffness, and the viscous damping effect of the eddy-current coupling between magnet and cup, leading to a second-order response of the form

$$\frac{\theta_o}{\omega_i}(D) = \frac{K}{D^2/\omega_n^2 + 2\zeta D/\omega_n + 1}$$

(4.109)

Nonlinearity of the order of 0.3 percent can be achieved in such units.

4.5 RELATIVE-ACCELERATION MEASUREMENTS

Transducers directly sensitive to relative acceleration are rare. Figure 4.70 shows one type commercially available.[1] When the shaft rotates at constant speed, eddy currents generated in the rotating conductive disk are constant and proportional to speed. These currents produce a magnetic field which links the two pickup coils, but no coil voltage is produced because the field is steady. When the shaft accelerates, a changing magnetic field causes coil output voltages which are pro-

---

[1] L. S. Hoodwin, Angular Accelerometer: Advantages and Limitations, *Inst. & Cont. Syst.*, p. 129, April 1967; Hoodwin Instruments, Sawyer, Mich.

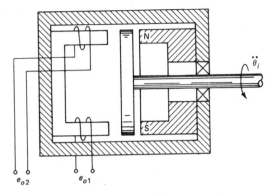

Figure 4.70 Relative-acceleration pickup.

portional to acceleration. Speed of response is limited by both electric resonance in the coils and torsional mechanical resonance. Connecting coils in series increases sensitivity by a factor of 4 (compared to parallel) but reduces electric resonant frequency by about the same amount. Sensitivities range from about 0.05 to 10.0 mV/(rad/s^2) with frequency response flat from dc to 200 to 1000 Hz for various models. Accelerations to thousands of radians per second squared and over 5,000-r/min shaft speed are possible.

Other possibilities for relative-acceleration measurement include graphical or electrical differentiation of displacement or velocity signals or records. The double differentiation required of displacement records or voltages can rarely be performed accurately except with very smooth signals. Single differentiation of velocity signals may be practical in some instances.

4.6 SEISMIC- (ABSOLUTE-) DISPLACEMENT PICKUPS

While the most general motion of a rigid body involves three-dimensional translation and rotation, the total vector quantities usually are not amenable to direct measurement. So we employ an array of uniaxial transducers oriented along selected axes to measure orthogonal components of the vectors, which are then combined by calculation to define the total vector magnitudes and directions.

Figure 4.71 shows the general construction of a seismic-displacement pickup for uniaxial translatory or rotary motions. These devices are used almost exclusively for measurement of vibratory displacement in those (many) cases where a fixed reference for relative-displacement measurement is not available. That is, the vibration of a body can be measured with any of the relative-motion transducers discussed earlier in this chapter, but only if one end of the transducer can be attached to a stationary reference. For measurements on moving vehicles, such references are not generally available, and in many other situations measurement

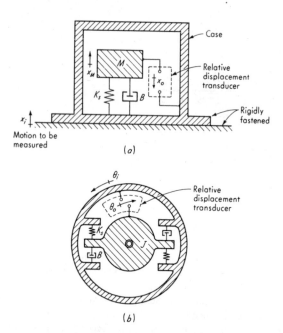

Figure 4.71 Translational and rotational seismic pickups.

of absolute motion is easier and more desirable. The basic principle of seismic-(absolute-) displacement pickups is simply to measure (with any convenient relative-motion transducer) the relative displacement of a mass connected by a soft spring to the vibrating body. For frequencies above the natural frequency, this relative displacement is also very nearly the absolute displacement since the mass tends to stand still.

To obtain a quantitative measure of performance for such systems, we analyze the configuration of Fig. 4.71a. The rotational configuration is completely analogous. Newton's law may be applied to the mass M as follows:

$$K_s x_o + B\dot{x}_o = M\ddot{x}_M = M(\ddot{x}_i - \ddot{x}_o) \tag{4.110}$$

where x_i and x_M are the absolute displacements and we have chosen our reference for x_o such that x_o is zero when the gravity force (weight of M) is acting along the x axis statically. Manipulation gives

$$\frac{x_o}{x_i}(D) = \frac{D^2/\omega_n^2}{D^2/\omega_n^2 + 2\zeta D/\omega_n + 1} \tag{4.111}$$

where $\omega_n \triangleq \sqrt{\dfrac{K_s}{M}}$

$\zeta \triangleq \dfrac{B}{2\sqrt{K_s M}}$

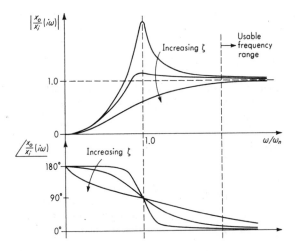

Figure 4.72 Seismic-displacement-pickup frequency response.

Since the pickup is intended mainly as a vibration sensor, the frequency response is of prime interest:

$$\frac{x_o}{x_i}(i\omega) = \frac{(i\omega)^2/\omega_n^2}{(i\omega/\omega_n)^2 + 2\zeta i\omega/\omega_n + 1} \tag{4.112}$$

This is graphed in Fig. 4.72. We note that there is no response to static displacement inputs and that ω_n should be much less than the lowest vibration frequency ω for accurate displacement measurement. For frequencies much above ω_n, $(x_o/x_i)(i\omega) \rightarrow 1 \angle 0°$, indicating perfect measurement. The characteristics of the relative-displacement transducer in converting x_o to a voltage e_o also must be considered. Since the force in spring K_s is directly proportional to x_o, if strain gages are used, they can be applied directly to this spring, which may be in the form of a cantilever beam. Since a low ω_n is desired, either a large mass or a soft spring (or both) is necessary. To keep size (and thereby loading on the measured system) to a minimum, soft springs are preferred to large masses. Intentional damping in the range $\zeta = 0.6$ to 0.7 is often employed to minimize resonant response to slow transients.

4.7 SEISMIC- (ABSOLUTE-) VELOCITY PICKUPS

Again the application here is limited to vibratory velocities, and the basic configuration is exactly the same as in Fig. 4.71. To measure velocity $\dot{x}_i$ rather than displacement x_i, three possibilities are considered. First, a voltage signal from a displacement pickup may be sent to an electric differentiation circuit. Second (and this is the most practical), the relative-displacement transducer of Fig. 4.71 is replaced by a relative-velocity transducer (usually a moving-coil pickup). Then, since $e_o = K_e \dot{x}_o$, we have

$$\frac{e_o}{\dot{x}_i}(D) = \frac{K_e D^2/\omega_n^2}{D^2/\omega_n^2 + 2\zeta D/\omega_n + 1} \tag{4.113}$$

and accurate velocity measurement is possible if $\omega \gg \omega_n$. Signals from such pickups may be readily integrated electrically to get displacement information. The third possibility is revealed by rewriting Eq. (4.111) as

$$\frac{x_o}{Dx_i}(D) = \frac{D}{D^2 + 2\zeta\omega_n D + \omega_n^2} \tag{4.114}$$

$$\frac{x_o}{\dot{x}_i}(i\omega) = \frac{1}{2\zeta\omega_n - i[(\omega_n^2 - \omega^2)/\omega]} \tag{4.115}$$

Now if we wish x_o to be a measure of $\dot{x}_i$, then $(x_o/\dot{x}_i)(i\omega) \approx$ constant. From Eq. (4.115) we see that this will be the case if $(\omega_n^2 - \omega^2)/\omega \approx 0$, since then

$$\frac{x_o}{\dot{x}_i}(i\omega) \approx \frac{1}{2\zeta\omega_n} \tag{4.116}$$

Now $(\omega_n^2 - \omega^2)/\omega \approx 0$ if $\omega \approx \omega_n$. This would allow measurement only at frequency ω_n. However, if ζ is made very large, the range of frequencies around ω_n for which $(\omega_n^2 - \omega^2)/\omega$ is negligible compared with $2\zeta\omega_n$ is fairly broad. Figure 4.73 shows this graphically. While a possibility, this approach is rarely employed in practice. The required large value of ζ reduces the sensitivity [see Eq. (4.116)], and *both* very low and very high frequencies are not accurately measured.

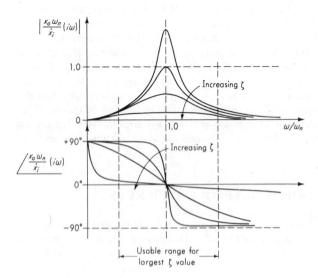

Figure 4.73 Seismic-velocity-pickup frequency response.

4.8 SEISMIC- (ABSOLUTE-) ACCELERATION PICKUPS (ACCELEROMETERS)

The most important pickup for vibration, shock, and general-purpose absolute-motion measurement is the accelerometer. This instrument is commercially available in a wide variety of types and ranges to meet correspondingly diverse application requirements. The basis for this popularity lies in the following features:

1. Frequency response is from zero to some high limiting value. Steady accelerations can be measured (except in piezoelectric types).
2. Displacement and velocity can be easily obtained by electrical integration, which is much preferred to differentiation.
3. Measurement of transient (shock) motions is more readily achieved than with displacement or velocity pickups.
4. Destructive forces in machinery, etc., often are related more closely to acceleration than to velocity or displacement.

The basic accelerometer configuration is again that of Fig. 4.71. The operating principle is as follows: Suppose the acceleration $\ddot{x}_i$ to be measured is constant. Then, in steady state, the mass M will be at rest relative to the case, and thus its absolute acceleration will also be $\ddot{x}_i$. If mass M is accelerating at $\ddot{x}_n$ there must be some force to cause this acceleration, and if M is not moving relative to the case, this force can come only from the spring. Since spring deflection x_o is proportional to force, which in turn is proportional to acceleration, x_o is a measure of acceleration $\ddot{x}_i$. Thus absolute-acceleration measurement is reduced to the measurement of the force required to accelerate a known mass (sometimes called the "proof" mass). This dependence on mass leads to problems (mainly in inertial guidance systems, not in vibration measurement) since a mass also experiences forces due to gravitational fields. Thus an accelerometer cannot distinguish between a force due to acceleration and a force due to gravity.

The majority of accelerometers may be classified as either deflection type or null-balance type. Those used for vibration and shock measurement are usually the deflection type whereas those used for measurement of gross motions of vehicles (submarines, aircraft, spacecraft, etc.) may be either type, with the null-balance being used when extreme accuracy is needed.

Deflection-Type Accelerometers

A large number of practical accelerometers have the configuration of Fig. 4.71 and differ only in details, such as the spring element used, relative-motion transducer employed, and type of damping provided. Since the desired input is now $\ddot{x}_i$, we can rewrite Eq. (4.111) as

$$\frac{x_o}{D^2 x_i}(D) = \frac{x_o}{\ddot{x}_i}(D) = \frac{K}{D^2/\omega_n^2 + 2\zeta D/\omega_n + 1} \qquad (4.117)$$

where $$K \triangleq \frac{1}{\omega_n^2} \quad \text{cm/(cm/s}^2) \tag{4.118}$$

Since output voltage $e_o = K_e x_o$ for many motion transducers, Eq. (4.117) has the correct form for the acceleration-to-voltage transfer function also. We see that the accelerometer is an ordinary second-order instrument; thus all our previous work on this type is immediately applicable. The frequency response extends from zero to some fraction of ω_n, depending on the accuracy required and the damping. Because sensitivity $K = 1/\omega_n^2$, high-frequency response must be traded for sensitivity. Since the dynamic characteristics of second-order instruments have been thoroughly discussed, here we discuss mainly the specific characteristics of commercially available instruments.

Accelerometers using resistive potentiometers as their motion pickup are intended mainly for slowly varying accelerations and low-frequency vibration. A typical family[1] of such instruments offers nine models covering the range of $\pm 1g$ full scale to $\pm 50g$ full scale. The natural frequencies range from 12 to 86 Hz, and ζ is 0.5 to 0.8 over the temperature range -65 to $+165°$F, using a temperature-compensated liquid damping arrangement. Potentiometer resistance may be selected in the range 1,000 to 10,000 Ω, with corresponding resolution of 0.45 to 0.25 percent of full scale. The potentiometer power rating is 0.5 W at $+165°$F. The sensitivity to acceleration at right angles to the desired axis (cross-axis sensitivity) is less than ± 1 percent of the sensitivity to the desired direction. The operating life is 2,000,000 cycles. Overall inaccuracy is ± 1 percent of full scale or less at room temperature. This increases to ± 1.8 percent if the temperature is allowed to vary over the design range of -65 to $+165°$F. Size is about a 2-in cube; weight is about 1 lb.

Unbonded-strain-gage accelerometers use the strain wires as the spring element and as the motion transducer. They are useful for general-purpose motion measurement and for vibration up to relatively high frequencies. They are available in a wide range of characteristics, typical[2] values including ± 0.5 to $\pm 200g$ full scale, natural frequency 17 to 800 Hz, excitation voltage about 10 V alternating or direct current, full-scale output ± 20 to ± 50 mV, resolution less than 0.1 percent, inaccuracy 1 percent of full scale or less, cross-axis sensitivity less than 2 percent, and damping ratio (using silicone-oil damping) of 0.6 to 0.8 at room temperature. (Temperature-compensated models are also available.) These instruments can be made quite small and light, a typical size being $\frac{1}{2} \times \frac{1}{2} \times 2$ in, with a weight of 26 g.

Bonded-strain-gage accelerometers generally use a mass supported by thin flexure beams, with strain gages cemented to the beam so as to achieve maximum sensitivity, temperature compensation, and insensitivity to cross-axis and angular acceleration. Their characteristics are similar to those of unbonded-gage accelerometers except that size and weight tend to be greater. Silicone-oil damping is

[1] Bourns, Inc., Riverside, Calif.
[2] Gould Inc., Oxnard, Calif.

again widely used. Semiconductor strain gages are widely used as strain sensors in cantilever-beam/mass types of accelerometers, allowing high outputs (0.2 to 0.5 V full scale) and miniaturized designs. A $\pm 25g$ unit has flat response from 0 to 750 Hz, damping ratio 0.7, mass 28 g, and compensated temperature range -18 to $+93°C$.[1] A triaxial $\pm 20,000g$ model[1] has flat response from 0 to 15,000 Hz, damping ratio 0.01, compensated temperature range 0 to 45°C, 13 × 10 × 13 mm size, and 10-g mass.

A family of liquid-damped differential-transformer accelerometers[2] exhibits the following characteristics: full-scale ranges from ± 2 to $\pm 700g$, natural frequency from 35 to 620 Hz, nonlinearity 1 percent of full scale, full-scale output about 1 V with excitation of 10 V at 2,000 Hz, damping ratio 0.6 to 0.7 at 70°F, residual voltage at null less than 1 percent of full scale, and hysteresis less than 1 percent of full scale. The size is about a 2-in cube, with a weight of 4 oz.

Piezoelectric accelerometers are in wide use for shock and vibration measurements. In general, they do not give an output for constant acceleration because of the basic characteristics of piezoelectric motion transducers. But they do have large output-voltage signals, small size, and can have very high natural frequencies, which are necessary for accurate shock measurements. No intentional damping is provided, with material hysteresis being the only source of energy loss. This results in a very low (about 0.01) damping ratio, but this is acceptable because of the very high natural frequency. The transfer function is a combination of Eqs. (4.70) and (4.117):

$$\frac{e_o}{\ddot{x}_i}(D) = \frac{[K_q/(C\omega_n^2)]\tau D}{(\tau D + 1)(D^2/\omega_n^2 + 2\zeta D/\omega_n + 1)} \tag{4.119}$$

The low-frequency response is limited by the piezoelectric characteristic $\tau D/(\tau D + 1)$ while the high-frequency response is limited by mechanical resonance. The damping ratio ζ of piezoelectric accelerometers is not usually quoted by the manufacturer, but can be taken as zero for most practical purposes. The accurate (5 percent high at the high-frequency end and 5 percent low at the low-frequency end) frequency range of such an accelerometer is $3/\tau < \omega < 0.2\omega_n$. Accurate low-frequency response requires large τ, which is usually achieved by use of high-impedance voltage amplifiers or charge amplifiers. Systems designed for response below about 1.0 Hz and subject to temperature transients may exhibit errors because of the pyroelectric effect present in most piezoelectric materials. Here a charge output is produced in response to a temperature input. For systems with negligible response at low frequencies, these temperature-induced signals (because they are "slow" transients) produce little output. But serious errors may occur if τ has been made large to measure low-frequency accelerations and if the accelerometer has not been designed to minimize thermal effects.

The design details of piezoelectric accelerometers can be varied to emphasize

[1] Endevco Corp., San Juan Capistrano, Calif.
[2] Schaevitz Engineering, Camden, N. J.

selected features of performance desired for particular applications. No single configuration is ideal for all situations since tradeoffs exist here just as in all engineering design. Figure 4.74 shows several designs from a single manufacturer. The basic compression design (Fig. 4.74a) is the simplest and most rugged and has the best mass/sensitivity ratio; but because the housing acts as an integral part of the spring/mass system, this type is more sensitive to spurious inputs. [For piezoelectric accelerometers these include temperature,[1] acoustic noise, "base bending"[2] (surface strains caused by bending at the mounting surface), cross-axis motion, and magnetic fields.] The spring is generally preloaded to work the piezoelectric material at a more linear portion of its charge-strain curve. This preload also allows measurement of both positive and negative accelerations without putting the piezo material in tension. That is, the initial preload causes an output voltage of a certain polarity. However, this immediately leaks off, and the polarity of subsequent acceleration-induced voltages will follow the direction of the motion since the charge polarity depends on the *change* in strain, not its total value. The preload is made large enough that it is never relaxed by even the largest input acceleration.

Reduced response to spurious inputs is achieved in the "single-ended compression" or center-mounted designs of Fig. 4.74b and c. The inverted design of Fig. 4.74c is especially tolerant of base bending strains, as is the shear design of Fig. 4.74d. Shear designs using bolted stacks of flat plate elements have been introduced recently to gain further improvements in performance. The Delta Shear unit of Fig. 4.74 is a variation which uses a shrink-fitted collar, rather than bolts, to hold the elements fast.

Microcircuit electronics has allowed the design of piezoelectric accelerometers with charge amplifiers (see Chap. 10) built into the instrument housing. A single two-conductor cable which carries both amplifier power and the measurement signal connects the instrument to a simple constant-current power supply, which also provides a high-level (a few volts) output signal directly to oscilloscopes or signal analyzers. This arrangement allows greater sensitivity with a smaller, higher-frequency accelerometer, reduces cable-noise effects and length limitations, and lowers costs. These advantages are traded for a reduced temperature range (the microcircuit electronics is more temperature-limited than the accelerometer itself) and less versatile signal conditioning (the built-in amplifiers allow little or no adjustment).

Piezoelectric accelerometers are available in a very wide range of specifications from a large number of manufacturers. The sensitivity-frequency response tradeoff is apparent in typical specifications; a shock accelerometer may have 0.004 pC/g and natural freuqency of 250,000 Hz, while a unit designed for low-level seismic measurements has 1,000 pC/g and 7,000 Hz. The smallest units

[1] C. F. Vezzeti and P. S. Lederer, An Experimental Technique for the Evaluation of Thermal Transient Effects on Piezoelectric Accelerometers, *NBS Tech. Note* 855, January 1975.
[2] Tech. Data A510, 508A, 509A, Endevco Corp., San Juan Capistrano, Calif.

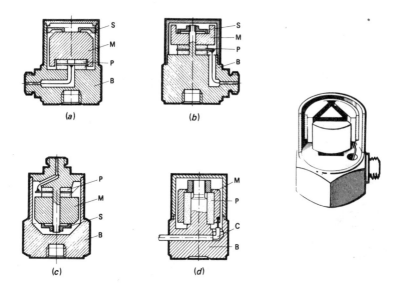

S: Spring   M: Mass   P: Piezoelectric element   B: Base   C: Cable

Figure 4.74 Piezoelectric accelerometer designs. (*a*) Peripheral mounted compression design. (*b*) Center mounted compression design. (*c*) Inverted center-mounted compression design. (*d*) Shear design. (*Courtesy Bruel and Kjaer Instruments, Marlboro, Mass.*)

(needed to reduce mass-loading errors in measurements on light structures) are about 3 × 3 mm with 0.5-g mass including cable. Response to spurious thermal and base-bending inputs for the best isolated-shear designs may be 200 times less than for designs not optimized in this respect. Triaxial units as small as a 7-mm cube with 1-g mass are available. Uncooled instruments usable over a temperature range of -450 to $+1500°F$ typically exhibit a ± 10 percent sensitivity change from -100 to $+1500°F$ and may be designed to survive radiation environments typical of nuclear reactors.

Piezoelectric accelerometers tend to have somewhat larger cross-axis sensitivities than other types; however, this is typically held to about 2 to 4 percent and usually is not a critical factor. Some manufacturers indicate the location of the cross axis of *least* transverse sensitivity, allowing the user to orient the instrument so as to minimize this effect in each installation. Accelerometers may be mounted with threaded studs (the preferred method), with cement or wax adhesives, or with magnetic holders. The main effect of the various mounting methods is a reduction of the natural frequency below that of the unmounted accelerometer itself, as a result of elastic and inertial characteristics of the mounting. The

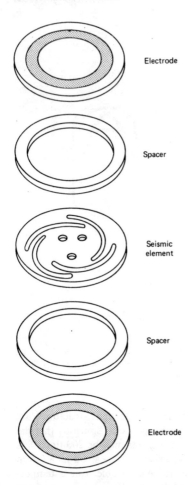

Figure 4.75 Capacitive accelerometer design. (*Courtesy Setra Systems, Natick, Mass.*)

unmounted resonant frequency is measured by suspending the accelerometer freely from its cable, exciting the piezo element electrically (sinusoidally), and finding the frequency at which current and voltage are in phase. Mounted natural frequency is the frequency of peak voltage output when the accelerometer is mechanically vibrated sinusoidally. For a solid steel stud (the optimum mounting), the mounted natural frequency is typically about 60 percent of the unmounted. Experience has shown that for measurements above 5 kHz, the steel-stud mounting is significantly improved when a drop of light oil is applied between the mating surfaces before torquing the connection tight. Apparently a stiffer elastic coupling is achieved by the oil filling microscopic voids at the interface. When electrically insulated studs are used to reduce electric noise problems arising from ground loops, the lower stiffness of the insulating material causes reduced frequency response. Data on this and other mounting effects are available in the literature.[1]

Figure 4.75 shows an accelerometer design using the capacitance displacement transducer of Fig. 4.37. A thin diaphragm with spiral flexures provides the spring, proof mass, and moving plate of the differential capacitor. Plate motion between the electrodes "pumps" air parallel to the plate surface and through holes in the plate to provide squeeze-film damping. Since air viscosity is less temperature-sensitive than oil, the desired 0.7 damping ratio changes only about 15 percent per 100°F. A family of such instruments with full-scale ranges from $\pm 0.2g$ (4-Hz flat response) to $\pm 1,000g$ (3000 Hz), cross-axis sensitivity less than 1.0 percent, and full-scale output ± 1.5 V are available in a case size about a 1-in cube weighing 1.7 oz.

While angular accelerometers based on the configuration of Fig. 4.71b are available, the selection of manufacturers and models is much less extensive than for the translational units. Angular accelerometers of the servotype are available from several manufacturers and are discussed in the next section. It is also possible to use an array of translational accelerometers to measure rotational motions. Figure 4.76 shows the most general situation, where nine translational accelerometers are attached to a fixture which is itself then fastened to the rigid body whose rotational motion is to be measured. By suitable processing of the signals from the nine accelerometers, it has been shown[2] that one can compute the three perpendicular components of both angular velocity and angular acceleration for the body-fixed axes shown.

We conclude this section with Fig. 4.77, which provides guidance for accelerometer selection to meet dynamic-response requirements.

Null-Balance- (Servo-) Type Accelerometers[3]

So-called servoaccelerometers using the principle of feedback have been developed for applications requiring greater accuracy than is generally achieved with instruments using mechanical springs as the force-to-displacement transducer. In these null-balance instruments, the acceleration-sensitive mass is kept very close to the zero-displacement position by sensing this displacement and generating a magnetic force which is proportional to this displacement and which always opposes motion of the mass from neutral. This restoring force plays the same role

[1] G. K. Rasamen and B. M. Wigle, Accelerometer Mounting and Data Integrity, *Sound & Vib.*, pp. 8–15, November 1967.

[2] A. I. King, A. J. Padgaonkar, and K. W. Krieger, Measurement of Angular Acceleration of a Rigid Body Using Linear Accelerometers, *Biomechanics Res. Center Rep.*, Wayne State University, Detroit, Mich., December 6, 1974.

[3] E. J. Jacobs, New Developments in Servo Accelerometers, *Proc. Inst. of Environ. Sci., 14th Ann. Meeting*, April 29, 1968, St. Louis (Endevco Corp., Reprint); C. E. Bosson and D. W. Busse, Aotonetics Microminiature Digital Accelerometer, *Proc. Inst. of Navig., Natl. Space Nav. Mtg.*, Boston, April 21–22, 1966 (Autonetics Corp., Reprint); H. K. P. Neubert, "Instrument Transducers," chap. 5, Oxford University Press, London, 1963.

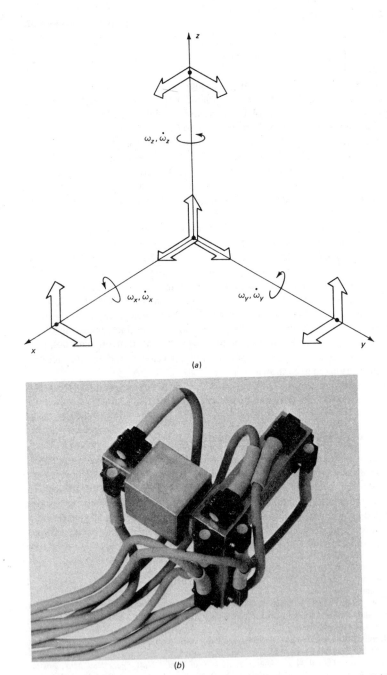

(a)

(b)

Figure 4.76 Nine-accelerometer array. (*Courtesy Endevco Corp., San Juan Capistrano, Calif.*)

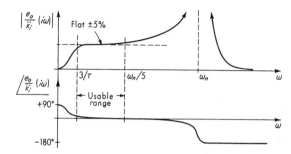

Piezoelectric accelerometer frequency response

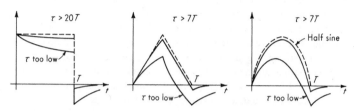

Requirements for accurate peak measurements ± 5 %

Low-frequency response problems (piezoelectric only)

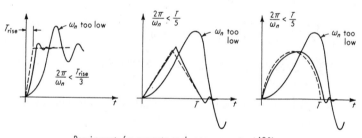

Requirements for accurate peak measurements ± 10 %

High-frequency response problems (all accelerometers)

Figure 4.77 Accelerometer selection criteria.

as the mechanical spring force in a conventional accelerometer. Thus we may consider the mechanical spring to have been replaced by an electrical "spring." The advantages derived from this approach are the greater linearity and lack of hysteresis of the electrical spring as compared with the mechanical one. Also, in some cases, electrical damping (which can often be made less temperature-sensitive than mechanical damping) may be employed. There is also the possibility of testing the static and dynamic performance of the device just prior to a test run by introducing electrically excited test forces into the system. This convenient and rapid remote self-checking feature can be quite important in complex and expensive tests where it is extremely important that all systems operate correctly before the test is commenced. These servoaccelerometers usually are used for general-purpose motion measurement and low-frequency vibration. Also they are particularly useful in acceleration-control systems since the desired value of acceleration can be put into the system by introducing a proportional current i_a from an external source.

Figure 4.78 illustrates in simplified fashion the operation of a typical instrument designed to measure a translational acceleration $\ddot{x}_i$. (Angular acceleration also can be measured by these techniques by using an obvious mechanical modification.) The acceleration $\ddot{x}_i$ of the instrument case causes an inertia force f_i on the sensitive mass M, tending to make it pivot in its bearings or flexure mount. The rotation θ from neutral is sensed by an inductive pickup and is amplified, demodulated, and filtered to produce a current i_a directly proportional to the motion from null. This current is passed through a precision stable resistor R to produce the output-voltage signal and is applied to a coil suspended in a magnetic field. The current through the coil produces a magnetic torque on the coil (and the attached mass M) which acts to return the mass to neutral. The current required to produce a coil magnetic torque that just balances the inertia torque due to $\ddot{x}_i$ is directly proportional to $\ddot{x}_i$; thus e_o is a measure of $\ddot{x}_i$. Since a nonzero displacement θ is necessary to produce a current i_a, the mass is not returned exactly to null, but comes very close because a high-gain amplifier is used. Analysis of the block diagram reveals the details of performance as follows:

$$\left(Mr\ddot{x}_i - \frac{e_o K_c}{R}\right)\frac{K_p K_a/K_s}{D^2/\omega_{n1}^2 + 2\zeta_1 D/\omega_{n1} + 1} = \frac{e_o}{R} \tag{4.120}$$

$$\left(\frac{D^2}{\omega_{n1}^2} + \frac{2\zeta_1 D}{\omega_{n1}} + 1 + \frac{K_c K_p K_a}{K_s}\right)e_o = \frac{MrRK_p K_a}{K_s}\ddot{x}_i \tag{4.121}$$

Now, by design, the amplifier gain K_a is made large enough so that $K_c K_p K_a/K_s \gg 1.0$; then

$$\frac{e_o}{\ddot{x}_i}(D) = \frac{K}{D^2/\omega_n^2 + 2\zeta D/\omega_n + 1} \tag{4.122}$$

where
$$K \triangleq \frac{MrR}{K_c} \qquad V/(m/s^2) \tag{4.123}$$

$$\omega_n \triangleq \omega_{n1}\sqrt{\frac{K_p K_a K_c}{K_s}} \qquad rad/s \tag{4.124}$$

$$\zeta \triangleq \frac{\zeta_1}{\sqrt{K_p K_a K_c/K_s}} \tag{4.125}$$

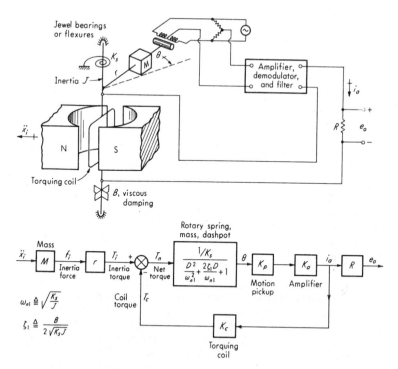

Figure 4.78 Servotype accelerometer.

Equation (4.123) shows that the sensitivity now depends on only M, r, R, and K_c, all of which can be made constant to a high degree. This demonstrates again the usefulness of high-gain feedback in shifting the requirements for accuracy and stability from many components to a chosen few where the requirements can be met. As in all feedback systems, the gain cannot be made arbitrarily high because of dynamic instability; however, a sufficiently high gain can be achieved to give excellent performance. Turning to Eq. (4.124), we see that ω_n is increased from the basic spring-mass frequency ω_{n1} by the factor $\sqrt{K_p K_a K_c / K_s}$, another benefit of high-gain feedback. However, ζ is decreased by the same factor, and so ζ_1 must be made sufficiently high to compensate for this.

A typical accelerometer[1] of this kind is available in full-scale ranges of ± 0.5 to $\pm 100g$, natural frequency of 50 to 250 Hz, damping ratio of 0.7 to 1.1, cross-axis sensitivity of 0.2 percent, combined (root-sum-square) nonlinearity, hysteresis, nonrepeatability and resolution of 0.2 percent in a 1.7-in-diameter, 1.4-in-long case, weighing 8 oz. An angular acceleration version has ranges from

[1] Columbia Research Labs Inc., Woodlyn, Pa.

100 to 1500 rad/s^2, natural frequencies from 25 to 200 Hz, and sensitivity to translational acceleration of 0.1 percent full scale per g.

Accelerometers for Inertial Navigation

Inertial navigation is accomplished in principle by measuring the absolute acceleration (usually in terms of three mutually perpendicular components of the total-acceleration vector) of the vehicle and then integrating these acceleration signals twice to obtain displacement from an initial known starting location. Thus instantaneous position is always known without the need for any communication with the world outside the vehicle. To keep the accelerometers' sensitive axes always oriented parallel to their original starting positions, elaborate stable platforms using gyroscopic references and feedback systems are necessary. Since the accelerometers are sensitive also to gravitational force, this force must be computed and corrections applied continuously. Since the inertial navigation system measures absolute motion, systems for navigation over the earth's surface (such as for submarines) must include means for compensating for the earth's own motions.

While accelerometers for such navigation systems must operate on essentially the same basic principles that we considered above, their extreme performance requirements and the desire to obtain integrals of the acceleration rather than acceleration itself lead to special techniques and configurations. The desire for compatibility with the required data-processing computers (often digital) also influences the designer's choice of alternatives. The details of these applications are beyond the scope of this book, but may be found in numerous references.[1]

Mechanical Loading of Accelerometers on the Test Object[2]

The attachment of an accelerometer to a vibrating system results in a change in the motion measured as compared with the undisturbed case. We can apply general impedance principles to this problem to calculate the significance of this effect in any particular instance. In doing so, a useful simplification, which is adequate in most cases, is to regard the entire accelerometer as one rigid mass equal to the total mass of the instrument. This approximation generally holds since accelerometers are used below their natural frequency and thus there is little relative motion of the proof mass and the instrument case.

---

[1] H. B. Sabin, 17 Ways to Measure Acceleration, *Contr. Eng.*, p. 106, February 1961; J. M. Slater and D. E. Wilcox, How Precise Are Inertial Components? *Contr. Eng.*, p. 86, July 1958; *Sperry Eng. Rev.* Spring 1964; J. M. Slater, Inertial Guidance Notes, North American Aviation Corp., Autonetics Div.; C. F. Savant et al., "Principles of Inertial Navigation," McGraw-Hill, New York, 1961; P. H. Savet (ed.), "Gyroscopes: Theory and Design," McGraw-Hill, New York, 1961.

[2] Weight Limitations of Accelerometers for High Frequency Vibration Measurements, Wilcoxon Research, Bethesda, Md., *Tech. Bull., No. 1*, August 1963.

4.9 CALIBRATION OF VIBRATION PICKUPS

While the response of vibration pickups to interfering and modifying inputs such as temperature, acoustic noise, and magnetic fields is often of interest, here we are concerned with the response to the desired input of displacement, velocity, or acceleration. An excellent reference giving a more complete treatment of this subject is available.[1] Here we briefly touch on some of the main points.

The calibration methods in wide use may be classified into three broad types: constant acceleration, sinusoidal motion, and transient motion. Constant-acceleration methods (which are suitable only for calibrating accelerometers) include the tilting-support method and the centrifuge. The tilting-support method utilizes the accelerometer's inherent sensitivity to gravity. Static "accelerations" over the range $\pm 1g$ may be accurately applied by fastening the accelerometer to a tilting support whose tilt angle from vertical is accurately measured. This method requires that the accelerometer respond to static accelerations; therefore most piezoelectric devices cannot be calibrated in this way. The accuracy of the method depends on the accuracy of angle measurement and the knowledge of local gravity. The accuracy is of the order of $\pm 0.0003g$. In the centrifuge method, the sensitive axis of the accelerometer is radially disposed on a rotating horizontal disk so that it experiences the normal acceleration of uniform circular motion. Static accelerations in the range 0 to $60,000g$ are achievable with an accuracy of ± 1 percent. The allowable weight of the pickup varies from 100 lb at $100g$ to 1 lb at $60,000g$.

The sinusoidal-motion method is exemplified by the calibration facility of the National Bureau of Standards.[2] This consists of a modified electrodynamic vibration shaker which has been carefully designed to provide uniaxial pure sinusoidal motion and which is equipped with an accurately calibrated moving-coil velocity pickup to measure its table motion. If a motion is known to be purely sinusoidal, knowledge of its velocity and frequency enables accurate calculation of the displacement and acceleration. (The motion frequency is easily obtained with high accuracy by electronic counters.) This technique is thus useful for displacement, velocity, or acceleration pickups. The particular equipment referred to above can calibrate pickups (obtaining both amplitude ratio and phase angle) of a weight up to 2 lb over the frequency range 8 to 2,000 Hz. The acceleration range available is 0 to $25g$, velocity range is 0 to 50 in/s, and displacement range is 0 to 0.5 in. Accuracy is ± 1 percent from 8 to 900 Hz and ± 2 percent from 900 to 2,000 Hz.

Calibration of primary standards such as the National Bureau of Standards moving-coil velocity pickup above is accomplished by the so-called absolute

[1] R. R. Bouche, Calibration of Shock and Vibration Measuring Transducers, Shock and Vibration Info. Ctr., Naval Res. Lab., Code 8404, Washington, 1979.
[2] R. R. Bouche, Improved Standard for the Calibration of Vibration Pickups, *Exp. Mech.*, April 1961.

method, with the reciprocity[1] and laser interferometer[2] techniques being the most common. These are both rather involved procedures reserved for standards-laboratory use. Figure 4.79 shows typical apparatus for the laser method, as used to calibrate a high-precision piezoelectric accelerometer which is then utilized as a "transfer standard" to conveniently calibrate other accelerometers by a comparison method. The laser method actually measures displacement amplitude x_o and relies on the purity of the shaker's sinusoidal motion, plus accurate frequency measurement, to infer acceleration amplitude as $\omega^2 x_o$. The particular accelerometer shown is made with a mirror-finish surface to reflect the laser light; thus there is no need to attach a separate mirror when calibration is performed. When the accelerometer has a displacement amplitude x_o, the number of fringes counted by the phototransistor observing the interferometer output is $8x_o/\lambda$ for each complete vibration cycle, which gives 12.64 cycles/μm when a laser with $\lambda = 0.6328$ μm is used. For a typical calibration frequency of 160 Hz, a 10 m/s² peak acceleration corresponds to a displacement of 9.9 μm, which gives 125.1 fringes per cycle and 20.016-kHz fringe frequency. To "automate" the needed division of the 20.016-kHz fringe signal by the shaker frequency, a ratio-type counter is used as shown, and it gives a direct readout of the 125.1 motion-amplitude signal.

Once a precision standard accelerometer is calibrated by one of the absolute methods, it can serve as a comparison standard for calibrating "working" transducers. By providing a threaded mounting hole in each end of the standard accelerometer, we can attach it to the calibration shaker and then attach the transducer to be calibrated directly to the standard, ensuring that both will feel the same motion when they are shaken sinusoidally over the desired frequency range. This is the most common method of calibration for working transducers.

Transient-motion calibration methods include the physical pendulum, ballistic pendulum, and drop-test techniques.[3] These techniques are somewhat specialized and are not pursued further here.

4.10 JERK PICKUPS

In some measurement and control applications, the rate of change of acceleration, or *jerk*, d^3x/dt^3, must be measured. An obvious approach is to apply the electrical output from an accelerometer to a differentiating circuit. A more subtle technique which avoids the noise-accentuating problems of differentiating circuits is applied in the Donner Jerkmeter.[4] By ingenious use of feedback principles, this null-balance instrument provides both acceleration and jerk signals of good quality. The physical configuration is essentially that of Fig. 4.78 with the addition of an electronic integrator. The resulting block diagram is shown in Fig. 4.80.

---

[1] Bouche, Calibration of Shock and Vibration Measuring Transducers, pp. 95–109.

[2] Accelerometer Calibration, BR 0173, Bruel and Kjaer Instruments, Marlboro, Mass.

[3] Harris and Crede, "Shock and Vibration Handbook," vol. 1, chap. 18.

[4] Systron Donner Corp., Concord, Calif.

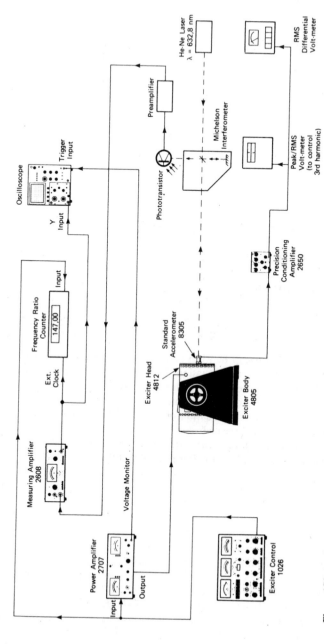

Figure 4.79 Accelerometer calibration by laser interferometer. (*Courtesy Bruel and Kjaer Instruments, Marlboro, Mass.*)

331

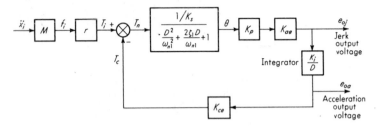

Figure 4.80 Jerkmeter block diagram.

Analysis gives

$$\left(MrD^2 x_i - \frac{K_i K_{ce} e_{oj}}{D} \right) \frac{K_p K_{ae}/K_s}{D^2/\omega_{n1}^2 + 2\zeta_1 D/\omega_{n1} + 1} = e_{oj} \qquad (4.126)$$

which leads to the differential equation

$$\left(\frac{K_s}{K_i K_p K_{ae} K_{ce} \omega_{n1}^2} D^3 + \frac{2\zeta_1 K_s}{K_i K_p K_{ae} K_{ce} \omega_{n1}} D^2 \right.$$

$$\left. + \frac{K_s}{K_i K_p K_{ae} K_{ce}} D + 1 \right) e_{oj} = \frac{Mr}{K_i K_{ce}} D^3 x_i \qquad (4.127)$$

We note that the relationship between jerk input $D^3 x_i$ and voltage output e_{oj} is that of a third-order differential equation. The static sensivity is easily seen to be $Mr/(K_i K_{ce})$ V/(cm/s^3); however, the dynamic behavior is not obvious since we have not considered third-order instruments. For any given set of numerical values, the cubic characteristic equation of Eq. (4.127) will have three numerical roots. These will be either three real roots or one real root plus a pair of complex conjugates. In an actual design, the latter situation often prevails. This means that the transfer function of Eq. (4.127) is generally of the form

$$\frac{e_{oj}}{D^3 x_i}(D) = \frac{K}{(\tau D + 1)(D^2/\omega_n^2 + 2\zeta D/\omega_n + 1)} \qquad (4.128)$$

where $$K \triangleq \frac{Mr}{K_i K_{ce}}$$

and τ, ζ, and ω_n can be found if numerical values for all the constants are given. The frequency response for Eq. (4.128) is easily plotted by using logarithmic methods. Actually, the frequency response for Eq. (4.127) is quite revealing. It is clear that, for sufficiently low frequencies, $e_{oj} \approx [Mr/(K_i K_{ce})]\dddot{x}_i$ for *any* values of the system constants since the first three terms involve ω^3, ω^2, and ω as factors and thus go to zero as $\omega \to 0$. To increase the usable frequency range, the *coefficients* of the ω^3, ω^2, and ω terms must be made small. This can be done by making $K_s/(K_i K_p K_{ae} K_{ce})$ small, which corresponds to making the gain of the feedback loop large. A limit is placed on the gain, however, since excessive gain will cause dynamic instability and resultant destruction of the instrument.

The Routh stability criterion[1] shows that for a cubic characteristic equation of the form $a_3 D^3 + a_2 D^2 + a_1 D + a_0 = 0$, stability is ensured if all the a's are positive and $a_0 a_3 < a_1 a_2$. In our case this becomes

$$\frac{K_s}{K_i K_p K_{ae} K_{ce} \omega_{n1}^2} < \frac{2\zeta_1 K_s^2}{(K_i K_p K_{ae} K_{ce})^2 \omega_{n1}} \tag{4.129}$$

or

$$\zeta_1 > \frac{K_i K_p K_{ae} K_{ce}}{2 K_s \omega_{n1}} \tag{4.130}$$

This shows that if $K_s/(K_i K_p K_{ae} K_{ce})$ is made small to increase the usable frequency range, then the damping ζ_1 must be correspondingly increased to retain adequate stability. This required damping effect also can be obtained by adding proper electrical compensating networks to the circuit. This approach is often used. No matter how the damping is achieved, however, a tradeoff will always be necessary between frequency range and stability.

4.11 PENDULOUS (GRAVITY-REFERENCED) ANGULAR-DISPLACEMENT SENSORS

In a number of applications, the measurement of angular displacement relative to the local vertical (gravity vector) is a useful technique. Examples include sensing elements for control systems of road-paving and scraping machines; drainage-tile-laying machines; alignment of construction forms, piles and bridges; and attitude control of vehicles (such as submarines) and torpedoes or missiles. These relatively simple instruments (basically plumb bobs with electrical output) can sometimes replace more complex and expensive gyroscopic instruments which perform similar functions. Their main disadvantages relative to gyros are their sensitivity to interfering translatory acceleration inputs and their dependence on a gravity field. (They do not work in essentially gravity-free space.)

Figure 4.81 shows a typical configuration of a single-axis pendulum-type sensor. The desired input to be measured is the case rotation angle θ_c. Most commercial sensors do not include the springs K_s; we include them here because their presence makes possible interesting and potentially useful dynamic behavior. For the usual case of no springs, K_s is simply set equal to zero. The damping effect is not essential to the theoretical operation of the device, but is included in most practical instruments to reduce oscillations at pendulum frequency caused by transient interfering inputs. A variety of electrical displacement transducers may be employed, depending on the required characteristics; a potentiometer is shown for simplicity. The following assumptions are justifiable for most purposes

---

[1] Doebelin, "Dynamic Analysis and Feedback Control," p. 175.

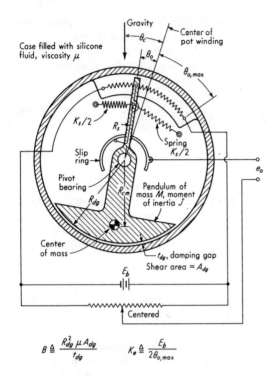

$$B \triangleq \frac{R_{dg}^2 \mu A_{dg}}{t_{dg}} \qquad K_e \triangleq \frac{E_b}{2\theta_{o,max}}$$

Figure 4.81 Pendulum-displacement sensor.

in simplifying the analysis:

1. Angles are small enough that the sine and the angle are nearly equal and the cosine is nearly 1.
2. The inertia effect of the fluid on the pendulum motion is negligible.
3. The damping effect of the fluid is limited to the damping gap.
4. All dry-friction effects in pot wipers, bearings, and slip rings may be neglected for dynamic analysis.
5. The buoyant force on the pendulum is negligible.
6. The springs provide a linear restoring torque.

The analysis is left for the problems at the end of this chapter. However, the results are as follows: When the springs are present ($K_s \neq 0$), we have

$$\frac{e_o}{\theta_c}(D) = \frac{K(D^2/\omega_{n1}^2 + 1)}{D^2/\omega_{n2}^2 + 2\zeta D/\omega_{n2} + 1} \qquad (4.131)$$

where

$$K \triangleq \frac{MgR_{cm}K_e}{R_s^2 K_s + MgR_{cm}} \qquad (4.132)$$

$$\omega_{n1} \triangleq \sqrt{\frac{MgR_{cm}}{J}} \tag{4.133}$$

$$\omega_{n2} \triangleq \sqrt{\frac{R_s^2 K_s + MgR_{cm}}{J}} \tag{4.134}$$

$$\zeta \triangleq \frac{B}{2\sqrt{J(R_s^2 K_s + MgR_{cm})}} \tag{4.135}$$

The frequency response of this system is shown in Fig. 4.82a. Note the "notch-filter" effect at ω_{n1} followed by a resonant peak near ω_{n2}. If the springs are removed (the usual case), we get $\omega_{n1} = \omega_{n2} = \omega_n$ and $K_e = K$, which yields

$$\frac{e_o}{\theta_c}(D) = \frac{K(D^2/\omega_n^2 + 1)}{D^2/\omega_n^2 + 2\zeta D/\omega_n + 1} \tag{4.136}$$

and the frequency response of Fig. 4.82b.

The pendulum sensor is unfortunately sensitive to horizontal accelerations, and so its application is ruled out where such accelerations are large enough to cause a significant output signal. A simple analysis shows that a steady horizontal acceleration A_x will cause an output voltage $K_e A_x/g$ for an instrument with no springs.

A commercial pendulum[1] using a potentiometer motion pickup has a full-scale range of $\pm 8°$, natural frequency of 2 Hz, damping ratio of 0.6, and resolution of $0.1°$ ($0.05°$ if vibration is present). Pendulums are also available for measuring rotation about two mutually perpendicular axes.

Closely related to the pendulum is the Tiltmeter[2] sensor of Fig. 4.83,[3] based on the ancient carpenter's spirit level. A cylindrical glass cavity has a spherical inner top surface plated with four equally spaced platinum strips which act as electrodes; the flat bottom surface has a single central pin electrode. The cavity is partially filled with a conductive liquid, which leaves at the top a bubble that covers equal areas of the four top electrodes when the device is level. Opposite electrodes are connected to form from the fluid body two adjacent legs of a bridge, while two matched resistors complete the bridge which is ac-excited by applying a voltage between the bottom electrode and the common termination point of the resistors. As the case is titled about one axis, the bubble, which remains fixed with respect to the gravity vector, moves to partially uncover one cap electrode and cover the opposite. Thus the bridge is unbalanced as a result of the resistance changes of the fluid bodies caused by their changed contact with the upper electrodes. Two such bridges are formed to resolve a biaxial tilt into two perpendicular components. Analysis[4] shows each axis of the device responds

---

[1] Honeywell Inc., Minneapolis, Minn.

[2] Rockwell International, Autonetics Group, Anaheim, Calif.

[3] C. G. Kirkpatrick, A New Biaxial Tiltmeter, Paper X76-894/201, *ISA-CIM Proc.*, August 1976.

[4] Ibid.

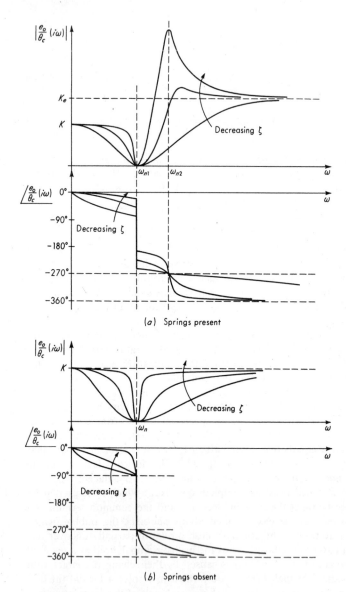

Figure 4.82 Frequency response of pendulum sensor.

essentially as an overdamped simple pendulum to desired tilt input and spurious inputs of horizontal displacement and acceleration, with the typical frequency response ($f_n \approx 1$ Hz) being as shown in Fig. 4.83. Sensitivity is extremely high;

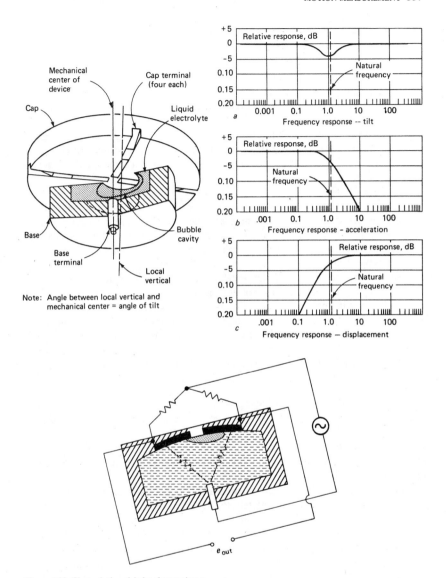

Figure 4.83 Electrolytic spirit-level transducer.

angles the order of 10^{-6} degree can be measured. To prevent overranging the device, it must be level within about 0.05°. If angular motions larger than this are expected, we can build a servocontrolled tilt table, mount the sensor on this table, and use it as the servo's null detector.

4.12 GYROSCOPIC (ABSOLUTE) ANGULAR-DISPLACEMENT AND VELOCITY SENSORS

While gyroscopic instruments have been used in limited numbers and applications (gyrocompasses in ships and aircraft, turn-and-bank indicators in aircraft, etc.) since around World War I, developments during and after World War II brought them to an extreme degree of refinement, and their use is now common in military and commercial applications.[1] In most of these applications, the gyro is used to measure the motion of some vehicle, either as a permanent part of a vehicle measurement-control system or as a test instrument temporarily attached to gather data.

Perhaps the simplest gyro configuration is the free gyro shown in Fig. 4.84. These instruments are used to measure the absolute angular displacement of the vehicle to which the instrument frame is attached. A single free gyro can measure rotation about two perpendicular axes, such as the angles θ and ϕ. This can be accomplished because the axis of the spinning gyro wheel remains fixed in space (if the gimbal bearings are frictionless) and thus provides a reference for the relative-motion transducers. If the angles to be measured do not exceed about 10°, the readings of the relative-displacement transducers give directly the absolute rotations with good accuracy. For larger rotations of both axes, however, there is an interaction effect between the two angular motions, and the transducer readings do *not* accurately represent the absolute motions of the vehicle. The free gyro is also limited to relatively short-time applications (less than about 5 min) since gimbal-bearing friction causes gradual drift (loss of initial reference) of the gyro spin axis. A constant friction torque T_f causes a drift (precession) of angular velocity ω_d given by

$$\omega_d = \frac{T_f}{H_s} \tag{4.137}$$

where H_s is the angular momentum of the spinning wheel. It is clear that a high angular momentum is desirable in reducing drift. A typical drift rate is about 0.5 degree/min for each axis.

Rather than using free gyros to measure two angles in one gyro (thus requiring two gyros to define completely the required three axes of motion), most recent high-performance systems utilize the so-called single-axis or constrained gyros. Here a single gyro measures a single angle (or angular rate); therefore three gyros are required to define the three axes. This approach avoids the coupling or interaction problems of free gyros, and the constrained (rate-integrating) gyros can be constructed with exceedingly small drift. We consider here two common types of constrained gyros: the rate gyro and the rate-integrating gyro. The rate gyro measures absolute angular velocity and is widely used to generate stabilizing signals in vehicle control systems. The rate-integrating gyro measures absolute angular displacement and thus is utilized as a

[1] Sidney Lees (ed.), "Air, Space, and Instruments," p. 32, McGraw-Hill, New York, 1963.

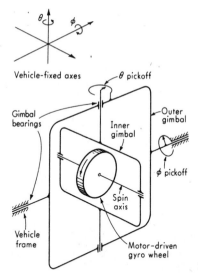

Figure 4.84 Free gyroscope (two-axis position gyro).

fixed reference in navigation and attitude-control systems. The configuration of a rate gyro is shown in Fig. 4.85, the rate-integrating gyro is functionally identical except that it has no spring restraint.

While a general analysis of gyroscopes is exceedingly complex, useful results for many purposes may be obtained relatively simply by considering small angles only. This assumption is satisfied in many practical systems. Figure 4.86 shows a gyro whose gimbals (and thus the angular-momentum vector of the spinning

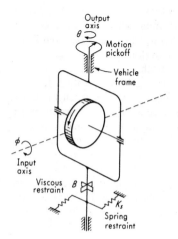

Figure 4.85 Single-axis restrained gyro.

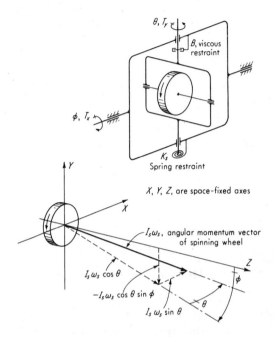

θ, T_y

B, viscous restraint

ϕ, T_x

K_s
Spring restraint

$X, Y, Z,$ are space-fixed axes

$I_s \omega_s$, angular momentum vector of spinning wheel

$I_s \omega_s \cos \theta$

$-I_s \omega_s \cos \theta \sin \phi$

$I_s \omega_s \sin \theta$

Figure 4.86 Gyro analysis.

wheel) have been displaced through small angles θ and ϕ. We apply Newton's law

$$\sum \text{torques} = \frac{d}{dt} (\text{angular momentum}) \qquad (4.138)$$

to the x and y axis components of angular momentum. This angular momentum is made up two parts, one part due to the spinning wheel and another part (due to the motion of the wheel, case, gimbals, etc.) which would exist even if the wheel were not spinning. The latter part depends on (for the x axis) the angular velocity $d\phi/dt$ and the moment of inertia I_x of everything that rotates when the outer gimbal turns in its bearing. For the y axis, the angular momentum depends on $d\theta/dt$ and the moment of inertia I_y of everything that rotates when the inner gimbal turns in its bearing. The external applied torques T_x and T_y are included to allow for the possibility of bearing friction and for intentionally applied torques from small electromagnetic "torquers" which are used in some systems to cause desired precessions for control or correction purposes. The inertias I_x and I_y (which are about *space-fixed* axes) actually change when θ and ϕ change, but this effect is negligible for small angles. Also, the exact equations would contain terms in the *products* of inertia as well as the moments of inertia, but these are again negligible because of the small angles and the inherent symmetry of gyro structures. With the above qualifications, we may write for the x axis

$$T_x = \frac{d}{dt}\left(H_s \sin\theta + I_x \frac{d\phi}{dt}\right) \qquad (4.139)$$

and for the y axis

$$T_y - B\frac{d\theta}{dt} - K_s\theta = \frac{d}{dt}\left(-H_s \cos\theta \sin\phi + I_y \frac{d\theta}{dt}\right) \qquad (4.140)$$

We now assume H_s is a constant (the gyro wheel is driven by a constant-speed motor) and $\cos\theta = 1$, $\sin\theta = \theta$, $\sin\phi = \phi$ to get

$$T_x = H_s \frac{d\theta}{dt} + I_x \frac{d^2\phi}{dt^2} \qquad (4.141)$$

$$T_y - B\frac{d\theta}{dt} - K_s\theta = -H_s\frac{d\phi}{dt} + I_y\frac{d^2\theta}{dt^2} \qquad (4.142)$$

These are two simultaneous linear differential equations with constant coefficients relating the two inputs T_x and T_y to the two outputs θ and ϕ. Writing these equations in operator form, we can treat them as algebraic equations to solve for ϕ and θ as desired. For ϕ we get

$$\phi = \frac{(I_y D^2 + BD + K_s)T_x - (H_s D)T_y}{D^2(I_x I_y D^2 + BI_x D + H_s^2 + I_x K_s)} \qquad (4.143)$$

Since ϕ depends on both T_x and T_y, transfer functions can be obtained by considering each input separately and then using superposition. Letting $T_y = 0$, we get

$$\frac{\phi}{T_x}(D) = \frac{I_y D^2 + BD + K_s}{D^2(I_x I_y D^2 + BI_x D + H_s^2 + I_x K_s)} \triangleq G_1(D) \qquad (4.144)$$

and letting $T_x = 0$ gives

$$\frac{\phi}{T_y}(D) = -\frac{H_s}{D(I_x I_y D^2 + BI_x D + H_s^2 + I_x K_s)} \triangleq G_2(D) \qquad (4.145)$$

This leads to the block diagram of Fig. 4.87a. Similar analysis for θ gives

$$\frac{\theta}{T_x}(D) = \frac{H_s}{D(I_x I_y D^2 + BI_x D + H_s^2 + I_x K_s)} \triangleq G_3(D) = -G_2(D) \qquad (4.146)$$

and

$$\frac{\theta}{T_y}(D) = \frac{I_x}{I_x I_y D^2 + BI_x D + H_s^2 + I_x K_s} \triangleq G_4(D) \qquad (4.147)$$

which leads to the block diagram of Fig. 4.87b. An overall block diagram may then be constructed as in Fig. 4.87c.

The above results are of quite general applicability to gyro systems of various

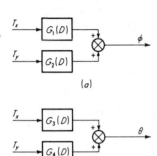

(a)

(b)

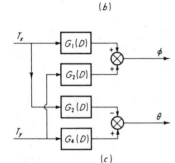

(c)

Figure 4.87 Gyro block diagrams.

configurations as long as the small-angle requirement is met. For single-axis rate and rate-integrating gyros, considerable simplification of the above results is possible. In these applications (see Fig. 4.85) we consider the input to be the *motion* ϕ. A torque T_x also exists and would be felt by the vehicle; however, it does not enter our calculations since we assume the motion ϕ is given. The angle θ is an indication of the angle ϕ (rate-integrating gyro) or angular velocity $\dot{\phi}$ (rate gyro); thus we should like to have transfer functions relating θ to ϕ. The torque T_y (neglecting bearing friction) is zero for this application unless a torquer is used for some special purpose. Then the desired θ-ϕ relation may easily be obtained by solving Eqs. (4.144) and (4.146) for T_x and setting them equal. The result is

$$\frac{\theta}{\phi}(D) = \frac{H_s D}{I_y D^2 + BD + K_s} \tag{4.148}$$

For a rate gyro, then, we have the second-order response

$$\frac{\theta}{D\phi}(D) = \frac{\theta}{\dot{\phi}}(D) = \frac{K}{D^2/\omega_n^2 + 2\zeta D/\omega_n + 1} \tag{4.149}$$

where

$$K \triangleq \frac{H_s}{K_s} \qquad \text{rad/(rad/s)} \tag{4.150}$$

$$\omega_n \triangleq \sqrt{\frac{K_s}{I_y}} \quad \text{rad/s} \tag{4.151}$$

$$\zeta \triangleq \frac{B}{2\sqrt{I_y K_s}} \tag{4.152}$$

A high sensitivity is achieved by large angular momentum H_s and soft spring K_s, although low K_s gives a low ω_n. Natural frequencies of commercially available rate gyros are of the order of 10 to 100 Hz. The damping ratio is usually set at 0.3 to 0.7. Large angular momentum is obtained in small size by using high-speed (often 24,000 r/min) motors to spin the gyro wheel. Full-scale ranges of about ± 10 to ± 1000 degree/s are readily available. Resolution of a ± 10 degree/s range instrument is of the order of 0.005 degree/s. In some high-performance rate gyros, the mechanical spring is replaced by an "electrical spring" arrangement similar to that used in the servo accelerometer of Fig. 4.78.

To measure all three components (roll, pitch, and yaw) of angular velocity in a vehicle, an arrangement of three rate gyros, such as in Fig. 4.88 may be employed. It should be pointed out that in a rate gyro only the output angle θ must be kept small. This requires use of a very sensitive motion pickoff, but such are available. The input angle ϕ may be indefinitely large since no matter how large it gets the spin angular-momentum vector is always perpendicular (except for a small error resulting from nonzero θ) to the input angular-velocity vector.

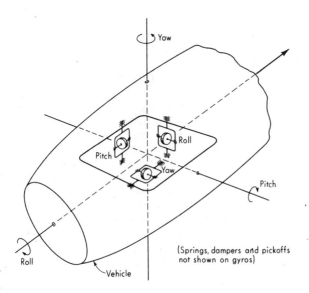

Figure 4.88 Three-axis rate-gyro package.

Thus a roll-rate gyro in an aircraft will function correctly even if the aircraft rolls completely over. Of course, the angular-velocity components measured are those about the *vehicle* axes rather than about space-fixed axes, but these are usually the velocities desired for stabilization signals in control systems.

To obtain a rate-integrating gyro, we merely remove the spring restraint from the configuration of Fig. 4.85. Equation (4.148) then becomes

$$\frac{\theta}{\phi}(D) = \frac{K}{\tau D + 1} \tag{4.153}$$

where

$$K \triangleq \frac{H_s}{B} \quad \text{rad/rad} \tag{4.154}$$

$$\tau \triangleq \frac{I_y}{B} \quad \text{s} \tag{4.155}$$

We see that the output angle θ is a direct indication of the input angle ϕ according to a standard first-order response form. High sensitivity again requires high H_s, a typical value being 5×10^5 g · cm^2/s. Low damping B also increases sensitivity, but at the expense of speed of response, as shown by Eq. (4.155). For a K of 15, a τ of 0.006 s is typical of high-performance units.[1] Increase of K to 440 raises τ to 0.17 s. The rate-integrating gyro is the basis of highly accurate inertial navigation systems, where it is used as a reference to maintain so-called stable platforms in a fixed attitude within a vehicle while the vehicle moves arbitrarily. This is done by using the motion signals from the gyros to drive servomechanisms which maintain the platform in a fixed angular orientation. The system accelerometers are mounted on this platform, and their double-integrated output is an accurate measure of vehicle motion along the three orthogonal axes since the platform always moves parallel to its initial orientation.

The rate-integrating gyro has been subjected to extremely intensive and extensive engineering development to bring its performance characteristics to remarkable levels while maintaining small size and weight and great resistance to rugged environments. This has been accomplished by painstaking attention to minute mechanical, thermal, and electrical details which would be completely negligible in a less sophisticated device. Figure 4.89 shows a floated rate-integrating gyro which incorporates a number of clever design features necessary to high performance. The inner gimbal is in the form of a hollow cylinder containing the gyro wheel and is mounted in precision pivot jewel bearings. A small radial gap between this cylinder and its housing is filled with damping fluid to form the damper B. Temperature feedback-control systems regulate electric heaters to maintain damping-fluid viscosity constant. This heating must be carefully applied so as to minimize convection currents in the damping fluid which would exert spurious torques on the inner gimbal. The density of the damping fluid (which completely surrounds the hollow cylinder) is adjusted so as to obtain

---

[1] General Precision Inc., Kearfott Div., Little Falls, N. J.

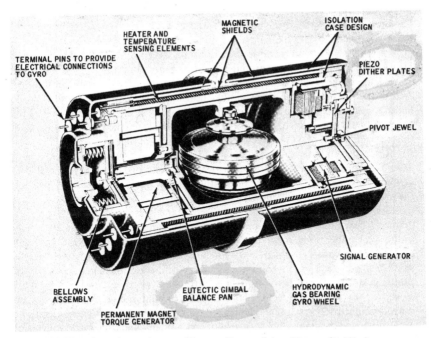

Figure 4.89 Floated, rate-integrating gyro. (*Courtesy Honeywell, Inc., Minneapolis, Minn.*)

neutral buoyancy (this explains the "floated gyro" terminology), thereby re-moving the radial load from the jewel bearings and reducing their already-low friction even more. A further 10-to-1 bearing-friction reduction is achieved by vibrating ("dithering") the pivot bearings at low amplitude and high frequency by means of electrically driven piezo crystals. A further advantage of the floated construction is an increase in shock and vibration tolerance since case shocks are transmitted to the inner gimbal through the damping fluid rather than the rather fragile jewel bearings.

At original assembly, the inner gimbal is very carefully statically balanced since any unbalance creates a spurious gravity torque. When such a gyro is built into, say, an ICBM, it may be inactive for long periods, and various environmental and aging effects can cause a loss of balance. To correct for such unbalance without removing the gyro from the missile or even disturbing its hermetic seals, the eutectic gimbal balance pan is provided. If the gyro is activated by running its wheel up to speed, the gyro becomes a very sensitive detector of unbalance since the inner gimbal will slowly precess under the action of the unbalanced gravity torque and this motion will be detected by the θ displacement transducer (signal generator). To rebalance the gimbal, an electric current is applied to melt a eutectic alloy in the balance pan so that the molten metal migrates to move its mass center and thereby regain balance. When gimbal precession stops, rebalance has been obtained and the metal is allowed to solidify. Gyro wheel bearings of the hydrodynamic gas type contribute to long life and low noise since they ride on a film of gas with no metal-to-metal contact, except momentarily when starting or stopping. The permanent-magnet torque generator allows precise electrical application of inner gimbal torques for self-testing and correction procedures. An "isoelastic" design for the inner gimbal bearing suspension ensures that elastic deflections caused by case linear accelerations will be in the same direction as the acceleration. This means that the inertia torque will not have a moment about the center of mass, which would cause a spurious gyro output.

Recent developments in absolute angular-motion sensors have included both exotic versions of the classical "spinning wheel" concept and totally new devices devoid of gyroscopic effects but often still called "gyros" because they perform the same measurement function as the traditional mechanical instrument. Electrostatic gyros[1] utilize feedback-type electrostatic systems to levitate a 1-cm-diameter beryllium sphere spinning at 146,000 r/min without physical contact inside a chamber evacuated to 10^{-7} torr. The gyro is a free-rotor type with two-axis angle-measuring capability for large angles, which makes it especially suitable for strapdown navigation systems. A deliberate mass unbalance in the spinning sphere causes motions related to sphere angular attitude which are picked up by a capacitance-displacement sensor that uses eight electrodes arranged octahedrally over the spherical housing's inner surface. These electrodes are timeshared between the suspension servosystem and the angle-measuring system, with 20 percent of the duty cycle being devoted to angle measurement. Extreme accuracy is obtained by calibrating each gyro to obtain numerical values for a sophisticated error model containing 56 sources of error. This model is stored by the navigation computer and used to process the gyro angle data. Here again we see the use of computing power (made feasible by microcomputer technology) to overcome basic electromechanical hardware limitations in instrument design.

Optical (ring-laser) "gyros"[2] do not use gyroscopic principles at all. Rather, they measure phase shifts between two laser beams directed around a loop by mirrors fastened to the object whose rotation is to be measured. One beam travels with the rotation while the other travels against it, causing a phase shift proportional to rotation when the beams are recombined and compared. The practical realization of this basic concept uses two active optical oscillators internal to the loop, rather than an external light source. These oscillators (lasers) are arranged so that rotation produces a frequency difference between them. In a typical instrument with a triangular optical path of 8.4-in. perimeter and laser wavelength of 0.6328 μm, a 1 rad/s rotation rate produces a 65-kHz output frequency. Since the time integral of the output-pulse rate is proportional to input

[1] R. R. Duncan, "Micron—A Strapdown Inertial Navigator Using Miniature Electrostatic Gyros," Autonetics Div., Rockwell Intl., Anaheim, Calif., 1973.

[2] J. A. Tekiela, Technology Assessment of Ring Laser Gyroscopes, GACIAC TA-79-01, ITT Res. Inst., Chicago, July 1979.

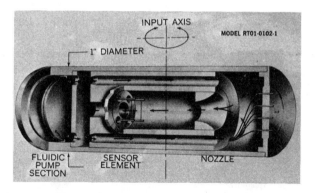

Figure 4.90 Hot-wire jet-deflection sensor. Model RT01-0102-1.

angle, a simple summing of pulses gives a rate-integrating-gyro type of response with a resolution of 1 count for 3.1 seconds of arc of rotation. The development of the no-moving-parts ring-laser gyro appears to be synergistic with that of the microcomputer since the two together make the use of so-called strapdown guidance systems very attractive. These systems, rather than using complex servo-systems to maintain the gyro mounting platform in a fixed attitude, strap the gyro directly to the vehicle frame and compute, from the gyro signals, the axis-transformation data required to convert from the vehicle reference frame to the necessary inertial coordinates. The low-cost, compact microcomputers thus allow us to substitute computing power for electromechanical hardware.

A recent variation[1] on the laser gyro principle utilizes a long optical fiber wound in a multiturn coil to replace the mirrors. Angular-rate sensors utilizing fluid (nongyroscopic) principles include the fluidic vortex device[2] and hot-wire jet-deflection sensor.[3] Figure 4.90[4] shows the latter device, which utilizes vibrating piezo crystals to circulate a steady laminar flow of gas in a sealed chamber. The gas jet passes over two hot-wire flow sensors (see hot-wire anemometers in Chap. 7) connected differentially in a bridge circuit. Rotation of the instrument case about the input axis causes a lateral deflection of the gas stream, differential cooling of the wires, and a bridge-unbalance voltage proportional to angular rate.

In this short section we can only indicate a few major concepts while many significant details are neglected. A few samples[5] of the voluminous literature are mentioned here while more are found in the Bibliography of this chapter.

[1] *Invention Rep.* 30-3873/NPO-14258, Jet Propulsion Lab., California Institute of Technology, Pasadena, October 1978.

[2] G. P. Wachtell, Fluidic Vortex Angular Rate Sensor, *USAAVLABS Rep.* 70-25, Fort Eustis, Va., 1970.

[3] *Eng. Bull.* 1075, Humphrey Inc., San Diego.

[4] Ibid.

[5] B. Lichtenstein, "Technical Information for the Engineer, Gyros," General Precision Inc., Kearfott Div., Little Falls, N. J., Savet "Gyroscopes."

PROBLEMS

4.1 Derive an equation analogous to (4.1) for the system of Fig. 4.5b.

4.2 Derive Eq. (4.1).

4.3 The output of a potentiometer is to be read by a recorder of 10,000-Ω input resistance. Nonlinearity must be held to 1 percent. A family of potentiometers having a thermal rating of 5 W and resistances ranging from 100 to 10,000 Ω in 100-Ω steps are available. Choose from this family the potentiometer that has the greatest possible sensitivity and meets the other requirements. What is this sensitivity if the potentiometers are single-turn (360°) units?

4.4 If a potentiometer changes resistance because of temperature changes, what effect does this have on motion measurements?

4.5 A 10-in-stroke wirewound translational potentiometer is excited with 100 V. The output is read on an oscilloscope with a "sensitivity" of 0.5 mV/cm. It would appear that measurements to the nearest 0.0001 in are easily possible. Explain why this is not so.

4.6 In Fig. P4.1 a potentiometer whose moving part weighs 0.01 lbf measures the displacement of a spring-mass system subjected to a step input. The measured natural frequency is 30 Hz. If the spring constant and mass M of the system are unknown, can the true natural frequency be deduced from the above data? Suppose an additional 0.01-lbf weight is attached to the potentiometer and the test repeated, giving a 25-Hz frequency. Calculate the true natural frequency of the system, that is, the frequency before the potentiometer was attached.

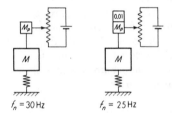

$f_n = 30\,\text{Hz}$ $f_n = 25\,\text{Hz}$ **Figure P4.1**

4.7 Explain why increasing the cross-sectional area of the end loops in foil-type strain gages reduces transverse sensitivity.

4.8 If, in the discussion following Eq. (4.21), dynamic strains only in the range 0 to 10,000 Hz need to be measured, explain how and to what extent the noise voltage may be reduced.

4.9 In a Wheatstone bridge, leg 1 is an active strain gage of Advance alloy and 120-Ω resistance, leg 4 is a similar dummy gage for temperature compensation, and legs 2 and 3 are fixed 120-Ω resistors. The maximum gage current is to be 0.030 A.

(a) What is the maximum permissible dc bridge excitation voltage? (Use this value in the remaining parts of this problem.)

(b) If the active gage is on a steel member, what is the bridge output voltage per 1,000 lb/in^2 of stress?

(c) If temperature compensation were *not* used, what bridge output would be caused by the active gage increasing temperature by 100F° if the gage were bonded to steel? What stress value would be represented by this voltage? Thermal-expansion coefficients of steel and Advance alloy are 6.5×10^{-6} and 14.9×10^{-6} in/(in · F°), respectively. The temperature coefficient of resistance of Advance is 6×10^{-6} $\Omega/(\Omega \cdot F°)$.

(d) Compute the value of a shunt calibrating resistor that would give the same bridge output as 10,000 lb/in^2 stress in a steel member.

4.10 From Eq. (4.31), find an expression for the frequency at which zero phase shift occurs.

4.11 Perform an analysis similar to that leading to Eq. (4.31), assuming output loaded with R_m, for (a) the circuit of Fig. 4.19a, (b) the circuit of Fig. 4.19b, (c) the circuit of Fig. 4.19c, (d) the circuit of Fig. 4.19d.

4.12 In Fig. 4.21c, let x_i be a periodic motion with a significant frequency content up to 500 Hz, and let the excitation frequency be 10,000 Hz. The output voltage e_o is connected to an oscillograph galvanometer, which is a second-order system with $\zeta = 0.65$ and a natural frequency of 1,000 Hz. Will this combination result in a satisfactory measurement system? Justify your answer with numerical results.

4.13 In Eq. (4.48), suppose a flat amplitude ratio within 5 percent down to 20 Hz is required. What is the minimum allowable τ? If $A = 0.5$ in^2 and x_0 is 0.005 in, what value of R is needed?

4.14 A piezoelectric transducer has a capacitance of 1,000 pF and K_q of 10^{-5} C/in. The connecting cable has a capacitance of 300 pF while the oscilloscope used for readout has an input impedance of 1 MΩ paralleled with 50 pF.

    (a) What is the sensitivity (V/in) of the transducer alone?

    (b) What is the high-frequency sensitivity (V/in) of the entire measuring system?

    (c) What is the lowest frequency that can be measured with 5 percent amplitude error by the entire system?

    (d) What value of C must be connected in parallel to extend the range of 5 percent error down to 10 Hz?

    (e) If the C value of part d is used, what will the system high-frequency sensitivity be?

4.15 A piezoelectric transducer has an input

$$x_i = \begin{cases} At & 0 \leq t < T \\ 0 & T < t < \infty \end{cases}$$

Solve the differential equation to find e_o. For $t = T^-$, find the error [(ideal value of e_o) − (actual value of e_o)]. Approximate this error by using the truncated series

$$e^{-T/\tau} \approx 1 - \frac{T}{\tau} + \frac{1}{2}\left(\frac{T}{\tau}\right)^2$$

Express this approximate error as a percentage of the ideal value of e_o. What must T/τ be if the error is to be 5 percent? For this value of T/τ, evaluate the error caused by truncating the series. (Use the theorem on the remainder of an alternating series.)

4.16 Analyze the nozzle-flapper displacement pickup of Fig. P4.2, using the simple incompressible relations. Explain the advantages of this configuration.

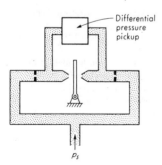

p_s **Figure P4.2**

4.17 Prove Eqs. (4.105) and (4.106).

4.18 In the variable-capacitance velocity pickup shown in Fig. P4.3, prove that the current i is directly proportional to the angular velocity $d\theta/dt$. Since voltage signals are more readily manipulated, how might the current signal be transduced to a proportional voltage? Does your method of doing this affect the basic operation? What must be required if the basic operation is to be only slightly affected?

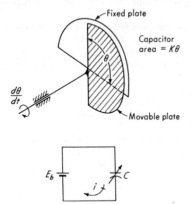

Figure P4.3

4.19 Construct logarithmic frequency-response curves for a seismic-displacement pickup with $\zeta = 0.3$ and $\omega_n/2\pi = 10$ Hz. For what range of frequencies is the amplitude ratio flat within 1 db?

4.20 Make a comprehensive list and explain the action of modifying and/or interfering inputs for the systems of the following:

(a) Fig. 4.4	(g) Fig. 4.55
(b) Fig. 4.11	(h) Fig. 4.65
(c) Fig. 4.17	(i) Fig. 4.66
(d) Fig. 4.33	(j) Fig. 4.69
(e) Fig. 4.40	(k) Fig. 4.78
(f) Fig. 4.54	(l) Fig. 4.85

4.21 Derive $(\theta_o/\theta_i)(D)$ for Fig. 4.71b.

4.22 Construct logarithmic frequency-response curves for a piezoelectric accelerometer with $K_q/(C\omega_n^2) = 0.001$ V/(in/s^2), $\tau = 0.10$ s, $\omega_n/(2\pi) = 10,000$ Hz, and $\zeta = 0$. Will this accelerometer be satisfactory for shock measurements of half-sine pulses with a duration of 0.05 s? If not, suggest needed changes.

4.23 Explain, giving a sketch, how the principle of the system of Fig. 4.78 can be adapted to the measurement of angular acceleration. Your device must *not* be sensitive to translational acceleration.

4.24 In the system of Fig. 4.78, if K_s and B are made zero, $(\theta/T_n)(D) = 1/(JD^2)$. Obtain $(e_o/\ddot{x}_i)(D)$ for this situation. What is the defect in this system? To remedy this, electric "damping" may be introduced by adding a circuit with a transfer function as shown in Fig. P4.4. Obtain $(e_o/\ddot{x}_i)(D)$ for this arrangement. Why must $\tau_1 > \tau_2$?

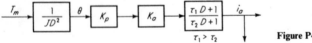

Figure P4.4

4.25 When a seismic-displacement pickup is used in its proper frequency range, what is an adequate mechanical model (masses, springs, dashpots) for its loading effect on the measured system?

4.26 In the Jerkmeter of Fig. 4.80, replace ω_{n1} and ζ_1 by their values in terms of J, B, and K_s, thus rewriting Eq. (4.127). Suppose a sensitivity $Mr/(K_i K_{ce}) = 0.05$ V/(ft/s^3) is required. Assume temporarily that $B = K_s = 0$, and neglect the fact that this system would be unstable. Assume also that J is due mainly to M; thus take $J = Mr^2$.

(a) Find the numerical value of $K_p K_{ae}/r$ needed to give a flat amplitude ratio for (e_o/D^3x_i) $(i\omega)$ within 5 percent over the range 0 to 5 Hz.

(b) If $r = 0.1$ ft and $K_p = 57.3$ V/rad, find K_{ae}.

(c) Suppose now that $M = 0.01$ lbm. Find $K_i K_{ce}$.

(d) If B and K_s are not zero, find the value of BK_s needed to put the system just on the margin of instability. (Use the Routh criterion.)

(e) Let the design value of BK_s be 10 times the value of part d. If $K_s = 0.275$ ft·lbf/rad, what is B?

(f) With the above values of B and K_s, recheck the amplitude ratio at 5 Hz. Does it meet the 5 percent requirement?

(g) To correct the situation found in part f, make $K_{ae} = 2,490$. Recheck the amplitude ratio at 5 Hz. Recheck the stability.

(h) To regain the stability lost in part g, reduce M to 0.001 lbm. Recheck the amplitude ratio at 5 Hz.

(i) Recheck the overall system static sensitivity. How much external amplification is now needed to return to the required 0.05 V/(ft/s^3)?

4.27 Derive Eqs. (4.131) to (4.135). Give a physical explanation of the "notch-filter" effect. Explain the apparent discontinuity in phase angle.

4.28 Find the steady-state response of the system of Fig. 4.81 to a constant horizontal acceleration.

4.29 A commercial version of the system of Fig. 4.81 has a pendulum made as shown in Fig. P4.5. Where is the center of buoyancy of this pendulum? Where is the center of mass? Will the buoyant force tend to cause an output? Why? Derive a relation showing the requirements for completely unloading the pivot bearing. (This "floating" reduces bearing friction and thus system threshold.)

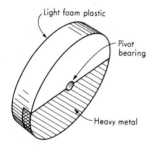

Light foam plastic

Pivot bearing

Heavy metal

Completely immersed in liquid   **Figure P4.5**

4.30 Equation (4.145) can be written as

$$(I_x I_y D^2 + B I_x D + H_s^2 + I_x K_s)\phi = -\frac{H_s}{D} T_y$$

For a gyro with no spring or damping, suppose T_y is a unit impulse. Solve Eqs. (4.145) and (4.147) for ϕ and θ. The combined sinusoidal motion of ϕ and θ causes the spin axis to rotate in space. This motion is called *nutation*. Describe it qualitatively and define its frequency. What is the effect on the above results if damping is present?

4.31 A rate gyro is mounted on a long, thin missile, which is quite flexible in bending, as in Fig. P4.6. Bending vibrations cause the slope at the gyro location to go through sinusoidal oscillations of 0.1° amplitude at 50 Hz. What maximum angular velocity will the gyro feel because of vibration? If the gross (rigid-body) rotation of the missile (which is what the gyro is *intended* to measure) is 10 rad/s, what percentage of the total gyro signal is due to the vibration? Where could one relocate the gyro to minimize this problem? If the fastest rigid-body motions expected are 1 Hz, what other solution is possible?

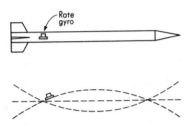

Mode shape of bending vibrations   **Figure P4.6**

4.32 In a rate gyro, the steady-state output/input ratio θ/ϕ is not strictly linear because the angular-momentum vector does not remain perpendicular to the angular-velocity vector ϕ when θ rotates away from zero.

    (a) What is the maximum allowable θ if this nonlinearity is not to exceed 1 percent?

    (b) If a rate gyro requires $\omega_n = 100$ rad/s and if $I_y = 0.0013$ in·lbf·s^2, what must the spring constant K_s be?

    (c) If the maximum input rate ϕ is 10 rad/s what must H_s be if θ_{max} is the value found in part a? Use K_s from part b.

    (d) If the spin motor runs at 24,000 r/min and if the wheel is a solid cylinder of a length equal to its diameter and is made of a material with a specific weight of 0.3 lbf/in^3, what are the required dimensions of the wheel? Use all necessary numbers from the previous parts of the problem.

4.33 The float-type wave height gage of Fig. P4.7 operates on the buoyant-force principle. The float has cross-sectional area A and mass M, and B represents the only significant frictional effect in the system. Derive $(x_o/h_i)(D)$. Suppose waves pass the gage at the rate of 0.2 waves per second and that

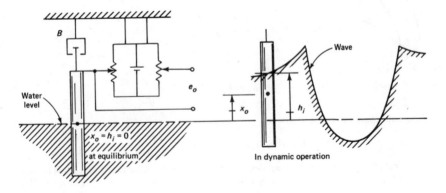

Figure P4.7

the wave profile contains significant Fourier harmonics up to the fifth. We wish to have 95 percent dynamic accuracy, based on the frequency-response-amplitude ratio. Get an expression for the maximum allowable float mass.

4.34 Modify the gyro analysis of Fig. 4.86 to account for a spurious input in the form of a vehicle rotation θ_f about the θ axis. Call the absolute gimbal rotation θ_g, and note that the θ pickoff output would now be proportional to $\theta_g - \theta_f$. Use these results to obtain disturbance-input transfer functions $(\theta/\theta_f)(D)$ for rate and rate-integrating gyros.

4.35 In a surface-roughness instrument (Fig. P4.8), the stylus of a displacement transducer is free to move vertically as the carriage is moved over the surface at constant velocity V. If V is too large, the stylus will bounce off the surface; too small V wastes time. To study this problem, consider the sinusoidal surface of wavelength L and amplitude A (Fig. P4.8b). Assume the weight of the stylus to be the only effect tending to maintain contact; neglect all friction. For a given A and L find the maximum allowable V to prevent bouncing. Find this V if $A = 2.5 \times 10^{-4}$ cm and $L = 2.5 \times 10^{-2}$ cm.

The configuration of Fig. P4.8c is suggested as a method of increasing the allowable V by a factor of about 6. Investigate this conjecture and list any drawbacks to this scheme.

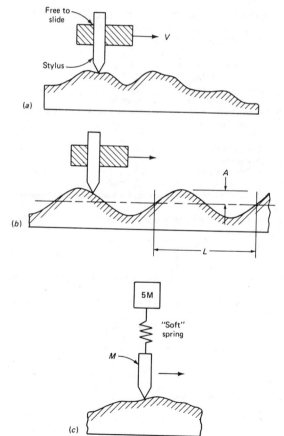

Figure P4.8

4.36 Perform a linear dynamic analysis on the pneumatic noncontact displacement gage of Fig. 4.55, using a sufficiently complete model to allow the possibility of instability. Then perform an accuracy-stability tradeoff study.

4.37 A servoaccelerometer (Fig. 4.78) is intended to respond to motion $\ddot{x}_i$ only; however, it may have some response to other motions.

(a) If $\ddot{x}_i \equiv 0$ but a cross-axis acceleration $\ddot{y}_i$ (perpendicular to $\ddot{x}_i$ and perpendicular to the paper in Fig. 4.78) occurs, will the instrument produce a spurious output? Explain.

(b) If a constant $\ddot{x}_i$ is simultaneously present with a constant $\ddot{y}_i$, will the instrument output have an error due to $\ddot{y}_i$? Calculate this error.

(c) If $\ddot{x}_i = \ddot{y}_i \equiv 0$ but the instrument frame experiences a steady angular acceleration $\ddot{\theta}_f$, will a spurious output occur? Derive $(e_o/\ddot{\theta}_f)(D)$.

BIBLIOGRAPHY

1. J. W. Dally and W. F. Riley: "Experimental Stress Analysis," 2d ed., McGraw-Hill, New York, 1978.
2. D. N. Keast: "Measurements in Mechanical Dynamics," McGraw-Hill, New York, 1970.
3. K. H. Hardman: Conductive Plastic Precision Pots, *Electromech. Des.*, p. 44, October 1963.
4. J. F. Blackburn (ed.): Potentiometers, "Components Handbook," chap. 8, McGraw-Hill, New York, 1948.
5. High Temperature Potentiometers, *Electromech. Des.*, p. 41, January 1960.
6. T. L. Foldvari and K. S. Lion: Capacitive Transducers, *Instr. Contr. Syst.*, p. 77, November 1964.
7. Electro-optical Moiré Fringe Transducer, *Electromech. Des.*, p. 26, December 1964.
8. W. H. Kliever: Measure Position Digitally, *Contr. Eng.*, p. 107, November 1956.
9. J. O. Morin: 6 Transducers for Precision Position Measurement, *Contr. Eng.*, p. 107, May 1960.
10. J. H. Brown: Measure Motion to 0.0001 In. without Friction or Wear, *Contr. Eng.*, p. 50, April 1955.
11. G. E. Bowie: Measurement of Displacements in Contact-Stress Experiments, *ASD Tech. Rep.* 61–450, Wright-Patterson Air Force Base, Ohio October 1961.
12. T. G. Baxter: Measurement of Pier Tilt with a Quartz Torsion Fiber Pendulum, *Jet Propulsion Lab. Tech. Rep.* 32–44, Jet Propulsion Laboratory, Pasadena, Calif., 1960.
13. W. H. Faulkner and J. G. Wood: Thickness Measuring Devices for Sheet and Web Materials, *Automation*, p. 67, July 1962.
14. K. V. Olsen: On the Standardization of Surface Roughness Measurements, *B & K Tech. Rev.*, B & K Instruments, Cleveland, Ohio, no. 3, 1961.
15. Length Calibration, National Bureau of Standards, *ISA J.*, p. 75, April 1963.
16. B. Sternlicht: An Indirect Method of Film-Thickness Measurement in Fluid-Film Bearings, *ISA Trans.*, vol. 2, no. 1, p. 28, January 1963.
17. A. L. Browne Fluid Film Thickness Measurement with Moiré Fringes, Gen'l Mtrs. Res. Lab. Publ. GMR-1186, March 16, 1972.
18. T. J. Rudd and E. L. Brandenburg: Inertial Profilometer as a Rail Surface Measuring Instrument, *ASME Paper* 73-ICT-102, 1973.
19. R. D. Young: The National Measurement System for Surface Finish, *NBS* IR75-927, March 1976.
20. R. W. Kern: Optical Inspection of Ground Metal Surfaces, *Electro-optical Syst. Des.*, pp. 20–27, July 1978.
21. J. B. Bryan and G. I. Boyadjieff: Measuring Surface Finish, A State-of-the-Art Report, *Mech. Eng.*, p. 42, December 1963.
22. F. Farago: Measuring the Critical Profile of Barrel Roller Bearings with Microinch Sensitivity, *Gen. Motors Eng. J.*, p. 17, January–February–March 1964.
23. R. Zito: Nuclear-Resonance Sensing of Mechanical Motion, *Electro-Technol.*, p. 43, June 1964.
24. K. G. Overbury: Temperature in Length Measurement, Sandia Corp., Albuquerque, N. Mex., February 1962.

25. D. H. Parkes: The Application of Microwave Techniques to Noncontact Precision Measurement, *ASME Paper* 63-WA-346, 1963.
26. D. R. Johnson: An Approach to Large-Scale Non-Contact Coordinate Measurements, *Hewlett-Packard J.*, pp. 3–20, September 1980.
27. J. G. Collier and G. F. Hewitt: Film-Thickness Measurements, *ASME Paper* 64-WA/HT-41, 1964.
28. D. E. Smith: Electronic Distance Measurement for Industrial and Scientific Applications, *Hewlett-Packard J.*, pp. 3–19, June 1980.
29. R. Zito: Velocity Sensing for Spacecraft Docking, *Space/Aeron.*, p. 90, December 1963.
30. J. Frey: The AC Tachometer, *Electro-Technol.*, p. 88, August 1963.
31. R. L. Pike: Measurement of Low Angular Rates, *Instr. Contr. Syst.*, p. 83, December 1962.
32. A. L. Fisher: Ball and Disk Read Angular Velocity Directly, *Contr. Eng.*, p. 125, November 1962.
33. L. E. Bollinger and K. E. Kissell: Measurement of Detonation-Wave Velocities, *ISA J.*, p. 170, May 1957.
34. Shaoul Ezekiel: Towards a Low-Level Accelerometer, *NASA*, CR-56941, 1964.
35. P. K. Chapman: A Cryogenic Test-Mass Suspension for a Sensitive Accelerometer, *NASA*, N64-27883, 1964.
36. Pressure Sensitivity of Accelerometers and Cables, Wilcoxon Research, Bethesda, Md.
37. J. T. Broch: Mechanical Vibration and Shock Measurements, Brüel and Kjaer Instruments, Marlboro, Mass., 1972.
38. H. R. Judge: Performance of Donner Linear Accelerometer Model 4310, *Space Technol. Lab. Tech. Note* 60-0000-09117, Space Technology Laboratories, Los Angeles, 1960.
39. A. Castle: Accelerometer Scribes Vector-Force Signatures, *Contr. Eng.*, p. 105, March 1965.
40. C. M. Harris and C. E. Crede (eds.): "Shock and Vibration Handbook," McGraw-Hill, New York, 1961.
41. C. K. Stedman: Some Characteristics of Gas Damped Accelerometers, Statham Instruments Inc., Los Angeles, 1958.
42. H. B. Sabin: 17 Ways to Measure Acceleration, *Contr. Eng.*, p. 106, February 1961.
43. R. P. Bowen: Calibrating Vibration Pickup Calibrators, *ISA J.*, p. 58, March 1951.
44. V. B. Corey: Measuring Angular Acceleration with Linear Accelerometers, *Contr. Eng.*, p. 79, March 1962.
45. K. E. Pope: A New Double-Integrating Accelerometer, *Contr. Eng.*, p. 97, November 1958.
46. K. N. Sergeyev: Investigation of Acceleration Pickup Having Filtering Properties, *NASA*, N64-23543, 1964.
47. D. K. Phillips: Balanced Beam Improves Angular Accelerometer, *Contr. Eng.*, p. 91, July 1964.
48. S. Rubin: Design of Accelerometers for Transient Measurement, *J. Appl. Mech.*, p. 509, December 1958.
49. A. Degenholtz: Optical-Wedge Technique for Measuring Angular Vibration, *Mach. Des.*, p. 167, March 12, 1964.
50. Design and Construction of a Lunar Seismograph, *NASA* N63-18290, 1963.
51. G. B. Foster: Non-Contacting Self-Calibrating Vibration Transducer, *Instr. Contr. Syst.*, p. 83, December 1963.
52. B. D. Van Deusen: Analysis of Vehicle Vibration, *ISA Trans.*, vol. 3, no. 2, p. 138, April 1964.
53. D. F. Wilkes and C. E. Kreitler: The Long Period Horizontal Air Bearing Seismometer, Sandia Corp., Albuquerque, N. Mex., SCTM74-62(13), 1962.
54. J. M. Slater: Exotic Gyros, *Contr. Eng.*, p. 92, November 1962.
55. Traverse Meter Uses Gyroscope Sensor, *Mach. Des.*, p. 163, April 13, 1961.
56. Intermediate Accuracy Gyroscope Design Criteria, Monograph, A. D. Little, Inc., Cambridge, Mass., June 1970.
57. A. W. Lane et al.: Achieving Extremely Accurate Non-Floated Gyros, *Aero/Space Eng.*, p. 43, January 1959.
58. H. Stern: Which Rate Gyro to Use, *Contr. Eng.*, p. 79, February 1958.
59. C. S. Draper et al.: The Floating Integrating Gyro, *Aeron. Eng. Rev.*, p. 46, June 1956.
60. Inertial Gyro Test, *Electromech. Des.*, p. 18, November 1960.

61. R. E. Barnaby et al.: Control of Thermal Drift in Floated Gyroscopes, *Sperry Eng. Rev.*, Sperry Gyroscope Co., Great Neck, N.Y., p. 36, September 1961.

62. W. G. Wing: Fluid Rotor Gyros, *Contr. Eng.*, p. 105, March 1963.

63. G. C. Newton: Vibratory Rate Gyros, *Contr. Eng.*, p. 95, June 1963.

64. E. H. Ernst: Basic Theory-Particle Gyroscope, *Electro-Technol.*, p. 12, August 1963.

65. H. L. Kreitzburg: Compensating Gyro Drifts, *Contr. Eng.*, p. 113, November 1963.

66. H. W. Knoebel: The Electric Vacuum Gyro, *Contr. Eng.*, p. 70, February 1964.

67. Gimbal-less Gyro, *Mech. Eng.*, p. 59, May 1964.

68. S. Redner and F. Zandman: Experimental Stress Analysis, *Ind. Res.*, p. 67, May 1965.

69. L. H. Ravitch: Some Applications of Stress Analysis Techniques in Improving Casting Designs, *Gen. Motors Eng. J.*, p. 22, October–November–December 1958.

70. S. S. Manson: Thermal Stresses, Measurements by Photoelasticity, *Mach. Des.*, p. 143, November 26, 1959.

71. F. Zandman: Stress Analysis with a Photoelastic Coating, *Metal Progr.*, p. 111, November 1960.

72. F. B. Stern: Strain Sensitive Ceramic Base Brittle Coatings, *Mach. Des.*, p. 147, May 29, 1958.

73. G. Gerard and H. Tramposch: Photothermoelastic Investigation of Transient Thermal Stresses in a Multiweb Wing Structure, *J. Aerosp. Sci.*, p. 783, December 1959.

74. A. F. Rudl and M. J. Ward: Laser Transducer Systems for High-Accuracy Machine Positioning, *Hewlett-Packard J.*, pp. 2–18, February 1976.

75. P. K. Stein: Pulsing Strain-Gage Circuits, *Instr. Contr. Syst.*, p. 128, February 1965.

76. P. K. Stein: Strain Gages, "Measurement Engineering," Stein Engineering Services, Inc., Phoenix, Ariz.

77. S. S. Manson: Thermal Stresses in Design-Strain-Gage Applications, *Mach. Des.*, p. 683, November 12, 1959.

78. A. Kaufman: Performance of Electrical-Resistance Strain Gages at Cryogenic Temperatures, *NASA, Tech. Note*, D-1663, 1963.

79. R. H. Kemp et al.: Application of a High-Temperature Static Strain Gage to the Measurement of Thermal Stresses in a Turbine Stator Vane, *NACA, Tech. Note* 4215, 1958.

80. S. S. Manson: Thermal Stresses in Design-Strain-Gage Measurements, *Mach. Des.*, p. 109, October 29, 1959.

81. Typical Examples of Measurement Problems and Their Solutions, Hottinger Baldwin Measurements Inc., Natick, Mass., 1978.

82. J. Gunn and E. Billinghurst: Magnetic Fields Affect Strain Gages, *Contr. Eng.*, p. 109, August 1957.

83. "Semiconductor Strain Gage Handbook," Baldwin-Lima-Hamilton Corp., Waltham, Mass.

84. R. J. Chaka: The Design and Development of a Highway Speed Road Profilometer, *SAE Paper* 780064, 1978.

85. R. Shiver and W. Putman: Measuring Dynamic Strain on High-Speed Turbine Wheels, *Instr. Contr. Syst.*, p. 118, September 1962.

86. A. J. Bush: Soldered-Cap Strain Gages, *Mach. Des.*, p. 163, November 8, 1962.

87. C. E. Mathewson: The "Dimensionless" Strain Gage, *Instr. Contr. Syst.*, p. 1870, October 1961.

88. D. Post: The Moiré Grid-Analyzer Method for Strain Analysis, *Exp. Mech.*, pp. 368–377, November 1965.

89. G. R. Sarna: Extending the Range of Seismic Transducers, *Inst. & Cont. Syst.*, February 1972.

90. W. J. Pastorius: Vibration Analysis by Holography, *Mech. Eng.*, June 1972.

91. J. K. S. Walter: Motion Sensor Development, Sandia Labs Rep. SC-DR-710910, Albuquerque, N. Mex., 1972.

92. Y. T. Li: Air Damped Capacitance Accelerometers and Velocimeters, *Trans. of IECI Group of IEEE Sec. Trans. Conf.*, May 4, 1970.

93. S. Y. Lee: Signal Transduction with Differential Pulse Width Modulation (Capacitance Transducers), *Trans. of IECI Group of IEEE Sec. Trans. Conf.*, May 4, 1970.

94. D. Olshove: Velocity Measurement with a Piezoresistive Acceleration-Integration System, "Inst. in the Aerospace Ind., Vol. 17," Inst. Soc. of Am., Pittsburgh, 1971.

95. R. L. Gates: Reviewing the Status of Inertial Sensors, *Contr. Eng.*, March 1971.

96. J. K. Emery: Roundness Measurement, *Mech. Eng.*, October 1969.
97. R. F. Hill and M. Skunda: Measuring Tool Wear Radiometrically, *Mech. Eng.*, February 1972.
98. J. B. Bryan et al.: Thermal Effects in Dimensional Metrology, *ASME Paper* 65-*Prod*-13, 1965.
99. R. Iltis: Solid-State Sensors: Strain Gages, *Contr. Eng.*, January 1970.
100. R. E. Herzog: Forecasting Failures with Acoustic Emission, *Mach. Des.*, June 14, 1973.
101. F. H. Middleton, L. R. Le Blanc, and M. F. Czarnecki: Spectral Tuning and Calibration of a Wave Follower Buoy, Eighth Annual Offshore Tech. Conf., Houston, May 3–6, 1976.
102. D. G. Fleming et al. (eds.): "Handbook of Engineering in Medicine and Biology," sec. B, Instruments and Measurements, CRC Press, Boca Raton, Fla., 1978.

FIVE

FORCE, TORQUE, AND SHAFT POWER MEASUREMENT

5.1 STANDARDS AND CALIBRATION

Force is defined by the equation $F = MA$; thus a standard for force depends on standards for mass and acceleration. Mass is considered a fundamental quantity, and its standard is a cylinder of platinum-iridium, called the International Kilogram, kept in a vault at Sèvres, France. Other masses (such as national standards) may be compared with this standard by means of an equal-arm balance, with a precision of a few parts in 10^9 for masses of about 1 kg. Tolerances on various classes of standard masses available from the National Bureau of Standards may be found in its publications.[1]

Acceleration is not a fundamental quantity, but rather is derived from length and time, two fundamental quantities whose standards are discussed in Chap. 4. The acceleration of gravity, g, is a convenient standard which can be determined with an accuracy of about 1 part in 10^6 by measuring the period and effective length of a pendulum or by determining the change with time of the speed of a freely falling body. The actual value of g varies with location and also slightly with time (in a periodic predictable fashion) at a given location. It also may change (slightly) unpredictably because of local geological activity. The so-called standard value of g refers to the value at sea level and 45° latitude and is numerically 980.665 cm/s^2. The value at any latitude ϕ degrees may be computed

---

[1] T. W. Lashof and L. B. Macurdy, Precision Laboratory Standards of Mass and Laboratory Weights, *Natl. Bur. Std. (U.S.), Circ.* 547, sec. 1, 1954; P. E. Pontius, Mass and Mass Values, *Natl. Bur. Std. (U.S.), Monograph* 133, 1974.

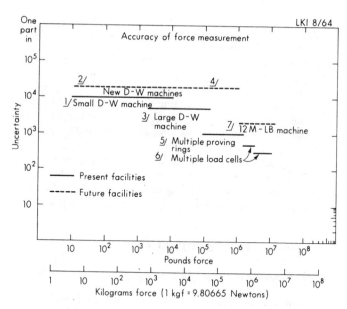

Figure 5.1 Force standards.

from

$$g = 978.049(1 + 0.0052884 \sin^2 \phi - 0.0000059 \sin^2 2\phi) \quad \text{cm/s}^2 \quad (5.1)$$

while the correction for altitude h in meters above sea level is

$$\text{Correction} = -(0.00030855 + 0.00000022 \cos 2\phi)h$$

$$+ 0.000072 \left(\frac{h}{1,000}\right)^2 \quad \text{cm/s}^2 \quad (5.2)$$

Local values of g also may be obtained from the National Ocean Survey, National Oceanic and Atmospheric Administration.

When the numerical value of g has been determined at a particular locality, the gravitational force (weight) on accurately known standard masses may be computed to establish a standard of force. This is the basis of the "deadweight" calibration of force-measuring systems. The National Bureau of Standards capability (Fig. 5.1[1]) for such calibrations is an inaccuracy of about 1 part in 5,000 for the range of 10 to 1 million lbf. Above this range, direct deadweight calibration is not presently available. Rather, proving rings[2] or load cells of a capacity of 1 million lbf or less are calibrated against deadweights, and then the unknown force

---

[1] Accuracy in Measurements and Calibrations, *Natl. Bur. Std. (U.S.), Tech. Note* 262, 1965.
[2] Proving Rings for Calibrating Testing Machines, *Natl. Bur. Std. (U.S.), Circ.* C454, 1946.

is applied to a multiple array of these in parallel. The range 1 to 10 million lbf is covered by such arrangements with somewhat reduced accuracy. At the low-force end of the scale, the accuracy of standard masses ranges from about 1 percent for a mass of 10^{-5} lbm to 0.0001 percent for the 0.1 to 10 lbm range to 0.001 percent for a 100-lb mass. The accuracy of *force* calibrations using these masses must be somewhat less than the quoted figures because of error sources in the experimental procedure.[1] A commercially available[2] calibrating machine using deadweights, knife edges, and levers covers the range of 0 to 10,000 lbf (or 0 to 50 kN) with an accuracy of ± 0.005 percent of applied load and a resolution of ± 0.0062 percent of applied load.

The measurement of torque is intimately related to force measurement; thus torque standards as such are not necessary, since force and length are sufficient to define torque. The power transmitted by a rotating shaft is the product of torque and angular velocity. Angular-velocity measurement was treated in Chap. 4.

5.2 BASIC METHODS OF FORCE MEASUREMENT

An unknown force may be measured by the following means:

1. Balancing it against the known gravitational force on a standard mass, either directly or through a system of levers
2. Measuring the acceleration of a body of known mass to which the unknown force is applied
3. Balancing it against a magnetic force developed by interaction of a current-carrying coil and a magnet
4. Transducing the force to a fluid pressure and then measuring the pressure
5. Applying the force to some elastic member and measuring the resulting deflection
6. Measuring the change in precession of a gyroscope caused by an applied torque related to the measured force
7. Measuring the change in natural frequency of a wire tensioned by the force

In Fig. 5.2, method 1 is illustrated by the analytical balance, the pendulum scale, and the platform scale. The analytical balance, while simple in principle, requires careful design and operation to realize its maximum performance.[3] The beam is designed so that the center of mass is only slightly (a few thousandths of an inch) below the knife-edge pivot and thus barely in stable equilibrium. This makes the beam deflection (which in sensitive instruments is read with an optical

---

[1] Calibration of Force-Measuring Instruments for Verifying the Load Indication of Testing Machines, *ASTM Std.* E-74, 1974.

[2] W. C. Dillon Co., Van Nuys, Calif.

[3] L. B. Macurdy, Performance Tests for Balances, *Inst. & Cont. Syst.*, pp. 127–133, September 1965.

micrometer) a very sensitive indicator of unbalance. For the low end of a particular instrument's range, often the beam deflection is used as the output reading rather than attempting to null by adding masses or adjusting the arm length of a poise weight. This approach is faster than nulling but requires that the deflection-angle unbalance relation be accurately known and stable. This relation tends to vary with the load on the balance, because of deformation of knife edges, etc., but

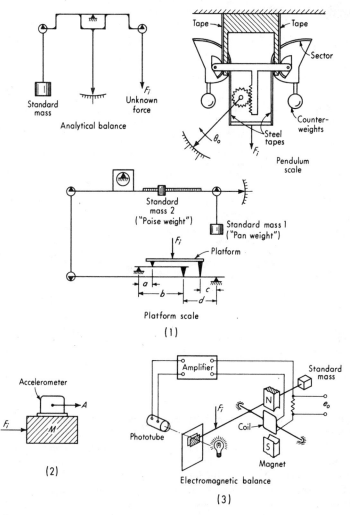

Figure 5.2 Basic force-measurement methods.

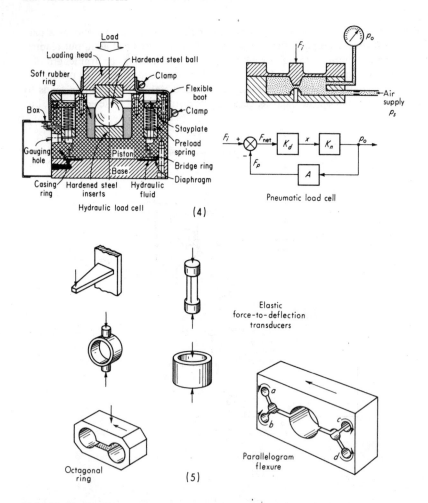

Hydraulic load cell

(4)

Pneumatic load cell

Elastic
force-to-deflection
transducers

Octagonal
ring

(5)

Parallelogram
flexure

Figure 5.2 (*Continued*)

careful design can keep this to a minimum. For highly accurate measurements, the buoyant force due to the immersion of the standard mass in air must be taken into account. Also, the most sensitive balances must be installed in temperature-controlled chambers and manipulated by remote control to reduce the effects of the operator's body heat and convection currents. Typically, a temperature difference of $1/20C°$ between the two arms of a balance can cause an arm-length ratio change of 1 ppm, significant in some applications. Commercially available analytical balances may be classified as follows:[1]

Description	Range, g	Resolution, g
Macro analytical	200–1,000	10^{-4}
Semimicro analytical	50–100	10^{-5}
Micro analytical	10–20	10^{-6}
Micro balance	less than 1	10^{-6}
Ultramicro balance	less than 0.01	10^{-7}

The pendulum scale is a deflection-type instrument in which the unknown force is converted to a torque that is then balanced by the torque of a fixed standard mass arranged as a pendulum. The practical version of this principle utilizes specially shaped sectors and steel tapes to linearize the inherently nonlinear torque-angle relation of a pendulum. The unknown force F_i may be applied directly as in Fig. 5.2 or through a system of levers, such as that shown for the platform scale, to extend the range. An electrical signal proportional to force is easily obtained from any angular-displacement transducer attached to measure the angle θ.

The platform scale utilizes a system of levers to allow measurement of large forces in terms of much smaller standard weights. The beam is brought to null by a proper combination of pan weights and adjustment of the poise-weight lever arm along its calibrated scale. The scale can be made self-balancing by adding an electrical displacement pickup for null detection and an amplifier-motor system to position the poise weight to achieve null. Another interesting feature is that if $a/b = c/d$, the reading of the scale is independent of the location of F_i on the platform. Since this is quite convenient, most commercial scales provide this feature by use of the suspension system shown or others that allow similar results.

While analytical balances are used almost exclusively for "weighing" (really determining the *mass* of) objects or chemical samples, platform and pendulum scales are employed also for force measurements, such as those involved in shaft power determinations with dynamometers. All three instruments are intended mainly for static force measurements.

Method 2, the use of an accelerometer for force measurement, is of somewhat limited application since the force determined is the *resultant* force on the mass. Often *several* unknown forces are acting, and they cannot be separately measured by this method.

The electromagnetic balance[2] (method 3) utilizes a photoelectric (or other displacement sensor) null detector, an amplifier, and a torquing coil in a servosystem to balance the difference between the unknown force F_i and the gravity force on a standard mass. Its advantages relative to mechanical balances are ease

[1] F. Baur, The Analytical Balance, *Ind. Res.*, p. 64, July–August 1964.

[2] L. Cahn, Electromagnetic Weighing, *Instrum. Contr. Syst.*, p. 107, September 1962; Cahn Instrument Div., Cerritos, Calif.

of use, less sensitivity to environment, faster response, smaller size, and ease of remote operation. Also, the electric output signal is convenient for continuous recording and/or automatic-control applications. Balances with built-in microprocessors[1] allow even greater convenience, versatility, and speed of use by automating many routine procedures and providing features not formerly feasible. Automatic tare-weight systems subtract container weight from total weight to give net weight when material is placed in the container. Statistical routines allow immediate calculation of mean and standard deviation for a series of weighings. "Counting" of small parts by weighing is speeded by programming the microprocessor to read out the parts count directly, rather than the weight. Accurate weighing of live laboratory animals (difficult on an ordinary balance because of animal motion) is facilitated by averaging scale readings over a preselected time. Interfacing the balance to (external or built-in) printers for permanent recording also is eased by the microprocessor.

Method 4 is illustrated in Fig. 5.2 by hydraulic[2] and pneumatic load cells. Hydraulic cells are completely filled with oil and usually have a preload pressure of the order of 30 lb/in^2. Application of load increases the oil pressure, which is read on an accurate gage. Electrical pressure transducers can be used to obtain an electrical signal. The cells are very stiff, deflecting only a few thousandths of an inch under full load. Capacities to 100,000 lbf are available as standard while special units up to 10 million lbf are obtainable. Accuracy is of the order of 0.1 percent of full scale; resolution is about 0.02 percent. A hydraulic totalizer[3] is available to produce a single pressure equal to the sum of up to 10 individual pressures in multiple-cell systems used for tank weighing, etc. (see Chap. 10).

The pneumatic load cell shown uses a nozzle-flapper transducer as a high-gain amplifier in a servoloop. Application of force F_i causes a diaphragm deflection x, which in turn causes an increase in pressure p_o since the nozzle is more nearly shut off. This increase in pressure acting on the diaphragm area A produces an effective force F_p that tends to return the diaphragm to its former position. For any constant F_i, the system will come to equilibrium at a specific nozzle opening and corresponding pressure p_o. The static behavior is given by

$$(F_i - p_o A)K_d K_n = p_o \tag{5.3}$$

where
$$K_d \triangleq \text{diaphragm compliance, in/lbf} \tag{5.4}$$

$$K_n \triangleq \text{nozzle-flapper gain, (lb/in}^2\text{)/in} \tag{5.5}$$

Solving for p_o, we get

$$p_o = \frac{F_i}{1/(K_d K_n) + A} \tag{5.6}$$

Now K_n is not strictly constant, but varies somewhat with x, leading to a nonlin-

---

[1] B. Ludewig, Microprocessor Balance, *Am. Lab.*, pp. 81–83, May 1979.
[2] A. H. Emery Co., New Canaan, Conn.
[3] Ibid.

earity between x and p_o. However, in practice, the product $K_d K_n$ is very large, so that $1/(K_d K_n)$ is made negligible compared with A, which gives

$$p_o = \frac{F_i}{A} \tag{5.7}$$

which is linear since A is constant. As in any feedback system, dynamic instability limits the amount of gain that actually can be used. A typical supply pressure p_s is 60 lb/in^2, and since the maximum value of p_o cannot exceed p_s, this limits F_i to somewhat less than $60A$. A line of commercial pneumatic weighing systems[1] using similar principles (combined with lever/knife-edge methods) is available in standard ranges to 110,000 lbf.

While all the previously described force-measuring devices are intended mainly for static or slowly varying loads, the elastic deflection transducers of method 5 are widely used for both static and dynamic loads of frequency content up to many thousand hertz. While all are essentially spring-mass systems with (intentional or unintentional) damping, they differ mainly in the geometric form of "spring" employed and in the displacement transducer used to obtain an electrical signal. The displacement sensed may be a gross motion, or strain gages may be judiciously located to sense force in terms of strain. Bonded strain gages have been found particularly useful in force measurements with elastic elements. In addition to serving as force-to-deflection transducers, some elastic elements perform the function of resolving vector forces or moments into rectangular components. An example, the parallelogram flexure of Fig. 5.2 is extremely rigid (insensitive) to all applied forces and moments except in the direction shown by the arrow. A displacement transducer arranged to measure motion in the sensitive direction thus will measure only that component of an applied vector force which lies along the sensitive axis. Perhaps the action of this flexure may be most easily visualized by considering it as a four-bar linkage with flexure hinges at the thin sections a, b, c, and d.

Because of the importance of elastic force transducers in modern dynamic measurements, we devote a considerable portion of this chapter to their consideration. Although they may differ widely in detail construction, their dynamic-response form is generally the same, and so we treat an idealized model representative of all such transducers in the next section. Discussion of methods 6 and 7 is deferred to the end of the chapter since they are not as common as methods 1 through 5.

5.3 CHARACTERISTICS OF ELASTIC FORCE TRANSDUCERS

Figure 5.3 shows an idealized model of an elastic force transducer. The relationship between input force and output displacement is easily established as a simple

---

[1] An Introduction to the Darenth Gnu-Weigh Pneumatic Weighing System, Darenth Americas, Bridgeville, Del., 1980.

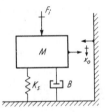

Figure 5.3 Elastic force transducer.

second-order form:

$$F_i - K_s x_o - B\dot{x}_o = M\ddot{x}_o \tag{5.8}$$

$$\frac{x_o}{F_i}(D) = \frac{K}{D^2/\omega_n^2 + 2\zeta D/\omega_n + 1} \tag{5.9}$$

where

$$\omega_n \triangleq \sqrt{\frac{K_s}{M}} \tag{5.10}$$

$$\zeta \triangleq \frac{B}{2\sqrt{K_s M}} \tag{5.11}$$

$$K \triangleq \frac{1}{K_s} \tag{5.12}$$

Note that devices of this type are also (unintentional) accelerometers and produce a spurious output in response to base vibration inputs (see Prob. 5.1).

For transducers that do not measure a gross displacement but rather use strain gages bonded to the "spring" K_s, the output strain ϵ may be substituted for x_o if K_s is reinterpreted as force per unit strain rather than force per unit deflection. In many transducers a distinct and separate "spring" and "mass" cannot be distinguished because the elasticity and inertia are distributed rather than lumped. In these cases, for design purposes the natural frequency must be calculated from the appropriate formulas[1] for the geometric shapes involved rather than by employing Eq. (5.10). Once the transducer is constructed, its lowest natural frequency generally can be found experimentally, as can ζ, which is usually small and difficult to calculate since B usually represents parasitic (rather than designed-in) friction effects. Generally the sensitivity K is available theoretically from strength-of-materials or elasticity formulas, whether it relates to a gross deflection or a local unit strain. Once the transducer is constructed, it should be given an overall calibration relating electrical output to force input since none of the theoretical formulas is sufficiently accurate for this purpose.

[1] C. M. Harris and C. E. Crede (eds.), "Shock and Vibration Handbook," vol. 1, chap. 7, McGraw-Hill, New York, 1961; J. P. Den Hartog, "Mechanical Vibrations," 4th ed., pp. 431–433, McGraw-Hill, New York, 1956; R. K. Mitchell, Some Considerations in the Design of Elastic Force Transducers, M.Sc. Thesis, The Ohio State University, Mechanical Engineering Department, 1965.

Since the dynamic response of second-order instruments has been fully discussed, we concentrate mainly on details peculiar to specific force transducers.

Bonded-Strain-Gage Transducers

A typical construction for a strain-gage load cell for measuring compressive forces is shown in Fig. 5.4. (Cells to measure both tension and compression require merely the addition of suitable mechanical fittings at the ends.) The load-sensing member is short enough to prevent column buckling under the rated load and is proportioned to develop about 1,500 $\mu\epsilon$ at full-scale load (typical design value for all forms of foil gage transducers). Materials used include SAE 4340 steel, 17-4 PH stainless steel, and 2024–T4 aluminum alloy, with the last being quite popular for "homemade" transducers. Foil-type metal gages are bonded on all four sides; gages 1 and 3 sense the direct stress due to F_i, and gages 2 and 4 the transverse stress due to Poisson's ratio μ. This arrangement gives a sensitivity $2(1 + \mu)$ times that achieved with a single active gage in the bridge. [See Eq. (10.8) for bridge-circuit behavior.] It also provides primary temperature compensation since all four gages are (at least for steady temperatures) at the same temperature. Furthermore, the arrangement is insensitive to

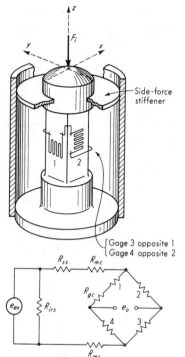

Figure 5.4 Strain-gage load cell.

bending stresses due to F_i being applied off center or at an angle. This can be seen by replacing an off-center force by an equivalent on-center force and a couple. The couple can be resolved into x and y components which cause bending stresses in the gages. If the gages are carefully placed so as to be symmetric, the bending stresses in gages 1 and 3 will be of opposite sign, and by the rules of bridge circuits the net output e_o due to bending will be zero. Similar arguments hold for gages 2 and 4 and for bending stresses due to F_i being at an angle. The side-force stiffener plate also reduces the effects of angular forces, since it is very stiff in the radial (x, y) direction but very soft in the z direction.

The deflection under full load of such load cells is of the order of 0.001 to 0.015 in, indicating their high stiffness. Often the natural frequency is not quoted since it is determined almost entirely by the mass of force-carrying elements external to the transducer. This is especially true in the many applications where the load cell is used for weighing purposes. The high stiffness also implies a low sensitivity. To increase sensitivity (in low-force cells where it is needed) without sacrificing column stability and surface area for mounting gages, a hollow (square on the outside, round on the inside) load-carrying member may be employed.

To achieve the high accuracy (0.3 to 0.1 percent of full scale) required in many applications, additional temperature compensation is needed.[1] This is accomplished by means of the temperature-sensitive resistors R_{gc} and R_{mc} shown in Fig. 5.4. These resistors are permanently attached internal to the load cell so as to assume the same temperature as the gages. The purpose of R_{gc} is to compensate for the slightly different temperature coefficients of resistance of the four gages. The purpose of R_{mc} is to compensate for the temperature dependence of the modulus of elasticity of the load-sensing member. That is, although we wish to measure force, the gages sense strain; thus any change in the modulus of elasticity will give a different strain (and thus a different e_o) even though the force is the same. Since all metals change modulus somewhat with temperature, this effect causes a sensitivity drift. The resistance R_{mc} compensates for this by changing the excitation voltage actually applied to the bridge by just the right amount to counteract the modulus effect.

Two additional (non-temperature-sensitive) resistors are often found in commercial load cells. They are R_{ss}, which is adjusted to standardize the sensitivity for a nominal e_{ex} to a desired value, and R_{irs}, which is used to adjust the input resistance to a desired value.

When adequate sensitivity cannot be achieved by use of tension/compression members, configurations employing bending stresses may be helpful. These generally provide more strain per unit applied force, but at the expense of reduced stiffness and thus natural frequency. Of the many possibilities, two are shown in Fig. 5.5. The cantilever-beam gage arrangement provides 4 times the sensitivity of

---

[1] J. Dorsey, Homegrown Strain-Gage Transducers, *Exp. Mech.*, pp. 255–260, July 1977; *VRE Tech. Ed. News*, no. 21, Vishay Research and Education, Malvern, Pa., February 1978; J. Dorsey, Linearization of Transducer Compensation, Vishay Intertechnology, Romulus, Mich., 1977; SR-4 Strain Gage Handbook, BLH Electronics, Waltham, Mass., 1980.

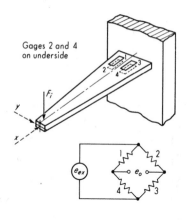

Gages 2 and 4
on underside

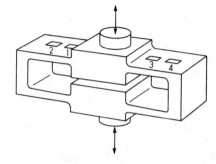

Figure 5.5 Strain-gage beam transducers.

a single gage, temperature compensation, and insensitivity to x and y components of force if identical gages and perfect symmetry are assumed.

Transducers using shear loading[1] can be designed to be very compact (in the load application direction), have little sensitivity to off-axis forces and moments, good symmetry for tension/compression, long fatigue life, simple overload protection, and high stiffness. Figure 5.6 shows one of several possible designs, where four gages (only two show) oriented at 45° with the load axis are cemented to the shear webs between the holes to pick up the tension and compression strains produced by the applied shear stress. With two gages in compression and two in tension, a full-bridge circuit can be used. The nature of the overload stop allows "stacking" of a high-range (say 100,000 lbf) and a low-range (say 10,000 lbf) cell to provide a dual-range transducer whose accuracy is maintained at loads below 10 percent of high range. That is, at loads below 10,000 lbf we use the output of the low-range cell, but as load goes above 10,000 lbf, this cell "bottoms out" (safely) and we take our readings from the high-range cell.

---

[1] A. Umit Kutsay, Flat Load Cells, *Inst. & Cont. Syst.*, pp. 123–125, February 1966; Interface Load Cell—A New Dimension in Force Measurement, Interface Inc., Scottsdale, Ariz., 1980.

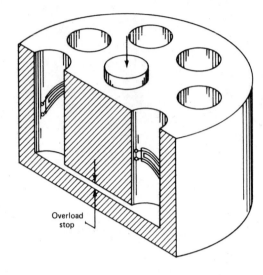

Figure 5.6 Shear-web force transducer.

Commercial strain-gage force transducers of the various types are available for full-scale loads of a few pounds to hundreds of thousands of pounds and in several accuracy (and price) grades. The lowest-accuracy grade typically has an overall (combined nonlinearity, hysteresis, nonrepeatability, etc.) error of about 1 percent of full scale, temperature effects of ± 0.005 percent of full scale/°F on zero shift and ± 0.01 percent of reading/°F on span. Corresponding figures for the highest-accuracy grade are about 0.15 percent full-scale overall error, ± 0.0015 percent full-scale/°F zero shift, and ± 0.008 percent of reading/°F on span.

When maximum output is desired for any strain-gage transducer, one should consider the possible use of low-modulus materials (such as aluminum) to increase strain per unit force, several gages (or one high-resistance gage) per bridge leg (if space allows), the intentional introduction of stress concentrations at the gage locations, and/or use of semiconductor gages.[1] However, such techniques also present associated problems. Low modulus reduces stiffness and natural frequency, and some low-modulus materials have excessive hysteresis and low fatigue life. Stress concentrations also lower fatigue life, and their effect may be difficult to calculate for design purposes. Semiconductor gages present installation and temperature compensation problems.

Differential-Transformer Transducers

Figure 5.7 shows the design of a family of LVDT load cells available in ranges from 10 g to 1000 lbf which use two helical flexures (each machined from one

---

[1] M. Lebow and R. Caris, Semiconductor Gages or Foil Gages for Transducers, Lebow Associates Inc., Troy, Mich., 1965.

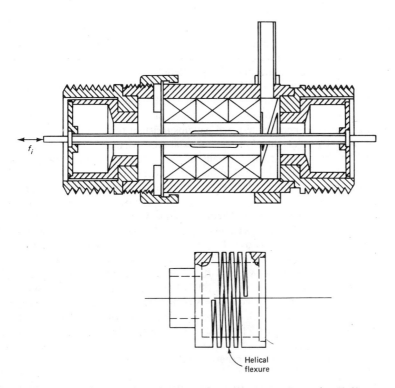

Figure 5.7 LVDT force transducer. (*Courtesy Schaevitz Engineering, Pennsauken, N. J.*)

solid piece) as the elastic element. An external threaded ring allows convenient
zero adjustment by changing the axial position of the spring-loaded LVDT coil
form. Full-scale deflections are about 0.020 to 0.030 in, nonlinearity is better than
0.2 percent full scale, and operating temperature range is -65 to $+200°F$.

Piezoelectric Transducers

These force transducers have the same form of transfer function as piezoelectric
accelerometers. They are intended for dynamic force measurement only, although
some types (quartz pickup with electrometer charge amplifier) have sufficiently
large τ to allow short-term measurement of static forces and static calibration.
Just as in the piezoelectric accelerometers of Chap. 4, force transducers are available with or without built-in microelectronics. Typical units have about 1 percent
nonlinearity and very high stiffness (10^6 to 10^8 lbf/in) and natural frequency
(10,000 to 300,000 Hz), though, as with any force transducer, the *system* natural
frequency will be lower because of the unavoidable mass of necessary attached
components. A single transducer often is useful over a very wide range of forces

and preloads because the 1 percent nonlinearity applies to any calibrated range and the natural "leak-off" of output voltage for steady loads gives a convenient zero reading. Useful temperature ranges and sensitivity to temperature inputs depend on piezoelectric material used, design details, and whether built-in electronics are employed. Piezoelectric force pickups tend to be sensitive to side loading, and most manufacturers recommend special precautions to minimize this, but do not always quote numerical values of cross-axis sensitivity. One manufacturer[1] offering specially designed pickups resistant to side loading quotes transverse sensitivity as less than 7 percent of axial sensitivity.

Figure 5.8a shows construction details of two piezoelectric load cells. The smaller one is permanently preloaded to measure in the range 1,000-N tension to 5,000-N compression without relaxing the built-in compressive preload. The larger unit comes with external preloading nuts which may be adjusted to cover the range 4,000-N tension to 16,000-N compression. An analysis[2] by the manufacturer suggests that such transducers may be modeled as a spring (piezoelectric elements) sandwiched between two end masses. Figure 5.8b shows such a transducer being used to measure the force applied by a vibration shaker to some structure being vibration-tested (K_{m_1} and K_{m_2} represent the stiffness of mounting screws). If the impedance of the structure (including K_{m_2}) is called Z_s, then since $Z_s \triangleq f_s/v_s$, the force actually applied to the structure is $Z_s v_s$. The force F_m *measured*, however, is that in K_t, which is proportional to the relative displacement of M_{t_1} and M_{t_2} since this is the deflection of the piezoelectric element. Under dynamic conditions F_m is not necessarily equal to F_s, and a dynamic analysis as shown below is useful in developing criteria for accurate measurement:

$$F_m - v_s Z_s = M_{t_2} D v_s \qquad \frac{F_m}{F_s}(D) = \frac{M_{t_2} D}{Z_s(D)} + 1 \qquad (5.13)$$

If $M_{t_2} \equiv 0$, there would be no error no matter what Z_s might be; thus transducers with small M_{t_2} are clearly preferable (the small unit of Fig. 5.8a has $M_{t_2} \approx 3g$). For a "springlike" structure, $Z_s(D) = K_s/D$, and

$$\frac{F_m}{F_s}(i\omega) = 1 - \frac{M_{t_2}}{K_s}\omega^2 \qquad (5.14)$$

which shows how accuracy varies with frequency ω. Other assumed forms of Z_s are easily investigated for accuracy criteria.

Studies of structural impedance actually are quite useful and common in dynamic analysis of mechanical systems.[3] Thus manufacturers have developed the *impedance head*, a dual sensor which combines a separate load cell and an accelerometer into a single, compact package (see Fig. 5.9). While impedance basically involves force and *velocity*, generally accelerometers are used since they are more convenient and the integration necessary to obtain velocity is easily

[1] Wilcoxon Research, Bethesda, Md.

[2] W. Braender, High-Frequency Response of Force Transducers, *B & K Tech. Rev.*, no. 3, Bruel and Kjaer Instruments, Marlboro, Mass., 1972.

[3] E. O. Doebelin, "System Modeling and Response," Wiley, New York, 1980.

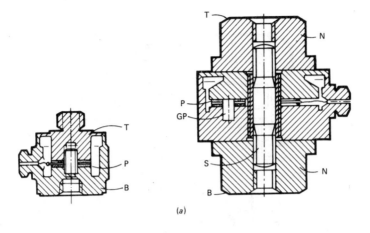

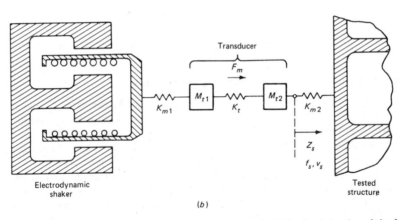

Figure 5.8 Piezoelectric load cells and dynamic error analysis. (*a*) Sectional drawing of the force transducer. T = top; P = piezoelectric disks; GP = guide pin; S = preloading screw; N = preloading nut; B = base. (*Courtesy Bruel and Kjaer Instruments, Marlboro, Mass.*)

accomplished electronically or numerically. Again, low mass between the force-sensing crystals and the driving point is desired, but we now also strive for high *stiffness* between the accelerometer base and the driving point, so that the measured acceleration accurately reflects that of the driving point. When this acceleration measurement is accurate (by using either an impedance head or *separate* force and acceleration sensors), an accuracy-improving trick called *mass cancellation* is possible. Here the "true" value of $F_s = v_s Z_s$ is calculated from Eq. (5.13) by electrically adding to the directly measured F_m a value $M_{t2} \dot{v}_s$, using the known value of M_{t2} and the measured acceleration $\dot{v}_s$.

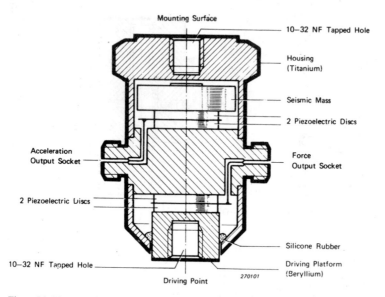

Mounting Surface

10–32 NF Tapped Hole

Housing
(Titanium)

Seismic Mass

2 Piezoelectric Discs

Acceleration
Output Socket

Force
Output Socket

2 Piezoelectric Discs

Silicone Rubber

10–32 NF Tapped Hole

Driving Platform
(Beryllium)

Driving Point

270101

Figure 5.9 Piezoelectric impedance head. (*Courtesy Bruel and Kjaer Instruments, Marlboro, Mass.*)

Variable-Reluctance/FM-Oscillator Digital Systems

While the electrical signal from any force transducer can be converted to digital form by suitable equipment (see Chap. 10), some types of pickups are specifically designed with this in mind. We here discuss briefly one such type which has been used in rocket-engine testing where digital data provide advantages of accuracy and ease of performing computations automatically. The elastic member is a proving ring with a two-arm variable-reluctance bridge displacement transducer. The signal from this transducer is used to change the frequency of a frequency-modulated (FM) oscillator. The frequency change (from some base value) of this oscillator is directly proportional to displacement (and thus to force). A digital reading of force is accomplished by applying the output signal of the oscillator to an electronic counter over a known time interval. The total number of pulses accumulated is thus a digital measure of force. In a typical unit, a change of force from minus full scale to plus full scale causes the frequency to change from 10,000 to 12,500 Hz. For a 1-s counting period, a full-scale force will thus cause a counter reading 1,250 above the base value of 11,250.

Computing advantages of such a system arise from the desire to know the total impulse of the rocket engine and to be able to add and subtract various forces in multicomponent test stands. The total impulse is the integral of force with respect to time; this is simply the total number of counts over the desired integrating time. When several forces must be added or subtracted, each has its own force pickup and 10,000 to 12,500 Hz oscillator. The oscillator output sig-

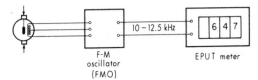

Force measurement

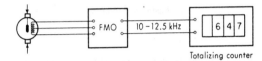

Total impulse measurement

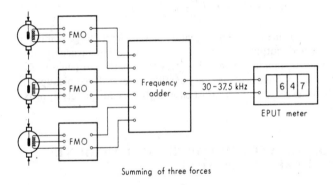

Summing of three forces

Figure 5.10 Digital force measurement, integration, and summing.

nals are combined in an electronic adder unit, which produces a single output whose frequency is the sum (or difference) of the input frequencies. The output of the adder unit is a signal with a frequency of 30,000 to 37,500 Hz; counting this over a timed interval gives the algebraic sum of the measured forces. Integration of this sum signal can be easily accomplished by the same method used for a single signal. Figure 5.10 shows block diagrams of these systems. By applying the dc output voltage to a voltage-to-frequency converter (see Chap. 10), these digital methods can be extended to any load cell with dc output.

Loading Effects

Since force is an effort variable, the pertinent loading parameter is either stiffness or impedance, and the associated flow variable is either displacement or velocity. Stiffness is perhaps more convenient for elastic force sensors since impedance

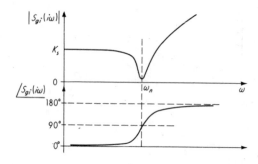

Figure 5.11 Stiffness of force transducers.

would be infinite for the static case. The generalized input stiffness S_{gi} of the system of Fig. 5.3 is given by

$$S_{gi}(D) = \frac{F_i}{x}(D) = K_s \left(\frac{D^2}{\omega_n^2} + \frac{2\zeta D}{\omega_n} + 1 \right) \qquad (5.15)$$

Recall that for small error due to loading, S_{gi} must be sufficiently large compared with S_{go}, the generalized output stiffness of the system being measured. The frequency characteristic $S_{gi}(i\omega)$ is shown in Fig. 5.11 for a particular (small) value of ζ. Note that, near ω_n, $S_{gi}(i\omega)$ becomes very small. However, force pickups are generally used only for $\omega \ll \omega_n$; thus in most cases $S_{gi}(i\omega)$ is adequately approximated as simply K_s.

5.4 RESOLUTION OF VECTOR FORCES AND MOMENTS INTO RECTANGULAR COMPONENTS

In a number of important practical applications, the force or moment to be measured is not only unknown in magnitude but also of unknown and/or variable direction. Outstanding examples of such situations are "balances" for measuring forces on wind-tunnel models,[1] dynamometers (force gages) for measuring cutting forces in machine tools,[2] and thrust stands[3] for determining forces of rocket engines. Elastic force transducers of either the bonded-strain-gage or gross-deflection variety are employed in these applications. Ingenious use of various types of flexures for isolating and measuring different force components

[1] P. K. Stein, "Measurement Engineering," vol. 1, p. 431, Stein Engineering Services, Inc., Phoenix, Ariz., 1964; C. C. Perry and H. R. Lissner, "The Strain Gage Primer," p. 212, McGraw-Hill, New York, 1955; A. Pope and K. L. Goin, "High Speed Wind Tunnel Testing," chap. 7, R. E. Krieger Publ. Co., Huntington, N. Y., 1978; L. Bernstein, Force Measurements in Short-Duration Hypersonic Facilities, *AGARD*-AG-214 (AGARD ograph no. 214).

[2] E. G. Loewen, E. R. Marshall, and M. C. Shaw, Electric Strain Gage Tool Dynamometers, *Proc. Soc. Exp. Stress Anal.*, vol. 8, no. 2, 1951.

[3] The Design of High-Accuracy Thrust Stands and Calibrators, WEC-BA-7D, Daystrom-Wiancko Co., Pasadena, Calif., 1961.

Flexure
(Stiff axially, soft in all other directions)

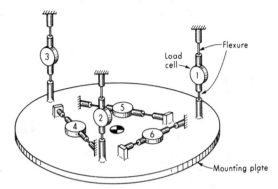

Flexure

Load cell

Load cell

Flexure

Mounting plate

ꟷꟷꟷ Denotes common rigid foundation

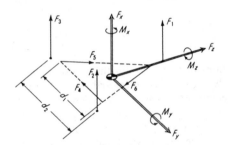

Figure 5.12 Six-component load frame.

characterizes the design of these devices. Depending on the degree to which the force or moment direction is unknown, force-resolving systems of varying degrees of complexity may be devised. The most general situation (measurement of three mutually perpendicular force components and three mutually perpendicular moment components) is regularly accomplished with high accuracy.

Figure 5.12 shows a six-component thrust stand[1] used in testing rocket engines. Load cells 1, 2, and 3 are mounted at the corners of an equilateral triangle, and load cells 4, 5, and 6 are in the sides of a concentric, smaller equilateral triangle. The engine to be tested is rigidly fastened at the common center of both triangles and produces a force of unknown magnitude and direction (which can be expressed in terms of components F_x, F_y, and F_z) and a moment of unknown magnitude and direction (which can be expressed in terms of components M_x,

M_y, and M_z). The rocket forces are transmitted from the mounting plate to the rigid foundation through the six load cells and their associated flexures. The action of the suspension system is most clearly seen if we consider each flexure as a pin joint. Actually pin joints are not used because of their lost motion and friction. A static analysis of the force system gives the following results, which allow calculation of the rocket forces and moments from the measured load-cell forces and stand dimensions:

$$F_x = F_1 + F_2 + F_3 \qquad (5.16)$$

$$F_y = \frac{F_5 - F_4 - (F_4 - F_6)}{2} \qquad (5.17)$$

$$F_z = \frac{\sqrt{3}\,(F_5 - F_6)}{2} \qquad (5.18)$$

$$M_x = \frac{-d_1(F_4 + F_5 + F_6)}{2\sqrt{3}} \qquad (5.19)$$

$$M_y = d_2 \frac{F_1 - F_2 - (F_3 - F_1)}{2\sqrt{3}} \qquad (5.20)$$

$$M_z = d_2 \frac{(F_3 - F_2)}{2} \qquad (5.21)$$

The indicated additions and subtractions of forces are (in the thrust stand described above) performed automatically by digital counters and adders since the load cells used are the variable-reluctance/FM oscillator type described in the previous section. For load cells with dc voltage output, we could, of course, use analog/digital converters to digitize the force signals and then send them to a suitably programmed digital computer for the necessary calculations.

A combination of bonded strain gages, Wheatstone-bridge circuits, and flexible elements of various geometries has proved a versatile tool in the development of multicomponent-force pickups of small size and high natural frequency. Figure 5.13 shows a beam with three separate bridge circuits of gages arranged to measure the three rectangular components of an applied force. All bridge circuits are temperature-compensated and respond only to the intended component of force; however, the point of application of the force must be at the center of the beam cross section. An eccentricity in the y direction, for example, would give F_z a moment arm, causing bending stresses in the y direction that would be indistinguishable from those due to F_y. If side loads F_x and F_y are present, the end of the beam will deflect, causing just such eccentricities; thus beam stiffness must be adequate to keep deflection sufficiently low. This stiffness tends, of course, to reduce sensitivity. While many strain-gage pickup configurations are possible and have been used, the octagonal ring[1] of Fig. 5.2 has particularly interesting properties. The reference gives information on its design and use.

[1] N. H. Cook and E. Rabinowicz, "Physical Measurement and Analysis," p. 162, Addison-Wesley, Reading, Mass., 1963.

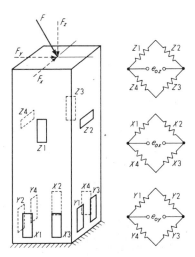

Figure 5.13 Resolution of vector forces.

Piezoelectric techniques also are well adapted to the problem of component resolution of vector forces and moments,[1] as briefly shown earlier (Fig. 4.43), where the x, y, and z components of a vector force were measured. The three-component load cell shown there can be used also as a building block in constructing measuring systems for specific purposes, such as the three-component dynamometer for machining research in milling and grinding operations illustrated in Fig. 5.14a. (*Dynamometer* is another word used for force transducer, particularly in Europe.) Four individual three-component cells share the total load, with all four x-component crystals paralleled electrically to give a single x output signal, and similarly for y and z. This arrangement produces an output which measures the force components correctly even when the force application point moves (as it *would*, in a milling operation). The reading of *each* of the four cells varies with force position, but their *sum* (which is what is produced by the electric parallel connection) gives the correct total force, irrespective of position. This scheme is carried even one step further in the milling operation of Fig. 5.14b, where two complete dynamometers support the workpiece and are electrically paralleled (externally) to achieve the desired force measurement. The individual sensing elements can be constructed from single disks of quartz or rings of disks, depending on the space available and other requirements (Fig. 5.15). When a torque must be measured, a ring of shear-sensitive disks oriented as in Fig. 5.15b can be used. Multicomponent devices combine disks or rings of the desired sensitivity in a "sandwich" fashion under sufficient compressive preload so that shear loads are transmitted reliably by friction. Here the "leakage" of piezoelectric voltages is an advantage since the preloads produce no electrical zero shift.

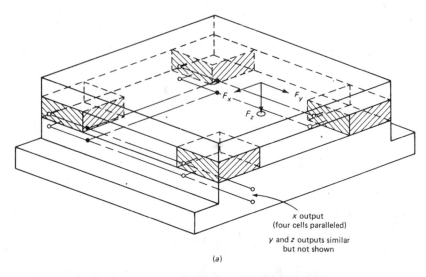

Figure 5.14 Multicomponent force transducer for machining studies. (*Courtesy Kistler Instruments, Amherst, N. Y.*)

When the coordinates of a moving force's point of application in a plane must be found, platforms supported by multicomponent load cells, as above, must be augmented with analog or digital computing elements to extract the needed data from the load-cell signals. A good example is a biomechanics platform[1] employed in sports medicine, orthopedics, posture control studies, etc.

[1] Model 9261A, Kistler Instrument Corp., Amherst, N. Y.

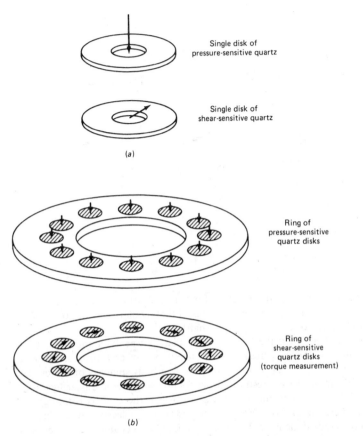

Single disk of
pressure-sensitive quartz

Single disk of
shear-sensitive quartz

(a)

Ring of
pressure-sensitive
quartz disks

Ring of
shear-sensitive
quartz disks
(torque measurement)

(b)

Figure 5.15 Component separation with quartz wafers. (*Courtesy Kistler Instruments, Amherst, N. Y.*)

Here, four three-component cells are utilized in the platform, together with two summing amplifiers and an analog divider, to obtain x, y, z force components, moment about a vertical axis through the force-application point, and x, y coordinates of this point for the force of a human foot acting on the platform's flat surface.

5.5 TORQUE MEASUREMENT ON ROTATING SHAFTS

Measurement of the torque carried by a rotating shaft is of considerable interest for its own sake and as a necessary part of shaft power measurements. Torque transmission through a rotating shaft generally involves both a source of power and a sink (power absorber or dissipator), as in Fig. 5.16. Torque measurement may be accomplished by mounting either the source or the sink in bearings

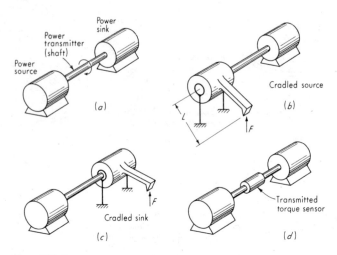

Figure 5.16 Torque measurement of rotating machines.

("cradling") and measuring the reaction force F and arm length L; or else the torque in the shaft itself is measured in terms of the angular twist or strain of the shaft (or a torque sensor coupled into the shaft).

The cradling concept is the basis of most shaft power dynamometers. These are utilized mainly for measurements of steady power and torque, by using pendulum or platform scales to measure F. A free-body analysis of the cradled member reveals error sources resulting from friction in the cradle bearings, static unbalance of the cradled member, windage torque (if the shaft is rotating), and forces due to bending and/or stretching of power lines (electric, hydraulic, etc.) attached to the cradled member. To reduce frictional effects and to make possible dynamic torque measurements, the cradle-bearing arrangement may be replaced by a flexure pivot with strain gages to sense torque,[1] as in Fig. 5.17. The crossing point of the flexure plates defines the effective axis of rotation of the flexure pivot. Angular deflection under full load is typically less than 0.5°. This type of cross-spring flexure pivot is relatively very stiff in all directions other than the rotational one desired, just as in an ordinary bearing. The strain-gage bridge arrangement also is such as to reduce the effect of all forces other than those related to the torque being measured. Speed-torque curves for motors may be obtained quickly and automatically with such a torque sensor by letting the motor under test accelerate an inertia from zero speed up to maximum while measuring speed with a dc tachometer.[2] The torque and speed signals are applied to an XY recorder to give automatically the desired curves.

---

[1] Lebow Associates, Troy, Mich.

[2] B. Hall, Motor Tests Using X-Y Recorders, *Electro-Technol.*, p. 116, May 1964.

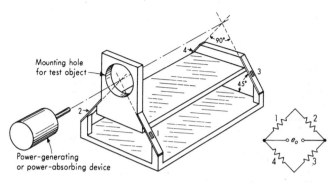

Figure 5.17 Strain-gage torque table.

Another variation (see Fig. 5.18) on the cradle principle is found in a null-balance torquemeter using feedback principles to measure small torques in the range 0 to 10 oz · in. In this device the test object is mounted on a hydrostatic air-bearing table to reduce bearing friction to exceedingly small values. Any

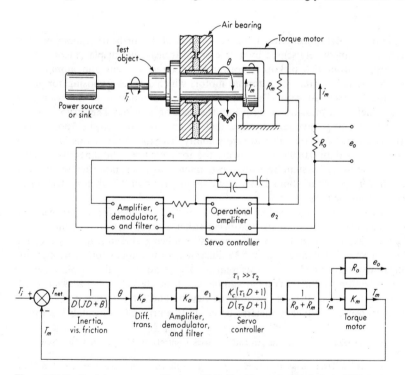

Figure 5.18 Feedback torque sensor. (*Courtesy McFadden Electronics, South Gate, Calif.*)

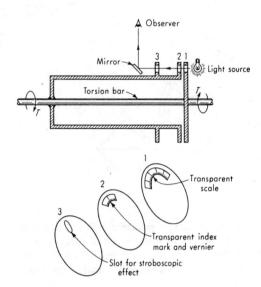

Figure 5.19 Torsion-bar dynamometer.

torque on the test object tends to cause rotation of the air-bearing table, but this rotation is immediately sensed by a differential-transformer displacement pickup. The output from this pickup is converted to direct current and amplified to provide the coil current of a torque motor, which applies opposing torque to keep displacement at zero. The amount of current required to maintain zero displacement is a measure of torque and is read on a meter. The servoloop uses integral control[1] to give zero displacement for any constant torque. Approximate derivative control is used also to give stability. The threshold of this air-bearing system is less than 0.0005 oz · in while the torque/current nonlinearity is 0.001 oz · in. The overall system behaves approximately as a second-order system with a natural frequency of about 10 Hz and damping ratio of 0.7 when no test object is present on the table.

The use of elastic deflection of the transmitting member for torque measurement may be accomplished by measuring either a gross motion or a unit strain. In either case, a main difficulty is the necessity of being able to read the deflection while the shaft is rotating. Figure 5.19 illustrates a torsion-bar torquemeter, using optical methods of deflection measurement. The relative angular displacement of the two sections of the torsion bar can be read from the calibrated scales because of the stroboscopic effect of intermittent viewing and the persistence of vision. A line of transmission dynamometers[2] based on this principle is available in ranges

[1] E. O. Doebelin, "Dynamic Analysis and Feedback Control," p. 223, McGraw-Hill, New York, 1962.
[2] Transmission Dynamometers, *Mech. Eng.*, p. 62, February, 1971; Torquemeters Ltd., Ravensthorpe, Northampton NN68EH, England.

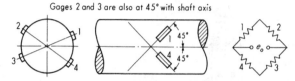

Gages 2 and 3 are also at 45° with shaft axis

Figure 5.20 Strain-gage torque measurement.

up to 50,000 ft · lbf and 50,000 r/min, with error of ±0.25 percent. By replacing the scales on disks 1 and 2 with sectored disks (alternate clear and opaque sectors) and employing an electro-optical transducer in place of the human eye, a version[1] with electrical output may be obtained. For zero torque, the sectored disks are positioned to give a 50 percent light transmission area. Then positive torque increases the area proportionally while negative torque decreases it, giving a linear and direction-sensitive electrical output.

Even though they require additional equipment to transmit power and signal between rotating shaft and stationary readout, strain-gage torque sensors are very widely used. Figure 5.20 shows the basic principle. (Chapter 11 gives data on slip rings, rotary transformers, and radiotelemetry systems to deal with the data transmission problem.) This arrangement (given accurate gage placement and matched gage characteristics) is temperature-compensated and insensitive to bending or axial stresses. The gages must be precisely at 45° with the shaft axis, and gages 1 and 3 must be diametrically opposite, as must gages 2 and 4. Accurate gage placement is facilitated by the availability of special rosettes in which two gages are precisely oriented on one sheet of backing material. In some cases the shaft already present in the machine to be tested may be fitted with strain gages. In other cases a different shaft or a commercial torquemeter must be used to get the desired sensitivity or other properties. Of the various configurations shown in Fig. 5.21, one manufacturer[2] uses the hollow cruciform for low-range units and the solid, square shaft for high-range ones. Placement of the gages on a square, rather than round, cross section of the shaft has some advantages. The gages are more easily and accurately located and more firmly bonded on a flat surface. Also, the corners of a square section in torsion are stress-free and thus provide a good location for solder joints between lead wires and gages. These joints are often a source of fatigue failure if located in a high-stress region. Also, for equivalent strain/torque sensitivity, a square shaft is much stiffer in bending than a round one, thus reducing effects of bending forces and raising shaft natural frequencies.

The torque of many machines, such as reciprocating engines, is not smooth even when the machine is running under "steady-state" conditions. If we wish to

[1] R. Foskett, Background Information on Torque Measuring Transducers, Vibrac Corp., Amherst, N. Y., 1968.

[2] Lebow Associates, Inc., Troy, Mich.

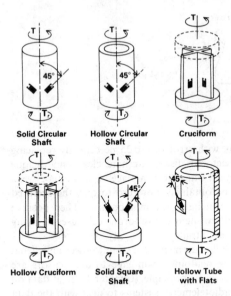

Solid Circular Shaft **Hollow Circular Shaft** **Cruciform**

Hollow Cruciform **Solid Square Shaft** **Hollow Tube with Flats**

Figure 5.21 Various torque sensor designs. (*Courtesy Lebow Associates, Inc., Troy, Mich., Torque Sensor and Dynamometer Catalog, no. 250, 1979.*)

measure the average torque so as to calculate power, the higher frequency response of strain-gage torque pickups may be somewhat of a liability since the output voltage will follow the cyclic pulsations and some sort of averaging process must be performed to obtain average torque. If exceptional accuracy is not needed, the low-pass filtering effect of a dc meter used to read e_o may be sufficient for this purpose. In the cradled arrangements of Fig. 5.16 (employed in many commercial dynamometers for engine testing, etc.), the inertia of the cradled member and the low-frequency response of the platform or pendulum scales used to measure F perform the same averaging function.

Commercial strain-gage torque sensors are available with built-in slip rings and speed sensors. A family[1] of such devices covers the range 10 oz · in to 3×10^6 in · lbf with full-scale output of about 40 mV. The smaller units may be used at speeds up to 24,000 r/min, the largest to 350 r/min. Torsional stiffness of the 10 oz · in unit is 112 in · lbf/rad while a 600,000 in · lbf unit has 4.0 × 10^6 in · lbf/rad. Nonlinearity is 0.1 percent of full scale while temperature effect on zero is 0.002 percent of full scale/F° and temperature effect on sensitivity is 0.002 percent/F° over the range 70 to 170°F.

The dynamic response of elastic deflection torque transducers is essentially slightly damped second-order, with the natural frequency usually determined by the stiffness of the transducer and the inertia of the parts connected at either end. Damping of the transducers themselves usually is not attempted, and any damping present is due to bearing friction, windage, etc., of the complete test setup.

[1] Lebow Associates, Inc., Troy, Mich.

A good example of the application of torque measurement (and several other measurement system concepts of general interest) is found in a dynamic test system used in the design and development of front-wheel-drive components for automobiles (see Figs. 5.22 through 5.24).[1] An alternative to conventional dynamometer loading systems, called the *four-square principle*, is utilized here to conserve energy and allow fast-response testing. Rather than connecting the shaft test specimens between a power source and a power sink, as in Fig. 5.16 (an arrangement which wastes all the shaft-transmitted power and couples large, slow-responding inertias to the specimen), the four-square principle (Fig. 5.22a) induces a "locked-in" torque into the system by counter-twisting the flanges of a pre-loading coupler at assembly. Now the geared assembly can be driven at any desired speed by, say, a dc motor drive, and the shaft specimens will be subjected to the desired torque/speed conditions, but the drive supplies *only* friction losses (plus accelerating power if speed is changed), *not* the shaft-transmitted power (product of locked-in torque and shaft speed).

The basic four-square principle just described has been widely utilized for some time; however, several new features were needed in the present application. It was desired to be able to maintain a certain level of locked-in torque in the face of inevitable component wear, creep, slippage, or fatigue, which all cause a progressive "relaxation" of torque with time in the basic four-square system. Also, it was necessary to dynamically vary the locked-in torque in response to electrical commands. By replacing the mechanical preloading coupler with a rotary hydraulic actuator (Fig. 5.22b), measuring shaft torque with a rotating torque sensor, and driving the actuator from an electrohydraulic servovalve responding to the error between commanded and measured torque, the desired features are achieved.

To simulate actual driving conditions, the testing machine includes means for subjecting the front-wheel-drive shafts to "jounce" (up-and-down) motions and steering motions, all simultaneous with speed and torque variations (see Fig. 5.23). All these testing modes are implemented by using electrohydraulic or electromechanical feedback control systems,[2] each of which requires appropriate measuring instruments for the controlled variable.

The remote parameter control (RPC) feature of the system is a further refinement (see Fig. 5.24). Input commands to all the torque and motion control loops are derived from tape recordings obtained from instrumented automobiles driven over actual roads or test tracks. Since the various feedback control systems cannot be designed to have "perfect" dynamic response, inputting a certain command waveform for, say, shaft torque will *not* result in exactly that waveform in the actual torque. To compensate for these unavoidable system response deficiencies, the RPC method *measures* them and then *alters* the input command so that the desired response is obtained. This is accomplished by implementing

[1] D. J. Manduzzi and K. E. Reid, Ford's New Programmable Front Wheel Drive Half Shaft Test System, *Closed Loop*, vol. 10, no. 1, MTS Systems Corp., Minneapolis, Minn., June 1980.
[2] Doebelin, "Dynamic Analysis and Feedback Control."

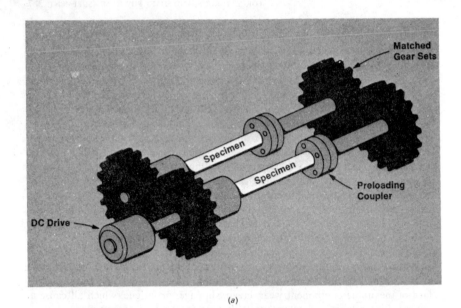

(a)

(b)

Figure 5.22 Four-square dynamometer principle.

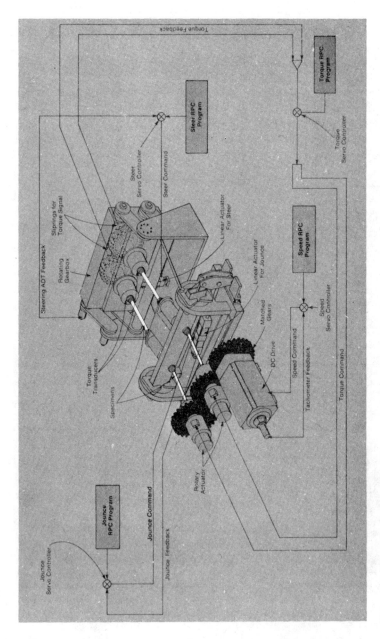

Figure 5.23 Closed-loop testing apparatus.

389

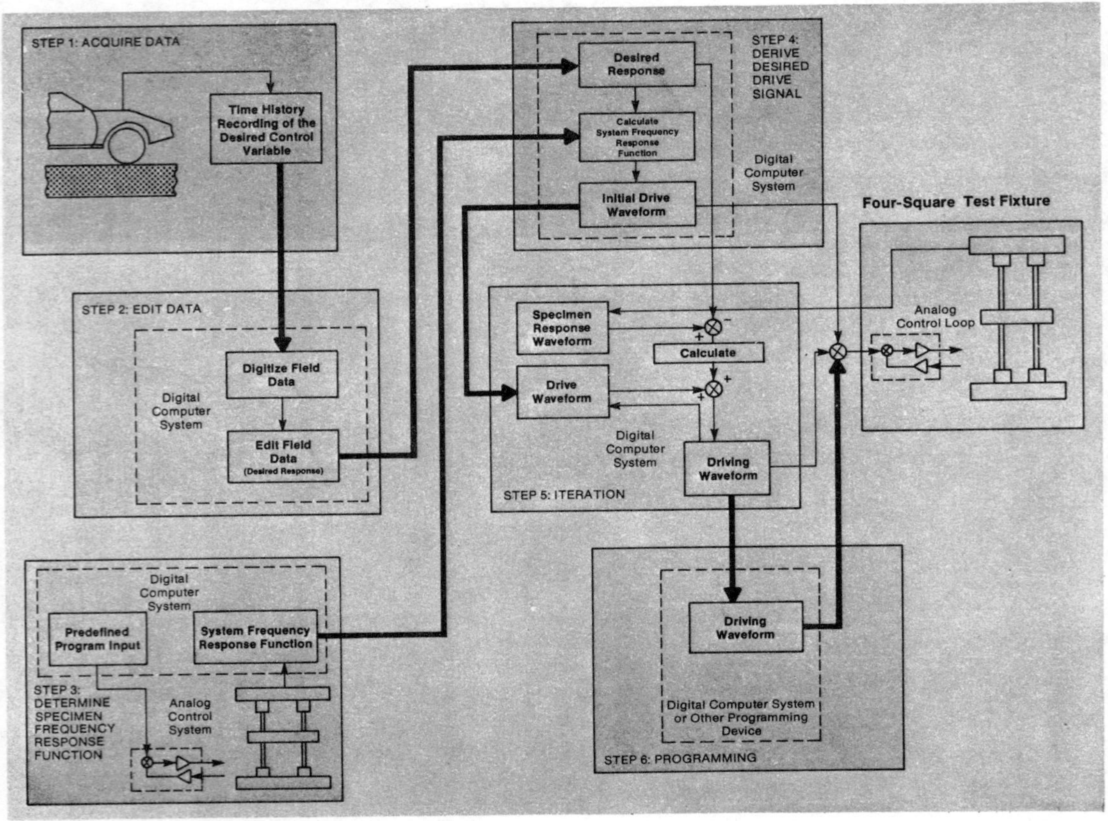

Figure 5.24 Remote parameter control (RPC) test technique.

frequency-domain methods (explained in Chap. 3) on a minicomputer, using fast Fourier transform (FFT) algorithms. Overall, this system is an excellent example of how concepts from the areas of measurement, system dynamics, automatic control, and computer-aided engineering are becoming a vital part of advanced design methods.

5.6 SHAFT POWER MEASUREMENT (DYNAMOMETERS)

The accurate measurement of the shaft power input or output of power-generating, -transmitting, and -absorbing machinery is of considerable interest. While the basic measurements—torque and speed—have already been discussed, their practical application to power measurement is considered briefly here. The term *dynamometer* generally is used to describe such power-measuring systems, although it is also used as a name for elastic force sensors.

The type of dynamometer employed depends somewhat on the nature of the machine to be tested. If the machine is a power generator, the dynamometer must be capable of absorbing its power. If the machine is a power absorber, the dynamometer must be capable of driving it. If the machine is a power transmitter or transformer, the dynamometer must provide both the power source and the load, unless a four-square method is feasible.

Perhaps the most versatile and accurate dynamometer is the dc electric type. Here a dc machine is mounted in low-friction trunnion bearings (see Fig. 5.16b) and provided with field and armature control circuits.[1] This machine can be coupled to either power-absorbing or power-generating devices since it may be connected as either a motor or a generator. When it is employed as a generator, the generated power is dissipated in resistance grids or recovered for use. The dc dynamometer can be adjusted to provide any torque from zero to the maximum design value for speeds from zero to the so-called base speed of the machine. At this speed the maximum torque develops the maximum design horsepower. At speeds above base speed, torque must be progressively reduced so as to maintain horsepower less than the design maximum. The controllability of the dc dynamometer lends it particularly to modern automatic load and speed programming applications.

Figure 5.25 illustrates such a situation. Tape recordings of engine torque and speed measured under actual driving conditions for an automobile are utilized to reproduce these conditions in the laboratory engine test. Two feedback systems control engine speed and torque. A tachometer generator speed signal from the dynamometer is compared with the desired speed signal from the tape recorder; if the two are different, the dynamometer control is automatically adjusted to change speed until agreement is reached. Actual engine torque is obtained from a

---

[1] P. S. Potts and P. T. Schuerman, How to Choose Electric Dynamometers, *Mach. Des.*, p. 102, June 27, 1957; R. F. Knudsen, A Discussion of Present-Day Dynamometers, *Gen. Motors Eng. J.*, p. 18, October–November–December 1957.

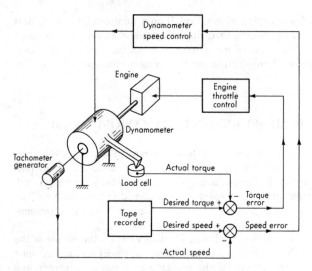

Figure 5.25 Servocontrolled dynamometer.

load cell on the dynamometer and compared with the desired torque from the tape recorder. If these do not agree, the error signal actuates the engine throttle control in the proper direction. Both systems operate simultaneously and continuously to force engine speed and torque to follow the tape-recorder commands.

By adapting recent developments in variable-frequency ac motor control, a trunnion-mounted induction motor becomes a versatile dynamometer[1] for both driving and absorbing applications. Features include fast response, flexible control, simplified maintenance, and energy conservation, since up to 85 percent of the energy absorbed is returned to the ac power line.

Dynamometers capable only of absorbing power (see Fig. 5.26) include the eddy-current brake[2] (inductor dynamometer) and various mechanical brakes employing dry friction (Prony brake) and fluid friction (air and water brakes). The eddy-current brake is easily controllable by varying a dc input, but it cannot produce any torque at zero speed and only small torque at low speeds. However, it is capable of higher power and speed than a dc dynamometer. The power absorbed is carried away by cooling water circulated through the air gap between rotor and stator. The Prony brake is a simple mechanical brake in which friction torque is manually adjusted by varying the normal force with a handwheel. Torque is available at zero speed, but operation may be jerky because of the

[1] C. S. Lassen, Conserving Energy in Engine Testing with Adjustable Frequency Regenerative Dynamometers, Eaton Corp., Kenosha, Wis., 1981.

[2] J. B. Winther, Dynamometer Handbook of Basic Theory and Applications, Eaton Corp., Kenosha, Wis., 1974.

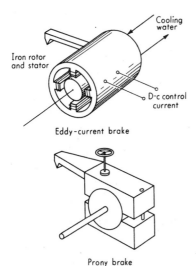

Eddy-current brake

Prony brake   **Figure 5.26** Absorption dynamometers.

basic nature of dry friction. Water and air brakes utilize the churning action of paddle wheels or vanes rotating inside a fluid-filled casing to absorb power. A flow of air or water through the device is maintained for cooling purposes. No torque is available at zero speed, and only small torques are available at low speeds. High speed and high power can be handled well, however, with some water brakes being rated at 10,000 hp at 30,000 r/min.

In all the above power measurement applications, torque and speed are measured separately and then power is calculated manually. This calculation can be performed automatically in a number of ways since the basic operation (multiplication) can be accomplished physically in various ways. An interesting scheme using the properties of bridge circuits is shown in Fig. 5.27. Speed is measured with a dc tachometer generator, and this voltage is applied as the excitation of a strain-gage load cell used to measure torque. Since bridge output is directly proportional to excitation voltage and directly proportional to torque, the voltage e_o is actually an instantaneous power signal.

Another ingenious solution of this problem is shown in Fig. 5.28. A torsion bar carrying the torque to be measured has a permanent-magnet ac generator (alternator) coupled at either end. Each alternator puts out an ac voltage of amplitude proportional to shaft speed and of frequency equal to shaft speed. When the torsion bar is unloaded, the two alternator rotor-stator positions are adjusted mechanically so that the ac output voltages are exactly out of phase. If the two alternators are connected electrically in series now, the net output will be zero at any shaft speed as long as no torque is present. When the shaft carries torque, it twists, causing a phase shift between the two alternators and a net output voltage whose amplitude is proportional to the product of torque and

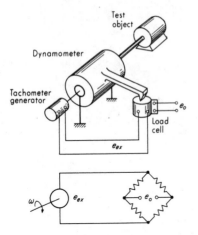

Figure 5.27 Instantaneous power measurement.

speed. This may be seen from the following analysis:

$$\text{Alternator 1 output} = K_\omega \omega \sin \omega t \tag{5.22}$$

$$\text{Alternator 2 output} = K_\omega \omega \sin (\omega t + \phi) \tag{5.23}$$

where

$$K_\omega \triangleq \text{peak volts/(rad/s)} \tag{5.24}$$

$$\phi = K_t T \tag{5.25}$$

$$K_t \triangleq \text{rad/(in} \cdot \text{lbf)} \tag{5.26}$$

$$T \triangleq \text{torque} \tag{5.27}$$

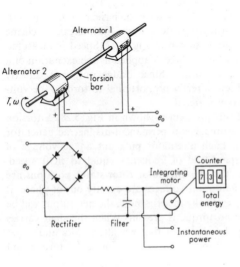

Figure 5.28 Alternator power measurement.

The net output of the series-connected alternators is

$$\text{Net output} = e_o = K_\omega \omega \left[\sin \omega t - \sin (\omega t + K_t T) \right] \tag{5.28}$$

$$e_o = K_\omega \omega \left[\sin \omega t - (\sin \omega t \cos K_t T + \cos \omega t \sin K_t T) \right] \tag{5.29}$$

The twist angle $\phi = K_t T$ is very small, and so $\cos K_t T \approx 1.0$ and $\sin K_t T \approx K_t T$. Thus

$$e_o = -K_\omega \omega K_t T \cos \omega t \tag{5.30}$$

and e_o is a sine wave of amplitude proportional to ωT and thus to power. In the instrument described above, this ac voltage is rectified and filtered to produce a proportional dc value of 20 V full scale. Also, if total energy over a time period is desired, an integrator is available to integrate the dc voltage. This is in the form of a precise dc motor whose speed is directly proportional to the voltage applied to it from the horsepower meter. If speed is proportional to horsepower, total revolutions (read by an ordinary mechanical counter) give total energy.

5.7 GYROSCOPIC FORCE AND TORQUE MEASUREMENT

The use of gyroscopic principles (method 6) in the measurement of force and torque is not widespread; however, some practical applications do exist. The method is particularly useful if you wish to obtain the time integral of a force or torque since the gyro supplies a mechanical rotation proportional to it. This concept has found practical application in a gyroscopic integrating mass flow-meter, which is discussed in more detail in the chapter on flow measurement.

From Chap. 4, the transfer function relating an applied torque T_x to an output axis rotation θ is

$$\frac{\theta}{T_x}(D) = \frac{H_s/I_x}{D(I_y D^2 + BD + H_s^2/I_x + K_s)}$$

For a free gyro, as in Fig. 5.29a, B and K_s are effectively zero, which gives

$$\frac{\dot{\theta}}{T_x}(D) = \frac{K}{D^2/\omega_n^2 + 1} \tag{5.31}$$

where

$$K \triangleq \frac{1}{H_s} \quad \text{(rad/s)/(in} \cdot \text{lbf)} \tag{5.32}$$

$$\omega_n \triangleq \sqrt{\frac{H_s^2}{I_x I_y}} \tag{5.33}$$

While the second-order term of Eq. (5.31) is undamped, small damping (friction) in bearings, etc., is always present to prevent sustained oscillation. Also, if T_x is applied gradually, oscillations will not be started. Thus a constant torque T_x will produce a precessional angular velocity $\dot{\theta}$ in direct proportion according to $\dot{\theta} =$

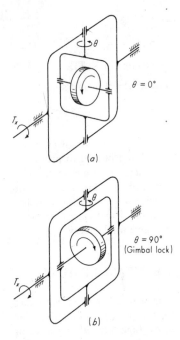

$\theta = 0°$

(a)

$\theta = 90°$
(Gimbal lock)

(b)

Figure 5.29 Gyroscopic torque measurement.

KT_x. This will be true only as long as θ is small, however, and since a constant torque produces a constant *velocity*, eventually θ must become large. This will ultimately (when θ reaches 90°) lead to what is called "gimbal lock." The gyroscopic precession actually depends on the component of the torque vector that is perpendicular to the spin angular-momentum vector. For small angles, this component is directly proportional to torque. For large angles, it is proportional to the product of torque and cos θ; this becomes smaller as θ increases and disappears completely at $\theta = 90°$. Thus the precession θ produced by a constant torque T_x gets smaller and smaller as θ approaches 90°. At 90° a torque T_x produces no precession at all; rather, the inner and outer gimbals both rotate together about the x axis.

To prevent gimbal lock and thus achieve a useful torque-sensing instrument, a simple mechanical solution is available. The requirement that torque vector and spin angular-momentum vector always be perpendicular is met by the configuration of Fig. 5.30. The equation $\dot{\theta} = KT_x$ now holds for *any* angle θ, and we can measure torque by measuring $\dot{\theta}$ or measure the time integral of torque by means of a simple revolution counter attached to the θ shaft. A commercial weighing system based on the above principles (with refinements) is available.[1]

[1] B. Wasko, The Gyroscope as a Force Measuring Device, Voland Corp., New Rochelle, N.Y., 1978.

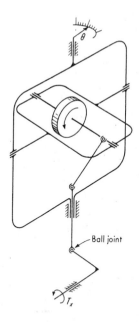

Figure 5.30 Solution of gimbal-lock problem.

5.8 VIBRATING-WIRE FORCE TRANSDUCERS

Method 7 in the list of basic force-sensing schemes is based on the well-known formula from classical physics for the first natural frequency ω of a string of length L and mass per unit length m_1, which is tensioned by the force F to be measured:

$$\omega = \frac{1}{2L} \sqrt{\frac{F}{m_1}} \qquad (5.34)$$

Since ω varies smoothly with F, the measuring principle is basically analog; however, the frequency is easily measured with conventional digital counters, so the transducer is sometimes described as a digital device. Temperature sensitivity (variation in L) and the nonlinear ω/F relation cause some inconvenience in practical use.[1]

The temperature and linearity problems above have been overcome by clever system design in a sensor used as part of industrial material feeding controllers[2] (Fig. 5.31). While the wire-tensioning force *determines* the natural frequency of vibration, some vibration *exciting system* must be provided in any vibrating-wire sensor to make up frictional losses and to maintain wire vibration at a fixed amplitude. The "wires" (actually thin tapes of electrically conducting but non-

[1] K. S. Lion, "Instrumentation in Scientific Research," pp. 83–85, McGraw-Hill, New York, 1959.

[2] M. Gallo and G. Bussian, The Benefits of Mass Measurement for Gravimetric Feeding of Bulk Solids in Industrial Environments, K-Tron Corp., Glassboro, N.J.

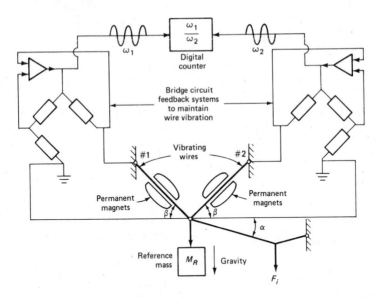

Figure 5.31 Vibrating-wire force transducer.

magnetic metal) are placed in the fields of permanent magnets. In this way, they serve as both velocity-to-voltage transducers (used in one leg of a bridge circuit to provide an oscillatory input voltage to an amplifier) and, current-to-force transducers, which accept an oscillatory output current from the amplifier and produce vibration-exciting transverse forces on the wires.

By proper design, these feedback systems maintain constant-amplitude wire vibrations at a frequency determined by the force carried by each wire. When the measured force is zero, a fixed reference mass M_R applies a preload tension F_o which is equal in each wire, causing both wires to vibrate at the same frequency and making the frequency ratio $\omega_1/\omega_2 = 1.0$. If temperature changes, both wires experience the *same* length and frequency changes, but the frequency *ratio* is unaffected, which provides temperature compensation. To make the output signal ω_1/ω_2 linear with F_i, a force proportional to F_i is applied through the (nonvibrating) wire at an angle α such that an increase in F_i increases F_1 (the tension in wire 1) but decreases F_2. Choice of angles β and α allows us to adjust the relation between ΔF_1 and ΔF_2 over a wide range. Analysis shows that the formulas for F_1 and F_2 are of the form

$$F_1 = F_o + k_1 F_i \qquad F_2 = F_o - k_2 F_i \qquad (5.35)$$

where k_1 and k_2 can be adjusted by using α and/or β. Then we get

$$\frac{\omega_1}{\omega_2} = \sqrt{\frac{F_o + k_1 F_i}{F_o - k_2 F_i}} \qquad (5.36)$$

The function $(F_o + k_1 F_i)/(F_o - k_2 F_i)$ of F_i is nonlinear with an increasing slope, while the square root function has a decreasing slope; so the two non-linearities, while not canceling, do *oppose* each other. By proper choice of k_1 and k_2 it is found that ω_1/ω_2 becomes linear with F_i to a close approximation. [Measurement of the frequency ratio ω_1/ω_2 is a standard digital counter technique in which the ω_2 signal is substituted for the usual internal clock oscillator of the counter and the number of cycles of ω_1 occurring during a preselected (and fixed) number of cycles of ω_2 is measured.] The referenced instrument is quoted as having ± 0.003 percent repeatability, ± 0.03 percent linearity, and stability of ± 0.03 percent over 18 months.

PROBLEMS

5.1 For the force transducer of Fig. 5.3, let the "foundation" have an interfering input displacement $X_i(t)$. Obtain the necessary transfer functions for a block diagram, showing how instrument output is produced by the superimposed effects of desired input f_i and interfering input X_i. If an accelerometer sensitive to $\ddot{X}_i$ (either built into the force transducer or as a separate instrument) is present, show how a summing amplifier could be used to obtain an output signal compensated for the spurious vibration input.

5.2 The torque wrench in Fig. P5.1 is claimed to produce an output voltage e_o proportional to the torque applied by force f to the nut, *irrespective* of the point of force application L_f, as long as $L_f > 3L$. Investigate the validity of this claim.

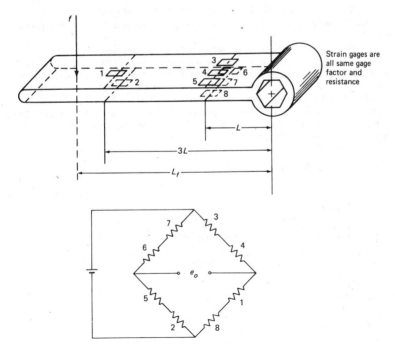

Strain gages are all same gage factor and resistance

Figure P5.1

5.3 An object with a volume of 10 in^3 is weighed on an equal-arm balance. The standard mass required for balance is 1 lbm and has a volume of 3 in^3. What is the value of the correction necessary for air buoyancy?

5.4 A brass balance beam has a length of exactly 1 m at 60°F, and the pivot is perfectly centered to give an equal-arm balance. If one end of the beam comes to 80°F and there is a uniform temperature gradient to the other end at 60°F, what inequality in arm length results?

5.5 What general form of dynamic response would you expect from the systems of Fig. 5.2-1? Why?

5.6 Prove that the reading is independent of location of F_i if $a/b = c/d$ in the platform scale of Fig. 5.2-1.

5.7 If, in Fig. 5.2-2, $M = 1$ lbm and $A = 20g$, what is the net force on M? If a friction force which is unknown but less than 1 lbf may be present, what error may be expected in F_i?

5.8 Carry out a simplified, linear dynamic analysis of the system of Fig. 5.2-3.

5.9 Carry out a linear dynamic analysis and an accuracy/stability tradeoff study of the pneumatic load cell of Fig. 5.2-4. Use $(x/F_{net})(D) = K_d/(D^2/\omega_n^2 + 2\zeta D/\omega_n + 1)$ and $(p_o/x)(D) = K_n/(\tau D + 1)$.

5.10 In Fig. 5.4 if F_i is eccentric and angularly misaligned, a torque is produced. Does this affect the bridge output? Explain.

5.11 A load cell deflects 0.005 in under its full-scale load of 1,000 lbf. It is used to measure force on a machine-tool slide which weighs 500 lbf. Estimate the highest frequency of force that may be accurately measured.

5.12 Derive Eqs. (5.16) to (5.21).

5.13 From the block diagram of Fig. 5.18, obtain the transfer function $(e_o/T_i)(D)$, assuming τ_2 is small enough to neglect. Investigate the effect of system parameters on dynamic accuracy and stability.

5.14 In Fig. 5.18 prove that $\theta = 0$ for any constant value of T_i.

5.15 Prove that the arrangement of Fig. 5.20 is insensitive to axial or bending stresses.

5.16 Prove that for equivalent strain/torque sensitivity, a square shaft is stiffer in bending than a round one.

5.17 A torque sensor with torsional stiffness of 1,000 in · lbf/rad is coupled between an electric motor and a hydraulic pump. The moments of inertia of the motor and pump are each 0.01398 in · lbf · s^2. If the motor has a small oscillatory torque component at 60 Hz, will this measuring system be satisfactory? Explain what torsional stiffness is needed if the response at 60 Hz is to be no more than 105 percent of the static response. The amount of damping is unknown.

5.18 Suppose the tachometer generator in the system of Fig. 5.27 puts out 6 V/1,000 r/min and the load cell produces 0.05 mV/(lbf · V excitation). What will be the power calibration factor for e_o in horsepower per millivolt if the arm length is 1 ft?

5.19 In the system of Fig. P5.2:

1. For $F = 0$ and heat off, $R_1 = R_2 = R_3 = R_4$ and $e_o = 0$.

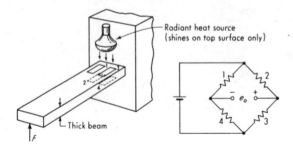

Radiant heat source
(shines on top surface only)

Thick beam

F

e_o

Figure P5.2

2. The gage factor of the gages is $+2.0$, and the temperature coefficient of resistance of the gages is positive.
3. The modulus of elasticity of the beam decreases with increased temperature.
4. The thermal-expansion coefficient of the gage is greater than that of the beam.
5. Assume the gage temperature is the same as that of the beam *immediately* beneath it.

At time $t = 0$, an upward force F is applied and maintained constant thereafter. After oscillations have died out, at a later time t_1 the radiant-heat source is turned on and left on thereafter. Sketch the general form of e_o versus t, justifying clearly by detailed reasoning the shape you give the curve.

5.20 It is necessary to design a strain-gage thrust transducer for small experimental rocket engines which are roughly in the shape of a cylinder 6 in in diameter by 12 in long. The following information is given:

1. Weight of motor and mounting bracket, 20 lbf.
2. Maximum steady thrust, 50 lbf.
3. Oscillating component of thrust, ± 10 lbf maximum.
4. Oscillating components of thrust up to 100 Hz must be measured with a flat amplitude ratio within ± 5 percent.
5. A recorder with a sensitivity of 0.1 V/in, frequency response flat to 120 Hz, and input resistance of 10,000 Ω is available.
6. Thrust changes of 0.5 lbf must be clearly detected.
7. Gages with a resistance of 120 Ω and a gage factor of 2.1 are available. They are 0.5×1.0 in in size.
8. An amplifier (to be placed between transducer and recorder) is available with a gain up to 1,000.

Design the transducer so as to require a minimum of amplifier gain. If damping is employed, calculate the required damping coefficient B, but do not design the damper. Use the cantilever-beam arrangement of Fig. P5.3.

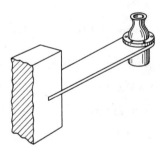

Figure P5.3

5.21 Repeat Prob. 5.20, using the configuration of Fig. P5.4.

Beam built in at both ends   **Figure P5.4**

5.22 Repeat Prob. 5.20 using the configuration of Fig. P5.5.

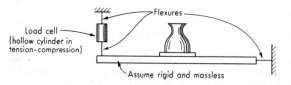

Load cell ─
(hollow cylinder in
tension-compression)

Flexures

Assume rigid and massless

Figure P5.5

BIBLIOGRAPHY

1. A. Krsek and M. Tiefermann: Optical Torquemeter for High Rotational Speeds, *NASA Tech. Note*, D-1437, October 1962.
2. H. E. Lockery: Applying the Strain-Gage Torque Transducer, *ISA J.*, p. 65, March 1962.
3. Torque-Gauge without Sliprings, *Electromech. Des.*, p. 6, November 1959.
4. J. Guthrie: Lever-Shaft Torque Measurement, *Instrum. Contr. Syst.*, p. 116, August 1964.
5. D. Ettleman and M. Hoberman: Torquemeters, *Mach. Des.*, p. 134, February 28, 1963.
6. O. Dahle: Heavy Industry Gets a New Load Cell, *ISA J.*, p. 32, August 1959.
7. F. M. Ryan: Automatic Weighing for Solids, *Contr. Eng.*, p. 103, September 1962.
8. D. W. Kennedy: Weighing Scales Couple to Computer, *Contr. Eng.*, p. 83, July 1962.
9. F. A. Ludewig: Digital Force Transducer, *Contr. Eng.*, p. 107, June 1961.
10. K. Harris: Servo-Balanced Supply Tank Measures Nozzle Thrust, *Contr. Eng.*, p. 115, February 1960.
11. Planetary Gearing in Torquemeter Does Away with Sliding Contacts, *Mach. Des.*, p. 135, September 1, 1960.
12. E. T. Gay: Precision Weighing with Platform Scales, *Tool Eng.*, June 1959.
13. S. Edwards: Dynamic Measurement of Vehicle Front Wheel Loads Using a Special Purpose Transducer, *Gen. Motors Eng. J.*, p. 15, October–November–December 1964.
14. S. Hejzlar: Backweighing Error in Scales, *Instrum. Contr. Syst.*, p. 95, February 1965.
15. Apparatus Measures Very Small Thrusts, *NASA Brief* 64-10284, 1964.
16. J. A. Bierlein: Methods of Measuring Thrust, *ARS J.*, p. 128, May–June 1953.
17. L. E. Stone: Criteria for Design and Use of an Internal Strain Gage Floating-Frame Balance, *ISA Trans.*, p. 152, April 1965.
18. R. L. Small: Belt Scales, *ISA J.*, p. 65, May 1965.
19. V. C. Plane: Total Impulse Measuring System for Solid-Propellant Rocket Engine, Rocketdyne Div., Canoga Park, Calif., *Rept.* R-5638, 1964.
20. R. W. Postma: Pulse Thrust Measuring Transducer (with Accelerometer Dynamic Compensation), Rocketdyne Div., Canoga Park, Calif., *Rept.* R-6044, 1965.
21. G. F. Malikov: Computation of Elastic Tensometric Elements (Load Cells), *NASA* TT F-513, 1968.
22. K. W. Stark: Design and Development of a Micropound Extended Range Thrust Stand, *NASA* TN D-7029, 1971.
23. J. Dimeff et al.: Characteristics of a New Type of Balance for Wind-Tunnel Models, *NASA* TM X-1278, August 1966.
24. S. P. Ragsdale: Alteration of Thrust Stand Dynamics Using Hydraulic Force Feedback, *AEDC-TR-67-232*, Arnold Air Force Station, Tenn., 1967.
25. J. E. Irby and J. C. Hung: Optimum Correction of Thrust Transient Measurements, *NASA* CR-269, July 1965.
26. F. L. Crosswy and H. T. Kalb: Dynamic Force Measurement Techniques, *Inst. & Cont. Syst.*, pp. 81–83, February 1970; pp. 117–121, March 1970.
27. J. D. Chang and J. Kukel: Advanced Torque Measurement System, *USAAMRDL Tech. Rept.* 73-54, Ft. Eustis, Va., June 1973.

28. J. M. Cummings: User's Experience with Instrumented Couplings for Continuous Measurement of Horsepower and Alignment on Large Turbomachinery Trains, *ASME Paper* 80-C2/DET-52, 1980.

29. M. L. Kuszewski, P. J. Mole, and S. A. Griffin: Study of Six-Component Internal Strain Gage Balances for Use in the Hirt Facility, *AEDC*-TR-75-63, Arnold Air Force Station, Tenn., July 1975.

30. I. A. Sutherland: Development of a Two-Component Force Dynamometer and Tool Control System for Dynamic Machine Tool Research, *NASA* TM X-64786, May 1973.

31. R. N. Zapata: Development of a Superconducting Electromagnetic Suspension and Balance System for Dynamic Stability Studies, *Rept.* ESS-4009-101-73U, University of Virginia, Charlottesville, February 1973.

32. J. Modransky et al.: Instrumented Locomotive Wheels for Continuous Measurements of Vertical and Lateral Loads, *ASME Paper* 79-RT-8, 1979.

SIX

PRESSURE AND SOUND MEASUREMENT

6.1 STANDARDS AND CALIBRATION

Pressure is not a fundamental quantity, but rather is derived from force and area, which in turn are derived from mass, length, and time, the latter three being fundamental quantities whose standards have been discussed earlier. Pressure "standards" in the form of very accurate instruments are available, though, for calibration of less accurate instruments. However, these "standards" depend ultimately on the fundamental standards for their accuracy. The basic standards[1] for pressures ranging from medium vacuum (about 10^{-1} mmHg) up to several hundred thousand pounds per square inch are in the form of precision mercury columns (manometers) and deadweight piston gages. For pressures in the range 10^{-1} to 10^{-3} mmHg, the McLeod vacuum gage is considered the standard. For pressures below 10^{-3} mmHg, a pressure-dividing technique allows flow through a succession of accurate orifices to relate the low downstream pressure to a higher upstream pressure (which is accurately measured with a McLeod gage).[2]

This technique can be further improved by substituting a Schulz hot-cathode or radioactive ionization vacuum gage for the McLeod gage. Each of these must

---

[1] D. P. Johnson and D. H. Newhall, The Piston Gage as a Precise Pressure-Measuring Instrument, *Instrum. Contr. Syst.*, p. 120, April 1962; Errors in Mercury Barometers and Manometers, *Instrum. Contr. Syst.*, p. 121, March 1962; 2″ Range Hg Manometer, *Instrum. Contr. Syst.*, p. 152, September 1962.

[2] J. R. Roehrig and J. C. Simons, Calibrating Vacuum Gages to 10^{-9} Torr, *Instrum. Contr. Syst.*, p. 107, April 1963.

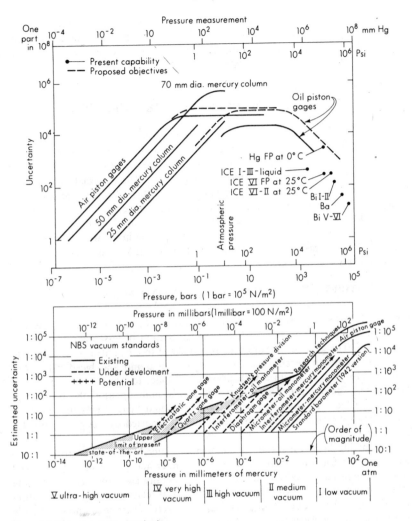

Figure 6.1 Pressure/vacuum standards.

be calibrated against a McLeod gage at one point (about 9×10^{-2} mmHg), but their known linearity is then used to extend their accurate range to much lower pressures.[1]

The inaccuracies of the above-mentioned pressure standards are summarized

---

[1] J. C. Simons, On Uncertainties in Calibration of Vacuum Gages and the Problem of Traceability, "Transactions of 10th National Vacuum Symposium," p. 246, Macmillan, New York, 1963.

graphically in Fig. 6.1.[1] Since the above-mentioned pressure standards are also pressure-measuring instruments (of the highest quality and used under carefully controlled conditions), their operating principles and characteristics are not discussed here since they are adequately covered later.

6.2 BASIC METHODS OF PRESSURE MEASUREMENT

Since pressure usually can be easily transduced to force by allowing it to act on a known area, the basic methods of measuring force and pressure are essentially the same, except for the high-vacuum region where a variety of special methods not directly related to force measurement are necessary. These special methods are described in the section on vacuum measurement. Other than the special vacuum techniques, most pressure measurement is based on comparison with known deadweights acting on known areas or on the deflection of elastic elements subjected to the unknown pressure. The deadweight methods are exemplified by manometers and piston gages while the elastic deflection devices take many different forms.

6.3 DEADWEIGHT GAGES AND MANOMETERS

Figure 6.2 shows the basic elements of a deadweight or piston gage. Such devices are employed mainly as standards for the calibration of less accurate gages or transducers. The gage to be calibrated is connected to a chamber filled with fluid whose pressure can be adjusted by some type of pump and bleed valve. The chamber also connects with a vertical piston-cylinder to which various standard weights may be applied. The pressure is slowly built up until the piston and weights are seen to "float," at which point the fluid "gage" pressure (pressure above atmosphere) must equal the deadweight supported by the piston, divided by the piston area.

For highly accurate results, a number of refinements and corrections are necessary. The frictional force between the cylinder and piston must be reduced to a minimum and/or corrected for. This is generally accomplished by rotating either the piston or the cylinder. If there is no axial relative motion, this rotation should reduce the axial effects of dry friction to zero. There must, however, be a small clearance between the piston and the cylinder and thus an axial flow of fluid from the high-pressure end to the low-pressure end. This flow produces a viscous shear force tending to support part of the deadweight. This effect can be estimated from theoretical calculations.[2] However, it varies somewhat with pressure since the piston and cylinder deform under pressure, thereby changing the clearance. The clearance between the piston and cylinder also raises the question

---

[1] Accuracy in Measurements and Calibrations, *NBS Tech. Note* 262, 1965.
[2] R. J. Sweeney, "Measurement Techniques in Mechanical Engineering," p. 104, Wiley, New York, 1953.

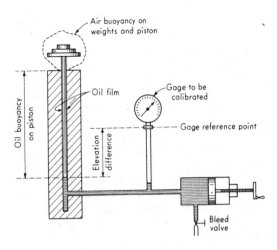

Air buoyancy on weights and piston

Oil buoyancy on piston

Oil film

Gage to be calibrated

Gage reference point

Elevation difference

Bleed valve

Figure 6.2 Deadweight gage calibrator.

of which area is to be used in computing pressure. The effective area generally is taken as the average of the piston and cylinder areas. Further corrections are needed for temperature effects on areas of piston and cylinder, air and pressure-medium buoyancy effects, local gravity conditions, and height differences between the lower end of the piston and the reference point for the gage being calibrated. Special designs and techniques allow use of deadweight gages for pressures up to several hundred thousand pounds per square inch. An improved design, the controlled-clearance piston gage,[1] employs a separately pressurized cylinder jacket to maintain the effective area constant (and thus achieve greater accuracy) at high pressures, which cause expansion and error in uncompensated gages.

Since the piston assembly itself has weight, conventional deadweight gages are not capable of measuring pressures lower than the piston weight/area ratio ("tare" pressure). This difficulty is overcome by the tilting-piston gage[2] in which the cylinder and piston can be tilted from vertical through an accurately measured angle, thus giving a continuously adjustable pressure from 0 lb/in^2 gage up to the tare pressure. The described gage uses nitrogen or other inert gas as the pressure medium and covers the range 0 to 600 lb/in^2 gage, having two interchangeable piston-cylinders and 14 weights. The accuracy is 0.01 percent of reading in the range 0.3 to 15 lb/in^2 gage and 0.015 percent of reading in the range 2 to 600 lb/in^2 gage. The tilting feature is used for the ranges 0 to 0.3 and 0 to 2.0 lb/in^2 gage; higher pressures are obtained in increments by the addition of discrete weights.

[1] D. H. Newhall and L. H. Abbot, Controlled-Clearance Piston Gage, *Meas. & Data*, January–February 1970.

[2] Ruska Instrument Corp., Houston, Tex.

A very convenient pressure standard[1] (although not really a deadweight gage) combines a precision piston gage with a magnetic null-balance laboratory scale[2] (Fig. 6.3). The gas or liquid pressure to be measured (generated and regulated by a system external to the pressure standard) is applied to one end of a rotating piston; the other end of the piston is supported by the "weighing platform" of the laboratory scale, which measures the pressure force and gives a digital readout of 40,000 counts full scale. Tungsten carbide piston-cylinders allow clearances less than 1 μm; the high hardness and elastic modulus maintain precision in the face of potential wear and pressure expansion. Piston-cylinder pairs are easily interchanged to give five full-scale ranges from 80 to 1,200 lb/in². Since deadweights are *not* utilized to measure the pressure force, periodic recalibration against a set of four precision masses is required. However, this is made quick and easy by using the scale's auto-tare feature and a simple screwdriver span adjustment. Instrument uncertainty on the 80 lb/in² range (other ranges are proportional) is $\pm(0.004 + 10^{-4}p)$ lb/in², where p is the actual pressure in pounds per square inch and repeatability is ± 1 count.

Deadweight gages may be employed for absolute- rather than gage-pressure measurement by placing them inside an evacuated enclosure at (ideally) 0 lb/in² absolute pressure. Since the degree of vacuum (absolute pressure) inside the enclosure must be known, this really requires an additional independent measurement of absolute pressure.

The manometer in its various forms is closely related to the piston gage, since both are based on the comparison of the unknown pressure force with the gravity force on a known mass. The manometer differs, however, in that it is self-balancing, is a deflection rather than a null instrument, and has continuous rather than stepwise output. The accuracies of deadweight gages and manometers of similar ranges are quite comparable; however, manometers become unwieldy at high pressures because of the long liquid columns involved. The U-tube manometer of Fig. 6.4 usually is considered the basic form and has the following relation between input and output for static conditions:

$$h = \frac{p_1 - p_2}{\rho g} \tag{6.1}$$

where $g \triangleq$ local gravity and $\rho \triangleq$ mass density of manometer fluid. If p_2 is atmospheric pressure, then h is a direct measure of p_1 as a gage pressure. Note that the cross-sectional area of the tubing (even if not uniform) has no effect. At a given location (given value of g) the sensitivity depends on only the density of the manometer fluid. Water and mercury are the most commonly used fluids. To realize the high accuracy possible with manometers, often a number of corrections must be applied. When visual reading of the height h is employed, the engraved-scale's temperature expansion must be considered. The variation of ρ

---

[1] Model 20400, DH Instruments, Inc., Pittsburgh, Pa.; P. Delajoud and M. Girard, The Development of a Digital Read-Out Primary Pressure Standard, DH Instruments, Pittsburgh, Pa., 1981.

[2] Mettler Instrument Corp., Hightstown, N.J.

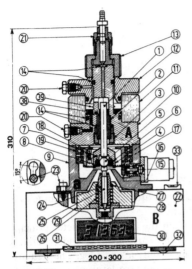

Figure 6.3 Pressure standard using electromagnetic balance. The digital standard is made up of a piston-cylinder measuring element (A + a) and an electronic dynamometer (B) *manufactured Mettler Instrument.*

Measuring element (A + a):

1. Piston in tungsten carbide
2. Cylinder in tungsten carbide
3. Cylinder retaining nut
4. Piston head
5. Ball in tungsten carbide
6. Ball bearing to center the ball (5)
7. Drive bearing
8. Retaining ring for ball (5)
9. Rotation mechanism housing
10. Piston-cylinder housing
11. Acrylic sight glass
12. Cover
13. Retaining nut
14. O-ring seals
15. Electric drive motor
16. Drive pinion
17. Toothed drive wheel
18. Drive bearing pin
19. Toothed wheel bearings
20. Purge screws
21. Quick-connect system (standard threads available)

Electronic dynamometer B:

22. Housing
23. 3 pins giving a quick release facility for the measuring assembly (A + a).
    (*After 15° rotation on the pins a locking mechanism secures the measuring element to the dynamométer.*)
24. Force limiting guide
25. Coupling rod
26. Force limiting spring
27-28. 2 vibration dampers
29. Force receiving plate
30. 40000 points 5 digits display
31. Auto-zero bar
32. 2 leveling screws
33. Bubble level

with temperature for the manometer fluid used must be corrected and the local value of g determined. Additional sources of error are found in the nonverticality of the tubes and the difficulty in reading h because of the meniscus formed by

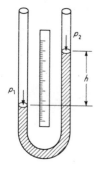

Figure 6.4 U-tube manometer.

capillarity. Considerable care must be exercised in order to keep inaccuracies as small as 0.01 mmHg for the overall measurement.[1]

A number of practically useful variations on the basic manometer principle are shown in Fig. 6.5. The *cistern* or *well-type manometer* is widely utilized because of its convenience in requiring reading of only a single leg. The well area is made very large compared with the tube; thus the zero level moves very little when pressure is applied. Even this small error is compensated by suitably distorting the length scale. However, such an arrangement, unlike a U tube, is sensitive to nonuniformity of the tube cross-sectional area and thus is considered somewhat less accurate.

Given that manometers inherently measure the pressure *difference* between the two ends of the liquid column, if one end is at zero absolute pressure, then h is an indication of absolute pressure. This is the principle of the *barometer* of Fig. 6.5. Although it is a "single-leg" instrument, high accuracy is achieved by setting the zero level of the well at the zero level of the scale before each reading is taken. The pressure in the evacuated portion of the barometer is not really absolute zero, but rather the vapor pressure of the filling fluid, mercury, at ambient temperature. This is about 10^{-4} lb/in^2 absolute at 70°F and usually is negligible as a correction.

To increase sensitivity, the manometer may be tilted with respect to gravity, thus giving a greater motion of liquid along the tube for a given vertical-height change. The *inclined manometer* (draft gage) of Fig. 6.5 exemplifies this principle. Since this is a single-leg device, the calibrated scale is corrected for the slight changes in well level so that rezeroing of the scale for each reading is not required.

The accurate measurement of extremely small pressure differences is accomplished with the *micromanometer*, a variation on the inclined-manometer principle. In Fig. 6.5 the instrument is initially adjusted so that when $p_1 = p_2$, the meniscus in the inclined tube is located at a reference point given by a fixed

[1] A. J. Eberlein, Laboratory Pressure Measurement Requirements for Evaluating the Air Data Computer, *Aeronaut. Eng. Rev.*, p. 53, April 1958.

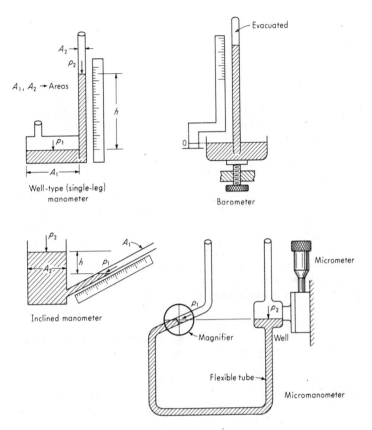

Figure 6.5 Various forms of manometers.

hairline viewed through a magnifier. The reading of the micrometer used to adjust well height is now noted. Application of the unknown pressure difference causes the meniscus to move off the hairline, but it can be restored to its initial position by raising or lowering the well with the micrometer. The difference in initial and final micrometer readings gives the height change h and thus the pressure. Instruments using water as the working fluid and having a range of either 10 or 20 in of water can be read to about 0.001 in of water.[1] In another instrument in which the inclined tube (rather than the well) is moved and which uses butyl alcohol as the working fluid, the range is 2 in of alcohol, and readability is 0.0002 in. This corresponds to a resolution of 6×10^{-6} lb/in^2.

While manometers usually are read visually by a human operator, various

[1] Meriam Instrument Co., Cleveland, Ohio.

schemes for rapid and accurate automatic readout are available, mainly for calibration and standards work using gaseous media. The sonar manometer[1] employs a piezoelectric transducer at the bottom of each 1.5-in-diameter glass tube to launch ultrasonic pulses, which travel up through the mercury columns, are reflected at the meniscus, and return to the bottom to be received by the transducers. The pulse from the shorter column turns on a digital counter, while that from the longer one turns it off. Thus a digital reading is obtained that is proportional to the difference in column height and thus to pressure. Resolution is 0.0003 in Hg, and accuracy is 0.001 inHg or 0.003 percent of reading, whichever is greater. Since temperature effects on sonic velocity and column length cause additive errors, a feedback control system keeps instrument temperature at $95 \pm 0.05°$F.

Another instrument[2] employs two large mercury cisterns (one fixed, one vertically movable by an electromechanical servosystem) connected by flexible tubing to create a U-tube manometer. Each cistern has a capacitor formed by a metal plate, the mercury surface, and a small air gap between them. The two capacitors are connected in an electric circuit which exhibits a null reading when the air gaps in both cisterns are equal and produces an error voltage when they are not. This error voltage causes the servosystem to drive the movable cistern to an elevation where balance is again achieved. A digital counter on the servosystem motor shaft reads out position to the nearest 0.0001 inHg. System accuracy is ± 0.0003 inHg ± 0.003 percent of reading. For manometers such as the two above, accessory automatic systems for generating and regulating the pressures of the gaseous (usually air or nitrogen) calibration media usually are available.

Manometer Dynamics

While manometers are utilized mainly for static measurements, their dynamic response is sometimes of interest. The general problem of the oscillations of liquid columns is an interesting (and rather difficult) question in fluid mechanics and has received considerable attention in the literature.[3] Here we take a considerably simplified view of the problem which nonetheless gives results of practical interest. In the U-tube configuration of Fig. 6.6a, the unknown pressures p_1 and p_2 are exerted by a gas whose inertia and viscosity may be considered negligible compared with those of the manometer liquid. If the pressures vary with time, the reading of the manometer varies with time; we are interested in the fidelity with which the manometer reading follows the pressure variation. The motion of the manometer liquid in the tube is caused by the action of various

---

[1] D. E. Van Dyck, Sonic Measurement of Manometer Column Height, Wallace & Tiernan Div., Belleville, N. J.

[2] Schwien Engineering, Pomona, Calif.

[3] J. F. Ury, Viscous Damping in Oscillating Liquid Columns, Its Magnitude and Limits, *Int. J. Mech. Sci.*, vol. 4, p. 349, 1962; P. D. Richardson, Comments on Viscous Damping in Oscillating Liquid Columns, *Int. J. Mech. Sci.*, vol. 5, p. 415, 1963.

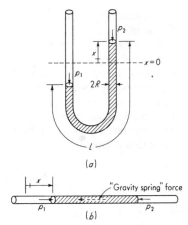

Figure 6.6 Manometer model.

forces. If we consider the manometer liquid in its entirety as a free body and search for forces acting on it, the following forces come to mind:

1. The gravity force (weight) distributed uniformly over the whole body of fluid
2. A drag force due to motion of the fluid within the tube and related to the wall shearing stress
3. The forces on the two ends of the free body due to pressures p_1 and p_2
4. Distributed normal pressure of the tube on the fluid
5. Surface-tension effects at the two ends of the body of fluid

A detailed analysis of all these effects would lead to rather complex and unwieldy mathematics. Fortunately, useful results may be obtained by a simplified analysis. The initial step is the assumption that the system shown in Fig. 6.6b is dynamically equivalent to that of Fig. 6.6a. The "gravity spring" force of Fig. 6.6b is explained as follows: In Fig. 6.6a, whenever $x \neq 0$, there is an unbalanced gravity force acting on the liquid column tending to restore the level to $x = 0$. The magnitude of this force is $-2\pi R^2 x \gamma$, where $\gamma \triangleq$ manometer-fluid specific weight, in pounds force per cubic inch. We see that this force is proportional to the displacement x and always opposes it; thus it has all the characteristics of a spring force. When the liquid column is "straightened out" in Fig. 6.6b, we must include this "gravity spring" force in our equivalent system if we are to preserve the analogy of the two configurations. In comparing Fig. 6.6b with Fig. 6.6a, we note that any effects of flow curvature in the 180° bend are lost. But these will probably be small if the diameter of the bend is large compared with the inside diameter of the manometer tube and if the total length L is large compared with the bend length. We also neglect any surface-tension effects at the ends of the column. This is usually a good assumption if the column is long relative to its diameter.

In addition to the gravity-spring force and the pressure forces due to p_1 and p_2, the liquid column is subjected to a drag force at the interface between the liquid and the wall of the tube. This drag force is equal to the wall shearing stress times the surface area of the liquid column. The motion of the liquid in the tube may be thought of as an unsteady pipe flow. We assume that at any instant of time the wall shearing stress can be computed from the instantaneous velocity of the liquid by *using the formulas commonly used for steady pipe flows.*

The flow of liquid in the tube may occur in the laminar, transition, or turbulent regimes. First let us assume laminar flow prevails. The pressure drop Δp due to pipe friction for both laminar and turbulent flow is given by

$$\Delta p = f \frac{\gamma L V_{\text{av}}^2}{2gd} = f \frac{L}{d} \frac{\rho V_{\text{av}}^2}{2g_0} \tag{6.2}$$

where $g_0 \triangleq$ mass unit conversion factor, lbm/slug
$\rho \triangleq$ fluid density, lbm/ft^3
$f \triangleq$ friction factor
$\gamma \triangleq$ fluid specific weight, local, lbf/ft^3
$V_{\text{av}} \triangleq$ average velocity
$g \triangleq$ local gravity acceleration, ft/s^2
$d \triangleq$ diameter of pipe
$L \triangleq$ pipe length

The wall shearing stress τ_0 is given by

$$\tau_0 = \Delta p \frac{d}{4l} \tag{6.3}$$

Thus
$$\tau_0 = f \frac{\gamma V_{\text{av}}^2}{8g} = f \frac{\rho V_{\text{av}}^2}{8g_0} \tag{6.4}$$

For laminar flow the friction factor is given by

$$f = \frac{64\mu g}{d\gamma V_{\text{av}}} = \frac{64}{dV_{\text{av}} \rho/(\mu g_0)} \tag{6.5}$$

so that
$$\tau_0 = \frac{4\mu V_{\text{av}}}{R} \tag{6.6}$$

where $R \triangleq d/2$.

This result also can be obtained directly from the laminar velocity distribution

$$V = V_c \left[1 - \left(\frac{r}{R} \right)^2 \right] \tag{6.7}$$

where $V \triangleq$ velocity at radius r and $V_c \triangleq$ center-line velocity.

The velocity gradient is

$$\frac{dV}{dr} = -\frac{V_c}{R^2}(2r) \tag{6.8}$$

which becomes at the wall

$$\left.\frac{dV}{dr}\right|_{r=R} = -\frac{2V_c}{R} \tag{6.9}$$

Shearing stress is given by

$$\tau = \mu\frac{dV}{dr} \tag{6.10}$$

and so the magnitude of the wall shearing stress τ_0 is

$$\tau_0 = \frac{4\mu V_{av}}{R}$$

since $V_c = 2V_{av}$ for laminar flow in circular pipes.

We are now in a position to apply Newton's law to the system of Fig. 6.6b. The average flow velocity V_{av} corresponds to $\dot{x}$, the first derivative of x with respect to time. Considering the entire body of liquid as a free body and taking the effective mass of the moving liquid as four-thirds of its actual mass, based on the kinetic energy of steady laminar flow, we can write for motion in the x direction

$$\pi R^2(p_1 - p_2) - 2\pi R^2\gamma x - 2\pi RL\frac{4\mu\dot{x}}{R} = \frac{4}{3}\frac{\pi R^2 L\gamma}{g}\ddot{x} \tag{6.11}$$

This reduces to

$$\frac{2\ddot{x}}{3g/L} + \frac{4\mu L}{R^2\gamma}\dot{x} + x = \frac{1}{2\gamma}p \tag{6.12}$$

where we have defined $p \triangleq p_1 - p_2$. In operator form, this becomes

$$\left(\frac{2D^2}{3g/L} + \frac{4\mu L}{R^2\gamma}D + 1\right)x = \frac{1}{2\gamma}p \tag{6.13}$$

The operational transfer function relating output x to input p is

$$\frac{x}{p}(D) = \frac{1/(2\gamma)}{2D^2/(3g/L) + [4\mu L/(R^2\gamma)]D + 1} \tag{6.14}$$

which is of the form $\quad \dfrac{x}{p}(D) = \dfrac{K}{D^2/\omega_n^2 + 2\zeta D/\omega_n + 1} \tag{6.15}$

where
$$K \triangleq \frac{1}{2\gamma} \quad \text{in/(lb/in}^2) \tag{6.16}$$

$$\omega_n \triangleq \sqrt{\frac{3g}{2L}} \quad \text{rad/s} \tag{6.17}$$

$$\zeta \triangleq 2.45\mu \frac{\sqrt{gL}}{R^2\gamma} \tag{6.18}$$

We note from the above that the manometer is a second-order instrument. The numerical values of the parameters are usually such that $\zeta < 1.0$; that is, the instrument is underdamped.

Since laminar flow was assumed in carrying out the above analysis, we should try to estimate a typical Reynolds number to see under what conditions laminar flow occurs. As a numerical example, we take a mercury manometer with

$$L = 26.5 \text{ in}$$

$$R = 0.13 \text{ in}$$

$$\mu = 2.18 \times 10^{-7} \text{ lbf} \cdot \text{s/in}^2$$

$$\gamma = 0.491 \text{ lbf/in}^3$$

Suppose that we wish to check our theoretical results by measuring ζ and ω_n experimentally for step-function input. Computing ζ and ω_n from Eqs. (6.17) and (6.18) gives $\zeta = 0.007$ and $\omega_n = 4.7$ rad/s. Since $\zeta = 0.007$ represents a *very lightly* damped system, we can estimate the maximum flow velocity by assuming no damping at all. A second-order system with no damping executes pure sinusoidal oscillations when subjected to a step-function input. Thus its motion would be given by

$$x = X \sin \omega_n t \tag{6.19}$$

where X is the size of the step function. The velocity $\dot{x}$, which is the same as the average flow velocity, would be

$$\dot{x} = \omega_n X \cos \omega_n t \tag{6.20}$$

and its maximum value would thus be $\omega_n X$. The Reynolds number for steady pipe flow is given by

$$N_R = \frac{2\gamma R V_{av}}{g\mu} \tag{6.21}$$

and the critical value for transition from laminar to turbulent flow is 2,100. Since $V_{av} = \omega_n X$, it should be clear that there is a maximum-size step function that can be used without exceeding $N_R = 2,100$. This limiting value X_m is given by

$$2,100 = \frac{2\gamma R \omega_n X_m}{g\mu} \tag{6.22}$$

which in this example gives

$$X_m = \frac{(2{,}100)(386)(2.18 \times 10^{-7})}{(2)(0.491)(0.13)(4.7)} = 0.30 \text{ in} \qquad (6.23)$$

Thus, to ensure laminar flow at all times during the oscillation, the step input can be no larger than 0.30 in. Suppose we wish to measure ζ and ω_n by simple visual methods—ζ from the size of the first overshoot and ω_n by counting and timing cycles. This requires much larger step inputs for reasonable accuracy. Therefore we must investigate the effect of the presence of turbulent flow on our analysis.

Suppose that we decide that a step input X_m of 5 inHg will be sufficiently large to allow accurate measurements of ζ and ω_n. The maximum Reynolds number then would be

$$N_R = \frac{5}{0.30}(2{,}100) = 35{,}000 \qquad (6.24)$$

For steady turbulent flow in smooth pipes with $3{,}000 < N_R < 100{,}000$, the Blasius equation for friction factor is

$$f = \frac{0.316}{(N_R)^{0.25}} \qquad (6.25)$$

The turbulent wall shearing stress is then given by Eq. (6.4) as

$$\tau_0 = \frac{0.0378\gamma^{0.75}\mu^{0.25}V_{av}^{1.75}}{g^{0.75}R^{0.25}} \qquad (6.26)$$

Using the numerical values of this particular example, we get

$$\tau_0 = 9.18 \times 10^{-6}V_{av}^{1.75} \qquad \text{lbf/in}^2 \qquad (6.27)$$

For laminar flow the comparable expression is

$$\tau_0 = 6.71 \times 10^{-6}V_{av} \qquad \text{lbf/in}^2 \qquad (6.28)$$

The most significant difference between Eqs. (6.27) and (6.28) is that (6.27) represents a nonlinear relation between shear stress and velocity whereas (6.28) is linear. This means that when the force due to wall shearing stress is substituted into Newton's law, the result is a nonlinear differential equation because of the term $(\dot{x})^{1.75}$. This nonlinear equation cannot be solved analytically, so the presence of turbulent flow leads to mathematical difficulties.

In working with oscillations of systems with nonlinear damping terms similar to the $(\dot{x})^{1.75}$ of this example, engineers have developed an approximate method of analysis which is quite useful. This approach is based on the observation that while the linear damping term $\dot{x}$ and the nonlinear term, such as $(\dot{x})^{1.75}$, are quite different mathematically, the general *form* of the oscillation in the two systems is not radically different in experimental tests. If the linear system is excited by a sinusoidal exciting force, it will respond with a sinusoidal motion, whereas the nonlinear system's motion will not be purely sinusoidal. However, observation shows that the deviation from pure sinusoidal motion is usually quite small. Using these facts as a basis, then, we might reason as follows: If a system with nonlinear damping is executing steady oscillations of fixed amplitude, during each cycle the damping force will dissipate a certain amount of energy. If we know from experience that the waveform of the nonlinear oscillation is nearly sinusoidal, we can compute approximately the energy dissipation per cycle. This is done as follows: Suppose there exists a steady oscillation of amplitude Y and frequency ω. If we assume the waveform to be sinusoidal, we can write

$$y = Y \sin \omega t \tag{6.29}$$

and

$$\dot{y} = Y\omega \cos \omega t \tag{6.30}$$

Now, in general, the instantaneous power is the product of instantaneous force and instantaneous velocity. Thus the power dissipation due to damping is the product of velocity and damping force. If the damping force is a known function of velocity $f(\dot{y})$, we can write

$$\text{Instantaneous power dissipation} = \dot{y}f(\dot{y}) \tag{6.31}$$

and the energy dissipated per cycle will be given by

$$\int_{\text{one cycle}} \dot{y}f(\dot{y}) \, dt \tag{6.32}$$

For a linear damping the function $f(\dot{y})$ is just $B\dot{y}$, where B is a constant. So the energy dissipation per cycle is

$$\int_0^{2\pi/\omega} (Y\omega \cos \omega t)(BY\omega \cos \omega t) \, dt = \pi B\omega Y^2 \tag{6.33}$$

For the nonlinear damping due to turbulent flow, the function $f(\dot{y})$ is $C(\dot{y})^{1.75}$, where C is a constant. The energy dissipation per cycle for this nonlinear damping is

$$\int_0^{2\pi/\omega} (Y\omega \cos \omega t)C(Y\omega \cos \omega t)^{1.75} \, dt \tag{6.34}$$

This is equal to

$$C(Y\omega)^{2.75} \int_0^{2\pi/\omega} (\cos \omega t)^{2.75} \, dt \tag{6.35}$$

In evaluating the integral in (6.35), we must use physical reasoning, because when $\cos \omega t$ becomes *negative*, the quantity $(\cos \omega t)^{2.75}$ is not defined in terms of real numbers. Physical reasoning, however, tells us that the physical processes occurring during the first quarter-cycle $[0 \leq t \leq \pi/(2\omega)]$ give exactly the same energy dissipation as those occurring during the other three quarters of the cycle. Thus, we can integrate over only the first quarter and multiply by 4 to get the total energy dissipation. During the first quarter-cycle, $\cos \omega t$ is always positive, and so no mathematical difficulties arise. This amounts to saying that, to agree with the known physical facts, integral (6.34) should really be written as

$$\int_0^{2\pi/\omega} |Y\omega \cos \omega t| \, [C(Y\omega)^{1.75}|(\cos \omega t)|^{1.75}] \, dt \tag{6.36}$$

with the absolute-value signs as shown.

By defining $\theta \triangleq \omega t$, integral (6.36) can be written as

$$CY^{2.75}\omega^{1.75} \int_0^{2\pi} (\cos \theta)^{2.75} \, d\theta \tag{6.37}$$

Since this integral is not available analytically, we use CSMP to obtain it numerically, which gives

$$2.77C\omega^{1.75}Y^{2.75} \tag{6.38}$$

Having obtained the above results, we now define the *equivalent linear damping* as that linear damping which would dissipate exactly the same energy per cycle as the nonlinear damping at a given frequency and amplitude. Thus we set (6.33) equal to (6.38) and get

$$\pi B_e \omega Y^2 = 2.77C\omega^{1.75}Y^{2.75} \tag{6.39}$$

$$B_e \triangleq \text{equivalent linear damping}$$
$$B_e = 0.882C(\omega Y)^{0.75} \tag{6.40}$$

Since C is the constant that multiplies $(\dot{y})^{1.75}$, in the manometer example we have

$$\text{Damping force} = 2\pi RL\tau_0 = \frac{0.237R^{0.75}\gamma^{0.75}\mu^{0.25}L\dot{x}^{1.75}}{g^{0.75}} \tag{6.41}$$

and thus

$$C = \frac{0.237\gamma^{0.75}R^{0.75}L\mu^{0.25}}{g^{0.75}} \tag{6.42}$$

Now Eq. (6.18) can be written as

$$\zeta = \frac{2.45\mu L\sqrt{g}}{R^2\gamma\sqrt{L}} = \frac{0.0974B\sqrt{g}}{R^2\gamma\sqrt{L}} \tag{6.43}$$

since $B = 8\pi\mu L$ for the linear system. We can now define the equivalent linear damping ratio ζ_e by substituting B_e from Eqs. (6.40) and (6.42) in (6.43):

$$\zeta_e \triangleq \frac{0.0203\sqrt{L}[\mu/(\gamma g)]^{0.25}}{R^{1.25}} (\omega Y)^{0.75} \tag{6.44}$$

This shows clearly the dependence of ζ_e on the frequency and amplitude of the oscillation. For turbulent flow the value of ω_n is also somewhat different since the velocity profile tends to be more nearly square than parabolic. If the velocity is assumed uniform over the cross section, the effective mass becomes equal to the actual mass since all particles have the same velocity. If the turbulent damping, though larger than the laminar, is assumed to be still quite small, it is reasonable to expect that it will have little effect on the frequency. We therefore compute ω_n for turbulent flow by neglecting the nonlinear damping completely where it would appear in Eq. (6.11). We then get

$$\omega_n = \sqrt{\frac{2g}{L}} \qquad (6.45)$$

for turbulent flow.

Many assumptions were made in the above analysis. The formulas for steady pipe flow were employed for an unsteady situation. In an oscillating flow, velocity actually goes to zero twice each cycle, no matter how great the amplitude or frequency; thus one wonders whether part of such a cycle is turbulent and part laminar. In the analysis above, turbulent equations were used for the whole cycle. The nonlinear differential equation containing $(\dot{x})^{1.75}$ actually has no closed-form analytical solution. Thus what is the meaning of a ζ_e and an ω_n attached to such a process? Such questions and others may be at least partially resolved by more complex analyses or experimental studies. To provide some idea of the degree of validity of our simplified analysis, some experimental results are given. They were obtained at The Ohio State University by undergraduate students who study manometer dynamics in a simple experiment in a measurement course.

The experiment consists in part of suddenly releasing an air pressure applied to a mercury manometer and observing the resulting oscillations. The process is slow enough that ζ_e can be estimated from the size of the first overshoot and ω_n by counting and timing cycles with a stopwatch. The manometers used have the numerical values quoted earlier in this section. A step pressure input of 10 inHg ($x = 5$ in) is utilized; thus turbulent flow may be expected. If laminar flow were assumed, the theoretical values would be $\zeta = 0.007$ and $\omega_n = 4.7$ rad/s. For turbulent flow, ω_n becomes 5.4 rad/s. To calculate ζ_e from Eq. (6.44), the frequency and amplitude of the oscillation must be known. For a step input the frequency is the damped natural frequency rather than ω_n; however, these are practically identical for the small damping present, and so we use ω_n. We experimentally measure ζ_e from the first overshoot, and so the proper amplitude to employ in the theoretical calculation might be an average of the initial amplitude and of that at the first overshoot. Only the initial amplitude (5 in) is known, though, and so this is used. Equation (6.44) then gives $\zeta_e = 0.0905.$ By timing and counting cycles (about 8 cycles of the decaying oscillation can be measured easily) the experimental value of ω_n is 5.2 rad/s. This lies between the values calculated for turbulent and laminar flow and thus is not unreasonable, since several of the 8 cycles used were of quite low amplitude because of the decay of the oscillation. The first overshoot is about 4.05 in, giving an experimental value of ζ of 0.067. Therefore the theoretical estimate of 0.0905 is fairly good. If we now use the experimental value of ω (5.2) and the average amplitude of the first half-cycle (4.53) in Eq. (6.44), then the predicted ζ_e is 0.082, which compares even more favorably. Based on even these limited results, a certain amount of confidence in the theoretical predictions is established.

Using digital simulation (such as CSMP), we can easily "solve" the nonlinear manometer differential equation. For the numerical values given, by substituting

Eq. (6.27) in (6.11) and using the actual (rather than 4/3 times the actual) mass appropriate for turbulent flow, a CSMP program could be written as:

```
XADOT2 = − .1113*SW*N − 29.2*XA
SW = FCNSW(XADOT1, − 1.0,0.0,1.0)
N = (ABS(XADOT1))**1.75
XADOT1 = INTGRL(0.0,XADOT2)
XA = INTGRL(5.0,XADOT1)
XLINZD = CMPXPL(5.0,0.0,.0905,5.4,0.0)
```

Here XADOT2 is $\ddot{x}$, and SW uses the CSMP function switch FCNSW to get the correct sign (± 1) on the nonlinear damping term as velocity XADOT1 changes direction. We start XA(x) off at 5 in, as in the experimental test. For comparison we also run a linear second-order system with $\zeta = 0.0905$ and $\omega_n = 5.4$, using the single statement XLINZD=.... . Figure 6.7 shows that the linearized approximate model agrees almost perfectly with the nonlinear for the first cycle, but damps down too quickly later, as we would expect, since the nonlinear damping *decreases* with amplitude while the linear stays fixed. Since the first overshoot is modeled almost perfectly by the linearized system, the discrepancies noted earlier between experimental and linearized theoretical results are clearly *not* due to inadequacies in the linearization technique. Rather, they must be charged to

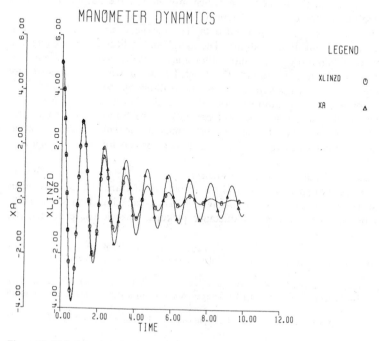

Figure 6.7 Digital simulation of manometer dynamics.

inadequacies in the original nonlinear model, inaccuracy in numerical values of parameters, and/or errors in the experimental results.

6.4 ELASTIC TRANSDUCERS

While a wide variety of flexible metallic elements conceivably might be used for pressure transducers, the vast majority of practical devices utilize one or another form of Bourdon tube,[1] diaphragm,[2] or bellows[3] as their sensitive element, as shown in Fig. 6.8. The gross deflection of these elements may directly actuate a pointer/scale readout through suitable linkages or gears, or the motion may be transduced to an electrical signal by one means or another. Strain gages bonded directly to diaphragms or to diaphragm-actuated beams are widely used to measure local strains that are directly related to pressure.

The Bourdon tube is the basis of many mechanical pressure gages and is also used in electrical transducers by measuring the output displacement with potentiometers, differential transformers, etc. The basic element in all the various forms is a tube of noncircular cross section. A pressure difference between the inside and outside of the tube (higher pressure inside) causes the tube to attempt to attain a circular cross section. This results in distortions which lead to a curvilinear translation of the free end in the C type and spiral and helical types and an angular rotation in the twisted type, which motions are the output. The theoretical analysis of these effects is difficult, and practical design at present still makes use of considerable empirical data. The C-type Bourdon tube has been utilized up to about 100,000 lb/in^2. The spiral and helical configurations are attempts to obtain more output motion for a given pressure and have been used mainly below about 1,000 lb/in^2. The twisted tube shown has a crossed-wire stabilizing device which is stiff in all radial directions but soft in rotation. This reduces spurious output motions from shock and vibration. Figure 6.9 shows construction of a higher accuracy (0.1 percent) C-type Bourdon test gage. Also shown is an optional bimetal temperature compensator which maintains accuracy over the range -25 to $+125°$F. This device corrects for both thermal zero shift and span shift.

Electrical output pressure transducers are available in forms combining various elastic elements with most of the displacement transducers described in Chap. 4, i.e., potentiometers, strain gages, LVDTs, capacitance pickups, eddy-current probes, reluctance and inductance pickups, piezoelectric elements, etc.

---

[1] D. M. Considine (ed.), "Process Instruments and Controls Handbook," sec. 3, McGraw-Hill, New York, 1957; R. W. Bradspies, Bourdon Tubes, *Giannini Tech. Notes*, Giannini Corp., Duarte, Calif., January–February 1961.

[2] Considine, ibid.; Pressure Capsule Design, *Giannini Tech. Notes*, November 1960; C. K. Stedman, The Characteristics of Flat Annular Diaphragms, *Statham Instrum. Notes*, Statham Instruments Inc., Los Angeles.

[3] Considine, op. cit.; R. Carey, Welded Diaphragm Metal Bellows, *Electromech. Des.*, p. 22, August 1963.

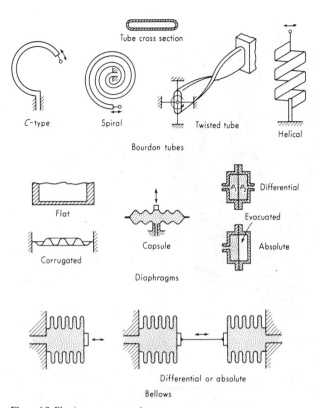

Figure 6.8 Elastic pressure transducers.

Space does not permit detailed discussion of all types, so we choose specific examples that illustrate features we feel to be particularly significant and interesting.

Figure 6.10 shows a sensor utilizing an infrared LED (light-emitting diode) and two photodiodes to optically measure displacement (0.020 in full scale) of the pressure-sensitive elastic element. The reference and measurement photodiodes are on the same chip and thus are equally affected by temperature changes. Changes in LED output due to temperature or age also cancel, since both diodes share the same illumination and a ratiometric integrating analog/digital converter is employed to obtain a digital output sensitive to only diode-illuminated areas A_R and A_X and pot settings α and β. While diode signals exhibit nonlinearity which varies from unit to unit, this nonlinearity and all others are linearized in the analog/digital converter using a look-up table resident in a pair of PROMs (programmable read-only memories). Thus each sensor has its individual nonlinearities linearized by programming its PROM at calibration time. This design is characteristic of recent trends which, rather than making a great effort

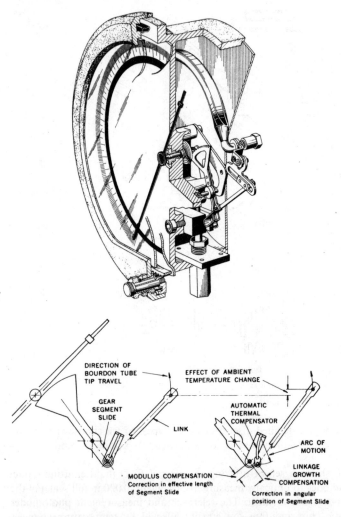

Figure 6.9 Bourdon-tube gage construction. (*Courtesy Heise Gage, Dresser Industries, Newton, Conn.*)

to obtain linearity in basic sensors, accept their nonlinearity (as long as it is repeatable) and then use programmable electronic compensation tailored to each unit to achieve overall linearity. The referenced unit is available in ranges of 0 to 50 in H_2O to 60,000 lb/in^2 gage and 5 lb/in^2 absolute to 60,000 lb/in^2 absolute with accuracy 0.1 percent of span and temperature effect on span of ± 0.004 percent/°F from 0 to 160°F. An automatic-zero feature in the analog/digital converter minimizes thermal zero shifts. Interestingly, the manufacturer also uses

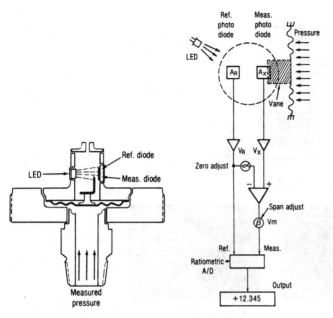

Figure 6.10 Electro-optic pressure transducer. (*Courtesy Heise Gage, Dresser Industries, Newton, Conn.*)

the same optical sensor as a null detector in a force-balance type of pressure transducer (see Fig. 6.18c). Here the diode nonlinearity need not be linearized since the servosystem always returns the sensor to the same balance point.

Piezoelectric pressure transducers share many common characteristics with piezoelectric accelerometers and force transducers, discussed earlier. Generally they have high natural frequencies and little response to spurious acceleration inputs [typically 0.02 $(lb/in^2)/g$]. However, transducers designed for low pressures and mounted on objects with severe vibration may require an acceleration-compensated design. Figure 6.11[1] shows how a compensating accelerometer is built right into the pressure transducer and its signal combined with that of the pressure-sensing crystals to cancel the vibration effect. About a 5:1 or 10:1 improvement is available in this way. A higher degree of compensation is technically possible, but is not attempted in practice since a *transverse* vibration sensitivity (which *cannot* be compensated) limits performance at this level. Figure 6.11b compares the output of a compensated (upper trace) and an uncompensated transducer when both are mounted on the same sinusoidal vibration shaker; Fig. 6.11c shows a similar comparison when the pressure transducers are each

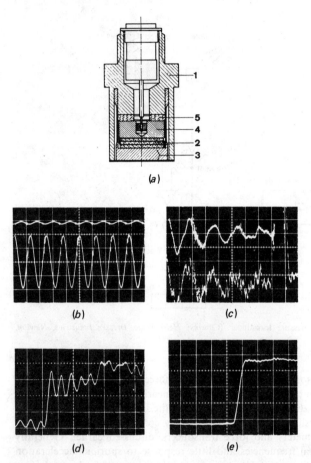

Figure 6.11 Piezoelectric transducer with acceleration compensation. (*a*) 1 = housing; 2 = pressure-sensitive quartz disks; 3 = diaphragm assembly; 4 = compensating mass; 5 = compensating quartz disk.

fastened to a strongly vibrating diesel-engine exhaust pipe to measure exhaust pressure. The compensation is clearly quite effective in these examples. However, it may *not* be effective in suppressing the pressure transducer's natural resonance if the accelerometer resonance does not adequately match the pressure transducer's, as can be seen in Fig. 6.11*d*, where a compensated transducer is mounted in the wall of an aerodynamic shock tube. The oscillatory signal at the far left actually occurs even *before* the air shock wave strikes the transducer diaphragm and is due to mechanical excitation of the transducer resonance caused by the *mechanical* shock wave in the tube wall, which travels *faster* than the airborne wave. To obtain the improved response of Fig. 6.11e, a transducer with about 4 times higher natural frequency was mounted with a plastic adapter between the wall and the transducer. The plastic adapter serves as a mechanical shock filter, while the higher-natural-frequency transducer is inherently less acceleration-sensitive.

Another solution to the transducer resonance phenomenon involves careful matching of the compensating accelerometer's resonance with that of the pressure elements. When this is successful, both acceleration effects and transducer "ringing" response to sharp pressure changes are reduced. Transducers[1] based on this approach may have 500-kHz resonant frequency, 1-μs rise time and less than 15 percent overshoot for step inputs, and acceleration sensitivity of 0.002 $(\text{lb/in}^2)/g$ for a 500 lb/in^2 full-scale unit with integral microelectronic amplifier (10-V full-scale output).

Pressure transducers based on foil-type metal strain gages may either apply the strain gages directly to a flat metal diaphragm or use a convoluted diaphragm to apply force to a strain-gage beam. Flat diaphragms exhibit nonlinearity at large deflections, where a stretching action adds to the basic bending, causing a stiffening effect. This nonlinearity in the stresses follows closely the nonlinearity in the center deflection y_c, for which the following theoretical result is available:

$$p = \frac{16Et^4}{3R^4(1 - v^2)} \left[\frac{y_c}{t} + 0.488\left(\frac{y_c}{t}\right)^3 \right] \qquad (6.46)$$

where $p \triangleq$ pressure difference across diaphragm
$\quad E \triangleq$ modulus of elasticity
$\quad t \triangleq$ diaphragm thickness
$\quad v \triangleq$ Poisson's ratio
$\quad R \triangleq$ diaphragm radius to clamped edge

By designing for a sufficiently small value of y_c/t, a desired nonlinearity may be achieved. However, note that small y_c/t also gives small strains and output voltage.

Such a diaphragm, clamped at the edges and subjected to a uniform pressure difference p, has at any point on the low-pressure surface a radial stress s_r and a tangential stress s_t given by the following formulas (see Fig. 6.12):

$$s_r = \frac{3pR^2v}{8t^2} \left[\left(\frac{1}{v} + 1\right) - \left(\frac{3}{v} + 1\right)\left(\frac{r}{R}\right)^2 \right] \qquad (6.47)$$

$$s_t = \frac{3pR^2v}{8t^2} \left[\left(\frac{1}{v} + 1\right) - \left(\frac{1}{v} + 3\right)\left(\frac{r}{R}\right)^2 \right] \qquad (6.48)$$

[1] PCB Piezotronics, Buffalo, N.Y.

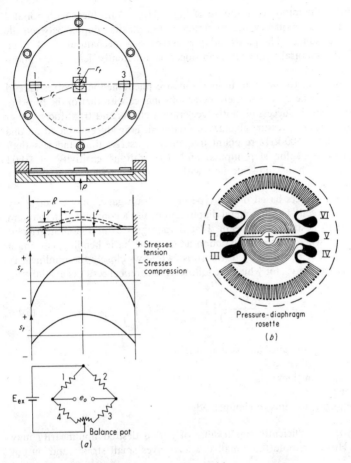

Figure 6.12 Diaphragm-type strain-gage pressure pickup.

where $v \triangleq$ Poisson's ratio. The deflection at any point is given by

$$y = \frac{3p(1 - v^2)(R^2 - r^2)^2}{16Et^3} \qquad (6.49)$$

Equations (6.47) to (6.49) all give linear relations between stress or deflection and pressure and are accurate only for sufficiently small pressures. Equation (6.46) may be used to estimate the degree of nonlinearity. The stress situation on the diaphragm surface is fortunate because both tension and compression stresses exist simultaneously. This allows use of a four-active-arm bridge in which all effects are additive (giving large output) and gives temperature compensation also. Gages 2 and 4 are placed as close to the center as possible and oriented to read tangential strain, since it is maximum (positive) at this point. Gages 1 and 3 are oriented to read radial strain and placed as close to the edge as possible, since radial strain has its maximum negative value at that point. The laws of bridge circuits show that the pressure effects on all four gages are additive. In computing the overall sensitivity, Eqs. (6.47) and (6.48) cannot be employed directly to determine the strains "seen" by the gages since the diaphragm surface is in a state of *biaxial* stress and *both* the radial and tangential stress contribute to the radial or tangential strain at any point. The general biaxial stress-strain relation gives

$$\epsilon_r = \frac{s_r - vs_t}{E} \tag{6.50}$$

$$\epsilon_t = \frac{s_t - vs_r}{E} \tag{6.51}$$

Once the gage strains are calculated, the individual gage ΔR's are obtained from the gage factors, and then e_o can be determined from the bridge-circuit sensitivity equations.

To facilitate construction and miniaturization of such pressure transducers, the discrete individual gages may be replaced by a rosette available in various sizes from several strain-gage manufacturers. Figure 6.12b^1 shows how such rosettes are configured to take advantage of radial strains at the diaphragm edge and tangential strains near the center, while the solder tabs are located in a low-strain region to increase reliability of the solder joints. For the referenced rosette, the manufacturer provides this design formula:

$$\frac{e_o}{E_{ex}} = 820 \frac{pR^2(1 - v^2)}{Et^2} \qquad \frac{mV}{V}$$

When such rosettes are utilized, the radius R can be made quite small and usually the diaphragm is "machined from solid," rather than using the "assembled from pieces" construction of Fig. 6.12a, thereby improving accuracy and hysteresis.

If the transducer is to be utilized for dynamic measurements, its natural frequency is of interest. A diaphragm has an infinite number of natural frequencies; however, the lowest is the only one of interest here. For a clamped-edge diaphragm vibrating in a vacuum (no fluid-inertia effects), the lowest natural frequency is given by

$$\omega_n = \frac{10.21}{R^2} \sqrt{\frac{Et^2}{12\rho(1 - v^2)}} \qquad \text{rad/s} \tag{6.52}$$

where $\rho \triangleq$ mass density of diaphragm material. A number of factors may make the actual operating value of ω_n different from the prediction of Eq. (6.52). The edge clamping is never perfectly rigid; any softness tends to lower ω_n. If the

[1] Design Considerations for Diaphragm Pressure Transducers, TN–129–3, Micro-Measurements, Romulus, Mich., 1974.

diaphragm is not perfectly flat, tightening the clamping bolts may cause a slight (perhaps imperceptible) "wrinkling," tending to stiffen the diaphragm and raise ω_n. If the diaphragm is used to measure liquid pressures, the inertia of the liquid tends to lower ω_n, especially if a small-diameter tube connects the pressure source to the diaphragm. When it is employed with gases, the volume of gas "trapped" behind the diaphragm may act as a stiffening spring, raising ω_n.

A pressure transducer of the discrete gage type constructed for use in a transducer research project at The Ohio State University serves as a numerical example for the above discussion. This transducer used a phosphor-bronze diaphragm with

$$E = 16 \times 10^6 \text{ lbf/in}^2 \qquad R = 1.830 \pm 0.002 \text{ in}$$

$$v = \tfrac{1}{3} \qquad r_t = 0.15 \pm 0.01 \text{ in}$$

$$\rho = 0.00083 \text{ lbf} \cdot \text{s}^2/\text{in}^4 \qquad r_\tau = 1.52 \pm 0.01 \text{ in}$$

$$t = 0.0454 \pm 0.0003 \text{ in}$$

The strain gages had a gage factor of 1.97 ± 2 percent and a resistance of $119.5 \pm 0.3 \; \Omega$. The bridge was excited with 7.5 V dc, and the transducer was designed for a full-scale range of 10 lb/in^2. The theoretically calculated sensitivity was 0.516 mV/(lb/in^2). Static calibration gave 0.513 mV/(lb/in^2) and indicated a maximum nonlinearity of 2 percent of full scale. The theoretically calculated natural frequency was 924 Hz, and the experimental value (with atmospheric air on both sides of the diaphragm) was 897 Hz.

Figure 6.13 shows construction details for a transducer in which a thin diaphragm (welded to the housing) serves as a seal and pressure-gathering member; however, the majority of the pressure force is taken by strain-gage bending beams in the form of spokes on a wheel. Nonlinearity as small as 0.05 to 0.1 percent can be achieved with designs of this type. Just as in the strain-gage

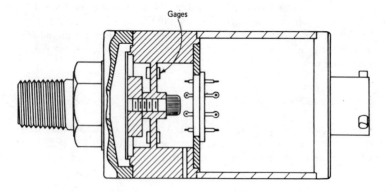

Figure 6.13 Bonded foil strain-gage pressure transducer. (*Courtesy Sensotec Inc., Columbus, Ohio.*)

force transducers of Chap. 5, various levels of temperature compensation can be provided, depending on application needs.

Figure 6.14 shows construction of a piezoresistive (semiconductor "strain gage") pressure transducer[1] of flush-diaphragm design with full-scale ranges from 0 to 2 to 0 to 200 bars. The measuring cell is a welded assembly of two silicon chips forming a sealed chamber containing the reference pressure (or vacuum for absolute pressure sensing). One chip is bored out on one side to form a diaphragm of thickness suited to the desired range, while the reverse side has a complete bridge of strain-sensitive regions diffused into the silicon surface. Use of constant-current bridge excitation (see figure) enhances primary temperature compensation, which together with external discrete compensating resistors brings the overall thermal zero shift to less than 0.01 percent of full-scale/°C and the sensitivity shift to less than 0.02 percent/°C. Pressure is transmitted from the thin metal diaphragm through an oil fill (which also provides damping) to the silicon capsule. Natural frequency is greater than 70 kHz, transducer compliance is 0.001 mm^3/bar, and acceleration sensitivity is 10^{-4} bar/g.

A line of piezoresistive pressure transducers[2] available in ranges from 2 to 50,000 lb/in^2 applies the fluid pressure *directly* to a silicon diaphragm with a complete Wheatstone bridge of strain elements diffused into the diaphragm. Chemical etching processes are employed to produce a "sculptured" diaphragm (rather than the usual flat configuration) with thick and thin sections carefully designed to concentrate strain at the locations where the gages will be diffused into the silicon. A further innovation employs the piezoresistive material in a transverse, rather than longitudinal, configuration, which gives improved linearity since the transverse gages exhibit compensating nonlinearities for the tension and compression legs of the bridge. While the transducers are specified for use with dry nonconductive gases, short-term applications in water[3] have been successful. A ±2 lb/in^2 gage unit is only 0.092 in in diameter; has 1 percent nonlinearity (3 percent at 3 times full-scale pressure), burst pressure of ±40 lb/in^2 gage, 157 mV/(lb/in^2) sensitivity, resonant frequency of 45,000 Hz, acceleration sensitivity of 0.0005 (lb/in^2)/g longitudinal (0.00003 lateral), compensated temperature range 0 to 200°F (3 percent zero shift, 4 percent sensitivity shift); and uses a 10-V constant-voltage excitation.

Diffused strain-gage pressure transducers intended for high-volume, low-cost applications such as appliances, automotive parts, etc., are available also. Figure 6.15a[4] shows the typical construction of the basic unit and an oil-filled version with a flexible silicone rubber barrier ("sock"), required for protection from certain working fluids. Often such transducers are designed to employ

[1] H. R. Winteler and G. H. Gautschi, Piezoresistive Pressure Transducers, Kistler Instruments, Amherst, N.Y., 1979.

[2] R. M. Whittier, Basic Advantages of the Anisotropic Etched, Transverse Gage Pressure Transducer, *Prod. Dev. News*, vol. 16, no. 3, Endevco Corp., San Juan Capistrano, Calif., 1980.

[3] K. Souter and H. E. Krachman, Measurement of Local Pressure Resulting from Hydrodynamic Impact, TP269, Endevco Corp., San Juan Capistrano, Calif., 1978.

[4] Pressure Transducer Handbook, National Semiconductor Corp., Santa Clara, Calif., 1977.

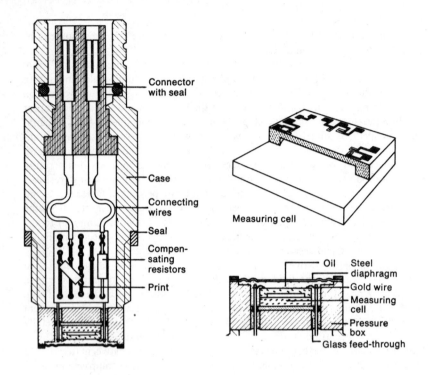

Connector
with seal

Case

Connecting
wires

Seal

Compen-
sating
resistors

Print

Measuring cell

Oil   Steel
     diaphragm

Gold wire

Measuring
cell

Pressure
box

Glass feed-through

Current excitation
The pressure transducers are excited by
constant current. The voltage rise due to
the increase in resistance with temperature
compensates for the decrease of the gage
factor with temperature. The graph shows
the typical relative changes of the
resistance R, the gage factor G and the
output voltage U_{out} in function of temperature.

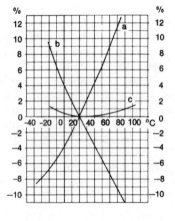

a) $\dfrac{\Delta R}{R}$

b) $\dfrac{\Delta G}{G} = \dfrac{\Delta U_{out}}{U_{out}}$ with constant voltage excitation

c) $\dfrac{\Delta U_{out}}{U_{out}}$ with constant current excitation

Figure 6.14 Diffused semiconductor strain-gage transducer. (*Courtesy Kistler Instruments, Amherst,
N. Y., model 4041.*)

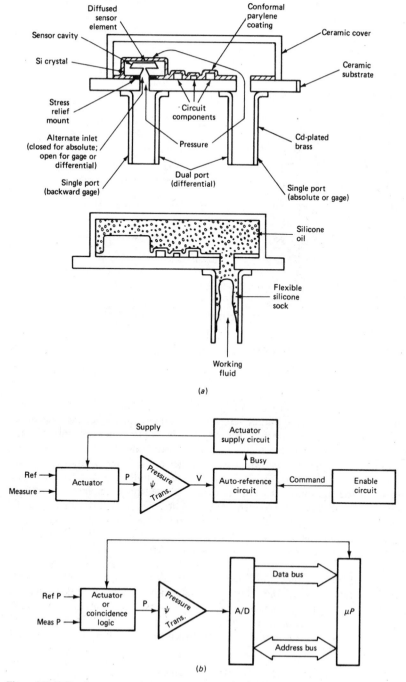

Figure 6.15 Diffused sensor transducers and auto-reference techniques.

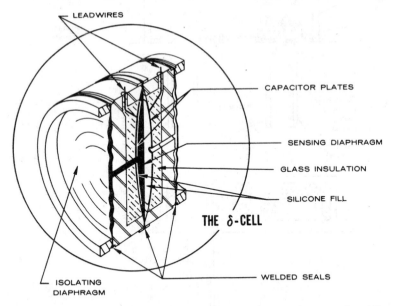

LEADWIRES

CAPACITOR PLATES

SENSING DIAPHRAGM

GLASS INSULATION

SILICONE FILL

THE δ-CELL

ISOLATING DIAPHRAGM

WELDED SEALS

Figure 6.16 Capacitive differential-pressure transmitter. (*Courtesy Rosemount Inc., Minneapolis, Minn.*)

auto-referencing[1] techniques in order to achieve good accuracy in the face of zero shifts resulting from time and/or temperature. This requires that the transducer periodically be exposed (momentarily) to some *known* reference pressure (perhaps the most common example would be a gage pressure unit, which could be vented to atmosphere, giving a known 0 lb/in² gage input). The automatic-reference circuit measures and "remembers" the voltage output for zero pressure input, and when the transducer is reconnected to its usual pressure signal, the circuit adds or subtracts this value, thereby correcting for bias errors from all sources. (If two different reference pressures can be employed, the technique can correct for slope errors too.) Figure 6.15b[2] shows block diagrams for "hardware" and "software" (microprocessor) versions of automatic-referencing (consult the reference for details).

Most designs of pressure transducers are available in gage (psig), absolute (psia), differential (psid), and sealed (psis) reference versions. A sealed reference transducer is similar to the absolute pressure version, except that the sealed reference chamber contains a nonzero pressure rather than 0 lb/in² absolute. Since this reference pressure will change with temperature, applications must take

---

[1] Ibid.; Auto Zeroing Circuits, *Bull.* 107-057, Data Instruments Inc., Lexington, Mass.
[2] Pressure Transducer Handbook, op. cit.

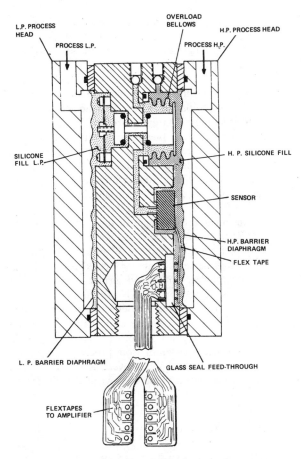

Figure 6.17 Diffused strain-gage differential-pressure transmitter. (*Courtesy Honeywell Inc., Ft. Washington, Pa.*)

this fact into account. Some transducer designs expose sensitive electrical elements to the measured fluid and thus may be limited to dry noncorrosive gases.

Differential-pressure transmitters are utilized widely in process instrumentation and control systems for flow, liquid level, density, viscosity, etc.[1] They present particularly difficult problems of transducer design since they often need to sense small differences (0 to 5 in H_2O) in large pressures, withstand full-line

---

[1] V. N. Lawford, Differential-Pressure Instruments: The Universal Measurement Tools, *Instrum. Tech.*, pp. 30–40, December 1974; M. Slomiana, Using Differential Pressure Sensors for Level, Density, Interface, and Viscosity Measurements, *In Tech*, pp. 63–68, September 1979.

pressure (perhaps 2,000 lb/in^2) overload, and interface with difficult process liquids. Mechanical, pneumatic, and electrical output devices are available.[1] Here we briefly describe some electrical output devices that provide the standard 4 to 20 mA current signal popular in process systems.[2]

Figure 6.16 shows a unit whose sensing diaphragm forms the moving plate of a differential capacitor; the motion is transduced to a proportional dc current by using an ac-excited bridge circuit, demodulation, and a feedback-type current regulator. Stainless-steel (316 SS) isolating diaphragms protect the sensor from process fluids while transmitting the measured pressures through silicone-fluid fills to the sensing diaphragm. Overpressure causes the sensing diaphragm to (safely) bottom out on the glass backup plates, causing less than 0.25 percent of span error when normal operation resumes after a 2,000 lb/in^2 overload.

Figure 6.17 shows a unit using diffused silicon "strain gage" technology. A single silicon crystal is cut to 1/4-in^2 by 1/10 in thick, and the center of one side is ground out and etched to form an integral diaphragm of thickness suitable for the desired pressure range (3 inH$_2$O to 10,000 lb/in^2). Then boron is diffused into the other side of the silicon chip to form four complete Wheatstone-bridge circuits, each having four strain-sensitive elements properly located on the diaphragm. A computerized test system checks each of the four bridges and selects for further processing only the best one. (This approach is typical of integrated-circuit manufacturing technology in which the most cost-effective approach often involves high-rate production with a certain percentage of defectives, which are later weeded out by automated test equipment.) Isolation diaphragms and silicone-fluid fills are employed to protect the sensor from process fluids. Overload protection is provided by valving (actuated by an overload bellows) which equalizes pressure on both sides of the silicon diaphragm when the differential pressure approaches unsafe values.

6.5 FORCE-BALANCE TRANSDUCERS

Feedback or null-balance principles may be applied to pressure measurement in a manner similar to that employed for force measurement. Figure 6.18a shows a pneumatic/mechanical type, and electromechanical methods are used in Fig. 6.18b. High loop gain in these servosystems gives good linearity and accuracy. The block diagrams give the static relations only; a dynamic analysis is necessary to determine the limit set on gain by stability requirements. The operation of these instruments is left for you to deduce from the schematic and block diagrams. Previous discussions of feedback-type instruments maybe helpful. In Fig. 6.18c[3], the optical displacement sensor of Fig. 6.10 is used as a null detector in the feedback system, giving an instrument with 0.05 percent overall accuracy.

---

[1] M. Slomiana, Selecting Differential Pressure Instrumentation, *In Tech*, pp. 32–40, August 1979.

[2] R. F. Wolny, Applying Electronic ΔP Transmitters, *Instrum. Tech.*, pp. 47–53, July 1978.

[3] Ashcroft Digigauge, Dresser Industries, Newton, Conn.

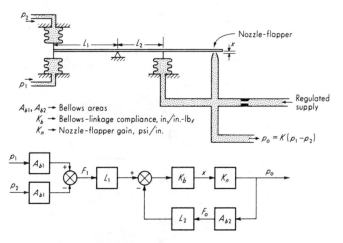

$A_{b1}, A_{b2} \rightarrow$ Bellows areas
$K_b \rightarrow$ Bellows-linkage compliance, in./in.-lb$_f$
$K_n \rightarrow$ Nozzle-flapper gain, psi/in.

Force-balance differential-pressure transducer

(a)

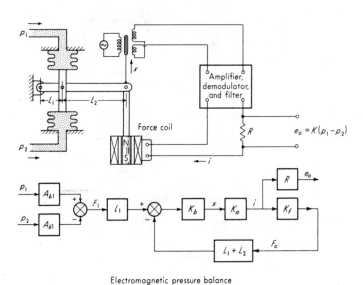

Electromagnetic pressure balance

(b)

Figure 6.18 Null-balance pressure sensors.

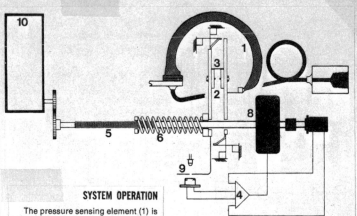

SYSTEM OPERATION

The pressure sensing element (1) is attached to the beam assembly (2) so that a variation in process pressure deflects the sensor and beam assembly. A slotted vane attached to the free end of the beam interrupts a ray of light impinging on a dual-element photoconductive cell (9). The photoconductive cell is in a bridge network, consequently any motion of the vane causes an unbalanced condition in the bridge. The differential signal of the bridge, amplified through solid-state circuitry, drives a servo motor with integral feedback tachometer generator, (8) and lead screw assembly (5). The feedback spring assembly (6) riding on the lead screw is also connected to the beam assembly. Motion of the lead screw moves the feedback spring to produce a force balancing that developed by the pressure sensor, thus returning the system to a null position.

ABSOLUTE PRESSURE

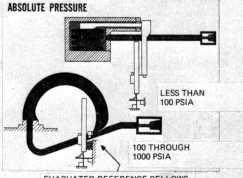

LESS THAN 100 PSIA

100 THROUGH 1000 PSIA

EVACUATED REFERENCE BELLOWS
BAROMETRIC COMPENSATION

DIFFERENTIAL PRESSURE

GAUGE PRESSURE

LESS THAN 100 PSIG lb/in^2

100 THROUGH 10,000 PSIG
lb/in^2 GAGE

(c)

Figure 6.18 (*Continued*)

438

6.6 DYNAMIC EFFECTS OF VOLUMES AND CONNECTING TUBING

We have mentioned the possible strong effect of fluid properties and "plumbing" configurations on the dynamic behavior of pressure-measuring systems. In this section some of these problems are investigated. First, it should be pointed out that if maximum dynamic performance is to be attained, then a flush-diaphragm transducer mounted directly at the point where a pressure measurement is wanted should be used if at all possible (see Fig. 6.19). Any connecting tubing or volume chambers will degrade performance to some extent. The fact that this degradation is studied in this section indicates that in many practical circumstances a flush-diaphragm transducer is not applicable.

Liquid Systems, Heavily Damped, Slow-Acting

In the system of Fig. 6.20, the spring-loaded piston represents the flexible element of the pressure pickup. For the present analysis, the only pertinent characteristic of the pressure pickup is its volume change per unit pressure change, the compliance C_{vp} [in^3/(lb/in^2)]. This can be calculated or measured experimentally. (The transducer of Fig. 6.14, for example, has $C_{vp} = 0.001$ mm^3/bar.) For systems that are heavily damped (a criterion for judging this is given shortly) or subjected to relatively slow pressure changes, the inertia effects of both the fluid and the moving parts of the pickup are negligible compared with viscous and spring forces. We show that under these conditions the measured pressure p_m follows the desired pressure p_i with a first-order lag. For steady laminar flow in the tube we have

$$p_i - p_m = \frac{32\mu L V_{t,\,av}}{d_t^2} \qquad (6.53)$$

where $\mu \triangleq$ fluid viscosity and $V_{t,\,av} \triangleq$ average flow velocity in tube. While this equation is exact only for steady flow, it holds quite closely for slowly varying

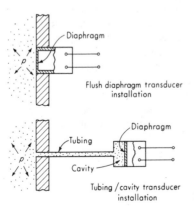

Figure 6.19 Transducer installation types.

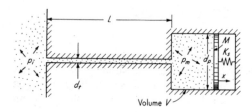

Volume V **Figure 6.20** Transducer/tubing model.

velocities. During a time interval dt, a quantity of liquid enters the chamber. This is given by

$$dV = \text{volume entering} = \frac{\pi d_t^2}{4} V_{t,\,av}\, dt = \frac{\pi d_t^4 (p_i - p_m)\, dt}{128 \mu L} \tag{6.54}$$

Any volume added or taken away results in a pressure change dp_m given by

$$dp_m = \frac{dV}{C_{vp}} = \frac{p_i - p_m}{\tau}\, dt \tag{6.55}$$

and thus

$$\tau \frac{dp_m}{dt} + p_m = p_i \tag{6.56}$$

where

$$\tau \triangleq \frac{128 \mu L C_{vp}}{\pi d_t^4} \tag{6.57}$$

Thus the response is seen to follow a standard first-order form. To keep τ small, the tubing length should be short, the diameter large, and C_{vp} small.

The viewpoint of this analysis is that any sudden changes in p_i will cause much more gradual changes in p_m, and thus the pickup spring-mass system will not be able to manifest its natural oscillatory tendencies. Under such conditions an overall first-order response from the tubing/transducer system may be expected. To obtain some numerical estimate of the conditions required for such behavior, we later carry out an analysis which includes inertia effects. This will lead to a second-order type of response, and when ζ of this model is greater than about 1.5, the simpler first-order model may be employed with fair accuracy, at least (in terms of frequency response) for frequencies less than ω_n.

The model used above predicts that, for a change in p_i, p_m starts to change immediately. This cannot be true since a pressure wave propagates through a fluid at finite speed (the velocity of sound) for small disturbances. There is thus a dead time τ_{dt} equal to the distance traversed divided by the speed of sound. For liquids and reasonable tube lengths usually this delay is small enough to ignore completely. The speed of sound v_s in a fluid contained in a nonrigid tube is given by

$$v_s = \sqrt{\frac{E_L}{\rho_L}} \sqrt{\frac{1}{1 + 2R_t E_L/(t E_t)}} \tag{6.58}$$

where $E_L \triangleq$ bulk modulus of fluid
$\quad \rho_L \triangleq$ mass density of fluid
$\quad R_t \triangleq$ tube inside radius
$\quad t \triangleq$ tube-wall thickness
$\quad E_t \triangleq$ tube-material modulus of elasticity

If this dead time is significant in a practical problem, it may be included in the transfer function as

$$\frac{p_m}{p_i}(D) = \frac{e^{-\tau_{dt}D}}{\tau D + 1} \tag{6.59}$$

with fair accuracy.

Finally, it should also be pointed out that the result [Eq. (6.56)] of this analysis also may be applied to systems using gases rather than liquids if the elastic pressure-sensing element is sufficiently soft that its volume change per unit pressure change is much larger than that due to gas compressibility.

Liquid Systems, Moderately Damped, Fast-Acting

When the motions of the liquid and the pickup elastic element are rapid, their inertia is no longer negligible. An analysis[1] of this situation using energy methods is available. In the system of Fig. 6.20, any change in pressure p_m must be accompanied by a volume change; this in turn requires an inflow or outflow of liquid through the tube. If the tube is of small diameter compared with the equivalent piston diameter of the pickup, the tube flow will be at a much higher velocity than the piston velocity, and the kinetic energy of the liquid in the tube may be a large (sometimes major) part of the total system kinetic energy. This increase in kinetic energy is equivalent to adding mass to the piston and ignoring the fluid inertia, and the analysis below calculates just how much mass should be added to give the same effect as the fluid inertia. This added mass lowers the system natural frequency and thereby degrades dynamic response.

To find the equivalent piston/spring configuration for a given transducer, the volume change per unit pressure change must be equal for both systems. This gives

Transducer volume change = equivalent piston volume change

$$pC_{vp} = \frac{\pi^2 d_p^4 p}{16 K_s} \tag{6.60}$$

$$\frac{d_p^4}{K_s} = \frac{16 C_{vp}}{\pi^2} \tag{6.61}$$

Thus d_p and K_s for the equivalent system can have any values that satisfy Eq.

---

[1] G. White, Liquid Filled Pressure Gage Systems, *Statham Instrum. Notes*, Statham Instruments Inc., Los Angeles, January–February 1949.

(6.61). Also, the natural frequencies of both systems with no fluid present must be equal. This gives

$$\omega_{n,t} = \sqrt{\frac{K_s}{M}} \tag{6.62}$$

$$\frac{K_s}{M} = \omega_{n,t}^2 \tag{6.63}$$

where $\omega_{n,t} \triangleq$ transducer natural frequency. Again, any values of K_s and M that satisfy Eq. (6.63) may be used. Thus, to define the equivalent system, only C_{vp} and $\omega_{n,t}$ need be known; they can be found from experiment if theoretical formulas are unavailable.

Since we have just shown how the equivalent system is related to the real system, now we can proceed with an analysis of the equivalent system. The volume change dV is related to the piston motion dx by

$$dV = \frac{\pi d_p^2\, dx}{4} \tag{6.64}$$

Thus

$$\frac{dV}{dt} = \frac{\pi d_p^2}{4} \frac{dx}{dt} \tag{6.65}$$

and

$$\frac{\pi}{4} d_t^2\, V_{t,\,av} = \frac{\pi d_p^2}{4} \frac{dx}{dt} \tag{6.66}$$

where $V_{t,\,av}$ = average flow velocity in tube. We then get

$$V_{t,\,av} = \left(\frac{d_p}{d_t}\right)^2 \frac{dx}{dt} \tag{6.67}$$

Next we assume laminar flow in the tube, with the parabolic velocity profile characteristic of steady flow:

$$V_t = V_{t,\,c}\left[1 - \left(\frac{r}{R}\right)^2\right] \tag{6.68}$$

where $V_t \triangleq$ velocity at radius r, $V_{t,\,c} \triangleq$ centerline velocity, and $R \triangleq$ tube inside radius. The kinetic energy of an annular element of fluid (density ρ) of thickness dr at radius r is

$$d(KE) = \frac{(2\pi r\, dr)L_\rho V_t^2}{2} \tag{6.69}$$

Substitution of Eq. (6.68) and integration give the fluid kinetic energy as

$$KE = \frac{\pi \rho L V_{t,\,av}^2\, d_t^2}{6} \tag{6.70}$$

For a square velocity profile the kinetic energy would be

$$KE_s = \frac{\pi \rho L V_{t,\,av}^2 \, d_t^2}{8} \tag{6.71}$$

The actual velocity profile will be somewhere between parabolic and square. Even if laminar flow exists, the velocity profile is nonparabolic except for steady flow.[1] Turbulent flow gives a rather square profile. We assume Eq. (6.70) to hold here; however, Eq. (6.71) can be carried through with equal ease to "bracket" the correct value. The rigid mass M_e, attached to M, which would have the same kinetic energy as the fluid, is given by

$$\frac{M_e}{2} \left(\frac{dx}{dt} \right)^2 = \frac{\pi \rho L d_p^4}{6 d_t^2} \left(\frac{dx}{dt} \right)^2 \tag{6.72}$$

$$M_e = \frac{\pi \rho L d_p^4}{3 d_t^2} \tag{6.73}$$

The natural frequency of the transducer/tubing system is then

$$\omega_n = \sqrt{\frac{K_s}{M + M_e}} = \sqrt{\frac{1}{M/K_s + M_e/K_s}} = \sqrt{\frac{1}{1/\omega_{n,\,t}^2 + 16\rho L C_{vp}/(3\pi d_t^2)}} \tag{6.74}$$

To keep ω_n as high as possible, L and C_{vp} must be as small as possible and d_t as large as possible. In many practical cases, $M_e \gg M$, which allows simplification of Eq. (6.74) to

$$\omega_n = \sqrt{\frac{3\pi d_t^2}{16\rho L C_{vp}}} \tag{6.75}$$

Next we calculate the damping ratio of the transducer/tubing system. The transducer itself is assumed to have negligible damping; thus the only damping is due to the fluid friction in the tube. Again we assume the validity of the steady laminar-flow relations to calculate the pressure drop due to fluid viscosity as $32\mu L V_{t,\,av}/d_t^2$. The force on the piston due to this pressure drop is the damping force $B(dx/dt)$; thus

$$\frac{\pi d_p^2}{4} \frac{32\mu L V_{t,\,av}}{d_t^2} = B \frac{dx}{dt} \tag{6.76}$$

and, since $V_{t,\,av} = (dx/dt)(d_p/d_t)^2$,

$$B = 8\pi\mu L \left(\frac{d_p}{d_t} \right)^4 \tag{6.77}$$

---

[1] C. K. Stedman, Alternating Flow of Fluid in Tubes, *Statham Instrum. Notes*, Statham Instruments Inc., Los Angeles, January 1956.

Then, using the general formula for the damping ratio of a spring-mass-dashpot system, we get

$$\zeta = \frac{B}{2\sqrt{K_s(M + M_e)}} = \frac{64\mu L C_{vp}}{\pi d_t^4 \sqrt{1/\omega_{n,t}^2 + 16\rho L C_{vp}/(3\pi d_t^2)}} \tag{6.78}$$

If $M_e \gg M$, this simplifies to

$$\zeta = \frac{16\sqrt{3/\pi}\,\mu\,\sqrt{L C_{vp}/\rho}}{d_t^3} \tag{6.79}$$

The above theory has been partially checked experimentally in an M.Sc. thesis by Fowler.[1] (A later, more detailed, investigation by NASA[2] confirmed the essential features of the theory and made useful extensions for those cases where the compliance of the liquid itself and/or that of entrained gas bubbles is not negligible compared with transducer compliance.) The transducer used was the diaphragm strain-gage instrument whose numerical parameters are presented on page 430. The liquid used was water, and ω_n and ζ were found from step-function tests using a bursting cellophane diaphragm to obtain a sudden 10 lb/in² release of pressure. Tubing of 0.042- to 1.022-in inside diameter and 4 to 32 in in length was studied. Conditions were such that turbulent flow probably existed much of the time; thus use of Eq. (6.71) was indicated. The value of C_{vp} was calculated from theory as 0.00441 in³/(lb/in²) while an experimental calibration gave 0.00500. For all cases in which oscillation occurred (and thus ω_n could be accurately measured), it was found that ω_n was accurately predicted within about 5 to 10 percent. To show the severity of the performance degradation, the natural frequency (which was 897 Hz with no tubing and in air) became about 60 Hz with a 1.022-in diameter tube 10 in in length and about 3 Hz with a 0.092-in tube 32 in long. However, the prediction of ζ was much less satisfactory, being invariably too low. (Fortunately, ζ is of little importance in those many practical applications which exhibit light damping.) The small-diameter long tubes (which encourage laminar flow) gave the best correlation with theory, but even there errors of 100 percent occurred. For example, the 0.042-in tube 10 in long had a ζ of about 1.0 while theory predicted 0.66. The poor agreement undoubtedly can be charged to turbulent flow and energy losses from expansion and contraction (end effects) at the tubing ends.

Fowler developed corrections for these effects that yielded much better agreement. However, the turbulent flow and end effects are nonlinear damping mechanisms; thus the meaning of ζ is confused for both theoretical predictions and experimental measurements. The use of smaller step functions should give lower velocities and thus more nearly laminar flow. This effect was checked for a

[1] R. L. Fowler, An Experimental Study of the Effects of Liquid Inertia and Viscosity on the Dynamic Response of Pressure Transducer-Tubing Systems, M.Sc. Thesis, The Ohio State University, Mechanical Engineering Department, 1963.

[2] R. C. Anderson and D. R. Englund, Jr., Liquid-Filled Transient Pressure Measuring Systems: A Method for Determining Frequency Response, *NASA* TN D-6603, December 1971.

0.092-in-diameter, 10-in-long tube. A 10 lb/in^2 step function gave $\zeta = 0.34$; a 5 lb/in^2 step function gave 0.22. The theoretical value was 0.065, suggesting that the theory might be quite accurate for sufficiently small inputs. Detailed discussion of these problems may be found in the reference.

When ζ calculated from Eq. (6.78) or (6.79) is 1.5 or greater, the simpler first-order model of Eq. (6.56) may be adequate. To show the relationships of these two analyses, recall that in the second-order form $K/[D^2/\omega_n^2 + 2\zeta D/\omega_n + 1]$, the inertia effect resides in the D^2 term. Neglecting this gives $K/(2\zeta D/\omega_n + 1)$. If we calculate $2\zeta/\omega_n$ from Eqs. (6.74) and (6.78), it is numerically equal to τ of Eq. (6.57).

Gas Systems with Tube Volume a Small Fraction of Chamber Volume

When, in the configuration of Fig. 6.20, a gas is the fluid medium, the compressibility of the gas in volume V becomes the major spring effect when the pressure pickup is at all stiff. It is reasonable, then, to assume that the volume V is enclosed by rigid walls ($C_{vp} = 0$). The majority of practical problems involve rather low frequencies. This allows treatment as a lumped-parameter system with fluid properties considered constant along the length of the tube for small disturbances. The validity of this viewpoint is shown by noting that pressure-wave propagation in the gas follows the general law of wave motion:

$$\lambda = \text{wavelength} = \frac{\text{velocity of propagation}}{\text{frequency}} = \frac{c}{f} \tag{6.80}$$

Our lumped-parameter assumption says that the gas in the tube moves as a unit, as opposed to wave motion *within* the tube. For any given tube length L, for sufficiently low frequencies, the lumped-parameter approach becomes valid. For example, the velocity of pressure waves in standard air is about 1,120 ft/s. If an oscillation of 10-Hz frequency exists, its wavelength must be 112 ft/cycle. This means that the spacewise variation of fluid pressure due to wave motion has a wavelength of 112 ft (see Fig. 6.21). For a tube of, say, 1-ft length, the variation in pressure (and thus in density, etc.) from one end of the tube to the other resulting from wave motion is very small. That is, there is negligible *relative* motion of

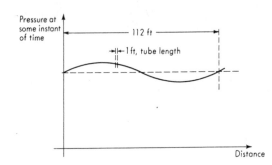

Pressure at some instant of time

112 ft

1 ft, tube length

Distance

Figure 6.21 Justification of lumped-parameter analysis.

particles in the tube; they all move together, as a unit. While the above aspect of wave motion is neglected, the dead time due to the finite speed of propagation can be relatively easily taken into account (approximately) by multiplying the second-order transfer function (which we obtain shortly) by $e^{-\tau_{dt}D}$, where

$$\tau_{dt} = \frac{L}{v_s} \tag{6.81}$$

$$\text{Sound velocity} = v_s = \sqrt{\frac{\gamma p}{\rho}} \tag{6.82}$$

where $\gamma \triangleq$ ratio of specific heats, $\rho \triangleq$ mass density, and $p \triangleq$ average pressure.

The analysis[1] we carry out below is valid only for small pressure changes; the system becomes nonlinear for large disturbances. Steady-laminar-flow formulas are used for calculating fluid friction. However, the effective mass is taken equal to the actual mass (rather than the four-thirds actual mass given by the parabolic velocity profile). We follow the reference in this respect; use of the $\frac{4}{3}$ factor is easily incorporated if we wish to bracket a more correct value. Numerically the effect is rather small in any case.

The analysis consists merely of applying Newton's law to the "slug" of gas in the tube. We assume that initially $p_i = p_m = p_0$ when p_i changes slightly in some way. From here on, the symbols p_i and p_m are taken to mean the *excess* pressures over and above p_0. The force due to the pressure p_i is $\pi p_i d_t^2/4$. The viscous force due to the wall shearing stress is $8\pi\mu L\dot{x}_t$, where x_t is the displacement of the slug of gas in the tube. If the slug of gas moves into the volume V an amount x_t, the pressure p_m will increase. We assume this compression occurs under adiabatic conditions. The adiabatic bulk modulus E_a of a gas is given by

$$E_a \triangleq -\frac{dp}{dV/V} = \gamma p \tag{6.83}$$

The displacement x_t causes a volume change $dV = \pi d_t^2 x_t/4$. This, in turn, causes a pressure excess $p_m = \pi E_a d_t^2 x_t/(4V)$. The force due to this pressure excess is $\pi^2 E_a d_t^4 x_t/(16V)$. Newton's law then gives

$$\frac{\pi p_i d_t^2}{4} - 8\pi\mu L\dot{x}_t - \frac{\pi^2 E_a d_t^4 x_t}{16V} = \frac{\pi d_t^2 L\rho}{4}\ddot{x}_t \tag{6.84}$$

and since $p_m = \pi E_a d_t^2 x_t/(4V)$,

$$\frac{4L\rho V}{\pi E_a d_t^2}\ddot{p}_m + \frac{128\mu L V}{\pi E_a d_t^4}\dot{p}_m + p_m = p_i \tag{6.85}$$

[1] G. J. Delio, G. V. Schwent, and R. S. Cesaro, Transient Behavior of Lumped-Constant Systems for Sensing Gas Pressures, *NACA, Tech. Note* 1988, 1949.

This is clearly the standard second-order form, and so we define

$$\omega_n \triangleq \frac{d_t}{2} \sqrt{\frac{\pi E_a}{L \rho V}} \tag{6.86}$$

$$\zeta \triangleq \frac{32 \mu}{d_t^3} \sqrt{\frac{VL}{\pi E_a \rho}} \tag{6.87}$$

Since E_a and ρ both vary during pressure changes, ζ and ω_n are not really constants; that is, the system is nonlinear. For small-percentage pressure variations around the original equilibrium pressure p_0, however, ζ and ω_n vary only slightly and the behavior is nearly linear. In calculating ζ and ω_n, E_a and ρ are computed by using pressure p_0.

When L becomes very short, Eq. (6.86) predicts a very large ω_n. In practice, this will not occur since, even when $L = 0$, there is some air (close to the opening in the volume) which has appreciable velocity and therefore kinetic energy. Theory shows that this end effect may be taken into account by using for L in Eq. (6.86) an effective length L_e given by

$$L_e = L \left(1 + \frac{8}{3\pi} \frac{d_t}{L} \right) \tag{6.88}$$

In most cases, the term $[8/(3\pi)](d_t/L)$ is completely negligible compared with 1.0. However, if $L = 0$ (tube degenerates into simply a hole in the side of volume V), we can still compute an ω_n since then $L_e = [8/(3\pi)]d_t$. The computation of damping for this case is not straightforward and is not discussed here.

Gas Systems with Tube Volume Comparable to Chamber Volume

When the volume of the tube becomes a significant part of the total volume of a system, compressibility effects are no longer restricted to the volume chamber alone and the above formulas become inaccurate. More refined analyses[1] give the following formulas for ζ and ω_n:

$$\omega_n = \frac{\sqrt{\gamma p/\rho}}{L \sqrt{1/2 + V/V_t}} \tag{6.89}$$

$$\zeta = \frac{16 \mu L}{d_t^2 \sqrt{\gamma p \rho}} \sqrt{\frac{1}{2} + \frac{V}{V_t}} \tag{6.90}$$

where $V_t \triangleq$ tubing volume. In these formulas, if $V/V_t \gg \frac{1}{2}$ (tubing volume negligible compared with chamber volume), the term $\frac{1}{2}$ may be neglected and the formulas become identical to (6.86) and (6.87).

[1] J. O. Hougen, O. R. Martin, and R. A. Walsh, Dynamics of Pneumatic Transmission Lines, *Contr. Eng.*, p. 114, September 1963.

While in general a flush-diaphragm installation is preferred, in some rocket-engine testing a short length of tubing purged with a steady helium flow has been found desirable to reduce fouling and temperature damage. A piezoelectric transducer[1] (resonant frequency 250,000 Hz) of this type using a tube of about 0.15-in diameter and 1-in length has flat response ± 10 percent to about 10,000 Hz. Helium is used since it has a value of γ/ρ about 9 times that of air, which gives faster response.

Conclusion

The results of this section are to be thought of as practical working relations. The general problem treated here is quite complex and has been the subject of many intricate analyses and experimental studies, some of which are found in the Bibliography of this chapter. Most of the difficulties encountered are in the area of very high frequencies, where the lumped-parameter models used in this section are inadequate and give faulty predictions.[2]

6.7 DYNAMIC TESTING OF PRESSURE-MEASURING SYSTEMS

To determine the regions of accuracy of theoretical predictions or to find accurate numerical values of system dynamic characteristics for critical applications, recourse must be made to experimental testing. This commonly takes the form of impulse, step, or frequency-response tests, with step-function tests being perhaps the most common. A comprehensive review[3] of this subject is available. Here we can mention only a few high points.

For step-function tests of systems in which natural frequencies are not greater than about 1,000 Hz, the bursting of a thin diaphragm subjected to gas pressure is often satisfactory. A general rule for step testing is that the rise time of the "step" function must be less than about one-fourth of the natural period of the system tested if it is to excite the natural oscillations. Thus a 1,000-Hz system requires a step with a rise time of 0.25 ms or less. Figure 6.22 shows schematically the principle of such devices. The pressures p_1 and p_2 are each individually adjustable. The volume containing p_2 is much smaller than that containing p_1; thus when the thin plastic diaphragm is ruptured by a solenoid-actuated knife, the pressure p_2 rises to p_1 very quickly. If a decreasing step function is wanted, p_2 can be made larger than p_1. Construction and operation of such devices are quite simple, and they have been utilized widely in their range of applicability.

[1] Helium-Bleed Rocket Probe, Kistler Instruments, Amherst, N.Y.

[2] T. W. Nyland et al., On the Dynamics of Short Pressure Probes, *NASA* TN D-6151, February 1971.

[3] J. L. Schweppe et al., Methods for the Dynamic Calibration of Pressure Transducers, *Natl. Bur. Std. (U.S.), Monograph* 67, 1963.

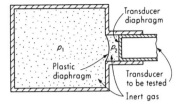

Figure 6.22 Step-test apparatus.

For pickups of natural frequency greater than 1,000 Hz, the simple burst-diaphragm testers are not capable of exciting the natural oscillations, and the pickup output is simply an accurate record of the terminated-ramp pressure input. To achieve sufficiently short pressure-rise times, the shock tube is widely used. Figure 6.23 shows a sketch of such a device. A thin diaphragm separates the high-pressure and low-pressure chambers, and the transducer to be tested is mounted flush with the end of the low-pressure chamber. When the diaphragm is caused to burst, a shock wave travels toward the low-pressure end at a speed that may greatly exceed the speed of sound (5,000 ft/s is not unusual). From one side of this shock front to the other, there is a pressure change of the order of 2 : 1 over a distance which may be of the order of 10^{-4} in. (At the same time, a rarefaction wave travels from the diaphragm toward the high-pressure end.) When the shock front reaches the end of the tube where the transducer is mounted, it is reflected as a shock wave with more than twice the pressure difference of the original shock wave. The transducer is thus exposed to a very sharp ($\sim 10^{-8}$ s) pressure rise which is maintained constant for a short interval before various reflected waves arrive to confuse the picture. The length of this interval may be controlled to a certain extent by proper proportioning and operation of the shock tube. Some numerical characteristics of a typical shock tube[1] are as follows: high-pressure chamber 7 ft long, low-pressure chamber 15 ft long, tubing inside dimensions 1.4 in^2 (wall thickness $\frac{1}{4}$ in), maximum high-pressure 600 lb/in^2, operating fluid air, maximum pressure step 350 lb/in^2, burst diaphragm 0.001- to 0.005-in-thick Mylar plastic, and duration of constant pressure 0.01 s. For a pressure pickup of, say, 100,000-Hz natural frequency, a pressure-rise time of less than 0.25×10^{-5} s is required. This is readily met by the tube described above. The 0.01-s step duration would give time for about 1,000 cycles of oscillation of a 100,000-Hz pickup—more than adequate to determine the dynamic characteristics.

A simple impulse-type test method applicable to flat, flush-diaphragm transducers utilizes a small steel ball dropped onto the diaphragm. The impact excites the natural oscillations, which are recorded and analyzed for natural frequency and damping ratio. Although this input is a concentrated force rather than a

[1] R. Bowersox, Calibration of High-Frequency-Response Pressure Transducers, *ISA J.*, p. 98, November 1958.

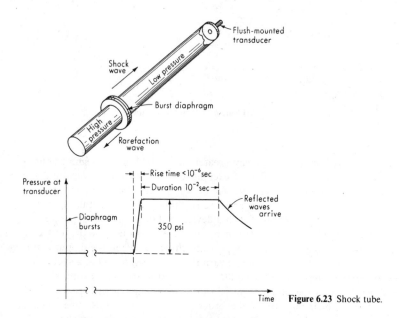

Figure 6.23 Shock tube.

uniform pressure, the results have been found[1] to correlate quite well with shock-tube tests. Another impulse method employs the shock wave created by discharging 25,000 V across a spark gap. The spark gap and transducer are located about 3 in apart in open air. A pressure impulse of 0.2-μs rise time and $100 \, \text{lb/in}^2$ peak can be obtained.

Figure 6.24a shows one method of constructing a frequency-response tester using liquid as the pressure medium. The vibration shaker applies a sinusoidal force of adjustable frequency and amplitude to the piston/diaphragm to create sinusoidal pressure in the liquid-filled chamber. Such vibration shakers are readily available in industry and cover a wide range of force and frequency. The average pressure about which oscillations take place may be adjusted by regulating the bias pressure on the air side of the cylinder. Since usually it is not possible to predict accurately and reproducibly the pressure actually produced by such an arrangement as frequency and/or amplitude is varied, it is customary to mount a reference transducer at a location where it will experience the same pressure as the transducer under test. The reference transducer must have a known flat frequency response beyond any frequencies to be tested. This can be determined by some independent method, such as a shock tube. In testing another transducer, one merely calculates the amplitude ratio and phase shift between the reference and test transducer to determine the test-transducer frequency response. A

[1] W. C. Bentley and J. J. Walter, Transient Pressure Measuring Methods Research, Princeton Univ. Aeron. Eng. Dept., *Rept.* 595g, p. 103, 1963.

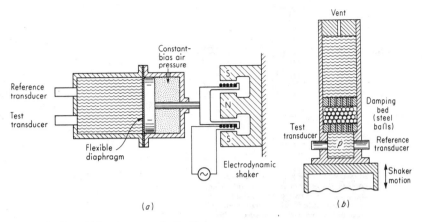

Figure 6.24 Sinusoidal test apparatus for liquid.

different version[1] of the shaker method uses an open, vertical fluid column (Fig. 6.24b) to generate the test pressure.

An interesting special-purpose version of this technique, which employs microprocessor technology, is the blood pressure systems analyzer of Fig. 6.25. At the dome where the pressure signal is generated, one can directly attach a reference pressure transducer and the catheter, tubing, and transducer to be dynamically tested. A microprocessor memory stores the 11 different waveforms and outputs the one selected through a digital/analog converter to a power amplifier and the shaker coil. By comparing the output traces of the reference and test transducers on a two-channel recorder, one can decide whether the tested system functions properly. The apparatus is useful for teaching purposes also since each waveform can be produced quickly for examination and discussion.

A different approach[2] to frequency testing, utilizing a flow-modulating principle and gas as the fluid medium, is shown in Fig. 6.26. A chamber is supplied with compressed gas from a constant-pressure source through a small inlet passage. The gas is exhausted to the atmosphere through an outlet passage whose area is modulated approximately sinusoidally with time. This is accomplished by rotating a disk containing holes in front of the exhaust port so that outflow periodically is cut on and off. This produces a periodic (nearly sinusoidal) variation in chamber pressure, which is measured by both a reference transducer and the test transducer. Varying the speed of the rotating disk changes the frequency. The amplitude of pressure oscillation of such a device drops off with frequency. For the system described, with helium gas at a supply pressure of 121 lb/in^2 absolute, the peak-to-peak pressure amplitude goes from about 15 lb/in^2 at

---

[1] C. F. Vezzetti et al., A New Dynamic Pressure Source for the Calibration of Pressure Transducers, *NBS Tech. Note* 914, June 1976.

[2] Bentley and Walter, op. cit., p. 63.

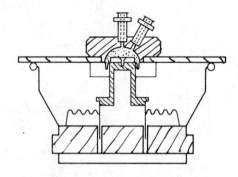

DOME VIEW: The pressure generator consists of a magnet and voice coil which applies pressure to the fluid through a piston and rubber seal. Current in the coil is accurately converted to pressure in the dome.

Waveform	Description	Output Level
	Aorta, Normal (BPM for all waveforms is 90 BPM unless indicated)	120/80mm Hg
	Aorta, Tachycardia (120 BPM)	120/80mm Hg
	Radial Artery, Typical	120/80mm Hg
	Radial Artery, Post Surgical	120/80mm Hg
	Right Ventricle	30/0mm Hg
	Pulmonary Artery	28/10mm Hg
	Pulmonary Artery, Wedge	15/10mm Hg
	Pulmonary Artery with Catheter Whip	28/8mm Hg
	Pulmonary Artery Wedge with large "v" wave	25/10mm Hg
	ECG Waveform	1mV@90BPM
	Square Wave (Rate 30 BPM; risetime 3m.)	
	Automatic Step Sequence	20, 0, 20, 40, 60, 80, 100, 120, 140, 160, 180, 200mm Hg

Figure 6.25 Microprocessor-based pressure calibrator. (*Courtesy Bio-Tek Instruments, Inc., Shelburne, Vt.*)

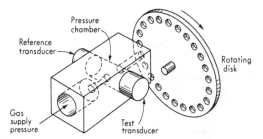

Figure 6.26 Sinusoidal test apparatus (siren) for gas.

1,000 Hz to about 2 lb/in^2 at 11,000 Hz. In addition to this reduction in amplitude, an increase in frequency also brings into play the resonant acoustical frequencies of the chamber. When these resonances occur, one cannot depend on the pressures being uniform throughout the chamber. This uniformity is a necessity when the method used is based on comparison of a reference transducer with the test transducer. The acoustic resonant frequencies depend on the chamber size (smaller chambers have higher frequencies) and the speed of sound in the gas (higher sound speed gives higher frequencies). The use of helium in the above example is based on this last consideration. A system of the above type was found to be usable for dynamic calibration up to about 10,000 Hz.

In summary, it should be pointed out that the dynamic testing of very-high-frequency pressure pickups involves a number of complicating factors. It has been found extremely difficult to generate high-frequency pressure sine waves that are of large enough amplitude to give a relatively noise-free transducer output signal. Small-amplitude pressure waves (such as are applicable to sound-measuring systems) can be produced relatively easily with loudspeaker-type systems, but their amplitude is far below the levels needed for pressure pickups whose full-scale range is tens, hundreds, or thousands of pounds per square inch. Step testing with a shock tube thus has been widely used, since the fast rise time and large pressure steps result in a transient input with strong high-frequency content. The pickups themselves present problems since at the high natural frequencies involved, many complex wave-propagation and reflection effects make the response deviate considerably from the simple second-order model. Also, these pickups generally have little damping ($\zeta \approx 0.01$ to 0.04), which makes them particularly prone to ringing at their natural frequency if any sharp transients occur.

6.8 HIGH-PRESSURE MEASUREMENT[1]

Pressures up to about 100,000 lb/in^2 can be measured fairly easily with strain-gage pressure cells or Bourdon tubes. Bourdon tubes for such high pressures have nearly circular cross sections and thus give little output motion per turn. To get a

[1] R. K. Kaminski, Measuring High Pressure above 20,000 PSIG, *Instrum. Tech.*, pp. 59–62, August 1968; W. H. Howe, The Present Status of High Pressure Measurement and Control, *ISA J.*, p. 77, March 1955; I. L. Spain and J. Paauwe (eds.), "High Pressure Technology," vol. 1, chap. 8, Marcel Dekker, New York, 1977.

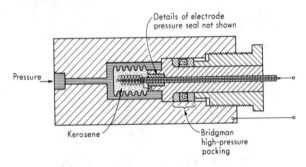

Figure 6.27 Very-high-pressure transducer.

measurable output, the helical form with many turns generally is used. Inaccuracy of the order of 1 percent of full scale may be expected with a temperature error of an additional 2 percent/100F°. Strain-gage pressure cells can be temperature-compensated to give 0.25 percent error over a large temperature range.

For fluid pressures above 100,000 lb/in², electrical gages based on the resistance change of Manganin or gold-chrome wire with hydrostatic pressure are generally utilized.[1] Figure 6.27 shows a typical gage. The sensitive wire is wound in a loose coil, one end of which is grounded to the cell body and the other end brought out through a suitable insulator. The coil is enclosed in a flexible, kerosene-filled bellows, which transmits the measured pressure to the coil. The resistance change, which is linear with pressure, is sensed by conventional Wheatstone-bridge methods. Pertinent characteristics of the common wire materials are as follows:

	Pressure sensitivity, $(\Omega/\Omega)/(\text{lb/in}^2)$	Temperature sensitivity, $(\Omega/\Omega)/F°$	Resistivity, $\Omega \cdot cm$
Manganin	1.69×10^{-7}	1.7×10^{-6}	45×10^{-6}
Gold chrome	0.673×10^{-7}	0.8×10^{-6}	2.4×10^{-6}

Although its pressure sensitivity is lower, gold chrome is preferred in many cases because of its much smaller temperature error. This is particularly significant since the kerosene used in the bellows will experience a transient temperature change when sudden pressure changes occur, because of adiabatic compression or expansion. The response of the wire resistance to pressure changes is practically instantaneous; however, the accompanying temperature change will cause a transient error if temperature sensitivity is too high. Gages of the above

[1] Howe, op. cit.

type are commercially available with full scale up to 200,000 lb/in^2 and inaccuracy of 0.1 to 0.5 percent. They also have been utilized successfully for much higher pressures on a special-application basis.

The measurement of local contact pressures between rolling elements in gears, cams, and bearings may be accomplished by depositing a thin strip of Manganin or gold chrome onto the surface as a pressure transducer. Studies[1] of such a technique, using a Manganin element 0.002 in wide and 3×10^{-6} in thick, have been reported.

6.9 LOW-PRESSURE (VACUUM) MEASUREMENT[2]

Two commonly employed units of vacuum measurement are the torr and the micrometer. One torr is a pressure equivalent to 1 mmHg at standard conditions; one micrometer is 10^{-3} torr. Manometers and bellows gages are usable to about 0.1 torr, Bourdon gages to 10 torr, and diaphragm gages to 10^{-3} torr. Below these ranges, other types of vacuum gages are necessary.

McLeod Gage

The McLeod gage is considered a vacuum standard since the pressure can be computed from the dimensions of the gage. It is not directly usable below about 10^{-4} torr; however, pressure-dividing techniques (see Sec. 6.1) allow its use as a calibration standard for considerably lower ranges. The multiple-compression technique[3] is also being studied to extend its range. The inaccuracy of McLeod gages is rarely less than 1 percent and may be much higher at the lowest pressures.

Of the many variations of McLeod gages, here we consider only the most basic. The principle of all McLeod gages is the compression of a sample of the low-pressure gas to a pressure sufficiently high to read with a simple manometer. Figure 6.28 shows the basic construction. By withdrawing the plunger, the mercury level is lowered to the position of Fig. 6.28a, admitting the gas at unknown pressure p_i. When the plunger is pushed in, the mercury level goes up, sealing off a gas sample of known volume V in the bulb and capillary tube A. Further motion of the plunger causes compression of this sample, and motion is continued until the mercury level in capillary B is at the zero mark. The unknown

[1] J. W. Kannel and T. A. Dow, The Evolution of Surface Pressure and Temperature Measurement Techniques for Use in the Study of Lubrication in Metal Rolling, *ASME Paper* 74-Lub5-7, 1974.

[2] S. Dushman, "Scientific Foundations of Vacuum Technique," chap. 6, Wiley, New York, 1949; L. G. Carpenter, "Vacuum Technology," chap. 4, American Elsevier, New York, 1970; J. H. Leck, "Pressure Measurement in Vacuum Systems," 2d ed., Chapman & Hall, London, 1964.

[3] W. Kreisman, Extension of the Low Pressure Limit of McLeod Gages, *NASA, CR*-52877.

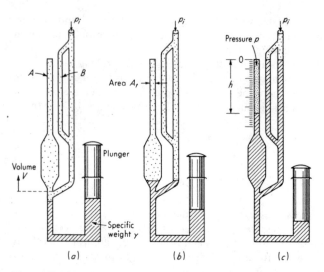

Figure 6.28 McLeod gage.

pressure is then calculated, by using Boyle's law, as follows:

$$p_i V = p A_t h \tag{6.91}$$

$$p = p_i + h\gamma \tag{6.92}$$

$$p_i = \frac{\gamma A_t h^2}{V - A_t h} \approx \frac{\gamma A_t h^2}{V} \quad \text{if } V \gg A_t h \tag{6.93}$$

In using a McLeod gage it is important to realize that if the measured gas contains any vapors that are condensed by the compression process, then the pressure will be in error. Except for this effect, the reading of the McLeod gage is not influenced by the composition of the gas. Only the Knudsen gage shares this desirable feature of composition insensitivity. The main drawbacks of the McLeod gage are the lack of a continuous output reading and the limitations on the lowest measurable pressures. When it is employed to calibrate other gages, a liquid-air cold trap should be used between the McLeod gage and the gage to be calibrated to prevent the passage of mercury vapor.

Knudsen Gage

Although the Knudsen gage is little utilized at present, we discuss it briefly since it is relatively insensitive to gas composition and thus gives promise of development into a standard for pressures too low for the McLeod gage. In Fig. 6.29 the unknown pressure p_i is admitted to a chamber containing fixed plates heated to absolute temperature T_f, which temperature must be measured, and a spring-

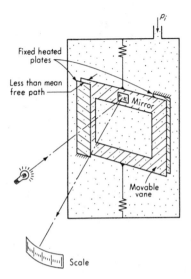

Fixed heated
plates

Less than mean
free path

Mirror

Movable
vane

Scale

Figure 6.29 Knudsen gage.

restrained movable vane whose temperature T_v also must be known. The spacing between the fixed and movable plates must be less than the mean free path of the gas whose pressure is being measured. The kinetic theory of gases shows that gas molecules rebound from the heated plates with greater momentum than from the cooler movable vane, thus giving a net force on the movable vane which is measured by the deflection of the spring suspension. Analysis shows that the force is directly proportional to pressure for a given T_f and T_v, following a law of the form

$$p_i = \frac{KF}{\sqrt{T_f/T_v - 1}}$$

where F is force and K is a constant. The Knudsen gage is insensitive to gas composition except for the variation of accommodation coefficient from one gas to another. The accommodation coefficient is a measure of the extent to which a rebounding molecule has attained the temperature of the surface. This effect results, for example, in a 15 percent change in sensitivity between helium and air. Knudsen gages at present cover the range from about 10^{-8} to 10^{-2} torr.

Momentum-Transfer (Viscosity) Gages

For pressures less than about 10^{-2} torr, the kinetic theory of gases predicts that the viscosity of a gas will be directly proportional to the pressure. The viscosity may be measured, for example, in terms of the torque required to rotate, at constant speed, one concentric cylinder within another. (For pressures greater than about 1 torr, the viscosity is independent of pressure.) The variation of viscosity with pressure is different for different gases; thus gages based on this principle must be calibrated for a specific gas. While gages based on viscosity principles can measure to about 10^{-7} torr, such ranges are characteristic of laboratory-type equipment requiring great care in its use.

A typical commercial gage, shown schematically in Fig. 6.30, is calibrated for dry air and covers the range from 0 to 20 torr. The range from 0 to 0.01 torr occupies about 10 percent of the total scale. The scale is nonlinearly calibrated because most of the range is above 10^{-2} torr and viscosity here is not proportional to pressure. To enable readings above 1 torr, where viscosity tends to become pressure-independent, bladed wheels rather than smooth concentric cylinders are used in this gage. These wheels cause a turbulent momentum exchange which is pressure-dependent above 1 torr, extending the useful range to 20 torr. To reach the quoted lower limit (10^{-7} torr) of viscosity gages, the construction of Fig. 6.30 is not employed. Rather, the rate of decay of delicately constructed vibrating systems subjected to the damping effects of the gas is determined and pressure inferred from this.

Thermal-Conductivity Gages

Just as for viscosity, when the pressure of a gas becomes low enough that the mean free path of molecules is large compared with the pertinent dimensions of the apparatus, a linear relation between pressure and thermal conductivity is predicted by the kinetic theory of gases. For a viscosity gage, the pertinent dimension is the spacing between the relatively moving surfaces. For a conductivity gage, it is the spacing between the hot and cold surfaces. Again, when the pressure is increased sufficiently, conductivity becomes independent of gas pressure. The transition region between dependence and nondependence of viscosity and thermal conductivity on pressure is approximately the range 10^{-2} to 1 torr for apparatus of a size convenient to construct.

The application of the thermal-conductivity principle is complicated by the

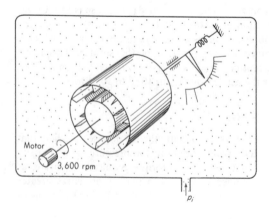

Motor

3,600 rpm

p_i

Figure 6.30 Momentum gage. (*Courtesy General Electric Co.*)

simultaneous presence of another mode of heat transfer between the hot and cold surfaces, namely, radiation. Most gages utilize a heated element supplied with a constant energy input. This element assumes an equilibrium temperature when heat input and losses by conduction and radiation are just balanced. The conduction losses vary with gas composition and gas pressure; thus for a given gas, the equilibrium temperature of the heated element becomes a measure of pressure, and this temperature is what is actually measured. If the radiation losses are a major part of the total, then pressure-induced conductivity changes will cause only a slight temperature change, giving poor sensitivity. Analysis shows that radiation losses may be minimized by using surfaces of low emissivity and by making the cold-surface temperature as low as practical. Since conduction and radiation losses depend on *both* the hot- and cold-surface temperatures, the cold surface may be maintained at a known constant temperature if overall accuracy warrants this measure. A further source of error is in the heat-conduction loss through any solid supports by which the heated element is mounted. The relative importance of the above-mentioned effects varies with the details of construction of the gage. The most common types of conductivity gages are the thermocouple, resistance thermometer (Pirani), and thermistor.

Figure 6.31 shows in schematic form the basic elements of a thermocouple

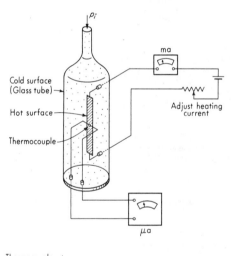

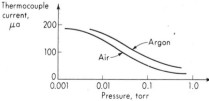

Figure 6.31 Thermocouple gage.

vacuum gage. The hot surface is a thin metal strip whose temperature may be varied by changing the current passing through it. For a given heating current and gas, the temperature assumed by the hot surface depends on pressure; this temperature is measured by a thermocouple welded to the hot surface. The cold surface here is the glass tube, which usually is near room temperature. Often the accuracy of such gages is not high enough to warrant measurement or correction for changes in room temperature. Thermocouple gages of one type or another are available to measure in the range 10^{-4} to 1 torr.

In the resistance-thermometer (Pirani) gage, the functions of heating and temperature measurement are combined in a single element. A typical construction is shown in Fig. 6.32. The resistance element is in the form of four coiled tungsten wires connected in parallel and supported inside a glass tube to which the gas is admitted. Again, the cold surface is the glass tube. Two identical tubes generally are connected in a bridge circuit, as shown. One of the tubes is evacuated to a very low pressure and then sealed off while the other has the gas admitted to it. The evacuated tube acts as a compensator to reduce the effect of bridge-excitation-voltage changes and temperature changes on the output

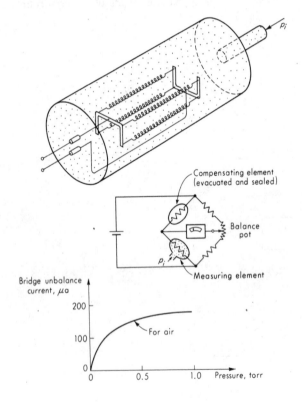

Figure 6.32 Pirani gage.

reading. Current flowing through the measuring element heats it to a temperature depending on the gas pressure. The electrical resistance of the element changes with temperature, and this resistance change causes a bridge unbalance. Generally the bridge is used as a deflection rather than a null device. To balance the bridge initially, the pressure in the measuring element is made very small and the balance pot set for zero output. Any changes in pressure will cause a bridge unbalance. Of course, the gage must be calibrated against some standard. Calibration is nonlinear and varies from one gas to another. Pirani gages cover the range from about 10^{-5} to 1 torr.

Thermistor vacuum gages operate on the same principle as the Pirani gage except that the resistance elements are temperature-sensitive semiconductor materials called thermistors, rather than metals such as tungsten or platinum. Thermistor gages are used in the range 10^{-4} to 1 torr.

Ionization Gages

An electron passing through a potential difference acquires a kinetic energy proportional to the potential difference. When this energy is large enough and the electron strikes a gas molecule, there is a definite probability that the electron will drive an electron out of the molecule, leaving it a positively charged ion. In an ionization gage, a stream of electrons is emitted from a cathode. Some of these strike gas molecules and knock out secondary electrons, leaving the molecules as positive ions. For normal operation of the gage, the secondary electrons are a negligible part of the total electron current; thus, for all practical purposes, electron current i_e is the same whether measured at the emitting point (cathode) or the collecting point (anode). The number of positive ions formed is directly proportional to i_e and directly proportional to the gas pressure. If i_e is held fixed (as in most gages), the rate of production of positive ions (ion current) is, for a given gas, a direct measure of the number of gas molecules per unit volume and thus of the pressure. The positive ions are attracted to a negatively charged electrode, which collects them and carries the ion current. The "sensitivity" S of an ionization gage is defined by

$$S \triangleq \frac{i_i}{p i_e} \tag{6.94}$$

where $i_i \triangleq$ ion current, gage output

$i_e \triangleq$ electron current

$p \triangleq$ gas pressure, gage input

According to our usual definition of sensitivity as output/input, the "sensitivity" would be $S i_e$ rather than S. But the definition of Eq. (6.94) makes "sensitivity" independent of i_e and dependent on only gage construction. This allows comparison of the "sensitivity" of different gages without reference to the particular i_e being used. A main advantage of ionization gages in general is their linearity; that is, the sensitivity S is constant for a given gas over a wide range of pressures.

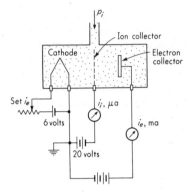

Figure 6.33 Ionization gage.

Figure 6.33 shows the basic elements of a hot-cathode ionization gage. The emission of electrons is due to the heating of the cathode. Some disadvantages of hot-cathode gages are filament burnout if exposed to air while hot, decomposition of some gases by the hot filament, and contamination of the measured gas by gases forced out of the hot filament. Hot-cathode gages cover the range from 10^{-10} to 1 torr.

The Philips cold-cathode gage[1] overcomes the problems associated with a high-temperature filament by the use of a cold cathode and a high accelerating potential (2,000 V). A superimposed magnetic field causes the electrons ejected from the cathode to travel in long helical paths to the anode. The long path results in more collisions with gas molecules and thus a greater ionization. Philips gages are used in the range 10^{-5} to 10^{-2} torr.

For the lowest pressures, hot-cathode and cold-cathode gages of the magnetron type[2] are available. They are useful down to about 10^{-13} torr. Mass spectrometers[3] are employed for even lower pressures; they allow identification of the partial pressures of components in gas mixtures.

Dual-Gage Technique[4]

When large amounts of water vapor (common during early stages of evacuation) or helium (from a leak-detection system) are present, simultaneous readings from an ionization gage and a thermocouple gage (both of which will read incorrectly) can be processed to not only get a correct vacuum reading but also determine the amount of moisture or helium present.

---

[1] I. M. Lafferty and T. A. Vanderslice, Vacuum Measurement by Ionization, *Instrum. Contr. Sys.*, p. 90, March 1963.

[2] Ibid.

[3] Ibid.; Carpenter, op. cit., pp. 62–65.

[4] C. A. Schalla, Making Accurate Vacuum Readings, *Mach. Des.*, pp. 122–124, February 26, 1981.

6.10 SOUND MEASUREMENT

The measurement of airborne and waterborne sound is of increasing interest to engineers. Airborne-sound measurements are important in the development of less noisy machinery and equipment, in diagnosis of vibration problems, and in the design and test of sound recording and reproducing equipment. In large rocket and jet engines, the sound pressures produced by the exhaust may be large enough to cause fatigue failure of metal panels because of vibration ("acoustic fatigue"). Waterborne sound has been applied in underwater direction and range-finding equipment (sonar). Since most sound transducers (microphones and hydrophones) are basically pressure-measuring devices, it is appropriate to consider them briefly in this chapter.

The basic definitions of sound are in terms of the magnitude of the fluctuating component of pressure in a fluid medium. The *sound pressure level* (SPL) is defined by

$$\text{SPL} \triangleq \text{sound pressure level} \triangleq 20 \log_{10} \frac{p}{0.0002} \quad \text{decibels (dB)} \qquad (6.95)$$

where $\quad p \triangleq$ root-mean-square (rms) sound pressure, μbar $\qquad (6.96)$

and $\quad 1 \ \mu\text{bar} = 1 \ \text{dyn/cm}^2 = 1.45 \times 10^{-5} \ \text{lb/in}^2 \qquad (6.97)$

The rms value of the fluctuating component of pressure is employed because most sounds are random signals rather than pure sine waves. The value 0.0002 μbar is an accepted standard reference value of pressure against which other pressures are compared by Eq. (6.95). Note that when $p = 0.0002 \ \mu$bar, the sound pressure level is 0 dB. This value was selected somewhat arbitrarily, but represents the average threshold of hearing for human beings if a 1,000-Hz tone is used. That is, the 0-dB level was selected as the lowest pressure fluctuation normally discernible by human beings. Since 0 dB is about $3 \times 10^{-9} \ \text{lb/in}^2$, the remarkable sensitivity of the human ear should be apparent. The decibel (logarithmic) scale is used as a convenience because of the great ranges of sound pressure level of interest in ordinary work. For example, an office with tabulating machines may have an SPL of 74 dB (1μbar). The average human threshold of pain is 144 dB. Sound pressure close to large rocket engines are the order of 170 dB (1 lb/in²). One atmosphere (14.7 lb/in²) is 194 dB. The span from the lowest to the highest pressures of interest is thus of the order of 10^{-9} to 1, a tremendous range.

Sound-Level Meter

The most commonly utilized instrument for routine sound measurements is the sound-level meter.[1] This is actually a measurement *system* made up of a number

[1] C. Thomsen, Sound Level Meters—Their Use and Abuse, *Sound & Vib.*, pp. 28–31, March 1979; J. R. Hassall and K. Zaveri, "Acoustic Noise Measurements," 4th ed., Bruel and Kjaer Instruments, Marlboro, Mass., 1979; Acoustics Handbook, *Appl. Note* 100, pp. 63–81, Hewlett-Packard Co., Palo Alto, Calif., 1968.

of interconnected components. Figure 6.34 shows a typical arrangement. The sound pressure p_i is transduced to a voltage by means of the microphone. Microphones generally employ a thin diaphragm to convert pressure to motion. The motion is then converted to voltage by some suitable transducer, usually a capacitance, piezoelectric, or moving-coil type. Microphones often have a "slow leak" (capillary tube) connecting the two sides of the diaphragm, to equalize the average pressure (atmospheric pressure) and prevent bursting of the diaphragm. This is necessary because the (slow) hour-to-hour and day-to-day changes in atmospheric pressure are much greater than the sound-pressure fluctuations to which the microphone must respond. (Note that the eustachian tube of the human ear serves a similar function.) The presence of this leak dictates that microphones will not respond to constant or slowly varying pressures. This is usually no problem since many measurements involve a human response to the sound, and this is known to extend down to only about 10 to 20 Hz. Thus the microphone frequency response need go only to this range, not to zero frequency.

The output voltage of the microphone generally is quite small and at a high impedance level; thus an amplifier of high input imedance and gain is used at the output of the microphone. This can be a relatively simple ac amplifier, since response to static or slowly varying voltages is not required. Capacitor microphones often use for the first stage a FET-input amplifier built right into the

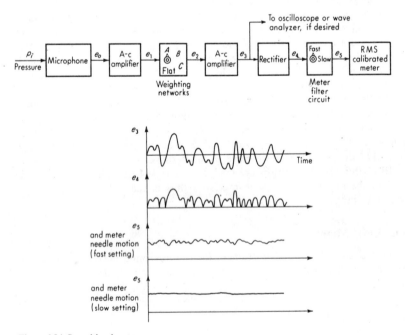

Figure 6.34 Sound-level meter.

microphone housing. This close coupling reduces stray capacitance effects by eliminating cables at the high-impedance end.

Following the first amplifier are the weighting networks. They are electrical filters whose frequency response is tailored to approximate the frequency response of the average human ear. Figure 6.35a[1] displays "equal-loudness contours" obtained from measurements on human beings, showing that the frequency response of the human ear is both "nonflat" and nonlinear. Each curve is labeled with a loudness unit called a *phon*, with 0 phon corresponding to the threshold of hearing. The ordinate (sound-pressure level in decibels) tells what pressure amplitude must be applied at any given frequency so that the human observer will perceive a sensation of equal loudness. For example, at a 50-phon loudness level, a 58-dB SPL at 100 Hz sounds as loud as a 50-dB SPL at 1,000 ·Hz, which demonstrates the nonflatness of the ear's frequency response. Its nonlinearity is manifested by the need for a *family* of curves for various loudness levels, rather than just a single curve. Since the main use of a sound-level meter is *not* the accurate measurement of pressure, but rather the determination of the loudness perceived by human beings, a flat instrument frequency response is not really wanted. The weighting networks of Fig. 6.34 are electrical filters designed to approximate the human ear's response at three different loudness levels, so that instrument readings will reflect perceived loudness. Usually three filters—A (approximates 40-phon ear response), B (70-phon), and C (100-phon)—are provided, and Fig. 6.35b shows the frequency response of these filters (dashed lines show tolerances allowed on "precision sound-level meters"). Some meters also provide a "flat" setting if true pressure measurements are wanted; if not, the C network is a good approximation. Actually, many practical measurements are made by employing the A scale since it is a simple approach which has given good results in many cases and has been written into many standards and codes. Readings taken with a weighting network are called *sound level* rather than *sound-pressure level*.

The output of the weighting network is further amplified and an output jack provided to lead this signal to an oscilloscope (if observation of the waveform is desired) or to a wave analyzer (if the frequency content of the sound is to be determined). If only the overall sound magnitude is desired, the rms value of e_3 must be found. While true rms voltmeters are available, their expense is justifiable only in the highest-grade sound-level meters. Rather, the *average* value of e_3 is determined by rectifying and filtering, and then the meter scale is *calibrated* to read rms values. This procedure is exact for pure sine waves since there is a precise relation between the average value and the rms value of a sine wave. For nonsinusoidal waves this is not true, but the error is generally small enough to be acceptable for relatively unsophisticated work. The filtering is accomplished by both a simple low-pass RC filter and the low-pass meter dynamics. Some meters have a slow-fast response switch which changes the filtering. The slow position gives a steady, easy-to-read needle position, but masks any short-term variations

[1] B. Katz, Primer on Sound Level Meters and Acoustical Calibration, Bruel & Kjaer Instruments, Marlboro, Mass.

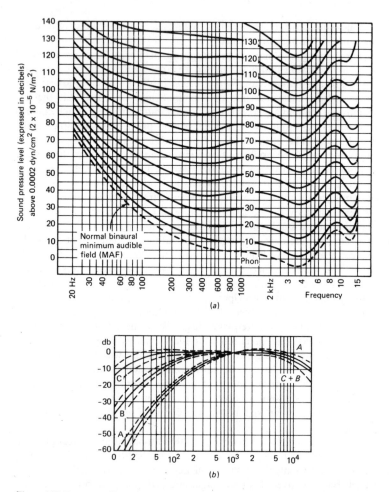

Figure 6.35 Response of human ear and weighting networks.

in the signal. If these short-term variations are of interest, they may be visually observed on the meter by switching it to fast response. While the meter is actually reading the rms value of e_3 (and thus of p_i), it is calibrated in decibels since Eq. (6.95) establishes a definite relation between sound pressure in microbars and decibels.

Microphones

While the design of microphones is a specialized and complex field with a large technical literature, here we can point out some of the main considerations. Frequency response is still of major interest; however, the effects on frequency response of sound wavelength and direction of propagation are aspects of dynamic behavior not regularly encountered in other measurements. The *pressure response* of a microphone refers to the frequency response relating a uniform sound pressure applied at the microphone diaphragm to the output voltage of the microphone. The pressure response of a given microphone may be estimated theoretically or measured experimentally by one of a number of accepted methods.[1]

What is usually desired is the *free-field response* of the microphone. That is, what is the relation between the microphone output voltage and the sound pressure that existed at the microphone location *before* the microphone was introduced into the sound field? The microphone distorts the pressure field because its acoustical impedance is radically different from that of the medium (air) in which it is immersed. In fact, for most purposes, the microphone (including its diaphragm) may be considered as a rigid body. Sound waves impinging on this body give rise to complex reflections that depend on the frequency, the direction of propagation of the sound wave, and the microphone size and shape. When the wavelength of the sound wave is very large compared with the microphone dimensions (low frequencies), the effect of reflections is negligible for any angle of incidence between the diaphragm and the wave-propagation direction, and the free-field response is the same as the pressure response. At very high frequencies, where the wavelength is much smaller than the microphone dimensions, the microphone acts as an infinite wall and the pressure at the microphone surface [for waves propagating perpendicular to the diaphragm (0° angle of incidence)] is twice what it would be if the microphone were not there. For waves propagating parallel to the diaphragm (90° incidence angle), the average pressure over the diaphragm surface is zero, giving no output voltage. Between the very low and very high frequencies, the effect of reflections is quite complicated and depends on sound wavelength (frequency), microphone size and shape, and angle of incidence.

For simple geometric shapes such as spheres and cylinders, theoretical results are available.[2] Experiments on actual microphone give results such as those shown in Fig. 6.36. Note that for sufficiently low frequencies (below a few thousand hertz) there is little change in pressure because of the presence of the microphone; also the angle of incidence has little effect. This flat frequency range can be extended by reducing the size of the microphone; however, smaller size tends to reduce sensitivity. The size effect is directly related to the relative size of the microphone and the wavelength of the sound. The wavelength λ of sound waves in air is roughly $13,000/f$ inches, where f is frequency in hertz. When λ becomes comparable to the microphone-diaphragm diameter, significant reflection effects can be expected. For example, a 1-in-diameter microphone would not be expected to have good response much above 13,000 Hz. (These limitations can be relaxed to some extent by clever use of acoustical mechanical techniques.[3])

[1] P. V. Bruel and G. Rasmussen, Free Field Response of Condenser Microphones, *B & K Tech. Rev.*, B & K Instruments Inc., Marlboro, Mass., no. 1, January 1959; no. 2, April 1959.

[2] L. Beranek, "Acoustic Measurements," chap. 3, Wiley, New York, 1949.

[3] Gunnar Rasmussen, Miniature Pressure Microphones, *B & K Tech. Rev.*, no. 1, B & K Instruments Inc., Marlboro, Mass., 1963.

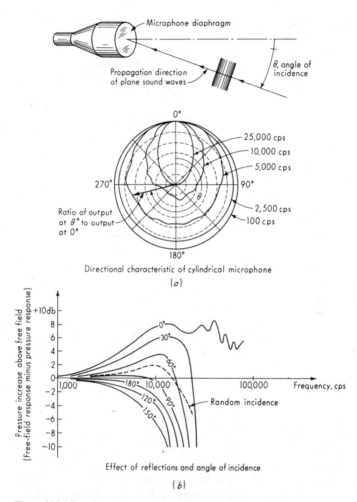

Figure 6.36 Microphone response characteristics.

The lower part of Fig. 6.36 shows a curve labeled "random incidence." This refers to the response to a diffuse sound field where the sound is equally likely to come to the microphone from any direction, the waves from all directions are equally strong, and the phase of the waves is random at the microphone position. Such a field may be approximated by constructing a room with highly irregular walls and placing reflecting objects of various sizes and shapes in it. A source of sound placed in such a room gives rise to a diffuse sound field at any point in the room. Microphones calibrated under such conditions are of interest because

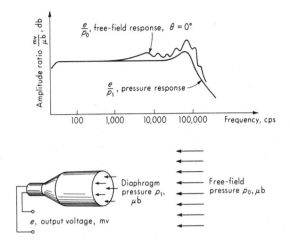

Figure 6.37 Free-field and pressure response.

many sound measurements take place in enclosures which, while not giving perfect random incidence, certainly do not give pure plane waves. Microphone calibrations may give the pressure response and the free-field response for selected incidence angles, usually 0 and 90°. Figure 6.37 shows typical curves.

Microphones used for engineering measurements are usually piezoelectric, capacitor, or electret types. Figure 6.38a[1] shows construction details of a piezoelectric unit which uses PZT (lead zirconate titanate) as a bending beam coupled to the center of a conical diaphragm of thin metal foil. An electret type is shown in Fig. 6.38b[2]. These are related to the capacitor types discussed in detail in the next section; however, they require no polarizing voltage since their charge is permanently "built into" the polymer film which forms the diaphragm. Since the unsupported polymer film would sag and creep excessively, a backup plate with "raised points" is used. Such microphones[3] are less expensive than the capacitor type, can be used under high-humidity conditions (where the capacitor type may arc over), and result in instruments of smaller size and power consumption. A version[4] which preserves the desirable features of an all-metal diaphragm also has been developed.

The selection and use of microphones for critical applications require some background in acoustics, which is beyond the scope of this text; fortunately

[1] Bruel & Kjaer Instruments, Marlboro, Mass.

[2] *GR Today*, p. 8, General Radio, Concord, Mass., Autumn 1972.

[3] R. W. Raymond and S. V. Djuric, The Latest in Instrumentation Quality Microphones, *Sound & Vib.*, pp. 4–6, May 1974.

[4] E. Frederiksen, N. Eirby, and H. Mathiasen, Prepolarized Condenser Microphones for Measurement Purposes, *B & K Tech. Rev.*, no. 4, pp. 3–25, B & K Instruments, Marlboro, Mass., 1979.

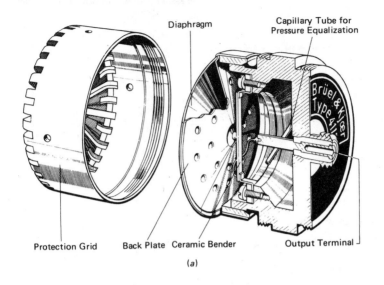

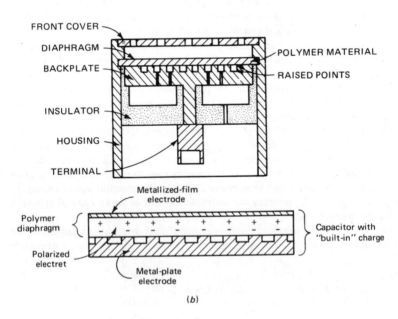

Figure 6.38 Piezoelectric and electret microphones.

useful references[1] are available. Figure 6.39[2] summarizes the main characteristics (size, frequency range, amplitude range) for a line of capacitor microphones from one manufacturer.

Pressure Response of a Capacitor Microphone

Of the several types of microphones in common use, generally the capacitor type is considered capable of the highest performance. Figure 6.40 shows in simplified fashion the construction of a typical capacitor microphone. The pressure response is found by assuming a uniform pressure p_i to exist all around the microphone at any instant of time. This is actually the case for sufficiently low sound frequencies, but reflection and diffraction effects distort this uniform field at higher frequencies, as pointed out earlier.

The diaphragm is generally a very thin metal membrane which is stretched by a suitable clamping arrangement. Diaphragm thickness ranges from about 0.0001 to 0.002 in. The diaphragm is deflected by the sound pressure and acts as the moving plate of a capacitance displacement transducer. The other plate of the capacitor is stationary and may contain properly designed damping holes. Motion of the diaphragm causes air flow through these holes with resulting fluid friction and energy dissipation. This damping effect is utilized to control the resonant peak of the diaphragm response. A diaphragm actually has many natural frequencies; however, only the lowest is of interest here. For frequencies near or below the lowest natural frequency, the diaphragm behaves essentially as a simple spring-mass-dashpot second-order system and may be analyzed as such.

A capillary air leak is provided to give equalization of steady (atmospheric) pressure on both sides of the diaphragm to prevent diaphragm bursting. For varying (sound) pressures the capillary-volume system results in the varying component of pressure acting *only* on the outside of the diaphragm and thus causing the desired diaphragm deflection.

The variable capacitor is connected into a simple series circuit with a high resistance R and "polarized" with a dc voltage E_b of about 200 V. This polarizing voltage acts as circuit excitation and determines the neutral (zero-pressure) diaphragm position because of the electrostatic attraction force between the capacitor plates. For a constant diaphragm deflection, no current flows through R and no output voltage e_o exists; thus there is no response to static pressure differences across the diaphragm. For dynamic pressure differences, a current *will* flow through R and an output voltage exists. The voltage e_o usually is applied to

[1] A. J. Schneider, Microphone Orientation in the Sound Field, *Sound & Vib.*, pp. 20–25, February 1970; W. R. Kundert, Everything You've Wanted to Know about Measurement Microphones, *Sound & Vib.*, pp. 10–26, March 1978; A. P. G. Peterson and E. E. Gross, Jr., Handbook of Noise Measurement, General Radio Corp., Concord, Mass., 1972; Microphone Calibration, Brochure BR-0092, B&K Instruments, Marlboro, Mass., 1980.

[2] Condenser Microphones and Microphone Preamplifiers, Theory and Application Handbook, B & K Instruments, Marlboro, Mass., 1976.

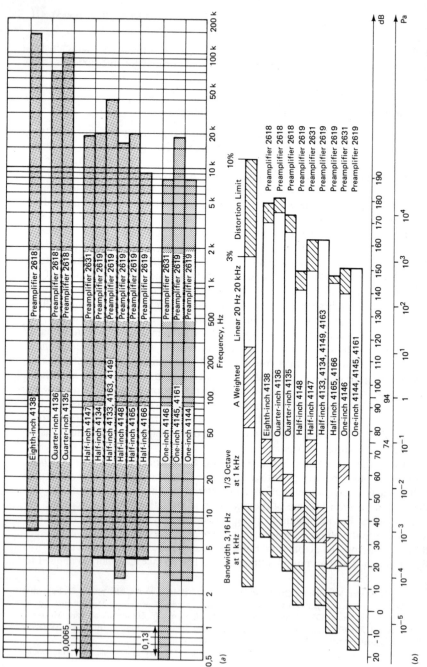

Figure 6.39 Performance of a typical microphone family. (a) Comparison of the frequency response ranges(± 2 dB) of recommended microphone and preamplifier combinations. (b) Comparison of the dynamic ranges (noise level to distortion limit) of recommended microphone and preamplifier combinations.

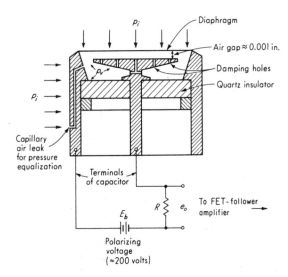

Figure 6.40 Capacitor microphone.

the input of a FET-follower amplifier, which always has a gain less than 1; thus the purpose of the amplifier is not to increase the voltage level. Rather, it has a high input impedance (> 1 GΩ) to prevent loading of the microphone, which has a high output impedance. Since the output impedance of the amplifier is low (< 100 Ω), its output signal may be coupled into long cables and low-impedance loads without loss of signal magnitude.

The first step in the analysis involves determination of the effective force tending to deflect the diaphragm in terms of the pressure p_i. The relation between p_i and the pressure p_v in the microphone internal volume V may be obtained from Eq. (6.85). We neglect the inertia term since in microphones the viscous effect predominates and the filtering effect of the capillary is significant only at low frequencies. Thus we get

$$\tau_l \dot{p}_v + p_v = p_i \tag{6.98}$$

where $$\tau_l \triangleq \text{leak time constant} \triangleq \frac{128 \,\mu L V}{\pi E_a d_t^4} \tag{6.99}$$

Now the deflection of the diaphragm is due to the *difference* between p_i and p_v. Operationally,

$$p_v = \frac{p_i}{\tau_l D + 1}$$

and thus $$p_i - p_v = p_i \left(1 - \frac{1}{\tau_l D + 1} \right) = \frac{\tau_l D}{\tau_l D + 1} p_i \tag{6.100}$$

The total force f_d on the diaphragm is $A_d(p_i - p_v)$, where A_d is the diaphragm area. Thus

$$\frac{f_d}{p_i}(D) = \frac{A_d \tau_l D}{\tau_l D + 1} \tag{6.101}$$

The frequency response of this shows clearly that $f_d \rightarrow 0$ as frequency $\rightarrow 0$; thus slow pressure changes do not result in forces tending to burst the diaphragm. However, the time constant τ_l must be small enough that $(f_d/p_i)(i\omega) \approx A_d$ for all frequencies above about 10 Hz, the lowest-frequency sound pressures usually of interest.

The next step requires study of the electromechanical-energy-conversion process in a moving-plate capacitor. Although the moving "plate" (diaphragm) of the microphone is not flat, we analyze the situation for a flat plate for simplicity. One can always find a flat-plate capacitor that is equivalent to the diaphragm capacitor in the sense that the linearized capacitance variation with plate separation is the same (at least for small motions) for both. (Analysis of a "dished"-shape capacitor is available.[1]) Considering Fig. 6.41, we recall that

$$\text{Energy stored by a capacitor} = \frac{q^2}{2C} = \frac{Ce^2}{2} \tag{6.102}$$

where $q \triangleq$ charge, $e \triangleq$ voltage, and $C \triangleq$ capacitance. We wish to show that the two plates attract each other with a force f. The capacitance of a parallel-plate capacitor whose area A is large compared with the plate separation x is given very closely by

$$C = \frac{\epsilon A}{x} \tag{6.103}$$

where $\epsilon \triangleq$ permittivity of material between plates
$= 8.86 \times 10^{-12} \text{F/m}$ for vacuum or dry air

We now suppose the capacitor is charged and then open-circuited so that q must remain constant. If the plates are separated an additional amount dx, we may write

$$\text{Original energy } (x = x_0) = \frac{q^2}{2C_0} = \frac{q^2 x_0}{2\epsilon A} \tag{6.104}$$

$$\text{Final energy } (x = x_0 + dx) = \frac{q^2}{2C_f} = \frac{q^2(x_0 + dx)}{2\epsilon A} \tag{6.105}$$

The energy change is thus $q^2 dx/(2\epsilon A)$. Since energy is conserved in this system, it must have required a force f on the plate to cause the motion dx, since then mechanical work $f\, dx$ would have been done and converted to electrical energy

---

[1] H. K. P. Neubert, "Instrument Transducers," pp. 274–278, Oxford University Press, London, 1963.

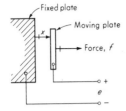

Figure 6.41 Moving-plate capacitor.

$(q^2/dx)/(2\epsilon A)$. The force f thus may be calculated from

$$\frac{q^2 dx}{2\epsilon A} = f\, dx$$

$$f = \frac{q^2}{2\epsilon A} = \frac{\epsilon A e^2}{2x^2} \tag{6.106}$$

For air, with e in volts and A and x in any consistent units, this becomes

$$f = 0.99 \times 10^{-12}\,\frac{Ae^2}{x^2}\quad \text{lbf} \tag{6.107}$$

As an example, if $e = 200$ V, $A = 1$ in^2, and $x = 0.001$ in, the force is 0.04 lbf.

If the capacitor is connected to an external circuit as in Fig. 6.42, we can show Eq. (6.106) still holds as follows: The work done in moving a charge dq through a potential difference e is $e\, dq$. Then, by conservation of energy,

$$f\, dx + e\, dq = d(\text{stored energy}) = d\left(\frac{Ce^2}{2}\right) \tag{6.108}$$

Then,
$$f = -e\frac{dq}{dx} + \frac{d}{dx}\left(\frac{Ce^2}{2}\right) = -e\frac{d}{dx}(Ce) + \frac{d}{dx}\left(\frac{Ce^2}{2}\right) \tag{6.109}$$

$$f = -e\left(C\frac{de}{dx} + e\frac{dC}{dx}\right) + Ce\frac{de}{dx} + \frac{e^2}{2}\frac{dC}{dx} \tag{6.110}$$

$$f = -\frac{e^2}{2}\frac{dC}{dx} = -\frac{e^2}{2}\left(-\frac{\epsilon A}{x^2}\right) = \frac{\epsilon A e^2}{2x^2} \tag{6.111}$$

Next we model the microphone as in Fig. 6.43. The mass M and spring K_s

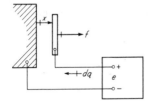

Figure 6.42 Capacitor with external circuit.

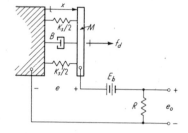

Figure 6.43 Microphone model.

must be such as to give the same natural frequency as the lowest natural frequency of the diaphragm. The dashpot B must be such as to give the same resonant peak as in the microphone's measured pressure response. The capacitor plate area and air gap (with no external forces acting) must be such as to give the same capacitance as is measured for the microphone under similar conditions. The spring constant K_s and capacitor dimensions also must be such that the force f_d causes a capacitance variation equal (at least for small motions) to that caused in the actual microphone by a pressure difference $p_i - p_v = f_d/A_d$. If all the above conditions are met, the simplified model of Fig. 6.43 will respond essentially in the same way as the microphone itself. While the equivalent system described is defined in terms of experimental measurements on an existing microphone, microphone designers have theoretical formulas for estimating these parameters *before* a new microphone is built.

Assuming that the equivalent system is a reasonable model, we can proceed with the analysis. With no force f_d applied and with the capacitor uncharged, the mass will assume an equilibrium position x_{fl}, where x_{fl} is the free length of the springs. If now the polarizing voltage E_b is applied, the moving plate will experience an attractive force and will move to a new position x_0 such that the spring force and electrostatic force just balance (see Fig. 6.44). Now, when pressure force f_d is applied, motion will take place around x_0 as an operating point. To find x_0, we can write

$$K_s(x_{fl} - x_0) = \frac{\epsilon A E_b^2}{2x_0^2} \tag{6.112}$$

This equation in x_0 has two positive solutions, x_0 and x_0', for a practical case. The solution (equilibrium position) x_0' is unstable in the sense that any slight motion away from x_0' results in *further* motion away from this point. The desired (stable) equilibrium position is x_0, where small disturbances from equilibrium give rise to forces tending to restore equilibrium. Thus the microphone must be designed to operate at x_0 rather than x_0'.

We apply Newton's law to the mass M to get

$$-B\frac{dx}{dt} + K_s(x_{fl} - x) - \frac{\epsilon A e^2}{2x^2} + f_d = M\frac{d^2x}{dt^2} \tag{6.113}$$

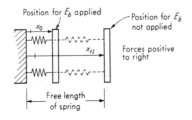

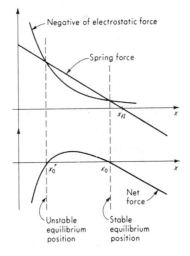

Figure 6.44 Determination of equilibrium points.

The electrostatic-force term makes this differential equation nonlinear. For small changes in e and x from the equilibrium operating point, this nonlinear term may be linearized approximately with good accuracy. This may be done by employing only the linear terms of a Taylor-series expansion of the nonlinear function. That is, if in general

$$z = z(x, y)$$

then

$$z \approx z(x_0, y_0) + \frac{\partial z}{\partial x}\bigg|_{\substack{x = x_0 \\ y = y_0}} (x - x_0) + \frac{\partial z}{\partial y}\bigg|_{\substack{x = x_0 \\ y = y_0}} (y - y_0) \qquad (6.114)$$

In this specific case, the nonlinear function is e^2/x^2; thus

$$\frac{e^2}{x^2} \approx \frac{E_b^2}{x_0^2} + E_b^2\left(-\frac{2}{x_0^3}\right)(x - x_0) + \frac{1}{x_0^2}\, 2E_b(e - E_b) \qquad (6.115)$$

We now define $x_1 \triangleq x - x_0$ and $e_o \triangleq e - E_b$ to get

$$\frac{e^2}{x_2} \approx \frac{E_b^2}{x_0^2} - \frac{2E_b^2}{x_0^3} x_1 + \frac{2E_b}{x_0^2} e_o \qquad (6.116)$$

Also
$$K_s(x_{fl} - x) = K_s(x_{fl} - x_1 - x_0) = -K_s x_1 + \frac{\epsilon A E_b^2}{2x_0^2} \qquad (6.117)$$

Now Eq. (6.113) may be written as

$$-B\frac{dx_1}{dt} - K_s x_1 + \frac{\epsilon A E_b^2}{2x_0^2} - \frac{\epsilon A}{2}\left(\frac{E_b^2}{x_0^2} + \frac{2E_b}{x_0^2} e_o - \frac{2E_b^2}{x_0^3} x_1\right)$$

$$+ f_d = M\frac{d^2 x_1}{dt^2} \qquad (6.118)$$

Bringing in Eq. (6.101), we may write

$$\left(MD^2 + BD + K_s - \frac{\epsilon A E_b^2}{x_0^3}\right) x_1 + \frac{\epsilon A E_b}{x_0^2} e_o = \frac{A_d \tau_l D}{\tau_l D + 1} p_i \qquad (6.119)$$

This equation contains two unknowns, x_1 and e_o; thus an additional equation must be found before a solution can be reached. This can be found from an analysis of the circuit of Fig. 6.45 as follows:

$$e_o = e - E_b = iR = -\frac{dq}{dt} R \qquad (6.120)$$

$$q = Ce = \frac{\epsilon A e}{x} \qquad (6.121)$$

Equation (6.121) may be linearized as

$$q \approx \frac{\epsilon A E_b}{x_0} - \frac{\epsilon A E_b}{x_0^2} x_1 + \frac{\epsilon A}{x_0} e_o \qquad (6.122)$$

Then, approximately,

$$\frac{dq}{dt} = -\frac{\epsilon A E_b}{x_0^2}\frac{dx_1}{dt} + \frac{\epsilon A}{x_0}\frac{de_o}{dt} \qquad (6.123)$$

and
$$\frac{e_o}{R} = -\frac{dq}{dt} = \frac{\epsilon A E_b}{x_0^2}\frac{dx_1}{dt} - \frac{\epsilon A}{x_0}\frac{de_o}{dt} \qquad (6.124)$$

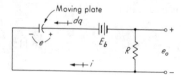

Figure 6.45 Circuit analysis.

thus finally giving $\quad -\dfrac{\epsilon A E_b R}{x_0^2} D x_1 + \left(1 + \dfrac{\epsilon A R}{x_0} D\right) e_o = 0 \qquad (6.125)$

Since we are primarily interested in e_o rather than x_1, Eq. (6.125) may be combined with (6.119) to eliminate x_1 and get

$$\left[\frac{M\tau_e}{K_e} D^3 + \left(\frac{M}{K_e} + \frac{B\tau_e}{K_e}\right) D^2 + \left(\frac{B}{K_e} + \tau_e + \frac{\tau_e^2 E_b^2}{x_0^2 R K_e}\right) D + 1\right] e_o$$

$$= \frac{A_d E_b \tau_e}{K_e x_0} \frac{\tau_l D^2}{\tau_l D + 1} p_i \quad (6.126)$$

where $\qquad\qquad K_e \triangleq K_s - \dfrac{\epsilon A E_b^2}{x_0^3} \qquad\qquad (6.127)$

$$\tau_e \triangleq \frac{\epsilon A R}{x_0} \qquad\qquad (6.128)$$

The cubic left-hand side is not readily factored until numerical values are known. In general, one gets two complex roots and one real root. This leads to a transfer function of the form

$$\frac{e_o}{p_i}(D) = \frac{K D^2}{(\tau_l D + 1)(\tau D + 1)(D^2/\omega_n^2 + 2\zeta D/\omega_n + 1)} \qquad (6.129)$$

The frequency response of the microphone is then as shown in Fig. 6.46. The sensitivity in the flat range is typically of the order of 1 to 5 mV/μbar, while the low-frequency cutoff is about 1 to 10 Hz, though lower values are possible.

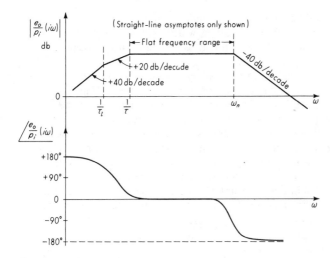

Figure 6.46 Microphone frequency response.

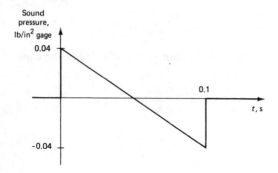

Sound
pressure,
lb/in² gage

Figure 6.47 Sonic boom pressure
signature.

The upper limit of frequency can be extended well beyond the range of human hearing; 100,000 Hz is not unattainable.

When microphones are employed to measure relatively long-duration transient sounds, such as the "sonic boom" (Fig. 6.47) caused by the overflight of supersonic aircraft, adequate response at low frequencies becomes critical.[1] This problem is conveniently studied by using digital simulation, which gives us another opportunity to demonstrate the utility of CSMP. If transfer function (6.129) is multiplied out and put in the form of a differential equation, we get

$$\frac{\tau\tau_l}{\omega_n^2}\overset{....}{e_o} + \left(\frac{\tau + \tau_l}{\omega_n^2} + \frac{2\zeta\tau\tau_l}{\omega_n}\right)\overset{...}{e_o} + \left[\frac{1}{\omega_n^2} + \frac{2\zeta(\tau + \tau_l)}{\omega_n} + \tau\tau_l\right]\overset{..}{e_o}$$

$$+ \left(\tau + \tau_l + \frac{2\zeta}{\omega_n}\right)\dot{e}_o + e_o = K\ddot{p}_i \quad (6.130)$$

To avoid the need to differentiate p_i, we employ the usual device of integrating the entire equation twice before simulation, which leads directly to the CSMP program of Fig. 6.48. The symbology should be self-explanatory except that τ is called TAUE.

In the first run, a "good" measurement system ($\omega_n = 2,000.$ rad/s, $\tau = \tau_l = 1.0$ s, $\zeta = 0.6$, $K = 2.5$) gives the almost-perfect results shown in Fig. 6.49. Note that we plot $(\tau\tau_l/K)e_o = $ EPLOT for direct comparison with p_i since the system amplitude ratio is $K/(\tau\tau_l)$ in the flat region of frequency response. Those familiar with microphones will note that $\omega_n = 2,000$ is very low, many microphones being good to 15,000 Hz. Use of the more realistic large value unfortunately causes simulation problems associated with "stiff" mathematical models.[2] A stiff mathematical model has both some very slow components ($\tau, \tau_l \approx 1.0$ s) and some very fast ($\omega_n > $ about 10,000 in this problem) which create numerical problems in the integrating algorithm. Fortunately, if ω_n is "large enough" to accurately follow

[1] J. J. Van Houten and R. Brown, Investigation of the Calibration of Microphones for Sonic Boom Measurement, *NASA* CR-1075, June 1968.

[2] Doebelin, "System Modeling and Response," pp. 218–221.

```
$$$CONTINUOUS SYSTEM MODELING PROGRAM  III   V1M3   TRANSLATOR OUTPUT$$$
TITLE MICROPHONE SONIC BOOM
PARAM WN=2000.
PARAM Z=.6
PARAM TAUE=1.,TAUL=1.
PARAM K=2.5
INIT
      A5=TAUE*TAUL/2.5
      A4=(TAUE*TAUL)/(WN*WN)
      A3=(TAUE+TAUL)/(WN*WN)+2*Z*TAUE*TAUL/WN
      A2=1./(WN*WN)+(2.*Z/WN)*(TAUE+TAUL)+TAUE*TAUL
      A1=TAUE+TAUL+2.*Z/WN
DYNAM
      PI=.04*STEP(0.0)-.8*RAMP(0.0)+.8*RAMP(.1)+.04*STEP(.1)
      EODOT2=(K*PI-A3*EODOT1-A2*EO-A1*INT1EO-INT2EO)/A4
      EODOT1=INTGRL(0.0,EODOT2)
      EO=INTGRL(0.0,EODOT1)
      INT1EO=INTGRL(0.0,EO)
      INT2EO=INTGRL(0.0,INT1EO)
      EPLOT=EO*A5
TIMER FINTIM=.15,DELT=.00001,OUTDEL=.003
PAGE WIDTH=50
OUTPUT DELT
OUTPUT PI,EPLOT
PAGE GROUP
END
PARAM WN=200.
PARAM Z=.3
PARAM TAUL=.1
PARAM TAUE=.1
END
PARAM WN=2000.
END
```

Figure 6.48 Digital simulation program for microphone response.

p_i, then making it even larger has essentially no effect on the response. The value $\omega_n = 2,000$ is just a convenient value; anything in the range of 1,000 to 10,000 gives almost identical results.

To show the effect of ω_n and τ values which are too small, the problem was rerun (new PARAM cards after first END card), yielding the results of Fig. 6.50. Here the slow rise due to low ω_n and oscillation due to $\zeta = 0.3$ are apparent. Too small values for τ and τ_l cause an exponential-type decay rather than the linear ramp of p_i. This last effect is made more obvious by rerunning with $\omega_n = 2,000$, $\tau = \tau_l = 0.1$, $\zeta = 0.3$ (PARAM card after second END card changes only ω_n from the previous run), with the results of Fig. 6.51. Here ω_n is adequate, and even the low ζ does not cause any trouble, but the low τ's (electrical and pneumatic "leakage" effects) cause inaccurate tracking of p_i.

Acoustic Emission

Acoustic emission methods[1] are utilized in materials research to study deformation and fracture processes, such as dislocation pileup and crack initiation, and in nondestructive testing[2] to evaluate structural integrity, monitor pressure vessels for incipent failure, etc. Acoustic emission transducers (usually piezoelectric) are fastened to the specimen surface and detect high-frequency

---

[1] R. E. Herzog, Forecasting Failures with Acoustic Emission, *Mach. Des.*, pp. 132–137, July 14, 1973; Acoustic Emission, STP 505, *ASTM*, Philadelphia, Pa., 1972; Acoustic Emission Bibliography 1970–72, STP 571, *ASTM*, Philadelphia, Pa., 1975; T. Licht, Acoustic Emission, *B&K Tech. Rev.*, no. 2, B&K Instruments, Marlboro, Mass., 1979.

[2] A. Vary, Nondestructive Evaluation Technique Guide, *NASA* SP-3079, 1973.

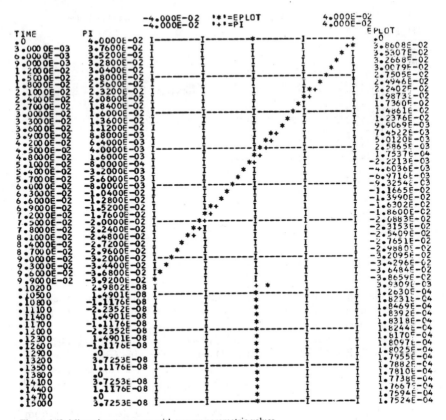

Figure 6.49 Microphone response with proper parametric values.

(20 kHz to 2 MHz) stress waves caused by the phenomena under study and propagated "acoustically" through the material. The above references should be sufficient to get the interested reader started on an in-depth study.

6.11 PRESSURE-SIGNAL MULTIPLEXING SYSTEMS

In certain applications, many channels of pressure data must be repetitively transduced to voltage signals. For example, when a pneumatic process control system is retrofitted for electronic computer control, there may be hundreds of 3 to 15 lb/in^2 gage air pressure signals to be interfaced to the computer. In wind-tunnel and fluid machinery testing,[1] hundreds of test points distributed over the

---

[1] J. C. Pemberton and G. O. Ellis, Flow Measurement in Rotating Machinery, *Inst. & Cont. Syst.*, March 1964.

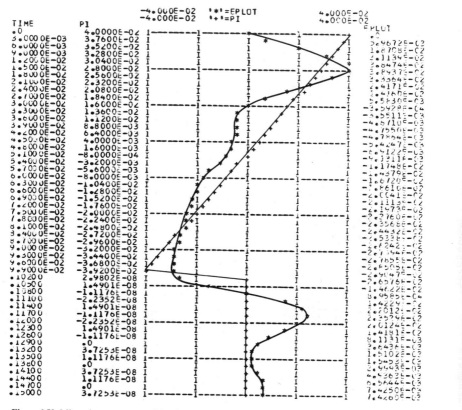

Figure 6.50 Microphone response with ω_n and τ too small.

surface of the model or machine may be instrumented for static or stagnation pressure sensing. When the needed data rates are sufficiently slow, a cost-effective solution timeshares a single transducer/amplifier by multiplexing the pressure lines using a pneumatic scanning device.

The classic device in this field is the Scanivalve (Fig. 6.52), a multiported rotating valve which sequentially connects each of many (up to 64) pressure lines distributed around its circumference to a single flush-diaphragm transducer. By keeping volume below 0.001 in^3, the response time is sufficiently fast to allow scan rates up to 10 or 20 channels per second. Unlimited numbers of channels can be accommodated by using as many Scanivalves as necessary; however, each requires its own pressure transducer. One 12-port model is designed with integral gearing to allow convenient interconnection of multivalve systems (24 valves and 288 channels are not unusual) which are driven by a single electric or pneumatic stepping motor. Often, two channels of a valve are used for calibration purposes

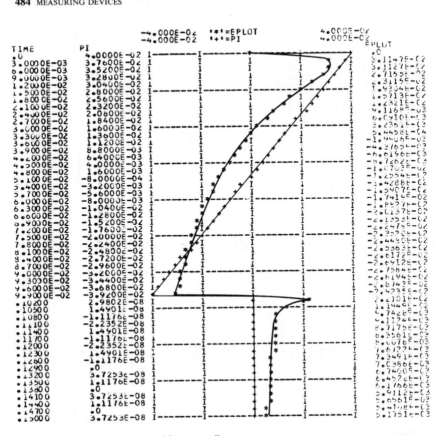

Figure 6.51 Microphone response with τ too small.

by leaving them connected to, respectively, zero pressure and a stable reference pressure. Modest computing capability in the electronics then allows periodic correction of transducer drifts. When multiple Scanivalves are employed, the output of each electrical transducer may be multiplexed electrically into the computer, thus combining pneumatic/mechanical scanning with electrical multiplexing and increasing the system scan rate in proportion to the number of valves used.

A different approach[1] to multichannel pressure scanning is somewhat more costly, but achieves a higher scan rate (64 channels per second) and produces an electrical output in digital form. Rather than using a single analog pressure transducer, as in the Scanivalve, each channel is provided with its own pressure

[1] Model SSP-D/64 MK4, Scanivalve Corp., San Diego, Calif.

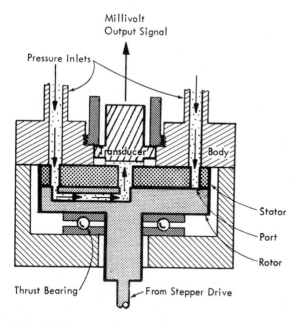

Figure 6.52 Pressure signal multiplexer. (*Courtesy Scanivalve Corp., San Diego, Calif.*)

comparator, a differential-pressure switch which provides a contact closure whenever the pressure differential across its diaphragm crosses zero (± 0.005 lb/in^2). For each channel, the unknown pressure is applied to one side of its diaphragm while the other side is connected to a chamber common to all the switches. In this chamber a pneumatic system produces a precision pressure ramp which rises linearly from zero to, say, 20 lb/in^2 gage in 0.800 s, discharges to zero, and then repeats this cycle once per second. A binary electronic counter is synchronized with this ramp; thus its contents at any time during the 0.800-s cycle are proportional to the ramp pressure. When the ramp pressure passes through any particular unknown pressure of the 64 channels, that particular pressure comparator switches, an event whose timing relative to the counter ramp measures the unknown pressure. Since all 64 pressures are within the range of the ramp, 64 individual digital pressure readings are obtained each cycle and put into a "two-page" memory, so that one page is being updated while the other can be asynchronously accessed by the computer. Again, 2 of the 64 channels may be utilized for calibration purposes by applying fixed precision pressures to them.

When required data rates are very high, systems using a separate analog transducer for each channel, with electronic multiplexing into a single analog/digital converter, have been available for some time, but at considerable expense. Recently silicon semiconductor pressure transducers with 32 individual sensors packaged in a 5-oz box of size $1 \times 2 \times 2.5$ in have become available[1] at reasonable cost. These transducers are presently available for gases only and do not meet as stringent accuracy specifications as those used in Scanivalve systems. However, the accuracy problem is overcome in complete measurement systems[2] by including a single precision digital quartz transducer as a reference and using microprocessor computing power to periodically recalibrate not only the transducers but also amplifiers and analog/digital converters against three precision pressure values. This three-point calibration corrects for transducer nonlinearity (which otherwise would be a problem in transducers of this type) and thermal zero and sensitivity shifts. Systems of up to 512 channels with acquisition rates to 14,000 measurements per second and throughput of 500 measurements per second are available.

PROBLEMS

6.1 For the system of Fig. 6.2:

(a) By what factor must the actual weight of steel weights be multiplied to correct for air buoyancy?

(b) What correction must be applied to the platform weight to account for oil buoyancy if the piston is immersed 5 in and has a diameter of 0.2 in? Take oil specific weight as 50 lbf/ft³.

(c) If, in part b, air, rather than oil, is the pressure medium, what would the correction be when the gage pressure is 100 lb/in² gage and temperature is 70°F? Make an estimate, assuming constant air temperature and pressure varying linearly from the high-pressure end of the piston to atmospheric at the top end.

6.2 A well-type mercury manometer employed to measure water flow rate is shown in Fig. P6.1 for zero flow rate $(p_1 = p_2)$. Derive a relation between $p_1 - p_2$ and h for this configuration.

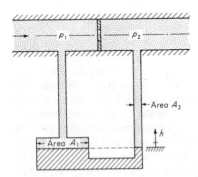

Figure P6.1

6.3 For the inclined manometer of Fig. 6.5, derive a relation between $p_1 - p_2$ and the displacement reading along the calibrated scale.

---

[1] Model ESP32, Pressure Systems Inc., Hampton, Va.
[2] Model 780B, Pressure Systems Inc., Hampton, Va.

6.4 Estimate the largest step change that will give linear behavior in a water manometer with $L = 26.5$ in and $R = 0.13$ in. What are ζ and ω_n for this manometer? If a step change 5 times the value found above is applied, estimate ζ_e and ω_n for this situation.

6.5 In Eq. (6.44), get an expression for $d\zeta_e/\zeta_e$ by taking the log of both sides and then differentiating. If L, μ, γ, g, R, ω, and Y are each in error by 1 percent, what is the percentage error in ζ_e?

6.6 From Eq. (6.46), plot p versus y_c/t for $0 \leq y_c/t \leq 1$ and $E = 25 \times 10^6$ lb/in^2, $t = 0.04$ in, $R = 4.0$ in, and Poisson's ratio $= 0.26$.

6.7 Design a pressure pickup and bridge circuit such as that in Fig. 6.12 to meet the following requirements:

Maximum pressure $= 100$ lb/in^2 gage
Natural frequency in vacuum $= 10$ Hz minimum
Maximum nonlinearity by Eq. (6.46) $= 3$ percent
Full-scale output $= 10$ mV minimum
Diaphragm material, stainless steel.
Strain gages with 350-Ω resistance, gage factor of 2, and size 0.3 by 0.3 in

6.8 A pressure pickup as in Fig. 6.12 has the following characteristics:

$$R = 3.0 \text{ in} \qquad E = 28 \times 10^6 \text{ lb/in}^2 \qquad \text{gage resistance} = 120 \ \Omega$$

$$r_r = 2.5 \text{ in} \qquad \mu = 0.26 \qquad \text{gage factor} = 2.0$$

$$r_t = 0.5 \text{ in} \qquad \gamma = 0.3 \text{ lbf/in}^3 \qquad \text{battery voltage} = 5.0 \text{ V}$$

$$t = 0.05 \text{ in}$$

(a) Calculate the sensitivity in mv/psi.

(b) What is the natural frequency in vacuum?

(c) Based on Eq. (6.46), what is the maximum allowable pressure for 2 percent nonlinearity? What is the voltage output at this point?

6.9 Explain in words the operation of the system of Fig. 6.18a. Derive an equation relating p_o to $p_1 - p_2$, and show how linearity is achieved even if K_n varies.

6.10 Explain in words the operation of the system of Fig. 6.18b. Derive an equation relating e_o to $p_1 - p_2$.

6.11 Perform a linearized dynamic analysis of the system of Fig. 6.18a, and discuss stability/accuracy tradeoff.

6.12 Perform a linearized dynamic analysis of the system of Fig. 6.18b, and discuss stability/accuracy tradeoff.

6.13 From Eq. (6.57), compute τ for a system with $C_{vp} = 0.4$ cm^3/(lb/in^2), $d_t = 0.10$ in, $L = 10$ ft, and $\mu = 0.001$ lbf $\cdot$ s/ft^2. Using Eq. (6.58), find the dead time associated with this system if the tube is of steel with a wall thickness of 0.02 in, $E_L = 100,000$ lb/in^2, and the fluid specific weight is 0.03 lbf/in^3. Is this dead time significant relative to τ?

6.14 A pressure transducer has a natural frequency in vacuum of 5,000 Hz and $C_{vp} = 0.0003$ in^3/(lb/in^2). It is used with a liquid of specific weight 0.04 lbf/in^3 and viscosity 0.0005 lbf $\cdot$ s/ft^2. The tubing inside diameter is 0.2 in, and its length is 5 ft. Find ω_n and ζ of the combined transducer/tubing system.

6.15 The pressure pickup of Prob. 6.14 has an internal volume of 0.004 in^3. If it is used with the same tubing as in Prob. 6.14 but the fluid medium is changed to air at 100 lb/in^2 absolute and 100°F, what are the values of ζ and ω_n?

6.16 If the transducer of Prob. 6.15 is used with tubing of 1-in length and 0.1-in inside diameter, what will ω_n and ζ be?

6.17 Compute the resistance change of 100-Ω coils of Manganin and gold chrome for 50,000 lb/in^2 pressure and 100°F temperature changes.

6.18 Design a capillary leak for a microphone such that frequencies of 10 Hz and above will be measured with an amplitude-ratio error of no more than 10 percent. Assume standard atmospheric air and a leak length L of 1 in. Find the required leak diameter d_l. The microphone has an internal volume of 0.5 in^3. What will be the amplitude ratio for atmospheric pressure drifts of 2 cycles/h frequency?

6.19 For a liquid-filled transducer/tubing system, modify the text analysis to get a formula for ω_n if the tubing is made of two sections of different lengths and diameters.

6.20 For a liquid-filled transducer/tubing system, modify the text analysis to account for a gas bubble of volume V_g trapped at the transducer diaphragm.

6.21 Derive formulas for the compliance C_{vp} of U-tube, well-type, and inclined manometers.

BIBLIOGRAPHY

1. J. B. Damrel: Quartz Bourdon Gage, *Instrum. Contr. Syst.*, p. 87, February 1963.
2. W. R. Myers: The Electromanometer, *Instr. Contr. Syst.*, p. 116, April 1962.
3. R. R. Koolman: Reference Pressure Cells, *Instrum. Contr. Syst.*, p. 123, February 1962.
4. Dead-weight Testers, *Instrum. Contr. Syst.*, p. 126, April 1962.
5. E. Moser: Automatic Pressure Calibration, *Meas. & Contr.*, December 1979.
6. H. Norville: A Dead Weight Pressure Balance with Extended Range to 5000 psig, *NASA*, N-64-33684, 1964.
7. S. Siegel: Pressure Calibration Circuits, *Instrum. Contr. Syst.*, p. 116, February 1965.
8. R. C. Cerni: Measuring 100 Pressures in 15 Seconds, *Contr. Eng.*, p. 89, July 1964.
9. R. I. Kreisler: Rocket Propellant Manometer System, *ISA J.*, p. 55, October 1962.
10. W. W. Willmarth: Wall Pressure Fluctuations in a Turbulent Boundary Layer, *NACA*, *Tech. Note* 4139, 1958.
11. P. K. Stein: Measuring Fluctuating Pressure, *Instrum. Contr. Syst.*, p. 156, September 1964.
12. E. J. Rogers: Semiconductor Pressure Transducer Features Mechanical Compensation, *Instrum. Contr. Syst.*, p. 128, April 1963.
13. Y. T. Li: Two-Cylinder Transducer Has Straight Line Response, *Contr. Eng.*, p. 151, April 1962.
14. Y. Kobashi et al.: Improvements of a Pressure Pickup for the Measurement of Turbulence Characteristics, *J. Aerosp. Sci.*, p. 149, February 1960.
15. T. Wrathall: Measuring Impact Pressures on Re-entering Missile Nose Cones. *ISA J.*, p. 54, October 1959.
16. G. E. Reis: Theoretical Examination of Variable Reluctance Diaphragm Gage, *SCR*-162, Sandia Corp., Albuquerque, N. Mex., 1960.
17. Capacitive Pressure Sensors, *Instrum. Contr. Syst.*, p. 119, May 1962.
18. P. Smelser: Pressure Measurements in Cryogenic Systems, National Bureau of Standards, Boulder, Colo.
19. H. Chelner: High Frequency Semiconductor Probe Pressure Transducer, *AIAA Paper* 64-508, 1964.
20. J. H. Thomson: Torsion Bar Pressure Transducer, *Electromech. Des.*, p. 46, June 1964.
21. D. S. Johnson: Design and Application of Piezoceramic Transducers to Transient Pressure Measurements, *NASA*, N-63-18139, 1962.
22. R. E. Engdahl: Pressure Measuring Systems for Closed Cycle Liquid Metal Facilities, *NASA*, CR-54140.
23. J. A. Haner: Pressure/Displacement Transducer, *Instrum. Contr. Syst.*, p. 107, November 1964.
24. D. D. Keough et al.: Piezoresistive Pressure Transducer, *ASME Paper* 64-WA/PT-5, 1964.
25. Pressure Transducer $\frac{3}{8}$-Inch in Size Can Be Faired into Surface, *NASA*, *Brief* 64-10021, 1964.
26. Welded Pressure Transducer Made as Small as $\frac{1}{8}$-Inch Diameter, *NASA*, *Brief* 63-10429, 1963.
27. Improved Variable-Reluctance Transducer Measures Transient Pressure, *NASA*, *Brief* 63-10321, 1963.

28. H. B. Jones et al.: Transient Pressure Measurements in Liquid Propellant Rocket Thrust Chambers, *ISA Trans.*, p. 117, April 1965.
29. R. L. Ledford and W. E. Smotherman: Miniature Transducers for Pressure and Heat Transfer Rate Measurements in Hypervelocity Wind Tunnels, *ISA Trans.*, p. 133, April 1965.
30. P. S. Lederer and R. O. Smith: An Experimental Technique for the Determination of the Fidelity of the Dynamic Response of Pressure Transducers, *Natl. Bur. Std. (U.S.), Rept.* 7862, 1963.
31. R. O. Smith: A Liquid-Medium Step-Function Pressure Calibrator, *ASME Paper* 63-WA-263, 1963.
32. J. L. Schweppe: Calibration of Pressure Transducers with Aperiodic Input-Function Generators, *ISA Trans.*, p. 72, January 1964.
33. W. C. Bentley and J. J. Walter: Dynamic Response Testing of Transient Pressure Transducers for Liquid Propellent Rocket Combustion Chambers, *NASA, CR*-51995, 1963.
34. W. E. Amend: Dynamic Performance of Pressure Transducers in Shock and Detonation Tubes, *NASA, N*-65-13313.
35. D. Baganoff: Pressure Gauge with One-tenth Microsecond Risetime for Shock Reflection Studies, *Rev. Sci. Instrum.*, p. 288, March 1964.
36. E. L. Davis: The Measurement of Unsteady Pressures in Wind Tunnels, *AGARD Rept.* 169, March 1958.
37. R. Oldenburger and R. E. Goodson: Hydraulic Line Dynamics, *NASA, CR*-52148, 1963.
38. T. R. Stalzer and G. J. Fiedler: Criteria for Validity of Lumped-Parameter Representation of Ducting Air-Flow Characteristics, *ASME Trans.*, p. 833, May 1957.
39. F. Nagao and M. Ikegami: Errors of an Indicator Due to a Connecting Passage, *Bull. JSME*, vol. 8, no. 29, 1965.
40. A. L. Ducoffe and F. M. White: The Problem of Pneumatic Pressure Lag, *ASME Trans.*, p. 234, June 1964.
41. A. S. Iberall: Attenuation of Oscillatory Pressures in Instrument Lines, *Natl. Bur. Std. (U.S.), Res. Paper* RP2115, July, *ASME Trans.*, 1950.
42. F. Nagao et al.: Influence of the Connecting Passage of a Low Pressure Indicator on Recording, *Bull. JSME*, vol. 6, no. 21, 1963.
43. C. B. Schuder and G. C. Blunck: The Driving Point Impedance of Fluid Process Lines, *ISA Trans.*, p. 39, January 1963.
44. R. P. Benedict: The Response of a Pressure-Sensing System, *ASME Paper* 59-A-289, 1959.
45. R. J. Martin and D. S. Moseley: Analysis of the Effect of Pulsations on the Response of Mercurial-Type Differential-Pressure Recorders, *ASME Trans.*, p. 1343, October 1958.
46. J. E. Broadwell and A. G. Hammitt: Transient Response of Fluid Systems, *J. Aerosp. Sci.*, July 1962.
47. A. F. D'Souza and R. Oldenburger: Dynamic Response of Fluid Lines, *ASME Paper* 63-WA-73, 1963.
48. R. Oldenburger and R. E. Goodson: Simplification of Hydraulic Line Dynamics by Use of Infinite Products, *ASME Paper* 62-WA-55, 1962.
49. J. D. Regetz: Experimental Determination of the Dynamic Response of a Long Hydraulic Line, *NASA, Tech. Note* D-576, 1960.
50. W. Lewis et al.: Study of the Effect of a Closed-End Side Branch on Sinusoidally Perturbed Flow of Liquid in a Line, *NASA, Tech. Note* D-1876, 1963.
51. I. Taback: The Response of Pressure Measuring Systems to Oscillating Pressures, *NACA, Tech. Note* 1819, 1949.
52. F. T. Brown: The Transient Response of Fluid Lines, *ASME Paper* 61-WA-143, 1961.
53. C. B. Schuder and R. C. Binder: The Response of Pneumatic Transmission Lines to Step Inputs, *ASME Trans.*, p. 578, December 1959.
54. High Pressure Measurement, *Mech. Eng.*, p. 76, February 1963.
55. D. H. Newhall and L. H. Abbott: High Pressure Measurement, *Instrum. Contr. Syst.*, p. 232, February 1961.
56. W. H. Howe: High-Pressure Measurement and Control, *Contr. Eng.*, p. 53, April 1955.
57. A. J. Yerman: The Tunnel Diode as an FM Hydrostatic Pressure Sensor, *ASME Paper* 63-WA-264, 1963.

58. R. J. Melling: Ionization Vacuum Gage Measures Absolute Pressures up to 1 mm Hg, *Instrum. Contr. Syst.*, p. 119, September 1964.
59. Vacuum Instrumentation, *Instrum. Contr. Syst.*, p. 110, September 1964.
60. J. M. Lafferty and T. A. Vanderslice, Vacuum Measurement by Ionization, *Instrum. Contr. Syst.*, p. 90, March 1963.
61. A. P. Flanick and J. Ainsworth: A Thermistor Pressure Gage, *NASA, Tech. Note* D-504, 1960.
62. J. M. Benson: Calibrating Thermal Conductivity Gauges, *Instrum. Contr. Syst.*, p. 115, September 1964.
63. J. P. Walsh: Molecular Vacuum Gages, *Instrum. Contr. Syst.*, p. 106, August 1963.
64. W. Kreisman: Extension of the Low Pressure Limit of McLeod Gages, *NASA, CR*-52877, 1963.
65. M. P. Hnilicka: Extreme High Vacuum, *Ind. Res.*, p. 36, September 1964.
66. J. M. Benson: Thermal Conductivity Vacuum Gages, *Instrum. Contr. Syst.*, p. 98, March 1963.
67. D. Alpert: Theoretical and Experimental Studies of the Underlying Processes and Techniques of Low Pressure Measurement, *NASA, N*-64-17582, 1964.
68. P. J. Bryant et al.: Extreme Vacuum Technology, *NASA, CR*-84, 1964.
69. Vacuum Instrumentation, *Instrum. Contr. Syst.*, p. 113, October 1964.
70. Precision Gage Measures Ultrahigh Vacuum Levels, *NASA, Brief* 63-10597, 1963.
71. Absolute Pressure Gage Feasibility Study, *NASA, CR*-58075, 1963.
72. J. Gavis: Vacuum Gage Systems, *NASA, N*-64-28208, 1964.
73. R. W. Roberts: Ultrahigh Vacuum Technology, General Electric Co., Schenectady, N. Y., *Rept.* 64-RL-3644C, 1964.
74. R. W. Roberts: An Outline of Vacuum Technology, General Electric Co., Schenectady, N.Y., *Rept.* 64-RL-3394C, 1964.
75. L. T. Melfi and P. R. Yeager: A Method for Calibration of Gas-Composition Sensitive Pressure Gages in Condensible Vapors, *NASA, Tech. Note* D-2567, 1965.
76. F. Feakes et al.: Gauge Calibration Study in Extreme High Vacuum, *NASA, CR*-167, 1965.
77. S. W. Athey: Acoustics Technology, A Survey, *NASA* SP-5093, 1970.
78. D. O. Conn, III: The Audio Dosimeter—A System for Measuring Personal Noise Exposure, *Sound & Vib.*, September 1972.
79. H. C. Sommer: Description and Use of a Measurement System for Air Bag Acoustic Transient Data Acquisition and Analysis, *Aero. Med. Res. Lab. Dept.* AMRL-TR-73-8, Wright Patterson AF Base, Ohio, 1973.
80. T. W. Nyland and R. E. Chase: High-Temperature Transient Pressure Transducer for Use in Liquid-Metal Systems, *NASA* TN D-5589, 1969.
81. J. D. Foote: A Complete Self-Contained Audio Measurement System, *Hewlett-Packard J.*, pp. 3–17, August 1980.
82. K. Chijiiwa and Y. Hatamura: Miniature Gages for Soil, Grains and Powders, *Bull. JSME*, vol. 15, no. 82, pp. 455–465, 1972.
83. J. S. Hilten et al.: Experimental Investigation of Means for Reducing the Response of Pressure Transducers to Thermal Transients, *NBS Tech. Note* 961, January 1978.
84. H. H. Taniguchi and G. Rasmussen: Selection and Use of Microphones for Engine and Aircraft Noise Measurements, *B&K Tech. Rev.*, no. 4, pp. 3–30, B&K Instruments, Marlboro, Mass., 1980.
85. D. S. Pallett and M. A. Cadoff: The National Measurement System for Acoustics, *Sound & Vib.*, pp. 20–31, October 1977.
86. D. G. Fleming (ed): "Indwelling and Implantable Pressure Transducers," CRC Press, Boca Raton, Fla., 1977.
87. A. Noordergraaf: "Circulatory System Dynamics," Academic, New York, 1978.
88. R. P. Benedict: "Fundamentals of Temperature, Pressure and Flow Measurement," 2d ed., Wiley, New York, 1977.
89. B. W. Spencer: Measurement of Fluctuating Pressure with Hot-Film Velocity Sensors, *Tech. Bull.* TB 34, Thermosystems, St. Paul, Minn., 1970.
90. G. S. Pick: A Study of Short-Time, Low-Pressure Response in a Transducer System, Naval Ship R&D Center, Washington, 1970.

SEVEN

FLOW MEASUREMENT

7.1 LOCAL FLOW VELOCITY, MAGNITUDE AND DIRECTION

In ma. ny experimental studies of fluid flow phenomena, it is necessary to determine the magnitude and/or direction of the flow-velocity at a point in the fluid and how this varies from point to point. That is, a description of the flow field is desired. Various methods of *flow visualization* allow us to gain an overall view of flow patterns. Sometimes the qualitative information available from direct visual observation is sufficient; however, most methods also allow quantitative analysis. Once (by flow visualization or other sources of information) localized regions of particular interest have been pinpointed, it may be necessary to insert *velocity probes* to obtain accurate point measurements. Such probes (pitot-static tubes, hot-wire anemometers, and laser-doppler velocimeters are the most common) always involve a sensing volume of finite size. Thus true "point" measurements are impossible. However, sensing volumes can be made sufficiently small to provide data of practical utility.

Flow Visualization[1]

The majority of flow visualization schemes are based on one of two basic principles: the introduction of tracer particles or the detection of flow-related changes in fluid optical properties. In liquids, colored dyes and gas bubbles are common

---

[1] W. Merzkirch, Making Fluid Flows Visible, *Am. Sci.*, vol. 67, pp. 330–336, May–June 1979; W. Merzkirch, "Flow Visualization," Academic, New York, 1974; T. Asanuma (ed), "Flow Visualization," Hemisphere, New York, 1979.

tracers. A line of hydrogen bubbles, for example, can be formed in water by applying a short electric pulse to a straight wire immersed in the flow. Photography with steady illumination shows the bubbles as short streaks whose length can be measured to obtain velocity data, while stroboscopic light gives a series of dots whose spacing gives similar information. For gas flows, smoke, helium-filled "soap" bubbles, or gas molecules made luminous by an ionizing electric spark have served as tracers.

Shadowgraph, schlieren, and interferometer techniques[1] employ, in different ways, the variation in refractive index of the flowing gas with density. For compressible flows (Mach number above about 0.3), density varies with velocity sufficiently to produce measurable effects. In shadowgraph and schlieren methods, light and dark patterns related to flow conditions are produced by the bending of light rays as they pass through a region of varying density. The optical apparatus for these two methods is relatively simple to use, and they are widely utilized, mostly for qualitative studies. For quantitative results, the more difficult interferometer approach may be necessary. Here the light/dark patterns are formed by interference effects resulting from phase shifts between a reference beam and the measuring beam. For no flow, a regular grid of light/dark fringes is present. When flow occurs, this grid is distorted and numerical values of density can be calculated from the fringe displacements.

Versions of the holographic techniques[2] discussed in Chap. 4 have been developed for flow measurements. Combinations[3] of several visualization methods are sometimes helpful in difficult problems.

Velocity Magnitude from Pitot-Static Tube

In some situations the direction of the velocity vector is known with sufficient accuracy without taking any measurements. If the direction is not known, it may be found in several ways discussed later. Let us assume the direction is known, so that a pitot-static tube may be properly aligned with this direction, as in Fig. 7.1. Assuming steady one-dimensional flow of an incompressible frictionless fluid, we can derive the well-known result

$$V = \sqrt{\frac{2(p_{stag} - p_{stat})}{\rho}} \tag{7.1}$$

where $V \triangleq$ flow velocity
$\rho \triangleq$ fluid mass density
$p_{stag} \triangleq$ stagnation or total pressure, free stream
$p_{stat} \triangleq$ static pressure, free stream

[1] A. Seiff, Shadow, Schlieren, and Interferometer Photographs, Ballistic Range Technology, chap. 8, *AGARDOGRAPH*-138, 1971.

[2] W. A. Benser, Holographic Flow Visualization within a Rotating Compressor Blade Row, *NASA* TMX-71788, August 1975.

[3] R. Sedney, C. W. Kitchens, Jr., and C. C. Bush, Combined Techniques for Flow Visualization, *AIAA Paper* 76-55, 1976.

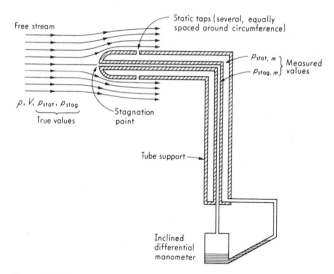

Figure 7.1 Pitot-static tube.

In an actual pitot-static tube, deviations from the ideal theoretical result of Eq. (7.1) arise from a number of sources. If ρ is accurately known, the errors can be traced to inaccurate measurement of p_{stag} and p_{stat}.

The static pressure is usually the more difficult to measure accurately. The difference between true (p_{stat}) and measured ($p_{stat, m}$) values of static pressure may be due to the following:

1. Misalignment of the tube axis and velocity vector. This exposes the static taps to some component of velocity.
2. Nonzero tube diameter. Streamlines next to the tube must be longer than those in undisturbed flow, which indicates an increase in velocity. This is accompanied by a decrease in static pressure, which makes the static taps read low. A similar (and possibly more severe) effect occurs if a tube is inserted in a duct whose cross-sectional area is not much larger than that of the tube.
3. Influence of stagnation point on the tube-support leading edge. This higher pressure causes the static pressure upstream of the leading edge also to be high. If the static taps are too close to the support, they will read high because of this effect. Note that this error and that of item 2 above tend to cancel. By proper design, effective cancellation may be achieved[1] (see also Prandtl pitot tube, Fig. 2.17b). Figure 7.2 shows the nature of both these errors as revealed by experimental tests.[2]

---

[1] V. S. Ritchie, Several Methods for Aerodynamic Reduction of Static-Pressure Sensing Errors for Aircraft at Subsonic, Near-Sonic, and Low Supersonic Speeds, *NASA, Tech. Rept.* R-18, 1959.

[2] R. G. Folsom, Review of the Pitot Tube, *Trans. ASME*, p. 1450, October 1956.

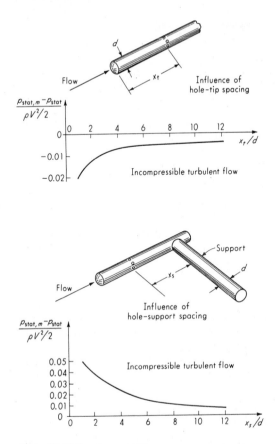

Figure 7.2 Static-pressure errors.

An important application of the pitot-static tube is found in aircraft and missiles.[1] Here the stagnation- and static-pressure readings of a tube fastened to a vehicle are used to determine the airspeed and Mach number while the static reading alone is utilized to measure altitude. If altitude is to be measured with an error of 100 ft, the static pressure must be accurate to 0.5 percent.[2] To achieve this accuracy, methods for compensating errors of the type mentioned in items 1, 2, and 3 of the above have been developed and reported.[3] An interesting and useful result of these studies is a simple method for reducing error from angular misalignment. It was found that by locating the static-pressure taps as in Fig. 7.3,

[1] W. Gracey, "Measurement of Aircraft Speed and Altitude," Wiley, New York, 1981.

[2] Ritchie, op. cit.

[3] Ibid.

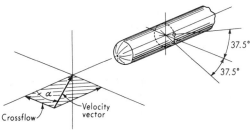

Figure **7.3** Probe insensitive to misalignment.

$\alpha \triangleq$ angle of attack

.the measurement was essentially insensitive to angle of attack for the range $-2° < \alpha < 12°$ and Mach numbers in the range 0.4 to 1.2. While this method, as shown, is effective only for misalignment in the particular plane shown, it can be extended to arbitrary directions of misalignment by designing the probe with a single vane to rotate about its longitudinal axis and automatically locate the taps $37.5°$ from the cross-flow stagnation point.[1] It is also possible to design multiple-vaned probes[2] mounted on a gimbal system with complete rotational freedom. These probes act in a fashion similar to a weather vane (except that they have angular freedom about two axes) and thus align themselves with the velocity vector. A conventional ring of evenly spaced static taps then gives accurate readings. By measuring the rotations of the gimbals, such probes also provide information on the *direction* of the velocity.

While errors in the stagnation pressure are likely to be smaller than those in the static pressure, several possible sources of error are present, namely:

1. Misalignment. This situation prevents formation of a true stagnation point at the measuring hole since the velocity is not zero. Tubes of special design have been developed which exhibit considerable tolerance to misalignment.[3] Figure 7.4 shows an example of such a tube which has an error less than 1 percent of the velocity pressure $\rho V^2/2$ for misalignments up to $\pm 38°$ for velocities from low subsonic to Mach 2. Conventional tubes not specifically designed for misalignment insensitivity may show 1 percent errors at only 5 or 10°.

2. Two- and three-dimensional velocity fields. When the velocity is not uniform, a probe of finite size intercepts streamlines of different velocities and the stagnation pressure measured corresponds to some sort of average velocity

---

[1] F. J. Capone, Wind-Tunnel Tests of Seven Static-Pressure Probes at Transonic Speeds, *NASA, Tech. Note* D-947, 1961.

[2] Ibid.

[3] W. Gracey, D. E. Coletti, and W. R. Russell, Wind-Tunnel Investigation of a Number of Total-Pressure Tubes at High Angles of Attack, Supersonic Speeds, *NACA, Tech. Note* 2261, January 1951; W. Gracey, W. Letko, and W. R. Russell, Wind-Tunnel Investigation of a Number of Total-Pressure Tubes at High Angles of Attack, Subsonic Speeds, *NACA, Tech. Note* 2331, April 1951.

(see Fig. 7.5a). For the two-dimensional situation of Fig. 7.5a, if we knew the displacement δ, we could assign the measured stagnation pressure (and thus velocity) to a specific point in the flow. Some limited data[1] on this problem are available.

3. Effect of viscosity. Equation (7.1) assumes the fluid to be frictionless. At sufficiently low Reynolds number, the viscosity of the fluid exerts a noticeable additional force at the stagnation hole, causing the stagnation pressure to be higher than predicted by Eq. (7.1). This effect can be taken into account by introducing a correction factor C as follows:

$$p_{\text{stag},\,m} = p_{\text{stat}} + \frac{C\rho V^2}{2} \tag{7.2}$$

For negligible viscosity effects, $C = 1.0$ and Eq. (7.2) is the same as (7.1). For a given probe, the factor C is a function of Reynolds number only and may be found theoretically for simple probe shapes such as spheres and cylinders. A typical result[2] for a cylindrical probe is

$$C = 1 + \frac{4}{N_R} \qquad 10 < N_R < 100 \tag{7.3}$$

where Reynolds number $\triangleq N_R \triangleq V\rho r/\mu$, $r \triangleq$ probe radius, and $\mu \triangleq$ fluid viscosity. Equation (7.3) shows that the effect is about 4 percent of the velocity pressure $\rho V^2/2$ at $N_R = 100$. At $N_R = 10$, however, the effect is 40 percent. Theory and experimental tests[3] show that viscosity corrections are rarely needed for $N_R > 500$, no matter what the shape of the probe.

When a pitot-static tube is employed in a compressible fluid, Eq. (7.1) no longer applies, although it may be sufficiently accurate if the Mach number is low enough. For subsonic flow (Mach number $N_M < 1$) the velocity is given by[4]

$$V = \sqrt{\frac{2k}{k-1}\frac{p_{\text{stat}}}{\rho_{\text{stat}}}\left[\left(\frac{p_{\text{stag}}}{p_{\text{stat}}}\right)^{(k-1)/k} - 1\right]} \tag{7.4}$$

where $\qquad k \triangleq \dfrac{\text{specific heat at constant pressure}}{\text{specific heat at constant volume}} = \dfrac{C_p}{C_v} \tag{7.5}$

Measurement of free-stream density ρ_{stat} requires knowledge of static temper-

[1] Folsom, op. cit., p. 1451.
[2] Ibid., p. 1453.
[3] Ibid.
[4] R. C. Binder, "Advanced Fluid Dynamics and Fluid Machinery," p. 51, Prentice-Hall, Englewood Cliffs, N.J., 1951.

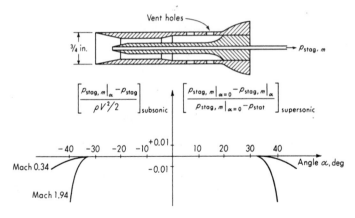

Figure 7.4 Special stagnation probe.

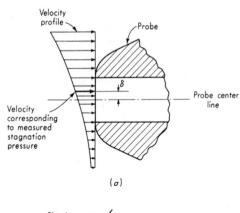

(a)

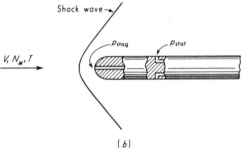

(b)

Figure 7.5 Nonuniform-velocity profile and supersonic probe.

ature, which may itself be a difficult measurement. Equation (7.4) may be rewritten as

$$p_{\text{stag}} = p_{\text{stat}} \left[1 + \frac{k-1}{2} \left(\frac{V}{c} \right)^2 \right]^{k/(k-1)} \tag{7.6}$$

where

$$c \triangleq \text{acoustic velocity} = \sqrt{\frac{k p_{\text{stat}}}{\rho_{\text{stat}}}} = \sqrt{kgRT} \tag{7.7}$$

and

$$g \triangleq \text{gravitational acceleration}$$
$$T \triangleq \text{free-stream static temperature}$$
$$R \triangleq \text{gas constant}$$

The right side of Eq. (7.6) may be expanded in a power series to give

$$p_{\text{stag}} = p_{\text{stat}} + \left(\rho_{\text{stat}} \frac{V^2}{2} \right) \left(1 + \frac{N_M^2}{4} + \frac{2-k}{24} N_M^4 + \cdots \right) \tag{7.8}$$

where

$$N_M \triangleq \frac{V}{c} \tag{7.9}$$

Since the Mach number of an incompressible fluid is zero, Eq. (7.8) shows that p_{stag} is higher for compressible than for incompressible flow. Also, if N_M is sufficiently small, Eq. (7.8) is closely approximated by Eq. (7.1).

For supersonic flow ($N_M > 1$), a compression shock wave forms ahead of the pitot tube. Between this shock wave and the tube end, the velocity is subsonic. This subsonic velocity is then reduced to zero at the tube stagnation point (see Fig. 7.5b). Analysis[1] gives the following formula for computing the free-stream Mach number and thus the velocity:

$$\frac{p_{\text{stag}}}{p_{\text{stat}}} = N_M^2 \left(\frac{k+1}{2} \right)^{k/(k-1)} \left[\frac{2kN_M - k + 1}{N_M^2(k+1)} \right]^{1 - 1/(k-1)} \tag{7.10}$$

The measurement of stagnation and static pressures may be combined in a single probe, as in Fig. 7.1, or two separate probes, one for stagnation and the other for static, may be employed. Figure 7.6 shows several examples[2] of commonly used forms. The wedge static-pressure probe of Fig. 7.6a also can be used to measure velocity direction in a single plane. When the two static taps read equal pressures, the wedge is aligned with the flow. This probe is usable for both subsonic and supersonic flow. At Mach 0.9 the sensitivity to misalignment is about 1.5 in of water per angular degree. The total (stagnation) probes of Fig. 7.6b and c also are intended for both sub- and supersonic flow. The simple tube is insensitive to misalignment up to about ±20° while the venturi shielded tube is good to ±50°. The boundary-layer probe is usable up to Mach 1.0 and is insensitive to misalignment up to ±5°. Boundary-layer thickness can be measured with such a probe with an error of the order of 0.002 in. The probe and

[1] Ibid., p. 52.
[2] Aero Research Instrument Co., Chicago.

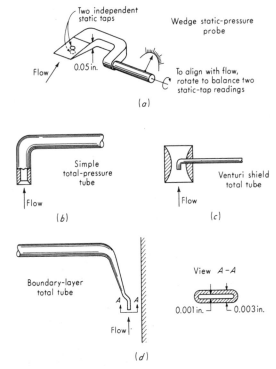

Figure 7.6 Pressure probes.

associated pressure-measuring equipment have a long time lag because of the small flow passage (0.001 in) at the probe tips.

At low velocities, where the pitot-static tube becomes insensitive, a larger Δp signal may be obtained by using a version called a "boost venturi" (Fig. 7.7). Here p_{stag} is the normal value; the measured p_{stat} is *not* the free-stream value, but instead is artificially lowered by locally accelerating the flow in a venturi section. Figure 7.7a and b shows two different designs while Fig. 7.7c combines them to get the maximum effect, about a 25 : 1 increase in Δp, compared with the value for a normal pitot-static tube. These instruments require some care in use since they must be individually calibrated and are sensitive to variations in Mach number and/or Reynolds number.

Velocity Direction from Yaw Tube, Pivoted Vane, and Servoed Sphere

In addition to laboratory studies of flow processes in fluid machinery, ducting, etc., flow-velocity direction information is of interest in flight vehicles[1] where

---

[1] H. H. Koelle, "Handbook of Astronautical Engineering," p. 13–33, McGraw-Hill, New York, 1961.

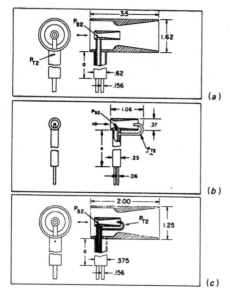

(a)

(b)

Figure 7.7 Boost-venturi pitot tubes. (*Courtesy United Sensor Control Corp., Watertown, Mass.*)

(c)

angle-of-attack measurements are utilized in attitude measurement and control, stability augmentation, and gust alleviation systems.

So-called yaw tubes[1] of one form or another conventionally are employed to determine the direction of local flow velocity. Perhaps the simplest form, useful for finding the angular inclination in one plane only, is shown in Fig. 7.8a. Taps 1 and 3 are connected to a differential-pressure instrument that reads zero when the tube is aligned with the flow. A central tap 2 is often included to read the stagnation pressure after alignment is attained (valid only if the angle of attack is zero). The claw tube of Fig. 7.8b operates on similar principles, but may be utilized in regions where the flow direction changes greatly, since its sensing holes may be located very close together. The two-axis probes of Fig. 7.8c and d conceivably could be designed to allow rotation about each axis; however, the complexity and size of such a design are generally prohibitive. Thus probe operation consists of rotation about the probe axis to balance taps 1 and 3. Then pressures 2 and 4 are each measured, and calibration charts give the angle of attack. Tap 5 does not read stagnation pressure directly; this can be obtained from calibration charts. Any of these probes may be made automatically self-aligning by using the pressure difference $p_1 - p_3$ as the error signal in a servosystem which rotates the probe until a null is achieved. The details of a system of this type are shown in Fig. 7.10.

Determination of angles of attack and yaw aboard flight vehicles is often

[1] Aero Research Instrument Co., Chicago.

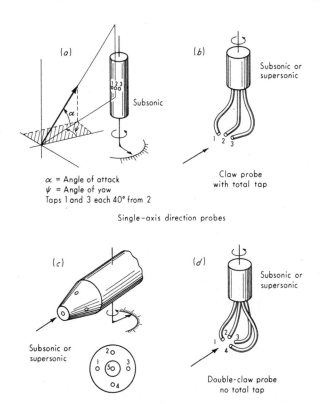

α = Angle of attack
ψ = Angle of yaw
Taps 1 and 3 each 40° from 2

Single–axis direction probes

Two–axis direction probes

Figure 7.8 Flow-direction probes.

accomplished with vane-type probes as in Fig. 7.9. These devices are essentially one- or two-axis weather vanes with suitable damping to reduce oscillation and with motion pickups to provide electrical angle signals. A dynamic analysis of these ·devices is available.[1] Limitations of this type of device for certain high-speed, high-altitude applications have led to the development of the servoed-sphere type of sensor shown in Fig. 7.10. The one shown was developed[2] for the X-15 rocket research aircraft. A servo-driven sphere is continuously and auto-matically aligned with the velocity vector by means of two independent servo-systems using the differential-pressure signals $p_1 - p_2$ and $p_3 - p_4$ as error sig-

[1] G. J. Friedman, Frequency Response Analysis of the Vane-Type Angle of Attack Transducer, *Aero/Space Eng.*, p. 69, March 1959; P. S. Barna and G. R. Crossman, Experimental Studies of Flow Direction Sensing Vanes, *NASA CR*-2683, May 1976.
[2] Northrop Corp., Nortronics Div., *Rept.* NORT60-46.

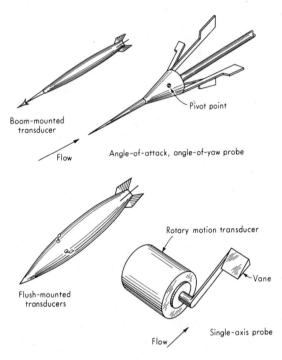

Boom-mounted transducer

Flow

Pivot point

Angle-of-attack, angle-of-yaw probe

Rotary motion transducer

Vane

Flush-mounted transducers

Flow

Single-axis probe

Figure 7.9 Vane-type probes.

nals. A fifth tap measures the stagnation pressure. Block diagrams and frequency response of a single axis are given in Fig. 7.10. The angle-of-attack axis is designed for the range -10 to $+40°$ while the sideslip axis covers $\pm 20°$. Static inaccuracy of angle measurement is 0.25°.

A five-tap sensor similar to that of Fig. 7.8c but requiring no rotation is available for aircraft applications (see Fig. 7.11[1]). The installation shown combines the five-tap angle sensor, integral pitot-static taps (P_{p2}, P_{s2}), and a separate side-mounted pitot-static tube for pilot instruments (P_{p1}, P_{s1}). Angle of attack α is related to pressure difference $P_{\alpha 1} - P_{\alpha 2}$; however, the relation varies with altitude and airspeed. Most of this variation can be accounted for by dividing $P_{\alpha 1} - P_{\alpha 2}$ by a normalizing factor $K_1[P_3 - P_{\beta 1} + (P_{\beta 1} - P_{\beta 2})/2]$, this quotient now being closely proportional to α. The "constant" K_1 still varies somewhat with Mach number (Fig. 7.11c); however, this correction can be included in the system air-data computer. Sideslip angle is measured by an identical scheme. Absolute- and differential-pressure transducers of the capacitance type are employed to perform the desired subtractions. Overall systems of this type, using aerodynamic compensation methods to optimize sensor design for a particular

---

[1] *Bull.* 1013, 1014, Rosemount Inc., Minneapolis, Minn.

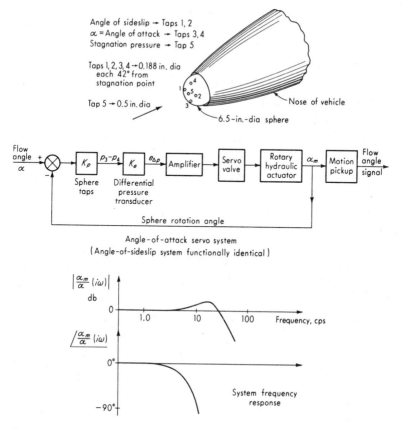

Figure 7.10 Servoed-sphere probe.

aircraft, can achieve errors as small as 0.25° for flow angles up to ±40° and wide ranges of subsonic and supersonic flight speeds. Additional information on the general class of five-tap sensors is available.[1]

Dynamic Wind-Vector Indicator

Figure 7.12 shows a transducer that measures the magnitude and direction of flow velocity in terms of the drag force exerted on a hollow sphere. The drag

---

[1] M. R. Hale, The Analysis and Calibration of the Five-Hole Spherical Pitot, *ASME Paper* 67-WA/FE-24, 1967; T. J. Dudzinski and L. N. Krause, Flow-Direction Measurement with Fixed-Position Probes, *NASA* TM X-1904, October 1969.

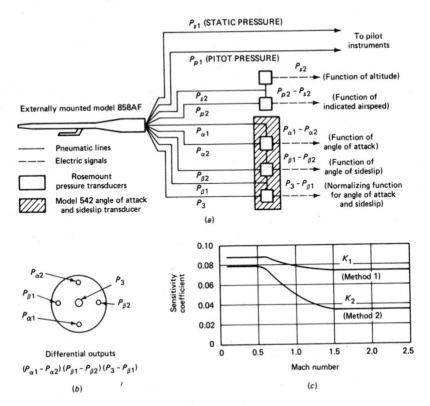

Figure 7.11 Five-tap pitot tube for aircraft instrumentation.

force F_d on a body is given by

$$F_d = C_d \frac{A\rho V^2}{2} \tag{7.11}$$

where $\qquad C_d \triangleq$ drag coefficient of body
$= 0.567$ for these transducers
$A \triangleq$ projected area of body

Clearly, if C_d, A, and ρ are known, then V may be found by measuring F_d. If all directional components of V are to be equally effective in producing drag force, a body with spherical symmetry must be employed. If this is done, measurement of the x, y, and z components of F_d completely defines the magnitude and direction of V. Since the drag coefficient of a smooth sphere is somewhat dependent on the Reynolds number (and thus V), wire roughening rings are attached to the sphere surface to ensure turbulence. The drag coefficient of the roughened sphere is

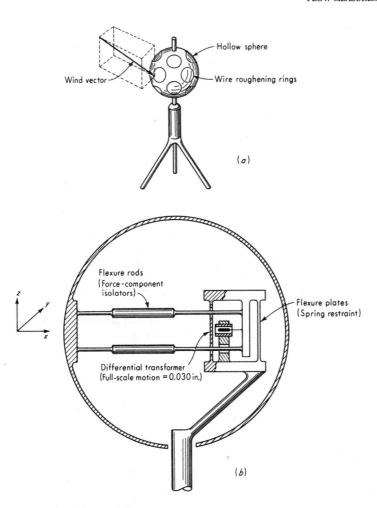

Figure 7.12 Wind-vector indicator.

constant over the entire design range of V for a given transducer. The separation of the total drag force into three rectangular components is accomplished by mounting the sphere on a force-resolving flexure assembly. Figure 7.12*b* shows the *x*-axis mechanism of this structure. (The *y* and *z* axes are identical except for their orientation.) The force components are applied to flexure-plate springs to produce proportional motions which are then measured by suitable displacement transducers. Sensors of this type have been used for both wind and ocean current measurement.

A miniaturized drag probe[1] for one-dimensional air velocity measurements, using a flat plate target 1.5×1.5 mm mounted on a commercially available semiconductor strain-gage beam, exhibited frequency response to 10 kHz. A two-dimensional version[2] utilized a hollow spherical glass target of 0.153-in diameter mounted on the stylus of a stereo phonograph pickup. Response to 26 kHz was estimated theoretically for 700 ft/s air flow.

Hot-Wire and Hot-Film Anemometers

Hot-wire anemometers commonly are made in two basic forms: the constant-current type and the constant-temperature type. Both utilize the same physical principle but in different ways. In the constant-current type, a fine resistance wire carrying a fixed current is exposed to the flow velocity. The wire attains an equilibrium temperature when the i^2R heat generated in it is just balanced by the convective heat loss from its surface. The circuit is designed so that the i^2R heat is essentially constant; thus the wire temperature must adjust itself to change the convective loss until equilibrium is reached. Since the convection film coefficient is a function of flow velocity, the equilibrium wire temperature is a measure of velocity. The wire temperature can be measured in terms of its electrical resistance. In the constant-temperature form, the current through the wire is adjusted to keep the wire temperature (as measured by its resistance) constant. The current required to do this then becomes a measure of flow velocity.

For equilibrium conditions we can write an energy balance for a hot wire as

$$I^2R_w = hA(T_w - T_f) \tag{7.12}$$

where $I \triangleq$ wire current
$R_w \triangleq$ wire resistance
$T_w \triangleq$ wire temperature
$T_f \triangleq$ temperature of flowing fluid
$h \triangleq$ film coefficient of heat transfer
$A \triangleq$ heat-transfer area

Now h is mainly a function of flow velocity for a given fluid density. For a range of velocities, this function (sometimes called *King's law*) has the general form,

$$h = C_0 + C_1\sqrt{V} \tag{7.13}$$

For the measurement of average (steady) velocities a "manual balance" constant-temperature mode of operation can be used. Figure 7.13 shows a possible circuit arrangement. For accurate work, a given hot-wire probe must be calibrated in the fluid in which it is to be used. That is, it is exposed to *known* velocities (measured accurately by some other means), and its output is recorded

---

[1] L. N. Krause and G. C. Fralick, Miniature Drag-Force Anemometer, *NASA* TM-3507, June 1977.

[2] D. Y. Cheng and P. Wang, Viscous Force Sensing Fluctuating Probe Technique, *AIAA Paper* 73-1044, 1973.

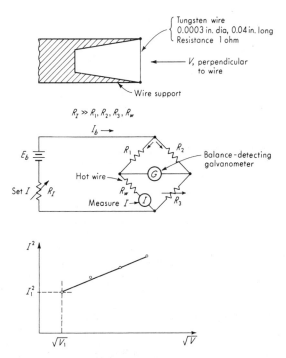

Figure 7.13 Hot-wire anemometer.

over a range of velocities. When velocities are too low to allow direct use of a pitot-static tube as the calibration standard, a number of alternative schemes are possible. By using a flow passage with two sections of widely differing areas connected in series, a pitot tube may be employed in the small-area (high-velocity) section, the hot wire in the large-area (low-velocity) section, and the known area ratio used to infer velocities at the hot wire from those measured at the pitot tube. Another approach[1] utilizes a choked-flow orifice (to establish a mass flow proportional to supply pressure) together with a flow passage with two areas of known ratio, providing a free jet whose velocity is closely proportional to supply pressure. For the lowest velocities, a rig which moves the probe through still fluid (a rotating arm[2] is perhaps most convenient) substitutes the (easy) measurement of solid-body velocity for the more difficult fluid-velocity measurement.

In the circuit of Fig. 7.13, the current through R_w stays essentially constant

[1] New Anemometer Calibration Equipment, *DISA Inform.*, no. 13, pp. 37–39, Disa Electronics, Franklin Lakes, N.J., May 1972.

[2] J. Anhalt, Device for In-Water Calibration of Hot-Wire and Hot-Film Probes, *DISA Inform.*, no. 15, pp. 25–26, Disa Electronics, Franklin Lakes, N.J., October 1973.

even when R_w changes because R_I is of the order of 2,000 Ω while R_1, R_2, R_3, and R_w are much less, of the order of 1 to 20 Ω. In calibration, V is set at some known value V_1. Then R_I is adjusted to set hot-wire current I at a value low enough to prevent wire burnout but high enough to give adequate sensitivity to velocity. The resistance R_w will come to a definite temperature and resistance. Then the resistor R_3 is adjusted to balance the bridge. This adjustment is essentially a measurement of wire temperature, which is held fixed at all velocities. The first point on the calibration curve is thus plotted as I_1^2, $\sqrt{V_1}$. Now V is changed to a new value, causing wire temperature and R_w to change and thus unbalancing the bridge. Then R_w, and thus wire temperature, is restored to its original value by adjusting I (by means of R_I) until bridge balance is restored (R_3 is *not* changed). The new current I and the corresponding V may be plotted on the calibration curve, and this procedure is repeated for as many velocities as desired.

Once calibrated, the probe can be employed to measure unknown velocities by adjusting R_I until bridge balance is achieved, reading I, and obtaining the corresponding V from the calibration curve. This assumes that the measured fluid is at the same temperature and pressure as for the calibration. Correction methods for varying temperature and pressure are fairly simple, but are not discussed here. For the above constant-temperature mode of operation, Eqs. (7.12) and (7.13) can be combined to give

$$I^2 = \frac{A(T_w - T_f)(C_0 + C_1\sqrt{V})}{R_w} \triangleq C_2 + C_3\sqrt{V} \tag{7.14}$$

indicating that the calibration curve of Fig. 7.13 should be essentially a straight line. This is borne out by experimental tests.

While the above described measurement of steady velocities is of some practical interest, perhaps the main application of hot-wire instruments is the measurement of rapidly fluctuating velocities, such as the turbulent components superimposed on the average velocity. Both constant-current and constant-temperature techniques are used; first we consider the constant-current operation. Figure 7.14 shows the basic arrangement. Again the current can be assumed constant at a value I even if R_w changes, since $R_I \gg R_w$. Let us suppose the velocity is constant at a value V_0. This will cause R_w to assume a constant value, say R_{w0}, and a voltage IR_{w0} will appear across R_w. Now, we let the velocity V fluctuate about the value V_0 so that $V = V_0 + v$, where v is the fluctuating component. This will result in R_w varying so that $R_w = R_{w0} + r_w$, where r_w is the varying component. Now, during a time interval dt, we may write for the wire (neglecting conduction and radiation effects)

Electrical energy generated − energy lost by convention

= energy stored in wire   (7.15)

The energy lost by convection is given by

$$A(T_w - T_f)(C_0 + C_1\sqrt{V})\,dt \tag{7.16}$$

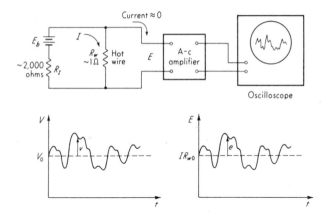

Figure 7.14 Velocity-fluctuation measurement.

while the wire temperature T_w may be related to its resistance by

$$T_w = K_{tr}(R_{w0} + r_w) \qquad (7.17)$$

where K_{tr} is the reciprocal of a temperature coefficient of resistance. The term $C_0 + C_1 \sqrt{V}$ may be approximately linearized for small changes in V with good accuracy as follows:

$$f(V) = C_0 + C_1 \sqrt{V} \approx (C_0 + C_1 \sqrt{V_0}) + \left.\frac{\partial f}{\partial V}\right|_{V=V_0} (V - V_0) \qquad (7.18)$$

$$C_0 + C_1 \sqrt{V} \approx (C_0 + C_1 \sqrt{V_0}) + K_v v \qquad (7.19)$$

Equation (7.15) then becomes

$$I^2(R_{w0} + r_w)\,dt - A(T_w - T_f)(C_0 + C_1 \sqrt{V_0} + K_v v)\,dt = MC\,dT_w \qquad (7.20)$$

where $M \triangleq$ mass of wire and $C \triangleq$ specific heat of wire.

Now,

$$I^2 R_{w0} + I^2 r_w - A[K_{tr}(R_{w0} + r_w) - T_f](C_0 + C_1 \sqrt{V_0}$$
$$+ K_v v) = MCK_{tr}\frac{dr_w}{dt} \qquad (7.21)$$

and since $\qquad I^2 R_{w0} - A(K_{tr}R_{w0} - T_f)(C_0 + C_1 \sqrt{V_0}) = 0 \qquad (7.22)$

because this represents the initial equilibrium state, we get

$$I^2 r_w - AK_{tr}r_w(C_0 + C_1 \sqrt{V_0}) - AK_{tr}r_w K_v v$$
$$- A(K_{tr}R_{w0} - T_f)K_v v = MCK_{tr}\frac{dr_w}{dt} \qquad (7.23)$$

The term $AK_{tr}K_v r_w v$ may be neglected relative to the other terms since it contains the product $r_w v$ of two small quantities. Now the voltage across R_w is $IR_w = I(R_{w0} + r_w)$. The fluctuating component of this is Ir_w, which we call e. Equation (7.23) then leads to

$$\frac{e}{v}(D) = \frac{K}{\tau D + 1} \tag{7.24}$$

where

$$K \triangleq \frac{-K_v AI(K_{tr} R_{w0} - T_f)}{K_{tr} A(C_0 + C_1 \sqrt{V_0}) - I^2} \qquad \text{V/(m/s)} \tag{7.25}$$

$$\tau \triangleq \frac{MCK_{tr}}{K_{tr} A(C_0 + C_1 \sqrt{V_0}) - I^2} \qquad \text{s} \tag{7.26}$$

We see that the voltage follows the flow velocity with a first-order lag. The time constant τ cannot be reduced much below 0.001 s in actual practice, which would limit the flat frequency response to less than 160 Hz. This is quite inadequate for turbulence studies since frequencies of 50,000 Hz and more are of interest. This limitation is overcome by use of electrical dynamic compensation (see Chap. 10 for a general discussion). Circuits whose frequency response just makes up the deficiency in the hot wire itself are employed, as in Fig. 7.15. The overall system then has a flat response to almost 100,000 Hz. The main difficulty in applying this technique is that the correct compensation depends on τ, whose value is not known and varies with flow conditions. The next paragraph explains a method of overcoming this difficulty.

The basic idea of the scheme is to force a square-wave current through the hot wire while it is exposed to the flow to be studied (see Fig. 7.16). We show that the output-voltage response to this *current* signal has exactly the same time constant as the response to the flow-*velocity* signal. Thus if the compensation can be adjusted to be correct for the current signal, it will be correct for the velocity input also. The "correctness" of the adjustment may be judged by the degree to which the output voltage corresponds to a square wave. In the circuit of Fig. 7.16, a good approximation to linear behavior may be expected for small input signals (current or velocity); thus the superposition principle will apply, and the effects of current and velocity inputs may be considered separately. If the square-wave current is turned off, R_I adjusted to give the desired hot-wire current I_0, and R_b adjusted to balance the bridge, then $R_a/R_{w0} = R_r/R_b$ and the voltage $E_{B1, B2} = 0$. Now we let the square-wave current i_1 be turned on, causing a current i, which we calculate to be $i_1(R_a + R_r)/(R_a + R_r + R_w + R_b)$, to flow through R_w and a current $i_2 = i_1(R_w + R_b)/(R_a + R_r + R_w + R_b)$ to flow through R_a and R_r. Equation (7.15) may be applied to this situation to give

$$(I_0 + i)^2(R_{w0} + r_w) - A[K_{tr}(R_{w0} + r_w) - T_f](C_0 + C_1 \sqrt{V_0}) = MCK_{tr}\frac{dr_w}{dt} \tag{7.27}$$

Now, since $I_0^2 R_{w0} = A(K_{tr} R_{w0} - T_f)(C_0 + C_1 \sqrt{V_0})$ and because we neglect

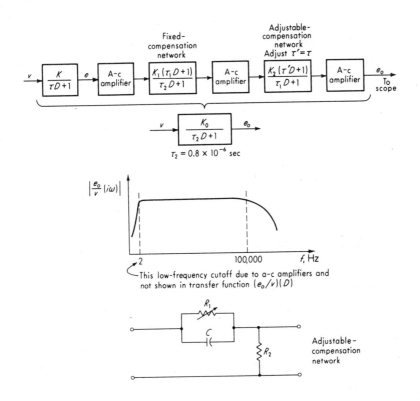

Figure 7.15 Dynamic compensation.

$2I_0 ir_w$ and $i^2(R_{w0} + r_w)$, since they are products of small quantities, Eq. (7.27) reduces to

$$\frac{r_w}{i}(D) = \frac{K_i}{\tau D + 1} \tag{7.28}$$

where

$$\tau \triangleq \frac{MCK_{tr}}{K_{tr}A(C_0 + C_1\sqrt{V_0}) - I_0^2} \quad \text{s} \tag{7.29}$$

$$K_i \triangleq \frac{2I_0 R_{\omega 0}}{K_{tr}A(C_0 + C_1\sqrt{V_0}) - I_0^2} \quad \Omega/\text{A} \tag{7.30}$$

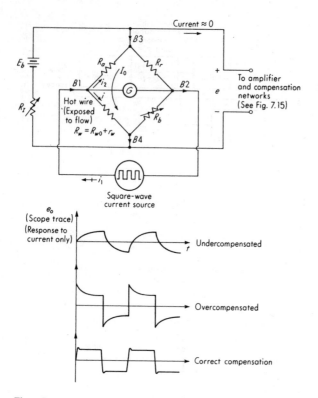

Figure 7.16 Compensation adjustment scheme.

Let us now calculate e, the varying component of the voltage appearing across $B3$ and $B4$, which will be the input to the amplifiers and compensating networks:

$$e = - R_a i_2 + (R_{w0} + r_w)i + I_0 r_w \approx - R_a i_2 + R_{w0} i + I_0 r_w \qquad i r_w \approx 0 \quad (7.31)$$

$$e = - R_a \frac{R_w + R_b}{R_a + R_r} i + R_{w0} i + I_0 r_w$$

$$= \frac{- R_a R_w - R_a R_b + R_{w0} R_a + R_{w0} R_r}{R_a + R_r} i + I_0 r_w \qquad (7.32)$$

Now $R_{w0} R_a \approx R_w R_a$ and $R_{w0} R_r = R_a R_b$ (balanced-bridge relation); thus

$$e \approx I_0 r_w \qquad (7.33)$$

Thus, finally, $$\frac{e}{i}(D) = \frac{K_e}{\tau D + 1} \qquad (7.34)$$

where $$K_e \triangleq I_0 K_i \qquad (7.35)$$

We see now that the response of the voltage e to impressed current signals has the identical time constant τ as the response to flow-velocity signals. Thus the compensating networks may be adjusted to optimize the response to current inputs and ensure optimum response for flow-velocity inputs. Since this adjustment is made while the probe is exposed to flow, the output will contain a superposition of current response and velocity response, resulting in a sometimes confusing picture, rather than the simple waveforms of Fig. 7.16. Usually, however, the compensation adjustment can be made satisfactorily.

The operation of the constant-temperature type of instrument for steady velocities is explained earlier in relation to Fig. 7.13. This mode of operation can be extended to measure both average and fluctuation components of velocity by making the bridge-balancing operation automatic, rather than manual, through the agency of a feedback arrangement. The advantages of this scheme are such that most instruments now employ it. A simplified functional schematic of such a system is shown in Fig. 7.17. With zero flow velocity and the bridge excitation shut off ($i_w = 0$), the hot wire assumes the fluid temperature. Then the variable resistor R_3 is adjusted manually so that $R_3 > R_w$, thereby unbalancing the bridge. When the excitation current is turned on, the unbalanced bridge produces

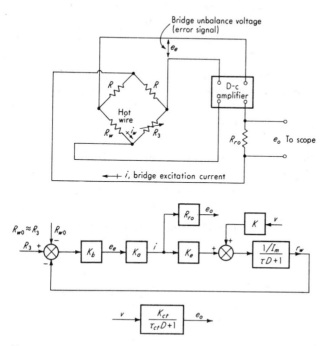

Figure 7.17 Constant-temperature anemometer.

an unbalance voltage e_e, which is applied to the input of a high-gain current amplifier supplying the bridge excitation current. The current now flowing through R_w increases its temperature and thus its resistance. As R_w increases, it approaches R_3 and the bridge-unbalance voltage e_e decreases. If the current amplifier had an infinite gain, the bridge current required to heat R_w to match R_3 precisely could be produced with an infinitesimally small error voltage e_e, and thus perfect bridge balance would be attained automatically. The current i (which is a measure of flow velocity V) produces an output voltage in passing through the readout resistor R_{ro}. Since an actual amplifier has finite gain, the bridge-unbalance voltage cannot go to zero; but it can be extremely small. This also means that R_w will be very close, but not exactly equal, to R_3.

The above-described self-adjusting equilibrium will come about at *any* steady-flow velocity, not just zero, with the equilibrium current in each case being a measure of velocity. If the velocity is fluctuating rapidly, perfect continuous bridge balance requires an infinite-gain amplifier. Lacking this, practical systems exhibit some lag between velocity and current. System dynamic response for small velocity fluctuations can be obtained by superimposing previously obtained results for response to velocity (wire-current constant) and response to current (flow-velocity constant), since in the constant-temperature mode both current and velocity are changing simultaneously. The effect of velocity on r_w (the varying component of R_w) is obtained from Eq. (7.24) as

$$\frac{r_w}{v}(D) = \frac{K/I_m}{\tau D + 1} \tag{7.36}$$

where I_m is the constant current about which fluctuations take place. From Eq. (7.34), the effect of current on r_w is given by

$$\frac{r_w}{i}(D) = \frac{K_e/I_m}{\tau D + 1} \tag{7.37}$$

The total effect of i and v on r_w is then, by superposition,

$$(\tau D + 1)r_w = \frac{Kv + K_e i}{I_m} \tag{7.38}$$

Now a change in r_w causes a bridge-unbalance voltage change according to

$$e_e = \frac{I_m R}{R + R_{w0}} r_w \triangleq -K_b r_w \tag{7.39}$$

We assume that the amplifier has no lag and produces output current proportional to input voltage as given by

$$i = K_a e_e \tag{7.40}$$

The block diagram of Fig. 7.17 embodies the above relations, which may now be manipulated to give the relation between input v and output e_o :

$$\left(\frac{K_e e_o}{R_{ro}} + Kv\right) \frac{(1/I_m)(-K_b K_a R_{ro})}{\tau D + 1} = e_o \tag{7.41}$$

This gives finally

$$\frac{e_o}{v}(D) = \frac{K_{ct}}{\tau_{ct} D + 1} \tag{7.42}$$

where

$$\tau_{ct} \triangleq \frac{\tau}{1 + K_e K_b K_a / I_m} \tag{7.43}$$

and

$$K_{ct} \triangleq \frac{-KK_b K_a R_{ro}/I_m}{1 + K_e K_b K_a / I_m} \tag{7.44}$$

Note that τ_{ct}, the time constant of the constant-temperature anemometer system, is always less than τ (the time constant of the wire itself) and in actual practice is *much* less, since a very high value of amplifier gain K_a is used. A more comprehensive analysis[1] of the system, which models the amplifier as a second-order dynamic system and adds a capacitor (used to "tune" system for optimum response) in parallel with R, leads to a third-order closed-loop differential equation. This allows the usual Routh-criterion analysis for stability and allowable loop gain. As in all feedback systems, too high a loop gain will cause instability; however, careful design allows sufficiently high gain to make τ_{ct} of the order of $\frac{1}{100}$ of τ or less. A typical instrument has flat (within 3 dB) frequency response to 17,000 Hz when the average flow velocity is 30 ft/s, 30,000 Hz for 100 ft/s, and 50,000 Hz at 300 ft/s.

The stated preference for the feedback-type constant-temperature instrument is based on a number of considerations. In the constant-current type, the current must be set high enough to heat the wire considerably above the fluid temperature for a given average velocity. If the flow suddenly drops to a much lower velocity or comes to rest, the hot wire will burn out since the convection loss cannot match the heat generation before the wire temperature reaches the melting point. The constant-temperature type does not have this drawback because the feedback system *automatically* sets wire current to maintain the desired (safe) wire temperature for every velocity.

A further advantage of the constant-temperature method lies in the nature of the dynamic compensation. In the constant-current method, the compensating network must be reset (using the square-wave current) whenever the average

---

[1] P. Freymuth, Feedback Control Theory for Constant-Temperature Hot-Wire Anemometers, *Rev. Sci. Instrum.*, vol. 38, no. 5, pp. 677–681, May 1967; J. A. Borgos, A Review of Electrical Testing of Hot-Wire and Hot-Film Anemometers, *TSI Quart.*, TSI Inc., St. Paul, Minn., August/September 1980.

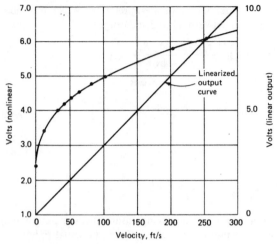

Calibration for a 0.002-in (0.051-mm) diameter hot-film sensor in atmospheric air; 0 − 300 ft/s (0 − 91m/s)

Figure 7.18 Nonlinear and linearized responses of hot-film sensor.

velocity changes appreciably. Furthermore, if velocity fluctuations about the average are large (more than, say, 5 percent of the average velocity), the dynamic compensation will not be complete since the value of τ varies with V and thus the compensating network time constant τ' should be continuously and instantaneously varied, which it is not. The feedback arrangement of the constant-temperature system provides more nearly correct compensation for large velocity fluctuations. While large velocity changes result in nonlinear response in both types of instrument, the nonlinearity of the constant-temperature feedback system appears algebraically on the output signal, where it is easily linearized with an external electronic linearizer. Usually these linearizers utilize a polynomial function through the fourth power, with the coefficient of each power individually adjustable to suit the probe and flow conditions. These units are fairly simple to adjust and are in wide use, but no similar apparatus has been devised for the constant-current type. Figure 7.18[1] shows system output voltage, both nonlinear and linearized, for a specific instrument, making clear the variation of sensitivity (high at low velocities, low at high velocities) typical of the basic thermal anemometer principle.

The hot-wire anemometer may be employed to measure the direction of the average flow velocity in several ways.[2] It has been found that a single wire, as in

[1] Hot Film and Hot Wire Anemometry: Theory and Application, *Bull.* TB5, TSI Inc., St. Paul, Minn.

[2] H. H. Lowell, Design and Applications of Hot-Wire Anemometers for Steady-State Measurements at Transonic and Supersonic Speeds, *NACA, Tech. Note* 2117, 1950.

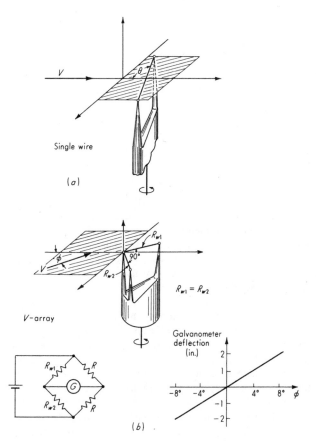

Figure 7.19 Flow-direction measurement.

Fig. 7.19a, responds essentially to the component of velocity perpendicular to it, if the angle between wire and velocity vector is between 90 and about 25°. For this range, then, the V in our derivations may be replaced by $V \sin \theta$. (For $\theta < 25°$ the heat loss is greater than predicted by $V \sin \theta$; for $\theta = 0$ it is about 55 percent of that for $\theta = 90°$.) With the arrangement of Fig. 7.13 and a rotatable probe as in Fig. 7.19a, the flow-direction angle (in a single plane) could be found by determining the probe-rotation angle which gives a maximum value of I. This method is quite inaccurate, however, since $\sin \theta$ changes very slowly with θ when θ is near 90°. A better procedure, if the flow angle is roughly known, is as follows: The wire is set at about 50° from the flow direction, and I is measured. Then the probe is rotated in the opposite direction until an angle is found at which the same I is measured as before. The bisector of the angle between the two locations then determines the flow direction. This method is more accurate since the rate of

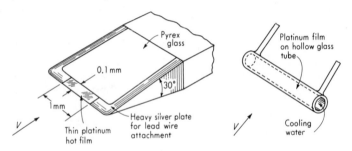

Figure 7.20 Hot-film sensors.

change of I with θ is a maximum near 50°. Even greater convenience and accuracy are achieved by use of a so-called V array of hot wires, as in Fig. 7.19b. The two hot wires R_{w1} and R_{w2} are assumed identical and form a V with an included angle, which is typically 90°. They are connected into a bridge as shown. When the probe is rotated, a bridge null occurs when the flow-velocity vector is aligned with the bisector of the V. In a specific case,[1] the sensitivity of this arrangement was sufficient to determine velocity direction within about 0.5°.

Practical problems in the application of hot-wire anemometers are found in the limited strength of the fine wires and the calibration changes caused by dirt accumulations. Unless the flow is very clean, significant calibration changes can occur in a relatively few minutes of operation. Larger dirt particles striking wires may actually break them. At high speeds, wires may vibrate because of aerodynamic loads and flutter effects. Applications are mainly in gas flows; however, special probes have been developed and successful results obtained in both electrically conducting and nonconducting liquids, such as water and oil. In compressible flows, the film coefficient h is really related to ρV; thus calibrations such as that of Fig. 7.13 must be adjusted when a probe calibrated at one density is employed at another. Further complications arise when fluid temperature varies, either as a steady change from calibration conditions or dynamically during measurement; however, various compensation techniques[2] are available.

A variation of the hot-wire anemometer intended to extend its utility is the hot-film transducer. Here the resistance element is a thin film of platinum deposited on a glass base. The film takes the place of the hot wire; the required circuitry is basically similar to that used in constant-temperature hot-wire approach. The film transducers have great mechanical strength and may be used at very high temperatures by constructing them with internal cooling-water passages. Various configurations of sensors are possible; Fig. 7.20 shows two possibilities.

[1] Ibid.
[2] Temperature Compensation of Thermal Sensors, *Tech. Bull.* 16, TSI Inc., St. Paul, Minn.; R. E. Drubka, J. Tan-atichat, and H. M. Nagib, Analysis of Temperature Compensating Circuits for Hot-Wires and Hot-Films, *DISA Inform.*, no. 22, pp. 5–14, Disa Electronics, Franklin Lakes, N.J., December 1977.

In addition to measurement of velocity magnitude and direction, hot-wire and hot-film instruments may be adapted to measurements of fluid temperatures, turbulent shear stresses, and concentrations of individual gases in mixtures.[1] By using a probe with two wires or films in an X configuration and two channels of electronics, velocity fluctuations in two perpendicular directions and the directional correlation (Reynolds stress) can be measured. For three-dimensional flows where flow direction is completely unknown, a system[2] employing six film sensors on three mutually perpendicular rods obtains velocity magnitude and direction in all eight octants with no ambiguity.

While the author is unaware of any texts devoted entirely to thermal anemometry, a voluminous literature of papers exists and has been documented in an excellent bibliography[3] organized to facilitate the reader's search for any particular topic.

Hot-Film Shock-Tube Velocity Sensors

In shock-tube experiments, the propagation velocity of the shock wave down the tube often must be measured. Of the various means available for making this measurement, hot-film temperature sensors are in wide use. The passage of the shock wave past a particular section of the tube is accompanied by a step change in gas temperature. By locating thin resistance films flush with the inside of the tube, the instant of wavefront passage may be detected as a temperature (and therefore resistance) change. If two such film sensors are mounted a known distance apart, the average wave velocity may be computed from the time interval between the two sensor responses. The films are operated at constant current by the simple circuit of Fig. 7.21, and a differentiating circuit is employed to sharpen the pulses for greater timing accuracy. With systems of this type,[4] shock-wave velocities have been measured with an accuracy of 1 percent for shock Mach numbers as high as 7.5 to 10.

Laser Doppler Velocimeter[5]

In the area of flow measurement, the most important recent advance is probably the *laser doppler velocimeter*. This instrument measures local flow velocity and is thus in competition with pitot tubes and hot-wire anemometers. Its main advan-

---

[1] S. Corrsin, Extended Applications of the Hot-Wire Anemometer, *NACA, Tech. Note* 1864, 1949.

[2] System 1080, TSI Inc., St. Paul, Minn.

[3] P. Freymuth, A Bibliography of Thermal Anemometry, *TSI Quart.*, TSI Inc., St. Paul, Minn., vol. 4, no. 4, November/December 1978; vol. 5, no. 1, February/March 1979; vol. 5, no. 4, November/December 1979.

[4] Shock Tubes, Handbook of Supersonic Aerodynamics, sec. 18 *NAVORD Rept.* 1488, vol. 6, pp. 543, 558.

[5] L. E. Drain, "The Laser Doppler Technique," Wiley-Interscience, New York, 1980; T. S. Durrani and C. A. Greated, "Laser Systems in Flow Measurement," Plenum, New York, 1976; H. D. Thompson and W. H. Stevenson, "Laser Velocimetry and Particle Sizing," Hemisphere, New York, 1979; F. Durst, A. Melling, and J. H. Whitelaw, "Principles and Practice of Laser Anemometry," Academic, New York, 1976.

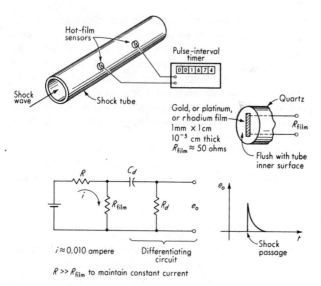

Figure 7.21 Thin-film velocity sensor.

tages relative to these more well-established techniques include:

1. Measurement of velocity is direct rather than by inference from pressure (pitot tube) or heat-transfer coefficient (hot-wire).
2. No "physical object" need be inserted into the flow; thus flow is undisturbed by measurement.
3. Sensing volume can be very small (a cube 0.2 mm on a side is not unusual).
4. Very-high-frequency response (to megahertz range) is possible.

Disadvantages involve the need for transparent flow channels, the necessity for tracer particles in the fluid, and the cost and complexity of the apparatus. In brief, the operating principle involves focusing laser beams at the point where velocity is to be measured and then sensing with a photodetector the light scattered by tiny particles carried along with the fluid as it passes through the laser focal point. The velocity of the particles (assumed to be identical to the fluid velocity) causes a doppler shift of the scattered light's frequency and produces a photodetector signal directly related to velocity.

Actually, *artificial* tracer particles are not always necessary; the microscopic particles normally present in liquids may suffice; however, gas flows often need to be seeded. Under extreme conditions, particles may not perfectly follow the flow, but studies have shown highly accurate following in many practical cases. A simple analysis[1] accurate for many gas flows, which models the particle as a

---

[1] Drain, op. cit., p. 183.

spherical body of radius r and density ρ immersed in a fluid of viscosity μ, uses Stokes' law for the viscous force to obtain the differential equation

$$\frac{4}{3} \pi r^3 \rho \frac{dv_p}{dt} = 6\pi\mu r(v_f - v_p) \tag{7.45}$$

This gives a first-order dynamic response with time constant $\tau \triangleq 2\rho r^2/(9\mu)$ relating particle velocity v_p to fluid velocity v_f. For water droplets of (typical) 1-μm diameter in air, we get $\tau \approx 3$ μs, which allows flat (within 1 percent) frequency response to about 7,400 Hz. More correct (and complex) models for particle motion, needed when Eq. (7.45) becomes inaccurate, are available.[1]

Laser doppler velocimeters (LDVs) have been operated in several different configurations; Figs. 7.22 to 7.24[2] give some details on the popular dual-beam (differential Doppler) or "fringe" mode. While a careful physical explanation of the device rests on a detailed study of the doppler-shift effect, many workers in the field find the interference-fringe explanation of Fig. 7.23 useful. Coated optical flats (2a, 2b) cause laser beam 1 to be split into two equal-intensity, parallel beams, 3 and 4. Lens 5 causes these beams to cross and focus at common point F. Light 6, scattered from particles moving through the fringe pattern, is selectively collected by the lens/pinhole aperture combination 7 and 8 and then detected by photosensor 9. Light intensity versus time is displayed on scope 10 and simultaneously processed by doppler burst signal processor 11 to derive flow-velocity data. The frequency f of the electrical signals (shown in Fig. 7.24) produced by a particle moving across the dark and light fringe pattern with a velocity component V normal to the fringes is given by

$$f = \frac{2V \sin(\theta/2)}{\lambda} \tag{7.46}$$

For typical laser-light wavelength $\lambda = 5 \times 10^{-5}$ cm and $\theta = 30°$, we get $f = 10,340$ V hertz, where V is in centimeters per second. Note that the method measures the velocity *component* perpendicular to the fringe pattern. We can rotate the fringe pattern 90° to measure the other component of a two-dimensional velocity vector.

In addition to the dual-beam forward-scatter ("fringe") mode of Fig. 7.22, equipment is commercially available[3] to implement other modes of application suited to particular measurement problems. The *reference beam* mode requires less critical optical alignment, but has difficulty handling low particle concentrations. Various *back-scattering* schemes require optical access from only one

[1] A. Melling and J. H. Whitelaw, Seeding of Gas Flows for Laser Anemometry, *DISA Inform.*, Disa Electronics, Franklin Lakes, N. J., no. 15, October 1973; N. S. Berman, Particle-Fluid Interaction Corrections for Flow Measurements with a Laser Doppler Flowmeter, *Rept.* for contract NAS 8-21397, Eng. Res. Ctr., Arizona State University, Tempe.

[2] D. B. Brayton et al., Project Squid Conf. on Laser Doppler Velocimeter, Purdue Univ., pp. 52–100, March 9–10, 1972.

[3] Laser Anemometer Systems, TSI 900-275, TSI Inc., St. Paul, Minn.; Laser Doppler Equipment Catalog, Publ. 8208E, Disa Electronics, Franklin Lakes, N. J.

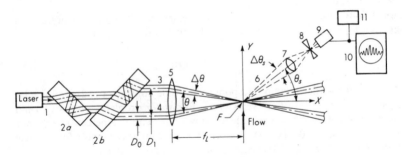

Figure 7.22 Layout of laser doppler velocimeter.

side of the flow, but again require high particle concentrations. Simultaneous measurement of several velocity components at a point may be achieved by either *polarization* schemes using a single laser or *two-color* systems employing two lasers of different wavelength. When velocity components actually change sign (reverse), ambiguities present in "ordinary" systems may be resolved by using *frequency-shifting* schemes employing acousto-optic (Bragg) cells. Two main types of signal processors are utilized. *Frequency trackers* work best with high particle concentrations, while *counter-type* instruments may be needed for high-velocity flows, which generally have few particles in the measurement volume.

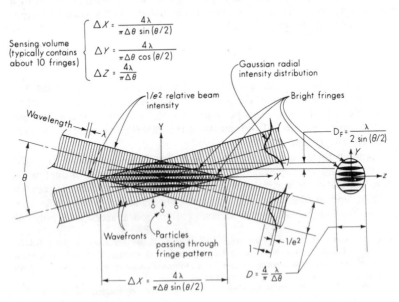

Figure 7.23 Interference fringes in sensing volume.

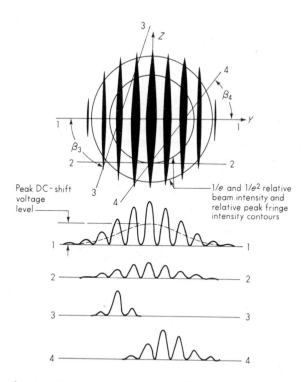

Signal amplitude vs. particle position near $X = 0$ for a number of particle trajectories. The indicated fringe width is proportional to the local peak fringe intensity.

Figure 7.24 Electronic signals from velocimeter.

7.2 GROSS VOLUME FLOW RATE

The total flow rate through a duct or pipe often must be measured or controlled. Many instruments (flowmeters) have been developed for this purpose. They may be categorized in various ways; a useful overall classification divides devices into those which measure volume flow rate [(ft^3 or m^3)/time] and those which measure mass flow rate [(lbm or kg)/time]. The total flow occurring during a given time interval also may be of interest. This measurement requires integration with respect to time of the instantaneous flow rate. The integrating function may be performed in various ways, sometimes being an integral part of the flowmetering concept and other times being performed by a general-purpose integrator more or less remote from the flowmeter.

Calibration and Standards

Flow-rate calibration depends on standards of volume (length) and time or mass and time. Primary calibration,[1] in general is based on the establishment of steady flow through the flowmeter to be calibrated and subsequent measurement of the volume or mass of flowing fluid that passes through in an accurately timed interval. If steady flow exists, the volume or mass flow rate may be inferred from such a procedure. Any stable and precise flowmeter calibrated by such primary methods then itself becomes a secondary flow-rate standard against which other (less accurate) flowmeters may be calibrated conveniently. As in any other calibration, significant deviations of the conditions of use from those at calibration will invalidate the calibration. Possible sources of error in flowmeters include variations in fluid properties (density, viscosity, and temperature), orientation of the meter, pressure level, and particularly flow disturbances (such as elbows, tees, valves, etc.) upstream (and to a lesser extent downstream) of the meter. A commercial calibrator for precise primary calibration of flowmeters using liquids is shown in Fig. 7.25.[2] These units use a convenient dynamic weighing[3] scheme, are available in models to cover the range 0.5 to 150,000 lbm/h, and have an overall accuracy of $\pm$ 0.1 percent.

Another type of calibrator, called a *ballistic flow prover*[4], is particularly useful for fast-response, high-resolution flowmeters such as turbine, positive-displacement, vortex-shedding, etc., types, where steady state can be achieved quickly and the integration of the flow rate to get total flow is accomplished accurately by accumulating the meter pulse-rate output in a counter. (The integration gives accurate total flow even if the flow rate is not perfectly steady.) One such calibrator[5] uses a Teflon-sealed air-driven free piston traveling down a precision-honed tube to dispense a precise volume of calibration fluid through the attached flowmeter to be calibrated. Precise time and displacement measurements on the moving piston are utilized in a microprocessor data-reduction system to achieve a claimed flow-rate accuracy of $\pm$ 0.02 percent. The small flow volume (about 5 gal for a 700 gal/min full-scale unit) involved in checking a single flow rate allows rapid calibration, 20 repeats of a single point typically being achieved in 8 minutes for a 300 gal/min flow rate. Units are available with maximum flow rates of 100 to 6,000 gal/min, each unit being usable over about a 1,000 : 1 range.

The calibration of flowmeters to be used with gases often can be carried out with liquids as long as the pertinent similarity relations (Reynolds number) are

---

[1] M. R. Shafer and F. W. Ruegg, Liquid-flowmeter Calibration Techniques, *Trans. ASME*, p. 1369, October 1958.

[2] *Bull.* CA6600, Cox Instrument, Detroit, Mich., 1978.

[3] L. O. Olsen, Introduction to Liquid Flow Metering and Calibration of Liquid Flowmeters, *NBS Tech. Note* 831, p. 30, 1974.

[4] Ibid., p. 35.

[5] Ballistic Flow Prover, May 1980; Ballistic Flow Calibrators, TD-016; The Flow Factor, vol. 6, no. 2, Flow Technology Inc., Phoenix, Ariz., August 1978.

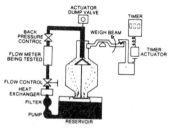

Running operation before test—Fluid contained in the reservoir is pumped through a closed hydraulic circuit. First, it enters the filter and then the heat exchange equipment, which controls temperature within ±1°F. It then passes through the control valves, the meter under test, the back-pressure valve, the weigh tank, then back into the reservoir.

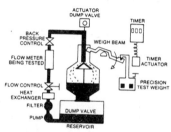

Weighing cycle in operation—The weighing cycle is continued as a precision weight is placed on the weigh pan, again deflecting the beam. The uniquely designed cone-shaped deflector at the inlet of the weigh tank permits the even distribution of the metered fluid.

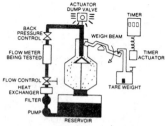

Start of preliminary fill (Tare Time)—When the control valves have been adjusted for desired flow, a tare weight is placed on the weigh pan. Then the cycle start button is pushed, resetting the timer, closing the dump valve which starts the filling of the weigh tank.

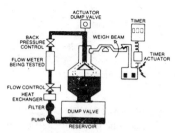

End of weighing cycle—As the tank fills, the weigh pan rises, until it again trips the timer actuator, stopping the timer and indicating the time within a thousandth of a second. By combining the precision test weight with the timed interval, the actual flow rate in pounds/hour is easily and accurately determined. From these basic mass units, other flow units can be accurately calculated.

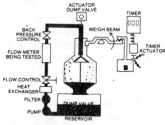

End of pre-fill, start of weighing cycle—As the weigh tank fills, the weigh pan rises, tripping the timer actuator, and the electronic timer begins counting in milliseconds, starting the actual weighing cycle. The preliminary fill, balanced out by the tare weight before actual weighing begins, permits a net measurement of the new fluid added after preliminary fill. The preliminary fill method permits measurement of only a portion of the cycle, eliminating the mechanical errors in the start and stop portions and allowing dynamic errors to be self-cancelling.

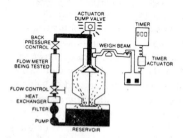

Emptying for recycling—After the beam movement trips the timer, the weigh tank automatically empties in less than 25 seconds, even at maximum flow. The calibrator is now ready for the next flow setting. This fast recycling cuts total calibrating time as much as 50%.

Figure 7.25 Flow-rate calibration by dynamic weighing.

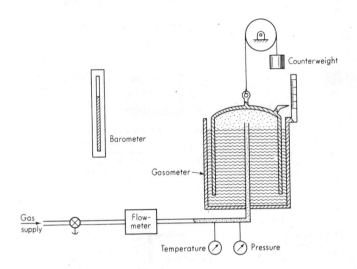

Figure 7.26 Gas-flow calibration.

maintained and theoretical density and expansion corrections are applied. If this procedure is felt to be of insufficient accuracy, a direct calibration with the actual gas to be employed can be carried out by means of the *gasometer* system of Fig. 7.26. Here the gas flowing through the flowmeter during a timed interval is trapped in the gasometer bell, and its volume is measured. Temperature and pressure measurements allow calculation of mass and conversion of volume to any desired standard conditions. By filling the bell with gas, raising it to the top, and adding appropriate weights, such a system may be used as a gas *supply* to drive gas through a flowmeter as the bell gradually drops at a measured rate. By using a precision analytical balance to measure the mass accumulated in a storage vessel over time, accuracies of ± 0.02 percent were obtained[1] for flow rates up to 9 kg/s (20 lbm/s). A volumetric method (corrected for pressure and temperature) is also explained in this reference. The size of the equipment makes it more economical for large flow rates; however, the accuracy is less (±0.08 percent).

When the above primary calibration methods cannot be justified, comparison with a secondary standard flowmeter connected in series with the meter to be calibrated may be sufficiently accurate. Turbine flowmeters and their associated digital counting equipment have been found particularly suitable for such secondary standards. With attention to detail, such standards can closely approach the accuracy of the primary methods themselves. The Navy Primary Standards Laboratory at Pensacola, Florida, has such a system[2] with an inaccuracy of the order of 0.2 percent. For gas flow, a flow nozzle discharging air to the atmosphere can

[1] B. T. Arnberg and C. L. Britton, Two Primary Methods of Proving Gas Flow Meters, *NASA CR*-72896, 1971.
[2] R. P. Bowen, Designing Portability into a Flow Standard, *ISA J.*, p. 40, May 1961.

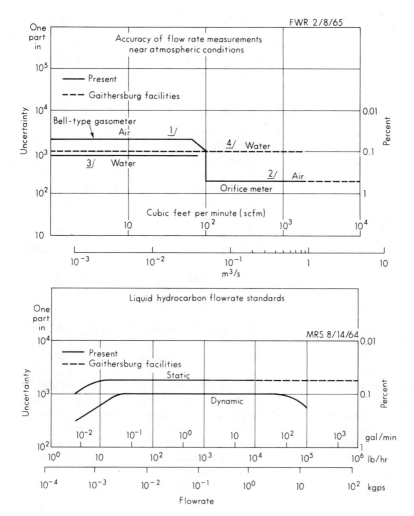

Figure 7.27 Flow-rate calibration capabilities.

be very accurately calibrated[1] for mass flow rate by means of pitot-tube traverses at the discharge. Then this nozzle can be connected in series with any flowmeter to be calibrated and used as an accurate standard. This can be done with very small error up to Mach number about 0.9.

Figure 7.27[2] summarizes the flow-rate calibration capabilities of the National Bureau of Standards.

---

[1] R. J. Sweeney, "Measurement Techniques in Mechanical Engineering," p. 220, Wiley, New York, 1953.

[2] Accuracy in Measurements and Calibrations, *NBS Tech. Note* 262, 1965.

Constant-Area, Variable-Pressure-Drop Meters ("Obstruction" Meters)

Perhaps the most widely used flowmetering principle involves placing a fixed-area flow restriction of some type in the pipe or duct carrying the fluid. This flow restriction causes a pressure drop which varies with the flow rate; thus measurement of the pressure drop by means of a suitable differential-pressure pickup allows flow-rate measurement. In this section we briefly discuss the most common practical devices that utilize this principle: the orifice, the flow nozzle, the venturi tube, the Dall flow tube, and the laminar-flow element.

The *sharp-edge orifice* is undoubtedly the most widely employed flowmetering element, mainly because of its simplicity, its low cost, and the great volume of research data available for predicting its behavior. A typical flowmetering setup is shown in Fig. 7.28. If one-dimensional flow of an incompressible frictionless fluid without work, heat transfer, or elevation change is assumed, theory gives the volume flow rate Q_t (m^3/s) as

$$Q_t = \frac{A_{2f}}{\sqrt{1 - (A_{2f}/A_{1f})^2}} \sqrt{\frac{2(p_1 - p_2)}{\rho}} \tag{7.47}$$

where $A_{1f}, A_{2f} \triangleq$ cross-section flow areas where p_1 and p_2 are measured, m^2
$\rho \triangleq$ fluid mass density, kg/m^3
$p_1, p_2 \triangleq$ static pressures, Pa

We see that measurement of Q requires knowledge of A_{1f}, A_{2f}, and ρ and measurement of the pressure differential $p_1 - p_2$. Actually, the real situation deviates from the assumptions of the theoretical model sufficiently to require experimental correction factors if acceptable flowmetering accuracy is to be attained. For example, A_{1f} and A_{2f} are areas of the actual flow cross section, which are *not*, in general, the same as those corresponding to the pipe and orifice diameters, which are the ones susceptible to practical measurement. Furthermore, A_{1f} and A_{2f} may change with flow rate because of flow geometry changes. Also, there are frictional losses that affect the measured pressure drop and lead to a permanent pressure loss. To take these factors into account, an experimental calibration to determine the actual flow rate Q_a by methods such as that of Fig. 7.25 is necessary. A discharge coefficient C_d may be defined by

$$C_d \triangleq \frac{Q_a}{Q_c} \tag{7.48}$$

and thus

$$Q_a = \frac{C_d A_2}{\sqrt{1 - (A_2/A_1)^2}} \sqrt{\frac{2(p_1 - p_2)}{\rho}} \tag{7.49}$$

where $A_1 \triangleq$ pipe cross-section area, $A_2 \triangleq$ orifice cross-section area, and Q_c is defined by Eqs. (7.48) and (7.49).

The discharge coefficient of a given installation varies mainly with the Reynolds number N_R at the orifice. Thus the calibration can be performed with a single fluid, such as water, and the results used for any other fluid as long as the Reynolds numbers are the same. Variation of C_d with N_R follows typically the

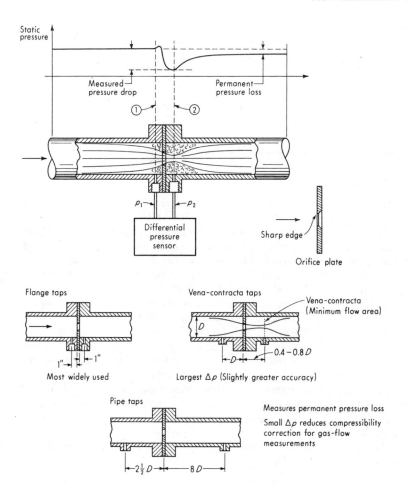

Figure 7.28 Orifice flowmetering.

trend[1] of Fig. 7.29. While the above discussion would seem to indicate that each installation must be individually calibrated, fortunately this is not the case. If one is willing to construct the orifice according to certain standard dimensions and to locate the pressure taps at specific points, then quite accurate (about 0.4 to 0.8 percent error) values of C_d may be obtained[2] based on many past experiments. Such data are available for pipe diameters 2 in and greater, β ratios (see

---

[1] "Handbook of Measurement and Control," p. 96, Instruments Publishing, Pittsburgh, Pa., 1954.

[2] "Fluid Meters, Their Theory and Application" 6th ed., American Society of Mechanical Engineers, New York, 1971; P. S. Starrett and P. F. Halfpenny, The Effect of Non-standard Approach Sections on Orifices and Venturi Meters, Lockheed California Co., LR17905, 1964. The ISO-ASME Orifice Coefficient Equation, *Mech. Eng.*, pp. 44–45, July 1981.

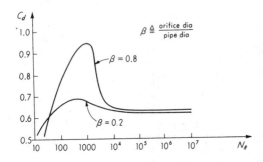

Figure 7.29 Variation in discharge coefficient.

Fig. 7.29) of 0.2 to 0.7, and Reynolds numbers above 10,000. Installations exceeding these limits should be calibrated individually if high accuracy is required. It also should be noted that the standard calibration data assume no significant flow disturbances such as elbows, bends, tees, valves, etc., for certain minimum distance upstream of the orifice. The presence of such disturbances close to the orifice can invalidate the standard data, causing errors of as much as 15 percent. Information on the minimum distances is available.[1] Figure 7.30 shows a typical example. If the minimum distances are not feasible, straightening vanes[1] may be introduced ahead of the flowmeter to smooth out the flow.

Since flow rate is proportional to $\sqrt{\Delta p}$, a 10 : 1 change in Δp corresponds to only about a 3 : 1 change in flow rate. Since a given Δp-measuring instrument becomes quite inaccurate below about 10 percent of its full-scale reading, this nonlinearity typical of all obstruction meters (other than the laminar-flow element) restricts the accurate range of flow measurement to about 3 : 1. That is, a meter of this type cannot be used accurately below about 30 percent of its maximum flow rating. The square-root nonlinearity also causes difficulties in pulsating-flow measurement,[2] where the average flow rate (the rate to be measured) has a fluctuating component superimposed on it. Let us consider, as a simple example, a flow Q where

$$Q = Q_{av} + Q_p \sin \omega t \qquad Q_p < Q_{av} \qquad (7.50)$$

and a flowmeter such that $\Delta p = KQ^2$. The Δp presented as input to the pressure-measuring system is then

$$\Delta p = K(Q_{av}^2 + 2Q_{av} Q_p \sin \omega t + Q_p^2 \sin^2 \omega t) \qquad (7.51)$$

If the Δp instrument has a low-pass filtering characteristic, it will tend to read the

---

[1] "Handbook of Measurement and Control", p. 96; "Fluid Meters," op. cit.; Starrett and Halfpenny, op. cit.
[2] A. K. Oppenheim and E. G. Chilton, Pulsating Flow Measurement—A Literature Survey, *Trans. ASME*, p. 231, February 1955; T. Isobe and H. Hattori, A New Flowmeter for Pulsating Gas Flow, *ISA J.*, p. 38, December 1959; H. J. Sauer, Jr., P. D. Smith, and L. V. Field, Metering Pulsating Flow in Orifice Installations, *Inst. Tech.*, pp. 41–44, March 1969.

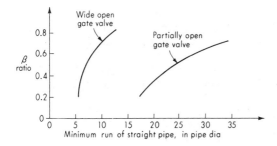

Figure **7.30** Effect of upstream disturbances.

average value of Δp. This is seen to be

$$\Delta p_{av} = K\left(Q_{av}^2 + \frac{Q_p^2}{2} \right) \qquad (7.52)$$

Thus if we take a measured Δp_{av} and compute the corresponding Q_{av} from it, using $Q_{av} = \sqrt{\Delta p_{av}/K}$, we get a flow rate *higher* than actually existed. A further difficulty caused by the nonlinearity occurs when flow rate must be integrated to get total flow during a given time interval. Then the square root of the Δp signal must be taken before integration or this compensation must be included in the integrating device.

The orifice has the largest permanent pressure loss of any of the obstruction meters (other than the laminar-flow element). This is one of its disadvantages since it represents a power loss[1] that must be replaced by whatever pumping machinery is causing the flow. The permanent pressure loss is given approximately by $\Delta p(1 - \beta^2)$, where Δp is the differential pressure used for flow measurement. Thus for the usual range of β (0.2 to 0.7), the permanent pressure loss ranges from 0.96 Δp to 0.51 Δp. The actual power loss, in fact, may be quite small since the Δp recommended[2] for conventional flowmetering of liquids is only 20 to 400 in of water (0.72 to 14.4 lb/in²).

Orifice discharge coefficients are quite sensitive to the condition of the upstream edge of the hole. The standard orifice design requires that this edge be very sharp and that the orifice plate be sufficiently thin relative to its diameter. Wear (rounding) of this sharp edge by long use or abrasive particles can cause significant changes in the discharge coefficient. Flows that contain suspended solids also cause difficulty since the solids tend to collect behind the "dam" formed by the orifice plate and cause irregular flow. Often this problem can be solved by use of an "eccentric" orifice in which the hole is at the bottom of the pipe rather than on the centerline. This allows the solids to be swept through

[1] W. M. Reese, Jr., Factor the Energy Costs of Flow Metering into Your Decisions, *InTech*, pp. 36–38, July 1980.

[2] L. K. Spink, "Principles and Practice of Flow Meter Engineering," The Foxboro Co., Foxboro, Mass., 1959; W. Buzzard, Flowmeter Orifice Sizing Handbook, 10B900, Fischer & Porter, Warminster, Pa., 1968.

continuously. Liquids containing traces of vapor or gas may be metered if the orifice is installed in a vertical run of pipe with the flow upward. Gases containing traces of liquid may be similarly handled except that the flow should be downward.

When compressible fluids are metered, Eq. (7.49) is no longer correct. By assuming an isentropic process between states 1 and 2, the following relation may be derived[1] for compressible fluids:

$$W = C_d A_2 \sqrt{\frac{2gkp_1}{(k-1)v_1}} \sqrt{\frac{(p_2/p_1)^{2/k} - (p_2/p_1)^{(k+1)/k}}{1 - \beta^4 (p_2/p_1)^{2/k}}} \tag{7.53}$$

where $W \triangleq$ weight flow rate, lbf/s
$k \triangleq$ ratio of specific heats (1.4 for air)
$g \triangleq$ local gravity
$v_1 \triangleq$ specific volume at state 1, ft^3/lbf

The discharge coefficient C_d is the same for liquids or gases as long as the Reynolds number is the same. In many practical gas-flow installations, the meter pressure drop is so small that the pressure ratio p_2/p_1 is 0.99 or greater. Under these conditions, the simpler incompressible relation of Eq. (7.49) may be used with an error less than 0.6 percent if, for example, $\beta = 0.5$ and $k = 1.4$. Modified to give weight rather than volume flow rate, this is

$$W = C_d A_2 \left[\sqrt{1 - \left(\frac{A_2}{A_1}\right)^2} \right]^{-1} \sqrt{\frac{2g(p_1 - p_2)}{v_1}} \tag{7.54}$$

For $p_2/p_1 < 0.99$, the error becomes greater, being 6 percent at $p_2/p_1 = 0.9$, 12 percent at 0.8, 19 percent at 0.7, and 26 percent at 0.6 (if $\beta = 0.5$ and $k = 1.4$). For such situations, the more complex compressible formula obviously must be used. In flow nozzles and venturis, the flow process is close enough to isentropic to allow theoretical calculation of compressibility effects from Eq. (7.53). In orifices, however, deviation from isentropic conditions is significant (greater turbulence), and an experimental compressibility factor Y is used in the equation:

$$W = Y \left\{ C_d A_2 \left[\sqrt{1 - \left(\frac{A_2}{A_1}\right)^2} \right]^{-1} \right\} \sqrt{\frac{2g(p_1 - p_2)}{v_1}} \tag{7.55}$$

For flange taps or vena-contracta taps[2]

$$Y = 1 - (0.41 + 0.35\beta^4) \frac{p_1 - p_2}{p_1} \frac{1}{k} \tag{7.56}$$

while for pipe taps[3]

$$Y = 1 - [0.333 + 1.145(\beta^2 + 0.7\beta^5 + 12\beta^{13})] \frac{p_1 - p_2}{p_1} \frac{1}{k} \tag{7.57}$$

---

[1] D. P. Eckman, "Industrial Instrumentation," p. 270, Wiley, New York, 1950.
[2] "Fluid Meters."
[3] Ibid.

These empirical formulas are accurate to ± 0.5 percent if $0.8 < p_2/p_1 < 1.0$ and the flowing fluid is a gas or vapor other than steam. For steam the accuracy is ± 1.0 percent. In sizing orifices for gas measurement, a useful rule of thumb is that the maximum Δp (in inches of water) should not exceed the upstream gage pressure in pounds per square inch. Handbooks[1] published by flowmeter manufacturers provide many practical guidelines and time-saving charts for orifice selection and application.

While most orifice applications involve essentially steady flow, dynamic response is sometimes of interest. The dynamic response of the differential-pressure sensor must be adequate, of course; the methods of Chap. 6 answer such questions. An analysis[2] is available that uses a nonlinear differential equation relating mass flow rate and the Δp signal and includes lumped fluid inertial and resistance effects. A linearized version defines a first-order dynamic system in which flow rate lags Δp with a time constant τ given by

$$\tau = \frac{C_d^2 A_o^2 \rho l}{w_o A_p} \tag{7.58}$$

where $C_d \triangleq$ discharge coefficient
  $A_o \triangleq$ orifice area
  $\rho \triangleq$ fluid density
  $l \triangleq$ length between pressure taps
  $w_o \triangleq$ fluid mass flow rate at linearization operating point
  $A_p \triangleq$ pipe area

The *flow nozzle*, *venturi tube*, and *Dall flow tube* (Fig. 7.31) all operate on exactly the same principle as the orifice; the significant differences lie in the numerical values of certain characteristics. Discharge coefficients of flow nozzles and venturis are larger than those for orifices and also exhibit an opposite trend with Reynolds numbers, varying from about 0.94 at $N_R = 10{,}000$ to 0.99 at $N_R = 10^6$. The Dall-flow-tube coefficient is more like that of an orifice; for $\beta = 0.7$, for example, it goes from about 0.68 at $N_R = 100{,}000$ to 0.66 at $N_R = 10^6$. Individual calibrations generally are needed on all these devices since their complicated shapes (compared with an orifice) make accurate reproduction difficult.

When comparing the permanent pressure losses of the various devices, you should require that each device be producing the same measured Δp, since this would keep the accuracy constant. On this basis, the permanent pressure loss of a flow nozzle is practically identical with that of an orifice, because, to get the same Δp, the flow nozzle must have a smaller β ratio, and losses increase with decreasing β ratio. The venturi tube also requires a smaller β for a given Δp, but because of its streamlined form, its losses are low and nearly independent of β. The permanent pressure loss is of the order of 10 to 15 percent of the measured Δp

[1] Spink, op. cit.; Buzzard, op. cit.; C. F. Cusick, Flow Meter Engineering Handbook, Minneapolis Honeywell Corp., Philadelphia, Pa., 1961.
[2] R. T. Lakey, Jr., and B. S. Shiralkar, Transient Flow Measurements with Sharp-Edged Orifices, *ASME Paper* 71-FE-30, 1971.

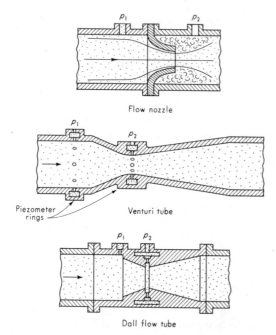

Flow nozzle

Piezometer rings

Venturi tube

Dall flow tube

Figure 7.31 Variable-pressure-drop flowmeters.

over the range $0.2 < \beta < 0.8$; thus a venturi gives a definite improvement in power losses over an orifice and often is indicated for measuring very large flow rates, where power losses, though a small *percentage*, become economically significant in absolute value. The initial higher cost of a venturi over an orifice thus may be offset by reduced operating costs. The Dall flow tube has the unexpected (though desirable) features of a high measured Δp (similar to an orifice) and a low permanent pressure loss (similar to, and sometimes better than, a venturi). These apparently inconsistent virtues have been checked experimentally, but are not fully explained theoretically.[1] The permanent pressure loss of a Dall tube is of the order of 50 percent or less of that of a venturi tube with the same Δp. Other factors to consider in choosing among the orifice, flow nozzle, venturi, and Dall tube include freedom from pressure-tap clogging resulting from suspended solids (venturi is best), loss of accuracy due to wear (venturi, flow nozzle, and Dall tube are better than an orifice), accuracy (venturi, when calibrated, is best), cost, and ease of changing the flow element to a different size.

Laminar-flow elements differ from the metering devices discussed above in that they are specifically designed to operate in the laminar-flow regime. Pipe flows generally are considered laminar if Reynolds number N_R is less than 2,000; however, in laminar-flow elements, often considerably lower values are designed for to ensure laminar conditions. The simplest form of laminar-flow element is

---

[1] I. O. Miner, The Dall Flow Tube, *Instrum. Eng.*, p. 45, April 1957.

merely a length of small-diameter (capillary) tubing.[1] For $N_R < 2,000$, the Hagen-Poiseuille viscous-flow relation gives for incompressible fluids

$$Q = \frac{\pi D^4}{128 \mu L} \Delta p \qquad (7.59)$$

where $Q \triangleq$ volume flow rate, m^3/s
     $D \triangleq$ tube inside diameter, m
     $\mu \triangleq$ fluid viscosity, $N \cdot s/m^2$
     $L \triangleq$ tube length between pressure taps, m
     $\Delta p \triangleq$ pressure drop, Pa

One usually designs for $N_R \gtrsim 1,000$ in such a device. Extremely small flows can be measured in this way; a 3-ft length of 0.004-in-diameter tubing measuring Δp with a 2-in water inclined manometer gives a threshold sensitivity of about 0.000175 in^3/h when hydrogen is flowing.[2]

A single capillary tube is capable of handling only small flow rates at laminar Reynolds numbers. To increase the capacity of laminar-flow elements, many capillaries in parallel (or their equivalent) may be employed. One commercially available variation[3] uses a large tube (about 1-in diameter) packed with small spheres. The passages between the spheres give the same effect as many capillary tubes. This particular instrument is designed to give a Reynolds number of 20 or less. A 5-in length of tube gives a Δp of 20 in of water for a $2\,cm^3/min$ flow rate of air at 14.7 lb/in^2 absolute and 70°F. Another approach[4] uses a "honeycomb" element (see Fig. 7.32) with triangular members a few thousandths of an inch on a side and a few inches long. These devices have been used mainly to measure flow of low-pressure air, and standard models are available in ranges from 0.1 to 2,000 ft^3/min at pressure drops of 4 to 8 in of water. All the above laminar elements have the advantages accruing from a linear (rather than square-root) relation between flow rate and pressure drop; these are principally a large accurate range of as much as 100 : 1 (compared with 3 : 1 or 4 : 1 for square-root devices), accurate measurement of average flow rates in pulsating flow, and ease of integrating Δp signals to compute total flow. The laminar elements also can measure reversed flows with no difficulty. They are usually less sensitive to upstream and downstream flow disturbances than the other devices discussed. Their disadvantages include clogging from dirty fluids, high cost, large size, and high pressure loss (*all* the measured Δp is lost).

Averaging Pitot Tubes

The pitot-static tube described earlier for point velocity measurements can be calibrated for volume flow rate when it is located at a specific point in a specific duct; however, such single-point flow-rate measurements become inaccurate

[1] L. M. Polentz, Capillary Flowmetering, *Instrum. Contr. Syst.*, p. 648, April 1961.
[2] Ibid.
[3] A. R. Hughes, New Laminar Flowmeter, *Instrum. Contr. Syst.*, p. 98, April 1962.
[4] Meriam Instrument Co., Cleveland, Ohio, *Tech. Note* 2A.

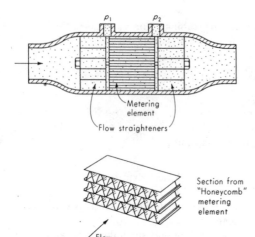

Figure 7.32 Laminar-flow element.

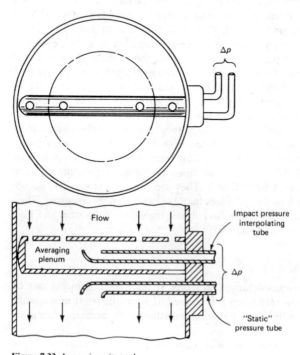

Figure 7.33 Averaging pitot tube.

when velocity profiles differ from calibration conditions. Division of the total flow area into several equal-area annuli, measurement of local velocity at the center of area of each annulus by moving the pitot tube to each such point, and summation of these individual flow rates to obtain the total are possible for laboratory applications but impractical for on-line process monitoring or control. By using a fixed probe with several spatially dispersed sensing ports, whose readings are averaged in a plenum to produce a single "impact" pressure and a single "static" pressure, a practical volume flowmeter is obtained which produces a Δp signal related to volume flow rate by a square-root relation similar to that of the "obstruction" meters. One manufacturer's[1] version (Fig. 7.33) employs an "interpolating tube" to sense at an optimum location the average of four impact ports positioned along a pipe diameter at the area centers of the respective flow areas. A single downstream-facing port reads a pressure somewhat below static. Advantages of such averaging pitot tubes include the possibility of installation in operating pipelines without shutdown (by using a sealable-drill technique), compatability with common $\sqrt{\Delta p}$-type instrumentation, and low permanent pressure losses (compared with those of orifices) leading to claimed[2] savings of \$1,000/year for a single steam line.

Constant-Pressure-Drop, Variable-Area Meters (Rotameters)[3]

A rotameter consists of a vertical tube with tapered bore in which a "float" assumes a vertical position corresponding to each flow rate through the tube (see Fig. 7.34). For a given flow rate, the float remains stationary since the vertical forces of differential pressure, gravity, viscosity, and buoyancy are balanced. This balance is self-maintaining since the meter flow area (annular area between the float and tube) varies continuously with vertical displacement; thus the device may be thought of as an orifice of adjustable area. The downward force (gravity minus buoyancy) is constant, and so the upward force (mainly the pressure drop times the float cross-section area) must be constant also. Since the float area is constant, the pressure drop must be constant. For a *fixed* flow area, Δp varies with the square of flow rate, and so to keep Δp *constant* for differing flow rates, the area must vary. The tapered tube provides this variable area. The float position is the output of the meter and can be made essentially linear with flow rate by making the tube area vary linearly with the vertical distance. Rotameters thus have an accurate range of about $10:1$, considerably better than square-root-type elements. Accuracy is typically ± 2 percent full scale (± 1 percent if calibrated) with repeatability about 0.25 percent of reading. Assuming incom-

[1] Annubar Flow Element, Dieterich Standard Corp., Boulder, Colo.

[2] Steam Measurement, *Appl. Note* 93-00-07-04 (580), Honeywell Process Control Div., Ft. Washington, Pa., 1980.

[3] Simplifying Flow Measurement and Control with Variable-Area Meters, Ametek Corp., Cornwells Heights, Pa., 1975; V. P. Head, Coefficients of Float-Type Variable-Area Flowmeters, *Trans. ASME*, pp. 851–862, August 1954.

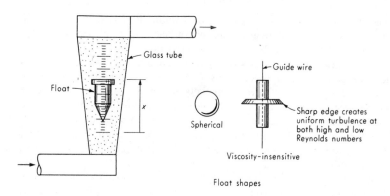

Float shapes

Figure 7.34 Rotameter.

pressible flow and the above described simplified model, we can derive the result

$$Q = \frac{C_d(A_t - A_f)}{\sqrt{1 - [(A_t - A_f)/A_t]^2}} \sqrt{2gV_f \frac{w_f - w_{ff}}{A_f w_{ff}}} \qquad (7.60)$$

where $Q \triangleq$ volume flow rate, ft³/s
$C_d \triangleq$ discharge coefficient
$A_t \triangleq$ area of tube, ft²
$A_f \triangleq$ area of float, ft²
$g \triangleq$ local gravity, ft/s²
$V_f \triangleq$ volume of float, ft³
$w_f \triangleq$ specific weight of float, lbf/ft³
$w_{ff} \triangleq$ specific weight of flowing fluid, lbf/ft³

If the variation of C_d with float position is slight and if $[(A_t - A_f)/A_t]^2$ is always much less than 1, then Eq. (7.60) has the form

$$Q = K(A_t - A_f) \qquad (7.61)$$

And if the tube is shaped so that A_t varies linearly with float position x, then $Q = K_1 + K_2 x$, a linear relation. The floats of rotameters may be made of various materials to obtain the desired density difference [$w_f - w_{ff}$ in Eq. (7.60)] for metering a particular liquid or gas. Some float shapes, such as spheres, require no guiding in the tube; others are kept central by guide wires or by internal ribs in the tube. Floats shaped to induce turbulence can give viscosity insensitivity over a 100 : 1 range. The tubes often are made of high-strength glass to allow direct observation of the float position. Where greater strength is required, metal tubes can be used and the float position detected magnetically through the metal wall. If a pneumatic or an electrical signal related to the flow rate is desired, the float motion can be measured with a suitable displacement transducer. Flow rates beyond the range of the largest rotameter may be measured by combining an orifice plate and a rotameter in a bypass arrangement.[1]

Turbine Meters

If a turbine wheel is placed in a pipe containing a flowing fluid, its rotary speed depends on the flow rate of the fluid. By reducing bearing friction and keeping other losses to a minimum, one can design a turbine whose speed varies linearly with flow rate; thus a speed measurement allows a flow-rate measurement. The speed can be measured simply and with great accuracy by counting the rate at which turbine blades pass a given point, using a magnetic proximity pickup to produce voltage pulses. By feeding these pulses to an electronic pulse-rate meter, one can measure flow rate; by accumulating the total number of pulses during a timed interval, the total flow is obtained. These measurements can be made very accurately because of their digital nature. If an analog voltage signal is desired, the pulses can be fed to a frequency-to-voltage converter. Figure 7.35 shows a flowmetering system of this type.

Dimensional analysis[2] of the turbine flowmeter shows that (if bearing friction and shaft power output are neglected) the following relation should hold:

$$\frac{Q}{nD^3} = \text{some function of } \frac{nD^2}{v} \qquad (7.62)$$

where $Q \triangleq$ volume flow rate, in³/s
$\quad n \triangleq$ rotor angular velocity, r/s
$\quad D \triangleq$ meter bore diameter, in
$\quad v \triangleq$ kinematic viscosity, in²/s

Actually, the effect of viscosity is limited mainly to low flow rates, with high flow rates being in the turbulent regime where viscosity effects are secondary. For negligible viscosity effects, a simplified analysis[3] based on strictly kinematic relationships gives the following result:

$$\frac{Q}{nD^3} = \frac{\pi L}{4D}\left[1 - \alpha^2 - \frac{2m(D_b - D_h)t}{\pi D^2}\sqrt{1 + \left(\frac{\pi D_b}{L}\right)^2}\right] \qquad (7.63)$$

---

[1] Guide to By-Pass Rotameter Application, *Tech. Bull.* T-023, Brooks Instrument Div., Hatfield, Pa., 1975.

[2] H. M. Hochreiter, Dimensionless Correlation of Coefficients of Turbine-Type Flowmeters, *Trans. ASME*, p. 1363, October 1958.

[3] Ibid.

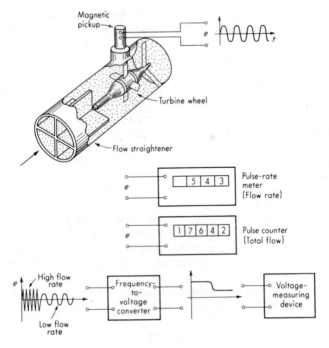

Figure 7.35 Turbine flowmeter.

where $L \triangleq$ rotor lead, in
  $\alpha \triangleq D_h/D$
  $m \triangleq$ number of blades
  $D_b \triangleq$ rotor-blade-tip diameter, in
  $D_h \triangleq$ rotor-hub diameter, in
  $t \triangleq$ rotor-blade thickness, in

Equation (7.63) gives $Q = Kn$, where K is a constant for any given meter and is independent of fluid properties. Thus this represents the ideal situation. Deviations from this ideal may be found from experimental calibrations,[1] such as are shown for a meter with $D = 1$ in in Fig. 7.36. We see for sufficiently high values of nD^2/v that $Q/(nD^3)$ becomes essentially constant, as predicted by Eq. (7.63). In the particular case shown, $Q/(nD^3) = 1.92$ for at least a 10 : 1 range of nD^2/v; thus this would be a useful linear operating range for this meter. The meter could be utilized at lower flow rates by applying corrections obtained from Fig. 7.36. However, usually this is not done since turbine meters are available in a wide range of sizes, each being linear over a different flow range. If the total flow range

[1] Ibid.

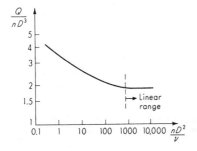

Figure 7.36 Turbine flowmeter characteristics.

to be accommodated is about 10 : 1 or less, one can usually select a turbine meter that is linear in the desired range. Linearity is particularly desirable if one is totalizing pulses to get a total flow over a timed interval during which flow rate fluctuates.

At low flow rates, linearity is degraded by both viscous effects and magnetic pickup drag. The latter can be reduced by using a modulated carrier type of pickoff. These are recommended[1] for all gas applications in meters of size 1 in or less and for any liquid applications where extended linear ranges are necessary. This technique makes the graph of pulse rate versus flow rate more nearly a straight line; however, this line does not go through the origin since the turbine will "stall" before flow goes to zero. By electronically adding[2] a bias frequency to the pickoff output, this defect can be corrected, which allows proper flow totalization over a wide range. While Fig. 7.36 shows linearity improving at high flows, this end of the range can be extended only at the expense of excessive pressure drop and reduced bearing life; but short-term overranging of 100 percent is usually possible without damage.[3]

Commercial turbine meters are available with full-scale flow rates ranging from about 0.01 to 30,000 gal/min for liquids and 0.01 to 15,000 ft³/min for air. Nonlinearity within the design range (usually about 10 : 1) can be as good as 0.05 percent in the larger sizes. The output voltage of the magnetic pickups is of the order of 10 mV rms at the low end of the flow range and 100 mV at the high. Pressure drop across the meter varies with the square of flow rate and is about 3 to 10 lb/in² at full flow. Turbine meters behave essentially as first-order dynamic systems for small changes about an operating point.[4] However, the time constant (typically 2 to 10 ms at maximum flow) is inversely proportional to the operating-point flow rate. If a frequency-to-voltage converter is used to get an analog

[1] R. Zimmerman and D. Deery, Turbine Flowmeter Handbook, Flow Technology Inc., Phoenix, Ariz., 1977.

[2] Ibid.

[3] Ibid.

[4] J. Grey, Transient Response of the Turbine Flowmeter, *Jet Propul.*, p. 98, February 1956; G. H. Stevens, Dynamic Calibration of Turbine Flowmeters by Means of Frequency Response Tests, *NASA* TM X-1736, 1969.

voltage output, however, its response speed may not be negligible since the operating frequencies of turbine meters are of the order of 100 to 2,000 Hz. That is, the frequency-to-voltage converter requires low-pass filtering, which rejects frequencies somewhat below the turbine operating frequency and is thus limited in transient response also. For example, if the turbine is putting out 500 Hz, the low-pass filter will have to cut off at *least* at 500 Hz and probably considerably lower. If a first-order filter is designed to attenuate 20 dB at 500 Hz, it will have a time constant of 0.0032 s, thus adding this much to the lag of the overall system.

Positive-Displacement Meters

These meters are actually positive-displacement fluid motors in which friction and inertia have been reduced to a minimum. The flow of a fluid through volume chambers of definite size causes rotation of an output shaft. Figure 7.37 shows a self-porting four-piston design with fluid entering at $P3$ and leaving at $E1$, causing a clockwise shaft rotation which sequentially brings pairs ($P4$-$E2$, $P1$-$E3$, $P2$-$E4$, $P3$-$E1$) of inlet/outlet ports into operation trhough the three-way valve action of each piston. Such meters exhibit little sensitivity to viscosity (accuracy actually improves with increased viscosity, because of decreased leakage) and can give high accuracy over wide flow ranges (up to 1,000 : 1, see Fig. 7.37b). Since the meters have a maximum allowable pressure drop [typically 20 to 40 lb/in^2 differential (140 to 290 kPa)], high viscosity reduces the maximum allowable flow rate for a given meter (Fig. 7.37c). As a result of precision clearances in the instrument, fluid filtration at the 10-μm level is recommended. Meter construction materials also limit the fluids that can be metered. To eliminate friction-producing fluid seals, the meter shaft is coupled to the speed sensor magnetically through a nonmagnetic wall. This "soft" coupling can introduce errors for unsteady flows. Since the crankshaft kinematics produces an epicycloidal speed variation for a steady flow rate, special analog tachometers that compensate for this fluctuation are available.

Digital (pulse-rate) output is also available that uses photo-optic techniques (Fig. 7.38) or rotary differential transformer methods. The latter method does not use the soft magnetic coupling and thus follows unsteady crankshaft motions faithfully. Both methods provide direction sensing, so that reverse flows actually subtract pulses, to give true net flow in totalizing systems using up-down counters. For larger flow rates and for food/beverage applications requiring a cleanable, "pocketless" design, a meter based on helical rotors (Fig. 7.39) is available with performance similar to that of the four-piston design. An advantage shared by positive-displacement meters in general is their insensitivity to distorted inlet/outlet flow profiles; thus flow straighteners and/or long runs of straight pipe upstream/downstream of the meter usually are not required. Dynamic response specifications are not usually quoted; however, the basic principle implies fast response since the output rotation is (ideally) "kinematically coupled" to the fluid flow rate. Deviations from ideal behavior which lead to response lags include fluid compressibility and meter leakage. Also, because of meter inertia and fric-

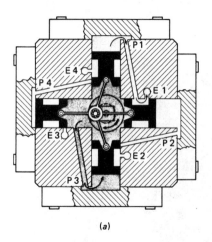

(a)

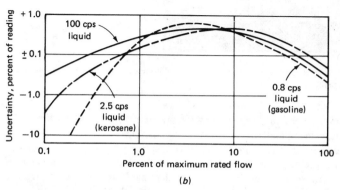

(b)

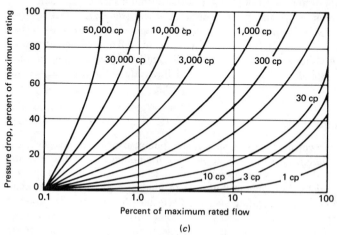

(c)

Figure 7.37 Piston-type positive-displacement flowmeter. (*Courtesy Fluidyne Instrumentation, Oakland, Calif.*)

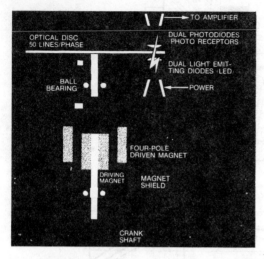

Figure 7.38 Output/drive details for positive-displacement flowmeter. (*Courtesy Fluidyne Instrumentation, Oakland, Calif.*)

tion, rapid flow changes can cause the meter pressure drop to exceed the maximum allowable, causing metering errors or actual damage.

An interesting variation[1] on the positive-displacement concept eliminates both the leakage present in all such meters and the loading effect of the meter on the measured flow process. Both effects are chargeable to the meter pressure drop, and would disappear if it were zero. This is accomplished by using an active feedback scheme in which the meter shaft actually is driven by an electric servomotor whose drive voltage is obtained from the output of a differential-pressure transducer connected across the meter. Whenever this differential pressure is not zero, the motor either speeds up or slows down until the pressure is again zero; thus the meter speed "tracks" the impressed flow rate without extracting any energy from the flow.

Metering Pumps

A variable-displacement positive-displacement pump, if properly designed, can serve both to *cause* a flow rate and to *measure* it simultaneously. The principle again is merely that a positive-displacement machine, except for leakage and compressibility, delivers a definite flow rate of fluid at a given speed. In many pumps of this kind, the operating speed is fixed and the flow rate is varied by changing pump displacement, usually with some form of mechanical linkage. Since these pumps are used often in automatic control systems, many are de-

[1] Laboratory Equipment Corp., Mooresville, Inc., model PLV.

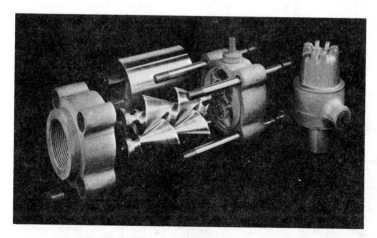

Figure 7.39 Helix-type positive-displacement flowmeter. (*Courtesy Fluidyne Instrumentation, Oakland, Calif.*)

signed to accept pneumatic or electrical input signals which adjust the pump displacement in a linear fashion. The flow rate of such a system can be set with an accuracy of the order of 1 percent.

Electromagnetic Flowmeters

The electromagnetic flowmeter[1] is an application of the principle of induction, shown in Fig. 7.40a. If a conductor of length l moves with a transverse velocity v across a magnetic field of intensity B, there will be forces on the charged particles of the conductor that will move the positive charges toward one end of the conductor and the negative charges to the other. Thus a potential gradient is set up along the conductor, and there is a voltage difference e between its two ends. The quantitative relation among the variables is given by the well-known equation

$$e = Blv \tag{7.64}$$

where $B \triangleq$ field flux density, $Wb/m^2 = v \cdot s/m^2$
$l \triangleq$ conductor length, m
$v \triangleq$ conductor velocity, m/s

If the ends of the conductor are connected to some external circuit that is stationary with respect to the magnetic field, the induced voltage, in general, will cause a current i to flow. This current flows through the moving conductor, which has a

[1] J. A. Shercliff, "The Theory of Electromagnetic Flow Measurement," Cambridge University Press, New York, 1962.

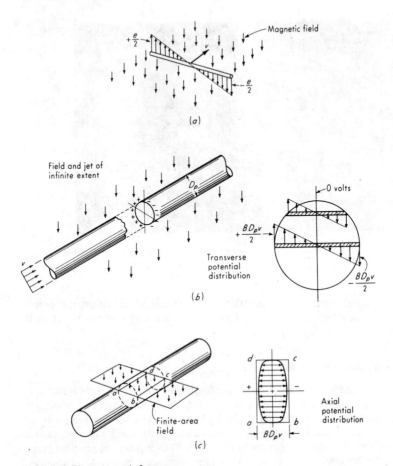

Figure 7.40 Electromagnetic flowmeter.

resistance R, causing an iR drop, so that the terminal voltage of the moving conductor becomes e-iR.

We consider now a cylindrical jet of conductive fluid with a uniform velocity profile, traversing a magnetic field as in Fig. 7.40b. In a liquid conductor the positive and negative ions are forced to opposite sides of the jet, giving a potential distribution as shown. The maximum voltage difference is found across the ends of a horizontal diameter and is $BD_p v$ in magnitude. In a practical situation, the magnetic field is of limited extent, as in Fig. 7.40c; thus no voltage is induced in that part of the jet outside the field. Since these parts of the fluid are, however, still conductive paths, they tend partially to "short-circuit" the voltages induced in the section exposed to the field; thus the voltage is reduced from the value

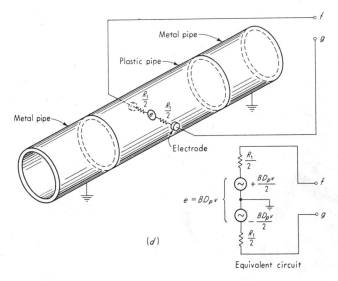

Equivalent circuit

Figure 7.40 (*Continued*)

$BD_p v$. If the field is sufficiently long, this effect will be slight at the center of the field length. A length of 3 diameters usually is sufficient.[1]

In a practical flowmeter (see Fig. 7.41[2]), the "jet" is contained within a stationary pipe. The pipe must be nonmagnetic to allow the field to penetrate the fluid and usually is nonconductive (plastic, for instance), so that it does not provide a short-circuit path between the positive and negative induced potentials at the fluid surface. This nonconductive pipe has two electrodes placed at the points of maximum potential difference. These electrodes then supply a signal voltage to external indicating or recording apparatus. While Fig. 7.40 shows a uniform velocity profile, it has been proved mathematically[3] that e corresponds to the average velocity of *any* profile which is symmetric about the pipe centerline. Because it is impractical to make the entire piping installation nonconductive, a short length (the flowmeter itself) of nonconductive pipe must be coupled into an ordinary metal-pipe installation. Since the fluid itself is conductive, this means that there is a conductive path between the electrodes. In Fig. 7.40d, this path is shown split into two equal parts $R_1/2$ and containing the signal source $e = BD_p v$. This resistance is not simple to calculate since it involves a

[1] T. C. Hutcheon, Electrical Characteristics of the Magnetic Flow Detector Head, *Instrum. Eng.*, p. 1, April 1964.

[2] *Tech. Bull.* 10D-14, Fischer and Porter Corp., Warminster, Pa., 1977.

[3] A. Kolin, An Alternating Field Induction Flowmeter of High Sensitivity, *Rev. Sci. Instrum.*, vol. 16, p. 109, May 1945; N. C. Wegner, Effect of Velocity Profile Distortion in Circular Transverse-Field Electromagnetic Flowmeters, *NASA* TN D-6454, 1971.

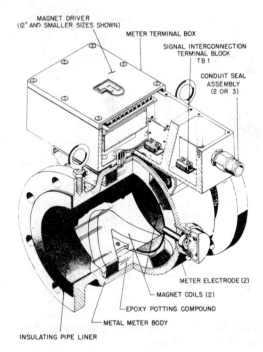

MAGNET DRIVER
(12" AND SMALLER SIZES SHOWN)

METER TERMINAL BOX

SIGNAL INTERCONNECTION
TERMINAL BLOCK
TB 1

CONDUIT SEAL
ASSEMBLY
(2 OR 3)

METER ELECTRODE (2)

MAGNET COILS (2)

EPOXY POTTING COMPOUND

METAL METER BODY

INSULATING PIPE LINER

Figure 7.41 Construction details of magnetic flowmeter.

continuous distribution of resistance over complex bodies. However, it can be estimated from theory, and once a device is built, it can be directly measured. The magnitude of this source resistance determines the loading effect of any external circuit connected to the electrodes.

The magnetic field used in such a flowmeter conceivably could be either constant[1] or alternating, giving rise to a dc or an ac output signal, respectively. For many years, ac systems (50 or 60 Hz) were most common, since they reduced polarization effects at the electrodes, did not cause flow-profile distortion from magnetohydrodynamic effects, could use high-pass filtering to eliminate slow, spurious voltage drifts resulting from thermocouple and galvanic actions, and allowed use of high-gain ac amplifiers whose drift was less than that of dc types of comparable gain. These advantages outweighed the major disadvantage that the powerful ac field coils induced spurious ac signals into the measurement circuit. To cancel this error, one had to periodically stop the flow to get a pipe-full, zero-velocity condition and then adjust a balance control to get a zero output reading.

About 1975, industrial meters utilizing an "interrupted dc" field became

[1] H. M. Hammac, An Application of the Electromagnetic Flowmeter for Analyzing Dynamic Flow Oscillations, *NASA* TM X-53570, 1967.

available, and currently the market is shared mainly by these two types. In the interrupted dc meter, a dc field is switched in a square-wave fashion between the working value and zero, at about 3 to 6 Hz. When the field is zero, any instrument output that appears is considered to be an error; thus by storing this zero error and subtracting it from the total instrument output obtained when the field is next applied, one implements an "automatic zero" feature which corrects for zero errors several times a second. Additional advantages include power savings of up to 75 percent and simpler wiring practices. A disadvantage in some applications is the slower response time of about 7 s (60-Hz systems have about 2 s).

For a 60-Hz ac system with tapwater flowing at 100 gal/min in a 3-in-diameter pipe, e is about 3 mV rms. The resistance between the electrodes is given by theory[1] as approximately $1/(\sigma d)$, where $\sigma \triangleq$ fluid conductivity and $d \triangleq$ electrode diameter. For tapwater, $\sigma \approx 200\ \mu S/cm$; so if $d \approx 0.64$ cm, there is a resistance of about 7,800 Ω as the internal resistance of the voltage source producing e, which requires the sensor amplifier to have an input impedance that is large relative to this value. Standard magnetic flowmeters accept fluids with conductivities as low as about 5 $\mu S/cm$, with special systems going down to about 0.1 $\mu S/cm$. Gasoline, with $10^{-8}\ \mu S/cm$, is definitely not measurable; alcohol, with 0.2 $\mu S/cm$, is just barely measurable; mercury (a liquid metal), with $10^{10}\ \mu S/cm$, presents no conductivity problem. While commercially available systems are limited to the 0.1 $\mu S/cm$ value, research devices which work with dielectric fluids such as liquid hydrogen have been demonstrated.[2]

A specialized but important application area which has received much attention is that of blood flow[3] in the vessels of living specimens. Miniaturized sensors allow measurements on vessels as small as 1 mm. To obtain dynamic flow capability, ac or switched dc ("square-wave") systems with 200 to 1,000 Hz frequency are used.

While ac and switched dc systems predominate, "pure" dc types have been employed in metering liquid metals, such as mercury. Here, no polarization problem exists. Also, an insulating pipe liner is not needed, since the conductivity of the liquid metal is very good relative to that of an ordinary metal (stainless-steel) pipe. This means that a metal pipe is not very effective as a "short circuit" for the voltage induced in a liquid-metal flow. Also, no special electrodes are necessary, the output voltage being tapped off the metal pipe itself at the points of maximum potential difference. Small, inexpensive pure dc meters using a permanent magnet field and special sintered silver-silver chloride electrodes, which greatly ease the classical electrode instability problems of dc meters, recently have appeared.[4] Originally developed for blood flow, they are usable with many

[1] V. P. Head, Electromagnetic Flowmeter Primary Elements, *Trans. ASME*, p. 662, Dec. 1959.

[2] V. Cushing, D. Reily, and G. Edmunds, Development of an Electromagnetic Induction Flowmeter for Cryogenic Fluids, NASA Lewis Res. Ctr. contract NASW-381, *Final Rept.*, 1964.

[3] R. S. C. Cobbold, "Transducers for Biomedical Measurements," chap. 8, Wiley-Interscience, New York, 1974.

[4] In Vivo Metric Systems, *Tech. Bull.* no. 1, Healdsburg, Calif., Dec. 1980.

liquids with conductivity greater than a few microsiemens per centimeter. Because of the permanent-magnet field and dc operation, the electronics can be very simple, giving a fast-response instrument of modest stability and accuracy useful in various research applications.

Some general features of electromagnetic flowmeters include the lack of any flow obstruction; ability to measure reverse flows; insensitivity to viscosity, density, and flow disturbances as long as the velocity profile is symmetrical; wide linear range; and rapid response to flow changes (instantaneous for a "pure" dc system; limited by the field frequency in an ac or switched dc system).

Drag-Force Flowmeters

A body immersed in a flowing fluid is subjected to a drag force F_d given by

$$F_d = \frac{C_d A \rho V^2}{2} \tag{7.65}$$

where $C_d \triangleq$ drag coefficient
$A \triangleq$ cross-section area, ft^2
$\rho \triangleq$ fluid mass density, slug/ft^2
$V \triangleq$ fluid velocity, ft/s

For sufficiently high Reynolds number and a properly shaped body, the drag coefficient is reasonably constant. Therefore, for a given density, F_d is proportional to V^2 and thus to the square of volume flow rate. Linearization can be achieved with standard analog electronic square-root elements. The drag force can be measured by attaching the drag-producing body to a strain-gage force-measuring transducer. One type[1] uses a cantilever beam with bonded strain gages (see Fig. 7.42). A hollow-tube arrangement with the gages on the outside serves to isolate the gages from the flowing fluid. If the drag body is made symmetric, reversed flows can be measured. A main advantage of this class of flowmeters is the high dynamic response. The type just described is basically second-order with a natural frequency of 70 to 200 Hz. The damping can be quite small; thus sharp transients may cause difficulty.

Ultrasonic Flowmeters

Small-magnitude pressure disturbances are propagated through a fluid at a definite velocity (the speed of sound) *relative to the fluid*. If the fluid also has a velocity, then the *absolute* velocity of pressure-disturbance propagation is the algebraic sum of the two. Since flow rate is related to fluid velocity, this effect may be used in several ways as the operating principle of an "ultrasonic" flowmeter.[2] The term *ultrasonic* refers to the fact that, in practice, the pressure

[1] Ramapo Instrument Co., Bloomingdale, N. J.

[2] L. C. Lynnworth, Ultrasonic Flowmeters, chap. 5, "Physical Acoustics," W. P. Mason and R. N. Thurston (eds.), Academic, N. Y., 1979.

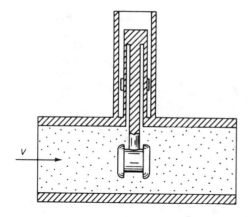

Figure 7.42 Drag-force flowmeter.

disturbances usually are short bursts of sine waves whose frequency is above the range audible to human hearing, about 20,000 Hz. A typical frequency might be 10 MHz.

The various methods of implementing the above phenomenon all depend on the existence of transmitters and receivers of acoustic energy. A common approach is to utilize piezoelectric crystal transducers for both functions. In a transmitter, electrical energy in the form of a short burst of high-frequency voltage is applied to a crystal, causing it to vibrate. If the crystal is in contact with the fluid, the vibration will be communicated to the fluid and propagated through it. The receiver crystal is exposed to these pressure fluctuations and responds by vibrating. The vibratory motion produces an electrical signal in proportion, according to the usual action of piezoelectric displacement transducers.

Figure 7.43a shows the most direct application of these principles. With zero flow velocity the transit time t_0 of pulses from the transmitter to the receiver is given by

$$t_0 = \frac{L}{c} \tag{7.66}$$

where $L \triangleq$ distance between transmitter and receiver and $c \triangleq$ acoustic velocity in fluid. For example, in water, $c \approx 5{,}000$ ft/s, and so if $L = 1$ ft, $t_0 = 0.0002$ s. If the fluid is moving at a velocity V, the transit time t becomes

$$t = \frac{L}{c + V} = L\left(\frac{1}{c} - \frac{V}{c^2} + \frac{V^2}{c^3} - \cdots\right) \approx \frac{L}{c}\left(1 - \frac{V}{c}\right) \tag{7.67}$$

and if we define $\Delta t \triangleq t_0 - t$, then

$$\Delta t \approx \frac{LV}{c^2} \tag{7.68}$$

Thus, if c and L are known, measurement of Δt allows calculation of V. While L may be taken as constant, c varies, for example, with temperature; and since c

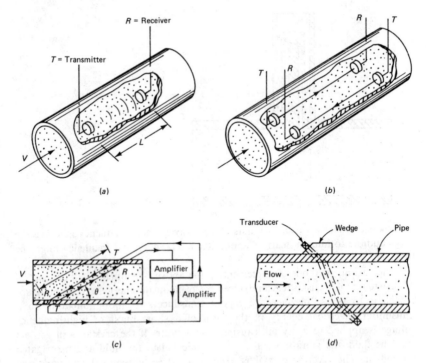

Figure 7.43 Ultrasonic flowmeters.

appears as c^2, the error caused may be significant. Also, Δt is quite small since V is a small fraction of c. For example, if $V = 10$ ft/s, $L = 1$ ft, and $c = 5,000$ ft/s, then $\Delta t = 0.4$ μs, a very short increment of time to measure accurately. Since the measurement of t_0 is not directly provided for in this arrangement, the modification of Fig. 7.43b may be preferable. If t_1 is the transit time with the flow and t_2 is the transit time against the flow, we get

$$\Delta t \triangleq t_2 - t_1 = \frac{2VL}{c^2 - V^2} \approx \frac{2VL}{c^2} \tag{7.69}$$

This Δt is twice as large as before and also is a time increment that physically exists and may be measured directly. However, the dependence on c^2 is still a drawback.

In Fig. 7.43c two self-excited oscillating systems are created by using the received pulses to trigger the transmitted pulses in a feedback arrangement. The pulse repetition frequency in the forward propagating loop is $1/t_1$ while that in the backward loop is $1/t_2$. The frequency difference is $\Delta f \triangleq 1/t_1 - 1/t_2$, and since $t_1 = L/(c + V \cos \theta)$ and $t_2 = L/(c - V \cos \theta)$, we get

$$\Delta f = \frac{2V \cos \theta}{L} \tag{7.70}$$

which is independent of c and thus not subject to errors due to changes in c. This technique is practical and forms the basis of many commercially available ultrasonic flowmeters. Two methods of reading out the frequency difference Δf are common: the "sing-around" and the up-down counter. In the sing-around method, the two signals of different frequency are multiplied, giving an output with sum and difference frequencies. Filtering then extracts the difference frequency. This approach gives fast response to flow changes, but requires pulses with well-defined leading edges,[1] which are not always available, particularly in the clamp-on type of meter, discussed below. The up-down counter scheme accumulates the two frequencies separately for 5 to 20 s and then subtracts them, giving an averaging effect which may reduce spurious noise problems.

To cut costs and reduce errors due to path-length changes caused by buildup of deposits on the transducer faces, most flowmeters of the above type now use two transducers, rather than the four shown in Fig. 7.43c. This is achieved by timesharing the single pair of transducers so it is used alternately for upstream and downstream propagation. High-speed electronic switching makes this possible, and since there is now only one value of L (rather than L_1 and L_2 when unequal deposit layers build up), we avoid the bias error ($\Delta f \neq 0$ when $V = 0$) and independence of c is retained. In the "clamp-on" type[2] of ultrasonic meter (Fig. 7.43d), the transducers are outside the pipe, which eliminates the fouling problems just described and gives an extremely convenient installation devoid of transducer/fluid compatibility problems. Generally, the existing pipe can be employed, and the transducers are attached by mechanical clamping or adhesive bonds, leaving the pipe intact and obstruction-free. Clamp-on meters exhibit their own problem areas, such as "acoustic short circuiting" (receiving transducer feels both liquid-path and pipe-path signals) and changes in beam path due to clamp slippage, temperature expansion, etc. However, these can often be overcome by proper design or use of compensation methods.

A problem common to both clamp-on and "wetted transducer" meters is sensitivity to flow velocity profile. Unlike electromagnetic meters, which read the correct average velocity as long as the flow profile is axisymmetric, ultrasonic meters of the type described above will give different readings for axisymmetric profiles of different shape but identical average velocity [Eq. (7.70) assumed a *uniform* profile]. The reason is that the pulse transit time is related to the integral of fluid velocity along the path. Thus 1 cm of path near the wall is weighted equally with 1 cm near the centerline, even though the contributions of the annular flow *areas* of these two regions would not be the same. If we defined a "meter coefficient" to be 1.0 for a uniform profile, this would drop to 0.75 for laminar flow ($N_R < 2,000$) and vary between 0.93 and 0.96 for turbulent flows with $4,000 < N_R < 10^7$. Rather than using the "tilted diameter" signal path assumed so far in our discussion, if one utilizes a *chordal* path, it has been shown[3]

---

[1] B. G. Liptak and R. K. Kaminski, Ultrasonic Instruments for Level and Flow, *Inst. Tech.*, p. 58, September 1974.

[2] L. C. Lynnworth, Clamp-on Ultrasonic Flowmeters: Limitations and Remedies, *Inst. Tech.*, pp. 37-44, September 1975; J. Baumoll, Letter to Editor, *Inst. Tech.*, p. 4, November 1975.

[3] Lynnworth, Ultrasonic Flowmeters, p. 430.

that the meter coefficient is nearly the same for both laminar and turbulent flow. Another approach is to employ several signal paths and average their results. Also, if flow conditions are reproducible, computed corrections always can be applied, even for a meter with a single diametral path.

The other main category of commercial ultrasonic meters uses the doppler principle. Whereas the "transit time" meters discussed above require relatively clean fluid to minimize signal attenuation and dispersion, doppler meters will not work at all unless sufficient reflecting particles and/or air bubbles are present. For example,[1] in an 8-in pipe, 10 percent by volume of 40-μm reflectors or 0.2 percent of 100-μm reflectors would be needed. Doppler meters usually employ a clamp-on configuration; Fig. 7.44 shows one of several possible arrangements. Here the transmitter propagates a continuous-wave (CW) ultrasonic (0.5 to 10 MHz) signal into the fluid (assumed to have a uniform-profile velocity V), whereupon particles (assumed to be moving at velocity V) reflect some of the energy to the receiver. Analysis[2] of a wetted-transducer design gives

$$\Delta f = f_t - f_r = \frac{2 f_t \cos \theta}{c} V \tag{7.71}$$

The apparent dependence on c is misleading for the clamp-on design of Fig. 7.44, since it can be shown[3] (1) that changes in c cause compensating changes in $\cos \theta$ and (2) that for such a design, one should interpret θ to be the transducer wedge angle and c to be the propagation velocity for the *wedge* material, not the fluid. This makes temperature effects manageable and contributes to the practicality of the doppler flowmeter.

The desired Δf signal can be obtained electronically from the f_t and f_r signals by methods already explained for transit-time meters. With regard to flow profile effects, the doppler meter "interrogates" only the sensing volume defined by the intersection of the transmitted and reflected beams; thus it measures an average velocity over this space (not over a tilted diametral path from wall to wall, as in a transit-time meter). This "point" velocity measurement can be calibrated, of course, to give total volume flow rate; however, the flow profile must be reproducible. (Also, any shift in the position of the sensing volume can cause errors even if the flow profile is reproducible.) If our goal is to measure "point" velocities rather than gross flow rates, this "problem" of the doppler meter actually becomes an advantage. A pulsed (rather than CW) system[4] using range-gating techniques actually has been utilized to get velocity profiles in blood vessels.

Because of the availability of meters for both clean and dirty fluids, little or no obstruction or pressure drop, convenient installation for clamp-on types, and

[1] J. M. Waller, Guidelines for Applying Doppler Acoustic Flowmeters, *InTech*, pp. 55–57, October 1980. T. R. Schmidt, What You Should Know about Clamp-on Ultrasonic Flowmeters, *InTech*, pp. 59–62, May 1981.

[2] Cobbold, op. cit., p. 281.

[3] Lynnworth, Ultrasonic Flowmeters, p. 444.

[4] Cobbold, op. cit., p. 286.

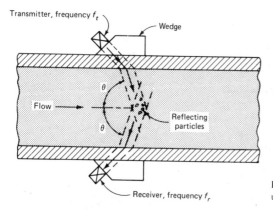

Transmitter, frequency f_t

Wedge

Flow

θ

θ

Reflecting particles

Receiver, frequency f_r

Figure 7.44 Clamp-on Doppler ultrasonic flowmeter.

good rangeability, ultrasonic meters are finding increasing application. Commercial meters are generally limited to liquids; however, ultrasonic principles have been applied successfully to gas flow in wind-measuring systems.[1]

Vortex-Shedding Flowmeters

The phenomenon of vortex shedding ("Karman vortex street") downstream of an immersed solid body of "blunt" shape when a steady flow impinges upstream is well known in fluid mechanics and is the basis of the vortex-shedding flowmeter. When the pipe Reynolds number N_R exceeds about 10,000, vortex shedding is reliable, and the shedding frequency f is given by

$$f = \frac{N_{st} V}{d} \qquad (7.72)$$

where $V \triangleq$ fluid velocity
$d \triangleq$ characteristic dimension of shedding body
$N_{st} \triangleq$ Strouhal number, an experimentally determined number, nearly constant in the useful flowmetering range (for example, 0.21 for cylinders)

By proper design of the shedding body shape, N_{st} can be kept nearly constant over a wide range of N_R (and thus flow rate), making f proportional to V and thus giving a "digital" flowmetering principle based on counting the vortex shedding rate (see Fig. 7.45). Various shedder shapes and frequency sensing schemes have been developed by several manufacturers. The vortices cause alternating forces or local pressures on the shedder; piezoelectric and strain-gage methods can be employed to detect these. Hot-film thermal anemometer sensors buried in the shedder can detect the periodic flow-velocity fluctuations. The interruption of ultrasonic beams by the passing vortices can be used to count

[1] Doppler-Acoustic Remote Sensor, Xonics Inc., Van Nuys, Calif., 1978.

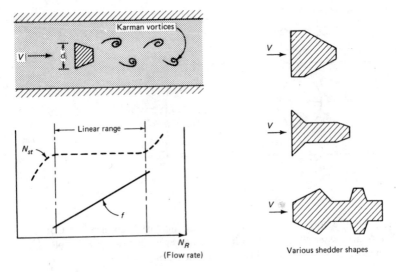

Various shedder shapes

Figure 7.45 Vortex-shedding flowmeter principles.

them. Vortex-induced differential pressures will cause oscillation of a small caged ball whose motion can be detected with a magnetic proximity pickup. Figure 7.46 shows yet another scheme which senses differential pressure with an elastic diaphragm.

A wide range of liquids and gases and steam may be metered. Linear ranges of 15:1 are common, with 200:1 sometimes possible. Vortex frequencies at maximum flow rate are the order of 200 to 500 Hz, and the frequency responds to changing flow rate within about 1 cycle. If a frequency-to-voltage converter is used to realize an analog output voltage, then its low-pass filter will determine overall system response, which can be quite fast at high flow rates but must become slower at low flow rates (because the filter time constant must be set longer to give acceptable ripple at the lowest frequency, typically 5 to 10 Hz). As with the ultrasonic flowmeters, desirable characteristics of vortex-shedding meters have allowed them to take over a portion of the flowmeter market from the more traditional types of instruments.

Miscellaneous Topics

We briefly mention here three topics that cannot be treated in depth because of space limitations but which are felt worthy of more than just a bibliographic entry. *Two-phase flow*[1] occurs in a number of practical applications and has been of particular interest for steam/water and air/water mixtures in steam power

[1] G. F. Hewitt, "Measurement of Two Phase Flow Parameters," Academic, New York, 1978.

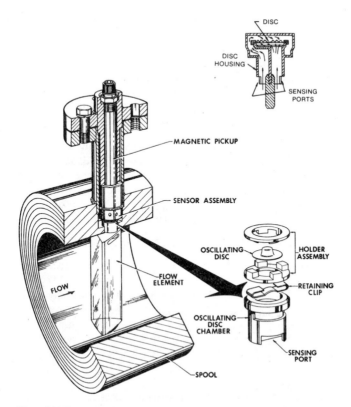

Figure 7.46 Vortex-shedding flowmeter details. (*Courtesy Neptune Eastech, Edison, N. J.*)

plants.[1] Severe environments and lack of well-established calibration standards make instrument development particularly difficult. Few commercial instruments are available, and special designs or adaptations of standard devices usually are necessary.

Metering of *cryogenic fluids* such as liquid oxygen[2] and hydrogen (used in rocket engines), liquefied natural gas, etc., presents problems of extreme low temperatures and explosion hazards. Use of safer liquid nitrogen as a surrogate fluid for calibration purposes[3] has been established as a valid procedure and is the basis of a calibration facility which provides a mass flow uncertainty of about

[1] Development of Instruments for Two-Phase Flow Measurements, *Rept.* ANCR-1181, Aerojet Nuclear Co., Idaho Falls, Idaho, 1974.

[2] D. B. Mann, ASRDI Oxygen Technology Survey, vol. 6, Flow Measurement Instrumentation, *NASA* SP-3084, 1974.

[3] J. A. Brennan et al., An Evaluation of Selected Angular Momentum, Vortex Shedding and Orifice Cryogenic Flowmeters, *NBS Tech. Note* 650, Cryogenics Div., Boulder, Colo., 1974.

± 0.2 percent over a flow range of 20 to 210 gal/min at temperatures of 70 to 90 K.

The use of *cross-correlation* as a basic flowmetering principle is being researched and may lead to a family of commercial instruments, since it can be implemented in several ways[1] by utilizing "tagging signals" already present in the flow or intentionally introduced. Briefly, the signals from two sensors separated axially in the pipe by a distance L and responding to the tagging signals are cross-correlated by using a range of delays. Since the tagging signals are swept from one sensor to the other at flow velocity V, the cross-correlation function should exhibit a peak at a delay equal to L/V seconds, which allows calculation of V since L is known. Turbulent eddies naturally present and detected by doppler ultrasonic methods, or thermal pulses intentionally injected and then sensed by thermocouples are only two of many possible schemes being considered. The practicality of the method hinges strongly on the availability of low-cost correlators based on integrated-circuit and microprocessor technology.

7.3 GROSS MASS FLOW RATE

In many applications of flow measurement, mass flow rate is actually more significant than volume flow rate. As an example, the range capability of an aircraft or liquid-fuel rocket is determined by the *mass* of fuel, not the volume. Thus flowmeters used in fueling such vehicles should indicate mass, not volume. In chemical process industries also, mass flow rate is often the significant quantity.

Two general approaches are employed to measure mass flow rate. One involves the use of some type of volume flowmeter, some means of density measurement, and some type of simple computer to compute mass flow rate. The other, more basic, approach is to find flowmetering concepts that are inherently sensitive to mass flow rate. Both methods are currently finding successful application in various forms.

Volume Flowmeter Plus Density Measurement

Some basic methods of fluid-density measurement are shown in Fig. 7.47. In Fig. 7.47*a* a portion of the flowing liquid is bypassed through a still well. The buoyant force on the float is directly related to density and may be measured in a number of ways, such as the strain-gage beam shown. In Fig. 7.47*b* a definite volume of flowing liquid contained within the U tube is continuously weighed by a spring and pneumatic displacement transducer. Flexible couplings isolate external forces from the U tube. A pneumatic force-balance feedback system also can be used to

---

[1] M. S. Beck, Correlation in Instruments: Cross Correlation Flowmeters, *J. Phys. E: Sci. Instrum.,* vol. 14, no. 1, 1981.

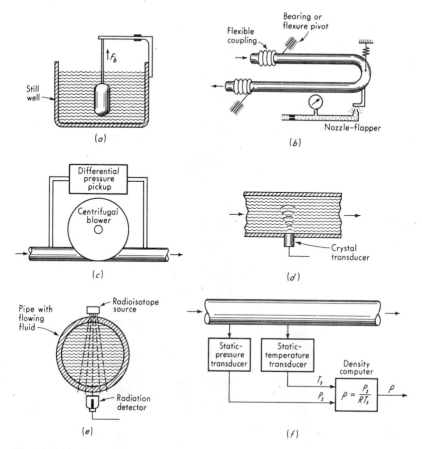

Figure 7.47 Fluid-density measurement.

measure the weight.[1] This minimizes deflection and thus reduces errors due to variable spring effects of flexible couplings and flexure pivots.

Figure 7.47c shows a method of measuring gas density by using a small centrifugal blower (run at constant speed) to pump continuously a sample of the flow. The pressure drop across such a blower is proportional to density and may be measured with a suitable differential-pressure pickup. Ultrasonic volume flowmeters often use an ultrasonic density-measuring technique when mass flow rate is wanted. In Fig. 7.47d the crystal transducer serves as an acoustic-impedance detector. Acoustic impedance depends on the product of density and speed of sound. Since a signal proportional to the speed of sound is available from the

---

[1] Halliburton Co., Duncan, Okla.

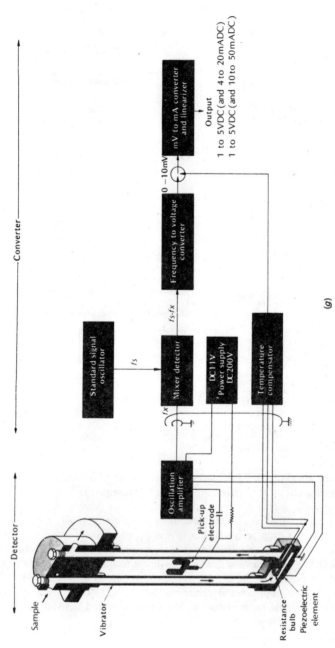

Figure 7-47 (*Continued*)

560

volume flowmeter, division of this signal into the acoustic-impedance signal gives a density signal.[1] The attenuation of radiation from a radioisotope source depends on the density of the material through which the radiation passes (see Fig. 7.47e). Over a limited (but generally adequate) density range, the output current of the radiation detector is nearly linear with density, for a given flowing fluid.[2] For gas flow, indirect measurement of density by means of computation from pressure and temperature signals (Fig. 7.47f) is also common.[3]

Densitometers based on the effect of density on the natural frequency of a vibrating system are available for both liquids and gases and are in wide use. These generally employ a positive-feedback electromechanical drive system fed from a vibration transducer, so as to maintain a constant amplitude of motion as the frequency changes because of density changes. Figure 7.47g[4] shows a design using piezoelectric drive elements. The sample fluid flows continuously through a metal U tube whose bending vibration frequency f_x is sensitive to the fluid mass in the tube and thus to fluid density. The density/frequency relation generally is nonlinear, but this can be linearized with suitable signal processing if desired.

In computing mass flow rate from volume flowmeter and densitometer (density-measuring device) signals, the necessary form of computer varies somewhat, depending on the type of flowmeter. For "head" meters (those producing a differential pressure or electrical signal proportional to ρV_{av}^2, such as an orifice) the computer multiplies the ρ signal by the ρV_{av}^2 to form $\rho^2 V^2$ and then takes the square root of this to get ρV, which is proportional to the mass flow rate. For "velocity" flowmeters, such as the turbine and electromagnetic types, the available signal is proportional to V_{av}; thus the computer simply must multiply this by the ρ signal, and the square-root operation is unnecessary.

Direct Mass Flowmeters

While the above indirect methods of mass-flow-rate measurement often are satisfactory and are in wide use, it is possible to find flowmetering concepts that are more directly sensitive to mass flow rate. These may have advantages with respect to accuracy, simplicity, cost, weight, space, etc., in certain applications. We discuss briefly some of the more common principles in terms of the practical hardware through which they have been realized.

A principle widely employed in aircraft fuel-flow measurement depends on the moment-of-momentum law of turbomachines. Fluid mechanics shows that, for one-dimensional, incompressible, lossless flow through a turbine or an impeller wheel, the torque T exerted by an impeller wheel on the fluid (minus sign) or on a turbine wheel by the fluid (plus sign) is given by

$$T = G(V_{ti} r_i - V_{to} r_o) \qquad (7.73)$$

[1] L. C. Lynnworth et al., Ultrasonic Mass Flowmeter for Army Aircraft Engine Diagnostics, *USAAMRDL Tech. Rept.* 72-66, Fort Eustis, Va., 1973.

[2] AccuRay Corp., Columbus, Ohio.

[3] Gas Mass Flow Computation, *Ref.* A0-09-01, Honeywell Inc., Fort Washington, Pa., 1977.

[4] Yokogawa Corp. of America, Elmsford, N.Y.

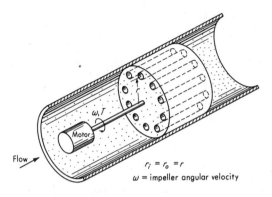

Flow

$r_i = r_o = r$
ω = impeller angular velocity

Figure 7.48 Angular-momentum element.

where $G \triangleq$ mass flow rate through wheel, slug/s (kg/s)
$V_{ti} \triangleq$ tangential velocity at inlet, ft/s (m/s)
$V_{to} \triangleq$ tangential velocity at outlet, ft/s (m/s)
$r_i \triangleq$ radius at inlet, ft (m)
$r_o \triangleq$ radius at outlet, ft (m)
$T \triangleq$ torque, ft · lbf (N · m)

Consider now the system of Fig. 7.48. The flow to be measured is directed through an impeller wheel which is motor-driven at constant speed. If the incoming flow has no rotational component ($V_{ti} = 0$); and if the axial length of the impeller is enough to make $V_{to} = r\omega$, the driving torque necessary on the impeller is

$$T = r^2\omega G \qquad (7.74)$$

Since r and ω are constant, the torque T (which could be measured in several ways) is a direct and linear measure of mass flow rate G. However, for $G = 0$, torque will *not* be zero, because of frictional effects; furthermore, viscosity changes also would cause this zero-flow torque to vary. A variation on this approach is to drive the impeller at constant *torque* (with some sort of slip clutch). Then, impeller *speed* is a measure of mass flow rate according to

$$\omega = \frac{T/r^2}{G} \qquad (7.75)$$

The speed ω is now nonlinear with G, but may be easier to measure than torque. If a magnetic proximity pickup is used for speed measurement, the time duration t between pulses is inversely related to ω; thus measurements of t would be linear with G.

A further variation, the basis of most commercial instruments,[1] is shown in Fig. 7.49. A constant-speed motor-driven impeller again imparts angular momentum to the fluid; however, no torque or speed measurements are made on this

---

[1] General Electric Aerospace Instruments Div., Wilmington, Mass.

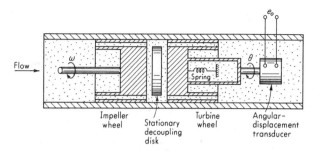

Figure 7.49 Angular-momentum mass flowmeter.

wheel. Closeby, downstream, a second ("turbine") wheel is held from turning by a spring restraint. For the impeller, $V_{to} = r\omega$; furthermore, this becomes V_{ti} for the turbine. Since the turbine cannot rotate, if it is long enough axially, the angular momentum is removed and V_{to} for the turbine is zero. Then the torque on the turbine is

$$T = r^2 \omega G \qquad (7.76)$$

If the spring restraint is linear, the deflection θ is a direct and linear measure of G and can be transduced to an electrical signal in a number of ways. The decoupling disk reduces the viscous coupling between the impeller and turbine so that, at zero flow rate, a minimum of viscous torque is exerted on the turbine wheel.

Figure 7.50 shows a version of Fig. 7.48 which embodies several advantageous features and is the basis of successful commercial instruments. Again a spring measures torque, but now both ends of the spring rotate at the same speed, and spring deflection is inferred from the time interval between pulses produced by two magnetic pickoffs. This makes the Δt signal independent of impeller speed ω, since a smaller ω gives a smaller torque (and spring deflection), but this is exactly compensated by the increase in Δt resulting from ω itself being lower. This independence of ω allows use of a drive (not shown in Fig. 7.50) employing a fluid turbine and thus eliminating the constant-speed electric motor drive completely. While the fluid drive need not maintain an exact speed, a simple internal valving scheme maintains a nominal speed for greater accuracy.

A number of other concepts using angular-momentum principles have been suggested. They include the vibrating gyroscope meter, the Coriolis meter, the rotating gyroscope meter, the twin-turbine meter, and the S-tube meter.[1] A Coriolis-type meter which has been successfully reduced to commercial practice[2] is shown in Fig. 7.51. Fluid flowing at mass flow rate G kilograms per second passes through a C-shaped pipe cantilevered from two supporting brackets. The

[1] C. M. Halsell, Mass Flowmeters, *ISA J.*, p. 49, June 1960; J. Haffner, A. Stone, and W. K. Genthe, Novel Mass Flowmeter, *Contr. Eng.*, p. 69, October 1962.

[2] K. O. Plache, Coriolis/Gyroscopic Flow Meter, *Mech. Eng.*, pp. 36–41, March 1979; Micro Motion Inc., Boulder, Colo.

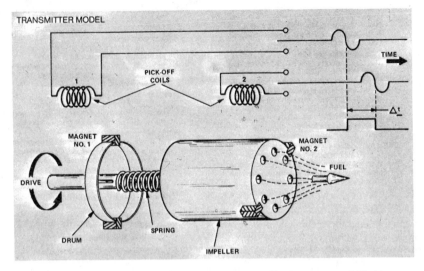

Figure 7.50 Flow-driven angular-momentum meter. (*Courtesy Eldec Corp., Lynnwood, Wash.*)

pipe is maintained in steady sinusoidal bending vibration (at its natural frequency, 50 to 80 Hz, as a cantilever beam) by an electromagnetic feedback system. This is a self-excited drive system which always runs at beam natural frequency (and thus minimum power), even though this frequency changes when fluid density changes. This is accomplished by deriving the force-motor drive signal from a velocity sensing coil wound on the same form as the driver coil (they share the same permanent-magnet core). Amplitude is stabilized by feedback which compares sense-coil voltage (velocity) with a fixed reference signal. The mechanical "tuning fork" configuration minimizes the vibratory force into the frame.

Coriolis-type meters require that the fluid experience an angular velocity ω whose vector is perpendicular to the fluid velocity V. In this example, ω is an oscillatory motion produced by the C tube bending about its cantilever supports. For the simplified analysis of Fig. 7.51b, ω is treated as a rigid-body rotation about a fixed axis, and fluid flow is represented by a single velocity V rather than by a velocity profile. The absolute acceleration $\ddot{r}$ of a point located by a vector ρ from the origin (located by a vector R from a fixed reference point) of a coordinate system rotating with vector angular velocity ω is given by[1]

$$\ddot{\mathbf{r}} = \ddot{\mathbf{R}} + \omega \times (\omega \times \rho) + \dot{\omega} \times \rho + \ddot{\rho}_r + 2\omega \times \dot{\rho}_r \qquad (7.77)$$

For our example $\mathbf{R} \equiv 0$, so $\ddot{\mathbf{R}} = 0$, and $\dot{\rho}_r = \mathbf{V}$. The flowmeter motion pickoffs sense the twist angle θ; thus we are interested in only those inertia forces which

[1] G. W. Housner and D. E. Hudson, "Applied Mechanics–Dynamics," p. 31, Van Nostrand, New York, 1950.

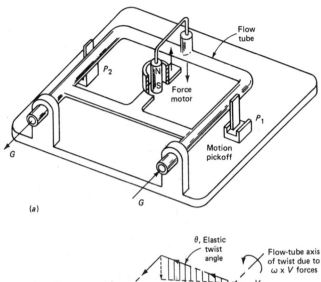

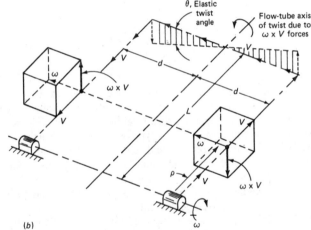

Figure 7.51 Coriolis-type mass flowmeter.

cause twist. In Eq. (7.77) the only term of this type is the Coriolis acceleration $2\omega \times V$ (see Fig. 7.51b).

An element of fluid mass dM at location ρ produces an inertia force of magnitude $(dM)(2\omega \times V)$ and direction opposite to $\omega \times V$. Since V changes sign from the right to the left leg of the C tube, a pair of right/left mass elements produces an inertia torque dT:

$$dT = 2(2\omega \times V)(dM)d = 2(2\omega \times V)\left(\frac{G}{V} d\rho\right)d \qquad (7.78)$$

$$T = \int_0^L dT = 4\omega Gd \int_0^L d\rho = 4Lr\omega G \qquad (7.79)$$

The angular velocity ω oscillates sinusoidally, so the torque T is sinusoidal also. It acts as a driving torque, tending to twist the C tube; and since the C-tube torsional natural frequency is well above this driving frequency, the torsional spring/mass system acts essentially as a spring of stiffness K_s, allowing calculation of the twist angle θ from

$$\theta = \frac{4Ld\omega}{K_s} G \tag{7.80}$$

The motion pickoffs P_1 and P_2 (both optical and magnetic types have been utilized) are located near the pipe neutral position and produce a switching action when the pipe passes the pickoff location; "proportional" sensors are not used. Because of twist θ, one of the pickoffs will be triggered a time interval Δt later than the other. If the average angular velocity over this Δt is ω_{av}, then

$$\theta = \frac{(L\omega_{av}) \Delta t}{2d} \approx \frac{L\omega \Delta t}{2d} \tag{7.81}$$

where the instantaneous value $\omega \approx \omega_{av}$ because the motion is sensed over a small fraction of the total cycle. Combining (7.80) and (7.81), we get

$$G = \frac{K_s}{8d^2} \Delta t \tag{7.82}$$

which shows that Δt is a linear measure of mass flow rate. In the actual system, the Δt measurement is implemented[1] as a pulse-width modulation scheme using a gated oscillator feeding an up/down counter. Also total flow over any time interval is obtained easily by digital integration. This type of meter is obstructionless; is essentially insensitive to viscosity, pressure, and temperature; and can handle clean liquids, mixtures, foams and slurries, and liquids with entrained gases. Since Δt is measured once per bending cycle, the meter can respond rapidly to changing flows; however, averaging over several cycles generally is used to improve accuracy for measurements of average flow rate.

An interesting mass flowmeter[2] for liquids, based on quite a different principle, is shown schematically in Fig. 7.52. For a given fluid, the pressure drop across an orifice is proportional to ρQ^2. In Fig. 7.52, four identical orifices are connected into a "bridge circuit." A positive-displacement pump of fixed displacement runs at constant speed and volume flow rate q. The pressure rise across this pump is Δp_{cb}, which pressure difference is the output signal of the flowmeter. The flow rates through the individual orifices must be as shown, because of symmetry. The output signal Δp_{cb} is

$$\Delta p_{cb} = K\rho \left(\frac{Q+q}{2}\right)^2 - K\rho \left(\frac{Q-q}{2}\right)^2 = Kq\rho Q = K_1 G \tag{7.83}$$

[1] Plache, op. cit.
[2] Flo-Tron Inc., Paterson, N.J.

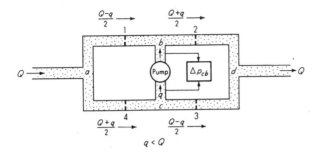

Figure 7.52 Bridge-circuit flowmeter.

Thus Δp_{cb} is linear with mass flow rate G. The "constant" K_1 includes the orifice discharge coefficient; thus all orifice flow rates must be maintained at high enough Reynolds numbers that the discharge coefficients of all orifices are equal and do not vary when Q varies. Various other arrangements of orifices, pumps, and pressure pickups have been devised; they give the same overall result, but have relative advantages and disadvantages in other respects.

Our final example of mass flowmeters uses heat-transfer principles. In Fig. 7.53 an electric heating coil is transferring heat to a fluid flowing inside a pipe. If the pipe wall is a good thermal conductor and quite thin and if heat losses are minimized by insulation, then the temperature drop across the boundary layer for turbulent flow is given by

$$\Delta T = \frac{K_1 P_h}{h} \tag{7.84}$$

where $K_1 \triangleq$ constant (conversion factor and heat-transfer area)
$P_h \triangleq$ heater power
$h \triangleq$ film conductance of boundary layer

For turbulent flow, the film conductance is given by

$$h = 0.023 \left(\frac{k^{0.6} c^{0.4}}{D^{0.2} \mu^{0.4}} \right) G^{0.8} \tag{7.85}$$

where $k \triangleq$ fluid thermal conductivity
$c \triangleq$ fluid specific heat
$D \triangleq$ pipe diameter
$\mu \triangleq$ fluid absolute viscosity
$G \triangleq$ mass flow rate

In Eq. (7.85), if the quantity in parentheses is constant or can be compensated for, then we see that h is given by $h = K_2 G^{0.8}$. Then from Eq. (7.84) we get

$$P_h = \frac{\Delta T \, K_2}{K_1} G^{0.8} \tag{7.86}$$

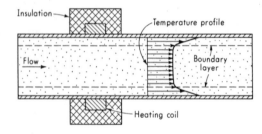

Figure 7.53 Boundary-layer principle.

If the heater power P_h is adjusted to keep ΔT always constant, then P_h becomes a direct (and almost linear) measure of mass flow rate G. This is the principle of the boundary-layer (electrocaloric) flowmeter.[1]

It would seem that to measure ΔT would require a temperature probe in the core of the flowing fluid. Fortunately, this complication is unnecessary; the pipe-wall temperature 3 to 5 in upstream of the heater is very close to the fluid core temperature downstream of the heater. The adjustment of heater power to maintain a constant ΔT (about 2°F is used) is accomplished continuously and automatically by a feedback system shown simplified in Fig. 7.54. A wattmeter reading P_h can be calibrated in mass-flow-rate units for a given fluid. Its reading is the output signal of the flowmeter. A typical value of P_h is about 40 W if 12,000 lb/h of water is flowing in a 1.5-diameter pipe.

The feedback system operation may be explained as follows: If the incoming fluid has a fixed temperature T_f, the resistance R_f will be constant. The bridge resistor R_{fb} also will be constant and made equal to R_f. Suppose that R_{wb} is set equal to R_w; then the bridge is balanced and no heater power is being supplied. If the temperature coefficient of R_w is known, the change in R_w corresponding to the desired ΔT can be calculated. Now R_{wb} is changed by this amount, unbalancing the bridge and thus turning on the heater power. This raises R_w toward the new value of R_{wb}, which tends to rebalance the bridge. Perfect balance ($R_w = R_{wb}$) cannot be achieved, since then the heater power would have to be zero. If the gain of the feedback system is high, however, nearly perfect bridge balance is possible without making the heater power zero. With the bridge in this condition, an increase, say, in mass flow rate G results in a momentary decrease in T_w (and thus R_w), unbalancing the bridge in the direction to increase heater power, increase T_w, and thus maintain the desired ΔT. A decrease in G results in the opposite action, again tending to maintain ΔT fixed. If ΔT is kept fixed for all flow rates, Eq. (7.86) shows that the wattmeter reading (system output signal) is a direct indication of mass flow rate.

Since the hot-wire and hot-film point velocity sensors discussed in Sec. 7.1 are really sensitive to the product ρV, rather than to V itself, such sensors can be mounted in a flow tube (usually of venturi shape) and calibrated for total mass

[1] J. H. Laub, Measuring Mass Flow with the Boundary-Layer Flowmeter, *Contr. Eng.*, p. 112, March 1957.

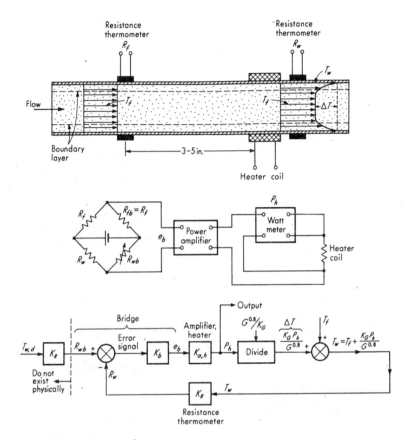

Figure 7.54 Boundary-layer flowmeter.

flow rate. Flowmeters of this type, complete with electronic linearization, are available for both liquids and gases and provide fast response, with time constants of 0.001 s being possible in some cases.

PROBLEMS

7.1 Water flows in a 1-in-diameter pipe at 10 ft/s. If a pitot-static tube of 0.5-in diameter is inserted, what velocity will be indicated? Assume one-dimensional frictionless flow. Find the pitot-static-tube diameter needed to reduce the above error to 1 percent.

7.2 For the system of Fig. 7.12, what diameter sphere is needed to obtain a 1-lb force from a 50 mi/h wind when atmospheric pressure is 14.7 lb/in^2 and temperature is 70°F? If the 50 mi/h wind is assumed to be the full-scale value and if 0.05-in full-scale deflection of the differential transformer is desired, what must be the flexure-plate spring constant? If the total moving mass is assumed to be due to the spherical shell alone and if the shell is made of $\frac{1}{16}$-in-thick aluminum, estimate the usable frequency range of this instrument.

7.3 If, in the system shown in Fig. 7.17, the amplifier lag is not neglected, so that (i/e_o) $(D) = K_a/[(\tau_1 D + 1)(\tau_2 D + 1)]$, find the value of K_a that will put the feedback system just on the margin of instability (use the Routh criterion). If $\tau = 0.001$ s and $\tau_1 = \tau_2 = 0.000001$, what is the maximum allowable value of $K_b K_a K_e/I_m$ for marginal stability? If a value of $K_b K_a K_e/I_m$ about one-fifth of that giving marginal stability can be used and if Eq. (7.43) is assumed applicable under these conditions, what percentage improvement of τ_{ct} as compared with τ may be achieved?

7.4 The frequency response of a hot-wire anemometer system also determines its ability to resolve *spacewise* variations in velocity. That is, at a given instant of time, if the velocity pattern in a flow were "frozen," the velocity component in a certain direction would be different at various stations along the line of travel of the gross flow. When this velocity structure is swept past a hot wire by the average velocity, it requires adequate frequency response to resolve the spacewise velocity variations. Consider a simple spacewise variation in which velocity deviation from average is given by $v = v_0 \sin(2\pi x/\lambda)$, where λ is the wavelength in inches and x is displacement in inches along the flow direction. If the average flow velocity is V_0 in/s find an expression for the smallest wavelength λ_{min} that can be resolved by a system whose flat frequency response extends to f_0 hertz. Plot λ_{min} versus V_0 for a system with f_0 of 10,000 Hz.

7.5 Analyze the error in flow-rate measurement caused by thermal expansion of an orifice plate.

7.6 A capillary-tube laminar-flow element is needed to measure water flow of 0.01 in³/min at 70°F. A flowmeter pressure drop of 3 in of water is desired. If the element is designed for a Reynolds number of 500, what length and diameter of tubing are needed?

7.7 Using the simplified model discussed in the text, derive Eq. (7.60).

7.8 Using the assumptions discussed following Eq. (7.60), one can write for the weight flow rate

$$W = K_1(A_t - A_f) \sqrt{w_{ff}(w_f - w_{ff})} \qquad \text{lbf/s}$$

where K_1 includes all the other constants in Eq. (7.60). To make the weight-flow indication relatively insensitive to changes in fluid density w_{ff}, the float density w_f should be twice the density of the flowing fluid. Show the truth of this statement. *Hint :* Set $\partial W/\partial w_{ff} = 0$.

7.9 A wind-velocity-measuring sphere (as in Fig. 7.12) is mounted on a buoy (as in Fig. P7.1) and has a diameter of 18 cm, a mass of 0.5 kg, $C_d = 0.567$, and the force transducer spring con-

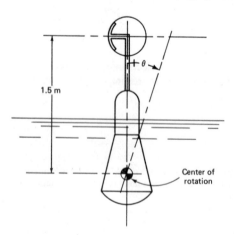

1.5 m

θ

Center of
rotation

Figure P7.1

stant = 3,600 N/m. Assume there is no wind, but the buoy is rocking about a single axis with θ = 0.2 sin $2t$ radians (t in seconds). This motion will cause a force transducer reading even though there is no wind. Analyze to find the wind measurement error caused by this spurious input. Then suggest several methods for reducing and/or compensating for this kind of error.

7.10 Outline the procedure you would use to design a drag-force flowmeter of the type shown in Fig. 7.42. The given specifications include static sensitivity, dynamic response, flow-velocity range to be covered, allowable size, and fluid-density range to be covered.

7.11 Perform a dynamic analysis on the system of Fig. 7.47a to obtain the transfer function relating fluid density as an input to strain-gage bridge voltage as an output. What is the effect of changes in liquid level in the still well on the output signal? What is the effect of thermal (volume) expansion of the float? If the entire assembly is aboard a vehicle that is accelerating vertically, explain the effect on the output.

7.12 Figure P7.2 shows a laminar-flow element and associated differential-pressure transducer. We wish to study the dynamic response of this flow measurement system. Model the laminar element as a lumped liquid inertance and resistance (no fluid compliance) and treat the pressure transducer/tubing system as heavily damped, slow-acting. Find $[(p_1 - p_2)/Q] (D)$.

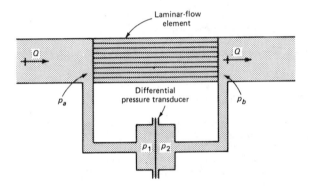

Figure P7.2

7.13 Perform a static analysis on the system of Fig. 7.47b to get a relation between fluid density as an input and nozzle-flapper pressure as output. List modifying and/or interfering inputs for this instrument.

7.14 Modify the system of Fig. 7.47b, using a feedback principle (null-balance system) to keep vertical deflection nearly zero for all densities. A bellows may be used to provide the rebalancing force.

7.15 For the system of Fig. 7.49, suppose that the full-scale flow rate is 10 lbm/s, an impeller speed of 100 r/min is used, and r = 1.0 in. If full-scale transducer rotation is to be $20°$, what torsional spring constant is required? If this spring constant must be increased to improve dynamic response, what design changes are possible to achieve this?

7.16 Intuitively, what would you expect the dynamic response of the system of Fig. 7.49 to be? What design parameters would have a major influence on this response and in what way? Devise an experimental technique for subjecting this instrument to an approximate step input.

7.17 List modifying and/or interfering inputs for the system of Fig. 7.54.

BIBLIOGRAPHY

1. "Flow: Its Measurement and Control in Science and Industry," vol. 2, ISA, Research Triangle Park, N. C., 1981.
2. "Flow: Its Measurement and Control in Science and Industry," vol. 1, ISA, Research Triangle Park, N. C., 1971.
3. D. R. Lynch: A Low-Conductivity Magnetic-Flowmeter, *Contr. Eng.*, p. 122, December 1959.
4. D. J. Lomas: Selecting the Right Flowmeter, *Inst. Tech.*, pp. 55–62, May 1977; pp. 71–77, June 1977.
5. J. A. Shercliff: "The Theory of Electromagnetic Flow Measurement," Cambridge University Press, New York, 1962.
6. B. E. Richards (ed): "Measurement of Unsteady Fluid Dynamic Phenomena," CRC Press, Boca Raton, Fla., 1977.
7. S. Blechman: Techniques for Measuring Low Flows, *Instrum. Contr. Syst.*, p. 82, October 1963.
8. E. W. Miller: Turbine Gas-Flow Sensor, *Instrum. Contr. Syst.*, p. 105, January 1962.
9. H. J. Evans: Turbine Flowmeter for Gases, *Instrum. Contr. Syst.*, p. 103, March 1964.
10. R. D. Wood: Steam Measurement by Orifice Meter, *Instrum. Contr. Syst.*, p. 135, April 1963.
11. R. B. Crawford: A Broad Look at Cryogenic Flow Measurement, *ISA J.*, p. 65, June 1963.
12. D. Shichman and B. S. Johnson: Tap Location for Segmental Orifices, *Instrum. Contr. Syst.*, p. 102, April 1962.
13. P. J. Klass: Laser Flowmeter, *Aviation Week*, January 11, 1965.
14. R. W. Henke: Positive Displacement Meters, *Contr. Eng.*, p. 56, May 1955.
15. T. Isobe and H. Hattori: A New Flowmeter for Pulsating Gas Flow, *ISA J.*, p. 38, December 1959.
16. A. K. Oppenheim and E. G. Chilton: Pulsating Flow Measurement—A Literature Survey, *ASME Trans.*, p. 231, February 1955.
17. J. R. Musham and B. G. Lewis: Direct Reading Flow Rate Meter for Low Flow Rates, *Contr. Eng.*, p. 115, December 1961.
18. L. Gess: Common Troubles with Head Flowmeters, *ISA J.*, p. 58, February 1958.
19. R. Shapcott: How to Select Flowmeters, *ISA J.*, p. 272, July 1957.
20. H. E. Wingo: Thermistors Measure Low Liquid Velocities, *Contr. Eng.*, p. 131, October 1959.
21. W. D. Hamilton: Flow Elements from Tubing Elbows, *ISA J.*, p. 61, July 1963.
22. E. G. Keshock: Comparison of Absolute- and Reference-System Methods of Measuring Containment Vessel Leakage Rates, *NASA, Tech. Note* D-1588, 1964.
23. Ball Bearing Used in Design of Rugged Flowmeter, *NASA, Brief* 64-10170, 1964.
24. Meter Accurately Measures Flow of Low-Conductivity Fluids, *NASA, Brief* 63-102080, 1963.
25. E. L. Upp: Flowmeters for High-Pressure Gas, *Instrum. Contr. Syst.*, p. 151, March 1965.
26. N. F. Cheremisinoff: "Applied Fluid Flow Measurement," Marcel Dekker, New York, 1979.
27. A. Haalman: Pulsation Errors in Turbine Flowmeters, *Contr. Eng.*, p. 89, May 1965.
28. R. Siev: Mass Flow Measurement, *Instrum. Contr. Syst.*, p. 966, June 1960.
29. Heat Transfer Flowmeter Has No Pressure Drop, *Space/Aeron.*, p. 259, January 1964.
30. Mass Flow by Temperature Measurement, *Instrum. Contr. Syst.*, p. 95, March 1964.
31. C. M. Holsell: Mass Flowmeters, *ISA J.*, p. 49, June 1960.
32. L. N. Mortenson: Mass Flowmeter Calibration, *Instrum. Contr. Syst.*, p. 133, March 1964.
33. J. Haffner et al.: Novel Mass Flowmeter, *Contr. Eng.*, p. 69, October 1962.
34. G. T. Gebhardt: What's Available for Measuring Mass Flow, *Contr. Eng.*, p. 90, February 1957.
35. G. F. Battista: The Use of Momentum Effects in Liquid Flow Measurement, U. S. Naval Air Turbine Test Station, Trenton, N. J., *NATTS-ATL-TN-*26, 1963.
36. G. Bloom: Low Flow Mass Flow Meter, *Instrum. Contr. Syst.*, p. 117, March 1965.
37. E. C. Evans and G. W. Ray: Gas Mass Flow Rate Measurement to 0.1%, *Instrum. Contr. Syst.*, p. 105, March 1965.
38. J. W. Freshour: Mass Flow Measurement of Cryogens, *Instrum. Contr. Syst.*, p. 97, March 1965.

39. J. C. Pemberton: Flow Measurement in Rotating Machinery, *Instrum. Contr. Syst.*, p. 105, March 1964.
40. T. J. Larson and L. D. Webb: Calibrations and Comparisons of Pressure-Type Airspeed-Altitude Systems of the X-15 Airplane from Subsonic to High Supersonic Speeds, *NASA, Tech. Note* D-1724, 1963.
41. W. Gracey et al.: Wind-Tunnel Investigation of a Number of Total-Pressure Tubes at High Angles of Attack, *NACA, Tech. Note* 2261, 1951.
42. F. J. Capone: Wind-Tunnel Tests of Seven Static-Pressure Probes at Transonic Speeds, *NASA, Tech. Note* D-947, 1961.
43. R. S. Ritchie: Several Methods for Aerodynamic Reduction of Static-Pressure Sensing Errors for Aircraft at Subsonic, Near-sonic, and Low Supersonic Speeds, *NASA, Tech. Rept.* R-18, 1959.
44. A. O. Pearson and H. A. Brown: Calibration of a Combined Pitot-Static Tube and Vane-Type Flow Angularity Indicator at Transonic Speeds and at Large Angles of Attack or Yaw, *NACA, RM*-L52F24, 1952.
45. J. M. Savino and A. J. Hilovsky: On the Use of Single Total- and Static-Pressure Probes to Measure the Average Mass Velocity in Thin Rectangular Channels, *NASA, Tech. Note* D-2212, 1964.
46. W. Gracey: Measurement of Static Pressure on Aircraft, *NASA Rept.* 1364, 1958.
47. W. H. Reed: Dynamic Response of Rising and Falling Balloon Wind Sensors with Application to Estimates of Wind Loads on Launch Vehicles, *NASA, Tech. Note* D-1821, 1963.
48. W. F. Van Tassell and C. E. Covert: Relaxation Effects on the Interpretation of Impact-Probe Measurements, *J. Aerosp. Sci.*, p. 147, February 1960.
49. F. S. Sherman: New Experiments on Impact-Pressure Interpretation in Supersonic Rarefied Air Streams, *NACA Tech. Note* 2995, 1953.
50. New Anemometer Has Fast Response, Measures Dynamic Pressure Directly, *NASA, Brief* 63-10530, 1963.
51. L. V. Baldwin and V. A. Sandborn: Hot-Wire Calorimetry: Theory and Application to Ion Rocket Research, *NASA, Tech. Rept.* R-98, 1961.
52. M. R. Neuman et al. (eds.): "Physical Sensors for Biomedical Applications," CRC Press, Boca Raton, Fla., 1980.
53. L. Kovasznay: Calibration and Measurement in Turbulence Research by Hot-Wire Method, *NACA, TM*-1130, 1947.
54. A. Noordergraaf: "Circulatory System Dynamics", Academic, New York, 1978.
55. W. G. Spangenberg: Heat-Loss Characteristics of Hot-Wire Anemometers at Various Densities in Transonic and Supersonic Flow, *NACA, Tech. Note* 3381, 1955.
56. L. Kovasznay: Development of Turbulence-Measuring Equipment, *NACA, Rept.* 1209, 1954.
57. V. A. Sandborn and J. C. Laurence: Heat Loss from Yawed Hot Wires at Subsonic Mach Numbers, *NACA, Tech. Note* 3563, 1955.
58. C. E. Shepard: A Self-excited, Alternating Current, Constant-Temperature Hot-Wire Anemometer, *NACA, Tech. Note* 3406, 1955.
59. W. G. Rose: Some Corrections to the Linearized Response of a Constant-Temperature Hot-Wire Aneomometer Operated in a Low-Speed Flow, *ASME Paper* 62-WA-11, 1962.
60. G. P. Katys: "Continuous Measurement of Unsteady Flow," Macmillan, New York, 1964.
61. R. R. Dowden: "Fluid Flow Measurement, A Bibliography," British Hydromechanics Res. Assoc. Cranfield, Bedford, England, 1971.
62. R. P. Benedict: "Fundamentals of Temperature Pressure and Flow Measurement," Wiley, New York, 1969
63. The Use of Laser Doppler Velocimeter for Flow Measurements, *Proc. of Project Squid Workshop*, Purdue University, 1971.
64. L. C. Lynnworth et al.: Ultrasonic Mass Flowmeter for Army Aircraft Engine Diagnostics, *USAA MRDL Tech. Rept.* 72-66, Fort Eustis, Va., 1973.
65. J. W. Tanney: Fluidic Velocity Sensor, *Inst. & Cont. Syst.*, June 1969.

66. Thin-Film Probes for Measurement of Instantaneous Blood Flow Velocity, *Disa Inform.*, Disa Electronics, Herlev, Denmark, October 1970.

67. M. R. Davis and P. O. A. L. Davies: The Physical Characteristics of Hot-Wire Anemometers, *Inst. Sound Vib., Rept.* no. 2, University of Southhampton, England, 1968.

68. F. J. Resch: Use of the Dual-Sensor Hot-Film Probe in Water Flow, *DISA Inform.*, Disa Electronics, Herlev, Denmark, March 1973.

69. H. A. Becker and A. P. G. Brown: Response of Pitot Probes in Turbulent Streams, *J. Fluid Mech.*, vol. 62, pt. 1, pp. 85–114, 1974.

70. J. A. Breman et al.: An Evaluation of Positive Displacement Cryogenic Volumetric Flowmeters, *NBS Tech. Note* 605, 1971.

EIGHT
TEMPERATURE AND HEAT-FLUX MEASUREMENT

8.1 STANDARDS AND CALIBRATION

The International Measuring System sets up independent standards for only four fundamental quantities: length, time, mass, and temperature. Standards for all other quantities are basically derived from these. We discussed previously the standards of length, time, and mass; we consider now the temperature standard.[1] It should be noted first that temperature is fundamentally different in nature from length, time, and mass in that it is an *intensive* quantity whereas the others are *extensive*. That is, if two bodies of like length are "combined," the total length is twice the original; the same is true for two time intervals or two masses. However, the combination of two bodies of the same temperature results in exactly the same temperature. Thus the idea of a standard unit of mass, length, or time that can be divided or multiplied indefinitely to generate any arbitrary magnitude of these quantities cannot be carried over to the concept of temperature. Also, even though statistical mechanics relates temperature to the mean kinetic energies of molecules, these kinetic energies (which are dependent on only mass, length, and time standards for their description) are not measurable at present. Thus an *independent* temperature standard is required.

The fundamental meaning of temperature, just as for all basic concepts of physics, is not given easily. For most purposes the zeroth law of thermodynamics

[1] R. P. Benedict, "Fundamentals of Temperature, Pressure, and Flow Measurements," 2d ed., Wiley, New York, 1977.

gives a useful concept: For two bodies to be said to have the same temperature, they must be in thermal equilibrium; that is, when thermal communication is possible between them, no change in the thermodynamic coordinates of either occurs. The zeroth law says that when two bodies are each in thermal equilibrium with a third body, they are in thermal equilibrium with each other. Then, by definition, the bodies are all at the same temperature. Thus if we can set up a reproducible means of establishing a range of temperatures, unknown temperatures of other bodies may be compared with the standard by subjecting any type of "thermometer" successively to the standard and to the unknown temperatures and allowing equilibrium to occur in each case. That is, the thermometer is calibrated against the standard and afterward may be used to read unknown temperatures.

In choosing the means of defining the standard temperature scale, conceivably we could employ any of the many physical properties of materials that vary reproducibly with temperature. For instance, the length of a metal rod varies with temperature. To define a temperature scale numerically, we must choose a reference temperature and state a rule for defining the difference between the reference and other temperatures. (Mass, length, and time measurements do *not* require universal agreement on a reference point at which each quantity is assumed to have a particular numerical value. Every centimeter, for example, in a meter is the same as every other centimeter.)

Suppose we take a copper rod 1 m long, place it in an ice-water bath which we have taken as our reference temperature source, and measure its length. Let us choose to call the ice-bath temperature 0°. We are now free to define any rule we wish to fix the numerical value to be assigned to all lower and higher temperatures. Suppose we decide that each additional 0.01 mm of expansion will correspond to $+1.0°$ on our temperature scale and each 0.01 mm of contraction to $-1.0°$. If the expansion phenomenon were reproducible, such a temperature scale would, in principle, be perfectly acceptable as long as everyone adhered to it. Would it be correct to say that each degree of temperature on this scale was "equal" to every other degree? That depends on what we mean by "equal." If "equal" means that each degree causes the same amount of expansion of the copper rod, then all degrees are equal. If, instead, we consider the expansion of, say, iron rods, then equal amounts of expansion would not, in general, be caused by a 1° (copper scale) change from -6 to $-5°$ as by a 1° change from 100 to 101°. Or, suppose we based our scale on conduction heat transfer in silver, for example. If a temperature difference from 100 to 200° causes a given heat-transfer rate, will the same rate be caused by a temperature difference from -50 to $+50°$? The answer is, in general, no.

The point of the above discussion is that, while our arbitrarily defined temperature scale is, in principle, as good as any other such scale based on some material property, its graduations have no particular significance with regard to physical laws *other* than the one used in the definition. We measure temperature for some *reason*, such as computing thermal expansion, heat-transfer rate, electrical conductivity, gas pressure, etc. The forms of the equations employed to make such calculations depend on the nature of the standard used to define temperature. A temperature scale that gives a simple form to thermal-expansion equations may give complex forms to all other physical relations involving temperature. Since this difficulty is common to *all* standards based on the properties of a particular substance, a way of defining a temperature scale independent of *any* substance is desirable.

The thermodynamic temperature scale[1] proposed by Lord Kelvin in 1848 provides the theoretical base for a temperature scale independent of any material property and is based on the Carnot cycle. Here a perfectly reversible heat engine transfers heat from a reservoir of infinite capacity at temperature T_2 to another such reservoir at T_1. If the heat taken from reservoir 2 is Q_2 and that supplied to reservoir 1 is Q_1, for a Carnot cycle $Q_2/Q_1 = T_2/T_1$; this may be taken as a *definition* of temperature ratio. If, also, a number is selected to describe the temperature of a chosen fixed point, then the temperature scale is completely defined. At present, the fixed point is taken as the triple point (the state at which solid, liquid, and vapor phases are in equilibrium) of water because this is the most reproducible state known. The number assigned to this point is 273.16 K since this makes the temperature interval from the ice point (273.15 K) to the steam point equal to 100 K. This would thus coincide with the previously established centigrade (now called Celsius) scale as a matter of convenience.[2]

While the Kelvin absolute thermodynamic scale is ideal in the sense that it is independent of any material properties, it is not physically realizable since it depends on an ideal Carnot cycle. Fortunately it can be shown[3] that a temperature scale defined by a constant-volume or constant-pressure gas thermometer using an ideal gas is *identical* to the thermodynamic scale. A constant-volume gas thermometer keeps a fixed mass of gas at constant volume and measures the pressure changes caused by temperature changes. The perfect-gas law then gives the fact that temperature ratios are identical to pressure ratios. The constant-pressure thermometer keeps mass and pressure constant and measures volume changes caused by temperature changes. Again, the perfect-gas law says that temperature ratios are identical to volume ratios. These ratios are identical to those of the thermodynamic scale; thus if the same fixed point (the triple point of water) is selected for the reference point, the two scales are numerically identical. However, now there is the problem that the ideal gas is a mathematical model, not a real substance, and therefore the gas thermometers described above cannot actually be built and operated.

To obtain a physically realizable temperature scale, *real* gases must be utilized in the gas thermometers; the readings must be corrected, as well as possible, for deviation from ideal-gas behavior; and then the resulting values are accepted as a definition of the temperature scale. The corrections for non-ideal-gas behavior are obtained for a constant-volume gas thermometer as follows: The ther-

---

[1] F. W. Sears, "Thermodynamics, Kinetic Theory and Statistical Mechanics," p. 116, Addison-Wesley, Reading, Mass., 1950.

[2] Ibid., p. 8.

[3] Ibid., p. 116.

mometer is filled with a certain mass of gas, and mercury is added until the desired volume is achieved (see Fig. 8.1). Suppose that this is done with the system at the ice-point temperature. The gas pressure is measured; let us call it p_{i1}. Then the system is raised to the steam-point temperature, causing volume expansion. By adding more mercury, however, the volume can be returned to the original value. The pressure will be higher now; we call it p_{s1}. For an ideal gas, the ratio of the steam-point and ice-point temperatures also would be given by the pressure ratio p_{s1}/p_{i1}. If we repeat this experiment but use a different mass of gas, thus giving different ice-point and steam-point pressures p_{i2} and p_{s2}, we find that $p_{s1}/p_{i1} \neq p_{s2}/p_{i2}$. This is a manifestation of the nonideal behavior of the gas; an ideal gas would have $p_{s1}/p_{i1} = p_{s2}/p_{i2}$.

Real gases approach ideal-gas behavior if their pressure is reduced to zero;

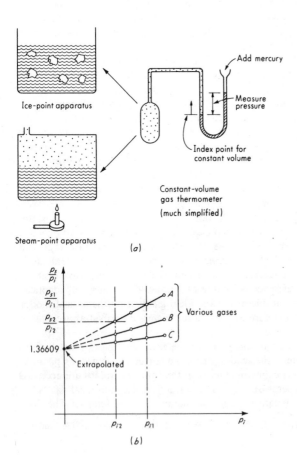

Figure 8.1 Gas-thermometer temperature scale.

thus we repeat the above experiment with successively smaller masses of gas, generating the curve A of Fig. 8.1b. Since we cannot use zero mass of gas, the zero-pressure point on this curve must be obtained by extrapolation. This zero-pressure point is taken as the true value of the pressure ratio corresponding to the steam-point/ice-point temperature ratio. If this experiment is repeated with *different* gases (B, C in Fig. 8.1b), all the curves intersect at the same point, showing that the procedure is independent of the type of gas used. Actual results give the numerical value $p_s/p_i = 1.36609 \pm 0.00004$. If we take $T_s/T_i = p_s/p_i$, the choice of a numerical value for any chosen reference point (such as calling $T_i = 273.15$ K) completely fixes the entire temperature scale. Such a scale, unfortunately, is not practical for day-to-day temperature measurements since the procedures involved are extremely tedious and time-consuming. Also, gas thermometers actually have a lower precision (repeatability) than some other temperature-measuring devices, such as resistance thermometers. This situation led to the acceptance in 1927 of the International Practical Temperature Scale which, with revisions in 1948, 1954, 1960, and 1968, is the temperature standard today.

The International Practical Temperature Scale is set up to conform as closely as practical with the thermodynamic scale. At the triple point of water, the two scales are in exact agreement, by definition. Five other primary fixed points are used. These are the boiling points of liquid oxygen ($-182.962°C$) and water ($100°C$) and the freezing points of zinc ($419.58°C$), silver ($961.93°C$), and gold ($1064.43°C$). Various secondary fixed points also are established, with the lowest being the triple point of hydrogen ($-259.34°C$), which also is the lowest value defined on the scale (the gold point $1064.43°C$ is the highest defined fixed point). In addition to the fixed points, the International Practical Temperature Scale also specifies certain instruments, equations, and procedures to be utilized to interpolate between the fixed points. From -259.34 to $630.74°C$, a platinum resistance thermometer is the interpolating instrument; however, since one equation will not serve the entire range, equations for several subranges are defined. Each equation contains certain constants whose values must be obtained from the readings of that particular thermometer at specific fixed points. Once these constants have been found, the equation allows us to calculate temperature values from thermometer readings at any point within the given subrange. A similar procedure is employed between 630.74 and $1064.43°C$, except now the interpolating instrument is a platinum/10% rhodium and platinum thermocouple.

Above the gold point, the International Practical Temperature Scale *is* defined and uses a narrow-band radiation pyrometer ("optical" pyrometer) and the Planck equation to establish temperatures. The formula is

$$\frac{J_t}{J_{Au}} = \frac{e^{1.4388/\lambda(t_{Au}+273.15)} - 1}{e^{1.4388/\lambda(t+273.15)} - 1} \tag{8.1}$$

where $t_{Au} \triangleq$ gold-point temperature, $1064.43°C$
$\lambda \triangleq$ effective wavelength of pyrometer, cm

The quantity J_t/J_{Au} is measurable with the pyrometer and is the ratio of the spectral radiance of a blackbody at temperature t to one at temperature t_{Au}. Since λ can be determined for a given pyrometer, Eq. (8.1) allows calculation of t when J_t/J_{Au} has been measured. In principle, this method can be applied to arbitrarily high temperatures, but in practice few reliable results above 4000°C are known.

The highest meaningful temperatures, existing in the interior of stars and for short times in atomic explosions, are inferred from kinetic theory to be in the range 10^7 to 10^9 K. Definition of temperature, much less measurement, is difficult at these extremes, although spectroscopic methods have given useful results. At the other extreme, temperatures of 10^{-6} K have been produced by using the concept of nuclear cooling. Magnetic susceptibility of certain materials has been employed to measure temperatures in the extremely low ranges.

The question of the accuracy of temperature standards may be considered from two viewpoints. First, how closely can the International Practical Temperature Scale be reproduced; second, how closely does it agree with the thermodynamic absolute scale? The highest reproducibility of the International Practical Temperature Scale occurs at the triple point of water, which can be realized with a precision of a few ten-thousandths of a degree, giving an accuracy of about 1 ppm. For either lower or higher temperatures the accuracy falls off. Figure 8.2[1] summarizes these data and shows calibration uncertainties for various instruments. The question of agreement among the various empirical scales (such as the International Practical Temperature Scale) and the absolute thermodynamic scale involves the fact that, in general, the thermodynamic scale is considerably less reproducible than the empirical scales. For example, the steam-point temperature is reproducible to 0.0005° with a platinum resistance thermometer, but to only 0.02° with a gas thermometer. The disagreement between the International Practical Temperature Scale and the absolute thermodynamic scale has been estimated in centigrade degrees as[2]

$$\frac{t}{100}\left(\frac{t}{100}-1\right)[0.04106 - 7.363(10^{-5})t] \qquad 0° < t < 444.6°C \qquad (8.2)$$

where t is the Celsius temperature. The error is seen to be zero at $t = 0$ and 100°C and has a maximum value of about 0.14C° near $t = 400$°C.

The International Practical Temperature Scale is being evaluated continuously, and a new version is expected[3] in about 1987 which may extend the low end to 0.5 K, replace the thermocouple interpolating instrument with a specially designed platinum resistance thermometer, and assign more closely thermodynamic values to fixed points. We should also make clear that although the present

[1] Accuracy in Measurements and Calibrations, *NBS Tech. Note* 262, 1965.

[2] "Temperature, Its Measurement and Control in Science and Industry," vol. 2, p. 93, Reinhold, New York, 1955.

[3] R. P. Hudson, Measurement of Temperature, *Rev. Sci. Instrum.*, vol. 51, no. 7, pp. 871–881, July 1980.

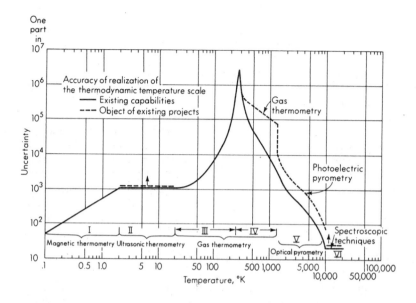

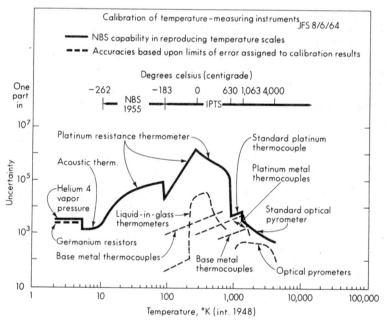

Figure 8.2 Temperature standards.

International Practical Temperature Scale is defined to only 13.81 K, measurements below this (and in fact below the 0.5 K mentioned above) are made regularly.[1]

Calibration of a given temperature-measuring device generally is accomplished by subjecting it to some established fixed-point environment, such as the melting and boiling points of standard substances, or by comparing its readings with those of some more accurate (secondary standard) temperature sensor which itself has been calibrated. The latter generally is accomplished by placing the two devices in intimate thermal contact in a constant-temperature-controlled bath. By varying the temperature of the bath over the desired range (allowing equilibrium at each point), the necessary corrections are determined. Accurate resistance thermometers,[2] thermocouples, or mercury-in-glass expansion thermometers generally are useful as secondary standards. Fixed-point standards using the melting points of various metals and the triple point of water are commercially available.

8.2 THERMAL-EXPANSION METHODS

A number of practically important temperature-sensing devices utilize the phenomenon of thermal expansion in one way or another. The expansion of solids is employed mainly in bimetallic elements by utilizing the differential expansion of bonded strips of two metals. Liquid expansion at essentially constant pressure is used in the common liquid-in-glass thermometers. Restrained expansion of liquids, gases, or vapors results in a pressure rise, which is the basis of pressure thermometers.

Bimetallic Thermometers

If two strips of metals A and B with different thermal-expansion coefficients α_A and α_B but at the same temperature (Fig. 8.3a) are firmly bonded together, a temperature change causes a differential expansion and the strip, if unrestrained, will deflect into a uniform circular arc. Analysis[3] gives the relation

$$\rho = \frac{t\{3(1+m)^2 + (1+mn)[m^2 + 1/(mn)]\}}{6(\alpha_A - \alpha_B)(T_2 - T_1)(1+m)^2} \tag{8.3}$$

where $\rho \triangleq$ radius of curvature
  $t \triangleq$ total strip thickness, 0.0005 in $< t <$ 0.125 in in practice
  $n \triangleq$ elastic modulus ratio, E_B/E_A
  $m \triangleq$ thickness ratio, t_B/t_A
$T_2 - T_1 \triangleq$ temperature rise

[1] Ibid.

[2] D. J. Curtis, Temperature Calibration and Interpolation Methods for Platinum Resistance Thermometers, *Rept.* 68023F, Rosemount Inc., Minneapolis, Minn., 1980.

[3] S. G. Eskin and J. R. Fritze, Thermostatic Bimetals, *Trans. ASME*, p. 433, July 1940.

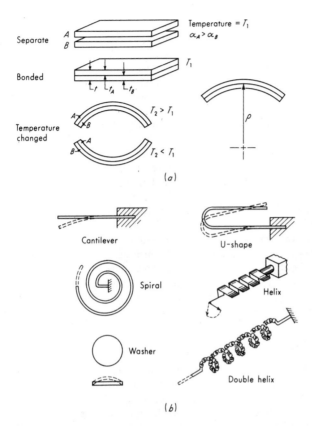

Figure 8.3 Bimetallic sensors.

In most practical cases, $t_B/t_A \approx 1$ and $n + 1/n \approx 2$, giving

$$\rho \approx \frac{2t}{3(\alpha_A - \alpha_B)(T_2 - T_1)} \tag{8.4}$$

Combination of this equation with appropriate strength-of-materials relations allows calculation of the deflections of various types of elements in practical use. The force developed by completely or partially restrained elements also can be calculated in this way. Accurate results require the use of experimentally determined factors[1] which are available from bimetal manufacturers.

Since there are no practically usable metals with negative thermal expansion, the B element is generally made of Invar, a nickel steel with a nearly zero

[1] General Plate Division, Attleboro, Mass., *Bull.* PR750.

$[1.7 \times 10^{-6} \text{ in}/(\text{in} \cdot \text{C}°)]$ expansion coefficient. While brass was employed originally, a variety of alloys are used now for the high-expansion strip, depending on the mechanical and electrical characteristics required. Details of materials and bonding processes are, in some cases, considered trade secrets. A wide range of configurations has been developed to meet application requirements (Fig. 8.3*b*).

Bimetallic devices are utilized for temperature measurement and very widely as combined sensing and control elements in temperature-control systems, mainly of the on-off type. Also they are used as overload cutout switches in electric apparatus by allowing the current to flow through the bimetal, heating and expanding it, and causing a switch to open when excessive current flows. Further applications are found as temperature-compensating devices[1] for various instruments that have temperature as a modifying or interfering input. The mechanical motion proportional to temperature is employed to generate an opposing compensating effect. The accuracy of bimetallic elements varies greatly, depending on the requirements of the application. Since the majority of control applications are not extremely critical, requirements can be satisfied with a rather low-cost device. For more critical applications, performance can be much improved. The working temperature range is about from -100 to $1000°F$. Inaccuracy of the order of 0.5 to 1 percent of scale range may be expected in bimetal thermometers of high quality. Recently, bimetal elements have been combined with conductive plastic potentiometers for automotive sensing applications.[2]

Liquid-in-Glass Thermometers[3]

The well-known liquid-in-glass thermometer is adaptable to a wide range of applications by varying the materials of construction and/or configuration. Mercury is the most common liquid utilized at intermediate and high temperatures; its freezing point of $-38°F$ limits its lower range. The upper limit is in the region of $1000°F$ and requires use of special glasses and an inert-gas fill in the capillary space above the mercury. Compression of the gas helps to prevent separation of the mercury thread and raises the liquid boiling point. For low temperatures, alcohol is usable to $-80°F$, toluol to $-130°F$, pentane to $-330°F$, and a mixture of propane and propylene giving the lower limit of $-360°F$.

Thermometers are commonly made in two types: total immersion and partial immersion. Total-immersion thermometers are calibrated to read correctly when the liquid column is completely immersed in the measured fluid. Since this may obscure the reading, a small portion of the column may be allowed to protrude with little error. Partial-immersion thermometers are calibrated to read correctly when immersed a definite amount and with the exposed portion at a definite

---

[1] R. Gitlin, How Temperature Effects Instrument Accuracy, *Contr. Eng.*, April, May, June 1955.

[2] P. J. Sacchetti and D. R. Phillips, A Ratiometric Temperature Sensor for High Temperature Applications, *SAE Paper* 80024, 1980.

[3] J. F. Swindells, Calibration of Liquid-in-Glass Thermometers, *Natl. Bur. Std. (U.S.)*, *Circ.* 600, 1959; M. F. Behar, "Handbook of Measurement and Control," p. 25, Instruments Publishing, Pittsburgh, Pa., 1954.

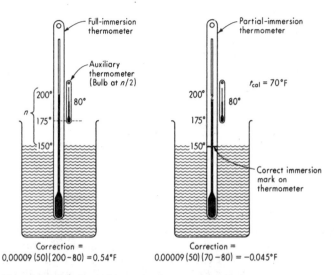

Correction =
0.00009 (50)(200 − 80) = 0.54°F

Correction =
0.00009 (50) (70 − 80) = −0.045°F

Figure 8.4 Full- and partial-immersion thermometers.

temperature. They are inherently less accurate than full-immersion types. If the exposed portion is at a temperature different from that at calibration, then a correction must be applied. Corrections for full- and partial-immersion thermometers used at conditions other than those intended are determined most accurately by the use of a special "faden" thermometer,[1] designed to measure the average temperature of the emergent stem. If such a thermometer is not available, the correction may be estimated by suspending a small auxiliary thermometer close to the stem of the thermometer to be corrected, as in Fig. 8.4. This auxiliary thermometer estimates the mean temperature of the emergent stem. When a partial-immersion thermometer is employed at correct immersion but with a surrounding air temperature different from that of its original calibration condition, the correction may be calculated from (for mercury-in-glass)

$$\text{Correction} = 0.00009n(t_{cal} - t_{act}) \qquad \text{F}° \qquad (8.5)$$

where $n \triangleq$ number of scale degrees equivalent to emergent stem length, F°

$t_{cal} \triangleq$ air temperature at calibration, F°

$t_{act} \triangleq$ actual air temperature at use, F° (from auxiliary thermometer)

When a total-immersion thermometer is utilized at partial immersion, the same formula may be used except that $t_{cal} - t_{act}$ is replaced by (main-thermometer reading) − (auxiliary-thermometer reading). For Celsius thermometers the constant 0.00009 becomes 0.00016.

The accuracy obtainable with liquid-in-glass thermometers depends on in-

---

[1] Swindells, ibid.

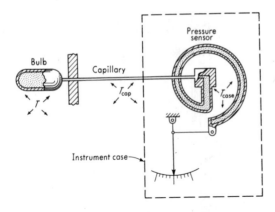

Figure 8.5 Pressure thermometer.

strument quality, temperature range, and type of immersion. For full-immersion thermometers the best instruments, when calibrated, are capable of errors as small as 0.4F° (range −328 to 32°F), 0.05F° (range −69 to 32°F), 0.04F° (range 32 to 212°F), 0.4F° (range 212 to 600°F), and 0.8F° (range 600 to 950°F). Errors in partial-immersion types may be several times larger, even after corrections have been applied for air-temperature variations. All the above figures refer to the ultimate performance attainable with the best instruments and great care in application. Errors in routine day-to-day measurements may be much larger.

Pressure Thermometers[1]

Pressure thermometers consist of a sensitive bulb, an interconnecting capillary tube, and a pressure-measuring device such as a Bourdon tube, bellows, or diaphragm (Fig. 8.5). When the system is completely filled with a liquid (mercury and xylene are common) under an initial pressure, the compressibility of the liquid is often small enough relative to the pressure gage $\Delta V/\Delta p$ that the measurement is essentially one of volume change. For gas or vapor systems, the reverse is true, and the basic effect is one of pressure change at constant volume.

Capillary tubes as long as 200 ft may be used for remote measurement. Temperature variations along the capillary and at the pressure-sensing device generally require compensation, except in the vapor-pressure type, where pressure depends on only the temperature at the liquid's free surface, located at the bulb. A common compensation scheme using an auxiliary pressure sensor and capillary is shown in Fig. 8.6. The motion of the compensating system is due to the interfering effects only and is subtracted from the total motion of the main system, resulting in an output dependent on only bulb temperature. The "trimming" capillary (which may be lengthened or shortened) allows the volume to be

[1] D. M. Considine (ed.), "Process Instruments and Controls Handbook," p. 68, McGraw-Hill, New York, 1957; Behar, op. cit., p. 29.

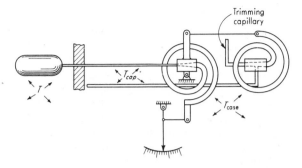

Figure 8.6 Compensation methods.

changed to attain accurate case compensation by experimental test. Bimetal elements also are used to obtain case and partial capillary compensation.

Liquid-filled systems cover the range -150 to $750°F$ with xylene or a similar liquid and -38 to $1100°F$ with mercury. Response is essentially linear over ranges up to about $300°F$ with xylene and $1000°F$ with mercury. Elevation differences between the bulb and pressure sensor different from those at calibration may cause slight errors. Gas-filled systems operate over the range -400 to $1200°F$ with linear ranges as great as $1000°F$; errors due to capillary temperature

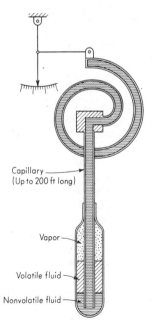

Figure 8.7 Vapor-pressure thermometer.

variations usually are small enough not to justify compensation. Case compensation is accomplished with bimetal elements. Vapor-pressure systems are usable in the range -40 to $600°F$. The calibration is strongly nonlinear; special linearizing linkages are needed if linear output is required. Characteristics of the system vary, depending on whether the bulb is hotter than, colder than, or equal in temperature to the rest of the system, since this determines where liquid and vapor will exist. The most versatile arrangement is shown in Fig. 8.7, where the volatile-liquid surface is *always* in the bulb. Capillary and case corrections are not needed in such a device since the vapor pressure of a liquid depends on only the temperature of its free surface. Commonly used volatile liquids include ethane (vapor pressure changes from 20 to 600 lb/in^2 gage for a temperature change from -100 to $+80°F$), ethyl chloride (0 to 600 lb/in^2 gage for 40 to 350°F), and chlorobenzene (0 to 60 lb/in^2 gage for 275 to 400°F). The accuracy of pressure thermometers under the best conditions is of the order ± 0.5 percent of the scale range. Adverse environmental conditions may increase this error considerably.

8.3 THERMOELECTRIC SENSORS (THERMOCOUPLES)[1]

If two wires of different materials A and B are connected in a circuit as shown in Fig. 8.8a, with one junction at temperature T_1 and the other at T_2, then an infinite-resistance voltmeter detects an electromotive force E, or if an ammeter is connected, a current I is measured. The magnitude of the voltage E depends on the materials and temperatures T_1 and T_2. The current I is simply E divided by the total resistance of the circuit, including the ammeter resistance. If current is allowed to flow, electric power is developed; this comes from a heat flow from the surroundings to the wires. Thus a direct conversion of heat energy to electric energy is obtained. The effect is reversible, so that forcing a current from an external source through a thermoelectric circuit will cause heat flows to and from the circuit. While we are concerned here with this thermoelectric effect only as a means of sensing temperatures, modern developments in materials have made the principle of practical application in electric power generation, heating, and cooling, though only on a small scale at present.

The overall relation between voltage E and temperatures T_1 and T_2, which is the basis of thermoelectric temperature measurement, is called the *Seebeck effect*. Temperatures T_1 and T_2 refer to the junctions themselves, whereas when using a thermocouple, we are trying to measure the temperature of some body in contact with the thermojunction. These two temperatures are not exactly the same if current is allowed to flow through the thermojunction, since then heat is generated or absorbed at the junction, which thus must be hotter or colder than the surrounding medium whose temperature is being measured. This heating and

---

[1] Manual on the Use of Thermocouples in Temperature Measurement, *ASTM* Spec. Pub. 470B, Philadelphia, Pa., 1981; D. D. Pollock, "The Theory and Properties of Thermocouple Elements," *ASTM* STP 492, Omega Press, Ithaca, N.Y., 1979.

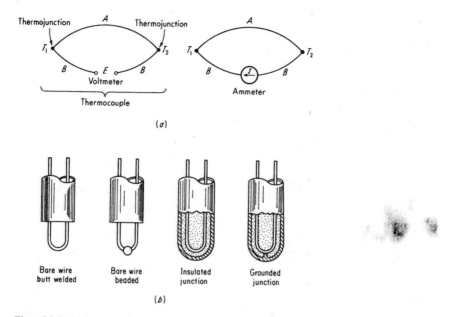

Figure 8.8 Basic thermocouple and junction types.

cooling process is related to the Peltier effect.[1] If the thermocouple voltage is measured with a potentiometer, no current flows and Peltier heating and cooling are not present. When a millivoltmeter is used, current flows and heat is absorbed at the hot junction (requiring it to become cooler than the surrounding medium) while heat is liberated at the cold junction, making it hotter than its surrounding medium. These heating and cooling effects are proportional to the current and fortunately are completely negligible[2] when the current is that produced by the thermocouple itself in a practical millivoltmeter circuit. These errors are even less when the thermocouple is connected (as is very common) to an instrumentation amplifier with high input impedance (1 to 1,000 MΩ).

Another reversible heat-flow effect, the *Thomson effect*,[3] influences the temperature of the conductors between the junctions rather than the junctions themselves. When current flows through a conductor having a temperature gradient (and thus a heat flow) along its length, heat is liberated at any point where the current flow is in the same direction as the heat flow, while heat is absorbed at any point where these are opposite. Since this effect also depends on current flow, it is not present if a potentiometer is used. Even if a millivoltmeter is employed, the effect of the heat flows on conductor temperature is completely negligible.

---

[1] P. H. Dike, "Thermoelectric Thermometry," Leeds & Northrup Co., Philadelphia, Pa., 1954.
[2] Ibid.
[3] Ibid.

Finally, it should be noted that in any current-carrying conductor, I^2R heat is generated, raising the circuit temperature above its local surroundings. Again, potentiometric and high-input-impedance voltage measurements give negligible error. Errors in millivoltmeter circuits usually are negligible also, but can be estimated if heat-transfer conditions are known.

The above physical effects can be analyzed[1] on a macroscopic scale by classical thermodynamics, with fewer assumptions by irreversible thermodynamics, and qualitatively on a microscopic basis by solid-state physics. Thermodynamic approaches are based on the two experimentally observed reversible energy-conversion processes, the Peltier and Thomson effects, and neither require nor give any explanation of the basic atomic mechanisms. The total emf produced is made up of a part due to the Peltier effect, which is localized at each junction, and a (usually much smaller) part caused by the Thomson effect, which is distributed along each conductor between the junctions. The Peltier emf's are assumed proportional to the junction temperature, while the Thomson emf's are proportional to the difference between the squares of the junction temperatures. For the total voltage, the equation takes the form

$$E = C_1(T_1 - T_2) + C_2(T_1^2 - T_2^2) \tag{8.6}$$

Copper/constantan thermocouples, for example, give

$$E = 62.1(T_1 - T_2) - 0.045(T_1^2 - T_2^2) \tag{8.7}$$

where $E \triangleq$ total voltage, μV
$T_1, T_2 \triangleq$ absolute junction temperatures, K

Unfortunately, the assumptions made in the analyses leading to Eq. (8.6) are not exactly satisfied in practice; thus equations such as (8.7) *cannot* usually be used to predict accurately temperatures from measured voltages. Rather, a given thermocouple material must be calibrated over the complete range of temperatures in which it is to be used. In this calibration only the overall voltage is of interest, and the separate contributions of Peltier and Thomson effects are not determined. Temperature measurement by thermoelectric means is thus based entirely on empirical calibrations and the application of so-called thermoelectric "laws" which experience has shown to hold. These laws, quoted below, are adequate for analysis of most practical thermocouple circuits. In those cases where the circuit configuration does not lend itself to direct application of these laws, alternative approaches[2] are available.

The laws of thermocouple behavior may be stated as follows:

1. The thermal emf of a thermocouple with junctions at T_1 and T_2 is totally unaffected by temperature elsewhere in the circuit if the two metals used are each homogeneous (Fig. 8.9a).

---

[1] R. R. Heikes and R. W. Ure, "Thermoelectricity," Interscience, New York, 1961.

[2] P. Stein, "Measurement Engineering," vol. 1, chap. 18, Stein Engineering Services, Inc., Phoenix, Ariz., 1964.

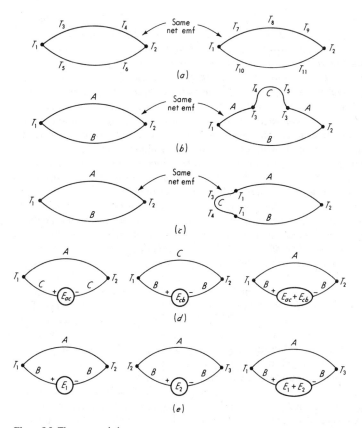

Figure 8.9 Thermocouple laws.

2. If a third homogeneous metal C is inserted into either A or B (see Fig. 8.9b), as long as the two new thermojunctions are at like temperatures, the net emf of the circuit is unchanged irrespective of the temperature of C away from the junctions.

3. If metal C is inserted between A and B at one of the junctions, the temperature of C at any point away from the AC and BC junctions is immaterial. As long as the junctions AC and BC are both at the temperature T_1, the net emf is the same as if C were not there (Fig. 8.9c).

4. If the thermal emf of metals A and C is E_{AC} and that of metals B and C is E_{CB}, then the thermal emf of metals A and B is $E_{AC} + E_{CB}$ (Fig. 8.9d).

5. If a thermocouple produces emf E_1 when its junctions are at T_1 and T_2, and E_2 when at T_2 and T_3, then it will produce $E_1 + E_2$ when the junctions are at T_1 and T_3 (Fig. 8.9e).

These laws are of great importance in the practical application of thermocouples. The first states that the lead wires connecting the two junctions may be safely exposed to an unknown and/or a varying temperature environment without affecting the voltage produced. Laws 2 and 3 make it possible to insert a voltage-measuring device into the circuit actually to measure the emf, rather than just talking about its existence. That is, the metal C represents the internal circuit (usually all copper in precise instruments) between the instrument binding posts. The instrument can be connected in two ways, as shown in Fig. 8.9b and c. Law 3 also shows that thermocouple junctions may be soldered or brazed (thereby introducing a third metal) without affecting the readings. Law 4 shows that all possible pairs of metals need not be calibrated since the individual metals can each be paired with *one* standard (platinum is used) and calibrated. Any other combinations then can be *calculated*; calibration is not necessary.

In considering the fifth law, we should note that, in using a thermocouple to measure an unknown temperature, the temperature of one of the thermojunctions (called the reference junction) must be known by some independent means. A voltage measurement then allows us to get the temperature of the other (measuring) junction from calibration tables. These calibration tables were obtained by maintaining the reference junction at a fixed known value (usually 32°F, the ice point), varying the measured junction over the desired range of temperatures (known by some independent means), and recording the resulting voltages. Thus most calibration tables are based on the reference junction being at the ice point. When a thermocouple is used, the reference junction may or may not be at the ice point. If it is, the calibration table may be employed directly to find the measuring-junction temperature. If it is not, the fifth law allows use of the standard table as follows: Suppose the reference junction is at 70°F and the voltage reading is 1.23 mV. In Fig. 8.9e we take $T_1 = 32°F$, $T_2 = 70°F$, and T_3 is unknown. We can look up E_1 directly in the table; suppose it is 0.71 mV. Now E_2 is the measured value 1.23 mV; thus $E_1 + E_2 = 1.94$ mV. The unknown temperature can be found by looking up the temperature value corresponding to 1.94 mV in the standard table; it is 100°F.

Common Thermocouples

Thermojunctions formed by welding, soldering, or merely pressing the two materials together give identical voltages. If current is allowed to flow, the currents may be different since the contact resistance differs for the various joining methods. Welding (either gas or electric) is used most widely although both silver solder and soft solder (low temperatures only) are used in copper/constantan couples. Special capacitor-discharge welding devices (particularly needed for very-fine-wire thermocouples) are available. Ready-made thermocouple pairs are, of course, available in a wide range of materials and wire sizes.

While many materials exhibit the thermoelectric effect to some degree, only a small number of pairs are in wide use. They are platinum/rhodium, Chromel/Alumel, copper/constantan, and iron/constantan. Each pair exhibits a combination of properties that suit it to a particular class of applications. Since the thermoelectric effect is somewhat nonlinear, the sensitivity varies with temperature. The maximum sensitivity of any of the above pairs is about 60 μV/C° for copper/constantan at 350°C. Platinum/platinum-rhodium is the least sensitive: about 6 μV/C° between 0 and 100°C.

The accuracy of the common thermocouples may be stated in two different ways. If you utilize standard thermocouple wire (which is *not* individually calibrated by the manufacturer) and make up a thermocouple to be used without calibration, you are relying on the wire manufacturer's quality control to limit deviations from the published calibration tables. These tables give the *average* characteristics, not those of a particular batch of wire. Platinum/platinum-rhodium is the most accurate; error is of the order of ± 0.25 percent of reading. Copper/constantan gives ± 0.5 percent or ± 1.5F° (whichever is larger) between -75 and 200°F and ± 0.75 percent between 200 and 700°F. Chromel/Alumel gives ± 5F° (32 to 660°F) and ± 0.75 percent (660 to 2300°F). Iron/constantan has ± 66 μV below 500°F and ± 1.0 percent from 500 to 1500°F. If higher accuracies are needed, the individual thermocouple may be calibrated. An indication of the achievable accuracy is available from National Bureau of Standards listings[1] of the results the Bureau will guarantee. At the actual calibration points, the error ranges from 0.05 to 0.5C°. Interpolated points are less accurate: 0.1 to 1.0C°, except platinum/platinum-rhodium at 1450°F, 2.0 to 3.0C°. The realization of this potential accuracy in applying such calibrated thermocouples to practical temperature measurement is, of course, dependent on the application conditions and is rarely possible.

Platinum/platinum-rhodium thermocouples are employed mainly in the range 0 to 1500°C. The main features of this combination are its chemical inertness and stability at high temperatures in oxidizing atmospheres. Reducing atmospheres cause rapid deterioration at high temperatures as the thermocouple metals are contaminated by absorbing small quantities of other metals from nearby objects (such as protecting tubes). This difficulty, causing loss of calibration, is unfortunately common to most thermocouple materials above 1000°C.

Chromel ($Ni_{90}Cr_{10}$)/Alumel ($Ni_{94}Mn_3Al_2Si_1$) couples are useful over the range -200 to $+1300$°C. Their main application, however, is from about 700 to 1200°C in nonreducing atmospheres. The temperature/voltage characteristic is quite linear for this combination (see Fig. 8.10).

Copper/constantan ($Cu_{57}Ni_{43}$) is used at temperatures as low as -200°C; its upper limit is about 350°C because of the oxidation of copper above this range. Iron/constantan is the most widely utilized thermocouple for industrial applications and covers the range -150 to $+1000$°C. It is usable in oxidizing atmospheres to about 760°C and reducing atmospheres to 1000°C.

Thermocouple manufacturers have a wealth of experience concerning the application of thermocouples to diverse temperature-measuring problems and should be consulted if special types of problems are foreseen in a particular case.

---

[1] Thermocouple Calibration, *Instrum. Contr. Syst.*, p. 1663, September 1961.

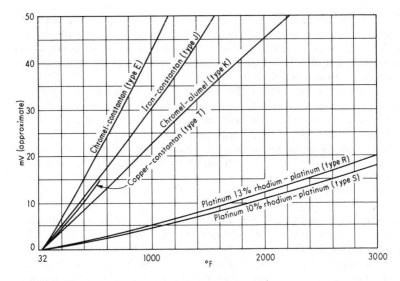

Figure 8.10 Thermocouple temperature/voltage curves.

Reference-Junction Considerations

For the most precise work, reference junctions should be kept in a triple-point-of-water apparatus whose temperature is $0.01 \pm 0.0005C°$. Such accuracy is rarely needed, and an ice bath is used much more commonly. A carefully made ice bath is reproducible to about $0.001C°$, but a poorly made one may have an error of $1C°$.[1] Figure 8.11 shows one method of constructing an ice-bath reference junction. The main sources of error are insufficient immersion length and an excessive amount of water in the bottom of the flask. Automatic ice baths that use the Peltier cooling effect as the refrigerator, rather than relying on externally supplied ice (which must be continually replenished), are available with an accuracy of $0.05C°$.[2] These systems use the expansion of freezing water in a sealed bellows as the temperature-sensing element that signals the Peltier refrigerator when to turn on or off by displacing a microswitch.

Since low-power heating is obtained more easily than low-power cooling, some reference junctions are designed to operate at a fixed temperature higher than any expected ambient. A feedback system operates an electric heating element to maintain a constant and known temperature in an enclosure containing the reference junctions. Since the reference junction is not at $32°F$, the thermocouple-circuit net voltage must be corrected by adding the reference-

[1] C. L. Feldman, Automatic Ice-Point Thermocouple Reference Junction, *Instrum. Contr. Syst.*, p. 101, January 1965.
[2] Ibid.

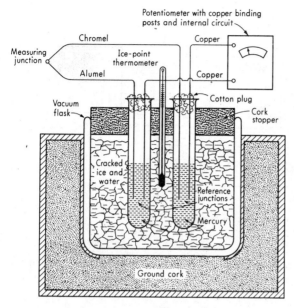

Figure 8.11 Ice-bath reference junction.

junction voltage before the measuring-junction temperature can be found. This correction is, however, a constant.

Figure 8.12 sketches a reference-junction technique widely utilized (in various versions) for digital thermometer instruments, data loggers, and data acquisition systems. Wires from the measuring junction are screwed directly to an isothermal block terminal strip. The temperature of this block (which has *no* active temperature control) drifts with ambient temperature; but because of careful thermal design, at all times it is *uniform* over its length (within about $\pm 0.05°C$). This reference temperature is measured by independent means [often a junction semiconductor sensor (Sec. 8.5)], and compensation circuitry[1] develops a voltage E_{comp} which is combined with that from the measuring junction so that the net voltage presented to the voltmeter represents T_{meas}. In multichannel instruments with microprocessor computing power, the isothermal block can accept many thermocouple pairs (of *mixed* types, if desired) since the T_{ref} sensor now sends its temperature data to the computer, which computes the needed voltage correction for each different thermocouple and applies it digitally as each channel is scanned. Such a "software" correction scheme, of course, must be intermittent (limiting response speed) because of the sampled nature of digital computation,

[1] D. H. Sheingold (ed.), Transducer Interfacing Handbook, p. 122, Analog Devices, Norwood, Mass., 1980.

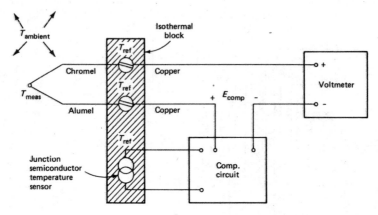

Figure 8.12 Isothermal block reference-junction technique.

while the "hardware" technique of Fig. 8.12 is continuous and does not significantly degrade sensor dynamics.

Special Materials, Configurations, and Techniques

Increasing interest in high-temperature processes in jet and rocket engines and nuclear reactors has led to requirements for reliable temperature sensors in the range 2000 to 4500°F. New thermocouples developed for these applications include rhodium-iridium/rhodium,[1] tungsten/rhenium,[2] and boron/graphite.[3] Rhodium-iridium is usable to about 4000°F under proper conditions and has a sensitivity of the order of 6 μV/C°. Various alloys of tungsten and rhenium may be utilized up to 5000°F under favorable conditions and have about the same sensitivity at the highest temperatures as rhodium-iridium. Boron/graphite has a high sensitivity (about 40 μV/C°) and is usable for short times up to 4500°F.

An alternative solution to high-temperature problems may be found in various cooling schemes. Two such[4] in actual use are shown in Fig. 8.13. In Fig. 8.13a the hot-gas flow whose temperature is to be measured impinges on a small tube carrying cooling water, causing a temperature rise of about 100°F. If heat-transfer coefficients are known, measurements of water flow rate, temperature, and temperature rise allow calculation of the hot-gas temperature. Figure 8.13b shows another approach in which the hot gas is aspirated through a heat exchanger, cooling it to about 1000°F. Knowledge of heat-transfer characteristics

---

[1] P. D. Freeze, Review of Recent Developments of High-Temperature Thermocouples, *ASME Paper* 63-WA-212, 1963.

[2] Ibid.

[3] Astro Industries Inc., Santa Barbara, Calif., *Bull.* BGT-1, 1963.

[4] *NASA, SP*-5015, p. 128, 1964.

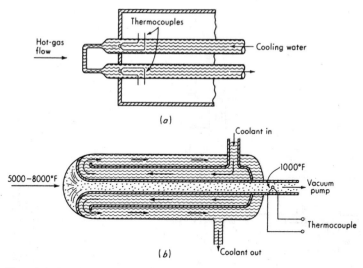

Figure 8.13 Cooled thermocouples.

and flow rates again allows calculation of the hot-gas temperature. Such methods have been used in the range 5000 to 8000°F.

Figure 8.14 shows in simplified fashion the principle of a pulse-cooling technique[1] which allows use of Chromel/Alumel thermocouples (melting point 2550°F) to measure temperatures up to 7000°F. (Further developments[2] have since appeared.) The measuring junction is kept at a low temperature by a cooling air flow. When this flow is shut off by a solenoid valve, the thermocouple starts to heat up, following the first-order equation

$$\tau \frac{dT_{tc}}{dt} + T_{tc} = T_{gas} \tag{8.8}$$

where $\tau \triangleq$ thermocouple time constant (assumed known)
$\quad T_{tc} \triangleq$ thermocouple temperature
$\quad T_{gas} \triangleq$ hot-gas temperature

This equation shows that T_{gas} can be *computed* any time after the cooling is shut off if dT_{tc}/dt is known. A voltage proportional to dT_{tc}/dt can be obtained by use of the differentiating circuit shown and, when it is summed with a voltage proportional to T_{tc}, provides a signal proportional to T_{gas}. Theoretically this signal is

[1] A. F. Wormser and R. A. Pfuntner, Pulse Technique Extends Range of Chromel-Alumel to 7000°F, *Instrum. Contr. Syst.*, p. 101, May 1964.

[2] D. Kretschmer, J. Odgers, and A. F. Schlader, The Pulsed Thermocouple for Gas Turbine Applications, *ASME Paper* 76-GT-1, 1976; G. E. Glawe, H. A. Will, and L. N. Krause, A New Approach to the Pulsed Thermocouple, *NASA* TM X-71883, 1976.

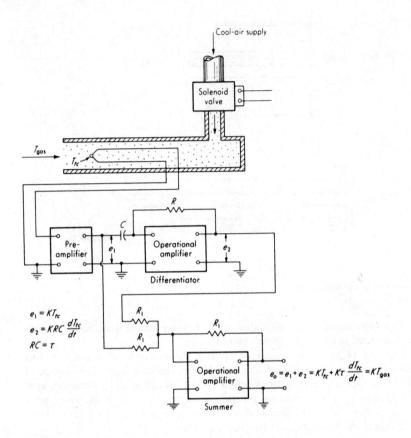

$$e_1 = KT_{tc}$$
$$e_2 = KRC \frac{dT_{tc}}{dt}$$
$$RC = \tau$$

$$e_o = e_1 + e_2 = KT_{tc} + K\tau \frac{dT_{tc}}{dt} = KT_{gas}$$

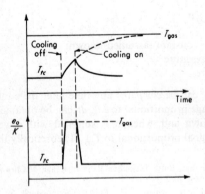

Figure 8.14 Pulsed-thermocouple technique.

available immediately after the cooling is shut off; in practice, however, the cooling is left off a finite interval during which the value of T_{gas} is recorded. The cooling is turned on again before the thermocouple is overheated. In the actual system of Ref. 1, additional computing elements also compute the numerical value of τ; thus this need not be known beforehand.

Still another approach to high-temperature gas measurements (particularly suited to dusty combustion gases because of its nonfouling characteristics) is the venturi pneumatic pyrometer of Fig. 8.15.[1] Here the hot gas at temperature T_h is aspirated through a water-cooled tube provided with two venturi flow-metering sections, one at the hot end and the other at the cold. The "cold" end (maximum temperature about 1600°C) has a platinum resistance thermometer (thermocouples also have been employed) to measure T_c, the temperature of the cooled gas. Two differential-pressure transducers measure ΔP_h and ΔP_c at the venturis. The two venturis carry the same mass flow rate, and the pressure drops are small enough to use incompressible flow relations, making $\Delta P_c/\Delta P_h \approx \rho_h/\rho_c$. Also, since $P_c \approx P_h$, the ideal-gas law gives $T_h/T_c \approx \rho_c/\rho_h$, which leads finally to

$$T_h = KT_c \frac{\Delta P_h}{\Delta P_c} \qquad (8.9)$$

where K is a calibration factor to account for various deviations from theory. Equation (8.9) is readily implemented in an analog multiplier/divider once the platinum thermometer signal is linearized, which gives a linear output of 40 $\mu V/°C$. Temperatures T_h up to about 2500°C can be measured with accuracy about ± 2 percent of reading.

The measurement of rapidly changing internal and surface temperatures of solid bodies may be accomplished with arrangements such as those in Fig. 8.16. The main requirements in such applications are that the thermojunction be of minimum size and be precisely located and that any materials placed into the wall have thermal properties identical with those of the wall, so that temperature distributions are not distorted. In Fig. 8.16a[2] the thermojunction is formed by drawing an abrasive tool, such as a file or an emery cloth, across the end of the sensing tip. This action flows metal from one thermocouple element to the other since the 0.0002-in mica insulation is easily bridged over, thus forming numerous microscopic hot-weld thermojunctions. Subsequent erosion or abrasive action forms new thermal junctions continuously as the tip wears away. Such thermocouples have time constants as small as 10^{-5} s and are available in materials usable to 5000°F and 10,000 lb/in² pressure. In Fig. 8.16b[3] two thermojunctions are formed by plating a thin rhodium film over the end of a coaxial pair of

[1] *Data Sheet* 39, Land Instruments Inc., Tullytown, Pa.; J. Chedaille and Y. Braud, "Measurements in Flames," pp. 35–48, Crane, Russak & Co., New York, 1972.

[2] Nanmac Corp., Framingham Centre, Mass.

[3] D. Bendersky, A Special Thermocouple for Measuring Transient Temperatures, *Mech. Eng.*, p. 117, February 1953.

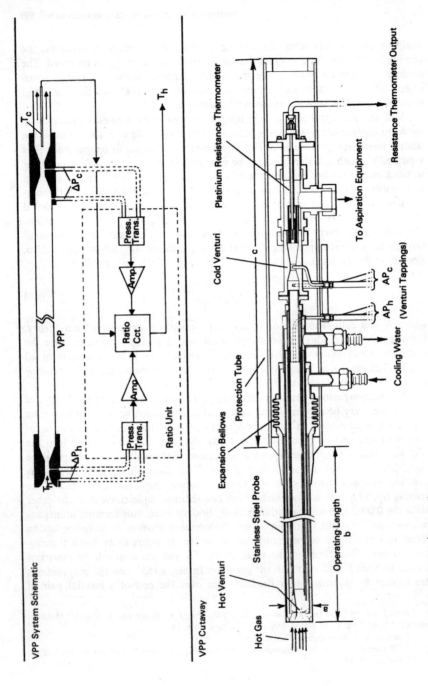

Figure 8.15 Venturi pneumatic pyrometer.

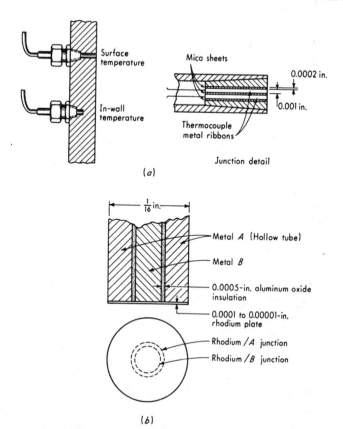

Figure 8.16 High-speed thermocouples.

thermocouple metals. Since the rhodium/metal A and rhodium/metal B junctions are at the same temperature, the third metal (rhodium) has no effect. The plating is performed by vacuum evaporation and results in a rhodium layer 10^{-4} to 10^{-5} in thick. Theoretical calculations indicate the time constant of such a probe is of the order of 0.3 μs.

When the surface whose temperature is to be measured is suitable as one member of a thermocouple pair, the *intrinsic thermocouple* of Fig. 8.17a may be used for fast response measurements. If this is not possible but the surface is an electric conductor, then the arrangement of Fig. 8.17b can be employed. The dynamic response for such systems has been studied[1] by using analytical solutions of partial-differential-equation models of various types, numerical me-

[1] N. R. Keltner, Heat Transfer in Intrinsic Thermocouples—Application to Transient Temperature Measurement Errors, SC-RR-72 0719, Sandia Corp., Albuquerque, N. Mex., 1973.

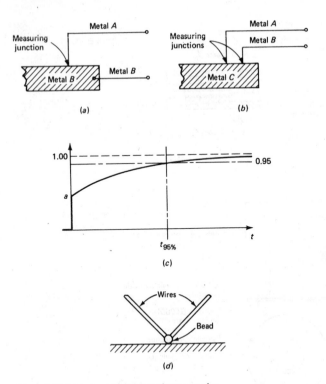

Figure 8.17 Intrinsic and bead-type thermocouples.

thods, and experimental tests. For the system of Fig. 8.17a, the response to a step change in surface temperature is given in Fig. 8.17c, where

$$a \triangleq \frac{1}{1 + \sqrt{(k\rho c)_w/(k\rho c)_s}} \qquad t_{95\%} \triangleq \frac{85\rho_s c_s R^2}{k_s} \left(\frac{k_w}{k_s}\right)^{1.08} \qquad (8.10)$$

where $R \triangleq$ wire radius, $k \triangleq$ thermal conductivity, $\rho \triangleq$ density, $c \triangleq$ specific heat, subscript $s \triangleq$ surface, and subscript $w \triangleq$ wire. For a constantan wire with $R = 0.001$ in and an iron surface, $a = 0.63$ and $t_{95\%} = 0.00092$ s. Figure 8.17d shows a related but different situation, in which an "ordinary" bead-welded thermocouple is fastened to the surface by either welding or adhesives. Response characteristics of this configuration also are available.[1] Thin-foil (0.0002 to 0.0005 in thick) thermocouples manufactured and attached like strain gages also are available[2] for surface-temperature measurement.

---

[1] K. Wally, The Transient Response of Beaded Thermocouples Mounted on the Surface of a Solid, *ISA Trans.*, vol. 17, no. 1, 1978.
[2] RdF Corp., Hudson, N.H.

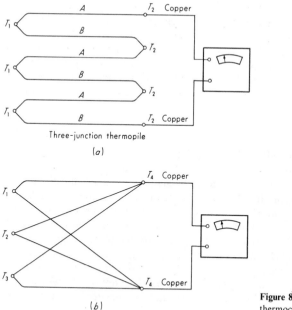

Three-junction thermopile

(a)

(b)

Figure 8.18 Multiple-junction thermocouples.

Thermocouples in common use are made from wires ranging from about 0.020 to 0.1 in in diameter; the larger diameters are required for long life in severe environments. Since speed of response, conduction and radiation errors, and precision of junction location all are improved by the use of smaller wire, very-fine-wire thermocouples are utilized in special applications requiring these attributes and where lack of ruggedness is not a serious drawback. Such couples are available ready-made[1] in most common materials and wire sizes from 0.0005 to 0.015 in diameter. The time constant of an iron/constantan couple of 0.0005-in-diameter wire for a step change from 200 to 100°F in still water is about 0.001 s.

Several thermocouples may be connected in series or parallel to achieve useful functions (Fig. 8.18). The series connection with all measuring junctions at one temperature and all reference junctions at another is used mainly as a means of increasing sensitivity.[2] Such an arrangement is called a thermopile and for n thermocouples gives an output n times as great as a single couple. A typical Chromel/constantan thermopile has 25 couples and produces about 1 mV/F°. Common potentiometers can resolve 1 μV, thus making such an arrangement sensitive to 0.001F°. The parallel combination generates the same voltage as a

[1] Omega Engineering Inc., Stamford, Conn.; Baldwin-Lima-Hamilton, Waltham, Mass., *Tech. Data* 4336-1.

[2] R. A. Schnurr, Thermopiles Aid in Measuring Heat Rejection, *Gen. Motors Eng. J.*, p. 8, April–May–June 1963.

single couple if all measuring and reference junctions are at the same temperature. If the measuring junctions are at different temperatures and the thermocouples are all the same resistance, the voltage measured is the average of the individual voltages. The temperature corresponding to this voltage is the average temperature only if the thermocouples are linear over the temperature range being measured. A 20-junction thermopile with the two sets of junctions built into two standard pipe fittings is available[1] for sensitive measurements of differential temperatures in flowing fluids.

8.4 ELECTRICAL-RESISTANCE SENSORS

The electrical resistance of various materials changes in a reproducible manner with temperature, thus forming the basis of a temperature-sensing method. Materials in actual use fall into two main classes: conductors (metals) and semiconductors. Conducting materials historically came first and traditionally have been called resistance thermometers. [More recently the terminology *resistance temperature detector* (RTD) has come into use.] Semiconductor types appeared later and have been given the generic name *thermistor*. Any of the various established techniques of resistance measurement may be employed to measure the resistance of these devices, with both bridge and "ohmmeter" methods being common.

Conductive Sensors (Resistance Thermometers)

The variation of resistance R with temperature T for most metallic materials can be represented by an equation of the form

$$R = R_0(1 + a_1 T + a_2 T^2 + \cdots + a_n T^n) \qquad (8.11)$$

where R_0 is the resistance at temperature $T = 0$ (see Fig. 8.19). The number of terms necessary depends on the material, the accuracy required, and the temperature range to be covered. Platinum, nickel, and copper are the most commonly used and generally require, respectively, two, three, and three of the a constants for a highly accurate representation. Tungsten and nickel/alloys are also in use. Only constant a_1 may often be used since quite respectable linearity may be achieved over limited ranges. Platinum,[2] for instance, is linear within ±0.4 percent over the ranges -300 to $-100°F$ and -100 to $+300°F$, ±0.3 percent from 0 to 300°F, ±0.25 percent from -300 to $-200°F$, ±0.2 percent from 0 to 200°F, and ±1.2 percent from 500 to 1500°F.

Sensing elements are made in a number of different forms (see Fig. 8.20). For measurement of fluid temperatures, the winding of resistance wire may be en-

---

[1] Delta-T Co., Santa Clara, Calif.
[2] Platinum Resistance Temperature Sensors, *Bull.* 9612, Rosemount Engineering Co., Minneapolis, Minn.

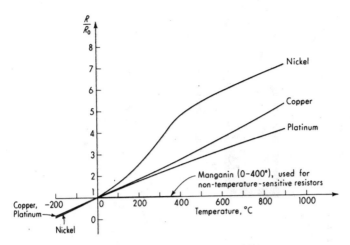

Figure 8.19 Resistance/temperature curves.

cased in a stainless-steel bulb to protect it from corrosive liquids or gases. Open-type pickups expose the resistance winding directly to the fluid (which must be noncorrosive) and give faster response. Various flat grid windings are available for measuring surface temperatures of solids. These may be clamped, welded, or cemented onto the surface. Thin deposited films of platinum also are used in place of wire windings. Surface-temperature transducers affixed to bodies may exhibit spurious output due to interfering strain inputs.[1] These strains may be due to loading of the structure or differential expansion. Careful design (Fig. 8.20b) can minimize strain effects.

Bridge circuits used with resistance temperature sensors may employ either the deflection mode of operation or the null (manually or automatically balanced) mode. If the null method is used, resistor R_4 in Fig. 8.21a is varied until balance is achieved. When the highest accuracy is required, the arrangement of Fig. 8.21b is preferred since the (variable and unknown) contact resistance in the adjustable resistor has no influence on the resistance of the bridge legs. If long lead wires subjected to temperature variations are unavoidable, errors due to their resistance changes may be canceled by use of the Siemens three-lead circuit of Fig. 8.21c. Three lead wires of identical length and material exhibit identical resistance variations, and since one of these leads is in each of legs 2 and 3, their resistance changes cancel. Resistance change in the third wire has no effect on bridge balance since it is in the null detector circuit for null-mode operation. For deflection operation, its effect is negligible if the indicating instrument draws little current.

While the resistance/temperature variation of the sensing element may be

[1] A. B. Kaufman, Bonded-Wire Temperature Sensors, *Instrum. Contr. Syst.*, p. 103, May 1963.

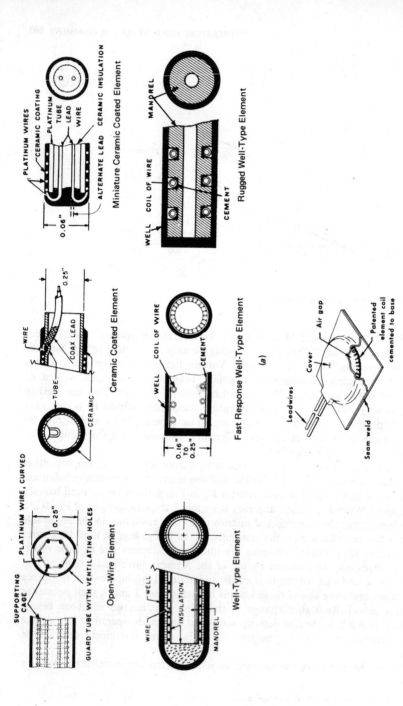

Figure 8.20 Resistance temperature detector (RTD) construction. (*Courtesy Rosemount Engineering, Minneapolis, Minn.*)

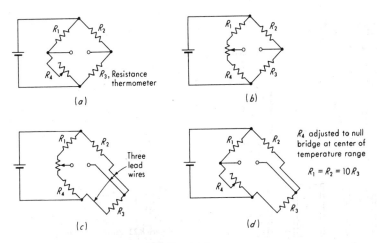

Figure 8.21 Resistance-thermometer bridge circuits.

quite linear, the output voltage signal of a bridge used in the deflection mode is not necessarily linear for large percentage changes in resistance. Unlike the case of strain gages, the resistance change of resistance thermometers for full-scale deflection may be quite large. Typically, a 500-Ω platinum element may exhibit a 100-Ω change over its design range. For a bridge with four equal arms, this would cause severe nonlinearity; however, by making the fixed arms R_1, R_2 of considerably higher resistance (say 10 : 1) than R_3 and R_4 and by balancing the bridge at the middle of the temperature range (rather than at one end), good linearity may be achieved (Fig. 8.21d). Typically,[1] a platinum element covering a range from 0 to 100°C using the 10 : 1 resistance ratio mentioned above gives a nonlinearity of only 0.5C°. For nickel elements, whose resistance/temperature variation is quite nonlinear, this nonlinearity and the bridge nonlinearity can be made to nearly cancel by proper design[2] since the two effects are of opposite directions. Commercially available[3] linear bridge modules requiring only ac power input and connection to a 100-Ω platinum probe (the most common resistance value) provide total accuracy (including linearity) of ±0.1 percent of span for standard spans of −100 to +500°F, 0 to +1000°F, −100 to +200°C, 0 to +200°C, and 0 to +500°C. A linear dc output of 1 mV/degree with 0 mV at 0°F or 0°C can be sent directly to data acquisition systems, reducing their computation load since no software linearization is required. Two such bridges also may be directly connected to measure differential temperature between two probes.

[1] Temperature Recording from Platinum Resistance Sensors, Brush Instruments, Cleveland, Ohio.
[2] D. R. Mack, Linearizing the Output of Resistance Temperature Gages, *Exp. Mech.*, p. 122, April 1961.
[3] Linear Bridge Model 414L, Rosemount Inc., Minneapolis, Minn.

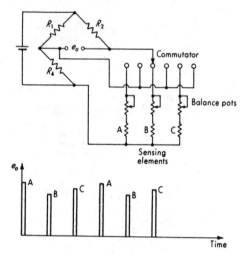

Figure 8.22 Pulse-excitation technique.

Resistance-thermometer bridges may be excited with either ac or dc voltages. The direct or rms alternating current through the thermometer is usually in the range 2 to 20 mA. This current causes an I^2R heating which raises the temperature of the thermometer above its surroundings, causing the so-called self-heating error. The magnitude of this error depends also on heat-transfer conditions and usually is quite small. A 450-Ω platinum element of open construction carrying 25-mA current has a self-heating error of 0.2F° when immersed in liquid oxygen. Actually, by using an unsymmetric pulse type of excitation voltage whose rms (heating) value is small compared with its peak value, quite large instantaneous currents (and thus large peak output voltages) may be obtained without significant self-heating. Such pulse-excitation voltages (Fig. 8.22) can be obtained by commutating a dc source; this also allows timesharing of the bridge among several resistance sensors. As much as a 5-V full-scale bridge output signal can be obtained from resistance sensors by using this technique or others.[1]

While bridge techniques are the classical method for conditioning RTD sensors, the four-wire "ohmmeter" technique of Fig. 8.23 is widely employed in digital thermometers and data acquisition systems, where sensor nonlinearity is corrected in the computer software. Since a precise *current* source (usually a few milliamperes) is utilized, resistance changes in these two lead wires have no effect on sensor current, while high voltmeter input impedance (typically 200 MΩ) makes current in its two lead wires (and thus these lead-wire resistance errors) negligible.

Resistance-thermometer elements range in resistance from about 10 to as high as 25,000 Ω. Higher-resistance elements are less affected by lead-wire and contact resistance variations, and since they generally produce large voltage signals, spurious thermoelectric emf's due to joining of dissimilar metals also are

[1] *Bull.* 9612, Rosemount Inc., op. cit.

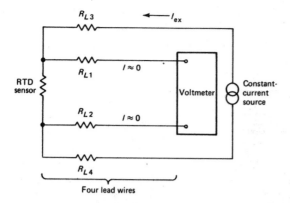

Figure 8.23 Four-wire ohmmeter technique.

usually negligible. Platinum is used from -450 to $1850°F$, copper from -320 to $500°F$, nickel from -320 to $800°F$, and tungsten from -450 to $2000°F$. Average temperatures may be measured by using resistance thermometers, as in Fig. 8.24a, while differential temperature is sensed by the arrangement of Fig. 8.24b.[1] Differential-temperature measurements to an accuracy of $0.05C°$ have been accomplished in a nuclear-reactor-coolant heat-rise application.[2]

Bulk Semiconductor Sensors (Thermistors)[3]

The earlier types of bulk semiconductor resistance temperature sensors were made of manganese, nickel, and cobalt oxides which were milled, mixed in proper proportions with binders, pressed into the desired shape, and sintered. These were given the name *thermistor* and are in wide use today. Compared with conductor-type sensors (which have a small positive temperature coefficient), thermistors have a very large negative coefficient. While some conductors (copper, platinum, tungsten) are quite linear, thermistors are very nonlinear. Their resistance/temperature relation is generally of the form

$$R = R_0 \, e^{\beta(1/T - 1/T_0)} \tag{8.12}$$

where $R \triangleq$ resistance at temperature T, Ω

$R_0 \triangleq$ resistance at temperature T_0, Ω

$\beta \triangleq$ constant, characteristic of material, K

$e \triangleq$ base of natural log

$T, T_0 \triangleq$ absolute temperatures, K

The reference temperature T_0 is generally taken as 298 K (25°C) while the constant β is of the order of 4,000. Computing $(dR/dT)/R$, we find the temperature

---

[1] Temperature Recording from Platinum Resistance Sensors, op. cit.

[2] B. G. Kitchen, Precise Measurement of Process Temperature Differences, *ISA J.*, p. 39, February 1959.

[3] H. B. Sachse, "Semiconductor Temperature Sensors," Wiley, New York, 1975.

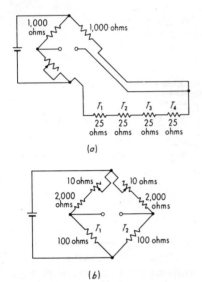

(a)

(b)

Figure 8.24 Average- and differential-temperature sensing.

coefficient of resistance to be given by $-\beta/T^2 \; \Omega/(\Omega \cdot \text{C}°)$. If β is taken as 4,000, the temperature coefficient at room temperature (25°C) is -0.045, compared with $+0.0036$ for platinum. While the exact resistance/temperature relation varies somewhat with the particular material used and the configuration of the resistance element, Fig. 8.25 shows the general type of curve to be expected.

Thermistors are available commercially in the form of beads, flakes, rods, and disks. However, most temperature-sensing applications will employ ready-made probes; some examples are given in Fig. 8.26. Resistance at 25°C can vary over a wide range, from 500 Ω to several megohms. The usable temperature range is from about -200 to $+1000$°C; however, a single thermistor cannot be used over such a large range. When the thermistor is utilized with a computerized data system, absolute temperature T (in kelvins) can be computed from measured resistance R (in ohms) from[1]

$$\frac{1}{T} = A + B \ln R + C \, (\ln R)^3 \tag{8.13}$$

Constants A, B, and C are gotten by solving three simultaneous equations obtained by substituting three pairs of known R, T points (at the low, middle, and high ends of the desired range) into Eq. (8.13). For ranges of 100°C or less, curve-fit accuracy is within about ±0.02°C. When a "hardware" type of linearization is desired, a linear bridge circuit (Fig. 8.27a) can be designed,[2] or

---

[1] Practical Temperature Measurements, *Appl. Note* 290, p. 20, Hewlett-Packard Corp., Palo Alto, Calif., 1980.

[2] Thermistors, *Appl. Note* AN-1A, Fenwal Electronics, Framingham, Mass.

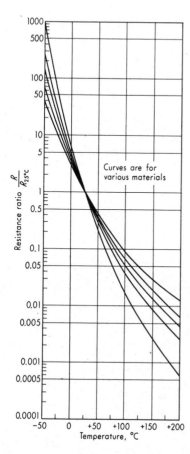

Figure 8.25 Thermistor resistance/temperature curves.

ready-made linear thermistor networks[1] can be employed in either a voltage or a resistance mode (Fig. 8.27b). The bridge of Fig. 8.27a covers the range -100 to $+100°F$ with nonlinearity $\pm 1°F$, while the referenced networks are available in three standard ranges: -5 to $+45°C$, -30 to $+50°C$, and 0 to $100°C$. Linear voltage/temperature outputs are provided with both positive (e_o^+) and negative (e_o^-) slopes, while the resistance mode gives a linear resistance/temperature relation of negative slope. Sensitivities are about ± 6 mV/°C and -20 to -150 Ω/°C, much greater than corresponding thermocouple or RTD values. Maximum nonlinearities are the order of ± 0.06 to $\pm 0.5°C$. Units can be packaged in various ways; one form is a $9 \times 9 \times 3$ mm epoxy molded unit.

Other bulk semiconductor temperature sensors include carbon resistors and

---

[1] Linear Thermistor Networks, Fenwal Electronics, Framingham, Mass.

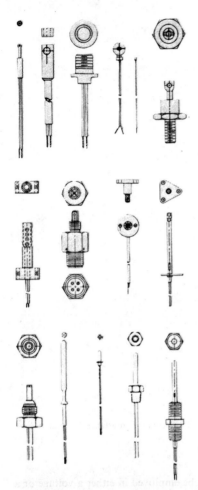

Figure 8.26 Thermistor probe designs. (*Courtesy Fenwal Electronics, Framingham, Mass.*)

silicon[1] and germanium[2] crystal elements. Carbon resistors are merely the commercial carbon-composition elements commonly used as resistance elements in radios and other electronic circuitry. The 0.1 to 1 W rated resistors with room-temperature resistance of 2 to 150 Ω are used widely for the measurement of cryogenic temperatures in the range 1 to 20 K. From about 20 K downward, these elements exhibit a large increase in resistance with a decrease in temper-

---

[1] J. R. Pies, A New Semiconductor for Temperature Measuring, *ISA J.*, p. 50, August 1959.

[2] J. S. Blakemore, Germanium for Low-Temp Resistance Thermometry, *Instrum. Contr. Syst.*, p. 94, May 1962.

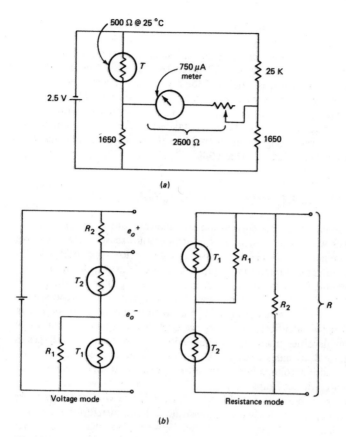

(a)

Voltage mode   Resistance mode

(b)

Figure 8.27 Thermistor linearization networks.

ature given by the relation[1]

$$\log R + \frac{K}{\log R} = A + \frac{B}{T} \tag{8.14}$$

where R is the resistance at Kelvin temperature T and A, B, and K are constants determined by calibration of the individual resistor. Reproducibility of the order of 0.2 percent is obtained in the range 1 to 20 K.

Silicon, with varying amounts of boron impurities, can be designed to have either a positive[2] or negative temperature coefficient over a particular temper-

---

[1] L. G. Rubin, Temperature, *Electron. Progr.*, p. 1, Raytheon Corp., Autumn 1963.
[2] W. Tarpley, Temperature-Resistance Characteristics of TSP102 Silicon Thermistor, *Bull.* CA-195, Texas Instruments, Dallas, 1978.

ature range. The resistance/temperature relation is quite nonlinear. A typical element shows a resistance change (from the nominal value at 25°C) of -80 percent at $-150°C$ to $+180$ percent at $+200°C$. The temperature coefficient near room temperature is of the order of $+0.7$ percent/C°. Germanium, doped with arsenic, gallium, or antimony, is employed for cryogenic temperatures, where it exhibits a large decrease in resistance with increasing temperature. The relation is quite nonlinear but very reproducible, giving precise measurements within 0.001 to 0.0001 K near 4 K when adequate care is taken in technique. Commercially available elements cover a range from about 0.5 to 100 K, a typical unit changing resistance from 7,000 Ω at 2 K to 6 Ω at 60 K.

8.5 JUNCTION SEMICONDUCTOR SENSORS

Junction semiconductor devices such as diodes[1] and transistors[2] exhibit temperature sensitivities which can be exploited as temperature sensors. Here we give details on a commercially available transistor-based device[3] (Fig. 8.28) that is produced as an integrated-circuit (IC) chip and available in chip form, in small metal can or ceramic flat-pack packages, or in ready-made probes. The metallization diagram (Fig. 8.28a) of the chip shows the small size of the basic sensor, while Fig. 8.28b shows internal details. It is a fundamental property of silicon transistors that if two identical transistors are operated at a constant ratio r of collector-current densities, then the difference in their base-emitter voltages is $(kT/q) \ln r$. Since Boltzman's constant k and the charge q of an electron are constants, the resulting voltage is directly proportional to absolute temperature T. In the referenced device, transistors Q_8 and Q_{11} produce the voltage, resistors R_5 and R_6 convert it to a current, and Q_{10} (whose collector current tracks those in Q_9 and Q_{11}) supplies all the bias and substrate leakage current for the rest of the circuit, forcing the total current (flowing between plus and minus power supply terminals) to be proportional to temperature. Resistors R_5 and R_6 are laser-trimmed on the wafer to calibrate each device to give 1 $\mu A/°C$ at 25°C.

The simplest application circuit would require only an unregulated supply of 4 to 30 V, a microammeter, and the device, all connected in series. (Figure 8.28c shows the insensitivity to supply-voltage drifts; between 5 and 15 V, the effect is only 0.2 $\mu A/V$.) For such a circuit and a design range of -55 to $+155°C$, the accuracy (including nonlinearity) of the medium-grade device (five accuracy/price grades are available) would be about $\pm 4.2°C$. This basic device accuracy can be improved by adding "trim" circuitry and calibrating at known temperatures.

[1] R. W. Treharne and J. A. Riley, A Linear-Response Diode Temperature Sensor, *Inst. Tech.*, pp. 59–61, June 1978; R. A. Pease, Using Semiconductor Sensors for Linear Thermometers, *Inst & Cont. Syst.*, pp. 80–81, June 1972; Sachse, op. cit.; M. F. Estes and D. Zimmer, Jr., New Temperature Probe Locates Circuit Hot Spots, *Hewlett-Packard J.*, pp. 30–32, March 1981.

[2] Sachse, op. cit.

[3] Two-Terminal IC Temperature Transducer AD590, *Publ.* C426C-5, Analog Devices, Norwood, Mass., 1979.

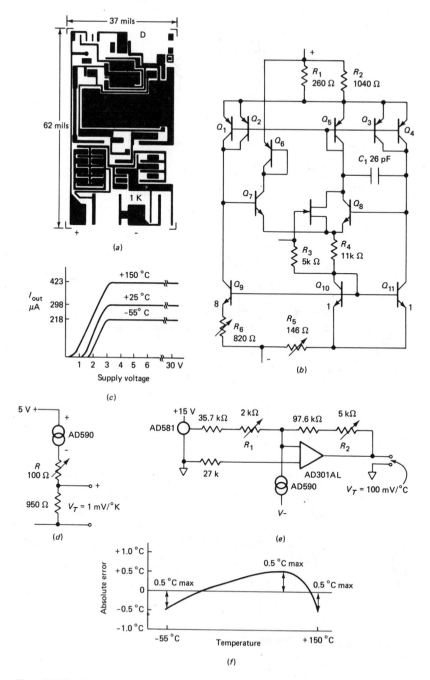

Figure 8.28 Junction semiconductor temperature sensor.

615

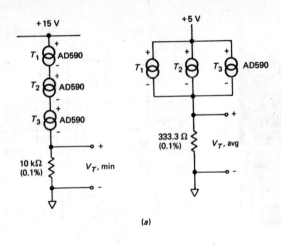

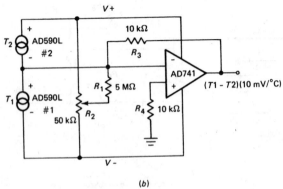

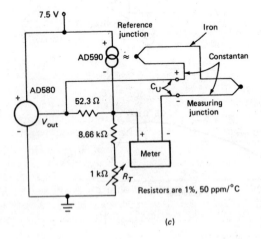

Figure 8.29 Junction semiconductor sensor applications.

Figure 8.28*d* shows a simple one-temperature trim circuit (which also gives a voltage, rather than current output) that improves accuracy to ±1.5°C. The two-temperature trim circuit of Fig. 8.28*e* (which also provides some amplification) gives the curve of Fig. 8.28*f* (accuracy ±0.5°C). (The AD581 is a precision reference voltage supply; AD301AL is an op-amp.) Just as in other electric sensors, self-heating effects are present and should be checked. For example, the metal-can package at 25°C in a stirred oil bath has a 0.06°C rise, while in still air this would be 0.72°C. Note that if the trim circuit techniques mentioned above are employed with calibration heat-transfer conditions the same as actual application conditions, self-heating causes no error.

The current-source nature of the device allows measurement of average temperature by using parallel connection of multiple sensors and selection of minimum temperature by utilizing series connection (Fig. 8.29*a*). The op-amp circuit of Fig. 8.29*b* gives a sensitive indication of differential temperature, while Fig. 8.29*c* shows application to a thermocouple reference-junction compensator of the "hardware" type (AD580 is a precision reference voltage). Junction semiconductor sensors also are widely utilized in "software" reference-junction compensation. Here the sensor measures reference-junction temperature and sends a proportional voltage to the data acquisition system computer, where measuring-junction temperature is corrected digitally.

The main advantages of junction semiconductor temperature sensors are linearity, simple external circuitry, and good sensitivity. However, their upper range is restricted to about 200°C by the damage limits of silicon transistors.

8.6 DIGITAL THERMOMETERS

In most applications, the nonlinearity of thermocouples and RTDs is sufficiently great that we must use calibration tables (or curve-fit formulas derived from them) to convert voltage or resistance measurements to the corresponding temperatures. When many measurements are to be made, this is inconvenient, time-consuming, and prone to error. Also, with thermocouples (TCs), the reference-junction temperature must be accounted for. Because of these considerations (which are *not* present in most other sensors), the fact that temperature is industry's most measured quantity, and the popularity of TCs and RTDs (market share in 1980: TCs, about 50 percent; RTDs, about 27 percent; thermistors, about 23 percent; junction semiconductors, less than 5 percent; radiation pyrometers, less than 5 percent,[1]) special-purpose digital voltmeters called *digital thermometers* have been developed and found a ready market.

Thermometers for RTDs generally use four-wire ohmmeter (rather than bridge) resistance measurement methods and provide the needed current source excitation (two- and three-wire sensors can be accommodated, but at reduced accuracy). Linearization methods for both TC and RTD thermometers can be

[1] D. M. Mackenzie and W. E. Kehret, Review of Temperature Measurement Techniques, *Inst. Tech.*, pp. 43–48, September 1976; pp. 49–54, November 1976.

strictly analog (see Fig. 10.54), nonmicroprocessor digital, or microprocessor digital (the most common). Simple units provide linearization for only a single type of TC (or RTD), while microprocessor linearization techniques allow switch selection from among six types and Fahrenheit or Celsius readout in more sophisticated instruments. Reference-junction compensation in TC thermometers usually uses an isothermal block whose temperature is monitored by a junction semiconductor sensor, which sends these data to the same microprocessor that does the linearization. Use of dual-slope analog/digital conversion for low noise and high resolution (RTDs: 0.01° low range, 0.1° high range; TCs: 0.1°) plus microprocessor computing lags, limits sampling rates to two or three readings per second. *Instrument* accuracies are consistent with the quoted resolutions; however, accuracy of the actual *temperature* measurement will be considerably less because of deviations of the individual TC or RTD from published tables and the usual problems of sensing any physical variable precisely. Sharing of a single digital thermometer among many TCs or RTDs is possible by using scanner accessories. Since digital information is already available, interfacing to printers, computers, etc., is also convenient.

8.7 RADIATION METHODS[1]

All the temperature-measuring methods discussed up to this point require that the "thermometer" be brought into physical contact with the body whose temperature is to be measured. Also, except for the pulsed thermocouple of Fig. 8.14, the temperature sensor generally is intended to assume the same temperature as the body being measured. This means that the thermometer must be capable of withstanding this temperature, which in the case of very hot bodies presents real problems, since the thermometer may actually melt at the high temperature required. Also, for bodies that are moving, a noncontacting means of temperature sensing is most convenient. Furthermore, if we wish to determine the temperature variations over the surface of an object, a noncontacting device can readily be "scanned" over the surface.

To solve problems of the type mentioned above, a variety of instruments based in one way or another on the sensing of radiation have been devised. These might, in general, be called radiometers; however, common usage employs terms such as *radiation pyrometer*, *radiation thermometer*, *optical pyrometer*, etc., to describe a particular type of instrument. Since this terminology is not standardized, you must inquire into the basic operating principle of a given instrument to be sure what its characteristics are, rather than relying on the name given to the instrument.

Other important applications of infrared radiation include missile guidance, satellite attitude sensing, and infrared spectroscopy. In missile guidance (the Sidewinder missile is an outstanding example) the missile is designed to "home" on

[1] R. Vanzetti, "Practical Applications of Infrared Techniques," Wiley, New York, 1972.

the infrared radiation emitted by the target, often the hot jet exhaust of the target aircraft's engine. A scanning system in the missile locates the target and produces error signals that steer the missile into the target. For satellite attitude sensing[1] the infrared sensors are able to distinguish the radiation from the earth, the moon, or a planet from the background of space and thus generate accurate orientation signals for control purposes. Infrared spectroscopy[2] involves the use of infrared principles for the analysis of gases, liquids, and solids to identify and determine the concentration of molecules or molecular groups.

Radiation Fundamentals

Radiation-temperature sensors operate with electromagnetic radiation whose wavelengths lie in the visible and infrared portions of the spectrum. The visible spectrum is quite narrow: 0.3 to 0.72 μm. The infrared spectrum generally is defined as the range from 0.72 to about 1,000 μm. Bordering the visible spectrum on the low-wavelength side are the ultraviolet rays, while microwaves border the infrared spectrum on the high side. Radiation-temperature sensing devices utilize mainly some part of the range 0.3 to 40 μm.

Physical bodies (solids, liquids, gases) may emit electromagnetic radiation or subatomic particles for a number of reasons. As far as temperature sensing is concerned, we need be concerned with only that part of the radiation caused solely by temperature. Every body above absolute zero in temperature emits radiation dependent on its temperature. The ideal thermal radiator is called a *blackbody*. Such a body would absorb completely any radiation falling on it, and for a given temperature, emit the maximum amount of thermal radiation possible. The law governing this ideal type of radiation is Planck's law, which states that

$$W_\lambda = \frac{C_1}{\lambda^5 (e^{C_2/(\lambda T)} - 1)} \tag{8.15}$$

where $W_\lambda \triangleq$ hemispherical spectral radiant intensity, W/(cm^2 · μm)
 $C_1 \triangleq 37,413$, W · μm^4/cm^2
 $C_2 \triangleq 14,388$, μm · K
 $\lambda \triangleq$ wavelength of radiation, μm
 $T \triangleq$ absolute temperature of blackbody, K

The quantity W_λ is the amount of radiation emitted from a flat surface into a hemisphere, per unit wavelength, at the wavelength λ. Equation (8.15) thus gives the distribution of radiant intensity with wavelength. That is, a blackbody at a certain temperature emits *some* radiation per unit wavelength at every wavelength from zero to infinity, but not the same amount at each wavelength. Figure 8.30a shows the curves obtained from Eq. (8.15) by fixing T at various values and

---

[1] Barnes Engineering Co., Stamford, Conn., *Bull.* 14-003 and 0-014, 1962.
[2] Considine, "Process Instruments and Controls Handbook," pp. 6–67.

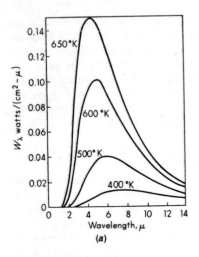

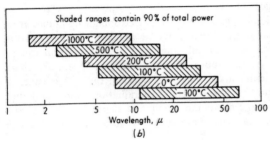

Figure 8.30 Blackbody radiation.

plotting W_λ versus λ. The curves exhibit peaks at particular wavelengths, and the peaks occur at longer wavelengths as the temperature decreases. The area under each curve is the total emitted power and increases rapidly with temperature. Equations giving the peak wavelength λ_p and the total power W_t are

$$\lambda_p = \frac{2,891}{T} \qquad \mu m \tag{8.16}$$

and

$$W_t = 5.67 \times 10^{-12} T^4 \qquad W/cm^2 \tag{8.17}$$

Figure 8.30b shows the wavelength range over which 90 percent of the total power is found for various temperatures. Note that lower temperatures require measurement out to longer wavelengths.

While the concept of a blackbody is a mathematical abstraction, real physical bodies can be constructed to approximate closely blackbody behavior. Such radiation sources are needed for calibration of radiation thermometers and generally take the form of a blackened conical cavity of about 15° cone angle. The temperature is adjustable, automatically controlled for constancy, and measured by some accurate sensor such as a platinum resistance thermometer. A typical unit[1] covers the range 500 to 1000 K with 1 K accuracy and emittance 0.99 ± 0.01 (blackbody has 1.00). While it is possible to construct a nearly perfect blackbody, the bodies whose temperatures are to be measured with some radiation-type instrument often deviate considerably from such ideal conditions. The deviation from blackbody radiation is expressed in terms of the emittance of the measured body.

Several types of emittance have been defined to suit particular applications. The most fundamental form of emittance is the hemispherical spectral emittance $\epsilon_{\lambda, T}$. Let us call $W_{\lambda a}$ the *actual* hemispherical spectral radiant intensity of a real body at temperature T, and let us assume it can be measured (by using optical bandpass filters). Then $\epsilon_{\lambda, T}$ is defined as

$$\epsilon_{\lambda, T} \triangleq \frac{W_{\lambda a}}{W_\lambda} \tag{8.18}$$

where W_λ is the blackbody intensity at temperature T. Thus emittance is dimensionless and always less than 1.0 for real bodies. In the most general case, it varies with both λ and T. With the definition of Eq. (8.18), the radiation from a real body may be written as

$$W_{\lambda a} = \frac{C_1 \epsilon_{\lambda, T}}{\lambda^5 (e^{C_2/(\lambda T)} - 1)} \tag{8.19}$$

Similarly, the total power W_{ta} of an actual body is given by

$$W_{ta} = C_1 \int_0^\infty \frac{\epsilon_{\lambda, T}}{\lambda^5 (e^{C_2/(\lambda T)} - 1)} \, d\lambda \tag{8.20}$$

and if we assume that W_{ta} can be measured experimentally, we may define the hemispherical total emittance $\epsilon_{t, T}$ by

$$\epsilon_{t, T} \triangleq \frac{W_{ta}}{W_t} \tag{8.21}$$

where W_t is the blackbody total power at temperature T. Thus if $\epsilon_{t, T}$ is known, the total power of a real body is given by

$$W_{ta} = 5.67 \times 10^{-12} \, \epsilon_{t, T} \, T^4 \qquad \text{W/cm}^2$$

If a body has $\epsilon_{\lambda, T}$ equal to a constant for all λ and at a given T, it is called a *graybody*. In this case we see that $\epsilon_{\lambda, T} \equiv \epsilon_{t, T}$. Also the curves of $W_{\lambda a}$ versus λ have exactly the same shape as for W_λ. Since many radiation thermometers operate in a restricted band of wavelengths, the hemispherical band emittance

---

[1] Infrared Industries Inc., Riverside, Calif.

$\epsilon_{b,\,T}$ has been defined by

$$\epsilon_{b,\,T} \triangleq \frac{\displaystyle\int_{\lambda_a}^{\lambda_b} \{\epsilon_{\lambda,\,T}/[\lambda^5(e^{C_2/(\lambda T)} - 1)]\}\,d\lambda}{\displaystyle\int_{\lambda_a}^{\lambda_b} \{1/[\lambda^5(e^{C_2/(\lambda T)} - 1)]\}\,d\lambda} \tag{8.22}$$

This is seen to be just the ratio of the total powers, actual and blackbody, within the wavelength interval λ_a to λ_b for bodies at temperature T. If the actual power can be measured directly, $\epsilon_{b,\,T}$ can be found without knowing $\epsilon_{\lambda,\,T}$. For a graybody, $\epsilon_{b,\,T} \equiv \epsilon_{\lambda,\,T}$.

If a radiation thermometer has been calibrated against a blackbody source, knowledge of the appropriate emittance value allows correction of its readings for nonblackbody measurements. Unfortunately, emittances are not simple material properties such as densities, but rather depend on size, shape, surface roughness, angle of viewing, etc. This leads to uncertainties in the numerical values of emittances, which are one of the main problems in radiation temperature measurement. Real-world objects are characterized by emittance ϵ, reflectance r, and transmittance t, which are related by

$$\epsilon = 1 - r - t \tag{8.23}$$

When r and/or t is not zero, we have nonblackbody behavior, and measurement errors as in Fig. 8.31 are possible. Commercial radiation thermometers usually include an emittance adjustment with a range from about 0.2 to 1.0. Thus if material emittance is known, it can be compensated for. The most reliable technique for determining emittance for such purposes requires that at some time we be able to measure specimen temperature (simultaneous with the radiation measurement) by some independent means, such as a thermocouple. The radiometer emittance dial is adjusted until the temperatures indicated by the two instruments agree. If there are no changes in conditions, this emittance setting can be used for any future radiation temperature measurements without, of course, requiring the presence of the thermocouple. Since emittance can vary with temperature, such an emittance calibration may be necessary over the desired temperature range. In such a case, a calibration in which the ϵ dial is left fixed at some arbitrary value, and a table of indicated versus true temperatures is developed, might be more useful for correction purposes.

Plastic film manufacture[1] makes wide use of radiation temperature measurement and is a good example for emittance considerations based on Eq. (8.23). Reflectance r is about 0.04 for all plastics in the infrared range and is independent of film thickness. Transmittance t is given for cellulose acetate film (as measured by a spectrophotometer) in Fig. 8.32. Note that in the range 8 to 10 μm, film of any thickness between 0.001 and 0.1 in behaves almost as a blackbody; thus we should design our radiometer for sensitivity to only this range.

---

[1] Plastic Film Measurement, TN100, Ircon Inc., Skokie, Ill., 1979.

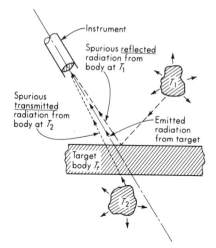

Figure 8.31 Measurement problems with reflective/translucent targets.

Another source of error is the losses of energy in transmitting the radiation from the measured object to the radiation detector. Generally the optical path consists of some gas (often atmospheric air) and various windows, lenses, or mirrors used to focus the radiation or protect sensitive elements from the environment. In atmospheric air the attenuation of radiation is due mainly to the resonance-absorption bands of water vapor, carbon dioxide, and ozone as well as the scattering effect of dust particles and water droplets. The combined absorption effect of H_2O, CO_2, and O_3 is roughly as shown in Fig. 8.33. Since the absorption varies with wavelength, a radiation thermometer can be designed to

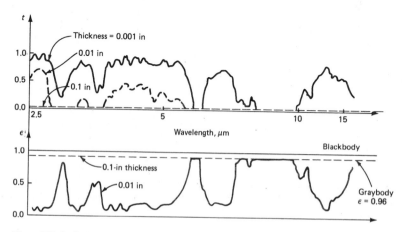

Figure 8.32 Radiation temperature measurement of plastic films.

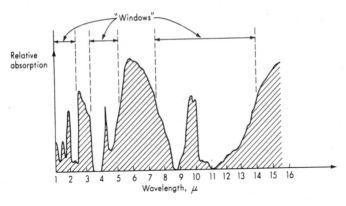

Figure 8.33 Atmospheric absorption.

respond only within one of the "windows" shown, thus making it insensitive to these effects. Since the absorption varies with the thickness of the gas traversed by the radiation, the effect is not an instrument constant and thus cannot be calibrated out. The lenses used in infrared instruments often must be made of special materials, since glasses normally utilized for the visible spectrum are almost opaque to radiation of wavelength longer than about 2 μm. Figure 8.34 shows the variation of transmission factor of various materials with wavelength. While infrared radiation follows the same optical laws used for lens and mirror design as visible light, some materials useful for infrared wavelengths (arsenic trisulfide, for example) are opaque to visible-light wavelengths.

To "see around corners," span interfering atmospheres between target and detector, and isolate system electronics from hostile sensing environments, flexible fiber optic[1] cables are available in lengths to 10 m to transmit infrared radiation from target to detector. Such systems work best at high temperatures (using silicon detectors) since fiber transmission is limited to the shorter wavelengths.

Radiation Detectors

In all radiation thermometers (other than the disappearing-filament optical pyrometer), the radiation from the measured body is focused on some sort of radiation detector which produces an electrical signal. Detectors may be classified as thermal detectors or photon detectors. Thermal detectors are blackened elements designed to absorb a maximum of the incoming radiation at all wavelengths. The absorbed radiation causes the temperature of the detector to rise until an equilibrium is reached with heat losses to the surroundings. Thermal

---

[1] O. W. Uguccini and F. G. Pollack, High-Resolution Surface Temperature Measurements on Rotating Turbine Blades with an Infrared Pyrometer, *NASA* TN D-8213, 1976.

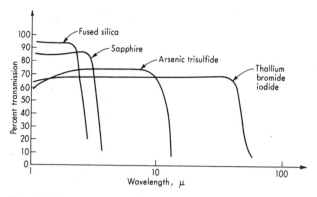

Figure 8.34 Spectral transmission of optical material.

detectors usually measure this temperature, using a resistance thermometer, thermistor, or thermocouple (thermopile) principle. (*Pyroelectric* thermal detectors operate on a different principle, as discussed below.)

Resistance-thermometer and thermistor elements are made in the form of thin films or flakes and are called *bolometers*. Performance criteria for both thermal and photon detectors include the time constant (most detectors behave roughly as first-order systems), the responsivity (volts of signal per watt of incident radiation), and the noise-equivalent power. The noise-equivalent power gives an indication of the smallest amount of radiation that can be detected, which is limited by the inherent electrical noise level of the detector. That is, with no radiation whatever coming in, the detector still puts out a small random voltage due to various electrical noise sources within the detector itself. The amount (watts) of incoming radiation required to produce a signal just equal in strength to the noise (signal-to-noise ratio of 1) is called the *noise-equivalent power*. Thus a low value of noise-equivalent power is desirable. An evaporated-nickel-film bolometer of about 35 mm^2 area has a resistance of about 100 Ω, a time constant of 0.004 s, responsivity of 0.4 V/W, and noise-equivalent power of 3×10^{-9} W. A thermistor bolometer of 0.5 mm^2 area might have 3-MΩ resistance, a time constant of 0.004 to 0.030 s, responsivity of 700 to 1,200 V/W, and noise-equivalent power of 2×10^{-10} W. Thermopiles of an area from 0.4 to 10 mm^2 have resistance of the order of 10 to 100 Ω, time constants in the range 0.005 to 0.3 s, responsivity of 3 to 90 V/W, and noise-equivalent power of 2×10^{-11} to 7×10^{-10}.

A class of thermal detectors of increasing importance since the mid-1960s is the pyroelectrics, useful over the spectral range 0.001 (soft x-rays) to 1,000 μm (far infrared). These devices can have a wide spectral response (similar to thermopiles and bolometers), but do not suffer from the slow response speed characteristic of other thermal detectors. The basis of the improved speed lies in the fact that the electrical output depends on the time rate of change in detector temperature

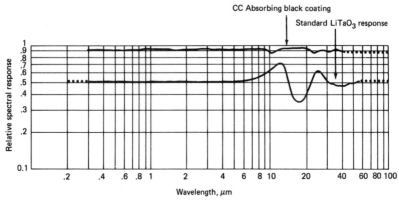

Figure 8.35 Wavelength response of pyroelectric detectors.

rather than on the detector temperature itself. Materials that exhibit the pyroelectric effect include lithium tantalate crystals, ceramic barium titanate, and polyvinyl fluoride plastic films.[1] Figure 8.35 shows spectral response for coated and uncoated lithium tantalate sensors.[2] Black absorbing coatings, which increase spectral uniformity and overall sensitivity, are used often with infrared radiation; however, they reduce speed of response by introducing a -10 dB/decade rolloff in amplitude ratio at high frequencies, with a typical breakpoint being at about 2,000 Hz.[3] When coatings are not employed, very fast response (rise times on the order of 170 ps)[4] are possible since the electrical output can be shown[5] to respond to the instantaneous volume-averaged detector temperature and thus does not have to "wait" for the high surface temperature created by a fast radiation pulse to diffuse throughout the detector mass (a relatively *slow* thermal process).

In Fig. 8.36*a* a pyroelectric crystal is shown with thin metal film electrodes, one of which faces the radiation (edge-type electrodes also are possible). In some cases face electrodes, while sometimes thin enough to be essentially transparent to the impinging radiation, may significantly effect spectral response and detector speed. Since pyroelectric materials are dielectrics, the electrode/crystal sandwich is a capacitor, in addition to being a radiation-sensitive charge generator. Pyro-

[1] J. Cohen and S. Edelman, Polymeric Pyroelectric Sensors for Fire Protection, *AFAPL Rept.* TR-74-16, Wright Patterson Air Force Base, Ohio, 1975.

[2] Molectron Corp., Sunnyvale, Calif.

[3] W. M. Doyle, A User's Guide to Pyroelectric Detection, *Electro-Opt. Syst. Des.*, pp. 12–17, November 1978; E. H. Putley, The Pyroelectric Detector, chap. 6, "Semiconductors and Semimetals," vol. 5, R. K. Willarson and A. C. Beer (eds.), Academic, New York, 1970; J. Cooper, A Fast Response Pyroelectric Thermal Detector, *J. Sci. Instrum.*, vol. 39, pp. 467–472, 1962.

[4] W. B. Tiffany, Commercial Applications of Pyroelectric Detectors, *Electro-Opt. Syst. Des.*, pp. 11–14, November 1976.

[5] P. A. Schlosser, Investigation of the Radiation Imaging Properties of Pyroelectric Detector Arrays, Ph.D. diss., Nuclear Eng. Dept., The Ohio State University, Columbus, 1972.

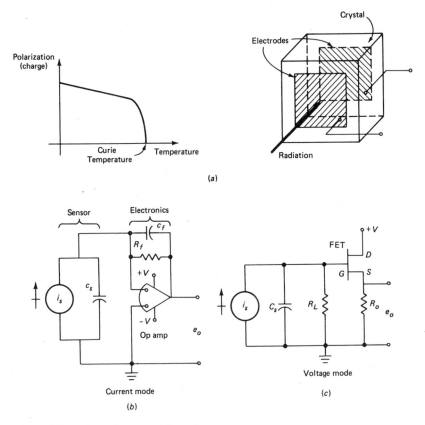

Figure 8.36 Pyroelectric detector and electronics.

electric materials exhibit a permanent electric polarization (charge) which is sensitive to temperature. Often a "poling" process which involves application of an external electric field at room or elevated temperature to obtain maximum alignment of the material's domains is employed to develop the strongest effect. Pyroelectric materials all exhibit a Curie temperature (610°C for lithium tantalate), above which the effect is lost and repolarization would be required.

Since connection of any measuring circuit to the sensor electrodes allows a flow of charge which ultimately neutralizes the polarization effect, the sensor, unlike thermopiles and bolometers, is not able to directly measure steady radiation (however, addition of a chopper system allows this capability). You may note that this behavior is analogous to that of the piezoelectric effect discussed in Chap. 4; and, in fact, one of the earliest pyroelectric materials studied (barium titanate) is employed also in piezoelectric applications. Both pyroelectric and piezoelectric behavior require the material to be ferroelectric; it must exhibit permanent electrical polarization. This piezoelectric response can cause spurious readings in pyroelectric detectors whose housings are subject to mechanical or acoustical excitation and even for nonvibrating housings when rapidly changing radiation input (such as short laser pulses) excites detector vibration.

For the linear portion of the polarization/temperature curve, charge q is proportional to volume-averaged detector temperature T, which gives sensor current i_s as

$$i_s = K_p \frac{dT}{dt} \tag{8.24}$$

where K_p is a sensitivity constant of the particular device. By modeling the thermal system as detector thermal capacitance C_t connected to a constant-temperature heat sink by a thermal resistance R_t, neglecting the coating dynamics mentioned earlier, calling the absorbed radiant power W (watts), and treating all variables as small perturbations away from an equilibrium operating point, conservation of energy gives

$$W \, dt - \frac{T - 0}{R_t} \, dt = C_t \, dT \tag{8.25}$$

$$(\tau_t D + 1)T = R_t W$$

where $\tau_t \triangleq$ thermal time constant $\triangleq R_t C_t$ $\qquad\qquad$ (8.26)

For the "current mode" circuit of Fig. 8.36b, we have

$$i_s = K_p D \left(\frac{R_t W}{\tau_t D + 1} \right) = - \frac{e_o}{R_f / (R_f C_f D + 1)} \tag{8.27}$$

$$\frac{e_o}{W} (D) = - \frac{K_e D}{(\tau_t D + 1)(\tau_e D + 1)} \tag{8.28}$$

where $K_e \triangleq K_p R_t R_f \triangleq$ sensitivity, V/(W/s) $\qquad\qquad$ (8.29)
$\qquad\qquad\quad \tau_e \triangleq$ electrical time constant $\triangleq R_f C_f$ $\qquad\qquad$ (8.30)

A typical system[1] might have $K_e = 50$, $\tau_t = 0.1$ s, and $\tau_e = 10^{-5}$ s. While the resistance R_f is an "intentional" component (typically 10^8 Ω), the capacitance C_f may represent parasitic effects (the order of 0.01 to 0.1 pF) associated with the wiring layout. When a smaller R_f (say 10^5 to 10^6 Ω) is used to extend the flat frequency range, an intentional C_f of 0.5 to 5.0 pF may need to be wired in to reduce oscillations caused by op-amp dynamics. Note that sensor capacitance C_s (typically 25 pF) does not enter into the system transfer function.

An alternative ("voltage mode") form of electronics commonly employed with pyroelectric detectors uses a field-effect transistor (FET) as in Fig. 8.36c. For a FET, point G (the gate) may be treated as essentially an open circuit, while the source (S) to drain (D) current, and thus e_o, is proportional to the voltage at G.

---

[1] PI-30 Detector/Op Amp, Molectron Corp., Sunnyvale, Calif.

Analysis gives

$$e_G = i_s \left(\frac{R_L}{R_L C_s D + 1} \right) = \frac{K_p R_t R_L D W}{(\tau_e D + 1)(\tau_t D + 1)} \tag{8.31}$$

And if $e_o = K_g e_G$ (with K_g typically 0.7 V/V), we get

$$\frac{e_o}{W}(D) = \frac{K_e D}{(\tau_t D + 1)(\tau_e D + 1)} \tag{8.32}$$

where
$$K_e \triangleq K_g K_p R_t R_L \triangleq \text{sensitivity, V/(W/s)} \tag{8.33}$$
$$\tau_e \triangleq \text{electrical time constant} \triangleq R_L C_s \tag{8.34}$$

Typical[1] values might be $R_L = 10^{11}$ Ω, $R_s = 10^4$ Ω, $C_s = 30$ pF, $\tau_t = 1$ s, and $K_e = 20,000$.

Equations (8.28) and (8.32) clearly show that the current-mode electronics exhibits the same form of transfer function as the voltage-mode, although each mode can be designed with widely varying numerical values. For the values given as typical above, Fig. 8.37 shows the frequency response (recall that an additional -10 dB/decade rolloff would be present at high frequency if an absorbing coating were present). The voltage-mode system of this example shows no range of flat response, although, of course, it could be obtained easily by changes in numerical values. Actually, a flat response is not always necessary or desirable. For example, to obtain response to *steady* radiation (which neither voltage mode nor current mode directly provides), a chopper is utilized to convert the steady radiation flux to a periodic variation. Since the chopping is at a fixed frequency, such an instrument need only have adequate amplitude ratio at that frequency, but there is no requirement for flat amplitude ratio. Also, some applications require measurement of the total energy of a radiation pulse. This requires integration, with respect to time, of the input signal W (which is power), and this is accomplished if the pulse-frequency content lies mainly in the downward-sloping (-20 dB/decade) region of the frequency response. Finally, if an overall system with flat amplitude ratio is desired in which the detector and preamp provide a -20 dB/decade slope, then the preamp can be followed with a dynamic compensator (approximate differentiator) to achieve this.

Our brief treatment of pyroelectric detectors has not included any discussion of electrical noise, always important in practical design studies. Useful data on this area may be found in the literature.[2]

In the various types of photon detectors, the incoming radiation (photons) frees electrons in the detector structure and produces a measurable electrical effect. These events occur on an atomic or a molecular time scale rather than on the gross time scale involved in the heating and cooling of thermal detectors. Thus a much higher response speed is possible. However, photon detectors have a sensitivity that varies with wavelength; thus incoming radiation of different

---

[1] Eltec Instruments Model 406, Cat. 8-78, 8-79, Daytona Beach, Fla.
[2] Doyle, op. cit.

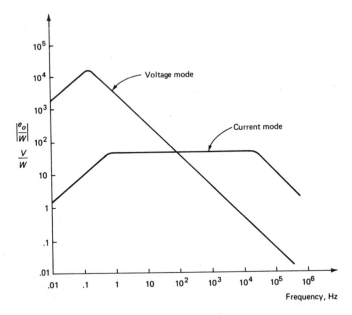

Figure 8.37 Pyroelectric-detector frequency response.

wavelengths is not equally treated. Typical spectral response of some common types is shown in Fig. 8.38. Photon detectors commonly in use operate in the photoconductive, photovoltaic, or photoelectromagnetic (PEM) modes.

Photoconductive types exhibit an electrical resistance that changes with the incoming radiation level. Photovoltaic cells, also called *barrier photocells*, employ a photosensitive barrier of high resistance, deposited between two layers of conducting material. A potential difference between these two layers is built up when the cell is exposed to radiation. In photoelectromagnetic detectors the Hall effect is utilized. A semiconductor crystal is subjected to a strong magnetic field, and radiation is applied to one side. A potential difference is developed across the ends of the crystal. Lead sulfide photoconductive cells are the most frequently used type by far, typical units of 1 to 35 mm^2 area having resistances of 10^5 to 2×10^6 Ω, time constants of 2 to 0.04 ms, responsivity of 5,000 to 150,000 V/W, and noise-equivalent power of 4×10^{-11} to 4×10^{-12} W. An indium antimonide photoelectromagnetic cell of 100-Ω resistance may have a time constant less than 1 μs, responsivity of 1 V/W, and noise-equivalent power of 10^{-9} W.

Some type of circuit must be employed to realize a usable electrical signal (generally a voltage) from a radiation detector. Thermopile devices generally work with an uninterrupted stream of radiation and require no circuitry other than the usual reference junction, which is commonly left at ambient temperature. Bolometers and photoconductive cells often employ a chopper to interrupt the

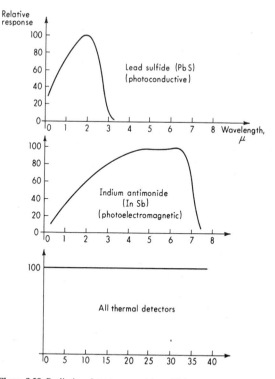

Figure 8.38 Radiation-detector spectral sensitivity.

radiation at a fixed rate of the order of several hundred hertz. This leads to an ac-type electric signal and allows use of high-gain ac amplifiers. Typical circuits for such arrangements are shown in Fig. 8.39.

Unchopped (DC) Broadband Radiation Thermometers

We begin our study of complete radiation-sensing instruments by consideration of the most common type used in day-to-day industrial applications.[1] These instruments employ a blackened thermopile as detector and focus the radiation by means of either lenses or mirrors. Figure 8.40 shows in simplified fashion the construction of this class of instruments. The reference of footnote 1 gives a very complete analysis of such devices.

Basically, for a given source temperature T_1, the incoming radiation heats the measuring junction until conduction, convection, and radiation losses just

[1] T. R. Harrison, "Radiation Pyrometry and Its Underlying Principles of Radiant Heat Transfer," Wiley, New York, 1960.

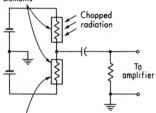

Two matched bolometer
elements

Chopped
radiation

To
amplifier

This element at same ambient temperature
but shielded from radiation to provide
ambient-temperature compensation

Simple circuit for a high-resistance bolometer

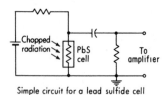

Chopped
radiation

PbS
cell

To
amplifier

Simple circuit for a lead sulfide cell

Figure 8.39 Basic detector circuits.

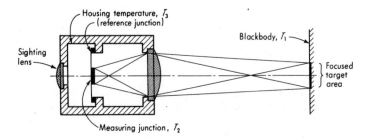

Housing temperature, T_3
(reference junction)

Blackbody, T_1

Sighting
lens

Focused
target
area

Measuring junction, T_2

Lens-type radiation thermometer

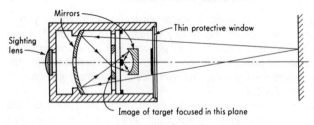

Mirrors

Thin protective window

Sighting
lens

Image of target focused in this plane

Mirror-type radiation thermometer

Figure 8.40 Lens- and mirror-type radiation thermometers.

632

balance the heat input. The measuring-junction temperature usually is less than 40°C above its surroundings even if the source is incandescent. An oversimplified analysis gives

$$\text{Heat loss} = \text{radiant heat input}$$

$$K_1(T_2 - T_3) = K_2 T_1^4 \tag{8.35}$$

If the thermocouple voltage is proportional to $T_2 - T_3$, the voltage output should be proportional to T_1^4. Figure 8.41 shows an actual calibration curve of such an instrument together with the ideal relationship. For high temperatures the agreement is quite close. The temperatures T_2 and T_3 are both influenced by the environmental temperature; thus compensation generally must be provided for this, particularly if the instrument is intended to measure low temperatures. This compensation may include a thermostatically controlled housing temperature.

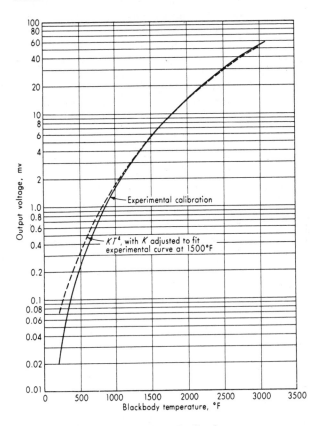

Figure 8.41 Theoretical and experimental calibration curves.

The thermopiles used may have from 1 or 2 to 20 or 30 junctions. A small number of junctions has less mass and thus faster response, but lower sensitivity limits application to high temperatures. Response is roughly that of a first-order system, with time constants ranging from about 0.1 (high-temperature systems) to 2 s (low-temperature systems). Instruments of this class are available to measure temperatures as low as 0°F; that is, actually the thermopile is cooler than the ambient temperature and reads a negative voltage. Theoretically there is no upper limit to the temperatures that can be measured in this way. Commercial instruments usable to 3200°F are readily available.

Conceivably, an instrument of the above type could be constructed with no focusing means, i.e., no lens or mirror. A simple diaphragm with a circular aperture (Fig. 8.42) would define the target from which radiation is received. To define smaller target areas for a given target distance, a smaller aperture could be utilized; however, there would be a proportionate loss in incoming radiation and thus sensitivity. The reading of such an instrument is independent of the distance between the target and the instrument, since the amount of radiation received is limited by the solid angle of the cone defined by the aperture and detector, and this is always the same. However, as the target distance increases, the target area necessary to fill the cone increases. If, because of nonuniformity of target temperature or small target size, we wish to restrict the target area to small values, a very small aperture is required, giving very low sensitivity.

The basic purpose of lens or mirror systems is to overcome this restriction,

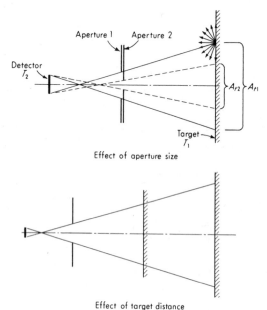

Aperture 1   Aperture 2

Detector
T_2

A_{t2} A_{t1}

Target
T_1

Effect of aperture size

Effect of target distance

Figure 8.42 Effect of aperture size and target distance.

thus allowing the resolution of small targets without loss of sensitivity. Thus, in Fig. 8.43, radiation emanating from the point P on the target is focused on the corresponding point P' of the detector. If a simple diaphragm, rather than a focused lens, had been used, the same amount of energy would be spread out over the area A_p, with the detector receiving only a fraction of the radiation. Commercial lens- or mirror-type instruments generally have parameters such that targets 2 ft or more distant are adequately focused with a fixed-focus lens, and the instrument calibration is independent of distance as long as the target fills the field of view. Minimum target size to fill the field of view is of the order of one-twentieth of the target distance for common instruments. For targets closer than 2 ft, focusing is necessary and may affect the calibration, depending on instrument construction and closeness of target. Target diameters of 0.1 to 0.3 in at target distances of 4 to 12 in are available. Since the focal length of a lens depends on the index of refraction, which in turn varies with wavelength, not all wavelengths are focused at the same point. In particular, if one focuses a lens visually (using visible light), the longer infrared wavelengths, which contribute a large portion of the total energy at lower temperatures, will be out of focus. Such an instrument may have to be focused by adjusting for maximum thermopile output rather than for sharpest visual definition. Another effect of lenses is selective transmission, as shown in Fig. 8.34. The use of mirrors rather than lenses is an attempt to alleviate some of these problems. However, instruments of both types are employed widely with success.

Chopped (AC) Broadband Radiation Thermometers

A number of advantages accrue when the radiation coming from the target to the detector is interrupted periodically (chopped) at a fixed frequency; therefore many infrared systems employ this technique. When high sensitivity is needed, amplification is required, and high-gain ac amplifiers are easier to construct than their dc counterparts. This is usually the main reason for using choppers. Additional benefits related to ambient-temperature compensation and reference-

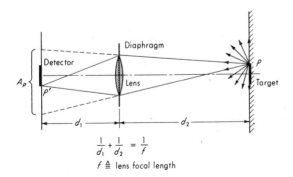

Figure 8.43 Advantages of focusing.

source comparison also may be obtained. Systems employing thermal (broadband) and photon (restricted-band) detectors and choppers are in common use. Here we consider those utilizing thermal detectors.

The time constants of adequately sensitive thermopile detectors are generally too long to allow efficient use of chopping; thus the faster bolometers, usually the thermistor type, are employed. We consider two specific forms of this class of instruments: the blackened-chopper type and the mirror-chopper type. Figure 8.44 shows the basic elements of a blackened-chopper radiometer. A mirror focuses the target radiation on the detector; however, this beam is interrupted periodically by the chopper rotating at constant speed. Thus the detector alternately "sees" radiation from the target and radiation from the chopper's blackened surface. For high target temperatures, sufficient accuracy may be achieved by leaving the chopper temperature at ambient. Higher accuracy, particularly at

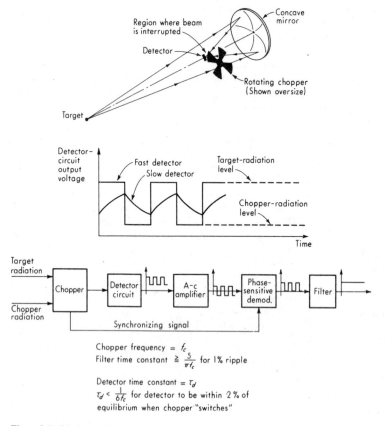

Chopper frequency = f_c

Filter time constant $\geqq \dfrac{5}{\pi f_c}$ for 1% ripple

Detector time constant = τ_d

$\tau_d < \dfrac{1}{6f_c}$ for detector to be within 2% of equilibrium when chopper "switches"

Figure 8.44 Blackened-chopper system.

low target temperatures, is obtained by thermostatically controlling the chopper temperature.

An arrangement similar to that of Fig. 8.39 is used for the detector circuit. The output voltage of the detector circuit is essentially as shown in Fig. 8.44. By amplifying this in an ac amplifier, the mean value (which is subject to drift) is discarded, and only the difference between the target and chopper radiation levels is amplified. If the chopper radiation level is considered as a known reference value, the target radiation and thus its temperature may be inferred. To provide a dc output signal related to target temperature and suitable for recording or control purposes, the ac amplifier is followed by a phase-sensitive demodulator and filter circuit. The necessary synchronizing signal for the demodulator may be generated by placing a magnetic proximity pickup near the chopper blades. While the response time of the detector itself may be of the order of a millisecond, the chopper frequency and necessary demodulator filter time constant greatly reduce the overall system speed. High chopper speeds allow faster overall response, but reduce sensitivity if the detector time constant is too large, since the detector does not have time to reach equilibrium during the time when either the target or the chopper is in view.

A typical instrument[1] uses a square thermistor detector, thus giving rise to a square field of view of size 1° by 1°. Sighting and focusing in the range 2 ft to infinity are accomplished with an attached optical telescope. The standard chopping frequency is 180 Hz, leading to an overall system time constant of about 0.008 s. Standard temperature range is from ambient to 1300°C.

In the mirror-chopper instrument[2] of Fig. 8.45, two thermistor detectors are utilized. Also, an accurate blackbody source whose temperature is controlled automatically and measured accurately (say by a thermocouple) is provided within the radiometer. A chopper operating at 77 Hz alternately exposes each of the two detectors to the target radiation and the blackbody source of known temperature. In Fig. 8.45 detector 1 is receiving target radiation while 2 is receiving blackbody. When the chopper disk rotates 90° so that a solid sector is in the line of sight, detector 1 receives blackbody radiation reflected from the rear mirror surface of the chopper while detector 2 receives target radiation reflected from the front surface. Circuitry similar to that of Fig. 8.44 can be employed to develop a dc output signal related to target temperature. Such a system can be utilized for very accurate static measurements by using it in a null method of operation. The blackbody source temperature is adjusted until no output is obtained. Then the target and blackbody are at identical temperatures, provided the target emittance is 1.0. If the emittance is not 1.0 but is known, a correction may be applied. Such a null method makes the reading independent of detector sensitivity (which may drift) and amplifier gain. A system of this type employing 1.5 × 1.5 mm thermistor detectors has a 0.5 by 0.5° field of view (1-in^2 target at 10 ft) and an overall system time constant of 0.016 s (bandwidth 10 Hz), and will

[1] Servo Corp. of America, New Hyde Park, N.Y.
[2] Barnes Engineering Co., Stamford, Conn.

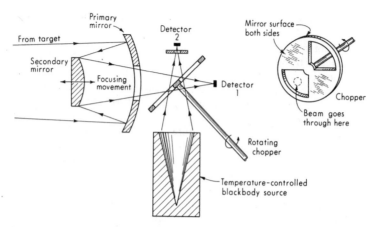

Figure 8.45 Mirror-chopper system.

detect (signal-to-noise ratio is 1.0) a target temperature change of 0.4C°. By sacrificing response speed for sensitivity, heavier filtering can be switched in to give a time constant of 0.16 or 1.6 s, with corresponding increase of resolution to 0.14 and 0.04C°. The instrument can be focused on targets from 4 ft to infinity.

While thermal detectors are sensitive to all wavelengths, overall instrument response can be limited to selected wavelength ranges by use of optical bandpass filters. This technique is quite common in instruments designed for a specific application, such as shown in Fig. 8.32.

Chopped (AC) Selective Band (Photon) Radiation Thermometers

The use of photon detectors allows faster response speeds and may reduce sensitivity to ambient temperature. Since such detectors respond *directly* to the incident photon flux, rather than detecting a temperature change, the disturbing influence of ambient-temperature changes is restricted mainly to changes in the responsivity of the detector. When radiation is expressed in terms of photon flux rather than watts, the formulas are somewhat modified. Equation (8.15) becomes

$$N_\lambda = \frac{2\pi c}{\lambda^4 (e^{C_2/(\lambda T)} - 1)} \tag{8.36}$$

where $N_\lambda \triangleq$ hemispherical spectral photon flux, photons/(cm$^2 \cdot$ s $\cdot \mu$m)
$c \triangleq$ speed of light, 3×10^{10} cm/s

The peak of the photon-flux curves occurs at a different wavelength from that of the radiant intensity. It is given by

$$\lambda_{p,\,p} = \frac{3,669}{T} \qquad \mu\text{m} \tag{8.37}$$

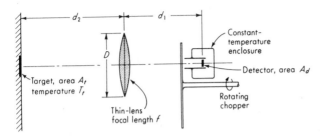

Figure 8.46 Photon-detector system.

The total photon flux for all wavelengths is

$$N_t = 1.52 \times 10^{11} T^3 \qquad \text{photons/(cm}^2 \cdot \text{s)} \qquad (8.38)$$

Figure 8.46 shows the basic arrangement of an instrument[1] using a photon detector and a chopper. Basic optics gives

$$\frac{1}{d_1} + \frac{1}{d_2} = \frac{1}{f} \qquad (8.39)$$

and

$$A_t = \frac{A_d d_2^2}{d_1^2} \qquad (8.40)$$

Combining these gives

$$A_t = \frac{A_d (d_2 - f)^2}{f^2} \qquad (8.41)$$

Now if $d_2 \gg f$, we get approximately

$$A_t = \frac{A_d d_2^2}{f^2} \qquad (8.42)$$

If the detector is a square with side L_d, the resolved target will be a square of side $L_d d_2/f$. For example, a 1 mm² detector used with a lens with a focal length of 75 mm requires a target of size $d_2/75$ to fill exactly the field of view.

For the general configuration of Fig. 8.47, basic radiation laws[2] give

$$\frac{\text{Radiation incident on } dA_2}{dA_2} = \frac{\cos \theta_1 \cos \theta_2}{\pi r^2} (\text{radiation emitted from } dA_1)$$

$$(8.43)$$

We can apply this to the configuration of Fig. 8.46 to find the radiation received over the area of the lens from the target. If d_2 is large compared with the size of

[1] Infrared Thermometry, Infrared Industries, Santa Barbara, Calif.

[2] A. I. Brown and S. M. Marco, "Introduction to Heat Transfer," p. 237, McGraw-Hill, New York, 1951.

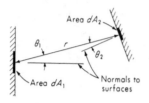

Figure 8.47 Basic radiation configuration.

target and lens (usually true), then $\cos \theta_1 \approx \cos \theta_2 \approx 1$, and dA_1 and dA_2 may be replaced by A_1 and A_2 in Eq. (8.43). To simplify the analysis, let us assume that the detector has uniform spectral response from $\lambda = 0$ to $\lambda = 6.9$ and zero response beyond and that the target is a graybody with emittance ϵ. The radiation emitted by the target is then

$$A_t \epsilon \int_0^{6.9} N_\lambda \, d\lambda = A_t \epsilon E(T_t) \qquad \text{photons/s} \tag{8.44}$$

where $$E(T_t) \triangleq \int_0^{6.9} N_\lambda \, d\lambda \qquad \text{a function of } T_t$$

The radiation received at the lens is

$$\frac{A_t \epsilon E(T_t)(\pi D^2/4)}{\pi d_2^2} = \frac{A_t \epsilon E(T_t)D^2}{4d_2^2} \tag{8.45}$$

If the lens transmits $100K_{tr}$ percent of the flux incident on it (perfect transmission has $K_{tr} = 1.0$) and if the system is focused so that the target image just fills the detector area, then the photon flux density at the detector is given by

$$N_d \triangleq \frac{K_{tr} A_t \epsilon E(T_t)D^2}{4d_2^2 A_d} \qquad \text{photons/(cm}^2 \cdot \text{s)} \tag{8.46}$$

For the indium antimonide photoelectromagnetic detector, the output voltage is proportional to N_d. Since for a focused target $A_t/(A_d d_2^2) = 1/f^2$, Eq. (8.46) can be written as

$$N_d = \frac{K_{tr} D^2}{4f^2} \epsilon E(T_t) \tag{8.47}$$

The output voltage of the overall instrument is $K_{dr} K_a N_d$, where $K_{dr} \triangleq$ detector responsivity, V/[photons/(cm$^2 \cdot$ s)], and K_a is the amplifier gain, V/V. Thus

$$\text{Instrument output voltage} \triangleq e_o = \frac{K_{dr} K_a K_{tr} D^2}{4f^2} \epsilon E(T_t) \tag{8.48}$$

Note that the first factor in Eq. (8.48) is a constant of the instrument while the second $[\epsilon E(T_t)]$ is a function of target temperature and emittance only. Also, as long as the target is focused, the reading is independent of the distance from

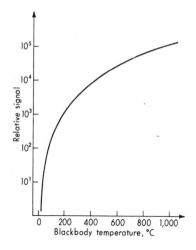

Figure 8.48 Indium antimonide system-response curve.

the target. The variation in $E(T_t)$ with T_t for a blackbody target can be found by experimental calibration of the overall instrument. Its general shape is shown in Fig. 8.48. Since Eq. (8.38) shows that the total photon flux varies as T^3 and since the detector output is roughly proportional to total flux, the instrument output signal varies approximately as T^3; thus, for a graybody

$$e_o \approx K \epsilon T^3 \qquad (8.49)$$

Since the instruments are calibrated against blackbody sources, for nonblack-bodies a value of ϵ must be known in order to find T. If the value of ϵ used is in error, an error in T will result. Because of the third power law, however, errors in ϵ do not cause proportionate errors in T. Rather

$$\frac{T_{\text{actual}}}{T_{\text{assumed}}} = \left(\frac{\epsilon_{\text{assumed}}}{\epsilon_{\text{actual}}}\right)^{1/3} \qquad (8.50)$$

Thus if we assume $\epsilon = 0.8$ when it is really 0.6, the temperature error is only 10 percent.

An instrument[1] of the above general class which uses mirror optics has a range from 100 to 2000°F (8000°F with calibrated aperture), focusing range 4 ft to infinity (18 to 60 in optional), 0.5° field of view, output signal 10 mV full scale, and an overall time constant (in chopped mode) of 2, 20, or 200 ms. For the study of very rapid transients, the chopper may be turned off and the system operated with just the detector and ac amplifier. Thus transients as brief as 10 μs may be measured. Other models of this manufacturer, using lens optics and lead sulfide detectors, have target sizes as small as 0.009 in at a target distance of 2.8 in.

---

[1] Infrared Industries, Santa Barbara, Calif.

Another instrument[1] of this class accepts radiation only in the wavelength band 4.8 to 5.6 μm. This band avoids the absorption bands of atmospheric water vapor and carbon dioxide, thus removing the effect of these variables on instrument response. The measurement of the surface temperature of glass is facilitated also since in this spectral range the emittance of glass is high and independent of thickness. This instrument has a range of 100 to 1000°F, target size equal to its distance divided by 57, time constant 0.2 to 0.5 s, focusing range 17 in to infinity, calibration accuracy 2 percent of range or 20°F (whichever is larger), resolution 0.25 percent of range or 3°F (whichever is larger), and repeatability of 0.5 percent of range or 10°F (whichever is larger).

While photon detectors are themselves selective with respect to wavelength, the use of optical bandpass filters provides a further (and more controllable) selectivity which is employed widely, as discussed in the previous section. Also, all types of radiation thermometers are inherently nonlinear because of the T^4 relation between temperature and radiant flux and the voltage/flux nonlinearities of the various detectors. The total instrument nonlinearity can be compensated electronically by either analog or digital computer means, and such linearization is widely available.

Automatic Null-Balance Radiation Thermometers

Figure 8.49 shows an interesting variation of the chopped, selective band type of instrument. Here the rotating chopper exposes the detector alternately (150 Hz) to radiant flux from the target and from a feedback-controlled reference source [an infrared light-emitting diode (LED)]. Thus the detector output signal is a square wave whose mean value is the average of the two flux levels and whose amplitude is their difference. The difference signal is extracted and amplified by an ac amplifier (acting as a high-pass filter), and the amplitude of this ac error signal is converted to a dc level (the feedback-system error signal) by a synchronous rectifier (phase-sensitive demodulator). Whenever the reference-source radiant flux is not equal to that from the measured target, the error signal drives the reference flux in the direction to null the error. By using sufficiently high gain in the feedback system, the reference flux almost perfectly tracks that from the measured target, and thus measurement of the LED drive signal is equivalent to measurement of the target flux. Such a system is largely unaffected by drifts in detector sensitivity because the detector is located in the forward path of the feedback system. The stability requirements, of course, are transferred from the detector to the LED reference source, but careful selection and aging give excellent stability here. The LED response is sufficiently fast that overall system dynamics is determined by the low-pass filter required in the demodulator, just as in conventional (nonfeedback) chopper instruments. (Figure 8.49 also shows linearizing, peak-holding, and current output features which are not necessary for operation of the basic feedback principle.)

[1] Ircon, Inc., Chicago.

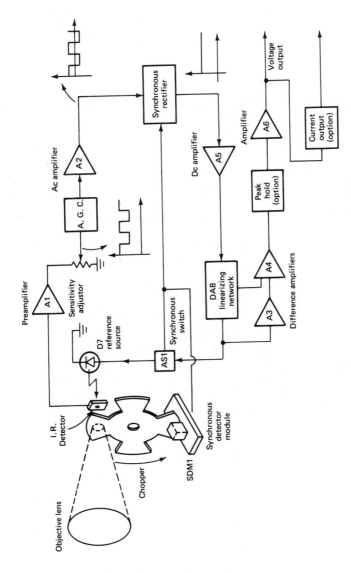

Figure 8.49 Automatic-null-balance radiation thermometer. (*Courtesy Williamson Corp., Concord, Mass.*)

Monochromatic-Brightness Radiation Thermometers (Optical Pyrometers)

The classical form of this type of instrument is the disappearing-filament optical pyrometer. It is the most accurate of all the radiation thermometers; however, it is limited to temperatures greater than about 700°C since it requires a visual brightness match by a human operator. This instrument is used to realize the International Practical Temperature Scale above 1063°C.

Monochromatic-brightness thermometers utilize the principle that, at a given wavelength λ, the radiant intensity ("brightness") varies with temperature as given by Eq. (8.15). In a disappearing-filament instrument (Fig. 8.50), an image of the target is superimposed on a heated tungsten filament. This tungsten lamp, which is very stable, has been calibrated previously, so that when the current through the filament is known, the brightness temperature of the filament is known. (Such a calibration is obtained by visually comparing the brightness of a blackbody source of known temperature with that of the tungsten lamp.) A red filter which passes only a narrow band of wavelengths around 0.65 μm is placed between the observer's eye and the tungsten lamp and target image. The observer controls the lamp current until the filament disappears in the superimposed target image. Then the brightnesses of the target and lamp are equal, and we can write

$$\frac{\epsilon_{\lambda_e} C_1}{\lambda_e^5(e^{C_2/(\lambda_e T_t)} - 1)} = \frac{C_1}{\lambda_e^5(e^{C_2/(\lambda_e T_L)} - 1)} \tag{8.51}$$

where $\epsilon_{\lambda_e} \triangleq$ emittance of target at wavelength λ_e

$\lambda_e \triangleq$ effective wavelength of filter, usually 0.65 μm

$T_t \triangleq$ target temperature

$T_L \triangleq$ lamp brightness temperature

For T less than about 4000°C, the terms $e^{C_2/(\lambda_e T)}$ are much greater than 1, which allows Eq. (8.51) to be simplified to

$$\frac{\epsilon_{\lambda_e}}{e^{C_2/(\lambda_e T_t)}} = \frac{1}{e^{C_2/(\lambda_e T_L)}} \tag{8.52}$$

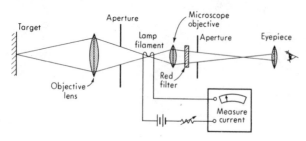

Figure 8.50 Disappearing-filament optical pyrometer.

Then
$$\epsilon_{\lambda_e} = e^{-(C_2/\lambda_e)(1/T_L - 1/T_t)} \tag{8.53}$$

and finally
$$\frac{1}{T_t} - \frac{1}{T_L} = \frac{\lambda_e \ln \epsilon_{\lambda_e}}{C_2} \tag{8.54}$$

If the target is a blackbody ($\epsilon_{\lambda_e} = 1.0$), there is no error since $\ln \epsilon_{\lambda_e} = 0$ and $T_t = T_L$. If ϵ_{λ_e} is not 1.0 but is known, Eq. (8.54) allows calculation of the needed correction. The errors caused by inexact knowledge of ϵ_{λ_e} for a particular target are not as great for an optical pyrometer as for an instrument sensitive to a wide band of wavelengths. The percentage error is given by

$$\frac{dT_t}{T_t} = -\frac{\lambda_e T_t}{C_2} \frac{d\epsilon_{\lambda_e}}{\epsilon_{\lambda_e}} \tag{8.55}$$

Thus, for a target at 1000 K, a 10 percent error in ϵ_{λ_e} results in only a 0.45 percent error in T_t. The use of a monochromatic red filter aids the operator in matching the brightness of target and of lamp since color effects are eliminated. Also, the target emittance need be known only at one wavelength. If ϵ_{λ_e} is exactly known, temperatures can be measured with optical pyrometers with errors on the order of 3C° at 1000°C, 6C° at 2000°C, and 40C° at 4000°C. With special optical systems, targets as small as 0.001 in can be measured at distances of 5 or 6 in.

Because of its manual null-balance principle, the optical pyrometer is not usable for continuous-recording or automatic-control applications. To overcome this drawback, automatic brightness pyrometers[1] have been developed. One model[2] of such a device uses a mirror-chopper arrangement to produce a square wave of radiation flux in which the target radiation and standard lamp radiation are alternately applied to a photomultiplier tube. The standard lamp is left at a fixed brightness; thus a null method is not used. A red filter with $\lambda_e = 0.653$ μm is employed. This instrument has a range of 700 to 3000°C (though not in a single instrument), an inaccuracy of 1 percent of span plus 0.3 percent of measured temperature, repeatability 0.3 percent of measured temperature, resolution 1°F, time constant 0.3 s, target size 0.4 in at 18 in, and a recorder output of 50 or 100 mV full scale.

Another class of brightness pyrometer which does not use a reference lamp source is available also. They are merely chopper-type radiation thermometers using photon detectors and narrow-band optical filters. One such instrument[3] has a range of 1400 to 8300°F, 0.5° field of view, focusing range 24 in to infinity, time constant 0.1 s (0.01 s optional), repeatability 0.25 percent of span, inaccuracy 1 percent of span, and recorder output 10 to 100 mV full scale. This instrument employs a filter centered at 0.80 μm with a bandwidth of 0.06 μm.

---

[1] S. Ackerman and J. S. Lord, Automatic Brightness Pyrometer Uses a Photomultiplier "Eye," *ISA J.*, p. 48, December 1960; J. S. Lord, Brightness Pyrometry, *Instrum. Contr. Syst.*, p. 109, February 1965; Instrument Development Laboratories, Attleboro, Mass., *Bull.* 614.

[2] *Bull.* 614, ibid.

[3] Infrared Industries, Santa Barbara, Calif.

Two-Color Radiation Thermometers

Since errors due to inaccurate values of emittance are a problem in all radiation-type temperature measurements, considerable attention has been paid to possible schemes for alleviating this difficulty. Although no universal solution has been found, the two-color concept[1] has met with some practical success. The basic concept requires that W_λ be determined at two different wavelengths and then the ratio of these two W_λ's be taken as a measure of temperature. For the usual conditions of practical application, the terms $e^{C_2/(\lambda T)}$ are much greater than 1.0, and we may write with close approximation

$$W_{\lambda 1} = \frac{\epsilon_{\lambda 1} C_1}{\lambda_1^5 e^{C_2/(\lambda_1 T)}} \tag{8.56}$$

$$W_{\lambda 2} = \frac{\epsilon_{\lambda 2} C_1}{\lambda_2^5 e^{C_2/(\lambda_2 T)}}$$

Then
$$\frac{W_{\lambda 1}}{W_{\lambda 2}} = \frac{\epsilon_{\lambda 1}}{\epsilon_{\lambda 2}} \left(\frac{\lambda_2}{\lambda_1}\right)^5 e^{(C_2/T)(1/\lambda_2 - 1/\lambda_1)} \tag{8.57}$$

For a graybody, $\epsilon_{\lambda 1} = \epsilon_{\lambda 2}$; thus

$$\frac{W_{\lambda 1}}{W_{\lambda 2}} = \left(\frac{\lambda_2}{\lambda_1}\right)^5 e^{(C_2/T)(1/\lambda_2 - 1/\lambda_1)} \tag{8.58}$$

and we see that the ratio $W_{\lambda 1}/W_{\lambda 2}$ is independent of emittance as long as it is numerically the same at λ_1 and λ_2. In one commercial instrument[2] the two filters ($\lambda = 0.71 \pm 0.02$, $\lambda_2 = 0.81 \pm 0.02$ μm) are mounted on a rotating wheel so that the incoming radiation passes alternately through each on its way to a silicon photon detector. Special electronic circuitry performs operations equivalent to taking the ratio of $W_{\lambda 1}$ and $W_{\lambda 2}$. Instruments covering the range 1500 to 4000°F are available as standard models.

Infrared Imaging Systems

Most of these systems have the function of providing a televisionlike visual display in which the shades of gray represent various temperature levels of the surface of some two-dimensional object (target) on which the infrared camera is focused. While these shades of gray are accurately related to infrared energy levels emitted from the target, recall that all infrared temperature sensing requires knowledge of the emittance of the target surface to convert the detector signal to degrees of temperature.

Figure 8.51 shows the arrangement of the camera for a unit of Swedish

---

[1] T. P. Murray and V. G. Shaw, Two-Color Pyrometry in the Steel Industry, *ISA J.*, p. 36, December 1958.

[2] A. S. Anderson, The Dual Wavelength Radiometer, Williamson Corp., Concord, Mass., 1980.

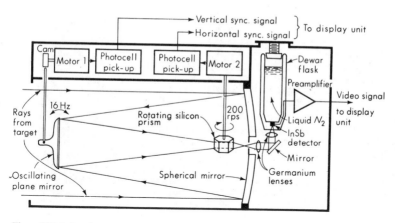

Figure 8.51 Infrared camera.

manufacture.[1] Scanning of the target surface is accomplished by focusing the radiation first on a plane mirror oscillating about a horizontal axis at 16 Hz. This scans the line of sight vertically over the target surface. The image from the plane mirror is focused on an eight-sided prism (rotating at 200 r/s) which provides the horizontal scanning and results in a "picture" with a frame rate of 16 frames/s and 100 lines per picture. An indium antimonide (InSb) detector (cooled by liquid nitrogen to reduce its noise level) produces an electrical signal proportional to the incident radiant flux. To produce a thermal image of the target on a TV picture tube, horizontal and vertical deflection signals are picked off the scanning-system motor shafts to position the electron beam, while the infrared detector signal (video signal) modulates the beam intensity. Thus the TV tube displays a 100-line picture whose local intensity (shades of gray) represents target temperature.

Using suitable optics, infrared scanners can produce full-scale thermal images of objects as small as 0.6 mm^2 at an 8-mm working distance with a "spot size" of 0.01 mm. Temperature differences as small as 0.1°C can be resolved for a target at 30°C. Of course, larger targets may be scanned by increasing the working distance; a typical 5° × 5° field of view gives a target size of 8.8 × 8.8 m at a range of 100 m. Figure 8.52 shows the displays produced by a commercial infrared microscanner[2] viewing a 1.5 mm^2 target area on a metallurgical sample. The right photo shows the "thermogram" with the calibrated gray scale at its left. Digits below give the "temperature" corresponding to black and the range of "temperature" from black to white on the gray scale. An oscilloscope-type trace of the temperature profile across any desired horizontal scan line also is shown. The black line on the thermogram shows its location on the target, while the

[1] AGA Corp., Secaucus, N.J.; Sven-Berta Borg, Thermal Imaging with Real-Time Picture Presentation, *Appl. Opt.*, vol. 7, p. 1697, September 1968.

[2] Model RM-50, Barnes Engineering, Stamford, Conn.

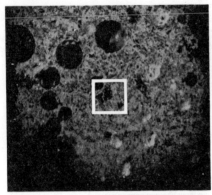

Photograph of metallurgical sample. Outline indicates area shown in thermogram.

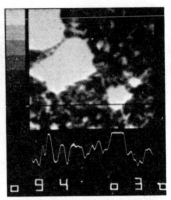

Thermogram (40 X objective, white: hot) of area outlined in photo shows infrared emissivity variation across surface.

Isometric display of metallurgical sample.

Figure 8.52 Displays from infrared microscanner.

white trace below displays the temperature profile. A manual knob allows positioning of this trace at any point on the target. The photo at the bottom shows an "isometric" display obtained by successively displacing horizontally each of the 64 temperature profiles produced by the scanning process. This gives a vivid three-dimensional picture of temperature variations over the surface. Another useful display mode (not shown) shows isotherms[1] (lines of constant temperature) as bright white lines on the thermogram. A further refinement[2] gives a 10-color display of isotherms, with each color being associated with a different known temperature.

The noncontacting nature of the sensing method, together with the display of the entire surface-temperature distribution over an object, gives infrared imaging

---

[1] AGA Thermovision, AGA Corp., Secaucus, N.J.
[2] Ibid.

systems unique application possibilities. These include medical diagnosis (cancer detection, peripheral circulation problems, etc.); hot-spot detection in electric power transmission equipment; surveys of earth and sea temperatures from aircraft and satellites; nondestructive testing of products for poor bonding, cracking, wear, heat generation, etc.; and biological studies of plant development and insect physiology.

Figure 8.53 shows the physical layout of the Barnes RM-50 Infrared Microscanner mentioned earlier. Basically the system is an infrared microscope with an optical chopper and two-axis scanning system inserted in the optical path. The chopper periodically interrupts the radiant beam so that the detector sees alternately the target and the blackened chopper vane. Since the chopper vane's temperature is known accurately (and controlled), the detector puts out an alternating electrical signal with an amplitude accurately related to target radiation. This signal is processed to form a thermal image on the cathode-ray tube screen. Target scanning uses two oscillating mirrors (x axis at 60 Hz, y axis at either 1 or 2 Hz) driven by small torque-motors to scan the beam over the target area. This produces an image composed of a raster of 64 horizontal lines at a rate of 1 frame/s when 2-Hz y scanning is used. To improve resolution (at the expense of frame rate), the 1-Hz scanning gives 128 lines at 2 s/frame.

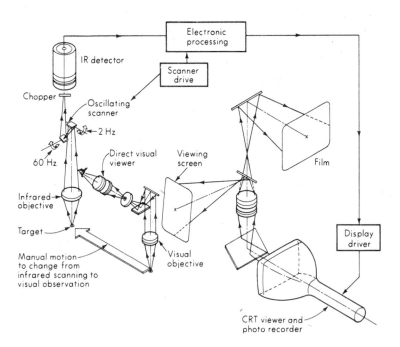

Figure 8.53 Layout of infrared microscanner.

Infrared versions of the detector arrays utilized for motion measurement (Fig. 4.51) are becoming available. The two-dimensional versions of these, using arrays of pyroelectric detectors,[1] make possible infrared imaging instruments with neither moving parts (for scanning), nor Dewar flasks for cooling.

8.8 TEMPERATURE-MEASURING PROBLEMS IN FLOWING FLUIDS

In attempting to measure the static temperature of flowing fluids (particularly high-speed gas flows), we encounter certain types of problems irrespective of the particular sensor being used. These have to do mainly with errors caused by heat transfer between the probe and its environment and the problem of measuring static temperature of a high-velocity flow with a stationary probe.

Conduction Error[2]

Let us consider the so-called conduction error first. Figure 8.54a shows a common situation. A probe has been inserted into a duct or other flow passage and is supported at a wall. In general, the wall will be hotter or colder than the flowing fluid; thus there will be heat transfer, and this leads to a probe temperature different from that of the fluid. We analyze a simplified model of this arrangement to find what measures can be taken to reduce and/or correct the error to be expected in such a case. Figure 8.54b shows the simplified model to be employed in analyzing this situation. A slender rod extends a distance L from the wall. We assume the rod temperature T_r is a function of x only; it does not vary with time or over the rod cross section at a given x. A fluid of constant and uniform temperature T_f completely surrounds the rod and exchanges heat with it by convection. For a steady-state situation

Heat in at x = (heat out at $x + dx$) + (heat loss at surface)

$$q_x = q_{x+dx} + q_l$$

One-dimensional conduction heat transfer gives

$$q_x = -kA \frac{dT_r}{dx} \tag{8.59}$$

where $k \triangleq$ thermal conductivity of rod and $A \triangleq$ cross-sectional area. Then

$$q_{x+dx} = q_x + \frac{d}{dx}(q_x)\, dx = -kA \frac{dT_r}{dx} + \frac{d}{dx}\left(-kA \frac{dT_r}{dx}\right) dx \tag{8.60}$$

---

[1] Pyroelectric Self-Scanning Infrared Detector Arrays, *Appl. Note* AN-10, Spiricon, Inc., Logan, Utah, 1981.

[2] R. P. Benedict, "Fundamentals of Temperature, Pressure, and Flow Measurements," Wiley, New York, 1977, chap. 12; F. D. Werner et al., Stem Conduction Error for Temperature Sensors, *Bull.* 9622, Rosemount Inc., Minneapolis, Minn.

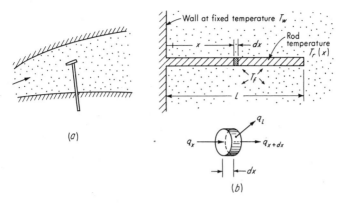

Figure 8.54 Probe configuration and conduction-error analysis.

Now if k and A are assumed constant,

$$q_{x+dx} = -kA \frac{dT_r}{dx} - kA \frac{d^2 T_r}{dx^2} dx \qquad (8.61)$$

We assume the heat loss by convection at the surface to be given by

$$q_l = h(C\ dx)(T_r - T_f) \qquad (8.62)$$

where $h \triangleq$ film coefficient of heat transfer
     $C \triangleq$ "circumference" of rod (it need not be circular)
  $C\ dx =$ surface area

So we have

$$\frac{d^2 T_r}{dx^2} - \frac{hC}{kA} T_r = -\frac{hC}{kA} T_f \qquad (8.63)$$

If we take h and C as being constants, then Eq. (8.63) is a linear differential equation with constant coefficients and is readily solved for T_r as a function of x. We need two boundary conditions to accomplish this. Clearly, $T_r = T_w$ at $x = 0$ is one such condition. The simplest assumption at $x = L$ is an insulated end; this gives $dT_r/dx = 0$ at $x = L$. Even if the end is not insulated, if L is quite large, we see intuitively that the variation of T_r with x must be as in Fig. 8.55; thus $dT_r/dx \approx 0$ for $x = L$. Using these two boundary conditions with Eq. (8.63) gives

$$T_r = T_f - (T_f - T_w)\left[\left(1 - \frac{e^{mL}}{2\cosh mL}\right)e^{mx} + \frac{e^{mL}}{2\cosh mL}e^{-mx}\right] \qquad (8.64)$$

where

$$m \triangleq \sqrt{\frac{hC}{kA}} \qquad (8.65)$$

Since generally the temperature-sensing element (thermocouple bead, thermistor, etc.) is located at $x = L$, we evaluate Eq. (8.64) there to get

$$\text{Temperature error} = T_r - T_f = \frac{T_w - T_f}{\cosh mL} \tag{8.66}$$

Equation (8.66) may be used in two ways: to indicate how to design a probe support to minimize the error and to allow us to calculate and correct for whatever error there might be. It is clear that the error is reduced if T_w is close to T_f. Insulating or actively controlling the temperature of the wall encourages this. The term $\cosh mL$ will be large if m and L are large. Thus the probe should be immersed (L is called the immersion length) as far as practical (see Fig. 8.56[1]). We see that to make m large, h should be large (high rate of convection heat transfer) and k should be small (the probe support made of insulating material). The term C/A depends on the shape of the rod. For the usual circular cross section, $C/A = 2/r$, where r is the rod radius. Thus we see that the rod should be of small cross section to reduce error.

If the boundary condition at $x = L$ is changed to a more realistic (and complicated) one in which there is convection heat transfer with a film coefficient h_e at the end, then at $x = L$

$$T_r - T_f = \frac{T_w - T_f}{\cosh mL + [h_e/(mk)] \sinh mL} \tag{8.67}$$

Since the error predicted by (8.67) is less than that of (8.66), the use of the simpler relation is conservative.

Radiation Error[2]

Additional error is caused by radiant-heat exchange between the temperature probe and its surroundings. This occurs simultaneously with the previously studied conduction losses, but here we consider it separately for simplicity. We also assume radiation exchange only between the probe and the surrounding walls, neglecting radiation of the gas itself or the absorption by the gas of radiation passing through it. Neglecting conduction losses, we may consider the probe as in Fig. 8.57a. For steady-state conditions

$$\text{Heat convected to probe} = \text{net heat radiated to wall} \tag{8.68}$$

$$hA_s(T_f - T_p) = 0.174\epsilon_p A_s \left[\left(\frac{T_p}{100} \right)^4 - \left(\frac{T_w}{100} \right)^4 \right] \tag{8.69}$$

[1] Werner et al., ibid.
[2] Benedict, op. cit.

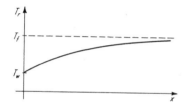

Figure 8.55 Temperature profile.

where $h \triangleq$ film coefficient at probe surface, Btu/(h · ft² · °F)

$A_s \triangleq$ probe surface area

$\epsilon_p \triangleq$ emittance of probe surface

$T_p \triangleq$ probe absolute temperature, °R

$T_w \triangleq$ wall absolute temperature, °R

Equation (8.69) assumes the radiation configuration described as "a small body

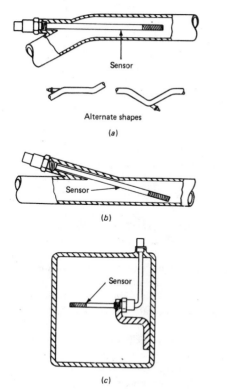

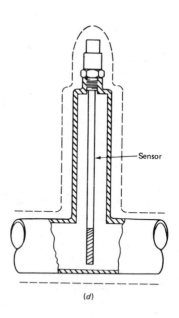

Figure 8.56 Schemes to reduce conduction error.

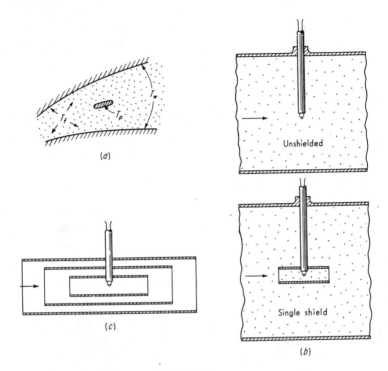

Figure 8.57 Radiation-error analysis and shielding.

completely enclosed by a larger one." The error due to radiation is given by

$$\text{Temperature error} = T_p - T_f = \frac{0.174\epsilon_p}{h}\frac{T_w^4 - T_p^4}{10^8} \qquad (8.70)$$

By insulating the wall or controlling its temperature, error can be reduced by making the difference between T_w and T_p as small as possible. A probe surface of low emittance ϵ_p (a shiny surface) will further reduce such errors, as will a high value of heat-transfer coefficient h. To obtain a high value of h when the fluid velocity is low, the aspirated type of probe may be utilized. Here a high local velocity is induced at the probe by connecting a vacuum pump to the probe tubing. Equation (8.70) may be used also to calculate corrections if numerical values of the needed quantities are available.

Probes with some form or another of radiation shield are employed widely to reduce radiation errors. Figure 8.57b shows a probe with a single shield. The principle of all radiation shields is to interpose between the probe and the wall a body (the shield) whose temperature is closer to the fluid temperature than is the wall. Thus the probe "sees" the shield rather than the wall, and if the temperature of the shield is close to fluid temperature, the probe radiation-heat loss will be small. For pure-radiation-heat transfer, it is easy to show that interposing a single screen between a body and its surroundings will reduce the heat loss to one-half the former value, since the screen comes to a temperature $T^4_{screen} = (T^4_{body} + T^4_{surroundings})/2$. For n screens the heat loss is reduced to $1/(n + 1)$ of the unscreened value. All these results are for the simplest case in which the screens and surroundings completely enclose the body. For actual probe shields, various geometric and emittance factors complicate the situation. Also, additional heating of the shield by convection raises its temperature and reduces probe error. Experimental tests[1] with concentric circular cylinder shields (Fig. 8.57c) have shown the following:

1. A significant decrease in error may be achieved by adding more shields, at least up to about four.
2. Little is gained by increasing the length/diameter ratio beyond 4 : 1 in attempting to reduce the unshielded angles at the open ends.
3. For multiple shields the spacing between shields must be sufficiently large to prevent excessive conduction heat transfer between shields and to allow high enough flow velocity for good convection from gas to shields. A double shield with only $\frac{1}{32}$-in spacing acted almost as a single shield.

To illustrate the need for shielding, an unshielded probe exposed to 1800°F gas flow at 270 lbm/(ft$^2 \cdot$ min) may exhibit an error of 160F°. A suitable quadruple shield can reduce this to about 20F°.

Another shielding technique employs an electrically heated shield with an additional temperature sensor fastened to the shield. The heat input to the shield is adjusted until the probe sensor and shield sensor register identical temperatures. At this point, probe and shield should both be at the fluid temperature, with the heat loss from the shield to the cooler wall being replaced by the shield heater.

Velocity Effects[2]

It is often necessary to determine the static temperature of a flowing gas, since its physical properties depend on this temperature. To measure this temperature directly with a probe, however, requires that the probe be stationary with respect to the fluid; thus it must be moving at the same velocity as the fluid. Since usually this is impractical, various indirect methods of measuring static temperatures are in use. If we can measure static pressure and density, sound velocity, or index of refraction, formulas allow calculation of static temperature. Experimental techniques based on each of these principles have been developed. However,

[1] W. J. King, Measurement of High Temperatures in High-Velocity Gas Streams, *Trans. ASME*, p. 421, July 1943.

[2] Benedict, op. cit., chap. 11; T. M. Stickney et al., Rosemount Total Temperature Sensors, *Tech. Rept.* 5755, Rosemount Inc., Minneapolis, Minn., 1975.

for routine measurements a different approach is employed. This involves placing a stationary probe in the stream and calculating the static temperature from the readings of this probe, by using suitable corrections. Ideally, if a perfect gas is decelerated from free-stream velocity to zero velocity adiabatically (not necessarily isentropically), the temperature rises from the free-stream static temperature T_{stat} to the so-called stagnation or total temperature T_{stag}, where

$$\frac{T_{stag}}{T_{stat}} = 1 + \frac{\gamma - 1}{2} N_m^2 \tag{8.71}$$

where
$\gamma \triangleq c_p/c_v$, ratio of specific heats
$N_m \triangleq$ Mach number
$T_{stag}, T_{stat} \triangleq$ absolute temperatures

This result holds for both subsonic and supersonic flows because the shock wave that forms ahead of a probe in supersonic flow affects only the entropy, not the total enthalpy of the gas. For air, Eq. (8.71) becomes

$$\frac{T_{stag}}{T_{stat}} = 1 + 0.2 N_m^2 \tag{8.72}$$

and if $N_m < 0.22$, T_{stag} is within 1 percent of T_{stat}. Thus for sufficiently low velocities, a stationary probe can be utilized to read static temperature directly. For higher Mach numbers, T_{stat} can be calculated from Eq. (8.71) if γ and N_m are known. Measurement of Mach numbers with a pitot-static tube is discussed in Chap. 7.

Unfortunately, real temperature probes do not attain the theoretical stagnation temperature predicted by Eq. (8.71). Even if the conduction and radiation errors discussed earlier in this section are corrected for, there remain further deviations of the actual situation from the assumed ideal. Correction for these effects generally is accomplished by experimental calibration to determine the recovery factor r of the particular probe. This is defined by

$$r \triangleq \frac{T_{stag, ind} - T_{stat}}{T_{stag} - T_{stat}} \tag{8.73}$$

where $T_{stag, ind} \triangleq$ temperature actually indicated by probe. If r is assumed to be a known number for a given probe, then combination of Eqs. (8.71) and (8.72) gives

$$T_{stat} = \frac{T_{stag, ind}}{1 + r[\gamma - 1)/2] N_m^2} \tag{8.74}$$

A probe that measures T_{stag} exactly would have a recovery factor of 1.0 while one that measures T_{stat} exactly would have $r = 0$.

A possible apparatus[1] for determination of r is shown in Fig. 8.58. The flow velocity in the stagnation chamber is $\frac{1}{100}$ of the nozzle flow velocity; thus measurement of tank temperature and pressure is carried out accurately under

[1] H. C. Hottel and A. Kalitinsky, Temperature Measurements in High-Velocity Air Streams, *J. Appl. Mech.*, p. A-25, March 1945.

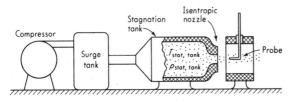

$T_{stat, tank}$ kept near room temperature to minimize
conduction and radiation errors

Figure 8.58 Recovery-factor calibration setup.

essentially zero-velocity conditions. By careful design to minimize friction and heat transfer, the nozzle can be made to provide an almost-perfect isentropic expansion. The validity of this assumption has been checked experimentally. For an isentropic process from tank to nozzle,

$$T_{stag, nozzle} = T_{stat, tank} \tag{8.75}$$

and

$$p_{stag, nozzle} = p_{stat, tank} \tag{8.76}$$

In a free jet

$$p_{stat, nozzle} = p_{atmosphere} \tag{8.77}$$

Now the nozzle Mach number N_m can be computed from the standard pitot-tube formulas since $p_{stat, nozzle}$ and $p_{stag, nozzle}$ are both known. However, the actual use of a pitot tube (with its attendant errors) is avoided since $p_{stag, nozzle}$ is obtained by measurement of $p_{stat, tank}$, and $p_{stat, nozzle}$ is obtained from a barometer reading of $p_{atmosphere}$. Once N_m is known, $T_{stat, nozzle}$ can be computed from

$$T_{stat, nozzle} = \frac{T_{stag, nozzle}}{1 + [(\gamma - 1)/2]N_m^2} = \frac{T_{stat, tank}}{1 + [(\gamma - 1)/2]N_m^2}. \tag{8.78}$$

The reading of the probe itself supplies $T_{stag, ind}$. Thus we can compute r from its definition

$$r = \frac{T_{stag, ind} - T_{stat, nozzle}}{T_{stag, nozzle} - T_{stat, nozzle}} \tag{8.79}$$

For bare thermocouple sensors, the numerical value of r usually lies in the range 0.6 to 0.9, depending on the form of the junction (butt-welded, twisted, or spherical bead) and the orientation (wire parallel to flow or transverse to flow). To get a high value of r and one relatively independent of flow conditions such as velocity magnitude and direction, sensors (usually thermocouples or resistance thermometers) are built into probes that have been specifically designed to approach ideal stagnation conditions. Figure 8.59 shows two examples[1] of such probes. Desirable characteristics of probes include the following:

1. Low heat capacity in the sensing element for a fast response.
2. Conduction loss of lead wires minimized by exposing enough length of the lead to the stagnation temperature.

[1] R. W. Ladenburg et al., "Physical Measurements in Gas Dynamics and Combustion," p. 186, Princeton University Press, Princeton, N.J., 1954.

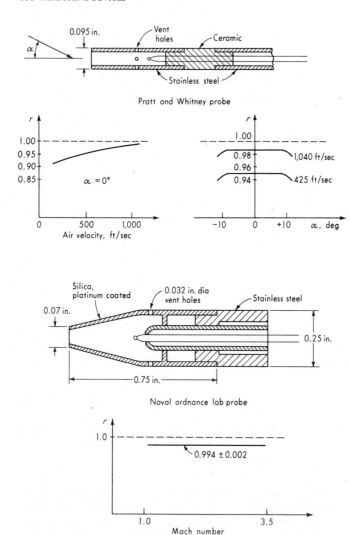

Figure 8.59 Stagnation-temperature probes.

3. Radiation shield of low thermal conductivity and low surface emittance.
4. Vent holes provided to replenish continuously the fluid in the stagnation chamber; otherwise, it would be cooled by conduction and radiation. This flow must be kept small enough, however, that stagnation conditions are essentially preserved. The increased convection coefficient caused by the flow speeds the response and reduces the radiation error.

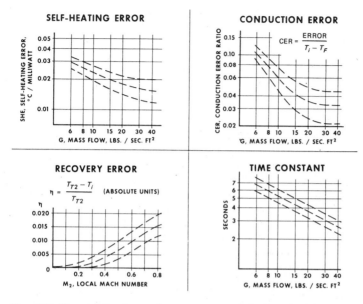

Figure 8.60 Characteristics of stagnation-temperature probe for jet-engine inlet.

5. Blunt shape causes formation of a normal shock wave in supersonic flow. This increases the temperature in the boundary layer and reduces the heat loss from the probe. The shock wave also reduces the influence of misalignment.

In addition to the laboratory-type probes of Fig. 8.59, permanently installed stagnation-temperature sensors for rugged environments, such as control-system sensing in aircraft gas-turbine engine inlets, are needed. Added to the problems discussed above, such probes must survive impacts by birds and ice balls and include provisions for deicing and inertial separation of water droplets. Figure 8.60[1] shows characteristics of sensors of this type which use platinum resistance elements.

8.9 DYNAMIC RESPONSE OF TEMPERATURE SENSORS

Since the conversion from sensing-element temperature to thermal expansion, thermoelectric voltage, or electrical resistance is essentially instantaneous, the dynamic characteristics of temperature sensors are related to the heat-transfer and -storage parameters that cause the sensing-element temperature to lag that of

---

[1] Product Data Sheet 2186, Rosemount Inc., Minneapolis, Minn.

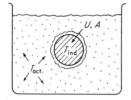

Figure 8.61 First-order sensor model.

the measured medium. When a fluid-immersed sensing element is used "bare" (not in a protective well), often the model of Fig. 8.61 is adequate (see Fig. 8.17 for sensors attached to solids). Here heat losses are neglected, resistance to heat transfer is lumped in a single element, and energy storage is lumped in a single element. Conservation of energy gives

$$UA(T_{\text{act}} - T_{\text{ind}}) \, dT = MC \, dT_{\text{ind}} \tag{8.80}$$

where $U \triangleq$ overall heat-transfer coefficient
$A \triangleq$ heat-transfer area
$T_{\text{act}} \triangleq$ actual temperature of surrounding fluid
$T_{\text{ind}} \triangleq$ temperature indicated by sensor
$M \triangleq$ mass of sensing element
$C \triangleq$ specific heat of sensing element

This leads to
$$\frac{T_{\text{ind}}}{T_{\text{act}}}(D) = \frac{1}{\tau D + 1} \tag{8.81}$$

where
$$\tau \triangleq \frac{MC}{UA} \tag{8.82}$$

Clearly, speed of response may be increased by decreasing M and C and/or increasing U and A. Since U, in general, depends on the surrounding fluid and its velocity, τ is not a constant for a given sensor, but rather varies with how it is employed.

Since temperature sensors often are enclosed in protective wells[1] or sheaths, a thermal model taking into account heat-transfer resistance and energy storage in the well is of practical interest. Figure 8.62 shows such a configuration. Analysis gives

$$\frac{T_{\text{ind}}}{T_{\text{act}}}(D) = \frac{1}{\tau_w \tau_s D^2 + [\tau_w + \tau_s + M_s C_s/(U_w A_w)]D + 1} \tag{8.83}$$

where $\tau_w \triangleq M_w C_w/(U_w A_w)$, time constant of well alone
$\tau_s \triangleq M_s C_s/(U_s A_s)$, time constant of sensor alone

We see that the addition of a well changes the form of response to second-order and increases the lag. The term $M_s C_s/(U_w A_w)$ is called the *coupling term* between

---

[1] D. W. Richmond, Selecting Thermowells for Accuracy and Endurance, *InTech.*, pp. 59–63, February 1980. J. A. Masek, Guide to Thermowells, Omega Engineering, Stamford, Conn., 1981.

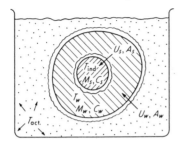

Figure 8.62 Second-order sensor model.

the well and the sensor. If it is small compared with $\tau_w + \tau_s$, we have approximately

$$\frac{T_{ind}}{T_{act}}(D) = \frac{1}{\tau_w \tau_s D^2 + (\tau_w + \tau_s)D + 1} = \frac{1}{\tau_w D + 1} \frac{1}{\tau_s D + 1} \qquad (8.84)$$

which is just a cascade combination of the sensor and well individual dynamics.

The accuracy of the theoretical model can be increased by increasing the number of "lumps" of heat-transfer resistance and energy storage employed, the ultimate limit being an infinite number corresponding to a distributed-parameter (partial differential equation) rather than a lumped-parameter (ordinary differential equation) approach. When temperature sensors are employed as measuring devices in feedback-control systems, usually they are allowed to contribute no more than 30° phase lag at the frequency where the entire open-loop lag is 180°. Under such conditions, usually they are modeled adequately as simple first-order systems. The best[1] time constant to utilize for such a model is determined by an experimental ramp-input test and is numerically the steady-state time lag observed in such a test.

When greater accuracy is needed in utilizing the results of experimental tests to determine sensor dynamics, a model[2] using three time constants and a dead time may be employed. The transfer function is then

$$\frac{T_{ind}}{T_{act}}(D) = \frac{e^{-\tau_{dt}D}}{(\tau_1 D + 1)(\tau_2 D + 1)(\tau_3 D + 1)} \qquad (8.85)$$

Numerical values of τ_1, τ_2, τ_3, and τ_{dt} may be obtained from step-function response tests. For example, a thermocouple used in a heavy-duty stainless-steel well had

$$\frac{T_{ind}}{T_{act}}(D) = \frac{e^{-2.6D}}{(21.6D + 1)(2.9D + 1)(2.1D + 1)} \qquad (8.86)$$

where the time constants are in seconds.

[1] G. A. Coon, Response of Temperature-Sensing-Element Analogs, *Trans. ASME*, p. 1857, November 1957.

[2] J. R. Louis and W. E. Hartman, The Determination and Compensation of Temperature-Sensor Transfer Functions, *ASME Paper* 64-WA/AUT-13.

When accurate numerical values are needed, generally experimental tests are required to determine temperature probe dynamics. For simple bare thermocouples, however, extensive research and testing have provided semiempirical formulas that allow calculation of the time constant with fair accuracy. One such relation[1] useful for temperatures from 160 to 1600°F, wire diameter 0.016 to 0.051 in, mass velocity 3 to 50 lbm/(ft^2 · s), and static pressure of 1 atm is

$$\tau = \frac{3,500 \rho c d^{1.25} G^{-15.8/\sqrt{T}}}{T} \quad \text{s} \tag{8.87}$$

where $\rho \triangleq$ average density of two thermocouple materials, lbm/ft^3

$c \triangleq$ average specific heat of two thermocouple materials, Btu/(lbm · F°)

$d \triangleq$ wire diameter, in

$G \triangleq$ flow mass velocity, lbm/ft^2 · s)

$T \triangleq$ stagnation temperature, °R

Within the above restrictions, this formula will predict τ for butt-welded junctions within about 10 percent. Another such result[2] based on tests for a Mach-number range of 0.1 to 0.9 and a Reynolds-number range of 250 to 30,000, gives

$$\tau = \frac{4.05 \rho c d^{1.50} \{1 + [\gamma - 1)/2] N_m^2\}^{0.25}}{p^{0.5} N_m^{0.5} T^{0.18}} \tag{8.88}$$

where $\gamma \triangleq$ ratio of specific heats

$N_m \triangleq$ Mach number

$p \triangleq$ static pressure, atm

This reference also presents a comprehensive analysis of conduction and radiation errors and the effects of differences in the thermal properties of the two metals used in a thermocouple.

Dynamic response tests of temperature sensors in the laboratory are not always accurate predictors of behavior when the sensor is installed in the process environment. Response degradations may be present in new sensors (if improperly installed) or may develop over time as a result of cracking of insulating cements or relocation of insulating powders used in probe construction. If actual installed response is critical to process safety and/or performance, *in situ* dynamic testing can provide the desired confidence. For both thermocouple and RTD sensors, the loop current step response (LCSR) technique[3] developed for nuclear power plant testing provides an *in situ* test method if the sensor lead wires may be disconnected for 15 to 60 min. The sensor circuit is subjected to a step increase (or decrease) of current, causing a heating (or cooling) effect, whose response can be analyzed to extract sensor time-constant information.

[1] R. J. Moffat, How to Specify Thermocouple Response, *ISA J.*, p. 219, June 1957.

[2] M. D. Scadron and I. Warshawsky, Experimental Determination of Time Constants and Nusselt Numbers for Bare-Wire Thermocouples in High-velocity Air Streams and Analytic Approximation of Conduction and Radiation Errors, *NACA, Tech. Note* 2599, 1952.

[3] T. W. Kerlin, H. M. Hashemian, and K. M. Peterson, Time Response of Temperature Sensors, *ISA Trans.*, vol. 20, no. 1, 1981.

Dynamic Compensation of Temperature Sensors

When environmental conditions require a rugged temperature sensor, the mass may be so high as to cause a sluggish response. It may be possible to obtain an improved overall measuring-system response by cascading an appropriate dynamic compensation device with the sensor. Such schemes have been applied in practice with considerable success and may be implemented in a number of ways.[1] An RC network such as that shown for hot-wire anemometer compensation in Chap. 7 may be used if the sensor is essentially first-order. Second-order sensors also may be compensated,[2] operational-amplifier networks provide a convenient means of implementation. Since the compensation is correct only for specific sensor dynamics, changes in numerical values or form of transfer function caused by changes in operating conditions can lead to loss of compensation. In general, increased speed of response is traded for overall sensitivity in such compensation schemes. This can be made up by additional amplification, but only up to a point; then the inherent noise level prevents further improvement. However, improvement of 100:1 or more is often possible.

8.10 HEAT–FLUX SENSORS

Requirements for measurement of local convective, radiative, or total heat-transfer rates have led to the development of several types of heat-flux sensors. Here we review briefly the operating principles and characteristics of the most common types.

Slug-Type Sensors

In Fig. 8.63a a slug of metal is buried in (but insulated from) the surface across which the heat-transfer rate is to be measured. Neglecting losses through the insulation and the thermocouple wires, we may write

$$\text{Heat transferred in} = \text{energy stored}$$

$$Aq \, dt = Mc \, dT \tag{8.89}$$

where $A \triangleq$ surface area of slug, cm^2

$q \triangleq$ local heat-transfer rate, w/cm^2

$M \triangleq$ mass of slug, kg

$c \triangleq$ specific heat of slug, $W \cdot s/(kg \cdot °C)$

$T \triangleq$ slug temperature, $°C$

---

[1] C. E. Shepard and I. Warshawsky, Electrical Techniques for Compensation of Thermal Time Lag of Thermocouples and Resistance Thermometer Elements, *NACA, Tech. Note* 2703, 1952; Louis and Hartman, op. cit.

[2] Louis and Hartman, op. cit.

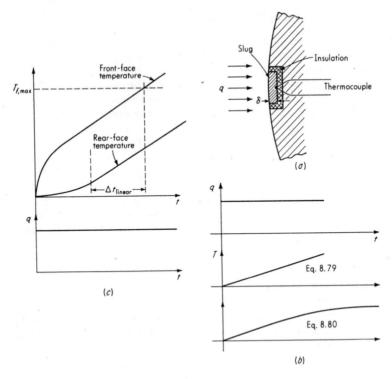

Figure 8.63 Slug-type heat-flux sensor.

Then

$$q = \frac{Mc}{A}\frac{dT}{dt}$$ (8.90)

and thus q may be determined by measuring dT/dt if Mc/A is known. Since the thermocouple reads T rather than dT/dt, a graphical, numerical, or electrical differentiation must be carried out to get q. For greater accuracy, the heat losses may be taken into account by modifying Eq. (8.90) to give

$$q = \frac{Mc}{A}\frac{dT}{dt} + K_l\,\Delta T$$ (8.91)

where $K_l \triangleq$ loss coefficient, W/(cm² · °C)

$\Delta T \triangleq$ temperature difference between slug and casing (usually taken as temperature rise of slug by assuming constant casing temperature)

The numerical values of Mc/A and K_l for a given sensor are determined by calibration and supplied by the manufacturer. Equation (8.90) predicts that, for a constant q, T increases linearly with time and without limit. Actually, the unavoidable heat losses eventually make dT/dt approach zero, as shown by the more correct Eq. (8.91).

The analysis of Eq. (8.89) assumes the slug is at all times at uniform temperature T throughout. This is not actually the case; thus there is a time-lag effect which has been evaluated[1] on the basis of a step input of q. A partial-differential-equation analysis leads to

$$q_m \approx q(1 - 2e^{-\pi^2\alpha t/\delta^2}) \tag{8.92}$$

where $q_m \triangleq$ measured flux, using temperature at back surface of slug
$\quad q \triangleq$ actual flux
$\quad \alpha \triangleq$ thermal diffusivity $= k/(\rho c)$
$\quad \delta \triangleq$ slug thickness
$\quad k \triangleq$ thermal conductivity
$\quad \rho \triangleq$ mass density

We see that a fast response requires a small value of $\delta^2 \rho c/k$.

Since the materials of which the sensor is made can withstand only a certain maximum temperature rise ΔT_{max}, a slug can be exposed to a given heat-transfer rate q for only a limited time t_{max}. Neglecting losses, we can integrate Eq. (8.90) to give the slug thickness δ required for a given q, ΔT_{max}, and t_{max} as

$$\delta = \frac{qt_{max}}{\rho c \Delta T_{max}} \tag{8.93}$$

For convective heat transfer from a gas of fixed temperature T_g, the heat flux into the slug is $h(T_g - T)$, where h is the film coefficient. Neglecting losses, we may write

$$h(T_g - T)\,dt = \rho c \delta\,dT \tag{8.94}$$

which leads to $\qquad \delta = \dfrac{ht_{max}}{\rho c \ln |1/[T - \Delta T_{max}/(T_g - T_i)]|} \tag{8.95}$

where T_i is the initial slug temperature.

A more refined analysis[2] for the case of constant q (which takes into account that the *front* surface will overheat before the back) shows that there is an optimum value of δ, in the sense that the linear (steady-state) part of the rear-surface response is the longest possible before the front surface overheats. This optimum value of δ is given by

$$\delta_{opt} = \frac{kT_{f,\,max}}{1.366q} \tag{8.96}$$

[1] Heat Technology Laboratory, Inc., Huntsville, Ala., *Rept. HTL-ER-4*, p. 4, 1962.

[2] R. H. Kirchhoff, Calorimetric Heating-Rate Probe for Maximum-Response-Time Interval, *AIAA J.*, p. 967, May 1964.

where $T_{f,\text{max}}$ is the maximum allowable front-surface temperature. The time interval of linear response is found to be

$$\Delta t_{\text{linear}} = \frac{0.366 k^2 T_{f,\text{max}}^2}{\alpha q^2} \tag{8.97}$$

Figure 8.63c illustrates these concepts.

Steady–State or Asymptotic Sensors (Gardon Gage)

Figure 8.64 shows the essential features of this type of sensor, which was first proposed by R. Gardon.[1] A thin constantan disk is connected at its edges to a large copper heat sink, while a very thin (<0.005-in-diameter) copper wire is fastened at the center of the disk. This forms a differential thermocouple between the disk center and its edges. When the disk is exposed to a constant heat flux, an equilibrium temperature difference is rapidly established which is proportional to the heat flux. Since the thermocouple signal is now directly proportional to the heat flux, no differentiating process (such as is required in a slug-type sensor) is necessary. Furthermore, loss corrections generally are not needed, nor is a thermocouple reference junction required. Instrument response[2] is approximately of the first-order type; thus

$$\frac{e_o}{q}(D) = \frac{K}{\tau D + 1} \tag{8.98}$$

where

$$K \triangleq \frac{d^2 K_e}{16 \delta k} \tag{8.99}$$

$$\tau \triangleq \frac{\rho c d^2}{16 k} \tag{8.100}$$

and $d \triangleq$ diameter of disk
$\delta \triangleq$ thickness of disk
$K_e \triangleq$ thermocouple sensitivity, mV/F°
$k \triangleq$ thermal conductivity of disk
$c \triangleq$ specific heat of disk

For copper/constantan, numerical values are

$$\frac{e_o}{q}(D) = \frac{0.0308(d^2/\delta)}{(5.96 d^2)D + 1} \tag{8.101}$$

where d and δ are in inches, e_o is in millivolts, q is in Btu/(s · ft²), and $5.96 d^2$ is in

---

[1] R. Gardon, An Instrument for the Direct Measurement of Intense Thermal Radiation, *Rev. Sci. Instrum.*, p. 366, May 1953.
[2] *Rept. HTL-ER-4*, op. cit.

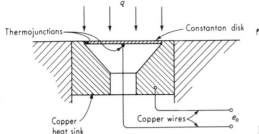

Figure 8.64 Gardon gage.

seconds. Typical commercial units are available for full-scale heat fluxes of 15 to 300 Btu/(s · ft^2), produce 10-mV full-scale output, and have time constants of 0.07 to 0.2 s.

Convenient foil-type heat-flow sensors which may be cemented directly to the measured surface and which use a differential thermopile to measure the ΔT across a polyamide plastic film are available.[1] The hot and cold junctions of the thermopile (Chromel/constantan or copper/constantan) are attached to opposite sides of the plastic (0.0005 to 0.01 in thick) to form a "sandwich." When the entire package (0.003 to 0.012 in thick, 0.3 × 0.5 in area) is cemented to the surface, any heat transferred must pass through the plastic film, causing a proportional ΔT, which is measured by the thermopile. If desired, a single thermocouple for measuring surface temperature itself may be included in the package. Maximum operating temperature is 500°F. In addition to the usual specifications, the manufacturer provides data on sensor thermal capacitance and resistance, allowing estimation of sensor disturbance effects.

Application Considerations

The introduction of the sensor into the wall locally alters the thermal properties of the wall and causes the measured heat flux to differ from that which would occur if the sensor were not present.[2] Thus it is desirable to match, insofar as feasible, the thermal properties of sensor and wall. For a Gardon gage, there is a radial temperature gradient over the disk which, if excessive, causes a variation in local convection coefficient and thus an error. By sacrificing sensitivity (and then recovering it by external amplication, if necessary) the temperature gradient and associated error may be reduced. When only the radiation component of the total flux is desired, the front of the sensor is covered with a thermally isolated sapphire window which passes the radiation flux but blocks the convective flux.

---

[1] Micro-Foil Heat Flow Sensors, RdF Corp., Hudson, N.H.

[2] D. R. Hornbaker and D. L. Rall, Thermal Perturbations Caused by Heat-Flux Transducers and Their Effect on the Accuracy of Heating-Rate Measurements, *ISA Trans.*, p. 123, April 1964.

PROBLEMS

8.1 An Invar/brass cantilever bimetal (see Fig. 8.3b) has $t = 0.05$ in, a length of 2 in, a width of 0.5 in, and $t_A = t_B$ and $n + 1/n \approx 2$. Estimate the end deflection for temperature changes of 30 and 60°C. If the end is held fixed, estimate the force developed for temperature changes of 30 and 60°C.

8.2 A thermometer is to have a sensitivity of 10 in/°C when mercury is used in the neighborhood of room temperature. Obtain an expression relating capillary cross-section area and bulb volume to meet this requirement. Obtain expressions for the time constant if the bulb is spherical and, if it is cylindrical with a length equal to 5 diameters. Also find the ratio of these two time constants.

8.3 Sketch and explain the operation of a bimetallic compensator to replace the auxiliary pressure sensor in Fig. 8.6. Can both case and capillary compensation be obtained? Explain.

8.4 Analyze the system of Fig. 8.13a to obtain a steady-state relation between hot-gas temperature as an input and thermocouple voltage as an output.

8.5 Develop equations to estimate the dynamic response of the system of Fig. 8.13a.

8.6 Repeat Prob. 8.4 for the system of Fig. 8.13b.

8.7 In Fig. 8.1, if $\tau = 2.8$ s, $T_{gas} = 5000°F$, and the thermocouple damage limit is 2000°F, how long can the cooling be left off if the steady-state cooled thermocouple temperature is 500°F?

8.8 A resistance-thermometer circuit as in Fig. 8.21d has $R_1 = R_2 = 10,000 \ \Omega$ and is to cover the temperature range 0 to 400°C. The thermometer element is platinum with a resistance of 1,000 Ω at 200°C. Plot the curve of bridge output voltage (open circuit) versus input temperature, using the data of Fig. 8.19. See Chap. 10 for bridge-circuit equations. Bridge excitation is 20 V.

8.9 A 500-Ω resistance thermometer carries 5-mA current. Its surface area is 0.5 in², and it is immersed in stagnant air, so that the heat-transfer coefficient is $U = 1.5$ Btu/(h · ft² · F°). Find its self-heating error. What would be the error in water with $U = 100$ Btu/(h · ft² · F°)?

8.10 A pulse-excited resistance thermometer (see Fig. 8.22) has an excitation voltage in the form of a rectangular pulse of 100-V height and 0.1-s duration. The pulse is on for 0.1 s and off for 0.9 s in a repetitive cycle. Compute the ratio of peak/rms voltage for this pulse. What average heating power would this voltage pulse produce in a 500-Ω resistor?

8.11 For blackbody radiation, what surface temperature is needed to radiate 1 hp/in²?

8.12 Estimate the percentage of total power found above 10 μm for blackbody radiation at $T = 400 \ K$.

8.13 Explain the disadvantage of a large time constant in a thermal-radiation detector using a chopper.

8.14 Derive Eq. (8.64).

8.15 Consider a subsonic air flow in a duct. Pitot-static-tube measurements give a static pressure of 100 lb/in² absolute and a stagnation pressure of 129.1 lb/in² absolute. A temperature probe with a recovery factor $r = 0.80$ extends from a 100°F wall a distance of 1 ft into the flow. The probe thermocouple reads 400°F. The probe support has a radius of 0.02 ft and a thermal conductivity of 100 Btu/(h · ft · F°). The end of the probe may be assumed insulated, and the surface convection coefficient is 10 Btu/(h · ft² · F°). Radiation effects are negligible. Calculate the static temperature of the flow.

8.16 Derive Eq. (8.83).

8.17 Make and analyze a third-order model of a temperature sensor analogous to the second-order model of Fig. 8.62.

8.18 A butt-welded 0.03-in bare-wire copper/constantan thermocouple is used to measure the temperature (near 100°F) of atmospheric air flowing at 100 ft/s. Estimate the time constant, using both formulas available.

8.19 A copper slug-type heat-flux sensor is 0.2 in thick. Plot its time response (q_m versus t) for a step change in q. Is this a first-order instrument? Explain. How long must one wait before q_m is 95 percent of q?

8.20 Derive Eq. (8.93).

8.21 Derive Eq. (8.95).

8.22 Compare the sensitivity and response speed of Gardon gages of like diameter and thickness but made of (a) constantan disk, copper heat sink; (b) copper disk, constantan heat sink; (c) iron disk, constantan heat sink; (d) constantan disk, iron heat sink.

8.23 Sketch and explain a test setup for evaluating the step-function response of temperature sensors exposed to air flows of different velocities.

8.24 Sketch and explain a test setup for static calibration of heat-flux sensors.

8.25 Sketch and explain a test setup for step-function testing of heat-flux sensors.

8.26 Perform a linearized dynamic analysis of the venturi pneumatic pyrometer (Fig. 8.15).

8.27 Explain the operation of the circuits of Fig. 8.29a.

8.28 Derive Eq. (8.31).

BIBLIOGRAPHY

1. H. F. Stimson: International Temperature Scale of 1948, Text Revision of 1960, *Natl. Bur. Std. (U.S.), Monograph* 37, 1961.
2. J. Nicol and C. J. Rauch: Below One Degree, *Ind. Res.*, p. 60, September 1964.
3. J. R. Van Orsdel et al.: Development of a Vapor-Pressure-Operated High Temperature Sensor Device, *NASA, CR*-50001, 1964.
4. C. F. Alban and C. C. Perry: Maximum-Work Bimetals, *Mach. Des.*, p. 143, April 16, 1959.
5. C. F. Alban and C. C. Perry: Adjusting Performance of Thermostatic Bimetals, *Mach. Des.*, p. 195, May 14, 1959.
6. C. F. Alban and C. C. Perry: Optimum Design of Thermostatic Bimetal Elements, *Mach. Des.*, p. 119, February 21, 1957.
7. J. M. Benson and R. Horne: Surface Temperature of Thin Sheets and Filaments, *Instrum. Contr. Syst.*, p. 115, October 1962.
8. C. E. Moeller: Do Shields Improve Thermocouple Response? *ISA J.*, p. 56, August 1960.
9. J. L. LeMay: More Accurate Thermocouples with Percussion Welding, *ISA J.*, p. 42, March 1959.
10. L. E. Bollinger: Thermocouple Measurements in an RF Field, *ISA J.*, p. 338, September 1955.
11. J. C. Lachman and F. W. Kuether: Stability of Rhenium/Tungsten Thermocouples in Hydrogen Atmospheres, *ISA J.*, p. 67, March 1960.
12. A. R. Driesner et al.: High Temperature W/W-25 Re Thermocouples, *Instrum. Contr. Syst.*, p. 105, May 1962.
13. F. W. Kuether and J. C. Lachman: How Reliable Are the Two New High-Temperature Thermocouples in Vacuum? *ISA J.*, p. 67, April 1960.
14. J. J. Van Drasek and B. A. Short: Conversion Formulas for Copper-Constantan Thermocouples, *Instrum. Contr. Syst.*, p. 106, February 1965.
15. G. E. Reis et al.: A Thermocouple Unit for Measuring Transient Temperatures at Specified Locations in Metal Bodies, Sandia Corp., Albuquerque, N. Mex., *SCR*-3, 1958.
16. H. C. Jordan: Welded Thermocouple Junctions, *Instrum. Contr. Syst.*, p. 988, June 1960.
17. Simple Circuit Continuously Monitors Thermocouple Sensor Continuity, *NASA, Brief* 63-10567, 1963.
18. Thermocouple Calibration, *Instrum. Contr. Syst.*, p. 1663, September 1961.
19. M. B. Dow: Comparison of Measurements of Internal Temperatures in Ablation Material by Various Thermocouple Configurations, *NASA, Tech. Note* D-2165, 1964.
20. R. Dutton and E. C. Lee: Surface-Temperature Measurement of Current-Carrying Objects, *ISA J.*, p. 49, December 1959.
21. J. Nanigian: Thermal Properties of Thermocouples, *Instrum. Contr. Syst.*, p. 87, October 1963.

22. Unusual Thermocouples and Accessories, *Instrum. Contr. Syst.*, p. 110, June 1963.
23. E. G. Weissenberger: Metal Sheathed Thermocouples, *Instrum. Contr. Syst.*, p. 109, May 1963.
24. C. E. Moeller: Special Thermocouple Solves Surface Temperature Problem, *ISA J.*, p. 47, June 1959.
25. C. M. Stover: Method of Butt Welding Small Thermocouples 0.001 to 0.01 Inch in Diameter, *Rev. Sci. Instrum.*, vol. 31, no. 6, p. 605, June 1950.
26. C. M. Stover: Method of Making Small Pointed Thermocouples, *Rev. Sci. Instrum.*, vol. 32, no. 3, p. 366, March 1961.
27. G. E. Reis: Temperature Measurements on High Speed Missiles, Sandia Corp., Albuquerque, N. Mex., *SCR*-73, 1956.
28. O. Schwelb and G. C. Temes: Thermistor-Resistor Temperature Sensing Networks, *Electro-Technol. (New York)*, p. 71, November 1961.
29. D. S. Saulson: The Thermistor Bridge, *Electro-Technol. (New York)*, p. 73, September 1961.
30. J. S. Blakemore: Germanium for Low-Temp Resistance Thermometry, *Instrum. Contr. Syst.*, p. 94, May 1962.
31. J. R. Pies: A New Semiconductor for Temperature Measuring, *ISA J.*, p. 50, August 1959.
32. J. M. Janicke: Direct-Reading Platinum Thermometer, *Instrum. Contr. Syst.*, p. 129, May 1965.
33. H. N. Norton: Resistance Elements for Missile Temperatures, *Instrum. Contr. Syst.*, p. 993, June 1960.
34. A. R. Anderson and T. M. Stickney: Ceramic Resistance Thermometers as Temperature Sensors above 2200°R, *Instrum. Contr. Syst.*, p. 1864, October 1961.
35. A. B. Kaufman: Cryogenic Characteristics of Alloy Wires, *Instrum. Contr. Syst.*, p. 119, March 1964.
36. H. C. Tsien: Piston Zone Temperature Measurement, *Instrum. Contr. Syst.*, p. 105, May 1964.
37. R. S. Benson: Measurement of Transient Exhaust Temperatures in I.C. Engines, *The Engineer*, February 28, 1964.
38. T. Coor and L. Szmanz: Digital Thermometer, *Instrum. Contr. Syst.*, p. 125, May 1965.
39. E. W. Jones: Calibration Techniques for Thermistors, *Instrum. Contr. Syst.*, p. 123, May 1965.
40. D. B. Schneider: The Thermistor Thermometer, *Instrum. Contr. Syst.*, p. 119, May 1965.
41. P. W. Montgomery and R. L. Lowery: Jet Temperature by IR Pyrometry, *ISA J.*, p. 61, April 1965.
42. R. J. Thorn and G. H. Winslow: Recent Developments in Optical Pyrometry, *ASME Paper* 63-WA-224, 1963.
43. R. H. Tourin: Recent Developments in Gas Pyrometry by Spectroscopic Methods, *ASME Paper* 63-WA-252, 1963.
44. P. S. Schmidt: Spectroscopic Temperature Measurements, *Inst. Tech.*, pp. 35–38, December 1975.
45. D. A. McGraw and R. G. Mathias: Application of Radiation Pyrometry to Glass-Forming Processes, *Ceram. Age*, August 1962.
46. R. A. Hanel: The Dielectric Bolometer, A New Type of Thermal Radiation Detector, *NASA, Tech. Note* D-500, 1960.
47. R. A. Hanel: A Low-Resolution Unchopped Radiometer for Satellites, *NASA, Tech. Note* D-485, 1961.
48. R. W. Reynolds: Infrared-Radiation Reference Sources, *Electro-Technol. (New York)*, p. 46, January 1963.
49. G. Conn and D. Avery: "Infrared Methods," Academic, New York, 1960.
50. F. Schwarz: Infrared Detectors, *Electro-Technol. (New York)*, p. 116, November 1963.
51. Calibration of Thermopiles, *NASA, N-64-28205*, 1964.
52. Pyroelectric Detection Techniques and Materials, *NASA, CR*-44, 1964.
53. D. Greenshields: Spectrometric Measurements of Gas Temperatures in Arc-Heated Jets and Tunnels, *NASA, Tech. Note* D-1960, 1963.
54. D. R. Buchele: Nonlinear-Averaging Errors in Radiation Pyrometry, *NASA, Tech. Note* D-2406, 1964.

55. M. Weiss: High Temperature Ultraviolet Radiometer, *Instrum. Contr. Syst.*, p. 95, May 1964.
56. E. W. Bivans: Measuring Infrared Detector Noise, *Electron. Des.*, August 2, 1962.
57. Reference Blackbody Is Compact, Convenient to Use, *NASA, Brief* 63-10004, 1963.
58. Lunar Surface Temperature Instrument, *NASA*, N-64-10097, 1964.
59. A. J. Metzler and J. R. Branstetter: Fast Response, Blackbody Furnace for Temperatures to 3000°K, *Rev. Sci. Instrum.*, vol. 34, no. 11, p. 1216, November 1963.
60. H. C. Ingrao et al.: Ferroelectric Bolometer for Space Research, *NASA*, CR-55542, 1964.
61. E. M. Wormser: Radiation Thermometer with In-Line Blackbody Reference, *Instrum. Contr. Syst.*, p. 101, December 1964.
62. J. Grey: Thermodynamic Methods of High-Temperature Measurement, *ISA Trans.*, vol. 4, no. 2, 1965.
63. Calibration of Optical Pyrometers, *Instrum. Contr. Syst.*, p. 84, May 1962.
64. B. Bernard: Flame Temperature Measurements, *Instrum. Contr. Syst.*, p. 113, May 1965.
65. A. G. Gaydon: "The Spectroscopy of Flame," Wiley, New York, 1957.
66. R. Looney: Method for Presenting the Response of Temperature Measuring Systems, *ASME Trans.*, p. 1851, November 1957.
67. W. J. King: Measurement of High Temperatures in High-Velocity Gas Streams, *ASME Trans.*, p. 421, July 1943.
68. W. M. Rohsenow and J. P. Hunsaker: Determination of the Thermal Correction for a Single-Shielded Thermocouple, *ASME Trans.*, p. 699, August 1947.
69. T. M. Stickney: Recovery and Time-Response Characteristics of Six Thermocouple Probes in Subsonic and Supersonic Flow, *NACA, Tech. Note* 3455, 1955.
70. R. C. Turner and G. D. Gordon: Thermocouple for Vacuum Tests Minimizes Error, *Space/ Aeron.*, p. 256, January 1964.
71. D. Wald: Measuring Temperature in Strong Fields, *Instrum. Contr. Syst.*, p. 100, May 1963.
72. L. M. K. Boelter et al.: Thermocouple Conduction Error Observed in Measuring Surface Temperatures, *NACA, Tech. Note* 2427, 1951; *Tech. Note* 1452, 1948.
73. M. Sibulkin: A Total-Temperature Probe for High-Temperature Boundary-Layer Measurements, *J, Aerosp. Sci.*, p. 458, July 1959.
74. J. C. Faul: Thermocouple Performance in Gas Streams, *Instrum. Contr. Syst.*, p. 104, December 1962.
75. I. Fruchtman: Temperature Measurement of Hot Gas Streams, *AIAA J.*, vol. 1, no. 8, p. 1909, 1963.
76. D. L. Goldstein and R. Scherrer: Design and Calibration of a Total-Temperature Probe for Use at Supersonic Speeds, *NACA, Tech. Note* 1885, 1949.
77. R. Sandri et al.: On the Measurement of the Average Temperature of a Fluid Stream in a Tube by Means of a Special Type of Resistance Thermometer, National Research Laboratory, Ottawa, Canada, *Mech. Eng. Div. Rept. M1-826*, April 1962.
78. R. D. Wood: A Heated Hypersonic Stagnation-Temperature Probe, *J. Aerosp. Sci.*, p. 556, July 1960.
79. T. R. Billeter: Using Microwave Techniques for High Temperature Measurement, *Inst. & Cont. Syst.*, pp. 107–109, February 1972.
80. R. V. DeLeo et al.: Measurement of Mean Temperature in a Duct, *Instrum. Contr. Syst.*, p. 1659, September 1961.
81. C. F. Hansen et al.: Investigation of Heat Conduction in Air, *NASA, Tech. Rept.* R-27 (Nickel Film Surface Temperature Detectors), 1959.
82. R. P. Benedict: High Response Aerosol Probe for Sensing Gaseous Temperature in a Two-Phase, Two-Component Flow, *ASME Paper* 62-WA-317, 1962.
83. I. Warshawsky: Measurements of Rocket Exhaust-Gas Temperatures, *ISA J.*, p. 91, November 1958.
84. M. G. Holland et al.: Temperature Measurement from 2°K–400°K, *Instrum. Contr. Syst.*, p. 89, May 1962.
85. J. Grey: Thermodynamic Methods of High-Temperature Measurement, *ISA Trans.*, p. 102, April 1965.

86. T. A. Perls and J. J. Hartog: Pyroelectric Transducers for Heat-Transfer Measurements, *ISA Trans.*, p. 21, January 1963.
87. D. L. Johnson: The Design and Application of a Steady-State Heat Flux Transducer for Aerodynamic Heat-Transfer Measurements, *ISA Trans.*, p. 46, January 1965.
88. Simple Transducer Measures Low Heat-Transfer Rates, *NASA, Brief* 64-10122, 1964.
89. F. C. Stempel and D. L. Rall: Direct Heat Transfer Measurements, *ISA J.*, p. 68, April 1964.
90. E. A. Laumann: A Steady-State Heat Meter for Determining the Heat-Transfer Rate to a Cooled Surface, *NASA, N-63-18868*, 1963.
91. L. R. Hunt and R. R. Howell: Experimental Technique for Measuring Total Aerodynamic Heating Rates to Bodies of Arbitrary Shape with Results to Mach 7, *NASA, Tech. Note* D-2446, 1964.
92. R. J. Conti: Heat-Transfer Measurements at a Mach Number of 2 in the Turbulent Boundary Layer on a Flat Plate Having a Stepwise Temperature Distribution, *NASA, Tech. Note* D-159, 1959.
93. R. A. Jones and J. L. Hunt: Use of Temperature-Sensitive Coatings for Obtaining Quantitative Aerodynamic Heat-Transfer Data, *AIAA J.*, p. 1354, July 1964.
94. P. H. Rose and J. O. Stankevics: Heat Transfer Measurements in Partially Ionized Gases, *NASA, CR-59768*, 1964.
95. C. H. Liebert et al.: Application of Various Techniques for Determining Local Heat-Transfer Coefficients in a Rocket Engine from Transient Experimental Data, *NASA, Tech. Note* D-277, 1960.
96. D. R. Beck and F. Kreith: A New Steady State Calorimeter for Measuring Heat Transfer through Cryogenic Insulation, *NASA, N-64-14283*, 1964.
97. L. Bogdon: High-Temperature, Thin-Film Resistance Thermometers for Heat Transfer Measurement, *NASA, CR-26*, 1964.
98. J. C. Cook and H. S. Levine: Calorimeter and Accessories for Very High Thermal Radiation Flux Measurements, *Rev. Sci. Instrum.*, October 1960.
99. L. Bogdan: Measurement of Radiative Heat Transfer with Thin-Film Resistance Thermometers, *NASA, CR-27*, 1964.
100. R. C. Bachmann et al.: Investigation of Surface Heat-Flux Measurements with Calorimeters, *ISA Trans.*, p. 143, April 1965.
101. Bibliography of Temp. Meas. 1953–1969, *NBS* SP-373, 1972.
102. S. S. Fam, L. C. Lynnworth, and E. H. Carnevale: Ultrasonic Thermometry, *Inst. & Cont. Syst.*, pp. 107–110, October 1969.
103. G. D. Nutter: Recent Advances and Trends in Radiation Thermometry, *ASME Paper* 71-WA/Temp-3, 1971.
104. J. Geist: Fundamental Principles of Absolute Radiometry and the Philosophy of This NBS Program (1968–1971), *NBS, Tech. Note* 594-1, 1972.
105. D. R. Buchele and D. J. Lesco: Pyrometer for Measurement of Surface Temperature Distribution on a Rotating Turbine Blade, *NASA TMX-68113*, 1972.
106. J. R. Branstetter: Some Practical Aspects of Surface Temperature Measurement by Optical and Ratio Pyrometers, *NASA TN D-3604*, 1966.
107. C. H. Liebert et al.: Turbine Blade Metal Temperature Measurement with a Sputtered Thin-Film Thermocouple, *NASA TM X-71844*, 1975.
108. R. P. Benedict and R. J. Russo: A Note on Grounded Thermocouple Circuits, *ASME Paper* 71-WA/Temp-1, 1971.
109. C. E. Moeller et al.: NASA Contributions to Development of Special-Purpose Thermocouples, *NASA* SP-5050, 1968.
110. G. Cataland and H. H. Plumb: Low Temperature Thermometry: Interim Report, *NBS, Tech Note* 765, 1973.
111. J. T. Hojnacki: Testing a Fluidic Temperature Sensor on a Subscale Ramjet Engine Combustion Chamber, *AFAPL-TR-75-1*, Air Force Aero. Prop. Lab, Wright-Patterson AFB, Ohio, 1975.
112. R. P. Shreeve and D. W. Peecher: Stagnation Temperature Measurement at High Mach Number Using Very Small Probes, *Boeing Res. Lab. Rept.* DI-82-0945, 1970.

113. D. J. Baines: Selecting Unsteady Heat Flux Sensors, *Inst. & Cont. Syst.*, May 1972.
114. An Oven for Many Thermocouple Reference Junctions, *NASA Tech. Briefs*, vol. 5, no. 4, FRC-10112, Winter 1980.
115. R. A. Gauthier: Surface Temperature Measurement, *InTech.*, pp. 57–60, February 1981.
116. M. C. Chuang: Transient Temperature Measurement in Moving Fluids, *ASME Paper* 77-WA/TM-3, 1977.

NINE

MISCELLANEOUS MEASUREMENTS

9.1 TIME, FREQUENCY, AND PHASE-ANGLE MEASUREMENT

The fundamental standard of time is discussed in Sec. 4.2. The United States Frequency Standard is a cesium-beam resonator whose precision is of the order of 1 part in 10^{11}. By radio-broadcasting signals related to the frequency of the standard, the National Bureau of Standards makes these frequency and time standards available to any other laboratory equipped to receive the signals. Stations WWVB and WWVL broadcast signals whose precision as received is about 1 part in 10^{10}.

Perhaps the most convenient and widely utilized instrument for accurate measurement of frequency and time interval is the electronic counter-timer. Figure 9.1 gives a block diagram showing the basic operation of such devices. The instrument's time and frequency standard is a piezoelectric crystal oscillator which generates a voltage whose frequency is very stable since the crystal is kept in a temperature-controlled oven. A typical frequency is 10^7 Hz, while the drift in frequency may be of the order of 3 parts in 10^7 per week. Over time, this gradual drift can cause errors; thus highly accurate measurements require periodic recalibration of the oscillator against a suitable standard such as the radio-broadcast signals. In Fig. 9.1a the instrument is set up for frequency measurement of a signal whose frequency is 6,843,169 Hz. This is accomplished by allowing the signal (suitably "shaped" to define each cycle more precisely) to go through a gating circuit to the decimal-counting units for a precisely timed interval. This interval may be selected in 10 : 1 steps from 10^{-7} to 1 s. Thus in the 1-s interval used in Fig. 9.1a the counters accumulate 6,843,169 pulses. This mode of operation is also called EPUT (events per unit time). To measure the frequency *ratio* of two signals *A* and *B*, use the scheme of Fig. 9.1*a*, connecting signal *A* as the "signal" but using signal *B* in place of the crystal oscillator (most counters provide for such an "external clock").

Sometimes it is more desirable to measure the period (rather than the frequency) of a signal. Figure 9.1*b* shows the arrangement used for this measurement. The trigger-level control is adjusted so that triggering occurs on the steepest part of the signal waveform to reduce error. There is usually provision for triggering on either a positive slope or a negative slope, as desired. Since there is an inherent potential error of ± 1 count in turning the gate on and off, frequency-mode measurements are more accurate for high-frequency signals whereas period-mode measurements are more accurate for low-frequency signals. For example, a 10-Hz signal measured in the frequency mode with the usual 1-s time interval gives only 10 counts; thus an error of ± 1 count is a 10 percent error. The same signal measured in the period mode with a 10-MHz counter gives 10^6 counts and an error of only 0.0001 percent. Thus for a given counter there is some frequency below which period measurements should be employed and above which frequency measurements should be utilized. For a 1-s sampling period, this frequency f_0 is given by

$$f_0 = \sqrt{f_c} \tag{9.1}$$

where $f_c \triangleq$ frequency of crystal oscillator ("clock" frequency). Thus a 10-MHz clock-frequency counter has $f_0 = 3,160$ Hz.

Measurements of the time interval between two events are very important in many experimental studies. The basic building blocks described above can be interconnected in a slightly different fashion, as in Fig. 9.1*c* and *d*, to accomplish this. In Fig. 9.1*c* two separate events have been transduced to electrical pulses; one event pulse is used to open the gate, and the other to close it, thereby timing the interval between them. Considerable versatility in triggering is obtained by providing trigger-level and slope controls on each input. By using the above arrangement but only one input signal (Fig. 9.1*d*), the widths of pulses may be determined. Sometimes additional circuits are provided to send to an oscilloscope pulses that show the exact point on the incoming signals at which triggering is initiated. These are helpful in adjusting the trigger-level and slope controls and in interpreting the resulting information.

Often measurement of the phase angle between two sinusoidal signals of the same frequency f_s is required. A general-purpose digital counter-timer can be employed for such measurements, as shown in Fig. 9.1*e*. To use this method, the amplitude of the two signals must be made equal and the triggering point of the two channels adjusted to be the same. Then the phase angle can be read directly with a resolution of 1° for the setup shown, or 0.1° if the reference frequency is set at $3,600 f_s$.

To prevent false counts as a result of the unavoidable noise on input signals, counter input circuits use a Schmitt trigger type of circuit with a built-in hysteresis effect (see Fig. 9.2). When frequency is measured, to cause one count in the

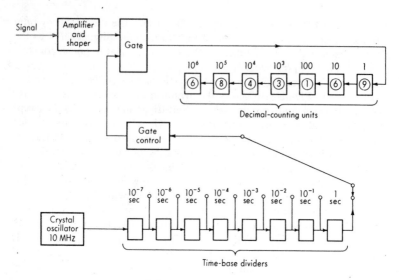

(a) Frequency measurement

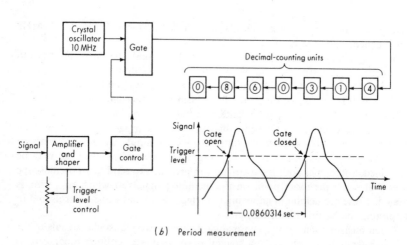

(b) Period measurement

Figure 9.1 Basic counter applications.

counting register, a signal must cross *both* the upper and the lower hysteresis levels. [For period or time-interval measurement the gate is opened (or closed) when the signal crosses a selected hysteresis level (either upper or lower).] The difference between the two hysteresis levels is called the *counter sensitivity*, and typically this might be 150 mV peak-to-peak. This is really a *threshold* sensitivity

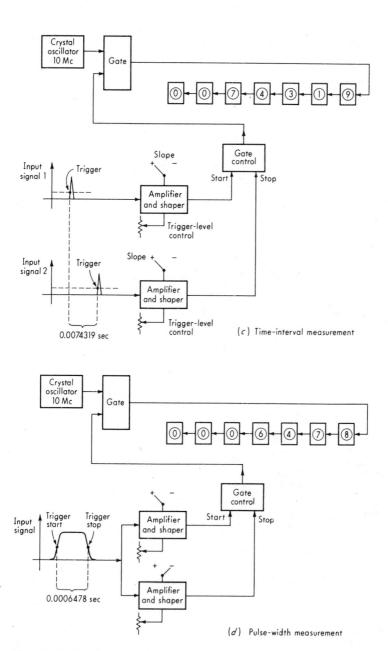

(c) Time-interval measurement

(d) Pulse-width measurement

Figure 9.1 (*Continued*)

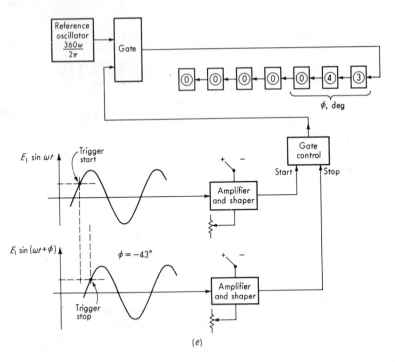

Figure 9.1 (*Continued*)

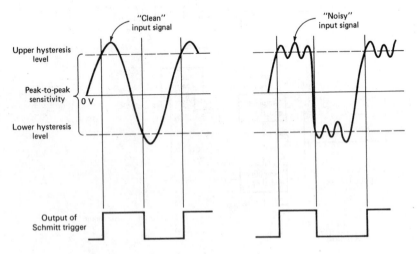

Figure 9.2 Use of counter hysteresis to reject noise.

since it fixes the amplitude of the smallest countable signal. But note that the optimum value is not necessarily the lowest, because of the false triggering problem mentioned earlier. The major error sources[1] in counter-timers are categorized as:

1. The ± 1 count ambiguity, a random error
2. The time-base (clock) error, resulting from temperature and line-voltage changes, aging, and short-term instability
3. The trigger error, a random error due to noise that causes early or late crossing of the hysteresis level
4. Channel-mismatch error, a systematic error present in two-channel measurements when input-circuit rise times and/or propagation delays are not identical for the two channels.

Frequency measurements are subject only to the ± 1 count and time-base errors. Period measurements have these plus the trigger error, while time-interval measurements suffer from all four. A high-performance instrument utilized within 1 week of time-base calibration and within the usual line-voltage (10 percent) and temperature (0 to 50°C) limits might have a time-base error of 1 part in 10^8. Channel-mismatch error might be 1 ns and could be reduced by calibration of this effect. Analysis[2] of the trigger error gives the standard deviation of the timing error as

$$\text{rms trigger error} = \sqrt{\frac{N_c^2 + N_{sa}^2}{\dot{V}_a^2} + \frac{N_c^2 + N_{sb}^2}{\dot{V}_b^2}}$$

where $N_c \triangleq$ counter noise, V rms (typically 0.2 to 2.0 mV)
$N_{sa} \triangleq$ start-signal noise, V rms
$N_{sb} \triangleq$ stop-signal noise, V rms
$\dot{V}_a \triangleq$ slew rate of start signal at trigger point, V/s
$\dot{V}_b \triangleq$ slew rate of stop signal at trigger point, V/s

The importance of using timing pulses with high slew rates is clear from this result.

Since even simple counter-timers allow time or period measurements of very high *resolution* (six or seven digits may be displayed), consideration of the above errors is very important, since total accuracy may turn out to be *much* less than resolution. This total error includes not only the contributions of the counter, but also any uncertainties in the sensors that transduce some physical event to the voltage which the timer accepts as input (for example, see Fig. 4.63). These sensor-related errors may completely overshadow those of the timer.

---

[1] Fundamentals of the Electronic Counters, *Appl. Note* 200, Hewlett-Packard Corp., Palo Alto, Calif., 1978.

[2] Ibid.; Understanding Frequency Counter Specifications, *Appl. Note* 200-4, Hewlett-Packard Corp., Palo Alto, Calif.

While the description of Fig. 9.1 explains the operation of most basic counter-timers, much more sophisticated instruments that employ microprocessor technology to extend capabilities, automate procedures, and provide calculating features are available.[1] The referenced instrument still uses a 10-MHz crystal clock as the basic timing source: however, an interpolation scheme increases time resolution from the 100 ns of the clock itself to about 1 ns for the instrument and removes the ± 1 count uncertainty always present in "ordinary" counters. Automatic trigger-setting circuits allow implementation of many useful features, such as 10 to 90 percent rise- (or fall-) time measurements and slew-rate measurements. Rise time is found by setting trigger levels at the 10 and 90 percent levels (this is automatic) and making a time-interval measurement, which is stored in a temporary register. Then the two trigger levels are measured automatically, and the microprocesser subtracts these and divides by the rise time to obtain and display slew rate in volts per second. Many other automatic features such as nonambiguous phase-angle measurement and various statistical calculations may be selected by push-button.

Another common method of phase-angle measurement involves cross-plotting the two sinusoidal signals against each other, by using an XY plotter for very low frequencies and an oscilloscope for high frequencies. The cross plot can be shown to be an ellipse, and suitable measurements on this ellipse give the phase angle (see Fig. 9.3a). We have

$$e_i = E_i \sin \omega t \tag{9.2}$$

and
$$e_o = E_o \sin (\omega t + \phi) \tag{9.3}$$

If we set $t = 0$, $e_i = 0$, and $e_o = E_o \sin \phi$, then

$$\sin \phi = \frac{e_o|_{e_i=0}}{E_o} \tag{9.4}$$

Since $e_o|_{e_i=0}$ has two values (plus and minus), the quadrant of ϕ is ambiguous; however, usually this can be resolved by visual observation of the two sine waves plotted against time (say on a dual-beam oscilloscope) or from knowledge of the system characteristics. The direction of travel of the "spot" as it plots the ellipse also resolves this difficulty, but may be hard to detect at high frequencies. An alternative method employing the same basic principle but a null technique is shown in Fig. 9.3b. Here the calibrated phase-shift circuit is adjusted until the ellipse degenerates into a straight line (0° phase shift). Then the phase angle ϕ is read directly from the phase-shifter dial. When the "sinusoidal" signals are noisy and/or distorted, special phase meters and tracking filters[2] may be necessary for accurate measurement. The FFT signal/system analyzers discussed in Chap. 10 can handle such problems also.

---

[1] G. D. Sasaki and R. C. Jensen, Automatic Measurements with a High-Performance Universal Counter, *Hewlett-Packard J.*, pp. 21–32, September 1980.

[2] E. O. Doebelin, "System Modeling and Response," p. 237, Wiley, New York, 1980.

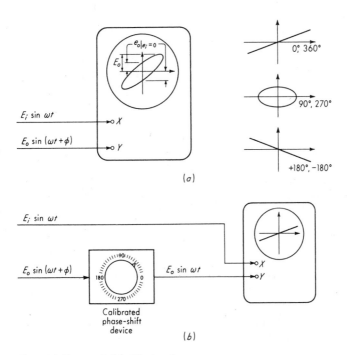

Figure 9.3 Phase angle from Lissajous figure.

9.2 LIQUID LEVEL

Measurement and/or control of liquid level in tanks is an important function in many industrial processes and in more exotic applications, such as the operation and fueling of large liquid-fuel rocket motors. Figure 9.4 illustrates some of the more common methods of accomplishing this measurement.

The simple float of Fig. 9.4a can be coupled to some suitable motion transducer to produce an electrical signal proportional to the liquid level. Figure 9.4b shows a "displacer" which has negligible motion and measures the liquid level in terms of buoyant force by means of a force transducer. Since hydrostatic pressure is related directly to liquid level, the pressure-sensing schemes of Fig. 9.4c and d allow measurement of the liquid level in open and pressure vessels, respectively. In the "bubbler" or purge system of Fig. 9.4e, the gas pressure downstream of the flow restriction is the same as the hydrostatic head above the bubble-tube end. The flow of gas is quite small; a bottle of nitrogen used as a source of pressurized gas may last six months or more.

Capacitance variation has been employed in various ways for level sensing. For essentially nonconducting liquids (conductivity less than 0.1 μS/cm^3), the bare-probe arrangement of Fig. 9.4f may be satisfactory since the liquid resistance

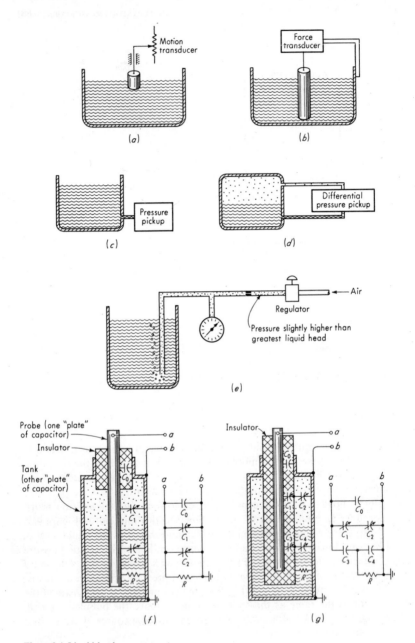

Figure 9.4 Liquid-level measurement.

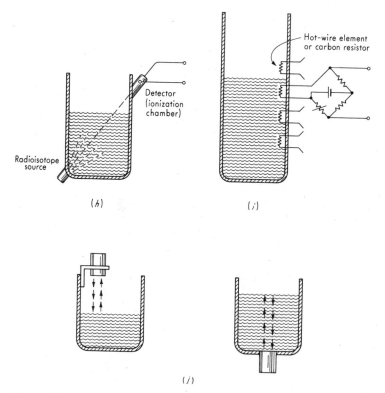

Figure 9.4 (*Continued*)

R is sufficiently high. For conductive liquids the probe must be insulated as in Fig. 9.4*g*, to prevent short-circuiting of the capacitance by the liquid resistance. The measurement of the capacitance between terminals *ab* may be accomplished in several ways. However, high-frequency ac (radio-frequency) methods offer significant advantages. Capacitance level-sensing techniques have been used with many common liquids, powdered or granular solids, liquid metals (high temperatures), liquefied gases (low temperatures), corrosive materials such as hydrofluoric acid, and in very-high-pressure processes.

Figure 9.4*h* illustrates the use of radioisotopes for level measurement. Since the absorption of beta-ray or gamma-ray radiation varies with the thickness of absorbing material between the source and the detector, a signal related to tank level may be developed. For analyzing such arrangements we may use the law

$$I = I_o e^{-\mu\rho x} \tag{9.5}$$

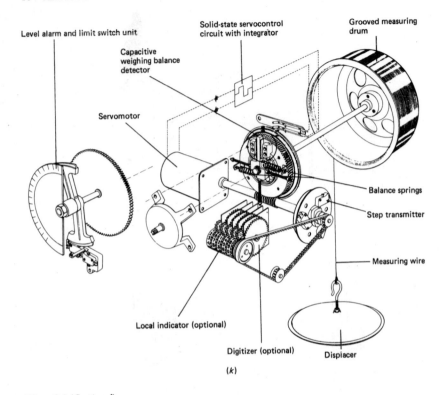

Figure 9.4 (*Continued*)

where $I \triangleq$ intensity of radiation falling on detector

$I_0 \triangleq$ intensity at detector with absorbing material not present

$e \triangleq$ base of natural logarithms

$\mu \triangleq$ mass absorption coefficient (constant for given source and absorbing material), cm^2/g

$\rho \triangleq$ mass density of absorbing material, g/cm^3

$x \triangleq$ thickness of absorbing material, cm

The gamma-ray source, cesium 137, has been used widely for liquid-level measurements and has a μ of 0.077 cm^2/g for water or oil, 0.074 for aluminum, 0.072 for steel, and 0.103 for lead. For gaging a tank of water, then, if a vertical radiation path (rather than the angled one of Fig. 9.4h) is assumed, the variation of I with liquid height h is given by (neglecting absorption of air path)

$$I = I_0 e^{-0.077h} \tag{9.6}$$

This exponential relation of I and h is nearly linear only for sufficiently small values of h. For h as large as, say, 100 cm, the nonlinearity is quite apparent. This can be overcome by using either a radiation source or a detector in the form of a strip oriented vertically rather than a "point" source or detector. Such arrangements are nonlinear for small values of $\mu\rho x_{max}$. Therefore point-to-point configurations are indicated for small ranges, whereas larger ranges require the more complex strip-to-point type. For a strip source (or detector), the strength (sensitivity for a detector) can be "tailored" to vary in just the right way with position along the strip to give a linear tank-level/detector-signal relation.

Figure 9.4i shows the method of using hot-wire or carbon resistor elements for the measurement of liquid level in discrete increments. The basic concept is that the heat-transfer coefficient at the surface of the resistance element changes radically when the liquid surface passes it. This changes its equilibrium temperature and thus its resistance, causing a change in bridge output voltage. By locating resistance elements at known height intervals, the tank level may be measured in discrete increments. Such arrangements have been used in filling fuel tanks of large rocket engines with cryogenic liquid fuels.

Ultrasonic "range-finding" techniques, discussed in Chap. 4, can be applied to liquid-level sensing, as in Fig. 9.4j.

An interesting variation on the classic float of Fig. 9.4a is found in the servopositioned level gage of Fig. 9.4k,[1] widely employed in deep tanks and underground caverns to 500-ft depth. The thin displacer has density (relative to liquid in tank) such that it would sink if it were not supported by the measuring wire, which applies just enough lifting force to support the displacer at a design immersion of 1 to 2 mm. The lifting force is maintained at this value (as the tank level rises and falls) by the servodrive, whose error signal comes from the elastic force sensor which deflects from neutral in proportion to the difference between lifting force and the design value set by adjustment of the balance springs. A capacitance displacement pickoff transduces elastic deflection to an electrical error signal for the servomotor. Competing (nonservo) gages use a mechanical "constant-force" spring to supply the lifting force, being otherwise similar except that displacer immersion is much larger, typically 10 mm. These simpler and cheaper all-mechanical devices suffer, however, from several error sources which are minimized in the servotype gage: increased friction, variation in constant-force spring lift, greater sensitivity to liquid-density change because of greater immersion, and greater cable-weight effect as a result of using a heavier tape (rather than a light wire) to suspend the displacer. Comparison of overall error for a 20-m-deep tank shows ± 13.5 mm for mechanical gages and ± 2.9 mm for servogages. Since 1 mm in a 100-ft-diameter tank is 4.6 bbl of product (at, say, \$40 per barrel), the economic importance of accuracy in a system used for custody transfer of product becomes obvious.

---

[1] Enraf-Nonius Service Corp., Bohemia, N.Y.

9.3 HUMIDITY[1]

Knowledge of the amount of water vapor in the air is important to the operation and/or automatic control of many industrial processes. This information may be gathered and presented in a number of ways, depending on the needs of the particular process and the measuring instrumentation utilized. In common use are the relative humidity (ratio of water partial pressure to saturation pressure), dew-point temperature, mixing ratio or specific humidity (mass of water per unit mass of dry gas), and volume ratio (parts of water vapor per million parts of air).

The ultimate standard for calibration of humidity-measuring devices is the National Bureau of Standards standard gravimetric hygrometer. This is a strictly laboratory apparatus in which the water vapor in an air sample is absorbed by suitable chemicals and then weighed very carefully. It directly determines the mixing ratio in grams per kilogram, covers the range 0.30 to 20.0, and has an uncertainty (systematic error plus 3 standard deviations) of about 0.1 percent of the reading. For less critical calibrations, the National Bureau of Standards uses its two-pressure humidity generator.[2] This equipment generates air/water mixtures in the dew-point range $-70°C$ (uncertainty $\pm 1.2C°$) to $+25°C$ ($\pm 0.1C°$), relative-humidity range 10 to 98 percent at temperatures ranging from $-55°C$ (relative-humidity uncertainty ± 2.5 percent) to $+40°C$ (relative-humidity uncertainty 0.5 percent), mixing ratio in the range 0.0013 g/kg (uncertainty ± 0.0003 g/kg) to 20 g/kg (uncertainty ± 0.5 percent), and volume ratio 2 ppm (± 0.5 ppm) to 30,000 ppm (± 0.5 percent).[3]

Classically, relative humidity has been found from psychrometric charts and the temperature readings of two thermometers. One, the dry-bulb thermometer, reads the ordinary air temperature; the other, the wet-bulb, is intended to read the temperature of adiabatic saturation. The latter measurement requires that the bulb be kept wet and a suitably high (about 1,000 ft/min) air velocity be maintained over the wet bulb. While these operations may be automated to a certain extent, the complexity of the calculations equivalent to the psychrometric chart hinders development of this technique into a continuous-reading instrument.

For continuous recording and/or control of relative humidity, electrical transducers of the Dunmore type are employed widely. These were first developed about 1944 by F. W. Dunmore of the National Bureau of Standards and are basically a resistance element that changes resistance with relative humidity. The resistance element is constructed of a dual winding of noble-metal wires on a plastic form with a definite spacing between them. When the windings are coated with a lithium chloride solution, a conducting path is formed between the windings. The electrical resistance of this path is found to vary reproducibly with the

[1] P. R. Wiederhold, Humidity Measurements, *Inst. Tech.*, pp. 31–37, June 1975; pp. 45–50, August 1975; A. Wexler, A Study of the National Humidity and Moisture Measurement System, *NBS* IR75-933, *Natl. Bur. Std. (U.S.)*, 1975.

[2] A. Wexler and R. D. Daniels, Pressure-Humidity Apparatus, *J. Res. Natl. Bur. Std.*, vol. 48, p. 269, 1952.

[3] Humidity Calibration Service, *Instrum. Contr. Syst.*, p. 123, November 1964.

relative humidity of the surrounding air and thus may be used as a sensing element. Bridge-type resistance-measuring circuitry with ac excitation is employed normally. The resistance/relative-humidity relation is quite nonlinear, and generally a single transducer can cover only a small range of the order of 10 percent relative humidity. Where large ranges (as great as 5 to 99 percent relative humidity) are needed, seven or eight of the transducers, each designed for a specific part of the total range, are combined in a single package. A single narrow-range sensing element may have an inaccuracy of the order of 1.5 percent relative humidity, resolution about 0.15 percent relative humidity, time constant as small as 3 s, and size 1 in in diameter by 2 in long. Since these units are sensitive to temperature also, some form of temperature compensation may be required. The sensors do not add or subtract moisture or heat from their environment in significant amounts and so may be utilized in sealed areas. Working temperatures in the range of -40 to $+150°F$ are possible.

Another electrical resistance type of sensor, the sulfonated polystyrene ion-exchange device called the *Pope cell*,[1] exhibits a nonlinear resistance change from a few megohms at 0 percent relative humidity to about 1,000 Ω at 100 percent, and a single sensor can cover the entire range. Accuracy is comparable to that of the Dunmore sensor. Proprietary semiconductor sensors,[2] highly porous at a molecular level, allow molecules of water vapor to drift freely in and out of the structure, changing device impedance and resistance when water dipoles interact with the lattice. A single sensor (and associated electronics) temperature-compensated from 0 to 50°C covers the range 0 to 100 percent relative humidity with a linearized output and ± 4 percent accuracy.

Dew-point temperature can be determined by noting the temperature of a polished metal surface (mirror) when the first traces of condensation ("fogging") appear. Commercial devices are available in which this operation has been completely automated by means of a feedback system. The mirror is cooled thermoelectrically (Peltier cooling), with the cooler's electrical input coming from an amplifier whose input is the error signal of the feedback system. Light-emitting diodes (LEDs) provide light to the mirror surface, where it is reflected to phototransistors that generate a feedback signal which is compared with a set reference voltage to generate the system error signal. The reference voltage is set at a value that just barely maintains condensation (fogging) on the mirror. Any deviation from this level of fogging causes an increase (or a decrease) in light reflected to the phototransistors and a proportional change in the cooling effort. Thus the mirror surface is always kept at the dew-point temperature, which is sensed with an embedded platinum RTD. For dew points below 0°C, the system tracks the frost point. Figure 9.5[3] shows a microprocessor-controlled version of this instrument which uses a dual-mirror system to correct continuously for mirror con-

---

[1] Wiederhold, op. cit.; G. Whitehaus, Linearizing Relative Humidity Measurements, *Inst. & Cont. Syst.*, pp. 72–73, September 1972.

[2] Thunder Scientific Corp., Albuquerque, N. Mex.

[3] *Bull.* 3–300, EG&G Environmental Equipment, Waltham, Mass.

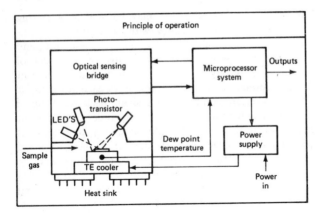

Figure 9.5 Microprocessor-based dew-point instrument.

tamination, which causes errors and requires periodic shutdown and/or cleaning in systems without this feature. The compensating mirror is controlled to stay "dry" at a precise temperature differential above the "wet" mirror, and both mirrors experience essentially the same contamination; so the dry mirror can be employed as a reference channel (phototransistor changes also are corrected by this scheme). Dew-point accuracy of $\pm 0.2°C$ over the range -75 to $+60°C$ is achieved. Addition of ambient temperature and/or pressure sensors allows microprocessor calculation of relative humidity, parts per million, grains, or pounds.

The lithium chloride sensor mentioned above under relative humidity can be modified to give a signal related to dew-point temperature.[1] The dual wire windings are supplied with ac power, causing a heating of the lithium chloride film. Lithium chloride shows a very sharp decrease in electrical resistance when *its* relative humidity increases above about 11 percent. Thus, when surrounded by moist air, the lithium chloride momentarily absorbs moisture, and its resistance drops, allowing more current to flow and more heat to be generated. This raises the temperature, driving off excess moisture, increasing the resistance, and reducing the heating. Thus the sensor regulates its own temperature, so that the relative humidity of the lithium chloride element stays near 11 percent no matter what the moisture content of the surrounding air is. It has been established that the temperature attained by the lithium chloride element, while not equal to the dew-point temperature, is directly related to it. Thus by measuring this temperature with some appropriate sensor, the dew-point temperature may be established. Probes of this type cover a dew-point range of -50 to $+160°F$ with an error of the order of 1 or $2F°$.

The final humidity sensor considered here is the electrolytic type. Here a continuous flow of sample gas (typically 100 cm^3/min, regulated ± 2 percent) is

---

[1] E. J. Amdur, Humidity Sensors, *Instrum. Contr. Syst.*, p. 93, June 1963.

passed through an analyzer tube. This tube has two platinum wires wound in a double helix on the inside of the tube. The space between the wires is coated with a strong desiccant (phosphorous pentoxide), and a dc potential is applied to the wire ends. When moisture in the sample gas is taken up by the P_2O_5, the water is electrolyzed into hydrogen and oxygen gas and a measurable electrolytic current flows. Such instruments have an inaccuracy of the order of 5 percent of full scale, typical range 0 to 2,000 ppm, resolution better than 1 ppm, and a time constant of the order of 30 s. They also have been adapted to the measurement of the water content of liquids and solids.

9.4 CHEMICAL COMPOSITION

In years past, the chemical composition of materials was necessarily determined by taking a sample to a laboratory and performing the required chemical tests, usually with somewhat tedious procedures. Today, many important measurements of this type are made on a relatively continuous and automatic basis without the need for a human operator. The need for such measuring systems is due largely to the desire to automatically control product quality directly in terms of its chemical composition, rather than inferring it from measurements of temperature, pressure, flow rate, etc. Even in manually controlled situations, the desire to increase production rates while also maintaining or improving quality leads to a need for rapid analysis methods. Rapid and accurate analysis techniques are very useful in research and development problems. These needs of industry have led to the development of a wide variety of instruments for measuring various aspects of chemical composition and related quantities.

Some examples of measurements of the above type include analysis of products of combustion, monitoring of the composition of dissolved gases in oil-well drilling mud, detection of alcohol contaminant in a heavy-hydrocarbon liquid stream, detection of explosive solvents in the atmosphere within a uranium-extraction kettle, measurement of pH of industrial-waste effluent to control river pollution, determination of alloying constituents in metals, outgassing of materials under high vacuum, use of rocketborne instruments for analyzing atmospheric gases at high altitudes, air-pollution studies, and analysis of anesthetic gases in blood. Most of the analyzers used for such measurements are rather complex systems, rather than simple sensors, and a discussion of these methods adequate for their selection and use is beyond the scope of this text. The references[1] serve this purpose for the interested reader.

There are, however, some simple sensors which do perform useful "chemical analysis," and now we describe one such from the important area of combustion control. Both industrial furnaces and automotive engine-control systems make

---

[1] D. M. Considine (ed.), "Process Instruments and Controls Handbook," sec. 6, McGraw-Hill, New York, 1957; J. M. Jutila, Guide to Selecting Gas and Liquid Chromatographs, *InTech*, pp. 24–31, August 1980.

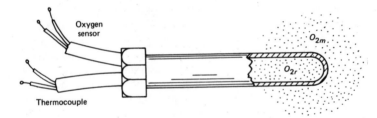

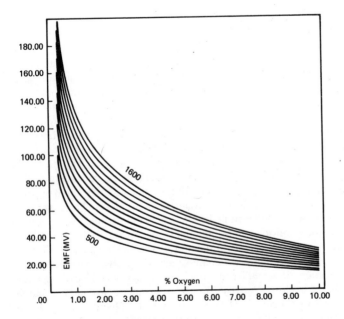

Figure 9.6 Zirconium-type oxygen sensor.

use of oxygen sensors of the zirconium oxide ceramic type.[1] The sensor consists of a closed-end hollow tube (similar in size and shape to a thermometer well) made of zirconium oxide ceramic with porous platinum electrodes coated on both the inner and outer surfaces. Above about 500°C the sensor becomes permeable to oxygen ions, and a solid electrolyte effect produces a voltage e_o milli-

[1] D. S. Howarth and R. V. Wilhelm, Jr., A Zirconia-Based Lean Air-Fuel Ratio Sensor, *SAE Paper* 780212, 1978; J. A. Brothers and F. Hirschfeld, Some New Applications for Zirconia Sensors, *Mech. Eng.*, pp. 35–37, November 1980; Oxygen Monitoring Systems, App. Eng. Handbook, Dynatron, Inc., Wallingford, Conn.

volts between the two electrodes, given by the Nernst equation

$$e_o = 0.0215T \ln \frac{O_{2r}}{O_{2m}} \tag{9.7}$$

where T is sensor temperature in kelvins, O_{2r} is the oxygen concentration in percent of a reference gas (often atmospheric air, $O_{2r} = 20.9$ percent) at the sensor inside surface, and O_{2m} is the oxygen concentration of the measured gas at the sensor outside surface. Sensors normally include a built-in RTD or platinum/rhodium thermocouple since T must be known to relate e_o to O_{2m}. Figure 9.6 graphs Eq. (9.7); accuracy is typically ± 0.1 percent of oxygen when the concentration is 2 percent, and response time is about 5 s.

PROBLEMS

9.1 Derive Eq. (9.1).

9.2 Prove that a plot of $A_o \sin(\omega t + \phi)$ against $A_i \sin \omega t$ is an ellipse.

9.3 Analyze the system of Fig. 9.4a to obtain the transfer function relating liquid level h_i to float motion x_o. Neglect dry-friction effects. If the liquid level increases very slowly and the float motion is subject to a dry-friction force F_f, develop a formula to estimate the maximum steady-state error.

9.4 Assume the force transducer in Fig. 9.4b is of the elastic deflection type, and obtain the transfer function relating liquid level h_i to force-transducer deflection x_o.

9.5 Discuss the effect of liquid density changes on the accuracy of liquid-level measurement in the systems of Fig. 9.4a and b.

9.6 For the system of Fig. 9.4a, discuss the effect of static and dynamic behavior of using a float that is a body of revolution but *not* a cylinder.

9.7 Repeat Prob. 9.6 for the system of Fig. 9.4b.

9.8 Discuss interfering and/or modifying inputs for the system of Fig. 9.4d. Assume the pressure pickup itself to be insensitive to such inputs.

9.9 Repeat Prob. 9.8 for the system of Fig. 9.4e.

9.10 For the system of Fig. 9.4k, make a study of the error caused by the weight of the measuring wire.

BIBLIOGRAPHY

1. S. J. Goldwater: Phase-Angle Measurement in Control Systems, *Trans. Soc. Instrum. Tech.* (*London*), p. 100, June 1960.
2. R. J. A. Paul and M. H. McFadden: Measurement of Phase and Amplitude at Low Frequencies, *Electron. Eng.*, vol. 31, no. 373, March 1959.
3. F. J. Huddleston: Frequency Response by Sum or Difference, *Contr. Eng.*, p. 113, October 1957.
4. Timers, *Electromech. Des.*, p. 51, March 1961.
5. Electric Timing Motors, *Electromech. Des.*, p. 59, May 1961.
6. Electronic Tuning Fork Beats Time for Accuracy, *Mach. Des.*, p. 30, October 27, 1960.
7. Time Interval Measurement, *Instrum. Contr. Syst.*, p. 125, September 1962.

8. P. Young: 1 Nanosecond Time Interval Counter, *Instrum. Contr. Syst.*, p. 105, January 1965.
9. A MacMullen: Sources of Error in Phase Measurement, *Instrum. Contr. Syst.*, p. 91, January 1965.
10. A New Approach to Precision Time Measurements, *Gen. Radio Exp.*, General Radio Co., West Concord, Mass., February–March 1965.
11. Correlating Time from Europe to Asia with Flying Clocks, *Hewlett-Packard J.*, Hewlett-Packard Co., Palo Alto, Calif., April 1965.
12. Level Measurement and Control, *Instrum. Contr. Syst.*, p. 148, March 1965.
13. N. Z. Alcock and S. K. Ghosh: Minimizing Measurement Errors in Nuclear Gages, *Contr. Eng.*, p. 87, May 1961.
14. F. W. Hannula: Use Capacitance for Accurate Level Measurement, *Contr. Eng.*, p. 104, November 1957.
15. R. C. Muhlenhaupt and P. Smelser: Carbon Resistors for Cryogenic Liquid Level Measurement, *NBS, Tech. Note* 200, 1963.
16. W. A. Olsen: An Integrated Hot Wire–Stillwell Liquid Level Sensor System for Liquid Hydrogen and Other Cryogenic Fluids, *NASA, Tech. Note* D-2074, 1963.
17. Liquid Hydrogen Level Sensors, *Instrum. Contr. Syst.*, p. 129, May 1964.
18. R. L. Rod: Propellant Gaging and Control, *Instrum. Contr. Syst.*, p. 119, October 1962.
19. G. H. Burger: Reliable Level Measurements for Liquid Metals, *Contr. Eng.*, p. 131, July 1959.
20. F. Marton: Level Measurement and Control, *Instrum. Contr. Syst.*, p. 107, January 1965.
21. E. Ulicki: Propellant Gaging System for Apollo Spacecraft, *Space/Aeron.*, p. 68, October 1964.
22. D. D. Kana: A Resistive Wheatstone Bridge Liquid Wave Height Transducer, *NASA, CR*-56551, 1964.
23. L. Siegel: Nuclear and Capacitance Techniques for Level Measurement, *Instrum. Contr. Syst.*, p. 129, July 1964.
24. N. H. Roos: Level Measurement in Pressurized Vessels, *ISA J.*, p. 55, May 1963.
25. Wire Matrix Gages Zero-g Liquids, *Mach. Des.*, p. 10, February 16, 1961.
26. C. L. Pleasance: Accurate Volume Measurement of Large Tanks, *ISA J.*, p. 56, May 1961.
27. Moisture and Humidity, *Instrum. Contr. Syst.*, p. 121, October 1964.
28. D. J. Fraade: Measuring Moisture in Gases, *Instrum. Contr. Syst.*, p. 100, April 1963.
29. R. E. Fishburn: Measurement and Control of Humidity, *Automation*, p. 61, January 1963.
30. R. M. Atkins: Wet/Dry Bulb Thermistor Hygrometer with Digital Indication, *Instrum. Contr. Syst.*, p. 111, April 1964.
31. H. Hellivig: Frequency Standards and Clocks: A Tutorial Introduction, *NBS, Tech. Note* 616, 1972.
32. E. H. Schulte: Carbon Resistors for Multipoint Level Sensing, *Cryog. Tech.*, September/October 1970.
33. B. E. Dozer: Liquid Level Measurement for Hostile Environment, *Inst. Tech.*, February 1967.
34. O. W. Schoen: A Continuously-Variable Humidity Reference, *Inst. & Cont. Syst.*, October 1972.
35. Survey of Humidity and Moisture Instrumentation, *Inst. & Cont. Syst.*, January 1972.
36. R. E. Ruskin (ed.): "Principles and Methods of Measuring Humidity in Gases," vol. 1 of "Humidity and Moisture," Reinhold, New York, 1965.
37. Dew Point Hygrometer Error Analysis, *Appl. Data* 3-051, EG & G Environmental Equip. Div., Waltham, Mass.
38. P. L. Mariam: Measuring Level in Hostile or Corrosive Environments, *Inst. Tech.*, pp. 45–47, April 1979.

THREE

MANIPULATION,
TRANSMISSION,
AND RECORDING
OF DATA

MANIPULATING, COMPUTING, AND COMPENSATING DEVICES

The information or data generated by a basic measuring device generally require "processing" or "conditioning" of one sort or another before they are presented to the observer as an indication or a record. Devices for accomplishing these operations may be specific to a certain class of measuring sensors, or they may be quite general-purpose. In this chapter we consider those devices most often needed in building up measurement systems.

10.1 BRIDGE CIRCUITS

Bridge circuits of various types are employed widely for the measurement of resistance, capacitance, and inductance. Since we have seen that many transducers convert some physical variable to a resistance, a capacitance, or an inductance change, bridge circuits are of considerable interest. While capacitance and inductance bridges are important, the simpler resistance bridge is in widest use, and so we concentrate on it here. Adequate technical literature on all types of bridge circuits is readily available.[1]

Figure 10.1 shows a purely resistive (Wheatstone) bridge in its simplest form. The excitation voltage E_{ex} may be either ac or dc; here we consider only dc. In measurement applications, one or more of the legs of the bridge is a resistive transducer such as a strain gage, resistance thermometer, or thermistor. The basic principle of the bridge may be applied in two different ways: the null method and

---

[1] E. Frank, "Electrical Measurement Analysis," chaps. 10 and 13, McGraw-Hill, New York, 1959.

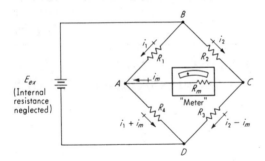

Figure 10.1 Basic Wheatstone bridge.

the deflection method. Let us assume that the resistances have been adjusted so that the bridge is balanced; that is, $e_{AC} = 0$. (It is easily shown that this requires $R_1/R_4 = R_2/R_3$.) Now we let one of the resistors, say R_1, change its resistance. This will unbalance the bridge, and a voltage will appear across AC, causing a meter reading. The meter reading is an indication of the change in R_1 and actually can be utilized to compute this change. This method of measuring the resistance change is called the *deflection method*, since the meter deflection indicates the resistance change. In the *null method*, one of the resistors is adjustable manually. Thus if R_1 changes, causing a meter deflection, R_2 can be adjusted manually until its effect just cancels that of R_1 and the bridge is returned to its balanced condition. The adjustment of R_2 is guided by the meter reading; R_2 is adjusted so that the meter returns to its null or zero position. In this case the numerical value of the change in R_1 is related directly to the change in R_2 required to effect balance.

Both the null and deflection methods are employed in practice. In the deflection method, a calibrated meter is needed, and if the excitation E_{ex} changes, an error is introduced, since the meter reading is changed by changes in E_{ex}. With the null method, a calibrated variable resistor is needed, and since there is no meter deflection when the final reading is made, no error is caused by changes in E_{ex}. The deflection method gives an output voltage across terminals AC that almost instantaneously follows the variations of R_1. This output voltage can be applied to an oscilloscope (rather than the meter shown in Fig. 10.1), and thus measurements of rapid dynamic phenomena are possible. The null method, however, requires that the balancing resistor be adjusted to null the meter before a reading can be taken. This adjustment takes considerable time if done manually; even when an instrument servomechanism makes the adjustment automatically, the time required is much longer than is allowable for measuring many rapidly changing variables. Thus the choice of the null or the deflection method in a given case depends on the speed of response, drift, etc., required by the particular application.

In order to obtain quantitative relations governing the operation of the bridge circuit, a circuit analysis is necessary. The following information is desired:

1. What relation exists among the resistances when the bridge is balanced ($e_{AC} = 0$)? The answer has been given as $R_1/R_4 = R_2/R_3$.
2. What is the sensitivity of the bridge? That is, how much does the output voltage e_{AC} change per unit change of resistance in one of the legs?
3. What is the effect of the meter internal resistance on the measurement?

We consider the question of bridge sensitivity first for the case where the "meter" has a very high internal resistance R_m, compared with the bridge resistances. If this is the case, the meter current i_m will be negligible compared with the currents in the legs. Often this situation is closely approximated in practice since most voltage-measuring instruments (digital voltmeters, oscilloscopes, chart recorders, etc.) have input amplifiers with 1-MΩ or more input resistance.

For $i_m = 0$ we have

$$i_1 = \frac{E_{ex}}{R_1 + R_4} \tag{10.1}$$

$$i_2 = \frac{E_{ex}}{R_2 + R_3} \tag{10.2}$$

$$e_{AB} = \text{voltage rise from } A \text{ to } B = i_1 R_1 = \frac{R_1}{R_1 + R_4} E_{ex} \tag{10.3}$$

$$e_{CB} = \frac{R_2}{R_2 + R_3} E_{ex} \tag{10.4}$$

and finally

$$e_{AC} = e_{AB} + e_{BC} = e_{AB} - e_{CB} = \left(\frac{R_1}{R_1 + R_4} - \frac{R_2}{R_2 + R_3} \right) E_{ex} \tag{10.5}$$

Thus we see that the output voltage is a linear function of the bridge excitation E_{ex} but, in general, a *nonlinear* function of resistances R_1, R_2, R_3, and R_4. If the bridge is balanced initially and then R_1, say, begins to change, the output voltage signal will *not* be directly proportional to the change in R_1. For certain practically important special cases, however, perfect linearity is possible. The best example is found in many strain-gage transducers in which, at the balanced condition, $R_1 = R_2 = R_3 = R_4 = R$. Also, the resistance changes are such that $+\Delta R_1 = -\Delta R_2 = +\Delta R_3 = -\Delta R_4$. Then we may write

$$e_{AC} = \left[\frac{R_1 + \Delta R_1}{(R_1 + \Delta R_1) + (R_4 + \Delta R_4)} - \frac{R_2 + \Delta R_2}{(R_2 + \Delta R_2) + (R_3 + \Delta R_3)} \right] E_{ex} \tag{10.6}$$

$$e_{AC} = \frac{\Delta R_1}{R} E_{ex} \tag{10.7}$$

Clearly, Eq. (10.7) shows a strictly linear relationship between e_{AC} and ΔR_1. Strict linearity also is obtained for R_2 and R_3 fixed and $\Delta R_1 = -\Delta R_4$.

Even when the above symmetry does not exist, the bridge response is very nearly linear as long as the ΔR's are small percentages of the R's. In strain gages, for example, the ΔR's rarely exceed 1 percent of the R's. Since the case of small ΔR's is of practical interest, we work out an expression for bridge sensitivity that is a good approximation for such a situation. From Eq. (10.5), $e_{AC} = f(R_1, R_2, R_3, R_4)$, and thus for small changes from the null condition we may write

$$\Delta e_{AC} = e_{AC} \approx \frac{\partial e_{AC}}{\partial R_1} \Delta R_1 + \frac{\partial e_{AC}}{\partial R_2} \Delta R_2 + \frac{\partial e_{AC}}{\partial R_3} \Delta R_3 + \frac{\partial e_{AC}}{\partial R_4} \Delta R_4 \qquad (10.8)$$

Now,
$$\frac{\partial e_{AC}}{\partial R_1} = E_{ex} \frac{R_4}{(R_1 + R_4)^2} \quad \text{V}/\Omega \qquad (10.9)$$

$$\frac{\partial e_{AC}}{\partial R_2} = -E_{ex} \frac{R_3}{(R_2 + R_3)^2} \qquad (10.10)$$

$$\frac{\partial e_{AC}}{\partial R_3} = E_{ex} \frac{R_2}{(R_2 + R_3)^2} \qquad (10.11)$$

$$\frac{\partial e_{AC}}{\partial R_4} = -E_{ex} \frac{R_1}{(R_1 + R_4)^2} \qquad (10.12)$$

The partial derivatives are taken as constants; thus Eq. (10.8) shows a linear relation between e_{AC} and the ΔR's.

We explained above, in a qualitative fashion, that if the meter resistance is "high enough," terminals AC may be thought of as an open circuit (no current i_m). It would be useful to have a more quantitative method of deciding whether the meter resistance was "high enough" and, if it were not, how to correct for it. We do this now.

By using Thévenin's theorem, the bridge circuit and the "meter" that loads it may be represented as in Fig. 10.2. Since we have been calling the bridge output voltage under assumed open-circuit conditions e_{AC}, this becomes the E_o of Fig. 3.22. Let us call the bridge output under the actual loaded condition e_{ACL}. Immediately we can write

$$i_m = \frac{e_{AC}}{R_{\text{total}}} = E_{ex} \frac{R_1/(R_1 + R_4) - R_2/(R_2 + R_3)}{R_m + R_1 R_4/(R_1 + R_4) + R_2 R_3/(R_2 + R_3)} \qquad (10.13)$$

Knowing i_m, we can compute the actual voltage e_{ACL} across the meter under the condition where the meter draws current, since the voltage across the meter will be the product of the current i_m and the meter resistance R_m. Carrying this out and simplifying, we get

$$e_{ACL} = \frac{E_{ex}(R_1 R_3 - R_2 R_4)}{(R_1 + R_4)(R_2 + R_3) + [(R_1 + R_4)R_2 R_3 + R_1 R_4(R_2 + R_3)]/R_m} \qquad (10.14)$$

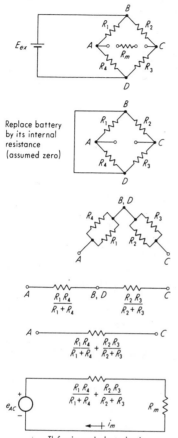

Replace battery
by its internal
resistance
(assumed zero)

Thévenin equivalent circuit

Figure 10.2 Thévenin analysis of bridge.

Now

$$e_{AC} = \frac{E_{ex}(R_1 R_3 - R_2 R_4)}{(R_1 + R_4)(R_2 + R_3)} \qquad (10.15)$$

and if we wish to display the effect of the meter resistance on the bridge output voltage, we can form the radio of e_{ACL} to e_{AC}. After some manipulation, this can be shown to be

$$\frac{e_{ACL}}{e_{AC}} = \frac{1}{1 + (1/R_m)[R_2 R_3/(R_2 + R_3) + R_1 R_4/(R_1 + R_4)]} \qquad (10.16)$$

Now we have a quantitative way of assessing the effect of the meter resistance R_m on the bridge output. We see that if $R_m = \infty$, then $e_{ACL} = e_{AC}$, as expected. If R_m is not infinite, there will be a *reduction* in the output signal, and the magnitude of this reduction depends on the relative values of R_m and the bridge "equivalent resistance" R_e, which is defined as

$$R_e \triangleq \frac{R_2 R_3}{R_2 + R_3} + \frac{R_1 R_4}{R_1 + R_4} \tag{10.17}$$

In terms of R_e, Eq. (10.16) becomes

$$\frac{e_{ACL}}{e_{AC}} = \frac{1}{1 + R_e/R_m} \tag{10.18}$$

Thus, if $R_m = 10R_e$,

$$\frac{e_{ACL}}{e_{AC}} = \frac{1}{1.1} = 0.91 \tag{10.19}$$

and there is a 9 percent loss in signal because of the noninfinite meter resistance. This type of loss usually is referred to as a *loading effect*; that is, the meter "loads down" the bridge and reduces its sensitivity.

The theory developed above is useful in assessing the effects of various parameters on the bridge sensitivity and actually could be utilized to compute the sensitivity if all quantities were known exactly. It is preferable, however, to calibrate the bridge directly by introducing a known resistance change and noting the effect on the bridge output. Often this known resistance change is introduced by means of the arrangement shown in Fig. 10.3. The resistance R_c of the calibrating resistor is known accurately. If the bridge is balanced originally with the switch open, when the switch is closed, the resistance in leg 1 will change and the bridge will be unbalanced. The output voltage e_{AC} is read on the meter, and the resistance change ΔR that caused this voltage is computed from

$$\Delta R = R_1 - \frac{R_1 R_c}{R_1 + R_c} \tag{10.20}$$

The bridge sensitivity is then

$$S \triangleq \frac{e_{AC}}{\Delta R} \quad V/\Omega \tag{10.21}$$

This procedure gives an overall calibration, since the values of all the resistors and the battery voltage are taken into account.

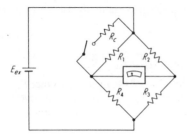

Figure 10.3 Shunt calibration method.

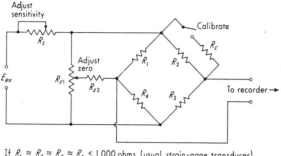

If $R_1 \approx R_2 \approx R_3 \approx R_4 < 1,000$ ohms (usual strain-gage transducer),
then $R_{z2} \approx 100 \ R_1$
$R_{z1} \approx 25,000$ ohms

Figure 10.4 Bridge with sensitivity, balance, and calibration features.

Figure 10.1 shows a bridge circuit with the bare essentials. Often additional features are necessary or desirable for the convenience of the user. Figure 10.4 shows a versatile arrangement providing the following capabilities:

1. Variation of overall sensitivity without the need to change E_{ex}
2. Provision for adjusting the output voltage to be precisely zero when the measured physical quantity is zero, even if the legs are not exactly matched
3. Shunt-resistor calibration

Commercial transducers also may include additional temperature-sensitive resistors to achieve temperature compensation (see Sec. 5.3).

10.2 AMPLIFIERS

Since the electrical signals produced by most transducers are at a low voltage and/or power level, often it is necessary to amplify them before they are suitable for transmission, further analog or digital processing, indication, or recording. While our discussion is aimed mainly at *users* (rather than designers) of amplifiers, the use of operational amplifiers in the construction of some simple "homemade" devices has become practical for nonspecialists in electronics, and we present some material along these lines.

Operational Amplifiers[1]

The operational amplifier (op-amp) is the most widely utilized analog electronic subassembly; it is the basis of instrumentation amplifiers, filters, and myriad analog and digital data processing equipment. We are not concerned with the

[1] G. B. Clayton, "Operational Amplifiers," 2d ed. Butterworth, London, 1979.

internal electronic details of the op-amp (these are of interest mainly to designers of op-amps); we simply accept on faith certain physical assumptions about their behavior, relying on op-amp designers for the validity of these assumptions. Details on op-amps are available from many sources; we draw heavily on manufacturers' handbooks and catalogs.[1] For the op-amp of Fig. 10.5a, e_A and e_B are the input voltages, i_A and i_B are called bias currents, Z_D is the differential input impedance, A is the open-loop gain (amplification), Z_o is the output impedance, V_{os} is the offset voltage, and $\pm V_s$ are the power-supply voltages. In establishing the basic operation of devices that employ op-amps, usually we utilize an ideal op-amp model which neglects certain real-world characteristics in the interest of simple analysis and understanding. Once this basic operation is established, the deviations from perfection can be studied by using a more correct (and more complicated) model. These ideal and real-world models are compared usefully as follows:

Characteristic	Ideal value	Typical real-world value
Open-loop gain A	∞	100,000 V/V
Offset voltage V_{os}	0	± 1 mV @ 25°C
Bias currents i_A, i_B	0	10^{-6} to 10^{-14}
Input impedance Z_D	∞	10^5 to 10^{11} Ω
Output impedance Z_o	0	1 to 10 Ω

Our ideal model also assumes instantaneous response (flat frequency response from dc to infinite frequency). Some op-amps are designed for low-frequency use only, others extend well into the megahertz range; so usually we can select a model whose dynamic behavior is close to the ideal, at least for the frequency range of interest. The op-amp simplified model of Fig. 10.5b assumes the ideal values for all the parameters and conventionally does not show the power supplies $\pm V_s$.

Assumption of the ideal behavior makes op-amp circuit analysis relatively simple, even for nonspecialists in electronics. Our first example, the *voltage-follower* of Fig. 10.5c, gives

$$e_o = (e_A - e_B - V_{os})A = (e_i - e_o - 0)\infty \tag{10.22}$$

$$\frac{e_o}{\infty} = 0 = e_i - e_o \qquad e_o = e_i \tag{10.23}$$

At first glance, a device that merely duplicates at its output a voltage which is applied to its input may not seem very useful. But note that the source (say a transducer) supplying e_i now works into an "infinite" impedance, and so no

[1] D. H. Sheingold (ed.), "Analog-Digital Conversion Handbook," pt. 3, chap. 1, Analog Devices, Norwood, Mass., 1972; Data Acquisition Components and Subsystems Catalog, Analog Devices, Norwood, Mass., 1980.

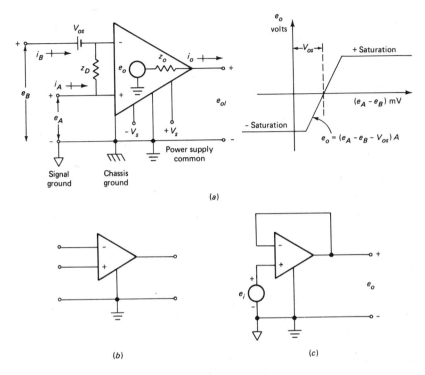

Figure 10.5 Op-amp terminology.

current or power is withdrawn from it. Yet e_o can be impressed across a load resistance (say a meter) which *does* draw current and consume power. Thus, while there is no voltage amplification, there is power amplification; such a device is sometimes called a *buffer amplifier*.

The output capabilities of a real op-amp are, of course, finite; typically, maximum e_o for proper operation might be specified at ± 10 V with ± 10-mA maximum current capability. Thus if we connected a resistive load of 1,000 Ω across e_o, both maximum voltage and current could be achieved. A 10,000-Ω load would reach 10 V at 1 mA, and a 100-Ω load would reach 10 mA at 1 V. Most op-amps are protected from short-circuits, so 0 Ω across e_o causes no damage; the output current merely saturates at its limiting value. To estimate the deviation of real-world behavior from the ideal, substitute the typical values from the table above into Eq. (10.22) to get $e_o = 0.9999e_i - 0.0009999$, clearly a very good approximation when $e_i \gg 1$ mV. When the output terminals at e_{ol} are open-circuit, the ideal output e_o and the "loaded" output voltage e_{ol} are identical, irrespective of the value of Z_o, since no output current flows. When e_{ol} is connected across a specific load impedance Z_l, then $e_{ol} = e_o - i_o Z_o$, causing another

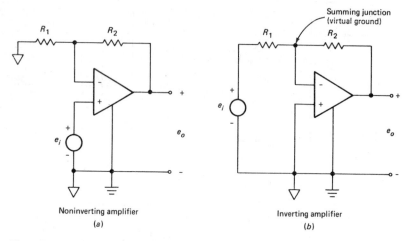

Figure 10.6 Noninverting and inverting op-amp configurations.

deviation from ideal behavior. For example, if e_i is 1.0 V, $Z_I = 1,000\ \Omega$, and $Z_o = 1\ \Omega$, then our earlier equation becomes

$$e_{ol} = (0.9999)(1.0) - 0.0009999 - \frac{0.9999 - 0.0009999}{1000 + 1}\ 1.0 = 0.9979\ \text{V} \quad (10.24)$$

which again compares favorably with the ideal value of 1.0.

When voltage amplification is desired, the inverting and noninverting configurations of Fig. 10.6 are basic. When the ideal model is employed, analysis gives for the noninverting amplifier

$$e_o = \left(\frac{R_2}{R_1} + 1\right)e_i \quad (10.25)$$

and for the inverting amplifier

$$e_o = -\frac{R_2}{R_1}\ e_i \quad (10.26)$$

Clearly, in both cases, e_o/e_i amplification (called *signal gain*) is determined entirely by the resistance ratio R_2/R_1. For the inverting amplifier, the input resistance presented to the source e_i is R_1, because e_i is effectively across R_1 (the summing junction is essentially at the same potential as the system ground). Analysis of the more correct model gives very nearly this same result. For the noninverting amplifier, input resistance is ideally infinite ($i_A \equiv 0$). However, analysis[1] of the more correct model gives

$$R_{in} = R_D\left(1 + A\ \frac{R_1}{R_1 + R_2}\ \frac{R_L}{R_o + R_L}\right) \quad (10.27)$$

[1] Clayton, op. cit., p. 22.

which can easily be $1{,}000R_D$, a very large value. In the noninverting case, since $e_A - e_B$ always must equal zero in the ideal model, the junction of R_1 and R_2 must be at voltage e_i at all times; in the inverting case, this junction (now called the *summing junction*) must be at 0 V since e_A is grounded, and for this reason it is called a *virtual ground*.

In Fig. 10.7*a*, the inverting configuration (generalized by addition of R_3 and R_4) is analyzed for voltage and current offset errors:

$$\frac{e_i - e_B}{R_1} + \frac{e_o - e_B}{R_2} - \frac{e_B}{R_4} - i_B = 0 \qquad (10.28)$$

$$e_B = e_A + V_{os} \qquad (10.29)$$

which lead to

$$e_o = -e_i \frac{R_2}{R_1} + (V_{os} + i_B R)\left(1 + \frac{R_2}{R_1} + \frac{R_2}{R_4}\right) \qquad \text{if } R_3 = 0 \qquad (10.30)$$

$$e_o = -e_i \frac{R_2}{R_1} + \underbrace{[V_{os} + (i_B - i_A)R]\left(1 + \frac{R_2}{R_1} + \frac{R_2}{R_4}\right)}_{\text{error}} \qquad \text{if } R_3 = R \quad (10.31)$$

$$\frac{1}{R} \triangleq \frac{1}{R_1} + \frac{1}{R_2} + \frac{1}{R_4} \qquad (10.32)$$

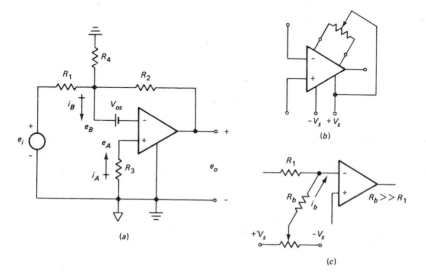

(a)

(b)

(c)

Figure 10.7 Offset-voltage and bias-current considerations.

Equation (10.31) shows why sometimes R_3 is added to the circuit: if R_3 is set equal to R, then the error due to bias currents is proportional to the *difference* current $i_B - i_A$, which often is about 10 times smaller than i_A or i_B. Resistance R_4 may or may not be present in a given application; it is included only for generality. While the addition of $R_3 = R$ may improve offset errors, many applications require additional efforts, and most op-amps provide a pair of terminals for connecting a manual balance pot (typically 10 kΩ), as in Fig. 10.7b. This adjustment nulls the effects of both offset voltages and currents, but the balance is gradually lost with time and/or temperature and must be retuned periodically.

When the offset is due mainly to bias current ($i_B R_1 > 4$ or 5 mV), the current injection bias circuit of Fig. 10.7c is preferred to the balance pot, since it causes less unbalance in the op-amp input stage. Some applications require use of both bias schemes.

Another basic configuration is the differential amplifier of Fig. 10.8a.[1] The input signal source in our example is a balanced grounded type such as found in a strain-gage bridge force transducer (Fig. 10.8b). At bridge null, e_x and e_y are both 5 V, and the amplifier output should be zero. When force application causes bridge unbalance, e_x goes to, say, 5.01 V and e_y to 4.99 V, and we want to amplify the 0.02-V difference. In the general circuit of Fig. 10.8a the original 5-V signal (common to both input paths) is called the *common-mode voltage e_{CM}*, while e_1 and e_2 represent the changes in e_x and e_y, respectively. Superimposing our earlier results for the inverting and noninverting configurations (allowed because of system linearity), we can write

$$e_o = -\frac{R_3}{R_1}(e_{CM} + e_1) + e_A \left(1 + \frac{R_3}{R_1}\right) \tag{10.33}$$

$$e_A = (e_{CM} + e_2)\frac{R_4}{R_2 + R_4} \tag{10.34}$$

which gives

$$e_o = -e_1 \frac{R_3}{R_1} + e_2 \frac{R_4}{R_2}\frac{R_3/R_1 + 1}{R_4/R_2 + 1} + e_{CM}\left(\frac{R_4}{R_2}\frac{R_3/R_1 + 1}{R_4/R_2 + 1} - \frac{R_3}{R_1}\right) \tag{10.35}$$

We see now that to eliminate common-mode errors due to e_{CM}, we must be careful to make $R_4/R_2 = R_3/R_1$, since then we achieve our original goal, i.e.,

$$e_o = \frac{R_3}{R_1}(e_2 - e_1) \tag{10.36}$$

Usually this balancing is done by making $R_1 = R_2$ and $R_3 = R_4$, where R_1 and R_2 must include not only the amplifier's resistors but also the source resistance (in our example, the strain-gage bridge resistance) of e_1 and e_2.

Equation (10.35) assumes the plus and minus amplifier input channels to

---

[1] Sheingold, op. cit., p. III-9.

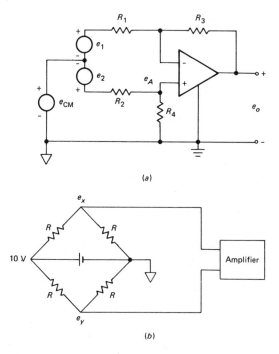

(a)

(b)

Figure 10.8 Differential input op-amp configuration.

have perfectly matched gains. A real amplifier deviates from this, and the degree of deviation is specified by the *common-mode rejection ratio* (CMRR):

$$\text{CMRR} \triangleq \frac{(e_{CM}) \text{ (differential gain)}}{e_o \text{ due to } e_{CM}} \tag{10.37}$$

For a CMRR of 80 dB (10,000), a differential gain [see Eq. (10.36)] of $R_3/R_1 = 100$, and an e_{CM} of 1 V, we would get a spurious output of 0.01 V even if $R_4/R_2 \equiv R_3/R_1$. Thus we get common-mode errors from *both* resistance mismatch and inherent amplifier-channel mismatch. These two effects may add or subtract, and sometimes intentional resistor mismatch is used to cancel amplifier common-mode error. Common-mode voltage e_{CM} is not limited to steady or intentionally applied effects like the bridge excitation of Fig. 10.8*b*; rather it is *any* signal that is applied equally to both differential input paths. Often it is a spurious noise voltage picked up by the circuitry from its surroundings. In any case, use of amplifiers with high common-mode rejection ratio and careful attention to symmetry in the lead-wire and amplifier impedances are helpful in reducing errors. The differential input resistance is $R_1 + R_2$, while the effective common-mode input resistances at e_1 and e_2 are $R_1 + R_3$ and $R_2 + R_4$, respectively; thus large R values are required for high input resistance. When these get

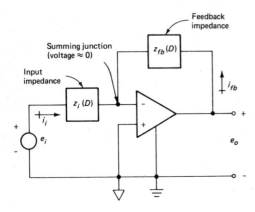

Figure 10.9 Op-amp generalized computing configuration.

too large, offset errors due to bias current and loss of common-mode rejection ratio at high frequencies (because of stray capacitance) limit use of this circuit; however, the *instrumentation amplifier* of the next section is available to meet such needs.

The inverting-type circuit with resistances replaced by generalized impedances $Z_i(D)$ and $Z_{fb}(D)$ is the basis of a wide variety of analog computing, filtering, and control devices (see Fig. 10.9):

$$i_i = \frac{e_i - o}{Z_i(D)} = -i_{fb} = -\frac{e_o - o}{Z_{fb}(D)} \tag{10.38}$$

$$\frac{e_o}{e_i}(D) = -\frac{Z_{fb}(D)}{Z_i(D)} \tag{10.39}$$

By proper choice of Z_{fb} and Z_i and by using combinations of resistors and capacitors (inductors are usually undesirable, not necessary, and rarely used), many practical e_o/e_i transfer functions may be synthesized (some are covered later in this chapter).

Instrumentation Amplifiers

Obviously, all kinds of amplifiers are utilized in instrumentation; however, a particular class and configuration conventionally is given the name *instrumentation amplifier*, mainly to distinguish it from the simpler op-amp circuits. Various designs[1] of instrumentation amplifier are in use; Fig. 10.10a shows a basic type[2] that utilizes three op-amps, while Fig. 10.10b shows a commercial unit together

---

[1] Sheingold, op. cit., p. III-33; Isolation and Instrumentation Amplifiers: Application, Theory, Selection, Analog Devices, Norwood, Mass.; Isolation and Instrumentation Amplifiers: Designer's Guide, Analog Devices, Norwood, Mass.

[2] Clayton, op. cit., p. 144.

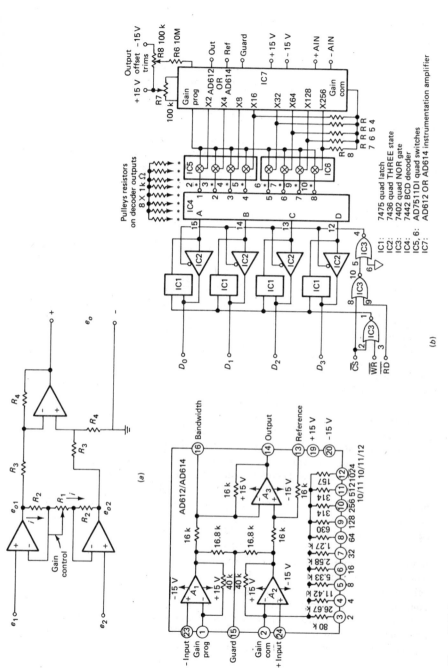

Figure 10.10 Instrumentation amplifiers.

with the circuitry needed to program its gain from a microprocessor.[1] You can construct instrumentation amplifiers, using op-amps as building blocks, or purchase them ready-made. Whereas op-amps are basic building blocks adaptable to a wide variety of unrelated uses, instrumentation amplifiers are "committed" devices of essentially fixed configuration intended for rather specific purposes. Their major characteristics include high common-mode rejection ratio and input impedance, low noise and drift, moderate bandwidth, and a limited range of gain (usually 1 to 1,000, programmable by a single resistor). In Fig. 10.10, since the op-amp inputs carry no current, a single current i flows from e_{o1} to e_{o2} through R_2, R_1, and R_2, giving

$$i = \frac{e_{o1} - e_1}{R_2} = \frac{e_1 - e_2}{R_1} = \frac{e_2 - e_{o2}}{R_2} \tag{10.40}$$

which leads to

$$e_{o1} - e_{o2} = (e_1 - e_2)\left(1 + 2\frac{R_2}{R_1}\right) \tag{10.41}$$

and finally

$$e_o = (e_2 - e_1)\frac{R_4}{R_3}\left(1 + \frac{2R_2}{R_1}\right) \tag{10.42}$$

Differential and common-mode input resistance are infinite ideally, with actual values of 10^9 Ω being readily available. A high common-mode rejection ratio (90 dB) is achieved even when source resistances are unbalanced by, say, 1,000 Ω.

Isolation amplifiers are a special subclass of instrumentation amplifiers intended for the most demanding applications, where low-level signals ride on top of high common-mode voltages; possibilities exist of troublesome ground disturbances and ground loops; processing circuitry must be protected from faults and power transients; interference from motors, power lines, etc., is heavy; and/or patient protection is important in biomedical applications. "Ordinary" instrumentation amplifiers require a return path for the bias currents. If this is not provided, these currents will charge stray capacitances, causing the output to drift excessively or to saturate. Therefore, when "floating" sources such as thermocouples are amplified, a connection to amplifier ground (Fig. 10.11a) must be provided; sometimes this leads to excessive noise problems. An isolation amplifier (Fig. 10.11b) does not require such a ground connection (the signal is now *isolated* from ground), and rejection of interfering noise may be much improved. A *two-port isolator* has its signal input circuit isolated from signal output and power input circuits. The ultimate in isolation is provided by *three-port isolators* which have, in addition, isolation between signal output and power input. To achieve isolation, two techniques are employed: transformer coupling and optical coupling. To allow signal transmission from input to output without a conductive connection, the transformer method uses a modulation/demodulation (carrier amplifier) scheme, in which the ac signals are transferred from input to output across a transformer. Opto-isolators transduce input voltage to proportional

---

[1] *Analog Dialogue*, vol. 15, no. 1, Analog Devices, Norwood, Mass., 1981.

light intensity by using LEDs. This light is transduced back to output voltage by light-sensitive diodes. Details of operation and comparison of the two methods are available in the literature.[1] Transformer isolated and "ordinary" instrumentation amplifiers may be compared[2] as follows:

Specification	Isolation amplifier	Instrumentation amplifier
Common-mode rejection ratio 5,000-Ω source unbalance, dc to 100 Hz	115 dB	80 dB
Common-mode voltage range	$\pm 2{,}500$ V dc ($\pm 7{,}500$ V peak)	± 10 V
Differential input voltage range	240 V rms ($\pm 6{,}500$ V peak)	± 10 V
Input-to-ground leakage	10^{11} Ω shunted by 10 pF	Feedback generated depends on linear circuit operation
Bias-current configuration	Single current; amplifier needs only two input conductors	Two currents; third wire needed for bias return
Small-signal bandwidth	dc to 2 kHz	dc to 1.5 MHz
Gain nonlinearity	0.05%	0.01%
Gain vs. temperature	$\pm 0.01\%/°C$	$\pm 0.0015\%/°C$
Offset vs. temperature	± 300 $\mu V/°C$	± 150 $\mu V/°C$

Figure 10.12[3] shows use of three isolation amplifiers (of a type that utilizes external modulator/demodulator drive signals) in a three-channel data acquisition system.

Noise Problems, Shielding, and Grounding[4]

In the progression from single-amplifier op-amp circuits to instrumentation and finally isolation amplifiers, capability for successfully dealing with noisy environments significantly improves, at the expense of greater cost, complexity, and size. Clearly, you should use any available methods to minimize noise effects so that the simplest and cheapest type of amplifier which provides satisfactory service can be utilized. Here we cover only briefly some of the main considerations,

[1] B. Morong, Isolator Stretches the Bandwidth of Two-Transformer Designs, *Electronics*, July 3, 1980.

[2] Isolation and Instrumentation Amplifiers: Designer's Guide, op. cit.

[3] Ibid.

[4] R. Morrison, "Grounding and Shielding Techniques in Instrumentation," Wiley, New York, 1977; D. H. Sheingold (ed.), "Transducer Interfacing Handbook," chap. 3, Analog Devices, Norwood, Mass., 1980; Elimination of Noise in Low-Level Circuits, Gould Inc., Cleveland, Ohio; Noise Control in Strain Gage Measurements, *TN*-501, Vishay Measurements Group, Raleigh, N.C., 1980.

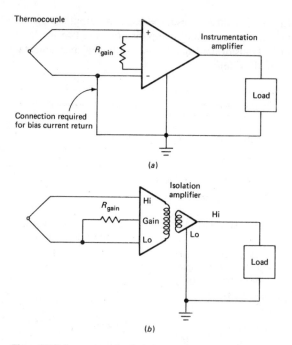

Figure 10.11 Instrumentation/isolation-amplifier comparison.

leaving details to the listed references. We begin by stating that transducers of low impedance generally cause less severe noise problems and should be employed in preference to high-impedance devices, when possible.

Gradual changes in amplifier offset voltages and bias currents resulting from temperature, time, and line voltage sometimes are called drift, rather than noise, but we include them here for completeness. If the temperature and/or line voltage can be kept more nearly constant, errors caused by these disturbances are, of course, reduced. Drift performance of amplifiers varies greatly from model to model; thus a close check of specifications allows a proper choice. Remember that any junction of dissimilar metals in a circuit is an unintentional thermocouple capable of introducing temperature-related errors. [An integrated-circuit (IC) amplifier, its socket, and a pair of binding posts connected to one intentional thermocouple can have 18 *unintentional* thermojunctions in series.] If frequency response to dc is not required, use of ac coupling can greatly reduce all drift effects. In digital data systems, often *automatic zero* techniques are utilized to compensate drift. Here, since data are already intermittent because of the sampling required in digital systems, the input can be shorted periodically (made zero) and the resulting output (which must be the total drift error) temporarily put in memory. Then the next data reading can be corrected for drift by adding the memorized error.

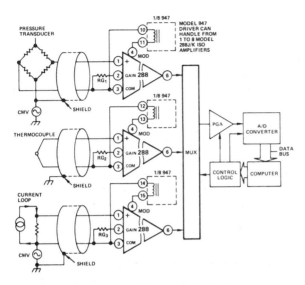

3 Channel Isolated Data Acquisition System

Figure 10.12 Isolation-amplifier application.

Inductive pickup, electrostatic pickup, and ground loops can cause large error voltages, often concentrated at the power-line frequency. The changing magnetic field that surrounds any conductor carrying ac current can link signal wires and produce large interfering voltages by inductive pickup. Because of the stray capacitance present between any adjacent conductors, a varying electric field can couple noise voltages to a signal circuit electrostatically. In solving (or preventing) such noise problems, start at the source by taking all measures possible to remove equipment such as power lines, motors, transformers, fluorescent lamps, relays, etc., from the near neighborhood of sensitive signal circuits, since the listed devices produce both inductive and electrostatic interference. If further measures are necessary, the use of an enclosing conductive shield (Fig. 10.13) can decrease electrostatic pickup. The shield functions by capturing charges that otherwise would reach the signal conductors. Once captured, these charges must be drained off to a satisfactory ground, or else they would be coupled to the signal conductors through the shield-to-cable capacitance.

Shielding of the above type is *not* practical for magnetically induced noise since magnetic shielding at low (60 Hz) frequencies requires thick (2.5 mm) shields of ferromagnetic metals. Rather, inductive noise is minimized by *twisting* the two signal conductors so that the "loop area" available for inducing error voltages is reduced and the mutual inductances between the noise source and each wire are balanced, to give a canceling effect. Commercially available cable provides twisted conductors, wrapped foil shields, and a grounding drain wire to meet the needs of both electrostatic and inductive noise reduction. Flexing of a cable can

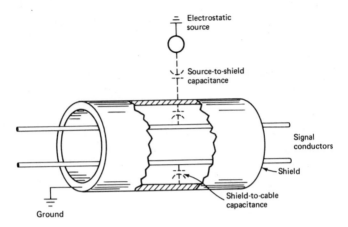

Figure 10.13 Electrostatic shielding.

generate noise as a result of rubbing friction within the cable itself (triboelectric effect), and so cables should be properly secured. When cable flexing is unavoidable, a construction employing a conductive tape between the primary insulation and the metal-foil shield reduces triboelectric noise considerably.

A *ground loop* is created by connecting a signal circuit to more than one ground. If the multiple "ground" points were truly at identical potentials, no problem would arise. However, the conductor that serves as ground (piece of heavy wire, metal chassis of instrument, ground plane of printed-circuit board, etc.) generally carries intentional or unintentional currents and has some resistance; thus two points some distance apart will *not* have identical potentials. If this potential difference produces current flows through the shield and/or signal circuit, then large noise voltages may occur. In Fig. 10.14a, two ground loops, one through the shield and the other through a signal wire, are caused by improper grounding practices. Current in the signal wire directly causes error voltages (because of wire resistance), while shield current couples voltages into the signal circuit through shield-to-cable capacitance. In Fig. 10.14b, the shield ground loop is broken simply by grounding the shield at only one point (the signal source), while use of a floating-input (isolated) amplifier breaks the other loop, thus greatly reducing noise pickup. If the amplifier has a floating internal shield ("guard shield" or "guard") surrounding its input section, then the cable shield is still grounded only at the signal source; but the amplifier end of the cable is connected to this guard, effectively extending it to the signal source. A final technique for noise control is filtering (see Sec. 10.3), but usually this can be utilized only if signal and noise occupy different frequency ranges.

Chopper, Chopper-Stabilized, and Carrier Amplifiers

In earlier generations of op-amps, chopper and chopper-stabilized designs were necessary when low drift was important, and these amplifier types were employed

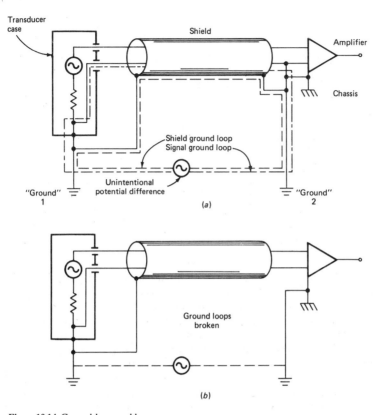

Figure 10.14 Ground-loop problems.

widely. Presently, the drift performance of "ordinary" op-amps has been much improved, and their simplicity, low cost, and wide bandwidth make them preferable to chopper-based designs in most cases. However, some extreme low-noise and low-drift applications still are best handled with chopper-type instruments; thus manufacturers continue to produce them, though in smaller quantities than earlier. A floating-input data amplifier design[1] based on a combination of a chopper-stabilized op-amp and an isolation amplifier with an op-amp front end has 200,000 gain, 100 nV/°C temperature drift, 5 μV/year time drift, 160 dB CMRR, and input noise of 1 μV, p-p in its design bandwidth of a few Hz.

Figure 10.15 shows the configuration of a chopper-type op-amp[2] designed for noninverting operation. The input signal is fed through a resistor to the MOSFET chopper (switch). When the MOSFET is off (high resistance), the error signal appears at the input to the ac-coupled amplifier. When the MOSFET

---

[1] Sheingold, ibid., pp. 226–228.
[2] Model 261, Analog Devices, Norwood, Mass.

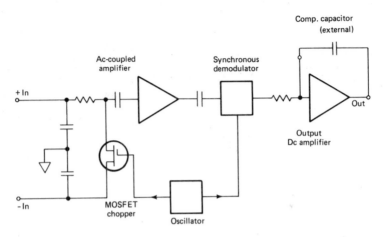

Figure 10.15 Chopper amplifier.

transistor is on (low resistance), the input to this amplifier is reduced to near zero. The difference between the on and off voltages at the amplifier is a square wave of amplitude slightly less than the error voltage. (A 3,500-Hz oscillator drives the chopper on and off and provides a synchronizing signal for the demodulator.) The ac-coupled amplifier amplifies the square-wave error signal; its output is ac-coupled to a synchronous demodulator which reconstructs the input signal. The drift of the input stage is not present in the demodulated signal, since it was not chopped by the input network. An output dc op-amp, connected as a low-pass filter with some additional gain, completes the system. For such an arrangement, offset and drift, referred to the *amplifier input*, are equal to the output dc amplifier stage input drift and offset divided by the ac-coupled amplifier gain. Typically, output-stage drift might be 100 μV/°C and ac gain is 1,000, which gives a system drift of 0.1 μV/°C, a very low value. As in any modulation/demodulation/filtering scheme, frequency response is limited by the modulation (chopping) frequency and the necessary low-pass output filter. Noninverting amplifiers built with this op-amp can have flat (-3 dB) frequency response to about 100 Hz, extremely high gain (100,000), and low noise (0.4 μV peak to peak, 0.01 to 1 Hz).

In a chopper-stabilized[1] (rather than chopper) op-amp, the input signal low-frequency components are diverted through a low-pass filter, amplified with a ("drift-free") chopper amplifier (gain $\approx$ 1,000), and then further amplified in an "ordinary" dc amplifier (gain $\approx$ 50,000). The unfiltered input signal is summed with the chopper amplifier output at the input of the final dc stage; thus low-frequency components go through both the chopper and dc amplifiers (total gain $\approx 5 \times 10^7$) while high-frequency go through only the dc (gain $\approx$ 50,000). Such a system combines the low-drift behavior of the chopper amplifier with the

[1] E. O. Doebelin, "Measurement Systems," Rev. Ed., pp. 619–621, McGraw-Hill, New York, 1975.

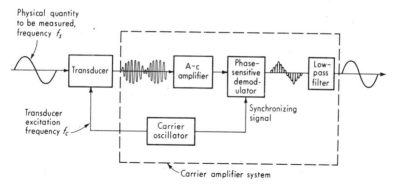

Figure 10.16 Carrier amplifier system.

wide bandwidth of the dc type, overcoming the frequency-response limitations of pure chopper amplifiers (response to several hundred kilohertz is possible).

When a transducer *requires* ac excitation (LVDTs are perhaps the most common example), the carrier amplifier system of Fig. 10.16 is still employed widely. When *either* ac or dc excitation is possible (strain-gage transducers, for example), the simplicity and bandwidth of dc amplifiers make dc excitation the preferred choice in most cases. However, there still are some applications in which the high gain, low drift, and good interference rejection of carrier systems are needed, so they are still available from several manufacturers. The final low-pass filter (needed to smooth the demodulator output), of course, sets the frequency-response limit of the entire system. A typical 5,000-Hz carrier system gives flat response to about 1,000 Hz.

Charge Amplifiers and Impedance Converters

Wide use of piezoelectric accelerometers, pressure pickups, and load cells has led to the development of an amplifier type that offers some advantages over the usual voltage amplifier in certain applications. Such a *charge amplifier* is shown connected to a piezoelectric transducer in Fig. 10.17. The idealized form is shown in Fig. 10.17a, where we note that a FET-input operational amplifier is used with a capacitor C_f in the feedback path. Assuming, as usual, that the input voltage e_{ai} and current i_a of the operational amplifier are small enough to take as zero, we get

$$K_q D x_i = -C_f D e_o \tag{10.43}$$

$$e_o = -\frac{K_q x_i}{C_f} \tag{10.44}$$

Equation (10.44) indicates that e_o would be instantaneously and linearly related to displacement x_i without the usual loss of steady-state response associated with piezoelectric transducers and voltage amplifiers. Unfortunately, this advantage is not realizable since a system constructed as in Fig. 10.17a would, because of nonzero op-amp bias current, exhibit a steady charging of C_f by the bias current until the amplifier saturated. To overcome this problem, in the practical circuit of Fig. 10.17b a feedback resistance R_f is included to prevent this small current from developing a significant charge on C_f. Analysis of this new circuit gives

$$\frac{e_o}{x_i}(D) = \frac{K\tau D}{\tau D + 1} \tag{10.45}$$

where
$$K \triangleq \frac{K_q}{C_f} \quad \text{V/in} \tag{10.46}$$

$$\tau \triangleq R_f C_f \quad \text{s} \tag{10.47}$$

Equation (10.45) is of identical form to the transfer function of a piezoelectric transducer and a *voltage* amplifier and exhibits the same loss of static and low-frequency response. The advantages of the charge amplifier are found in Eqs. (10.46) and (10.47). We note that both the sensitivity K and the time constant τ are now independent of the capacitance of the crystal itself and the connecting cable, whereas with a voltage amplifier neither of these advantages is obtained. Thus long cables (often several hundred feet in practical setups) do not result in a reduced sensitivity or a variation in frequency response. These advantages and others[1] are sufficient to make the charge amplifier of practical interest in many systems. Disadvantages[2] that may arise in certain applications include a possibly poorer signal-to-noise ratio and a reduction in natural frequency of the transducer because of a loss of stiffness, caused by what amounts to a short circuit across the crystal.

When utilized with quartz-crystal transducers,[3] the value of C_f is from 10 to 100,000, pF and R_f is 10^{10} to 10^{14} Ω. For $C_f = 100,000$ pF and $R_f = 10^{14}$ Ω, $\tau = 10^6$ s, showing that practically dc response, allowing static calibration and measurement, is possible under these conditions. For ceramic-type transducers, C_f is from 10 to 1,000 pF and R_f from 10^8 to 10^{10} Ω, making the maximum τ about 10 s and static measurements thus usually impractical.[4] Rather than the twisted, shielded pair recommended earlier for general-purpose low-level signal cables, usually piezoelectric transducers employ a miniature coaxial cable (Fig. 10.18a). Good noise rejection can be achieved by several means.[5] Figure 10.18b shows use of an insulated transducer mounting (to break the ground loop) with a case-grounded transducer and single-ended grounded amplifier. For accelerome-

[1] D. Pennington, Charge Amplifier Applications, Endevco Corp., Pasadena, Calif., April 1964.

[2] Wilcoxon Research, Bethesda, Md., *Res. Bull.* 5.

[3] Kistler Instrument Corp., Clarence, N.Y., *Tech. Notes* 133762 and 130662.

[4] Ibid.

[5] J. Wilson, Noise Suppression and Prevention in Piezoelectric Transducer Systems, *Sound & Vib.*, pp. 22–25, April 1979.

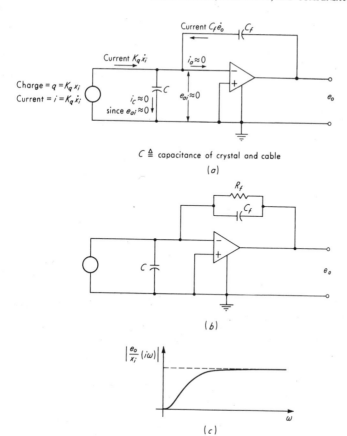

Figure 10.17 Charge amplifier.

ters, the insulated mounting (more compliant than "steel to steel") may not always be acceptable because of dynamic-response loss. Then a case-grounded transducer with grounded mounting may be necessary, and this requires a floating or differential charge amplifier if a noisy environment is present (Fig. 10.18c).

Another approach to piezoelectric signal conditioning is the *impedance converter* of Fig. 10.19. Here a field-effect transistor is connected directly across the crystal, presenting a very high impedance, but providing a large output (10 V) at a low impedance ($< 100 \ \Omega$). A single coaxial cable carries both power (from a 28-V battery) and readout signal. The FET device is small enough to be built into the transducer case, and such transducers are common where their lower allowable temperature is acceptable. No charge amplifier is needed; the output can go directly to conventional voltage amplifiers.

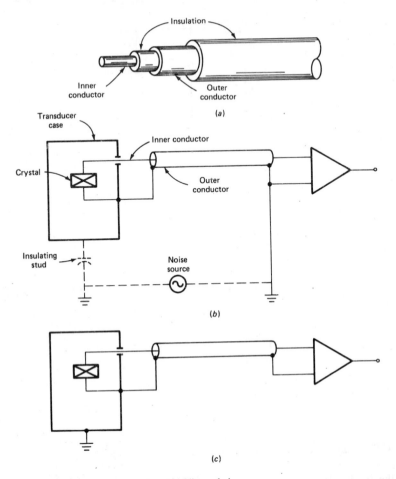

Figure 10.18 Piezoelectric transducer shielding techniques.

Concluding Remarks

The versatility of electronics ensures that most reasonable amplifier requirements can be met, though exotic conditions may require complex and expensive systems. Thus sensors, rather than electronics, usually limit overall system performance. The 5- or 10-mA current output of most instrumentation amplifiers is adequate except when high-current loads, such as oscillograph galvanometers, must be driven. Such needs ($\approx \pm 100$ mA) are met simply by adding a high-power output stage to the amplifier. Many amplifiers also include built-in filters with switch-selectable bandwidths, convenient for noise suppression.

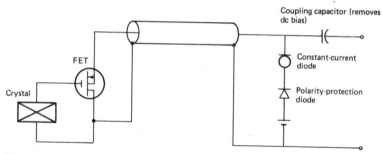

Figure 10.19 Piezoelectric transducer impedance converter.

10.3 FILTERS

The use of frequency-selective filters to pass the desired signals and reject spurious ones has been discussed. Figure 10.20 summarizes the most common frequency characteristics used. Filters may take many physical forms; however, the electrical form is most common and highly developed with regard to both theory and practical realization. By use of analogies, the material on electrical filters may suggest the configurations of mechanical, hydraulic, acoustical, etc., systems that will provide the desired filtering action in specific problems.

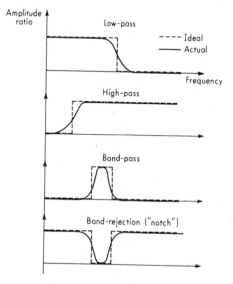

Figure 10.20 Basic filter characteristics.

Low-Pass Filters

The simplest low-pass filters commonly in use are shown in several different physical forms in Fig. 10.21. They all have identical transfer functions given by

$$\frac{e_o}{e_i}(D) = \frac{x_o}{x_i}(D) = \frac{p_o}{p_i}(D) = \frac{T_o}{T_i}(D) = \frac{1}{\tau D + 1} \qquad (10.48)$$

Since these are all simple first-order systems, the attenuation is quite gradual with frequency 6 dB/octave. This does not give a very sharp distinction between the frequencies that are passed and those that are rejected. By adding more "stages" (see Fig. 10.22a) the sharpness of cutoff may be increased. The use of inductance elements (Fig. 10.22b) also may lead to better filtering action. When a filter is inserted into a system, it is necessary to take into account possible loading effects by use of appropriate impedance analysis.

All the filters shown thus far are *passive* filters, since all the output energy must be taken from the input. Today many electrical filters are *active* devices

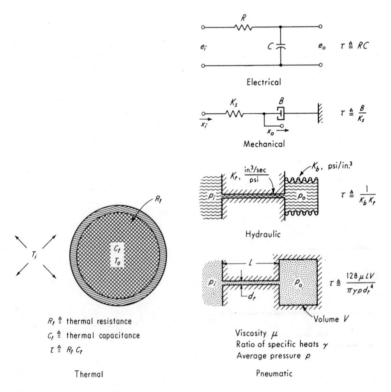

$R_t \triangleq$ thermal resistance
$C_t \triangleq$ thermal capacitance
$\tau \triangleq R_t C_t$

Thermal

Viscosity μ
Ratio of specific heats γ
Average pressure p

Pneumatic

Figure 10.21 Low-pass filters.

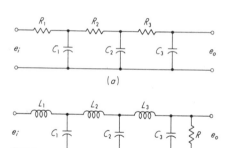

(a)

(b)

Figure 10.22 Sharper-cutoff low-pass filters.

based on op-amp technology. Whereas passive filters have very low noise, require no power supplies, and have a wide dynamic range, active filters[1] are much more adjustable and versatile, can cover very wide frequency ranges, have very high input and very low output impedances (which makes cascading and interconnection simple), and can be configured for simple switching from low-pass to high-pass and combination for band-pass or band-reject behavior.

Figure 10.23a shows an active second-order low-pass filter, while Fig. 10.23b illustrates a "state-variable filter" with electrically adjustable parameters.[2] The state-variable filter provides three simultaneous outputs: a low-pass, a high-pass, and a bandpass:

$$\frac{e_{lp}}{e_i}(D) = -\frac{1}{D^2/\omega_n^2 + 2\zeta D/\omega_n + 1} \tag{10.49}$$

$$\frac{e_{hp}}{e_i}(D) = -\frac{(100R^2C^2/V_c^2)D^2}{D^2/\omega_n^2 + 2\zeta D/\omega_n + 1} \tag{10.50}$$

$$\frac{e_{bp}}{e_i}(D) = -\frac{(20\zeta RC/V_c)D}{D^2/\omega_n^2 + 2\zeta D/\omega_n + 1} \tag{10.51}$$

These three outputs can be utilized individually or summed in various ways to obtain additional filtering effects. For example, if the low-pass and high-pass filters are summed (with appropriate coefficients), we can get a notch filter effect. The use of the multipliers (not necessary in an "ordinary" state-variable filter) allows adjustment of ω_n by simply applying the appropriate voltage at V_c. If digital control were desired, multiplying digital/analog converters could replace the analog multipliers.

Perhaps the most stringent requirements for sharp cutoff in low-pass filters are found in the antialiasing filters used in FFT signal and system analyzers (Sec.

[1] The Application of Filters to Analog and Digital Signal Processing, Rockland Systems Corp., Rockleigh, N.J., 1980; D. E. Johnson, J. R. Johnson, and H. P. Moore, "A Handbook of Active Filters," Prentice-Hall, Englewood Cliffs, N.J., 1980.

[2] D. H. Sheingold (ed.), Nonlinear Circuits Handbook, pp. 138–141, Analog Devices, Norwood, Mass., 1976.

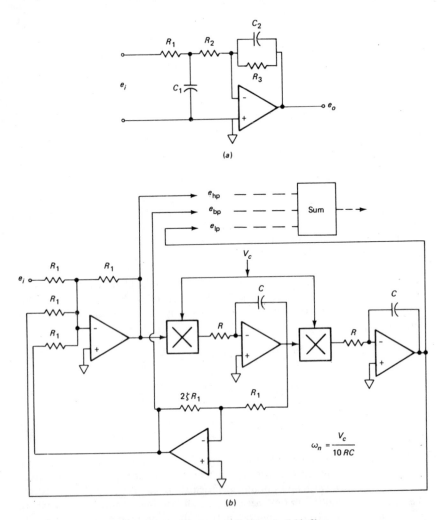

Figure 10.23 Active low-pass filter and voltage-controlled state-variable filter.

10.13). Here sophisticated active filters can provide 115 dB/octave rolloff, and pairs of filters [needed for two-channel (system) analyzers] are closely matched in phase ($\pm 1°$) and amplitude ratio (± 0.1 dB).

High-Pass Filters

Figure 10.24 shows the simplest passive high-pass filters, which all have the

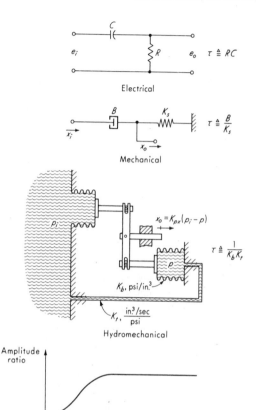

Figure 10.24 High-pass filters.

transfer function

$$\frac{e_o}{e_i}(D) = \frac{x_o}{x_i}(D) = \frac{x_o/K_{px}}{p_i}(D) = \frac{\tau D}{\tau D + 1} \qquad (10.52)$$

Again, the attenuation is quite gradual, and more complex passive or active configurations are needed to obtain a more sharply defined cutoff.

Bandpass Filters

By cascading a low-pass and a high-pass filter, we can obtain the bandpass characteristic (Fig. 10.25). To sharpen the rejection on either side of the passband, we can simply use the sharper low- and high-pass sections mentioned above or an active bandpass filter, such as that of Fig. 10.23b. Very sharply tuned

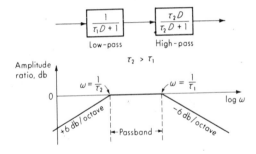

Figure 10.25 Bandpass filter.

bandpass filters (used in signal and system analyzers) are available by heterodyning or Fourier filtering.[1]

Band-Rejection Filters

A common application of a band-rejection filter is found in the input circuits of self-balancing potentiometer and XY recorders. These instruments are subject to interfering 60 Hz noise voltages. Since the frequency response of the overall recorder is good to only a few cycles per second, a band-rejection filter tuned to 60 Hz may be employed without distorting any desired signals. Such a filter prevents noise signals from saturating the recorder's amplifiers and distorting the proper amplification of the desired signals.

Passive networks commonly utilized for rejection of a band of frequencies include the bridged-T and the twin-T networks (Fig. 10.26). While the bridged-T does not completely reject any frequency, the twin-T can be so designed. Equations and charts for designing these filters are available.[2] One approach to active band-rejection filtering is discussed with Fig. 10.23b.

Digital Filters

As part of the general trend to replace analog electronics by digital wherever this is feasible and economic, digital filtering[3] is now quite common. Digital filtering can essentially duplicate any of the classical filter functions of Fig. 10.20 and can produce certain useful effects not possible in the analog domain. It also has the usual digital benefits of accuracy, stability, and adjustability by software (rather than hardware) changes. Digital filtering is an algorithm by which a sampled signal (or sequence of numbers), acting as an input, is transformed to a second sequence of numbers called the output. The algorithm may correspond to

[1] E. O. Doebelin, "System Modeling and Response," pp. 232–243, Wiley, New York, 1980.

[2] J. E. Gibson and F. B. Tuteur, "Control System Components," p. 43, McGraw-Hill, New York, 1958.

[3] R. E. Bogner and A. G. Constantinides (eds.), "Introduction to Digital Filtering," Wiley, New York, 1975.

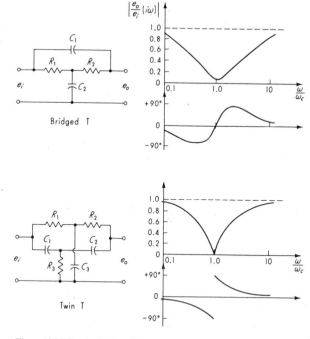

Figure 10.26 Band-rejection filters.

low-pass, high-pass, or other forms of filtering action. The output sequence can be employed in further digital processing or can be converted from digital to analog, producing a filtered version of the original analog signal. Since entire books are devoted to the theory of digital signal processing/filtering, here we give only a simple example.

A digital version of the simplest low-pass filter [Eq. (10.48)] can be found by converting the analog system's differential equation to a difference equation by using, say, rectangular numerical integration, as in Fig. 10.27:

$$N_o(nT) = N_o[(n-1)T] + \frac{T}{\tau}\{N_i[(n-1)T] - N_o[(n-1)T]\} \quad (10.52)$$

Here the sampling interval of the digital system is T seconds, and $N_o(nT)$ and $N_i(nT)$ are number sequences for the input and output, evaluated only at discrete-time points such as T, $2T$, $3T$, etc. Readers unfamiliar with difference equations may wish to do a sample calculation with Eq. (10.52) (use $T = 1$ s, $\tau = 5$ s) to verify the step-response behavior of Fig. 10.27. For digital filters the sampling frequency $1/T$ must be 2.5 or more times the highest frequency in the analog data to prevent aliasing problems.

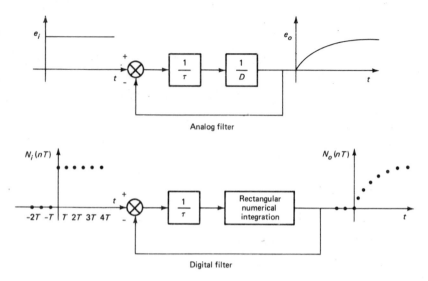

Analog filter

Digital filter

Figure 10.27 Digital low-pass filter.

A Hydraulic Bandpass Filter for an Oceanographic Transducer

While filtering with electrical networks is most common, in some instances other physical forms offer advantages. This section illustrates this idea with an example of a filter that has been successfully constructed and used.[1]

The Scripps Institution of Oceanography at La Jolla, California, employs pressure transducers in its studies of ocean-wave phenomena. A particular study required measurements of waves whose frequencies are lower than those of ordinary gravity waves (which one observes visually) and higher than those due to tides. These waves of intermediate frequency are of rather low amplitude relative to those due to tides and to gravity and thus are difficult to measure with a pressure pickup which treats all frequencies about equally. The bandpass filter and pressure pickup of Fig. 10.28 solves this problem since it is "tuned" to the frequency range of interest, which is about 0.001 Hz. Such low frequencies are very difficult to handle with electrical circuits, but the hydraulic filter shown gives very good results with quite simple and reliable components.

In use, the pressure transducer is located underwater, often buried in a foot of sand for temperature insulation, with a "snorkel" tube extending up through the sand to sense water pressure. This pressure is directly related to the height of the waves passing overhead; thus a record of pressure-transducer output voltage is a

[1] F. E. Snodgrass, Shore-Based Recorder of Low-Frequency Ocean Waves, *Trans. Am. Geophys. Union*, p. 109, February 1958.

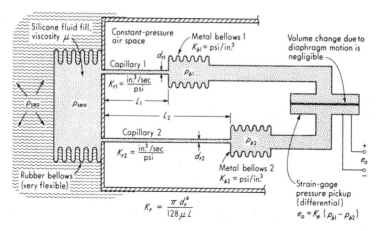

Figure 10.28 Hydraulic bandpass filter.

record of wave activity. Analysis of the system gives

$$\frac{e_o}{p_{sea}}(D) = \frac{K_e(\tau_2 - \tau_1) D}{(\tau_1 D + 1)(\tau_2 D + 1)} \tag{10.53}$$

where $K_e \triangleq$ sensitivity of differential pressure pickup [mV/(lb/in^2)] and

$$\tau_1 \triangleq \frac{1}{K_{t1} K_{b1}} \quad \text{s} \tag{10.54}$$

$$\tau_2 \triangleq \frac{1}{K_{t2} K_{b2}} \quad \text{s} \tag{10.55}$$

We can easily show that the frequency ω_p of peak response is given by

$$\omega_p = \sqrt{\frac{1}{\tau_1 \tau_2}} \quad \text{rad/s} \tag{10.56}$$

The amplitude ratio M_p at this frequency is

$$M_p = \frac{K_e(\tau_2 - \tau_1)}{\tau_2 + \tau_1} \tag{10.57}$$

and the phase angle at ω_p is zero.

Mechanical Filters for Accelerometers

While electrical filters at the *output* of piezoelectric accelerometers are in common use, mechanical filters at the *input* can protect the accelerometer from shock, prevent electrical saturation resulting from large, high-frequency acceleration

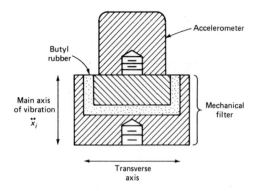

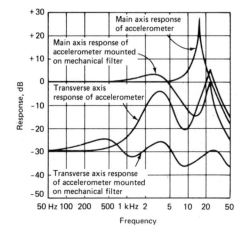

Figure 10.29 Mechanical filter for accelerometer.

when only low-frequency motions are of interest, and provide electrical isolation to break ground loops and thus reduce electrical noise pickup. Figure 10.29[1] shows construction (utilizing butyl rubber as both spring element and electrical insulation) and response characteristics of such filters.

Filtering by Statistical Averaging

All the filters mentioned above are of the frequency-selective type and, of course, require that the desired and spurious signals occupy different portions of the frequency spectrum. When signal and noise contain the same frequencies, such filters are useless. A basically different scheme may be employed usefully under

---

[1] B&K Instruments, Model UA 0559, Marlboro, Mass.

such circumstances if the following is true:

1. The noise is random.
2. The desired signal can be caused to repeat itself.

If these two conditions are fulfilled, it should be clear that if one adds up the ordinates of several samples of the total signal at like values of abscissa (time), the desired signal will reinforce itself while the random noise will gradually cancel itself. This will occur even if the frequency content of signal and noise occur in the same part of the frequency spectrum. It can be shown that the signal-to-noise ratio improves in proportion to the square root of the number of samples utilized. Thus theoretically the noise can be eliminated to any desired degree by adding a sufficiently large number of signals. In practice, various factors prevent realization of theoretically optimum performance.

10.4 INTEGRATION AND DIFFERENTIATION

Often in measurement systems it is necessary to obtain integrals and/or derivatives of signals with respect to time. Depending on the physical nature of the signal, various devices may be most appropriate. Generally accurate differentiation is harder to accomplish than integration, since differentiation tends to accentuate noise (which is usually high frequency) whereas integration tends to smooth noise. Thus second and higher integrals may be found easily while derivatives present real difficulties. For digital-computer signal processing, one of the many algorithms for numerical differentiation or integration (found in texts on numerical analysis) may be employed.

Integration. If the signal to be integrated is already a mechanical displacement or is easily transduced to one, the *ball-and-disk integrator* of Fig. 10.30 may be used. Assuming rigid bodies and no slippage, we can show that

$$\frac{\theta_o}{x_i}(D) = \frac{\omega_d}{R}\left(\frac{1}{D}\right) \qquad (10.58)$$

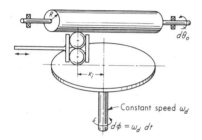

Figure 10.30 Ball-disk integrator.

and thus the rotation angle θ_o is proportional to the first time integral of the displacement x_i. A typical unit has a maximum ω_d of 500 r/min, $x_i = \pm 0.75$ in, maximum input force is 2 oz, output torque is 3 in · oz, reproducibility is 0.01 percent of full scale, accuracy is 0.05 percent of full scale, expected life is 10,000 h, and the unit uses a precision-lapped tungsten carbide disk with 1-μin surface finish and hardened-tool-steel roller and balls.

Two electromechanical means of obtaining integrals are shown in Fig. 10.31: the *integrating motor* and the *velocity-servo integrator*. Both accept electric signals as input and produce mechanical rotations in proportion to the time integral of the input voltage. The integrating motor is essentially a dc motor with permanent-magnet field in which friction, iron losses, and brush-contact voltage drop have been reduced to extremely low levels, resulting in an input-voltage/output-speed characteristic that is very linear over a wide range of input voltage. For a dc motor with constant field,

$$\text{Armature current} = i_a = \frac{e_i - e_m}{R} \tag{10.59}$$

where
$$e_m \triangleq \text{motor back emf} = K_e \dot{\theta}_o \tag{10.60}$$
$$R \triangleq \text{armature resistance} \tag{10.61}$$

Motor torque $T_m = K_{mt} i_a$, where K_{mt} is the motor-torque constant. Thus if rotor inertia is J, we have

$$T = J\ddot{\theta}_o \tag{10.62}$$

$$K_{mt} \frac{e_i - K_e \dot{\theta}_o}{R} = J\ddot{\theta}_o \tag{10.63}$$

$$\frac{\theta_o}{\int e_i \, dt}(D) = \frac{1/K_e}{\tau D + 1} \tag{10.64}$$

$$\tau \triangleq \frac{RJ}{K_{mt} K_e} \tag{10.65}$$

We see that the rotation angle θ_o (which can be counted by a simple mechanical counter) is a measure of the time integral of e_i with a first-order lag. A family of such instruments[1] has full-scale input voltage ranging from 1.5 to 24 V, R of 2.8 to 700 Ω, τ of about 0.01-s, full-scale speed of 1,885 to 1,260 r/min, and starting voltages of 4.2 to 79 mV. For a motor without any external load, the nonlinearity is better than 0.3 percent of full scale from 5 to 200 percent of full-scale voltage. These motors can be utilized only to drive very light loads, 1.8 to 12.4 g · cm at full voltage.

For greater accuracy and the ability to drive loads requiring greater power output, the velocity-servo integrator may be employed. Analysis of the block

---

[1] Electro Methods Ltd., Stevenage, England.

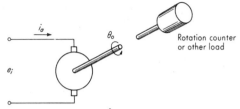

Integrating motor

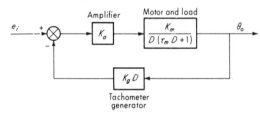

Velocity servo

Figure 10.31 Electromechanical integrators.

diagram of Fig. 10.31 gives

$$\frac{\theta_o}{\int e_i \, dt} (D) = \frac{1/K_g}{\tau D + 1} \tag{10.66}$$

where

$$\tau \triangleq \frac{\tau_m}{1 + K_a K_g K_m} \tag{10.67}$$

and $1/K_g \approx K_a K_m/(1 + K_a K_m K_g)$ since K_a is very large. Such integrators achieve accuracies of 0.1 percent and better.[1]

Figure 10.32 shows the op-amp type of integrator extensively employed in general-purpose electronic analog computers. By use of high-quality chopper-stabilized op-amps, an integrator of quite low drift can be constructed in this way. Accuracies of the order of 0.1 percent for short-term operation and 1 percent over 14 h are typical of high-quality electronic integrators of this type.[2] If higher integrals are desired, such units may be cascaded; however, drift becomes more troublesome. In addition to providing a closer approximation to true integration than the passive networks discussed in the following paragraph, the presence of the amplifier (with its own power supply) means that power can be supplied to the device following the integrator without taking any significant power from the device supplying the integrator. That is, op-amp circuits generally can have a high input impedance and low output impedance.

All the low-pass filters of Fig. 10.21 may be utilized as *approximate integra-*

---

[1] W. H. Barr, Integrators, *Electromech. Des.*, p. 57, October 1961.
[2] Ibid.

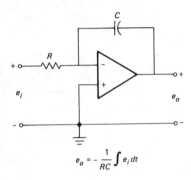

$$e_o = -\frac{1}{RC} \int e_i \, dt$$

Figure 10.32 Electronic integrator.

tors for input signals within a restricted frequency range. This can be shown as follows:

$$\frac{e_o}{e_i}(i\omega) = \frac{1}{i\omega\tau + 1} \tag{10.68}$$

Now if $\omega\tau \gg 1$,

$$\frac{e_o}{e_i}(i\omega) \approx \frac{1}{i\omega\tau} \tag{10.69}$$

and thus

$$\frac{e_o}{e_i}(D) \approx \frac{1}{\tau D} \tag{10.70}$$

$$e_o \approx \frac{1}{\tau} \int e_i \, dt \tag{10.71}$$

Thus, if the frequency spectrum of the input signal is such that $\omega\tau \gg 1$ for all significant frequencies, a good approximation to the desired integrating action is obtained. For a given τ, the approximation improves as ω increases. It appears as if any ω can be accommodated by choosing τ sufficiently large. However, large τ decreases the magnitude of the output; thus this can be carried only as far as the noise level of the system permits.

If a signal is available in digital form, it may be integrated by a general-purpose digital computer by programming it for one of the approximate numerical-integration schemes, such as Simpson's rule. Another type of *digital integration*, which can be carried out without use of a general-purpose computer, involves pulse-totalizing methods. Here the analog voltage signal is converted to a periodic voltage signal whose frequency is proportional to the input-signal amplitude (voltage-to-frequency converter). Then this periodic signal is applied to an electronic counter. Thus the reading of the counter at any time is proportional to the time integral of the input signal up to that time.

Differentiation. For mechanical displacement signals, the various velocity pickups, accelerometers, jerk pickup, tachometer generator, and rate gyro of Chap. 4 may be considered as differentiating devices.

All the high-pass filters of Fig. 10.24 may be employed as *approximate differentiators* for input signals within a restricted frequency range, as shown by the following analysis:

$$\frac{e_o}{e_i}(i\omega) = \frac{i\omega\tau}{i\omega\tau + 1} \tag{10.72}$$

Now if $\omega\tau \ll 1$,

$$\frac{e_o}{e_i}(i\omega) \approx i\omega\tau \tag{10.73}$$

$$\frac{e_o}{e_i}(D) \approx \tau D \tag{10.74}$$

$$e_o \approx \tau \frac{de_i}{dt} \tag{10.75}$$

We note here that for a given τ, the approximation improves for lower values of ω. Again τ may be reduced to extend accurate differentiation to higher frequencies. However, small τ reduces sensitivity; thus noise level is limiting, just as in the approximate integrators.

Use of op-amps results in both approximate and "exact" differentiators of improved performance relative to the passive high-pass filters discussed above. Figure 10.33 shows some of these circuits. In Fig. 10.33a, analysis of this "exact" differentiator gives

$$\frac{e_o}{e_i}(D) = -RCD \tag{10.76}$$

This circuit is rarely useful because the ever-present noise (generally of high frequency relative to the desired signal) will completely swamp the desired signal at the output. All exact differentiators must suffer from this problem. It can be alleviated only by shifting to approximate differentiators that include low-pass filters to take out the effects of high-frequency noise. Figure 10.33b shows a common scheme which, when analyzed, gives

$$\frac{e_o}{e_i}(D) = -\frac{R_2 CD}{R_1 CD + 1} \tag{10.77}$$

This gives an accurate derivative for frequencies such that $R_1 C\omega \ll 1$ and amplifies high-frequency noise only by an amount R_2/R_1. To attenuate noise, we must use a second-order type of low-pass filter, such as given by the circuit of Fig. 10.33c. Analysis yields

$$\frac{e_o}{e_i}(D) = -\frac{R_2 C_1 D}{(R_2 C_2 D + 1)(R_1 C_1 D + 1)} \tag{10.78}$$

Figure 10.33d shows an actual circuit[1] designed for measuring the rate of

[1] The Lightning Empiricist, Philbrick Researches Inc., Boston, October 1963.

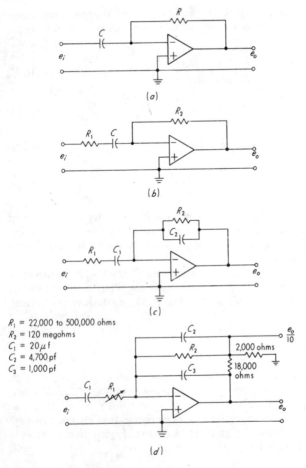

$R_1 = 22{,}000$ to $500{,}000$ ohms
$R_2 = 120$ megohms
$C_1 = 20\,\mu\mathrm{f}$
$C_2 = 4{,}700\,\mathrm{pf}$
$C_3 = 1{,}000\,\mathrm{pf}$

Figure 10.33 Electronic differentiators.

charging or discharging of batteries and using a solid-state operational amplifier. Analysis gives

$$\frac{e_o}{e_i}(D) = -\frac{10R_2 C_1 D}{(R_1 C_1 D + 1)[R_2(10C_3 + C_2)D + 1]} \qquad (10.79)$$

The output is read on a meter which may be connected to e_o or $e_o/10$, depending on the size of the output. For the numerical values given and $R_1 = 22{,}000$,

$$\frac{e_o}{e_i}(D) = -\frac{24{,}000D}{(0.44D + 1)(1.764D + 1)} \qquad (10.80)$$

We note that for $De_i = 10$ mV/min the output is 4 V. If, say, $\frac{1}{2}$ mV of 60-Hz noise is present at the input, then the output noise is only 41 mV, which is about 1 percent of the desired output.

At this point, we utilize a simple example to illustrate how "eyeball" design procedures can be combined with computer-aided analysis in the development of a practical differentiator. A spring-loaded rotary mechanism with dry friction has a motion whose general character is as shown in Fig. 10.34a. We are using a single-turn, 1,000-Ω, 1.25-W conductive plastic potentiometer to measure angular displacement θ, and we would like to employ a simple passive differentiator (Fig. 10.34b) to obtain a velocity signal. Both signals are recorded on an ink oscillo-graph with input resistance 1 MΩ and maximum sensitivity of 25 mV full scale. The differentiator gives a good approximation only for frequencies where $\omega\tau \ll 1$, so we must estimate the largest ω to be measured and choose τ accordingly.

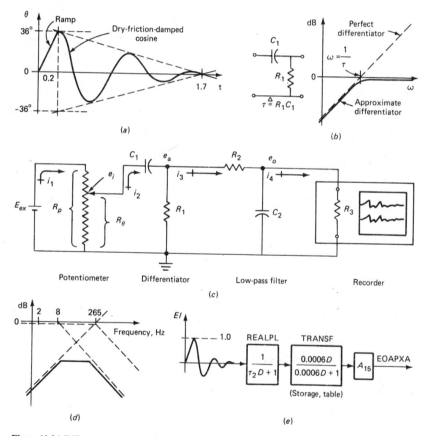

Figure 10.34 Differentiator design problem.

Rather than perform a Fourier transform to obtain a frequency spectrum of the signal to be measured, we use judgment to estimate that our signal will have little frequency content above about 2 Hz. Such judgment is necessary and desirable in preliminary design if we are to rapidly "rough out" an initial system, using models and methods simple enough to allow us to easily see the effects of parameter changes. Once rough numbers are available, computer-aided analysis can check performance accurately and quickly, providing data helpful in the final "trimming" of the design. Proceeding in this fashion, next we choose τ by deciding to make $\omega_{max} \tau = 0.01$ to satisfy the design requirement $\omega \tau \ll 1$. Again, note that this involves judgment, since a smaller τ would give a differentiator with a wider accurate frequency range but a *smaller* sensitivity. Also, while "good" accuracy in the frequency domain (amplitude ratio and phase close to the ideal) certainly implies "good" time-domain accuracy, no convenient *numerical* accuracy criteria are available.

Using $\omega_{max} = 12.6$ rad/s gives $\tau = 0.794$ ms, which can be achieved with an infinite number of combinations of R_1 and C_1. To choose specific values, we recall that good linearity in a potentiometer circuit requires the load resistance to be large compared with that of the potentiometer; a $10:1$ ratio gives about 1.5 percent nonlinearity. Again a judgment is necessary, and we choose $R_1 = 30,000$ Ω (a standard commercial value) to give nonlinearity of about 0.5 percent. Note that the high recorder resistance (1 MΩ) in parallel with R_1 makes the effective (parallel) resistance nearly R_1. Capacitance C_1 now can be calculated as 0.0265 μF, but we choose the closest commercial value, 0.02, making an adjusted value of $\tau_1 = 0.6$ ms.

An apparatus with these numerical values was assembled and run, and "visually" it seemed to behave properly (peaks of $\dot\theta$ occurred at zero crossings of θ, zero crossings of $\dot\theta$ occurred at peaks of θ). However, the $\dot\theta$ signal was excessively noisy, even though the θ signal looked perfectly smooth to the eye. The noise, caused by normal potentiometer roughness and accentuated by the differentiator, appeared to be of a "random" waveform with several hundred fluctuations in one cycle of the measured motion, which suggests that low-pass filtering could be employed, as in Fig. 10.34c. Design of this filter requires a compromise since a large filter time constant cuts noise strongly but also makes differentiating action less perfect. Again we assume no loading effects, to allow simple design of the filter as an isolated device, rather than analyzing the entire circuit, which is beginning to get somewhat complicated. From Fig. 10.34d we decide that a filter $\tau = R_2 C_2$ of about 0.02 s should not degrade differentiator performance too much at 2 Hz. To minimize loading, we want $R_2 \gg R_1$ and $R_3 \gg R_2$; and since R_3 is fixed at 1 MΩ and R_1 has been chosen as 30,000 Ω, we take $R_2 = 0.18$ MΩ and $C_2 = 0.10$ μF as acceptably close standard values.

The modified system was run, and again it appeared visually good, but now with practically no noise evident. To obtain quantitative data on dynamic accuracy, we should, of course, perform some kind of dynamic calibration, such as a frequency-response test with known sinusoidal variation of θ over a range of frequencies. In lieu of (or in addition to) this perhaps lengthy experimental effort, we could perform a theoretical analysis, using digital simulation to ease the task.

Such theoretical study could take several forms, depending on the level of complexity and accuracy felt necessary. Here we carry through several such studies, using the CSMP language introduced earlier. Since we wish to compare exact results with our approximate unloaded design model, case 3 in the program below (Fig. 10.35) handles this approximation as in Fig. 10.34e. Pot excitation EEX is taken as 25 V (a 1.25-W power rating allows as much as 35 V), and pot

```
$$$CONTINUOUS SYSTEM MODELING PROGRAM  III   V1M3   TRANSLATOR OUTPUT$$$
TITLE UNLOADED VERSUS EXACT DIFFERENTIATOR MODEL
METHOD RKSFX
PARAM EEX=25.
PARAM RP=1000.,RO=500.,RSP=100.
PARAM R1=30000.,C1=2.E-8,R2=180000.,C2=2.7E-7,R3=1.E6
INIT
      A1=C1*R2
      A2=(R1+R2)/(C1*R1*R2)
      A3=R2*C2
      A4=(R2+R3)/(C2*R2*R3)
      A5=(R1*R3*C1)/(R1+R2+R3)
      A6=(R1*R2*R3*C1*C2)/(R1+R2+R3)
      A7=((R1+R2)*R3*C2+(R2+R3)*R1*C1)/(R1+R2+R3)
      TAU2=R2*C2
      A14=C1*EEX*RO/RP
      A15=EEX*RSP/RP
DYNAM
*     INPUT SIGNAL
      EII=(COS(12.56*(TIME-.2)))*(1.-(TIME-.2)/1.5)*(STEP(.2)-STEP(1.7))
      EI=5.*RAMP(0.0)-5.*RAMP(.2)+EII-STEP(.2)
*     PERFECT DERIVATIVE
      E2=DERIV(0.0,EI)
      EODERV=A5*E2*A15
*     1 LINEAR AND UNLOADED POT, LOADED CIRCUIT (SIM. EQ.'S)
      INTEO=INTGRL(0.0,EO)
      EO=INTFA/A3-INTEO*A4
      INTEA=INTGRL(0.0,EA)
      EA=EI+INTEO/A1-INTEA*A2
      EOA=EO*A15
*     2 LINEAR AND UNLOADED POT, LOADED CIRCUIT (TRAN. FCN.'S)
      EXDT2=(EI-A7*EXDT1-EX)/A6
      EXDT1=INTGRL(0.0,EXDT2)
      EX=INTGRL(0.0,EXDT1)
      EOX=A5*EXDT1
      EOXA=EOX*A15
*     3 LINEAR AND UNLOADED POT, UNLOADED CIRCUIT (TRAN. FCN.'S)
      E1=REALPL(0.0,TAU2,EI)
      EOAPRX=TRANSF(1,B,1,A,E1)
STORAGE B(2),A(2)
TABLE B(1-2)=6.E-4,1.,A(1-2)=6.E-4,0.0
      EOAPXA=EOAPRX*A15
*     4 NONLINEAR AND LOADED POT, LOADED CIRCUIT (SIM. EQ.'S)
      RTHET=RO+RSP*EI
      A8=(R1*R1/(R1+R2))/A10
      A9=(R1/(C2*(R1+R2)))/A10
      A10=R1+RTHET
      A11=RTHET/(RP*A10)
      A12=(RTHET*RTHET)/(RP*A10)
      A13=1/(C1*A10)
      I2=(A9*INTI4-A9*INTI3+A11*EEX-A13*INTI2)/(1-A8-A12)
      INTI2=INTGRL(A14,I2)
      I1=(EEX+I2*RTHET)/RP
      I3=(R1*I2+INTI4/C2-INTI3/C2)/(R1+R2)
      INTI3=INTGRL(0.0,I3)
      I4=(INTI3-INTI4)/(C2*R3)
      INTI4=INTGRL(0.0,I4)
      EOACT=I4*R3
TIMER FINTIM=2.,DELT=.001,OUTDEL=.04
```

```
PAGE WIDTH=50
PAGE GROUP
OUTPUT EODERV,EOAPXA,EOACT
OUTPUT EODERV,EOA,EOAPXA
OUTPUT EODERV,EOA,EOXA
END
PARAM C2=1.E-7
END
```

Figure 10.35 Digital simulation program for differentiator analysis.

motion of $36°$ produces 2.5 V at $t = 0.2$ s. We assume that the pot starts at midrange ($R0 = 500\ \Omega$) and moves $100\ \Omega$ ($RSP = 100\ \Omega$) and $36°$ from $t = 0$ to $t = 0.2$ and that it produces an output voltage exactly proportional to its motion, with no loading effects. We ignore the initial steady voltage at e_i and work with only *changes* from this reference. The REALPL in CSMP models the first-order system; TRANSF, STORAGE, and TABLE (a general transfer-function module) are employed for the differentiator; and the Fortran multiplication by $A15 = (EEX)(RSP)/RP = 2.5$ gives the proper voltage scale. The EII and EI statements produce the desired input waveform.

Next we analyze an improved model that still assumes a perfect voltage at e_i (just as above) but treats the rest of the system "exactly." Using the current-node method of circuit analysis, we can write

$$C_1 D(e_a - e_i) + \frac{e_a}{R_1} + \frac{e_a - e_o}{R_2} = 0 \qquad (10.81)$$

$$\frac{e_o - e_a}{R_2} + C_2 De_o + \frac{e_o}{R_3} = 0 \qquad (10.82)$$

Since e_i is assumed a given input, these two equations in two unknowns e_a and e_o are solvable for the desired e_o. To avoid the need to differentiate, divide both equations by D (a standard trick of analog and digital simulation) to get the equations

$$EO = INTEA/A3 - INTEO*A4$$

$$EA = EI + INTEO/A1 - INTEA*A2$$

of case 1 in the program. Then use of the INTGRL statement gets us EA and EO.

Using the same model as in case 1, but manipulating Eqs. (10.81) and (10.82) analytically to eliminate e_a before simulation, leads to

$$\frac{e_o}{e_i}(D) = \frac{[(R_1 R_3 C_1)/(R_1 + R_2 + R_3)]\,D}{\dfrac{R_1 R_2 C_1 C_2}{R_1 + R_2 + R_3} D^2 + \dfrac{R_3 C_2(R_1 + R_2) + (R_1 C_1)(R_2 + R_3)}{R_1 + R_2 + R_3} D + 1} \qquad (10.83)$$

which is simulated in case 2 and should give results identical to those of case 1.

The most correct model, case 4, treats e_i as the unknown that it really is, since a given input motion θ tells us only the pot resistance R_θ (RTHET in the program). Our differential equations now have time-varying parameters and are analytically unsolvable. (Cases 1, 2, and 3 *could* have been solved analytically rather than numerically.) Simulation *can* be used to solve our new equations, but the model will be more complex. First we express R_θ, using the mathematical function EI, since it has the correct form and already is available in the program:

$$R_\theta = R_0 + R_{sp} e_i \qquad (10.84)$$

Be sure to note that e_i (EI) is *not* being utilized here as an input voltage, but just as a time-varying pure number useful in generating the R_θ variation we desire.

Employing Kirchhoff's voltage-loop law on the circuit of Fig. 10.34c yields

$$-E_{ex} + (R_p - R_\theta)i_1 + R_\theta(i_1 - i_2) = 0 \qquad (10.85)$$

$$\frac{1}{C_1} \int i_2 \, dt + R_1(i_2 - i_3) + R_\theta(i_2 - i_1) = 0 \qquad (10.86)$$

$$i_3 R_2 + \frac{1}{C_2} \int (i_3 - i_4) \, dt + R_1(i_3 - i_2) = 0 \qquad (10.87)$$

$$i_4 R_3 + \frac{1}{C_2} \int (i_4 - i_3) \, dt = 0 \qquad (10.88)$$

$$e_o = i_4 R_3 \qquad (10.89)$$

Usually analog or digital simulation works best (and involves less human effort) when you simulate the original set of basic equations by simply rearranging them so that the highest derivative of each unknown stands alone on the left side of one of the equations. This allows calculation of the highest derivative and successive use of the INTGRL statement then computes all lower derivatives, the unknown itself, and integrals of the unknown, if they appear in the equations. Occasionally this plan is frustrated by the appearance of the "highest derivative" of one of the *other* unknowns on the right-hand side of the equation for the highest derivative of an unknown, which creates a computing "loop" that prevents calculation. This problem occurs in our present example in the equation

$$I2 = (R1*I3 + RTHET*I1 - INTI2/C1)/(R1 + RTHET) \qquad (10.90)$$

where I1 and I3 are the "highest derivatives" of these two currents. The usual solution to this difficulty is to "manually" substitute for the offending terms (I1 and I3 here) from other equations in the set *before* simulation. This was done to get the equation for I2 in case 4. Then the rest of the equations in the set create no problems and are simulated as originally written. Note that INTI2 does *not* start at zero for $t = 0$ because

$$\frac{1}{C_1} \int i_2 \, dt \bigg|_{t=0} = E_{ex} \frac{R_o}{R_p} \qquad \int i_2 \, dt \bigg|_{t=0} = \frac{C_1 E_{ex} R_o}{R_p} \triangleq A14 \qquad (10.91)$$

Finally, to compare circuit model behavior with that of a *perfect* differentiator, we apply e_i as input to the CSMP statement DERIV, which takes the derivative. In both analog and digital simulation, great effort is exerted to use integration rather differentiation, and differentiation is rarely necessary. In this example, use of DERIV is convenient and justifiable since we are applying it to a *known* function (e_i) which is smooth and should cause no "noise" problems typical of differentiators. For the best comparison, the derivative signal is multiplied by A5, the steady-state gain of the actual circuit [see Eq. (10.83)]. This is appropriate since this scale-factor error could be corrected easily when the real system is used for measurements.

All the above models can be run simultaneously by using the CSMP program of Fig. 10.35. Most CSMP programs run well utilizing a variable-step-size integrating algorithm that is automatically invoked unless we ask for something different. In this example difficulties arose which were resolved by going to the Runge-Kutta fixed-step-size algorithm called by the card METHOD RKSFX. Cards between INIT and DYNAM are calculated only once at TIME = 0 and thus are used for constants in the differential equations. Cards between DYNAM and TIMER either have been explained already or can be understood by examining the basic equations. On the TIMER card, FINTIM = 2. lets TIME go from 0 to 2, DELT = .001 uses a computing increment of 0.001 s, and OUTDEL = .04 plots graphs with point spacing of 0.04 s. PAGE WIDTH = 50 selects a convenient graph size, while PAGE GROUP plots multicurve graphs all to the same scale, desirable in our problem for comparing various models. Each OUTPUT card lists variables to be plotted on a single graph, while END denotes the end of the first problem. After END, if we give new parameter values, the entire problem will be rerun using those new values. Note that the first problem used C2 = 2.7E-7, a value *larger* than our original design, while the rerun utilized C2 = 1.E-7, our design value. Once the basic program is written, changes in *any* system parameters and/or in the signal to be differentiated can be studied easily.

Because of space limitations and because we are more interested in showing the general simulation technique than in giving numerical results, only two graphs—Fig. 10.36 (C2 = 1.E-7) and Fig. 10.37 (C2 = 2.7E-7)—are given. Note that EOAPXA (not corrected for steady-state gain error resulting from loading) predicts values somewhat too large. Both EOAPXA and EOACT dynamically lag the perfect derivative, as we expect for any measurement system with flat amplitude ratio and linear phase shift. For C2 = 1.E-7, this delay is modest and peak values are nearly correct, while C2 = 2.7E-7 (too strong a filter) has departed sufficiently from flat amplitude ratio that peaks are attenuated and delay is more obvious because of greater phase lag. Whether C2 could be reduced even more to improve performance would depend on how much potentiometer noise were considered acceptable in the velocity signal. Overall, we see that the approximate, unloaded model employed for preliminary design is quite adequate for its purpose, allowing easy hand calculations and retaining a good feel for the effects of parameters, while the computer-aided analysis augments this capability by providing accurate results for models of almost any desired complexity.

A final example of a differentiator using nonelectrical methods is the aircraft rate-of-climb indicator shown in Fig. 10.38. Since atmospheric pressure varies with altitude, a device that measures rate of change of atmospheric pressure can indicate rate of climbing or diving. While actual design requires a more critical study,[1] here we consider a simplified linear analysis to show the main features. Static pressure p_s corresponding to aircraft altitude is fed from the vehicle's static pressure probe to the input tube of the rate-of-climb indicator. Leakage through a capillary tube into the chamber of volume V occurs at a mass flow rate assumed to be $K_c(p_s - p_c)$ lbm/s. Air in the chamber follows the perfect-gas law

[1] D. P. Johnson, Aircraft Rate-of-Climb Indicators, *NACA, Rept.* 666, 1939.

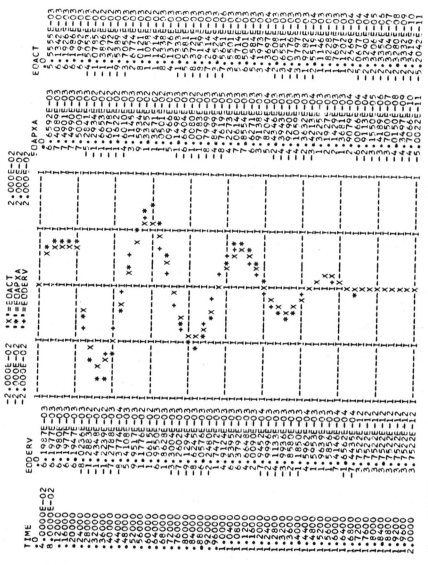

Figure 10.36 Differentiator performance.

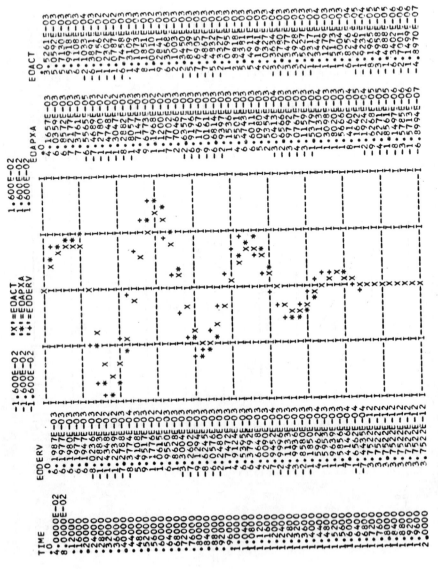

Figure 10.37 Differentiator performance.

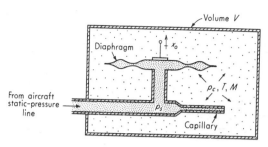

Figure 10.38 Rate-of-climb sensor.

$p_c V = MRT$. Motion of the output diaphragm is according to $x_o = K_d(p_s - p_c)$. By assuming K_c and T to be constant, analysis gives

$$\frac{x_o}{Dp_s}(D) = \frac{K}{\tau D + 1} \tag{10.92}$$

where

$$K \triangleq \frac{K_d V}{RTK_c} \quad \text{in/[(lb/in}^2)/\text{s]} \tag{10.93}$$

$$\tau \triangleq \frac{V}{RTK_c} \quad \text{s} \tag{10.94}$$

Thus x_o, which may be measured with any displacement transducer, is an indication of rate of change in p_s, and thereby a measure of rate of climb, if pressure is assumed to be a linear function of altitude. Since this is not exactly true, various compensating devices are needed in a practical instrument for this and other spurious effects.

10.5 DYNAMIC COMPENSATION

Sometimes it is not possible to obtain the desired behavior from a measuring device solely by adjusting its own parameters. To get fast response from a thermocouple, for example, very fine wire must be used. Perhaps the vibration and temperature environment might be so severe that such a fine-wire thermocouple would be destroyed before any readings could be obtained. For this and similar situations, dynamic compensation may provide a solution.

Figure 10.39 shows the general arrangement by which dynamic compensation may be employed. Ideally an instrument with transfer function $G(D)$ is cascaded with a compensator $K_1/G(D)$, and thus (if negligible loading is assumed) the overall system has *instantaneous response* since its transfer function is just the constant K_1. This result is, of course, too good to be true. The practical difficulty lies in the construction of the compensator $K_1/G(D)$, which generally is not realizable physically because of the need for perfect differentiating effects. While perfect compensation for instantaneous response is *not* possible, very great improvements *may* be achieved with the scheme of Fig. 10.39c. Here the undesirable

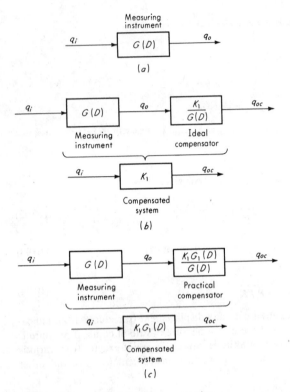

Figure 10.39 Generalized dynamic compensation.

dynamics $G(D)$ are *replaced* with more desirable ones, $G_1(D)$. This technique has been used with good success, for example, in speeding up the response of temperature-sensing elements and hot-wire anemometers. These are basically first-order instruments; thus $G(D) = K/(\tau D + 1)$. While the compensator, in general, can take any suitable physical form, because most sensors produce an electrical output, most compensators in use are electric circuits. The compensator generally utilized for first-order systems takes the form shown in Fig. 10.40. Note that for the passive version any increase in speed of response ($\tau_1 \ll \tau$) is paid for by a loss of sensitivity in direct proportion, since $K_1 = R_2/(R_1 + R_2) = \tau_1/\tau$. If this loss of sensitivity is not tolerable, additional amplification is needed. Usually it is placed between the sensor and the compensator because then it will also serve to unload the two circuits from each other. [The active (op-amp) version can *combine* compensation and amplification in one stage.] While several stages of such compensation may be used sometimes, such staging cannot be carried beyond a certain point because of additional noise introduced by the amplifiers and accentuated by the compensators. However, a response speedup of the order

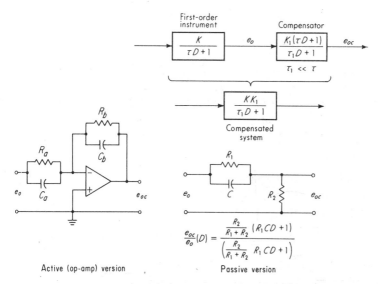

Figure 10.40 Dynamic compensation for first-order system.

of 100 : 1 is feasible and has been achieved for thermocouples and hot-wire anemometers.

Theoretically the concept of dynamic compensation is applicable to any-order system. A good example of a more complex application is found in the "equalization" of vibration shaker systems. Figure 10.41*a* shows the basic arrangement of a shaker system for sinusoidal vibration testing. Many tests involve "sweeping" the frequency of the oscillator through a certain range while maintaining the acceleration amplitude constant at the test object. While it is not difficult to maintain constant the amplitude of oscillator output voltage *e* while sweeping through the frequency range, the acceleration $\ddot{x}$ will not be constant since the transfer function $(\ddot{x}/e)$ $(i\omega)$ is not constant over this range. In fact, because of various resonances in the electromechanical shaker, test fixtures, and the test object itself, severely distorted frequency response is not uncommon (Fig. 10.41*b*). This difficulty can be overcome by use of a feedback scheme as in Fig. 10.41*c*. Here the actual acceleration $\ddot{x}$ is compared with the desired value $\ddot{x}_d$; if they differ, the amplitude of the oscillator is adjusted to obtain correspondence. This adjustment is performed automatically and continuously as the frequency range is swept.

When random vibration testing (see Fig. 10.42) rather than pure sinusoidal is desired, the above approach is not applicable directly since now the signal $\ddot{x}$ is random and the noise source cannot be adjusted in any simple fashion to force $\ddot{x}$ to have the desired frequency spectrum (the spectrum of *e*). One approach is to provide dynamic compensation such that the transfer function $(\ddot{x}/e)$ $(i\omega)$ *is* flat

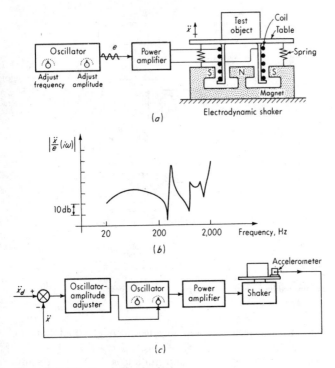

Figure 10.41 Vibration shaker systems.

over the desired frequency range. Then the spectrum at $\ddot{x}$ will be the same as that put in at e. The necessary dynamic compensation here is one which can put "peaks" where there are "notches" and notches where there are peaks in the curve of Fig. 10.41b. Thus the overall curve can be made relatively flat. A compensator that will provide one peak and one notch has the form

$$\frac{e_o}{e_i}(D) = \frac{D^2/\omega_{n,n}^2 + 2\zeta_n D/\omega_{n,n} + 1}{D^2/\omega_{n,p}^2 + 2\zeta_p D/\omega_{n,p} + 1} \tag{10.95}$$

Figure 10.42b shows the frequency response of such a compensator in which the peak occurs at a lower frequency than the notch. The reverse is also possible, if needed. Since ζ_n and ζ_p are adjustable, the compensator can be "tailored" to cancel exactly the undesired shaker-system dynamics. When several peaks and notches are present (as in Fig. 10.41b), several compensators are used; as many as 10 are not uncommon.

While the analog dynamic compensators described operate "online" in "real time," dynamic compensation also can be realized "offline" digitally, by using

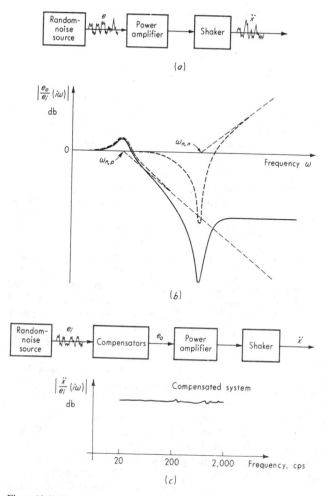

Figure 10.42 Dynamic compensation for vibration shaker.

frequency-spectrum or equivalent time-domain methods.[1] We merely compute $Q_o(i\omega)$ from the measured $q_o(t)$ (using, say, fast Fourier transform methods), multiply $Q_o(i\omega)$ by the (assumed known) (Q_o/Q_i) $(i\omega)$ of the instrument to get $Q_i(i\omega)$,

[1] G. L. Schulz, Method of Transfer Function Calculation and Distorted Data Correction, *Air Force Weapons Lab. Rept. AFWL-TR*-73-300, Kirtland AFB, N. Mex., March 1975; M. R. Weinberger, Flight Instrument and Telemetry Response and Its Inversion, *NASA* CR-1768, September 1971; L. W. Bickle, A Time Domain Deconvolution Technique for the Correction of Transient Measurements, *Rept.* SC-RR-71-0658, Sandia Labs, Albuquerque, N. Mex., 1971; H. Grote, Restoration of the Input Function in Measuring Instrumentation, US Army Electronics Command, *Rept.* ECOM-2662, Ft. Monmouth, N. J., 1966.

and then inverse-transform to get $q_i(t)$. Some FFT system analyzers (Sec. 10.13) provide this kind of capability almost in real time.

10.6 INSTRUMENT SERVOMECHANISMS

Measurement systems often require the conversion of a low-power electrical signal or a low-power mechanical motion to an accurately proportional and (relative to the input signal) high-power mechanical motion. When frequency response beyond about 5 Hz is not required, often this function may be performed by an appropriate instrument servomechanism. Figure 10.43 shows block diagrams of some of the most common types. The velocity servos generally are utilized to obtain the integral of the input signal, as explained in Sec. 10.4.

The dc-input position servo, while important in its own right, also is the basis of practically all self-balancing potentiometer recorders and XY plotters. Thus we explain its operation in greater detail. Some instrument servos utilize ac amplifiers and two-phase ac instrument servomotors even if the input signal is dc. The use of ac amplifiers is based on their freedom from drift and reasonable cost, while the use of two-phase ac motors relates to their low friction (no brushes are needed as in a dc motor) and controllability. Figure 10.44 briefly summarizes the operating characteristics of this type of motor. One of the phases is of fixed amplitude. The amplitude of the other phase (which must be displaced in phase by ± 90 electrical degrees from the fixed phase) controls the direction and amount of torque developed. When the controlled phase reverses polarity (goes from $+90$ to $-90°$ or vice versa), the torque reverses.

The schematic and graphs of Fig. 10.45 show how a dc signal is converted to ac by a chopper and how a reversal in polarity of the dc error signal e_{aa} results in a $180°$ phase shift (from $+90$ to $-90°$ or vice versa) in the motor-control phase voltage e_{dd}, thereby causing the required reversal in the direction of torque. We see that whenever $e_i \neq e_p$, there will be an error voltage e_{aa} which is converted to ac and amplified so that it tends to drive the motor in a direction to change e_p until it equals e_i. If e_i is changing and if the amplifier gain is high enough, e_p (and therefore output motion θ_o) will "track" e_i with very little error. Because of improvements in dc amplifier drift behavior, servos like those of Fig. 10.43a and b often use dc amplifiers and motors and no chopper today. Instrument servos regularly achieve high static accuracy, having errors as small as 0.1 to 0.2 percent of full scale in positioning θ_o as a linear function of e_i. However, their frequency response is limited by inertia of moving parts to about 5 Hz or less.

10.7 ADDITION AND SUBTRACTION

The addition or subtraction of mechanical-motion signals generally is accomplished by use of gear differentials or summing links (see Fig. 10.46a). Forces or pressures are summed and transduced to displacement by the schemes shown in

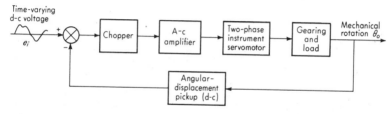

(a) D-c voltage-input position servo

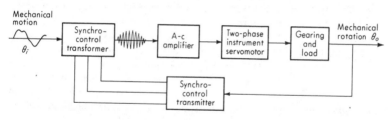

(b) D-c voltage-input velocity servo

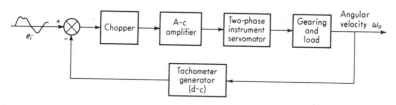

(c) Motion-input position servo (all a-c)

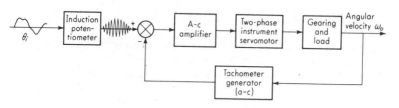

(d) Motion-input velocity servo (all a-c)

Figure 10.43 Instrument servomechanisms.

Fig. 10.46b. The spring restraints that transduce force to displacement may be removed if a feedback system using a null-balance force to return deflection to zero is employed. A totalizer for hydraulic load cells (Fig. 10.46c[1]) uses rolling

---

[1] A. H. Emery Co., New Canaan, Conn.

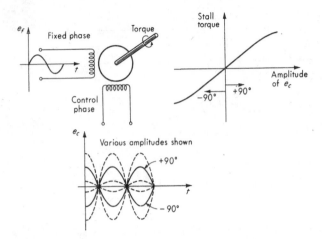

Figure 10.44 Two-phase servomotor.

diaphragm seals to achieve 0.0005 percent resolution. Summing of voltage signals is accomplished by the simple series circuit or the op amp circuit shown in Fig. 10.46d. Subtraction rather than addition in all the above devices is obtained by simply reversing the sense of the input to be subtracted. Addition is the basic operation of digital computers; thus addition or subtraction of digital signals in binary form is accomplished easily with such equipment. Equipment for interconverting numbers in binary and decimal form also is available. Digital signals in the form of pulse rates may be added and subtracted to obtain pulse rates which are the sum or difference of the input pulse rates.

10.8 MULTIPLICATION AND DIVISION

When data manipulation requires multiplication or division of two variable signals, a number of techniques are available, depending on the physical nature of the signals.[1] Mechanical, electromechanical, and pneumatic methods are described in the literature,[2] but are limited in application. General-purpose multiplication and division in data systems can be accomplished by analog, digital, or hybrid electronic means. For digital systems, these basic arithmetic operations present no problem for any size computer. Analog multiplication and division of 0.1 to 1 percent accuracy are readily accomplished in small IC devices at speeds up to several megahertz by using various electronic schemes,[3] with the "transconductance"[4] being perhaps the most common. Our interest is mainly in

[1] S. A. Davis, 31 Ways to Multiply, *Contr. Eng.*, p. 36, November 1954.
[2] Doebelin, "Measurement Systems," pp. 649–655.
[3] Multiplier Application Guide, Analog Devices, Norwood, Mass., 1980.
[4] Ibid.; Sheingold, "Nonlinear Circuits Handbook."

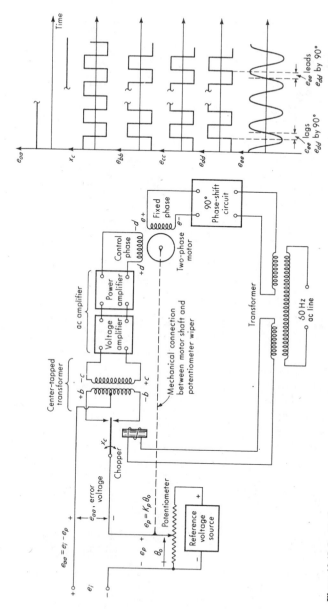

Figure 10.45 Position servo.

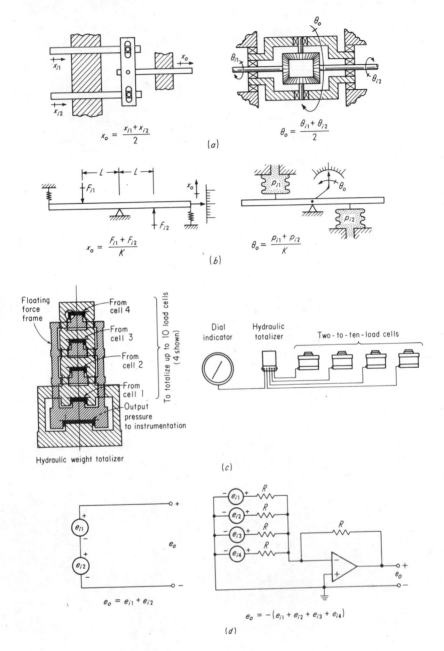

Figure 10.46 Addition and subtraction.

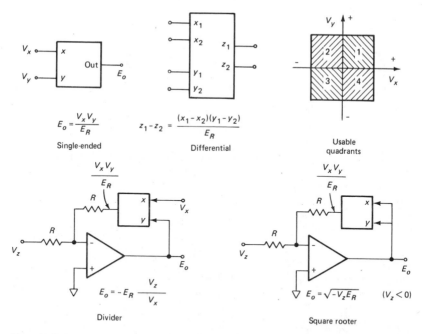

Figure 10.47 Electronic multiplier/divider.

the terminal characteristics of the devices (see Fig. 10.47). Multipliers are available as 1- to 4-quadrant devices; the 4-quadrant one is most versatile since V_x and V_y can assume any algebraic signs whatever. Differential-input types provide additional computing capability (additive constants or variables) and may allow better noise rejection. Usually the reference value E_R is 10 V, so that full-scale (± 10 V) inputs give full-scale (± 10 V) output. Division capability requires only the addition of a properly connected op-amp (included and switch-selectable in many multiplier/divider packages); but the denominator, of course, must remain of one polarity to avoid division by zero. Squaring is done easily by connecting the signal to both inputs, while taking the square root necessitates the op-amp feedback scheme of Fig. 10.47. For hybrid analog/digital applications, *multiplying digital/analog converters*[1] may be useful. Here a time-varying analog voltage becomes the reference voltage for a digital/analog converter, which makes the D/A output equal to the product of the analog input and the instantaneous value of the digital input to the digital/analog converter.

A Wheatstone bridge may be utilized for multiplication if one of the signals can be transduced to a voltage (which is used as the bridge-excitation voltage) and the other can be transduced to a resistance change (see Fig. 10.48). Then the

[1] Appl. Guide to CMOS Multiplying D/A Converters, Analog Devices, Norwood, Mass., 1978.

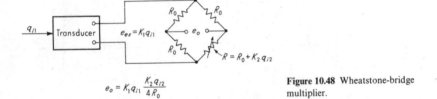

$$e_o = K_1 q_{i1} \frac{K_2 q_{i2}}{4 R_0}$$

Figure 10.48 Wheatstone-bridge multiplier.

bridge output is proportional to the product $q_{i1} q_{i2}$ for small-percentage resistance changes. While analog and/or digital electronics are the preferred schemes in general-purpose data systems, there is still room for mechanical ingenuity in special-purpose devices, such as the Btu meter of Fig. 10.49. This instrument/

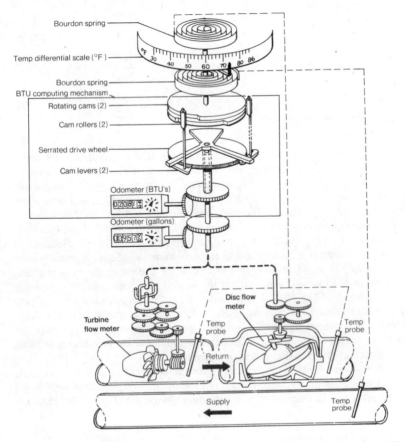

Figure 10.49 All-mechanical computing Btu meter. (*Courtesy Hersey Products, Spartanburg, S.C.*)

computer implements the relation

$$\text{Btu} = C \int \dot{M}(T_1 - T_2) \, dt \qquad (10.96)$$

to find the energy change in heated (or cooled) liquid flows in piping systems and requires subtraction, multiplication, and integration. Two pressure thermometers measure T_1 and T_2; the subtraction is effected by connecting one Bourdon tube to the pointer and the other tube to the scale, which produces a direct indication of $T_1 - T_2$. These same rotations also position two notched cams, forming a gap proportional to $T_1 - T_2$, which actuates a cam roller follower. The cam roller rotates continuously around the cams at a speed proportional to flow rate; this rotation is provided by either a positive-displacement or a turbine-type flowmeter (both are shown in Fig. 10.49, but one would be employed in an actual system). Two revolution counters (odometers), one reading Btu and the other reading gallons, are driven in intermittent fashion from a serrated drive wheel. As the rollers pass over and into the recess generated by the cams, two cam levers attached to the rollers engage the serrated circumference of the drive wheel. Thus each rotation of the cam roller carrier produces a partial rotation of the drive wheel that is proportional to $T_1 - T_2$. Integration is achieved simply by accumulating total rotations, since "average speed" is proportional to $\dot{M}(T_1 - T_2)$. The entire instrument requires no external power sources and achieves an accuracy of ± 1.5 percent for ΔT greater than 50 percent of rated.

10.9 FUNCTION GENERATION AND LINEARIZATION

When we need to generate a specific nonlinear function of a mechanical-motion signal, the use of cams, linkages, and noncircular gears allows great freedom since almost any reasonable function can be approximated adequately by one or a combination of these methods (see Fig. 10.50). The use of instrument servos also allows these methods to be employed with electrical signals, if, for some reason, all-electronic methods are not feasible.

Nonlinear potentiometers are used widely in function generation. They are constructed in basically the same manner as potentiometer displacement transducers except that a specific *nonlinear* relation between θ_i and e_o is wanted, rather than the linear relation desired for a motion transducer (see Fig. 10.51). A wide variety of functions are possible by distributing the resistance winding in a proper nonlinear fashion on the mandrel. Techniques also have been developed for constructing nonlinear potentiometers by using conducting plastic or deposited-film (rather than wirewound) resistance elements. While functions of rather arbitrary form are available as special items, certain basic functions are used so commonly that they are obtainable ready-made as stock items. These include sine and cosine over 360°, sine or cosine over 360°, 180° sine, 90° sine, $\pm 75°$ tangent, square function, and logarithmic function. The conformity of the voltage-output/rotation-input relation to the theoretical function is of the order of 0.3 to 2 percent of full scale, depending on the type of function and the instrument quality.

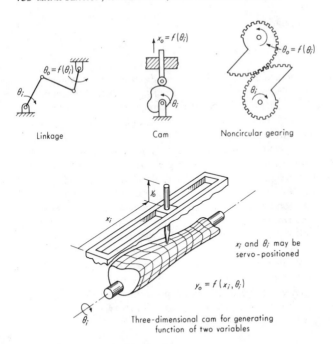

Linkage   Cam   Noncircular gearing

x_i and θ_i may be
servo-positioned

$y_o = f(x_i, \theta_i)$

Three-dimensional cam for generating
function of two variables

Figure 10.50 Mechanical function generation.

Potentiometer-type pressure transducers for flow rate, altitude, and airspeed applications are standard items.[1] Use of a computer-controlled mechanical milling process shapes the conductive-film resistance element of each transducer to give the desired voltage/pressure relation. This milling is done on each individual transducer with pressure and excitation voltage applied, thus compensating for any individual mechanical differences from unit to unit. The flow-rate transducer accepts a Δp signal from orifice-type flow elements and includes a square root function to develop an output linear with flow rate. Airspeed transducers are also differential-pressure units that accept stagnation and static pressures from aircraft pitot tubes and provide the proper nonlinear function to give an output linear with airspeed. A 0- to 1,000-kn unit conforms to airspeed linearity within ± 10 kn and can resolve ± 1.5 kn. Altitude transducers accept aircraft static pressure as input and linearize the U.S. standard atmosphere function with (for a 0- to 70,000-ft device) resolution of 10 ft.

When very accurate sine and/or cosine functions are needed (as in navigation and fire-control computers where resolution and composition of vectors must be performed), the use of resolvers rather than nonlinear potentiometers may be indicated. Resolvers are small ac rotating machines similar to synchros. In gener-

[1] CIC Vernitech, Deer Park, N.Y.

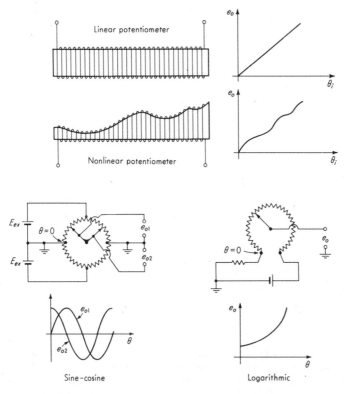

Figure 10.51 Nonlinear potentiometers.

al, they have two stator windings and two rotor windings (see Fig. 10.52). If one of the stator windings is excited with an ac signal of constant amplitude (60 or 400 Hz is employed commonly) and the other is short-circuited, rotation of the rotor through an angle θ_i from a null position gives at the two rotor windings ac signals whose amplitudes are proportional to $\sin \theta_i$ and $\cos \theta_i$, respectively. Other important computing functions[1] such as converting vehicle rotation angles to earth coordinates in navigation systems also can be performed by resolvers. A typical high-accuracy resolver has an excitation voltage of 26 V maximum at 400 Hz, open-circuit output voltage of 0 to 26 V, residual null voltage of 1 mV maximum, and a maximum deviation from the desired functional relation of 0.01 percent.

All the function generators shown so far involve moving parts and so are limited in speed. When the speed of all-electronic function generation is necessary or an electromechanical solution is unacceptable for other reasons, multi-

---

[1] Resolvers, Ford Instrument Co., Long Island City, N.Y.

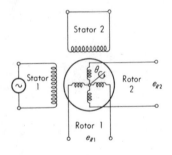

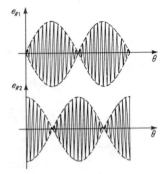

Figure 10.52 Resolver.

plier/dividers, multifunction modules, or diode function generators are available.[1] A cascade of multipliers with outputs scaled and summed with op-amp summers can generate a power series with adjustable coefficients. For example, two multipliers and one summer can generate $y = a_1 x + a_3 x^3$, which can fit $\sin x$ within ± 0.6 percent over the range $-\pi/2$ to $\pi/2$. Linearizers for hot-wire anemometers often utilize this sort of power-series scheme with terms up to the fourth power. Multifunction modules[2] provide the relation

$$E_o = \frac{10}{E_{\text{ref}}} V_y \left(\frac{V_z}{V_x} \right)^m \qquad (10.97)$$

where V_x, V_y, V_z are variable or constant input voltages (≥ 0), $E_{\text{ref}} \approx 9$ V, and m can be set in the range 0.2 to 5. This device, by itself or combined with others, provides very versatile function generation for linearization or other purposes.

Multiplier/dividers and multifunction modules provide "smooth" approximations to desired functions. Diode function generators, an alternative approach, give a *piecewise-linear* curve fit. Figure 10.53 conceptually illustrates the method, by assuming ideal performance for the diodes. The low cost and small size of IC op-amps today allow realization of essentially ideal diode behavior by utilizing one op-amp for each straight-line segment. Figure 10.54a[3] shows this approach

---

[1] Sheingold, "Nonlinear Circuits Handbook."
[2] Analog Devices Model 433, Norwood, Mass.
[3] Sheingold, "Nonlinear Circuits Handbook," pp. 92–97.

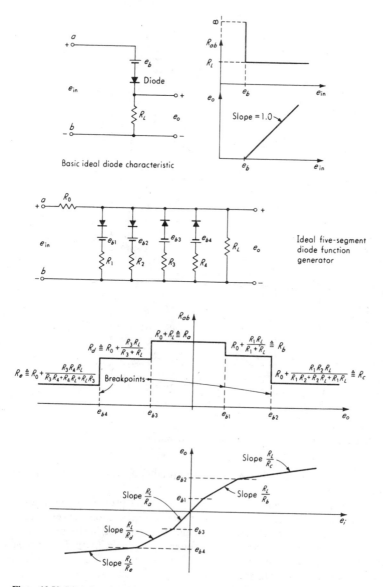

Figure 10.53 Diode function generator.

(see reference for details) practically applied to linearization of a Chromel-constantan thermocouple. Figure 10.54*b* shows an alternative scheme using a multifunction module, while Fig. 10.54*c* compares performance of the two meth-

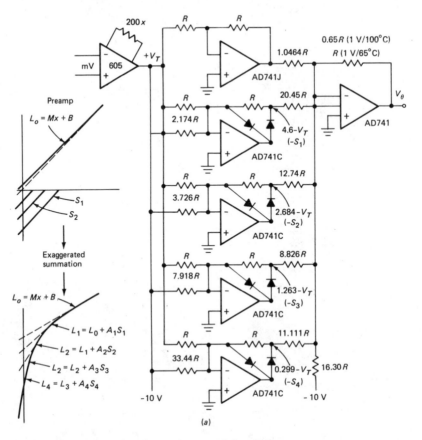

Figure 10.54 Piecewise, and continuous, thermocouple linearization.

ods. Both schemes are capable of response speed into kilohertz or even the megahertz range.

Because of their programmability, digital computers provide the most comprehensive, accurate, and versatile means for function generation; however, their speed generally cannot match that of all-electronic analog methods. Microcomputers have become small and cheap enough that units with fixed programs, dedicated to a particular operation, are found inside many instruments, such as digital thermometers and data loggers. Various approaches are employed. Perhaps the fastest (but one wasteful of memory) uses a "table lookup" procedure. Since the incoming data will have been digitized already with a certain resolution (say, 1 part in 1,000 for the full range), we can store in ROM (read-only memory) a table of 1,000 y values, with each corresponding to a particular X value in the desired function $y = f(X)$. A slower approach that utilizes less memory is to

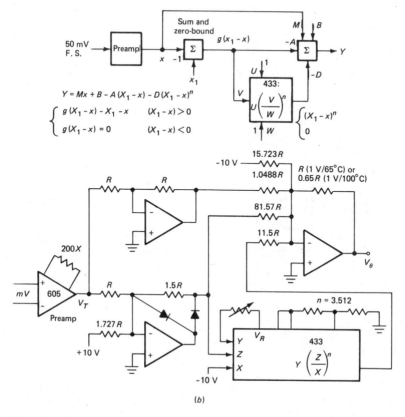

Figure 10.54 (*Continued*)

curve-fit the desired functional relation with analytic functions (such as power series) and then calculate y values from given X's. Frequently, breaking the complete range into smaller subranges (which allow simpler curve-fitting functions) can be used to increase speed.

10.10 AMPLITUDE MODULATION AND DEMODULATION

We have seen a number of examples of measurement systems in which interconversion between ac and dc signals was necessary and/or desirable. The conversion from dc to ac is a form of amplitude modulation, whereas conversion from ac to dc is called *demodulation*, or *detection*. In measurement systems, usually the demodulation must be phase-sensitive, so that the algebraic sign of the original time-varying dc signal is preserved. Modulation/demodulation may be accom-

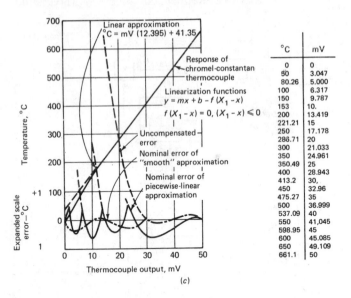

Figure 10.54 (*Continued*)

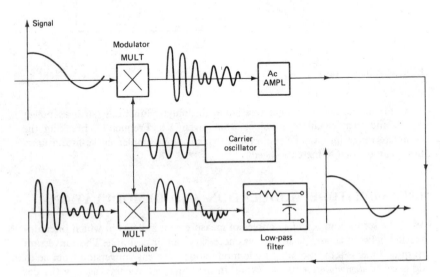

Figure 10.55 Multiplier-type modulator/demodulator system.

plished by assorted electronic schemes; the availability of small, accurate, and inexpensive IC analog multipliers makes the approach of Fig. 10.55 popular. There we show both modulation and synchronous demodulation, with an intermediate stage of ac amplification (since this is often the reason for using a modulation/demodulation scheme at all). The carrier oscillator can be either sine wave or square wave, and its frequency should be 5 to 10 times the highest signal frequency so that the low-pass filter can achieve a proper tradeoff between ripple and speed of response. Many infrared temperature instruments employ a scheme of this sort, except that the modulation is done mechanically with a rotating chopper wheel and a proximity pickup on this wheel generates the synchronizing waveform for the demodulator.

10.11 VOLTAGE-TO-FREQUENCY AND FREQUENCY-TO-VOLTAGE CONVERTERS

The conversion of a dc voltage input to a periodic-wave output whose frequency is proportional to the dc input may serve several useful functions in measurement systems. Such devices are employed widely in FM/FM telemetry ssytems, since the voltage-to-frequency conversion process is a form of frequency modulation. Also they are used in the integrating digital voltmeter where a dc signal is converted to a periodic wave of proportional frequency. Then this wave is applied to an electronic counter for a fixed time interval, giving a reading proportional to the average dc voltage over the time interval. The recording of dc voltages on magnetic tape recorders also is accomplished through the use of frequency modulation.

Figure 10.56a[1] shows a voltage-to-frequency (V/F) converter that utilizes one of the more popular operating principles. The input signal can be either current i_{in} (range 0 to 0.5 mA) or voltage e_{in} (0 to 10 V). The first op-amp serves as a buffer and provides a proportional voltage output (for either voltage or current input) to the second op-amp, which is connected as an integrator and charges at a rate proportional to the input signal. Integrator output is compared with a fixed threshold; whenever this threshold is crossed, a pulse of accurately known area is produced. This pulse serves both as output (via a buffer) and as a subtractive charge increment at the integrator input to return the integrator's net charge to zero, making it ready for another cycle to start. Thus a fixed input voltage produces an output pulse train of proportional frequency, with a scale factor (typically) of 10^4 Hz/V and ± 0.01 percent full-scale nonlinearity from 1 Hz to 100 kHz. For a $+10$-V step input, output settles within ± 0.01 percent in 3 output pulses plus 2 μs.

The frequency-to-voltage (F/V) converter of Fig. 10.56b accepts input signals of virtually any waveform and produces a time-varying dc output proportional to

[1] Data Acquisition Components and Subsystems Catalog, Analog Devices, Norwood, Mass., 1980.

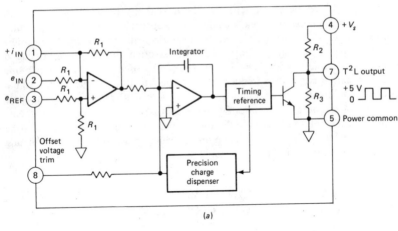

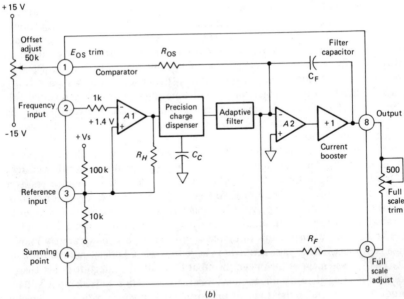

Figure 10.56 Voltage-to-frequency and frequency-to-voltage converters.

the rate at which the input signal crosses a fixed (but selectable) threshold. Input signals are applied directly to a comparator A1, which is set internally to provide a $+1.4$-V threshold with ± 100-mV hysteresis. (This threshold level offers excellent noise immunity for TTL input levels; for other types of signals, the threshold may be adjusted from 0 to ± 12 V.) Following the input comparator is a precision charge-dispensing circuit and output amplifier where the comparator signal is converted to a dc voltage. When the input comparator changes state, C_c is alternately charged from a precision voltage reference and discharged through the summing point of an output amplifier A2. A fixed amount of charge is controlled during each charge/discharge cycle. The higher the input frequency, the higher the average current into the summing point of A2. Then a current-to-voltage conversion is accomplished by R_F. The current pulses from the charge-dispensing circuit are integrated by C_F to reduce ripple. Added filtering for low-frequency input signals is provided by an adaptive filter at the output of the charge-dispensing circuit. For a unit with 0- to 100-kHz range, the "built-in" C_F gives a filter time constant of 24 μs, ripple of 35 mV rms at 100-kHz input, and 0.5 percent settling time (for a full-scale step-frequency change) of 0.8 ms. Full-scale output is 10 V with nonlinearity ± 0.01 percent from 1 Hz to 110 kHz. Frequency-to-voltage converters are utilized to obtain dc output signals from pulse-rate output devices, such as turbine and vortex flowmeters and pulse-type tachometers, and to recover the original signal in FM telemetry and tape recorders.

10.12 ANALOG-TO-DIGITAL AND DIGITAL-TO-ANALOG CONVERTERS; SAMPLE/HOLD AMPLIFIERS

Because most sensors have analog output while much data processing is accomplished with digital computers, devices for conversion between these two realms obviously play an important role. For motion inputs, the shaft-angle encoders of Chap. 4 provide analog-to-digital (A/D) conversion; here we concentrate on all-electronic devices whose analog input or output is a voltage.

We begin with digital-to-analog (D/A) converters since they are used as components in some A/D converters. The most fundamental property of any D/A or A/D converter is the number of bits for which it is designed, since this is a basic limit on resolution (see Fig. 10.57)[1]. Units with 8 to 12 bits are most common; however, resolution to about 18 bits is available, but special care must be taken in the application[2] (the least significant bit represents only 38 μV). Most D/A converters utilize a principle similar to that of Fig. 10.58, the so-called R-2R ladder network. Basic to the accuracy of such devices is the stability of the reference voltage and resistor values. Even with the best components, the highest-resolution converters need periodic calibration to remain within specification. To understand the operation of the 3-bit D/A circuit shown, note that the switches (called *current-steering switches*) are connected to "ground" irrespective of whether they are "on" (1) or "off" (0), because the op-amp negative input is a virtual ground. So we can redraw the circuit in this way and quickly see why the currents are as labeled. The op-amp merely sums the currents "steered" to it by the switches and produces an output voltage proportional to this sum.

[1] Design Engineers' Handbook and Selection Guide, A/D and D/A Converter Modules, BR-1021, Analogic Corp., Wakefield, Mass., 1980.

[2] Designer's Guide to High Resolution Products, Analog Devices, Norwood, Mass., 1981.

Binary Bits (n)	(2^n)	Equivalent percent or fraction of range of least-significant bit*		Residual $1 - \sum_{1}^{n} \left(\frac{1}{2^n}\right)$
		Percent	ppm	
1	2	50.0	500 000.	0.5
2	4	25.	250 000.	0.25
3	8	12.5	125 000.	0.125
4	16	6.25	62 500.	0.062 5
5	32	3.125	31 250.	0.031 25
6	64	1.562 5	15 625.	0.015 625
7	128	0.781 25	7 812.5	0.007 812 5
8	256	0.390 625	3 906.25	0.003 906 25
9	512	0.195 313	1 953.13	0.001 953 13
10	1 024	0.097 656	976.56	0.000 976 56
11	2 048	0.048 828	488.28	0.000 488 28
12	4 096	0.024 414	244.14	0.000 244 14
13	8 192	0.012 207	122.07	0.000 122 07
14	16 384	0.006 104	61.04	0.000 061 04
15	32 768	0.003 052	30.52	0.000 030 52
16	65 536	0.001 526	15.26	0.000 015 26
17	131 072	0.000 763	7.63	0.000 007 63
18	262 144	0.000 381	3.81	0.000 003 81
19	524 288	0.000 191	1.91	0.000 001 91
20	1 048 576	0.000 095	0.95	0.000 000 95
21	2 097 152	0.000 048	0.48	0.000 000 48
22	4 194 304	0.000 024	0.24	0.000 000 24
23	8 388 608	0.000 012	0.12	0.000 000 12
24	16 777 216	0.000 006	0.06	0.000 000 06

* May be limited by noise and other uncertainties in actual circuit.

Figure 10.57 Resolution of A/D and D/A converters.

A wide variety of D/A converters with a range of cost/performance and special features are available from many manufacturers: bipolar (± 10 V) outputs (the unit of Fig. 10.58 is unipolar, 0 to 10 V), current outputs (such as the process control 4- to 20-mA standard), multiplying types in which E_{ref} can be a dynamic variable, digitally buffered (data-bus-compatible) types for easy microprocessor interfacing, etc. Manufacturers' catalogs and handbooks provide comprehensive listings and discussions of specifications. In addition to resolution (mentioned earlier), here we mention only that settling times [within $\frac{1}{2}$ least-significant bit (LSB)] for full-scale input are in the range of 3 to 30 μs for "ordinary" units, while about 10 ns can be achieved by 8-bit "video" converters.

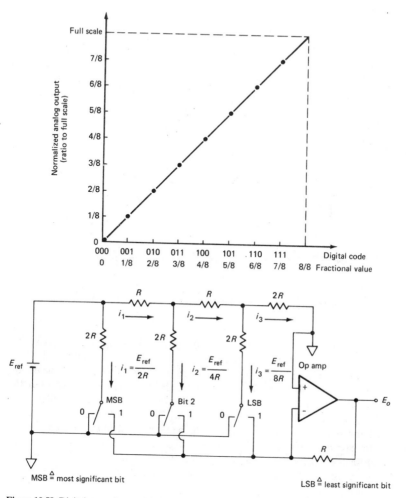

Figure 10.58 Digital-to-analog conversion (3 bits).

MSB $\overset{\triangle}{=}$ most significant bit

LSB $\overset{\triangle}{=}$ least significant bit

Let's turn now to A/D converters.[1] The majority of practical applications employ one of two different principles: successive approximation or dual slope (integrating). The dual-slope type is superior in most respects, except speed; however, its speed limitation rules out many applications, so successive-approximation types are very common. Figure 10.59 shows the ideal behavior of a 3-bit A/D converter. Note the inherent quantization uncertainty of $\pm\frac{1}{2}$ LSB. A

[1] B. M. Gordon, Linear Electronic Analog/Digital Conversion Architectures, Their Origins, Parameters, Limitations and Applications, *IEEE Trans.*, vol. CAS-25, no. 7, July 1978.

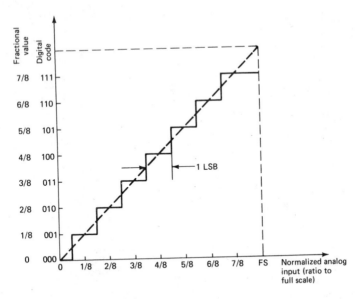

Figure 10.59 Analog-to-digital conversion (3 bits).

successive-approximation A/D converter operates according to the block diagram of Fig. 10.60.[1] When the "start conversion" command is applied, the D/A converter (which is built into the A/D device) outputs the most significant bit (MSB) for comparison with the analog input. If the input is greater than the MSB, the MSB remains on (1 in the output register) and the next smaller bit is tried; if the input is less than the MSB, the MSB is turned off (0 in the output register) and the next smaller bit is tried. If the second bit does not add enough weight to exceed the input, it is left on and the third bit is tried. If the second bit "tips the scales" too far, it is turned off and the third bit is tried. This process continues, in order of descending bit weight, until the last bit has been tried, at which point the status line changes state to indicate that the contents of the output register now constitute a valid conversion. (To force the transitions to occur at the ideal $\frac{1}{2}$ LSB points of Fig. 10.59, the comparator is biased by $\frac{1}{2}$ LSB.) This type of converter cannot tolerate much change in the analog input during the conversion process, so a sample/hold device must be used ahead of the converter if fast-changing analog signals are to be converted accurately. The status output of the converter can be employed to release the sample/hold from its hold mode at the end of conversion. Converters of this type are available to about 16 bits (conversion time ≈ 30 μs); 8-bit units can be as fast as 1 or 2 μs.

[1] D. H. Sheingold (ed.), Analog-Digital Conversion Notes, Analog Devices, Norwood, Mass., 1977.

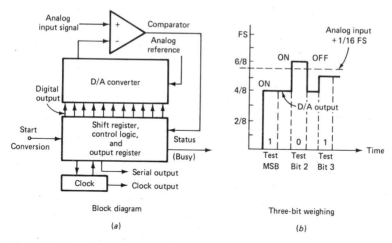

Figure 10.60 Successive-approximation A/D converter.

Several variations of the dual-slope principle are utilized; Fig. 10.61[1] shows a basic implementation. Conversion here is indirect since the analog input is first converted to a time interval which is then digitized by using a counter. The analog input V_{in} is applied to an integrator, and a counter (counting clock pulses) is started at the same time. After a preset number of counts (and thus a fixed time T), the input is disconnected and a reference voltage of opposite polarity is applied to the integrator. At this switching instant, integrator output is proportional to the average value of V_{in} over time interval T. The integral of V_{ref} is an opposite-going ramp of slope $V_{ref}/(RC)$. The counter, reset to zero at time T, now counts until the integrator output crosses zero, Δt seconds later. Therefore Δt (and thus counter output) will be proportional to the average value of V_{in} over time interval T. In Fig. 10.61, V_{in} has been offset by V_{ref} and divided by 2, which allows a bipolar analog input to produce an offset binary output suitable as input for computer systems.

Dual-slope converters have a number of advantages. Accuracy is unaffected by capacitor value or clock frequency, since these effect the up slope and down ramp equally. The integrating effect rejects high-frequency noise and averages changes in V_{in} during the integration period T. By choosing T as an integer multiple of the most prevalent noise signal's period (say, $T = \frac{1}{60}$ s for 60-Hz noise), theoretically perfect noise rejection at 60, 120, 180 Hz, etc., is obtained. Of course, this choice leads to the main disadvantage of dual-slope A/D converters: the slow conversion rate, usually less than 30 per second. A final advantage, widely implemented, is the easy inclusion of automatic zero capability. Here, before each measurement cycle the input is shorted; thus any output produced by

[1] Ibid.

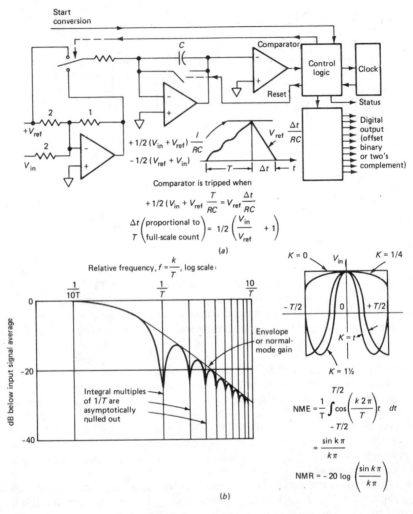

Figure 10.61 Dual-slope A/D converter.

the integration process over T must be zero-drift error, which is subsequently subtracted from the next reading, to give continuous zero correction. (Successive-approximation A/D devices do not use automatic zero because of the speed penalty.) Integrating A/D converters are available to about 17-bit resolution. Conversion rates are usually 3 to 30 per second; however, rates up to about 250 per second are available, although this sacrifices the line-frequency noise rejection.

We conclude this section with a brief discussion of sample/hold amplifiers

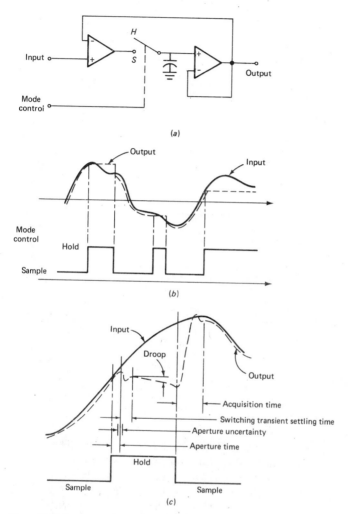

Figure 10.62 Sample-hold amplifier.

(SHAs),[1] often needed when successive-approximation A/D converters are employed with fast-changing inputs. Figure 10.62a shows the principle of the tracking-type[2] SHA we wish to discuss. When it is in tracking (sampling) mode, the switch is closed and the output follows the input, usually with a gain of +1. To engage the hold mold, the switch is opened, and ideally the output signal

---

[1] J. V. Wait, Sampled Data Reconstruction Errors, *Inst. & Cont. Syst.*, pp. 127–129, June 1970; L. W. Gardenhire, Selecting Sample Rates, *ISA J.*, p. 59, April 1964.

[2] Designer's Guide to High Resolution Products, Analog Devices.

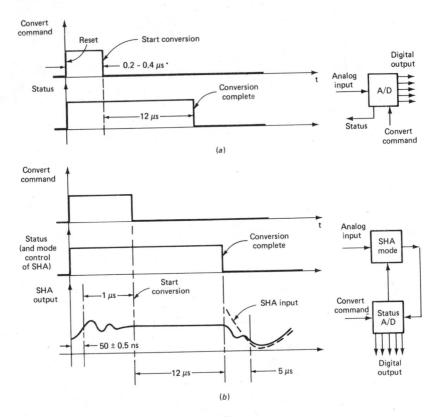

Figure 10.63 Effect of sample/hold on A/D conversion.

retains the last value it had when the hold command was given. This process continues until sampling (tracking) is again commanded, whereupon ideally the output jumps to the input value and follows it until hold is engaged again. In going from sample to hold, there is a delay called the *aperture time* (typically 50 ns) between the hold command and the actual opening of the switch. (Numerical values quoted here are for an SHA suitable for use with 14-bit A/D converters.) This delay is reproducible to within a random *aperture uncertainty* (typically 0.5 ns). Thus we can compensate for aperture delay by simply initiating the hold command early. After the switch opens, we must wait through the *switching transient settling time* (typically 1 μs) for the SHA output to settle within +0.003 percent before allowing the A/D device to start conversion. Once in hold, a real SHA will not hold perfectly steady, but will exhibit *droop*, which can be either positive or negative, typically 1 μV/s. In switching from hold to sample, an *acquisition time* (typically 5 μs to settle within ±0.01 percent after a 10-V step) delays return to accurate tracking of the input.

When a successive approximation A/D converter is used without an SHA, any input changes during the conversion period (typically 12 μs, see Fig. 10.63a) result in uncertainty in the digitized output. Take for example, a signal 10 sin $2\pi ft$ volts. Its maximum rate of change (*slew rate*) is $20\pi f$ volts per second. In deciding on how much uncertainty we can tolerate, a reasonable lower bound would be $\pm\frac{1}{2}$ LSB since the A/D converter's uncertainty is itself that great. For a 14-bit A/D device, $\pm\frac{1}{2}$ LSB is 610 μV (assuming 10 V full scale). Now we can calculate the highest allowable frequency of input such that the uncertainty of the digitized output is no greater than $\pm\frac{1}{2}$ LSB as 0.8 Hz! This severe frequency limitation is greatly relaxed when an SHA with properties as given above is connected at the A/D input. The sequence of events is now as in Fig. 10.63b, where we see that the total time for a conversion is somewhat longer. However, the uncertainty calculation now uses the SHA aperture uncertainty of 0.5 ns (in place of the A/D conversion time of 12 μs), which gives the highest allowable frequency as 19 kHz, a 24,000 : 1 improvement.

In addition to the above application, SHAs are utilized to store multiplexer outputs while the signal is being converted and the multiplexer is seeking the next signal to be converted; to determine peaks/valleys in analog data processing; to establish amplitudes in resolver-to-digital conversion; to facilitate analog computations involving signals obtained at different time instants; to hold converted data between updates in data distribution systems; and to perform "synchronous filtering." In synchronous filtering,[1] the source of the noise must be intermittent and predictable, such as the ignition spark of an oil-burner igniter. The same pulse that triggers the igniter can be used to put an SHA connected to the data signal into hold, protecting sensitive equipment from transient overload and slow recovery.

10.13 SIGNAL AND SYSTEM ANALYZERS

In the analysis and design of many devices and systems, it is necessary to have accurate knowledge about the characteristics of the inputs to the system. Once a system has been built, often its performance is checked by studying its output. Equipment for carrying out such studies may be characterized as *signal-analysis equipment*. Closely related to this is the problem of experimentally defining the characteristics (transfer function, frequency response, etc.) of a physical system which may be too complex to analyze accurately by theory alone. Equipment for such experimental modeling[2] investigations might be called *system-analysis equipment* and generally utilizes coordinated simultaneous measurements of both the system input and output signals, together with suitable data processing to obtain conveniently the desired system characterization.

Perhaps the most widely used signal-analysis equipment is that which mea-

---

[1] Sheingold, Transducer Interfacing Handbook, p. 64.
[2] Doebelin, "System Modeling and Response."

sures the frequency spectrum of a fluctuating physical quantity. The most common applications are in the field of sound and vibration, where the frequency spectrum of a sound pressure, stress, acceleration, etc., may be very useful in diagnosing faults in an operating machine or system. These faults can be traced to their origin by noting peaks in the frequency content at certain frequencies and then finding the machine parts that run at speeds which would produce such frequencies. The spatial variation ("mode shape") of resonant vibration at the various natural frequencies is also useful for guiding redesign efforts that redistribute material to achieve more optimum stress and deflection patterns and that alter the natural frequencies to avoid machine-excitation peaks. A number of commercially available computer-aided-design (CAD) systems include, in their methodology, the integration of dynamic test data of this type into the overall design-build-test-redesign sequence. Thus powerful theoretical methods (such as finite elements) can be employed when their accuracy and cost are appropriate, while automated dynamic test methods provide confidence in situations where theory is not yet completely reliable. The CAD system ties the two approaches together with a user-friendly interface.

While analog and hybrid analyzers[1] designed for sinusoidal testing are still in use, most current signal and system analyzers utilize the digital fast Fourier transform (FFT) approach,[2] which accommodates test signals of all types—sinusoidal, pulse, and random. These methods can be implemented offline on any general-purpose digital computer (by using tape-recorded data and appropriate software) or in essentially real time (by using special-purpose computers marketed as signal/system analyzers and including all necessary auxiliary equipment for rapid and convenient use). Two main types of instruments are available: single-channel devices useful for spectrum analysis of single signals and dual-channel units intended for system analysis that employ pairs of simultaneous input/output signals. The digital discrete Fourier transform process utilized by these instruments also can be thought of as a narrow-bandpass filtering operation, which makes the analyzers a form of digital filter.

To give some idea of the capabilities of this class of instrument, we describe briefly a recent dual-channel machine.[3] Two identical input channels accept analog signals in eight selectable full-scale ranges from ± 100 mV to ± 20 V, with either dc or ac coupling. The first operation, necessary to prevent aliasing (see Fig. 10.64) in the computed frequency spectrum, is a low-pass analog (antialiasing) filter which removes frequency components above the desired analysis range (selected from 16 full-scale frequency ranges extending from 1 Hz to 100 kHz). Selection of the desired frequency range, say 0 to 500 Hz, automatically chooses the correct antialiasing filter. Corresponding to each frequency range is a "time window" that defines the sample of incoming data to be digitally analyzed. For single-channel operation, a sample of 2,048 points can be employed, while

[1] Ibid., sec. 6.1.

[2] Ibid., secs. 6.1–6.3.

[3] Model 660A, Nicolet Scientific Corp., Northvale, N.J.

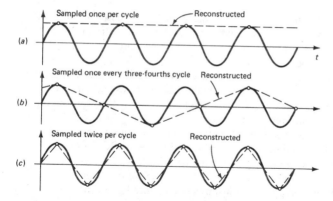

Figure 10.64 Aliasing caused by too sparse sampling.

dual-channel operation permits 1,024 points. For the 0- to 1-Hz range, a 400-s sample is required to get 1,024 points, while the 0- to 100-kHz range takes only 4 ms to gather the needed data. The sampling rate for each range is set so that 2.56 samples per cycle of the highest frequency are obtained. (For example, the range 0 to 1 Hz uses 2.56 samples per second, which explains the 400-s time window needed to accumulate 1,024 points.) The theoretical lower limit for such sampling, set by the Shannon sampling theorem, is 2 samples per cycle. If we think of the process as narrow-bandpass filtering with 400 filters covering the analysis range, the nominal bandwidth for the range 0 to 1 Hz is 0.0025 Hz, expanding proportionally to 250 Hz for the range 0 to 100 kHz. The computed frequency spectrum (Fourier transform) has 400 discrete values, spaced by the nominal bandwidths just quoted. These narrow bandwidths, particularly for the lowest frequency ranges, are impossible to obtain with analog approaches. In fact, "zoom"[1] options provide even greater frequency resolution, for example, 0.5 Hz for signals near 100 kHz.

The gathering of the desired 1,024-point digital sample by the 12-bit A/D converter requires different times, depending on the selected frequency range. But once the sample is in memory, the calculation of the discrete Fourier transform by the FFT algorithm always requires the same time—a fraction of a second. Thus (at least for the lower frequency ranges), all the incoming analog data can be analyzed without "gaps," since the next sample is being gathered while the present sample is being analyzed. This is called *real-time operation*, and it is another significant advantage over analog instruments, which take much longer to obtain a single spectrum and so have no chance of detecting and displaying spectrum variations, as they occur, in time-variant signals. The basic result of the

[1] N. Thrane, Zoom-FFT, *B&K Tech. Rev.*, no. 2, B&K Instruments Inc., Marlboro, Mass., 1980.

FFT calculation for an analog function sampled within the selected time window is its Fourier transform (magnitude and phase angle) in the form of 400 discrete values at 400 equally spaced frequencies in the selected range. In a dual-channel instrument analyzing the input and output signals of a "linear" physical system, the ratio of the two transforms, of course, gives the sinusoidal transfer function of the system. This is true irrespective of whether the input excitation is a swept-frequency sine wave, a single pulse, or a random signal of Gaussian or binary form, as long as the input-frequency content is adequate to "exercise" the system over the frequency range of interest. Actually, it has been found more accurate to compute the transfer function from the ratio of the cross spectrum of input and output signals and the power spectrum of the input. The cross spectrum is obtained by multiplying the output transform and the complex conjugate of the input transform, while the power spectrum is just the product of the input transform and its complex conjugate. Although this calculation gives results identical to those of the simple transform ratio when only a single record is analyzed, it gives improved accuracy when noise is present and averages of several records are used. The analyzer provides simple but versatile means for accumulating averages of various types.

In addition to the frequency-domain calculation of Fourier transform, power spectrum, cross spectrum, and transfer function, the analyzer provides the corresponding time-domain functions, such as autocorrelation function, cross-correlation function, and impulse response, all at the push of a button. Statistical functions such as probability density and cumulative distribution can be computed also. A built-in CRT display with full alphanumeric annotation displays all results. The results also can be read out for hard copy on external plotters or sent to external computers for further processing. A built-in microprocessor with simple front-panel programming allows selection of a wide variety of additional useful functions, such as time delays or advances, integration or differentiation, subtraction of background noise, correction for known transducer dynamics, recall of past setups for panel controls, etc. Addition of a special external mini-computer system gives computer-aided-design capabilities including animated-mode shape displays and finite-element analysis.

Single-channel units (spectrum analyzers) provide selectable narrow-band, one-third octave, and octave filtering capabilities which required separate analog machines in the past. The A-scale weighting function, much used in acoustics, is selectable. All the features mentioned for the dual-channel unit (except those peculiar to two-channel operations) are available.

10.14 MICROPROCESSOR APPLICATIONS

The integration of microprocessors into a wide variety of instruments is now well established, as you will note from the many specific applications mentioned throughout this text. The logical and computing power of microprocessors has extended the capabilities of many basic instruments, improving accuracy and efficiency of use. The development of microprocessor-based systems is, however, a specialized area beyond the scope of this text and one not practiced by most instrument users. So we cite some references[1] and move on to microprocessor applications which *are* accessible to a wide range of users who are not computer system designers. Since applications to specific instruments have been documented elsewhere in this text, here we want to indicate the availability of devices that make microprocessor computing power accessible in a more general way.

The main stumbling blocks which prevent a "casual" user from integrating a microprocessor into a measurement system are the complexities of interfacing; the need to learn a complicated, machine-dependent programming language; and possibly the need for an expensive microprocessor development system. All these problems are acceptable when you are developing a microprocessor system for an instrument that will sell thousands of units and these development costs can be spread out. Of course, you also get exactly the kind of performance desired since the system is designed at a rather basic level and can be optimized for the particular application. A control system engineer trying to implement some kind of improved measurement/control scheme for a process plant is in an entirely different situation, since perhaps only one or a few units are needed. Here the engineer would be looking for ready-made interfacing and a simple programming language and would be willing to pay considerably more than the cost of a "bare" microprocessor chip to gain these advantages.

Figure 10.65[2] shows the architecture of a process microcomputer designed for this type of situation and of a physical size ($3 \times 6 \times 20$ in) compatible with standard miniature panel instruments. It accepts any combination of eight frequency or analog inputs and four contact (digital) inputs, and it updates all results once per second, adequate for many process applications. Frequency/analog inputs are sampled at a rate of four frequency/voltage pairs per second. A simple programming language called FAPTRAN includes 43 program constants and 10 operators which, together with logic operations based on the four contact inputs, allow implementation of a wide variety of useful functions. Results of computations can be output through the four standard 4- to 20-mA current outputs provided and the one contact (logic) output. An LED numerical display can be set to provide a display of any of 8 measured or computed values (4 digits), any of 8 integrated flow totals (7 digits), a scan of all data and totals (each displayed for 5 s), and identification of up to 8 alarm conditions. Logging of all displayed data and integrated totals on an external teleprinter can be selected. Figure 10.66[3] shows a typical application to a boiler efficiency calculation.

The multifunction signal processor[4] of Fig. 10.67 is another example of mi-

---

[1] D. P. Burton and A. L. Dexter, Microprocessor Systems Handbook, Analog Devices, Norwood, Mass., 1977; P. H. Garrett, Analog Systems for Microprocessors and Minicomputers, Reston Publ., Reston, Va., 1978.

[2] Chameleon Process Microcomputer, Fischer and Porter Co., Warminster, Pa.

[3] Ibid.

[4] Model 7600, Oriel Corp., Stamford, Conn. T. C. O'Haver and A. Smith, A Microprocessor-Based Signal Processing Module, American Lab., February 1981, pp. 43–51.

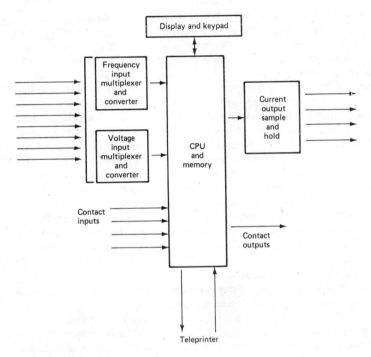

Figure 10.65 Process microcomputer architecture.

croprocessor computing power made available in a convenient package intended mainly for the analytical chemistry market but certainly useful elsewhere. Analytical instruments often require a variety of signal processing techniques to improve the signal-to-noise ratio and to increase the convenience and accuracy of measurement. Atomic absorption spectrometers utilize low-pass filters and running-mean devices for flame measurements, and integrators and peak detectors for furnace work. Ultraviolet-visible and fluorescent spectrophotometers may employ first- and second-derivative devices for improved resolution and specificity. While the latest top-grade instruments may have such capabilities built in, older or less expensive models do not and could benefit from a convenient "add-on" device, which would simply connect between the instrument output and recorder input. While all the listed functions can be (and are) performed with standard analog modules, the unit under discussion uses A/D conversion (10 bits), microprocessor calculation, and D/A conversion to achieve a strictly digital approach. Switch-selectable functions include test (data goes "straight through" without modification, convenient for checkout and setup), low-pass filter (time constant selectable from 0.002 to 32 s), first derivative, second derivative, integrator [scale factors 0.035 to 9.4 (V/s)/V], peak track and hold, and running average. The low-pass filter employs a special digital algorithm that improves on the ripple/response-speed tradeoff of analog RC filters. Noise-suppressing low-pass

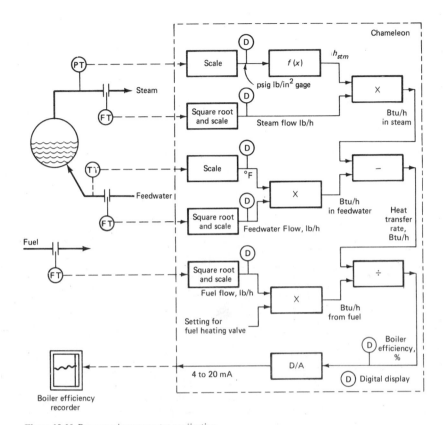

Figure 10.66 Process microcomputer application.

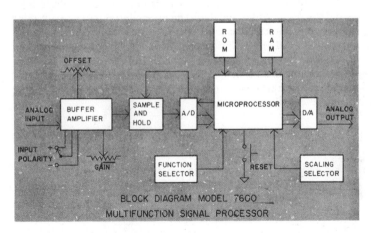

Figure 10.67 Multifunction signal-processor architecture.

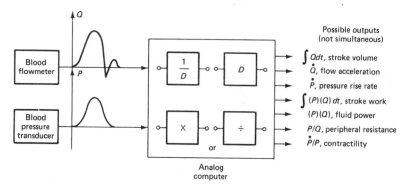

Figure 10.68 Blood-flow analog-computer system.

filters also are built into the differentiators. Digital integration provides the useful feature that when the output saturates (10 V), it "wraps around" back to zero and continues integrating. Running average is used to smooth noisy signals by taking readings at a 1-kHz rate until memory is full (256 entries) and then outputting the average. This is done over and over, with the average being augmented by each new set of data, until 65,536 readings have been taken, whereupon the system resets and starts fresh. The advantage of running average over a low-pass filter is that it is not necessary to judge beforehand what filter time constant will be appropriate for the noise in the signal. The main limitation on such a system is the modest processing speed of the microprocessor, which limits sampling rates to 1 kHz or less and gives the whole system a bandwidth less than 100 Hz; however, this speed is adequate for many applications.

Figure 10.68 shows, for comparison, that ready-to-use special-purpose analog processors are also available. The unit illustrated[1] is designed particularly for medical blood-flow studies and uses standard analog modules; however, all the details of power supplies, interconnections, range switching, etc., have been worked out, providing a ready-to-use instrument. The differentiator is an active (op-amp) type with two stages of low-pass filtering for noise suppression. The integrator (also a standard op-amp design) includes triggering and reset switching, so that integrals of flow rate (to get total flow) or power (to get total energy) may be computed for each individual cycle.

PROBLEMS

10.1 Derive the balanced-bridge relationship $R_1/R_4 = R_2/R_3$.

10.2 For a Wheatstone bridge, show that if $R_1 = R_2 = R_3 = R_4$ at balance, and if $\Delta R_2 = \Delta R_3 = 0$ and $\Delta R_1 = -\Delta R_4$, then the output voltage is a perfectly linear function of ΔR_1, no matter how large ΔR_1 gets.

---

[1] Model BL-620/622 Analog Computer, Biotronex Lab, Inc., Kensington, Md.

10.3 Discuss qualitatively the effect on bridge operation, for both the null method and the deflection method, of the excitation voltage source having an internal resistance.

10.4 In the system of Fig. 10.4, what considerations determine the numerical value of R_s?

10.5 In a Wheatstone bridge, $R_1 = 3,000$, $R_4 = 4,000$, $R_2 = 6,000$, and $R_3 = 8,000$ Ω at balance. Find the open-circuit output voltage if $\Delta R_1 = 30$, $\Delta R_2 = -20$, $\Delta R_3 = 40$, and $\Delta R_4 = -50$ Ω, and $E_{ex} = 50$ V. If the bridge output is connected to a meter of 20,000-Ω resistance, what will the output voltage now be?

10.6 Obtain $(e_o/e_i)(D)$ for the circuit of Fig. 10.23a.

10.7 Derive Eq. (10.45).

10.8 Why does a charge amplifier essentially amount to a short circuit across the crystal?

10.9 Derive the transfer functions of the circuits of (a) Fig. 10.22a, (b) Fig. 10.22b.

10.10 Derive the transfer function of the hydromechanical filter of Fig. 10.24.

10.11 In the circuit of Fig. 10.24, let e_i be supplied by a sinusoidal generator with an internal resistance of 1,000 Ω and an open-circuit voltage of 10 V peak to peak. Also let $C = 10$ μF and $R = 10$ Ω. If e_o is open-circuit, what voltage will actually appear at the e_i terminals for frequencies of 0, 100, 1,000, 10,000, and 100,000 Hz?

10.12 Derive Eq. (10.49).

10.13 Explain how a notch filter can be used in the feedback path of a high-gain feedback system to construct a bandpass filter.

10.14 Derive Eq. (10.50).

10.15 Derive Eq. (10.51).

10.16 Derive Eqs. (10.53) to (10.57). Show also that the phase angle at ω_p is zero.

10.17 Sketch the configuration of a hydraulic bandpass filter which has an amplitude-ratio attenuation of 40 dB/decade on either side of the passband. This is twice the attenuation rate of the system of Fig. 10.28. Only components of the type used in Fig. 10.28 are allowed. Short "transition" regions of slope ± 20 dB/decade are allowed between the flat response portion and ± 40 dB/decade portions. You must derive the transfer functions to prove your "invention" works as claimed.

10.18 Derive Eq. (10.58).

10.19 Derive Eq. (10.66).

10.20 Derive Eq. (10.76).

10.21 For the system of Fig. 10.33a, let $e_i = e_{signal} + e_{noise}$, where $e_{signal} = 10 \sin 20t$ and $e_{noise} = 0.1 \sin 377t$. What is the signal-to-noise ratio before and after the differentiation?

10.22 The input to a differentiator with transfer function $(e_o/e_i)(D) = D$ is a random signal with a constant mean-square spectral density of 0.001 V^2/Hz from 0 to 10,000 Hz and zero elsewhere. Calculate the rms voltage at the input and at the output of the differentiator.

10.23 Derive Eq. (10.78).

10.24 Derive Eq. (10.79).

10.25 Using the system of Eq. (10.80) and the e_i of Prob. 10.21, compute the signal-to-noise ratio at both the input and the output.

10.26 Derive Eq. (10.92).

10.27 Discuss sensitivity/response-speed tradeoffs in the system of Fig. 10.38.

10.28 Derive the transfer functions of the compensating circuits of Fig. 10.40.

10.29 Design a compensating network to speed up by a factor of 10 the response of a thermocouple with a time constant of 1 s. Thermocouple resistance is 10 Ω and full-scale output is 5 mV. The amplifier/recorder available has maximum full-scale sensitivity of 0.1 mV and an input resistance of 100,000 Ω.

10.30 List and explain the action of all effects that tend to degrade the static accuracy of the system in Fig. 10.45.

10.31 Derive the equation of the op-amp summing circuit of Fig. 10.46d.

10.32 Derive the operating equation of the mechanical filter of Fig. 10.29.

10.33 Discuss the dynamic response of the system of Fig. 10.49.

10.34 The "standard" op-amp integrating circuit, as used in analog-computer applications, becomes almost impossible to employ in those measurement instrumentation applications for which the signal is a short-duration (a few milliseconds) pulse. Since a "computing loop," as found in all analog-computer applications where a differential equation is being solved, is not present, drift is not self-correcting. Also, if input signal e_i is, say, 5 V but lasts only 1 ms, then $\int e_i \, dt$ is only 0.005 V · s, and we need a very large value of $1/(RC)$ to get an output voltage of convenient (volt level) size. For a typical $1/(RC) = 1,000$, if e_i (which comes from some sensor and/or amplifier) differs from zero (before the pulse to be integrated occurs) by only a small amount, say 0.005 V, then e_o will "drift" at the rate of 5 V/s, causing saturation (and disabling of the integrator) in just a few seconds. An approximate integrator which solves this drift problem is shown in Fig. P10.1.

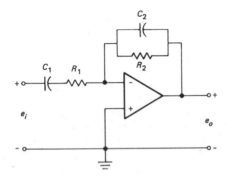

Figure P10.1

    (a) Find $(e_o/e_i)(D)$.

    (b) Take $R_1 C_1 = R_2 C_2$ and plot logarithmic frequency-response curves (dB and ϕ) for (e_o/e_i) $(i\omega)$. Superimpose on this graph the frequency response of a perfect integrator with $RC = R_1 C_2$. Discuss the nature of the approximation for the approximate integrator. Why is drift no longer a serious problem?

    (c) Take $R_1 = 10^6$ Ω, $R_2 = 10^9$ Ω, $C_1 = 10^{-6}$ F, and $C_2 = 10^{-9}$ F. Use CSMP (or other available digital simulation) to check performance of the approximate integrator if e_i is a rectangular pulse of height 5 V and duration 0.001 s. Would the integrator work well for short pulses of other waveforms? Why? What about "long-duration" pulses?

BIBLIOGRAPHY

1. R. L. Bannister: Signature Analysis of Turbomachinery, *Sound & Vib.*, September 1971.
2. P. R. Perino: The Effect of Transmission Line Resistance in the Shunt Calibration of Bridge Transducers, *Statham Instrum. Notes* 36, Statham Instruments Inc., Los Angeles, November 1950. (Statham Instruments is now Gould Measurement Systems, Oxnard, Calif.)
3. P. Pohl: Signal Conditioning for Semiconductor Strain Gages, *ISA J.*, p. 33, June 1962.
4. Another Look at the Wheatstone Bridge, *Electromech. Des.*, p. 36, February 1965.
5. A. Baracz: Graph Finds Temperature Sensing Bridge Response, *Contr. Eng.*, p. 85, October 1961.
6. R. B. F. Schumacher: Differential High-Resistance Bridge, *ISA J.*, p. 65, April 1965.
7. P. Perino: System Considerations for Bridge Circuit Transducers, *Statham Instrum. Notes* 37, Statham Instruments Inc., Los Angeles, September 1964.

8. G. White: Temperature Compensation of Bridge Type Transducers, *Statham Instrum. Notes* 5, Statham Instruments Inc., Los Angeles, October 1948.

9. B. B. Helfand: Summation and Averaging of Multiple Measurements by Parallel Transducer Operation, *Statham Instrum. Notes* 16, Statham Instruments Inc., Los Angeles, July 1950.

10. B. B. Helfand and J. Burns: Calibration of Resistance Bridge Transducer Circuits under Temperature Extremes, *Statham Instrum. Notes* 14, Statham Instruments Inc., Los Angeles.

11. H. E. Darling: Magnetic Amplifiers for Instrumentation, *ISA J.*, p. 58, January 1960.

12. J. J. Rado: Input Impedance of a Chopper-Modulated Amplifier, *Electro-Technol. (New York)*, p. 140, June 1962.

13. J. DiRocco: Potentiometric Amplifiers Improve Impedance Buffering, *Contr. Eng.*, p. 87, July 1962.

14. J. Minck and E. Smith: Noise Figure Measurement, *Instrum. Contr. Syst.*, p. 115, August 1963.

15. W. R. Williams and R. C. Hawes: Vibrating Reed Electrometer, *Instrum. Contr. Syst.*, p. 112, November 1963.

16. A. Pearlman: Selecting and Testing Solid-State Operational Amplifiers, *Instrum. Contr. Syst.*, p. 121, February 1965.

17. R. D. Moore: Lock-in Amplifiers for Signals Buried in Noise, *Electronics*, June 8, 1962.

18. C. T. Stelzried: Loaded Parallel-T *RC* Filters, *Contr. Eng.*, p. 113, May 1961.

19. G. Cocquyt: Evaluating Bridged-T Networks for AC Systems, *Contr. Eng.*, p. 77, December 1963.

20. A. I. Zverev: Introduction to Filters, *Electro-Technol. (New York)*, p. 61, June 1964.

21. W. Gile: Solid-State Low-Frequency Filter, *Electro-Technol (New York)*, p. 34, September 1964.

22. A. W. Langill: Designing Passive Compensators, *Electro-Technol (New York)*, p. 26, January 1965.

23. Miniature Servo Packages, *Electromech. Des.*, p. 70, June 1960.

24. J. B. Heaviside: Sources of Error in AC Servos, *Contr. Eng.*, p. 85, February 1964.

25. H. J. Huttenlocker et al.: Instrument Servomechanism Systems, *Electromech. Des.*, p. 37, July 1964.

26. Miniature Servo Packages, *Electromech. Des.*, p. 202, May 1962.

27. Specifying an Instrument Servomechanism, *Electromech. Des.*, p. 32, November 1959.

28. M. Richter: A Simplified Technique in Instrument Servo Analysis, *Electromech. Des.*, p. 36, February 1962.

29. A. Svoboda: "Computing Mechanisms and Linkages," McGraw-Hill, New York, 1948.

30. T. R. Fredriksen: A Way to Design Low-Loss Nonlinear Networks, *Contr. Eng.*, p. 117, June 1962.

31. A. J. Baracz: How to Design a Compensating Bridge, *Contr. Eng.*, p. 81, March 1965.

32. F. M. Ryan: Special Purpose Analog Computers, *Contr. Eng.*, p. 103, May 1963.

33. J. T. Nichols: Zener-Regulated Power Supplies, *Instrum. Contr. Syst.*, p. 2242, December 1961.

34. J. Nagy: Zener Diode Power Supplies, *ISA J.*, p. 65, July 1964.

35. G. A. Korn: "Minicomputers for Engineers and Scientists," McGraw-Hill, New York, 1973.

36. D. M. Auslander and P. Sague: "Microprocessors for Measurement and Control," Osborne/McGraw-Hill, Berkeley, Calif., 1981.

ELEVEN

DATA TRANSMISSION

When the components of a measurement system are located more or less remotely from one another, it becomes necessary to transmit information among them by some sort of communication channel. Also in some cases, even though components are close together, transmission problems arise because of relative motion of one part of the system with respect to another. We examine briefly questions of this sort and some of the equipment commonly used to solve such problems.

The transmission of information is amenable to mathematical analysis totally dissociated from any hardware considerations, and there is a large body of technical literature on this subject. This science of communication has been extremely useful in putting the design of hardware on a rational basis, showing the tradeoffs in competitive systems, and putting theoretical limits on what can be done. Its consideration, however, is beyond the scope of this text, and we restrict ourselves to rather qualitative, hardware-oriented discussions.

11.1 CABLE TRANSMISSION OF ANALOG VOLTAGE AND CURRENT SIGNALS

Perhaps the most common situation is that in which a simple cable is used to transmit an analog voltage signal from one location to another. The accurate analysis of a cable or transmission line involves the use of a distributed-parameter (partial differential equation) approach since the properties of resistance, inductance, and capacitance are not lumped or localized.[1] Figure 11.1a shows the model generally employed for such an analysis. An approximation

[1] H. H. Skilling, "Electric Transmission Lines," McGraw-Hill, New York, 1951.

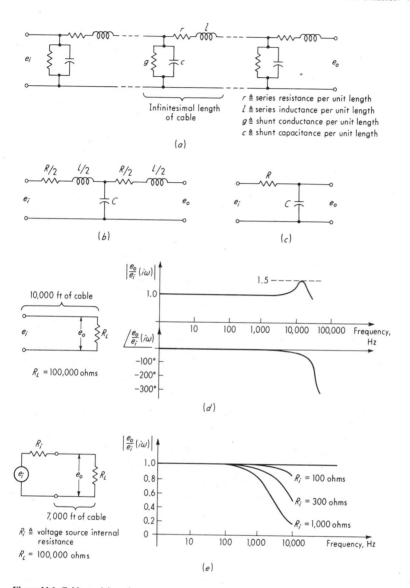

Figure 11.1 Cable models and response.

suitable for low frequencies is the lumped network of Fig. 11.1b. The shunt conductance has been neglected here since, in practice, generally it is negligible. If the line is not too long and the frequencies are not too high, the crude model of

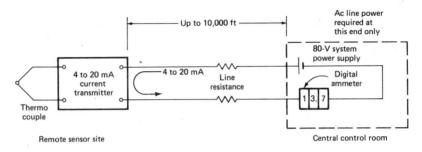

Figure 11.2 Current loop of 4 to 20 mA.

Fig. 11.1c may be adequate. The inductance has been totally neglected; R represents the total resistance of both conductors in the cable, while C represents the total capacitance between them. These can be numerically calculated from the values per foot of length given by cable manufacturers. Typically, resistance per foot might be of the order of 0.01 Ω while capacitance would be about 30 pF/ft. Thus a 1,000-ft length of two-conductor cable has $R = 20\ \Omega$ and $C = 30,000$ pF. The actual frequency response of a length of cable can be measured experimentally to get its exact characteristics. Figure 11.1d and e shows some typical results.[1] The use of cables up to 7,000 ft long to transmit low-level (± 10 mV full scale) data has been accomplished.[2] However, even short cables (less than 10 ft) can cause difficulties in high-impedance transducers such as the piezoelectric type. The use of charge amplifiers rather than voltage amplifiers may be helpful in such cases.

A hard-wired (cable) data-transmission technique widely utilized in the process industries is the so-called current loop (usually a 4- to 20-mA range). Transmitters are available which convert millivolt, thermocouple, RTD, frequency, slidewire potentiometer, or bridge-circuit inputs into a proportional output current. A zero input signal produces 4-mA current while full-scale input produces 20-mA. Such transmitters are available in several forms. The two-wire version of Fig. 11.2 is particularly convenient since the connection between the central control room and the remote sensor requires only two wires to transmit both power and signal. The voltage appearing across the transmitter output terminals (which will vary as output current changes) is actually the transmitter's power supply, but the transmitter operation is insensitive to changes in this voltage as long as it stays above some minimum, say 9 V. Thus for a system supply voltage of, say, 80 V, line resistance can be as high as $(80 - 9)/0.02 = 3,550\ \Omega$. The transmitter is a true current source, which makes the system relatively immune to induced noise voltages and line resistance changes (within the 3,550-Ω limit). While an actual transmitter is somewhat more complicated, the op-amp circuit of

---

[1] R. L. Smith, Transmission of Low-Level Voltage Over Long Telephone Cables, *NASA, Tech. Note* D-1320, p. 14, January, 1963.

[2] Ibid.

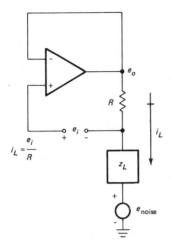

Figure 11.3 Basic op-amp current source.

Fig. 11.3 shows in principal how such current-source behavior may be achieved. Digital-to-analog converters with 4- to 20-mA current output are available. Figure 11.4a[1] shows one that takes all its power from the loop supply and provides a redundant analog output, which automatically takes over in the case of computer crashes. Figure 11.4b[2] shows a number of such D/A devices (all sharing a common loop supply) embedded in a computerized process-control system.

11.2 CABLE TRANSMISSION OF DIGITAL DATA

When data must be transmitted very long distances (100 mi is not unusual), analog signals tend to be corrupted by the response characteristics of the transmission line and the pickup of spurious noise voltages from a number of sources. Under such conditions, it may be desirable to convert the analog data to some digital form, transmit them in digital form, and then reconvert to analog form if desired. A time-honored example of digital transmission is the telegraph system, in which letters and numbers are represented by a system of coded pulses (dots and dashes).

In many cases the information can be transmitted over lines that were installed for other purposes. Electric power systems, for example, transmit information signals over their power transmission lines simultaneously with the transmission of 60-Hz power. It is simply necessary to keep the frequencies utilized sufficiently separated to allow easy filtering for elimination of unwanted signals.

---

[1] *Analog Dialogue*, vol. 14, no. 1, Analog Devices, Norwood, Mass., 1980.
[2] Ibid.

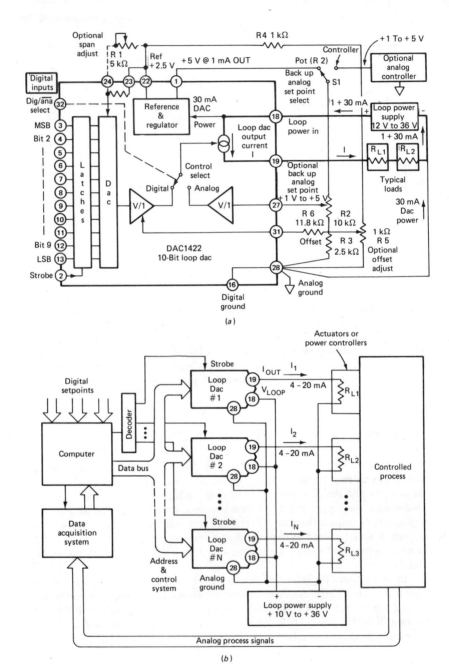

Figure 11.4 Digital-to-analog converter with current output of 4 to 20 mA.

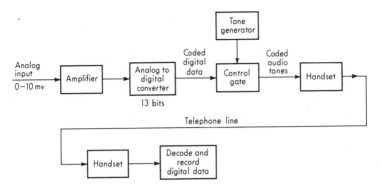

Figure 11.5 Telephone digital-data transmission.

Telephone lines are also in wide use for data transmission. An interesting system[1] shown in block-diagram form in Fig. 11.5 employs ordinary telephone handsets together with auxiliary equipment to transmit and receive data. The telephone line is not leased; *any* telephone can be used to call any other in the usual way. Once voice contact is established, the handsets are placed into an "acoustic coupler." Here tones at 1,400 Hz (representing binary 1) and 2,100 Hz (representing binary 0) produced by a tone generator actuated from the analog-to-digital converter are coupled into a loudspeaker which the telephone "hears." These sounds are transmitted over the telephone line in the usual way, to be received by a similar system (but working in reverse fashion) at the other end. Frequency response of such a system is quite low, but accuracy of transmission is the order of 0.1 percent. Maximum bit rates are currently[2] 1,200 Bd [1 baud (Bd) = 1 bit/s].

Digital communications, of course, are necessary over shorter distances also. The 20-mA current transmitters discussed in Sec. 11.1 can be designed to provide digital transmission at up to 9,600 Bd and distances up to 10,000 ft (see Fig. 11.6[3]). The standard ASCII RS-232C transmission is limited to about 50 ft at rates up to 9,600 Bd. Higher-speed interfaces (such as the various versions of IEEE-488) intended for computer based systems can handle 20,000 bytes/s or more, but the distance between instruments may be limited to 6 ft with total length no more than 60 ft. The most sophisticated systems can handle about 3×10^6 Bd over about 300 ft without use of repeaters.

The high accuracy of digital data transmission as compared with analog is due to the fact that the size or precise shape of a pulse in a digital system is not particularly important. Rather, the system operates on the presence or absence of some sort of pulse. Thus even rather severe degradation of pulse shape by the

[1] M. L. Klein, Telephonic Transmission of Data, *Instrum. Contr. Syst.*, p. 99, June 1962.
[2] T. J. McShane, Acoustic Coupling for Data Transmission, *Digital Design*, pp. 38–46, June 1980.
[3] *Publ.* C603b-15-1/81, Analog Devices, Norwood, Mass.

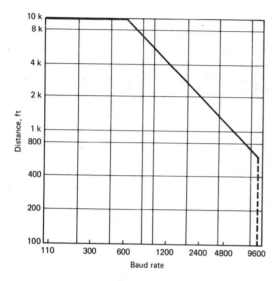

Figure 11.6 Distance/speed trade-off for current loop of 4 to 20 mA.

transmission medium does not affect the accuracy of a digital system *at all* as long as the presence or absence of a pulse can be detected.

11.3 FIBER OPTIC DATA TRANSMISSION

The use of optical (rather than electrical) means of data transmission is of increasing importance. Methods of transmitting both analog and digital information are available.[1] While communication over either an air path or a glass fiber is possible, we emphasize the fiber approach.[2] Basically, an electrically controllable light source (often an LED), an optical fiber, and a photodetector (often a silicon photodiode) are needed to make a complete system. Fiber optics offers wide bandwidths and high data rates while avoiding many of the interference and security constraints associated with traditional electronic communications. Glass cable can carry signals to or from instrumentation at high common-mode voltage levels without creating electrical paths between devices; that is, it provides high isolation. Optical fibers also are immune to electromagnetic interference from sources such as lightning or power switching, and provide good noise rejection. Glass cables produce no sparks; one application uses a fiber running inside a natural-gas pipeline. Data security is enhanced since tapping into fiber optic lines

---

[1] I. Math, Basic Optical Links, *Electro-Opt. Syst. Des.*, pp. 48–51, September 1977.

[2] W. F. Trover, Fiber Optic Communications for Data Acquisition and Control, *InTech*, pp. 33–41, April 1981; B. G. Mahrenholz and R. R. Little, Jr., A Fiber Optic Link for High-Speed Data Acquisition, *InTech*, pp. 43–46, April 1981; High Sensitivity–Low Speed Fiber Optic Transmitter and Receiver, PDS-408A, PDS-425, and AN-97, Burr-Brown Res. Corp., Tucson, Ariz.

is extremely difficult without causing fiber failure. Present applications range from distances of a few meters to 30 km (without repeaters) and data rates up to about 10 MHz.

11.4 FM/FM RADIO TELEMETRY

When interconnecting wires are not possible or desirable, data may be transmitted by radio. A number of different schemes[1] are in use; we consider first one that is widely employed. The word *telemetry* means simply measurement at a distance and includes all forms of such systems, irrespective of the means of transmission of physical nature of the hardware.

Radio telemetry probably received its greatest impetus from the requirements of aircraft and missile flight testing during and after World War II. Considerable standardization based on the requirements of such systems has been accomplished, and our discussion here reflects this emphasis. Figure 11.7 shows the widely utilized FM/FM system of radio telemetry. The symbol "FM/FM" refers to the fact that two frequency-modulation processes are employed. In the first process, time-varying dc voltages are converted to proportional frequencies, by using voltage-to-frequency converters as described in Sec. 10.11. When employed in FM/FM telemetry systems, these converters generally are called *subcarrier oscillators*. Instead of voltage-to-frequency converters, subcarrier oscillators may be inductance-controlled. The inductance of a variable-inductance transducer forms part of the oscillator circuit, and changes in inductance cause proportional changes in frequency from the center frequency.

The standard FM/FM system of Fig. 11.7 has 18 available channels; thus 18 different physical variables may be measured and transmitted simultaneously. Figure 11.8 shows the characteristics of these 18 channels. Note that the center frequencies range from 400 to 70,000 Hz. Such low frequencies cannot be practically transmitted by radio propagation since they would require antennas of immense size, because the size of an antenna must be of the order of the wavelength to be transmitted [wavelength in meters = 3×10^8/(frequency in hertz)]. Thus an additional frequency modulation to boost all frequencies into the radio-frequency (RF) range is employed. Rather than utilize a separate radio-frequency transmitter for each of the 18 channels (which is wasteful of the crowded RF spectrum and requires much more equipment), the 18 channels are "mixed" (added) and sent out together over one radio-frequency channel. Two such channels, 217.550 and 219.450 MHz, are available for radio telemetry. The frequency deviation caused by any of the subcarrier-oscillator frequency variations cannot exceed ± 125 kHz around either of these two frequencies. (Other radio frequencies in the range 216 to 235 MHz may be used if available.) When the

---

[1] C. M. Harris and C. E. Crede (eds.), "Shock and Vibration Handbook" vol. 1, pp. 19–76, McGraw-Hill, New York, 1961; M. H. Nichols and L. L. Rauch, "Radio Telemetry," Wiley, New York, 1956.

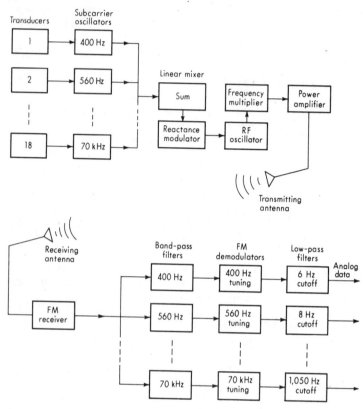

Figure 11.7 FM/FM radio telemetry.

radio signals are received, the 18 channels must be reseparated by suitable bandpass filters, FM-demodulated and low-pass-filtered to reconstruct the original analog data. We see from Fig. 11.8 that by using optional band *E*, a channel with frequency response as great as 0 to 2,100 Hz is available. This is adequate for many purposes. Systems of the above type have ranges up to about 400 mi and accuracies of the order of ± 2 percent of full scale. Digital forms of telemetry, pulse-code modulation[1] (PCM) being the most important, also are in wide use. The digital aspect here occurs before (and after) the hardwired or radio transmission link; the transmission itself is not "digital." For data of relatively low frequency content, any one of the channels may be timeshared by several transducers if a commutator is employed to sample each transducer periodically. If a

---

[1] C. S. Johnson, Telemetry Data Systems, *Instr. Tech.*, pp. 39–43, August, 1976; pp. 47–53, October 1976.

Band	Center frequency, Hz	± Full-scale frequency deviation, %	Overall frequency response, dc to Hz
1	400	7.5	6.0
2	560	7.5	8.4
3	730	7.5	11.0
4	960	7.5	14
5	1,300	7.5	20
6	1,700	7.5	25
7	2,300	7.5	35
8	3,000	7.5	45
9	3,900	7.5	59
10	5,400	7.5	81
11	7,350	7.5	110
12	10,500	7.5	160
13	14,500	7.5	220
14	22,000	7.5	330
15	30,000	7.5	450
16	40,000	7.5	600
17	52,500	7.5	790
18	70,000	7.5	1,050

Optional bands

	Band	Center frequency, Hz	± Full-scale frequency deviation, %	Overall frequency response, dc to Hz
Omit 15 and B	A	22,000	15.0	660
Omit 14, 16, A, C	B	30,000	15.0	900
Omit 15, 17, B, D	C	40,000	15.0	1,200
Omit 16, 18, C, E	D	52,500	15.0	1,600
Omit 17 and D	E	70,000	15.0	2,100

Figure 11.8 Telemetry channel characteristics.

very large number of low-frequency signals must be telemetered, other methods such as pulse-duration modulation (PDM/FM) may be indicated. More sophisticated telemetry systems can be used over ranges of millions of miles, as in the various space probes.

Radio telemetry is also useful over very short distances when the relative motion of the measuring device and the readout equipment prevents a suitable direct connection. Good examples of such situations are found in measurements on rotating machinery,[1] where slip-ring techniques are not feasible because of high speeds or inaccessibility, and in physiological measurements on test animals or human beings, in which restriction of motion due to connecting wires is not tolerable. Miniaturization of the telemetry components and improvement of shock resistance by use of IC semiconductor devices now make possible many such applications formerly not feasible.

[1] A. Adler, Telemetry for Turbomachinery, *Mech. Eng.*, pp. 30–35, March 1979; R. E. Kemp, Closed-Coupled Telemetry for Measurements on Gas Turbines, *Instr. Tech.*, pp. 105–112, September 1978; V. Donato and S. P. Davis, Radio Telemetry for Strain Measurements in Turbines, *Sound & Vib.*, pp. 28–34, April 1973.

11.5 PNEUMATIC TRANSMISSION

Transmission of pressure signals in industrial pneumatic control systems is accomplished regularly over distances of several hundred feet. Pneumatic-transmission-line dynamics is analogous to that of electric cables but, of course, at a much lower frequency. Adequate simplified models employing a dead time equal to the acoustic transmission time and either a first-order or second-order system are available and are discussed in Chap. 6.

11.6 SYNCHRO POSITION REPEATER SYSTEMS

Figure 11.9 illustrates synchro position repeater systems used for transmitting low-power mechanical motion over considerable distances with only a three-wire interconnecting cable. Whenever the two angles θ_i and θ_o are not identical, an electromagnetic torque is exerted on the rotor of *each* machine, which tends to bring the shafts into alignment. Thus, if the transmitter shaft θ_i is turned, the receiver shaft θ_o will follow accurately as long as there is no appreciable torque load on the θ_o shaft. The accuracy of such systems depends on the torque gradient (torque per unit error angle) of the transmitter/receiver system. A typical value might be 0.35 in · oz/degree. The electrical system serves only to transmit power from the θ_i to the θ_o shaft; all the mechanical work taken out at θ_o must be provided as mechanical work at θ_i. The torque gradient is reduced as the resistance of the connecting cable is increased; thus long cables result in reduced accuracy. In a typical unit, 10-Ω resistance in each of the three wires results in a 50 percent loss of torque. The dynamic response of these systems is essentially second-order; a mechanical analog is shown in Fig. 11.9. Sometimes a damper is put on the θ_o shaft to reduce oscillations since little inherent damping is present. When one transmitter drives several receivers (all units of identical size), the torque available at each receiver is $2/(N + 1)$ times the torque for a single pair, where N is the number of receivers. The synchro differential shown in Fig. 11.9 is useful for comparing two rotations at a location remote from either. Its static and dynamic behavior is essentially the same as for a transmitter/receiver pair.

11.7 SLIP RINGS AND ROTARY TRANSFORMERS

When transducers must be mounted on the rotating members of machines, some means must be provided to bring excitation power into the transducer and to take away the output signal. Some transducers (such as synchros) themselves are rotating "machines" in which such data and/or power transmission between a rotating and a stationary member is necessary. When only a small relative motion is involved, continuous flexible conductors (often in the form of light coil springs) can be employed. In some cases of limited rotation through a few revolutions, the connecting wires simply can be allowed to wind or unwind on the

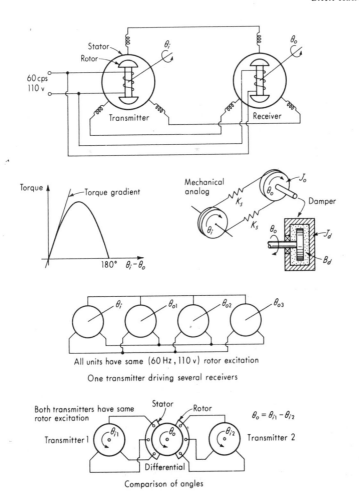

Figure 11.9 Torque-synchro angle transmission.

rotating shaft. However, continuous high-speed rotation requires slip rings, radio telemetry, or some form of magnetic coupling between rotating and stationary parts.

Figure 11.10 shows the common forms of slip rings.[1] Rings are made of coin gold, silver, or other noble metals and alloys. Block-type brushes often are sintered silver graphite while wire-type brushes are alloys of platinum, gold, etc. An important consideration in slip rings used to transmit low-level instrumentation signals is the electrical noise produced at the sliding contact. One component of

[1] A. J. Ferretti, Slip Rings, *Electromech. Des.*, p. 145, July 1964.

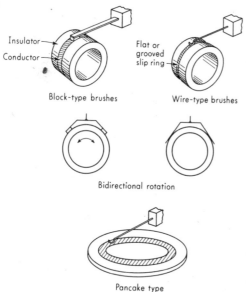

Insulator
Conductor
Flat or grooved slip ring

Block-type brushes   Wire-type brushes

Bidirectional rotation

Pancake type

Figure 11.10 Slip-ring configurations.

this noise is due to thermocouple action if the brush and ring are of different materials. The other main effect is a random variation of contact resistance from surface roughness, vibration, etc. If the contact carries current, a variation in contact resistance causes a noise voltage to appear at the contact. A high-quality miniature sliding slip ring may exhibit a contact-resistance variation of the order of 0.05 Ω peak to peak and 0.005 Ω rms.[1]

While slip rings have been operated successfully at about 100,000 r/min, applications above 10,000 r/min generally require extreme care because of heating and vibration problems. A particular slip-ring assembly[2] usable to 100,000 r/min and intended for strain-gage work had peak-to-peak noise voltage of 0.02 mV at 52,000 r/min and 0.40 mV at 100,000 r/min. This assembly used liquid cooling and lubrication of slip rings and bearings and gave a brush life of 30 h at 35,000 r/min. At 52,000 r/min the noise level in a typical strain-gage circuit gave a signal- to- noise ratio of about 150 : 1. Hard gold rings of $\frac{1}{4}$-in diameter were utilized with two cantilevered wire-tuft brushes per ring.

When slip rings are used with strain-gage circuits, particular care must be taken since the resistance variation of the sliding contact may be comparable with the small strain-gage resistance change to be measured. If possible, a full bridge on the rotating member should be employed so that the sliding contacts

[1] E. J. Devine, Rolling Element Slip Rings for Vacuum Application, *NASA, Tech. Note* D-2261, p. 11, 1964.
[2] Ferretti, op. cit., p. 159.

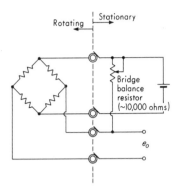

Figure 11.11 Bridge-circuit slip-ring configuration.

can be taken out of the bridge circuit. This arrangement (Fig. 11.11) greatly reduces the effects of slip-ring resistance variations. For the most demanding applications, more complex schemes are available[1] to reduce noise to even lower levels.

A rotating disk dipping into a mercury pool (see Fig. 11.12) can perform the same function as a conventional slip ring. A commercially available device[2] is usable from 0 to 10,000 r/min; has a contact resistance of 0.005 Ω, contact-resistance variation of ± 0.00025 Ω for 0 to 600 rpm and no measurable resistance variation from 600 to 10,000 rpm; is compensated for self-generated thermoelectric voltages; and can be made with 2 to 160 terminals.

Ordinary sliding slip rings may not operate properly in the high-vacuum environment of space. Preliminary research[3] using a thrust-type ball bearing as the signal-transfer mechanism indicates that rolling contact slip rings may provide a solution for such problems. A particular test at 2,000 r/min and vacuum of 2×10^{-9} torr gave operation for over 100 million revolutions at a resistance variation of 0.002 Ω rms.

Another alternative to slip rings is the rotary transformer. With this device, signal and/or power voltages are transferred through an annular air gap between concentrically rotatable primary and secondary transformer coils. Figure 11.13[4] shows two data systems (developed for jet-engine testing) which employ rotary transformers. One system is used with slowly changing thermocouple signals and utilizes an 8-bit A/D converter to digitize the temperature data, which are then transmitted serially (one bit at a time) through the rotary transformer. The other

[1] C. C. Perry and H. R. Lissner, "The Strain Gage Primer," p. 186, McGraw-Hill, New York, 1955; P. K. Stein, "Measurement Engineering" vol. 2, chap. 29, Stein Engineering Services, Inc., Phoenix, Ariz.

[2] Meridian Laboratory, Lake Geneva, Wis.

[3] Devine, op. cit.

[4] D. J. Lesco, J. C. Sturman, and W. C. Nieberding, On-the-Shaft Data Systems for Rotating Engine Components, *NASA* TM X-68112, 1972.

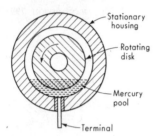

Figure 11.12 Mercury-pool slip ring.

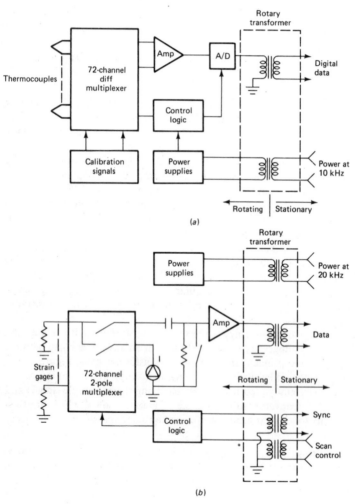

Figure 11.13 Rotary-transformer applications.

system handles dynamic strain-gage data of bandwidth 150 to 3,000 Hz, and since no dc data are present, this information can be directly coupled to the transformer.

PROBLEMS

11.1 Find a general expression for (e_o/e_i) (D) for the system of Fig. 11.1b. If $R = 20\ \Omega$, $C = 0.3\ \mu\text{F}$, and $L = 0.2$ mH, plot the logarithmic frequency-response curves.

11.2 A synchro repeater system has one transmitter and five receivers. The torque gradient of a single pair of devices with very short cable connections is 0.5 in · oz/degree, and 10 percent of this is lost for each ohm of cable resistance. Each receiver drives a dial with 0.05 in · oz of friction. If the allowable error is 0.5° and cable resistance is 0.05 Ω/ft., find the maximum allowable cable length.

11.3 Explain the operation of the system of Fig. 11.3.

BIBLIOGRAPHY

1. R. H. Cerni and L. E. Foster: "Instrumentation for Engineering Measurement," chap. 5, Wiley, N. Y., 1961.
2. J. D. Tate: Synchro Systems, *Mach. Des.* p. 150, June 8, 1961.
3. R. J. Barber: 21 Ways to Pick Data Off Moving Objects, *Contr. Eng.*, p. 82, October 1963; p. 61, January 1964.
4. F. W. Hannula: Transmitting Test Information, *Contr. Eng.*, p. 173, September 1959.
5. E. D. Lucas: Techniques for Radio Telemetry, *Contr. Eng.*, p. 71, December 1962.
6. E. A. Ragland and D. E. Wassall: The Digital Answer to Data Telemetering, *Contr. Eng.*, p. 95, August 1957.
7. E. H. Krause: Telemetering for Interplanetary Flight, *ISA J.*, p. 478, October 1957.
8. C. I. Cummings and A. W. Newberry: Radio Telemetry, *ARS J.*, p. 141, May–June 1953.
9. E. H. de Grey and J. G. Bayly: Measuring through Vessel Walls, *ISA J.*, p. 82, May 1963.
10. L. W. Gardenhire: Evolution of PCM Telemetry, *Instrum. Contr. Syst.*, p. 87, April 1965.
11. M. K. Stark: Short Range Telemetry System Provides Test Data on Rotating Parts, *Gen. Motors Eng. J.*, p. 23, January–February–March 1965.
12. J. Valentich: Simple Slip Rings for Strain Gage Measurement, *Mach. Des.*, p. 154, January 7, 1969.
13. J. Valentich: Broadcasting Power to Shaft-Mounted Sensors, *Mach., Des.*, August 9, 1973.
14. Liquid Metal Contacts Outrun Slip Rings, *Electromech. Des.*, November 1973.
15. G. M. Dick: The Lowly Modem, *Datamation*, pp. 69–73, March 1977.
16. R. T. Troutner: Acoustic Telemetry, *Oceanology Int.*, January 1970.
17. Medical and Biological Applications of Space Telemetry, *NASA* SP-5023, 1965.

TWELVE

VOLTAGE-INDICATING
AND -RECORDING DEVICES

The majority of signals in measurement systems ultimately appear as voltages. Since voltage cannot be seen, it must be transduced to a form intelligible to a human observer. The form in which the data are presented is generally that of a pointer moving over a scale, a pen writing on a chart (including light beams writing on photosensitive paper and electron beams writing on cathode-ray tubes), visual presentation of a set of ordered digits, or printout of digital data by a typewriter or similar device. We consider the most common types of such indicating and/or recording devices.

12.1 STANDARDS AND CALIBRATION

Figure 12.1[1] gives information on the primary standards for voltage, resistance, capacitance, and inductance. For most routine purposes in engineering laboratories, the secondary standards (voltage-calibration sources, precision resistor decades, etc.) available from many manufacturers are adequate.

12.2 ANALOG VOLTMETERS AND POTENTIOMETERS

While digital voltmeters are very popular, analog meters are still the preferred choice for certain applications.[2] The most widely employed meter movement for dc and (with rectifiers) ac measurement in electronics and instrumentation work is the classical D'Arsonval movement (see Fig. 12.2). This basically current-sensitive device is used to measure voltage by maintaining circuit resistance

---

[1] Accuracy in Measurements and Calibration, *NBS, Tech. Note* 262, 1965.

[2] J. Harte, Jr., Analog Panel Meters—Alive and Well! *Inst. & Cont. Syst.*, pp. 19–23, July 1979; J. Hayes, Digital or Moving-Coil Meters? *Mach. Des.*, pp. 113–115, September 5, 1974.

constant by means of compensating techniques (see Fig. 2.17a). Relatively recent improvements on this basic configuration include taut-band suspension (rather than pivot-and-jewel bearings), individually calibrated scale divisions, and expanded-scale instruments. Taut-band suspension completely eliminates bearing friction, reduces inertia and temperature effects, increases ruggedness, and results in less loading on the measured circuit since the reduced friction requires less power drain. The increased accuracy made possible by taut-band construction can be provided at reasonable cost through the use of automatic calibration systems, which print an individual scale for each and every instrument. Expanded-scale instruments use a precision voltage-suppression circuit to measure a small variation around a larger voltage. Thus if you need to measure a 100-V signal accurately, it is possible to do this with a meter whose scale goes from 99 to 101 V. Static inaccuracies of 0.1 percent are attainable in a rugged and portable instrument with these methods.

Transistorized voltmeters (see Fig. 3.23) still utilize the D'Arsonval meter movement, but precede it by amplifier circuits. These increase the input impedance and overall sensitivity. Such instruments generally accept a wide range of dc and ac input voltages and have static error of the order of 1 to 3 percent of full scale.

When ac (not necessarily sinusoidal) voltages are to be measured with a D'Arsonval movement, it is necessary to perform rectification. Depending on the circuitry used, a meter may be sensitive to the average, peak, or rms value of the input waveform. It is common practice to calibrate the scale of the meter to read rms value no matter what quantity is fundamentally sensed. This procedure is accurate only if pure sinusoidal waveforms are being measured since, in this case only, the peak, average, and rms values are all related by fixed constants and thus can be included in the scale calibration. For nonsinusoidal waveforms, peak- or average-sensing meters will not read the correct rms value. In some cases, peak or average value is actually what is wanted; however, rms is desired most often. A true rms voltmeter is complex and expensive; thus peak- and average-sensing meters calibrated to read rms are in wide use and are generally satisfactory, except in the most critical applications.

Figure 12.3 shows circuits for peak, average, and rms meters. In the peak circuit, the capacitor is charged to the peak value of a periodic input voltage. This charge cannot leak off rapidly because of the one-way conduction of the diodes and the high input impedance of the voltmeter (transistorized voltmeters often use peak sensing). The voltage across the meter thus stays near the peak value of the input with only slight fluctuations resulting from diode reverse leakage and meter noninfinite impedance. The meter reads the *largest* peak, whether positive or negative. In the average-reading circuit, the input is full-wave-rectified, and the low-pass filtering characteristic of the meter movement is employed to extract the average value. The rms-reading circuit[1] approximates the

---

[1] C. G. Wahrman, A True RMS Instrument, *B & K Tech. Rev.*, B & K Instruments, Marlboro, Mass.

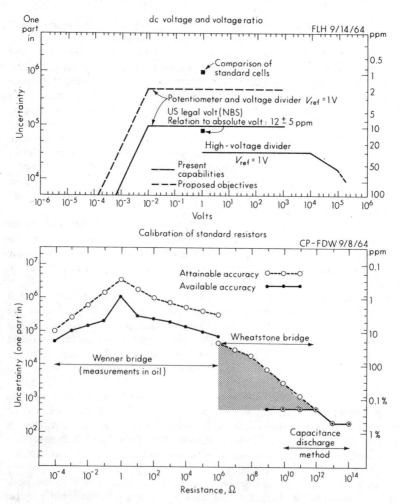

Figure 12.1 Electrical standards.

required square-law parabola with a few straight-line segments in the fashion of a diode function generator. The average voltage on the capacitor is utilized to provide a variable bias on the diodes in the function generator, thereby obtaining higher accuracy than possible in a fixed-bias unit using the same number of diodes. The averaging required in obtaining an rms value is performed by the meter's low-pass filtering characteristic while the square-root operation is obtained simply by meter-scale distortion.

When highly accurate rms measurements of nonsinusoidal signals are re-

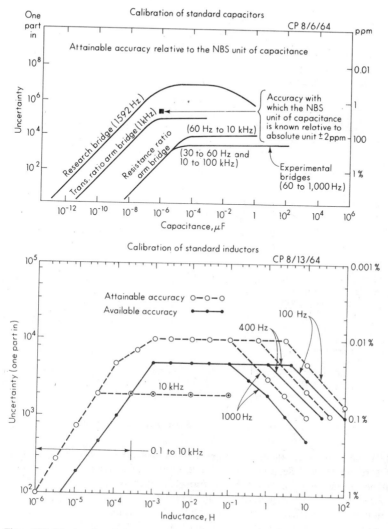

Figure 12.1 (*Continued*)

quired (random signals are a good example), sometimes methods based on the heating power of the waveform are employed since heating power is directly proportional to the mean-squared voltage. Voltmeters for random signals must be able to handle peaks that are large compared with the rms value. This is specified by the peak factor of the meter. Large peak factors (ratio of peak to rms value) are desirable; values up to about 10 are available.

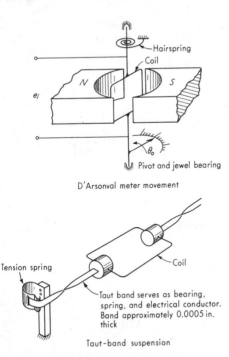

D'Arsonval meter movement

Tension spring

Coil

Taut band serves as bearing,
spring, and electrical conductor.
Band approximately 0.0005 in.
thick

Taut-band suspension

Expanded scale meter

Figure 12.2 DC analog meters.

When the most accurate measurements of dc voltage are needed, potentiometers rather than deflection meters are employed. The potentiometer is a
null-balance instrument in which the unknown voltage is compared with an
accurate reference voltage, which can be adjusted until the two are equal. Since,
at the null point, no current flows, errors due to IR drops in lead wires are
eliminated. Such IR drops are always present when a D'Arsonval-type meter is
used to measure voltage directly. Figure 12.4a shows the basic potentiometer
circuit. We see that a galvanometer (just a very sensitive D'Arsonval movement)
is utilized as a null detector. It detects the presence or absence of current by
deflecting whenever the unknown and reference voltages are unequal. However, it
need not be calibrated since it must indicate only the presence of current, not its
numerical value. The basic circuit of Fig. 12.4a is not practical since the accuracy
of the reference voltage picked off the slidewire is directly influenced by changes
in the dry-cell voltage. Since the dry cell supplies power to the slidewire, its

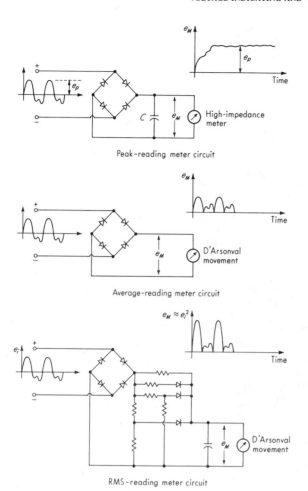

Figure 12.3 Peak, average, and rms circuits.

voltage is bound gradually to drop off. This problem is solved in the practical circuit of Fig. 12.4b by inclusion of an additional component, the standard cell.

Figure 12.4c shows the Weston cadmium saturated standard cell which is the basic working standard of voltage. Its terminal voltage is 1.018636 V and is reproducible to the order of 0.1 to 0.6 ppm. Its accuracy in terms of the fundamental mass, length, and time standards can be established to only about 10 ppm, however. Its temperature coefficient is $-40\ \mu V/C°$; thus close temperature control obviously must be employed in the most exacting situations. Since its accuracy is destroyed if any appreciable current is drawn from it over a time interval, a standard cell cannot be substituted for the dry cell of Fig. 12.4a. So the

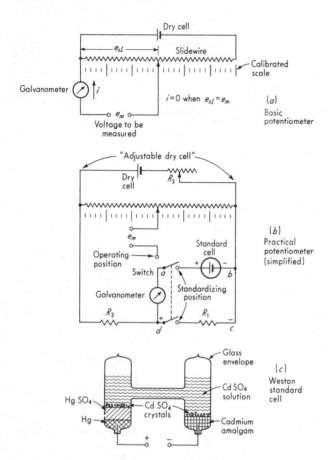

Figure 12.4 Manually balanced potentiometer.

standard cell must be utilized as an intermittent reference against which the slidewire excitation voltage can be checked whenever desired. The *unsaturated* Weston cell is employed in practical instruments since it is more portable. Its terminal voltage varies from one unit to another. However, its drift at constant temperature is only about -0.003 percent per year; thus it is perfectly adequate for most purposes. Its temperature coefficient is about -10 $\mu V/C°$.

The operation of the circuit of Fig. 12.4b is as follows: When the slidewire scale on the potentiometer was calibrated originally at the factory, slidewire excitation-adjusting resistor R_3 was set at a fixed value and resistor R_1 was adjusted until, when loop *abcd* was completed by the switch, no current flowed in the galvanometer. This means that the voltage drop across R_1 was just equal to the standard-cell voltage. Now the slidewire, R_1, and R_2 are all fixed and stable resistors; thus if the voltage across R_1 is at its calibration value, the slidewire excitation voltage also must be at its calibration value. Thus, whenever we wish to check the calibration (this is called standardization), we merely complete the loop *abcd* momentarily (so as not to draw much current from the standard cell) and note whether the galvanometer deflects. If it does, we adjust the slidewire excitation with R_3 until deflection ceases. Then we are assured that the slidewire excitation is at its original calibration value. The resistor R_2 is merely a current-limiting resistor to prevent drawing large current from the standard cell through the slidewire path, which is fairly low-resistance.

Fairly common and inexpensive potentionetemers which can be read to the nearest microvolt are in wide use. More sophisticated instruments intended for the most accurate calibration work provide greater accuracy and sensitivity. One such commercially available instrument[1] measures in three ranges: 0 to 1.611110 V in steps of 1.0 μV, 0 to 0.1611110 in steps of 0.1 μV, and 0 to 0.01611110 in steps of 0.01 μV. The total parasitic thermoelectric voltage is less than 0.1 μV. The limit of error on the high range is ± 0.003 percent of reading ± 0.1 μV, while on the medium and low ranges it is ± 0.005 percent of reading ± 0.1 μV. These values approach the level of the National Standards achieved by the National Bureau of Standards, which are about 0.001 percent from 0.01 to 1,000 V.

12.3 DIGITAL VOLTMETERS AND MULTIMETERS

While analog meters require no power supply, give a better visual indication of trends and changes, suffer less from electric noise and isolation problems, and are simple and inexpensive, digital meters offer higher accuracy and input impedance, unambiguous readings at greater viewing distances, smaller size, and a digital electrical output (for interfacing with external equipment) in addition to visual readout. The three major classes of digital meters are panel meters, bench-type meters, and systems meters. All employ some type of A/D converter [often the dual-slope integrating type (see Sec. 10.12)] and have a visible readout that displays the converter output. Usually panel meters are dedicated to a single function (and perhaps even a fixed range), while bench and system meters often are *multimeters*; that is, they can read ac and dc volts and amps as well as resistance over several ranges. The basic circuit is always dc volts; current is converted to volts by passing it through a precision low-resistance shunt while ac is converted to dc by employing rectifiers and filters. For ohms measurement, the meter includes a precision low-current source that is applied across the unknown resistor; again this gives a dc voltage which is digitized and read out as ohms (see Fig. 12.5[2]). Bench meters are intended mainly for stand-alone operation and

---

[1] Honeywell Inc., Philadelphia, Pa.
[2] Catalog and Buyers Guide, Keithley Instruments, Cleveland, Ohio, 1981.

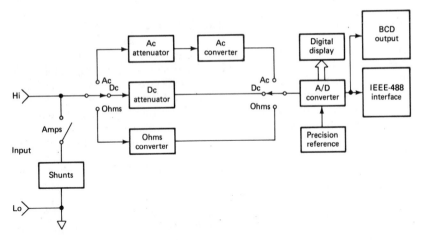

Figure 12.5 Digital multimeter configuration.

visual reading, while systems meters provide at least an electrical binary-coded decimal output (in "parallel" with the visual display) and perhaps sophisticated interconnection and control capability (such as the IEEE-488 interface of Fig. 12.5) or even microprocessor-based computing power.

Digital panel meters[1] are available in a very wide variety of special-purpose functions. Readouts range from basic 3-digit (999 counts, accuracy ± 0.1 percent of reading, ± 1 count) to high-precision $4\frac{3}{4}$-digit ones ($\pm 39,999$ counts, accuracy ± 0.005 percent of reading, ± 1 count). Units are available to accept inputs such as dc volts (microvolts to ± 20 V), ac volts (true rms measurement), line voltage, strain-gage bridges (meter provides bridge excitation), RTDs (meter provides sensor excitation), thermocouples of many types (meter provides cold-junction compensation and linearization), and frequency inputs such as pulse tachometers. Figure 12.6[2] shows some details for a high-precision unit with input resistance of 10^9 Ω, ± 0.0025 percent resolution (10 μV), and ± 0.005 percent of reading ± 1 count accuracy, which uses dual-slope A/D conversion with automatic zero. The reading rate is 2.5 per second when free-running and 10 per second maximum when externally triggered. These meters can be obtained with TRI-STATE binary-coded decimal outputs. TRI-STATE outputs provide a "disconnected" state in addition to the usual digital HI and LO. This facilitates interconnection to microcomputer data busses since any number of devices can be serviced by a single bus, one at a time, by "disconnecting" all but the two that are talking to each other.

[1] Digital Panel Instruments: Designer's Handbook and Catalog, Analogic Corp., Wakefield, Mass., 1980–1981.
[2] Ibid.

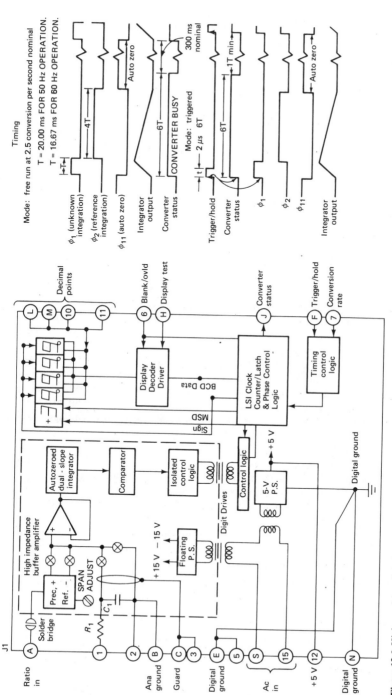

Figure 12.6 High-precision digital panel meter.

811

Bench-type meters range from inexpensive hand-held units with $3\frac{1}{2}$-digit readout and 0.5 percent accuracy to $5\frac{1}{2}$-digit (200,000 count) devices with 1-μV resolution. Instruments designed for extremely low voltages are called digital *nanovoltmeters*,[1] and they provide resolution down to about 10 nV (comparable analog meters go to about 1 nV). Digital *picoammeters*[2] measure very small currents and can resolve about 1 pA (analog instruments go to 3 fA). When extremely high input impedance is required for current, voltage, resistance, or charge measurements, an *electrometer*[3] type of instrument is employed. Digital electrometers can resolve 10^{-17} A, 10 μV, and 1 fC and measure resistance as high as 200 TΩ. Input impedance can be as high as 10,000 TΩ. Corresponding analog instruments achieve 10^{-15} A, 10 μV, and 1 fC and can measure resistance to 100 TΩ.

Systems-type digital voltmeters (DVMs) or multimeters (DMMs) are designed to provide the basic A/D conversion function in data systems assembled by interfacing various peripheral devices with the DVM. Capabilities and costs vary widely; here we describe briefly a meter[4] which utilizes a microprocessor to provide several useful mathematical functions in addition to managing meter operations. A modified dual-slope A/D converter is used with selectable integration times from 0.01 to 100 power-line cycles to trade off speed versus accuracy. At maximum speed (330 readings per second) accuracy is ± 0.1 percent, while 0.57 reading per second gives $6\frac{1}{2}$-digit resolution and 0.001 percent accuracy. Modes available are dc volts, ac volts, and resistance. Mathematical functions include the following: with NULL, the first reading is subtracted from each successive reading, and the difference is displayed (the first "reading" can be entered manually from a built-in keypad). The function STAT accumulates readings and calculates mean and variance. With PASS/FAIL, any readings falling outside user-entered upper and lower limits will display HI or LO and pull the service-request line. With dBm(R), the user enters resistance R, and then all readings are displayed as power dissipated in R in units of decibels referenced to 1 mW. With THMS°F (with the voltmeter in OHMS), the temperature of a thermistor probe is displayed in degrees Fahrenheit (or Celsius). The function $(X - Z)/Y$ provides offsetting and scaling with user-entered Z and Y constants (X is the reading). The function $100(X - Y)/Y$ finds the percentage deviation and 20 log (X/Y) displays X in decibels relative to the stored value Y. A built-in IEEE-488 type of interface allows versatile interconnection with other instruments and computers. An internal memory (RAM) can be used to store the results of measurements and programs for taking the measurements.

[1] Catalog and Buyers Guide, Keithley Instruments.

[2] Ibid.

[3] Ibid.

[4] L. T. Jones, J. J. Ressmeyer, and C. A. Clark, Precision DVM Has Wide Dynamic Range and High Systems Speed, *Hewlett-Packard J.*, pp. 23–32, Hewlett-Packard Corp., Palo Alto, Calif., April 1981.

12.4 ELECTROMECHANICAL SERVOTYPE XT AND XY RECORDERS

Figure 12.7 shows the principle of servotype XT recorders employed for obtaining indication and simultaneous recording of a voltage $e_i(X)$ against time (T). The instrument servomechanism is designed so that displacement x_o tracks the voltage e_i accurately over the design frequency range. Variations on this general principle include use of ac or dc amplifiers, ac or dc motors, rotary or translational motors, various mechanical-drive arrangements (piano wire and pulleys, etc.), assorted writing schemes, and different displacement transducers (potentiometers, RVDTs, capacitance, etc.). Adjustable chart drive speeds (these establish the time base) were obtained originally from constant-speed synchronous motors and pushbutton mechanical change gears; however, now the step-motor drive of Fig. 12.7 is more common.

One subclass of this type of recorder utilizes rather wide charts (10 in) and is intended for slowly changing inputs (<1 Hz for full-scale travel). Throw-away fiber ink pens are employed most, but some recorders use heated styli and heat-sensitive paper. With suitable preamps, most any desired voltage range can be accommodated, with resolution to as little as a few microvolts being common. Multichannel operation (a separate servo for each pen) is possible either reduced-width side by side or full-width overlapping (if pens are staggered to avoid mechanical interference); the limit is about six pens. Staggered pens give an undesirable chart-displacement error between channels. Some recorders[1] compensate for this with a microprocessor-implemented interchannel time delay. When inputs change very slowly, another approach to multichannel operation multiplexes the various input voltages into one servo and uses a numbered print wheel (or thermal dot matrix printhead) to print an identifying number or symbol for each channel of data (24 or 30 channels is a practical limit). Static accuracy of this entire subclass of recorders is about 0.1 percent full scale.

To cross-plot one variable against another, the XY configuration of Fig. 12.8a is available. Here the paper stands still (held down by vacuum or electrostatic attraction) while two independent servos move the pen horizontally and vertically. Standard sizes are $8\frac{1}{2} \times 11$ and 11×17 in (metric equivalents are available), and performance is very similar to the XT recorders just described since the mechanical motions are of a similar scale. Two Y pens on a single X carriage (pens staggered) also are obtainable. Many XY recorders include a ramp voltage generator, which can be applied as voltage input to one (or either) of the servos, creating a time base and allowing XT recording. Figure 12.8b[2] shows internal details of a typical XY recorder. Digital-input XY plotters for computer graphics work may employ open-loop linear step-motor drives with sophisticated

---

[1] Soltec Corp., Sun Valley, Calif., 1982.
[2] B & K Instruments, Marlboro, Mass.

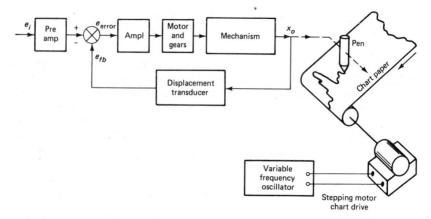

Figure 12.7 Servotype pen recorder.

piezoelectric accelerometer damping schemes[1] and provide multicolor plotting, using pens interchanged automatically under program control.

When higher response speeds (40 Hz at full-scale deflection, 150 Hz at 10 percent of full scale) are required, chart width must be reduced (5 or 10 cm is common). "Direct drive" (no gearing) limited-rotation motors (like Fig. 12.10 but without springs) are utilized, and sophisticated writing systems using special coated paper, pressurized ink systems and viscous inks, or fluid-dynamic compensation, such as the rubber accumulator[2] of Fig. 12.9, are necessary. Often such instruments are called direct-writing oscillographs, but we include them in this section because they are nothing but miniaturized and speeded-up versions of the basic servo principle of Fig. 12.7. Because of the high response speed and accumulation of large numbers of motion cycles, noncontacting displacement transducers such as RVDTs and rotary capacitance (Fig. 12.9) transducers are used. Multichannel instruments here are side by side, with about eight channels being a practical limit. Static accuracy is usually about ± 0.5 percent of full scale.

12.5 OPTICAL GALVANOMETER OSCILLOGRAPHS

While the D'Arsonval movement of Fig. 12.2 as applied to meters has a very limited frequency response (less than 1 Hz), it is possible to miniaturize it so as to

[1] J. A. Fenoglio, B. W. C. Chin, and T. R. Cobb, A High-Quality Digital X-Y Plotter Designed for Reliability, Flexibility and Low Cost; L. W. Tsai and R. L. Ciardella, Linear Step Motor Design Provides High Plotter Performance at Low Cost; J. A. Babiarz, Developing a Low-Cost Electrostatic Chart-Hold Table; W. G. Royce and P. Chu, Simple, Efficient Electronics for a Low-Cost X-Y Plotter; P. O. Maiorca and N. H. MacNeil, A Closed-Loop System for Smoothing and Matching Step Motor Responses—all in *Hewlett-Packard J.*, pp. 2–23, February 1979.

[2] L. Brunetti, A More Rugged, Cleaner Writing Oscillographic Ink Recorder, *Hewlett-Packard J.*, pp. 10–17, Hewlett-Packard Corp., Palo Alto, Calif., May 1973.

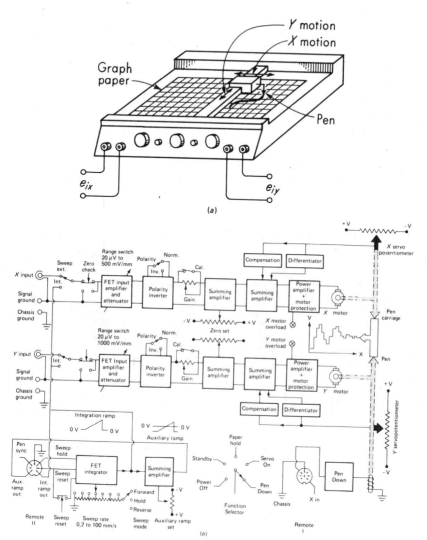

Figure 12.8 XY plotter.

obtain rotational natural frequencies of the order of 25 kHz. D'Arsonval movements of this type are called galvanometers and are the basic sensing elements of optical galvanometer oscillographs. We analyze the galvanometer to determine its performance characteristics.

Figure 12.10a shows schematically the construction of a typical galvanometer

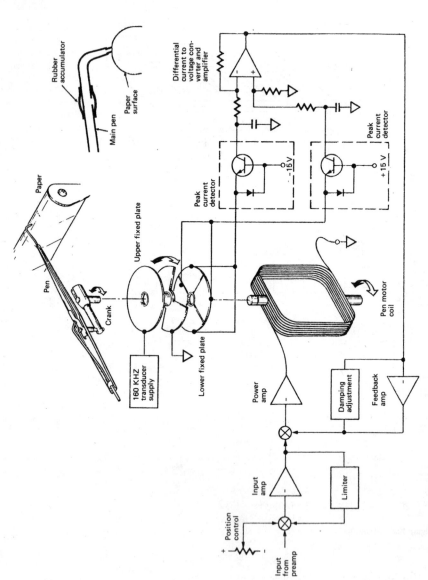

Figure 12.9 High-speed servorecorder (ink oscillograph).

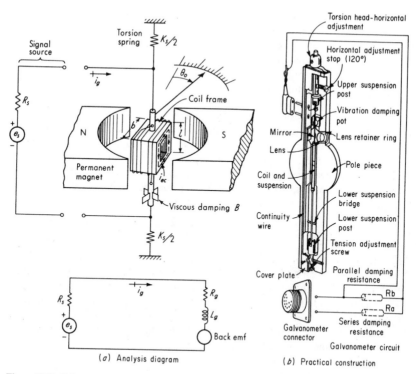

Figure 12.10 Galvanometer.

while Fig. 12.10b^1 gives practical details. Viscous damping B may or may not be present in a given design; we carry it along for generality. The galvanometer is an electromechanical transducer, and we write two equations, one electrical and one mechanical, to analyze its behavior. Basically, input voltage e_s from the signal source causes a current to flow in the coil. Then there is a current-carrying conductor in a magnetic field; thus the coil experiences an electromagnetic force which, since it has a lever arm, causes a torque. This torque tends to rotate the coil until it is just balanced by the restoring torque of the torsion springs. For a constant e_s the output pointer comes to rest at a definite value of θ_o. By proper design the static relation between θ_o and e_s can be made quite linear.

Writing a Kirchhoff voltage-loop law for the electric circuit, we have

$$i_g(R_s + R_g) + L_g \frac{di_g}{dt} + HNlb \frac{d\theta_o}{dt} - e_s = 0 \qquad (12.1)$$

[1] Honeywell Corp., Denver, Colo.

where $HNlb(d\theta_o/dt)$ = back emf of coil acting as generator

$\quad H \triangleq$ flux density

$\quad N \triangleq$ number of turns on coil

$\quad l \triangleq$ length of coil

$\quad b \triangleq$ breadth of coil

$\quad L_g \triangleq$ inductance of coil

$\quad R_g \triangleq$ resistance of coil

$\quad R_s \triangleq$ resistance of signal source

We may note that this analysis is comparable to that of a dc motor since the devices are basically the same. In fact, because this device uses one of the most basic electromechanical-energy-conversion principles, our analysis is useful for many other applications, such as electrohydraulic servovalves, gyro torquing coils, vibration shakers, and mirror scanners for lasers. Now we can apply Newton's law to the rotational motion of the coil:

$$\Sigma \text{ torques} = J \frac{d^2\theta_o}{dt^2}$$

The main electromagnetic torque is given by $HNlbi_g$; however, a more subtle effect also produces an additional torque. If the *frame* on which the coil is wound is itself a conductor, then a voltage $Hlb(d\theta_o/dt)$ will be induced in it (since it is a "one-turn coil") and an eddy current $i_{ec} = (Hlb/R_f)(d\theta_o/dt)$ will flow in the frame, where R_f is the resistance of the frame. This eddy current is also in the magnetic field, and so the frame feels an electromagnetic torque $- Hlbi_{ec}$. The minus sign is needed since such induced voltages always set up effects that oppose the original motion. Newton's law then reads

$$HNlbi_g - \frac{(Hlb)^2}{R_f} \frac{d\theta_o}{dt} - K_s \theta_o - B \frac{d\theta_o}{dt} = J \frac{d^2\theta_o}{dt^2} \tag{12.2}$$

where $B \triangleq$ viscous damping coefficient

$\quad K_s \triangleq$ torsional spring constant

$\quad J \triangleq$ moment of inertia of moving parts about axis of rotation

Equations (12.1) and (12.2) each contain two unknowns, i_g and θ_o, so they must be solved simultaneously. Since our main interest is the transfer function relating input e_s to output θ_o, i_g is of little interest, and we do not bother to solve for it. A simplification is possible in Eq. (12.1) since experience shows that, in terms of frequency response, within the useful operating frequency range of the galvanometer the effect of inductance is invariably negligible. Thus we drop this term and solve Eq. (12.1) for i_g:

$$i_g = \frac{e_s - HNlb(d\theta_o/dt)}{R_s + R_g} \tag{12.3}$$

Now this may be substituted in Eq. (12.2), which gives, after some manipulation,

$$\frac{\theta_o}{e_s}(D) = \frac{K}{D^2/\omega_n^2 + 2\zeta D/\omega_n + 1} \tag{12.4}$$

where
$$K \triangleq \frac{HNlb}{K_s(R_s + R_g)} \quad \text{rad/V} \tag{12.5}$$

$$\omega_n \triangleq \sqrt{\frac{K_s}{J}} \quad \text{rad/s} \tag{12.6}$$

$$\zeta \triangleq \frac{B + (Hlb)^2/R_f + (HNlb)^2/(R_s + R_g)}{2\sqrt{K_s J}} \tag{12.7}$$

We see that the galvanometer is a second-order instrument. We can increase sensitivity by increasing H, N, l, and b or by decreasing K_s, R_g, and R_s. The flux density H is usually at the maximum value with ordinary permanent magnets. Increases in N, l, and b result in increases in R_g since a longer total length of wire in the coil results; thus the net effect on sensitivity is not obvious. Decreases in K_s increase K directly; however, speed of response is lost since ω_n decreases. Note also that sensitivity varies with R_s; thus changing from one transducer to another with a different resistance results in a loss of calibration. This is due to the fact that the galvanometer is a current-sensitive device. Equation (12.7) shows that the mechanical viscous damping B can be eliminated completely and the system is still damped. The damping that remains is of electromagnetic origin and is caused by the back emf proportional to $d\theta_o/dt$ (mechanical-viscous-damping torque also is proportional to $d\theta_o/dt$). In low-frequency, high-sensitivity galvanometers the electrical damping is adequate to obtain the optimum ζ value of 0.65, and no intentional mechanical damping B is provided. Also such galvanometers may have the coil frame slotted so that $R_f = \infty$. Then

$$\zeta = \frac{(HNlb)^2}{2\sqrt{K_s J}\,(R_s + R_g)} \tag{12.8}$$

The most important feature of this result is that ζ depends on R_s, the source resistance. *Thus such a galvanometer can be properly damped for only one value of external resistance.* This does not mean it can be used only with transducers that have this value of R_s; but it does mean that a suitable resistance network may have to be interposed between the transducer and galvanometer. If the transducer resistance is too high, a shunt resistor is needed; if too low, a series resistor (see Fig. 12.11) is required. For high-frequency, low-sensitivity galvanometers the electromagnetic damping is inadequate, and intentional viscous damping is provided. The electromagnetic damping is a small percentage of the total, and so generally such galvanometers can be used with any external resistance in a wide range.

To construct a recording oscillograph, it is necessary to provide chart paper moving at a known speed, to give a time base, and a means of writing the galvanometer motion on this paper. To realize the high-frequency response (up to 25 kHz) mentioned earlier, a tiny mirror is rigidly fastened to the moving coil and a light beam reflected from it. When the coil turns, the light beam, which is focused as a spot on the moving chart paper, deflects over the paper, leaving a trace which develops under ordinary fluorescent room lighting. These records will

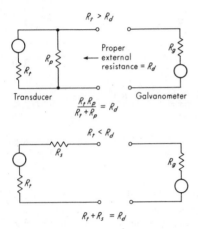

Figure 12.11 Damping networks.

last for years if kept away from sunlight and can be subjected to a simple liquid fixing process if absolute permanence is required.

Oscillographs utilizing light-beam galvanometers generally provide an entire family of interchangeable units covering the range from low frequency, high sensitivity to high frequency, low sensitivity. Figure 12.12 shows a typical

Undamped natural frequency, Hz	Flat ($\pm 5\%$) frequency response, Hz	External resistance for optimum damping, Ω	Coil resistance, Ω	Current sensitivity, in/ma	Maximum deflection for $\pm 2\%$ nonlinearity, in
		Electromagnetic damped types			
40	0–24	120	20	136	8
40	0–24	350	35	225	8
100	0–60	120	32	91	8
100	0–60	350	67	160	8
200	0–120	120	53	44	8
400	0–240	120	116	12	8
		Fluid damped types			
1,000	0–600	20–1,000	37	0.356	8
1,650	0–1,000	20–1,000	25	0.107	8
3,300	0–2,000	20–1,000	31	0.039	6
5,000	0–3,000	20–1,000	37	0.023	3.5
8,000	0–4,800	20–1,000	33	0.027	2.0

Figure 12.12 Galvanometer family.

selection.[1] The galvanometers can be removed easily and quickly from the magnet block and replaced by others suited to the particular job requirements. Since the light beams do not interfere with one another, as do mechanical arms, each channel of a multichannel instrument can employ the entire chart width. Optical galvanometers also allow greater deflections; full scale is 4 to 8 in in commercial instruments. Up to 42 channels may be recorded on one 12-in wide chart paper. Most instruments provide pushbutton selection of chart drive speeds; speeds up to about 200 in/s are available. The time base provided by the paper drive is usually of insufficient accuracy, so accurate time-line generators internal to the oscillograph print (optically) an accurate time grid at, say, 0.01-s intervals on the paper. A sequential trace interruption scheme allows identification of individual traces.

12.6 OPTICAL LIGHT-GATE ARRAY RECORDER

While optical galvanometer oscillographs have fewer mechanical moving parts than inking recorders, there is still incentive to go further in this direction. Figure 12.13a[2] shows the principle of a light-gate array recorder (based on a concept originated by Sandia Corp. to protect the pilot's vision from nuclear-explosion flashes) in which the only moving parts are the chart drive. A wide, flat beam of polarized light is directed toward an array of miniature light gates, which can be manufactured at a spacing of 80 per inch. Each individual gate can be opened or closed to the passage of light in about 1 μs by changing its optical polarization with an applied electrical signal. Thus if all gates but one are closed, a single spot of light gets through to the recording paper. For 12-in-wide paper there are 960 gates, and each is uniquely associated with one of 960 "time slots" of a 20-μs encoding ramp against which the analog input signal is compared (see Fig. 12.13b). When the analog signal crosses the ramp, a comparator switches the associated light gate open, plotting a discrete point at the correct location on the paper. At maximum paper speed (129 in/s) the paper moves a maximum of 0.0026 in in 20 μs. This, together with the gate spacing of 1/80 in and the light spot size, gives an essentially continuous trace. Instrument frequency response is quoted at 5,000 Hz. At this frequency and maximum paper speed, one cycle covers 0.026 in of paper and is made up of 10 "spots." Up to 28 channels of analog data may be accommodated, and time lines at six spacings from 100 to 0.001 s may be chosen.

12.7 FIBER OPTIC CRT RECORDING OSCILLOSCOPE

The familiar cathode-ray tube (CRT) oscilloscope has extremely high-frequency response, but permanent records normally require photography of the screen and

---

[1] Honeywell Corp., Denver, Colo.
[2] CEC Div., Bell and Howell, Pasadena, Calif.

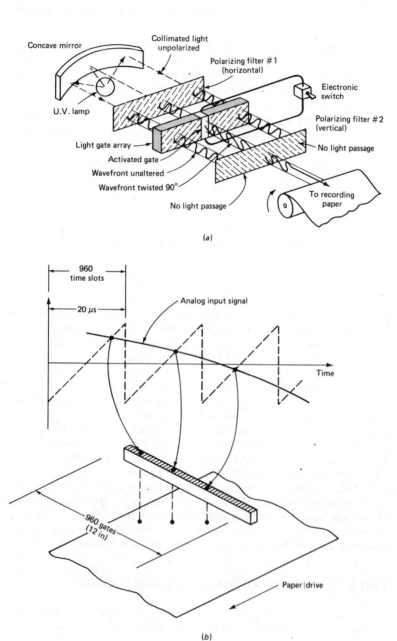

Figure 12.13 Light-gate array recorder.

822

are time limited to one sweep. By combining a special fiber optic CRT with an oscillograph-type paper drive (which passes the paper over the CRT face, where it is exposed by the light from the CRT phosphor), a recording oscilloscope[1] with uniquely useful characteristics is obtained. If the paper is held still, conventional oscilloscope operation allows single-sweep XT or XY recording of signals from dc to 1 MHz, much like the use of standard camera recording techniques, except that ordinary direct-print oscillograph paper is utilized. By employing both the CRT sweep and simultaneous paper drive (speeds up to about 100 in/s can be selected), the 1-MHz frequency response is retained; but now we can get permanent records of as many contiguous sweeps as we wish, because they are simply "stacked" one above the other on the recording paper. This technique gives an equivalent paper speed of up to 40,000 in/s. Time skew, resulting from paper motion during a single CRT sweep, is corrected electronically, but there will be small "gaps" in the data between sweeps because of the CRT sweep retrace time. Since the CRT allows beam intensity (Z axis) modulation, the instrument can produce gray-scale pictures such as video images. A multichannel version utilizes sampling techniques to obtain dc to 5,000-Hz response for up to 18 channels of data.

12.8 THERMAL AND ELECTROSTATIC ARRAY RECORDERS

With these recorders, heat-sensitive or xerographic chart paper is fed past a stationary printhead with an array of closely spaced (typically four per millimeter) heated or electrostatic elements. Analog input voltages activate the printing elements in a fashion similar to that of Fig. 12.13b. (Electrostatic instruments require a toner application similar to xerographic copying machines). Time lines, grid lines, and other annotation are combined with analog input-signal data, and a raster line which includes all these components is printed at one time. A typical thermal instrument[2] employs 6-in-wide paper and has 2 to 32 channels, static accuracy of ± 0.2 percent, and maximum paper speed of 50 mm/s. Dynamic response specification of such array recorders is complicated by the sampling process utilized; however, the maximum paper speed dictates that a 50-Hz signal could occupy only 1 mm per cycle. A 20-Hz sine wave actually is printed as 10 straight-line segments over a 25-mm length. Electrostatic instruments[3] exhibit similar behavior, except that the printing process is about 5 times as fast, allowing a maximum paper speed of 250 mm/s, and now a 100-Hz sine wave is printed as 10 segments over 2.5 mm.

[1] Honeywell Corp., Denver, Colo.
[2] Gould Inc., Cleveland, Ohio.
[3] Ibid.

12.9 CATHODE-RAY OSCILLOSCOPES AND GRAPHIC DISPLAYS

Figure 12.14 shows in simplified fashion the functional operation of a typical cathode-ray oscilloscope.[1] A focused narrow beam of electrons is projected from an electron gun through a set of horizontal and vertical deflection plates. Voltages applied to these plates create an electric field which deflects the electron beam and causes horizontal and vertical displacement of its point of impingement on the phosphorescent screen. By proper design this displacement can be made closely linear with deflection-plate voltage. The phosphorescent screen emits light which is visible to the eye and may be photographed for a permanent record.

The most common mode of operation is that in which a plot of the input signal against time is desired. This may be accomplished by driving the horizontal deflection plates with a ramp voltage, thus causing the spot to sweep from left to right at a constant speed. To ensure that the sweep and the input signal applied to the vertical deflection plates are synchronized properly, the triggering of the sweep can be initiated by energizing the trigger circuit from the leading edge of the input signal itself (see Fig. 12.14). This results in a loss of the first instants of the input signal on the screen, but generally this is not serious since only about 1 mm of deflection is needed to cause triggering. In those cases where this loss is objectionable, oscilloscopes with signal delay (Fig. 12.15) are available. These delay the application of the input signal to the vertical deflection plates so that the sweep starts *before* the rise of the input signal on the screen. Thus the complete input signal is recorded. Most oscilloscopes also provide for triggering from either positive-going or negative-going voltages (or both[2]), and the instant of triggering can be adjusted from the minimum 1-mm level upward to any point on the input waveform. Triggering also can be controlled from external signals or the 60-Hz power-line signal. When external trigger signals which are conveniently available occur somewhat before the input signal of interest, a sweep-delay feature may be useful; some instruments provide this capability. Since the deflection sensitivity of the cathode-ray tube itself is only of the order of 0.1 cm/V, oscilloscopes include amplifiers so that the instrument can directly handle input signals down to microvolts. Input amplifiers typically provide 1-MΩ input impedance, selectable single-ended or differential input, and selectable ac or dc coupling. Oscilloscopes are also useful for XY plotting. For such operation the horizontal deflection plates are merely disconnected from the sweep generator and connected to an amplifier identical to the vertical amplifier.

Cathode-ray tubes are obtainable with a number of different phosphors on the screen.[3] The choice of phosphor controls the intensity of light available for

---

[1] R. van Erk, "Oscilloscopes: Functional Operation and Measuring Examples," McGraw-Hill, New York, 1978.

[2] C. Baker, Digital Storage and Plug-in Versatility Distinguish New 10 MHz Oscilloscope, *Tekscope*, vol. 12, no. 14, Tektronix Corp., Beaverton, Ore., December 1980.

[3] R. A. Bell, CRT Phosphor Selection, *Inst. & Cont. Syst.*, pp. 86–90, March 1970.

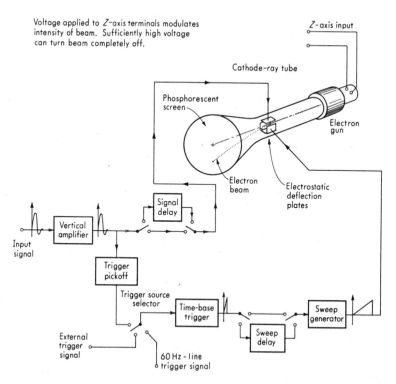

Figure 12.14 Cathode-ray oscilloscope.

visual observation or photographic recording, as well as the persistence of the trace after the electron beam has moved on. Both long- and short-persistence phosphors are available. Long-persistence phosphors are useful in visual observation of transients since the entire trace is visible long enough for an observer to note its characteristics. Persistence for several seconds is possible. When the moving-film method of trace photography is used, a very-short-persistence phosphor is necessary to prevent blurring. (In this method the electron beam is deflected vertically only, while the film is moved horizontally in front of the screen at a fixed, known velocity.) Persistence of less than 1 μs is available. Dual-persistence phosphors provide either a long or short persistence, depending on the color of the filter utilized over the scope screen. The most common method of photographing oscilloscope traces uses a still camera and the 10-s Polaroid[1] film process. A common method of photographing transients employs a double exposure to record both the trace and the grid lines. With the grid-line

[1] Polaroid Corp., Cambridge, Mass.

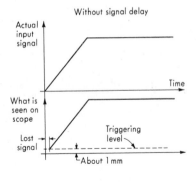

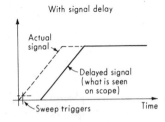

Figure 12.15 Signal delay.

illumination turned off and the camera shutter held open, the transient is triggered, thus recording its image on the film. Then the shutter is closed. Now the grid lines are turned on, and the shutter is snapped in the normal manner (say $\frac{1}{25}$ s at F : 16) to superimpose the grid lines on the picture. This procedure is necessary since the illumination from the grid lines is so great that it would completely fog the film if left on during the long time that the shutter is left open to catch the transient. Because of the rapidity and ease of making trial runs with Polaroid film, generally it is best to determine camera settings by trial and error rather than attempting to calculate them.

To obtain multichannel capability in oscilloscopes, several approaches[1] are taken. The dual-beam oscilloscope has two separate electron beams in one cathode-ray tube, with separate deflection plates and amplifiers for each beam. In some units both beams use the same sweep system; thus the two traces are plotted against the same time base. Completely independent beams allowing different time bases on each trace are available. The other approach (dual trace) uses a single-beam cathode-ray tube and a high-speed electronic switch to time-share the beam among several input signals. Such multitrace systems are available to give up to four traces on a single screen.

Versatility of operation is achieved in plug-in-type oscilloscopes by providing a wide variety of functional plug-in units for a single main frame. Typical plug-ins

[1] V. Lutheran and B. Floersch, Dual-Beam: An Often Misunderstood Type of Oscilloscope, *EDN*, August 5, 1974.

available include dual-trace and four-trace units, op-amps, carrier amplifiers, spectrum analyzers, high-gain amplifiers, and time bases with special features.

While most laboratory oscilloscopes have an 8 × 10 cm screen, larger screens up to about 21 in are available. These are useful for viewing by large groups, presentation of many channels of data by bar-graph types of displays, etc. Large-screen scopes usually cannot attain the high-frequency capability of the smaller types.

While the Polaroid photography process makes recording very convenient, a *permanent* retention of the trace on the scope screen has certain advantages. This feature is available in the various forms of storage oscilloscope.[1] Special cathode-ray tubes are obtainable that retain a trace for long periods until it is erased electrically by pushing a button. Oscilloscopes using such storage tubes cannot achieve as high-speed performance as conventional types; however, they allow one to examine traces visually with ease; when a desired trace is noted, it can be photographed, together with grid lines, in a single exposure.

Limited screen size and inherent nonlinearity of the CRT deflection system make the error of the basic oscilloscope (both voltage and time) about 2 to 5 percent. In dual-beam or dual-trace scopes, timing accuracy can be improved by applying an oscillatory signal of known frequency to one input as a reference. Voltage accuracy can be increased similarly in a dual-trace instrument by applying a reference square wave of accurately known amplitude to one input. Some scopes have built-in microprocessors and D/A converters[2] to increase accuracy and convenience of measurement. Digital storage scopes (see Sec. 12.10) achieve low errors since they do not depend on the CRT deflection system for accuracy. With respect to speed limitations, conventional scopes[3] can reach about 1 GHz while sampling scopes[4] go to about 20 GHz.

To improve productivity in oscilloscope applications to manufacturing, research and development, and automatic test systems, complete *oscilloscope measurement systems* incorporating microprocessors and using a computerlike bus architecture for easy interfacing are available (see Fig. 12.16[5]). Here, in addition to the usual signal trace, the CRT is used for alphanumeric messages, which can guide an operator quickly through a complex test sequence. (Simple CRT alphanumerics have been utilized on "ordinary" scopes for some time, for exam-

---

[1] New Instruments: 3 Kinds of Storage, *Tekscope*, Tektronix Corp., Beaverton, Ore., July 1972; J. Rogers and K. Hawken, A Big Step Forward for Direct-View Storage, *Tekscope*, vol. 9, no. 1, Tektronix Corp., Beaverton, Ore., 1977.

[2] W. A. Fischer and W. B. Risley, Improved Accuracy and Convenience in Oscilloscope Timing and Voltage Measurements, *Hewlett-Packard J.*, pp. 2–11, Hewlett-Packard Corp., Palo Alto, Calif., December 1974.

[3] H. Springer, 1 GHz at 10 mV in a General Purpose Plug-in Oscilloscope, *Tekscope*, Tektronix Inc., Beaverton, Ore.; A. G. Shephard, Pushing the Laboratory Scope up to 1 GHz, *Electro-Opt. Syst. Des.*, pp. 72–74, November 1976.

[4] J. A. Mulvey, The Simple Sampling Oscilloscope, *Tekscope*, Tektronix Inc., Beaverton, Ore., May/June 1973; Sampling Oscilloscopes, *Electronics Test*, pp. 56–60, March 1979.

[5] Model 1980 A/B Oscilloscope Measurement System, *Publ.* 5953-3869(D), Hewlett-Packard Corp., Palo Alto, Calif., 1980.

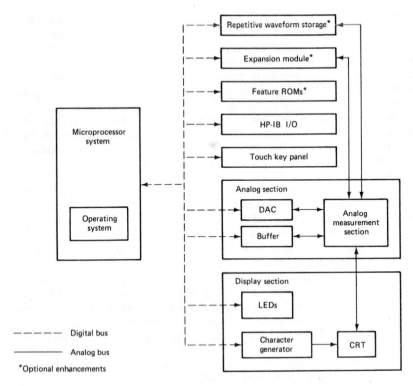

Figure 12.16 Architecture of oscilloscope measurement system.

ple, to include time and voltage scales in trace photographs.) Programming capability allows automatic measurement setup and readout. The microprocessor and digital readouts permit implementation of rapid methods for voltage and time measurements. Alphanumeric displays include menus for selection of basic tasks and advisory messages and error codes. Even the instrument's calibration procedure has been put in memory, allowing calibration without reference to a manual.

In addition to signal traces and alphanumerics, graphic displays [as used in computer-aided design (CAD) and manufacture (CAM)[1]] are often implemented using CRT technology. Monochromatic and full-color displays are employed widely as interactive devices between human operators and measurement/control systems. In aircraft cockpits,[2] CRT displays can reduce the confusing prolifer-

[1] C. Machover and R. E. Blauth, The CAD/CAM Handbook, Computervision Corp., Bedford, Mass., 1980; J. K. Krause, Graphic Terminals for CAD/CAM, *Mach. Des.*, pp. 95–100, August 6, 1981; Raster Graphics Handbook, Conrac Div., Covina, Calif., 1980.

[2] A. R. Tebo, Cockpit Displays—Works of Ingenuity and Splendor, *Electro-Opt. Syst. Des.*, pp. 31–43, July 1981.

ation of dials and pointers by displaying on the screen only the information needed at a particular time. The number of instruments has not been reduced; simply not all are visible simultaneously. While digital readouts might seem preferable at first, the trend information available from a "moving pointer" often proves advantageous; so CRT displays may be computer-generated views which mimic the face of a traditional mechanical instrument such as a gyro horizon. Sophisticated head-up and helmet-mounted displays allow pilots to view the "real world" out the cockpit window while at the same time observing a superimposed data display projected from a CRT. Many process plants[1] now employ color CRT displays to present flowcharts of operator-selected portions of the process, with superimposed digital/analog readouts of critical variables. Again, capability of the CRT for dynamically reconfiguring the display in a split second allows operators to "zero in" on selected portions of a large process without diverting their gaze over a large panel board. Operator interaction with CRT displays is available through both keyboards and light pens.[2] Although CRT systems (storage tubes, random scan, raster scan) dominate the field, they are being challenged by various "flat-panel" display technologies[3] (liquid crystal, plasma, etc.) which promise more compact configurations.

12.10 DIGITAL WAVEFORM RECORDERS AND DIGITAL STORAGE OSCILLOSCOPES

The digital-memory waveform recorder[4] provides capabilities difficult or impossible to achieve by earlier methods. Figure 12.17 shows a block diagram for a unit[5] capable of recording four independent signals simultaneously. The practical and economic implementation of this concept rests on the ready availability, small size, and reasonable cost of digital memories and associated digital-processing hardware brought about by explosive growth and intense competition in the computer field. Consider a single channel. The analog voltage input signal is digitized in a 10-bit A/D converter with a resolution of 0.1 percent (1 part in 1,024) and frequency response to 25 kHz. The total digital memory of 4,096 words can be used for a single channel, 2,048 can be employed for each of two channels, or 1,024 for each of four channels. The analog input voltage is sampled at adjustable rates (up to 100,000 samples per second), and the data points are read into the memory; 4,096 points maximum are storable in this particular instrument. (Sampling rate and memory size must be chosen to suit the duration

[1] P. Sniqier, Graphic Display Devices, *Digital Des.*, pp. 39–44, April 1980; Terminals: CRT, Graphic Display and Printing, *Digital Des.*, pp. 55–75, January 1978.
[2] R. Klatt, CRT Light Pen Technology and Design, *Digital Des.*, pp. 56–60, September 1979; F. J. Porter, Light Pen Aids User Interaction with Display, *Hewlett-Packard J.*, pp. 10–19, December 1980.
[3] L. Teschler, New Technology for Flat Displays, *Mach. Des.*, pp. 52–59, July 23, 1981.
[4] C. Somers, A New Way to Capture Elusive Signals, *Mach. Des.*, pp. 111–115, May 1981.
[5] Gould Inc., Biomation Div., Santa Clara, Calif.

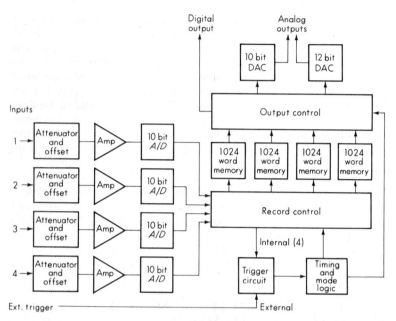

Figure 12.17 Digital-memory waveform recorder.

and waveform of the physical event being recorded. Some models have sampling rates to 200 MHz.)

Once the sampled record of the event is captured in the memory, many useful manipulations are possible, since the memory can be read out without erasing it (nondestructive readout). By reading the memory out *slowly* through a digital-to-analog converter (DAC), the original event (which could have been extremely rapid) is reproduced as a slowly changing voltage that is easy to record permanently in large size on a slow *XY* plotter. If the memory is read out *rapidly and repetitively*, an input event which was a "one-shot" transient becomes a repetitive waveform that now may be observed easily on an ordinary (not storage) oscilloscope. The digital memory also may be read out directly (without going through the DAC) to, say, a computer where a stored program can manipulate the data in almost any desired way.

"Pretrigger recording" allows the device to record the input signal *preceding* the trigger point, a unique and often useful capability. In "ordinary" triggering, the recording process is *started* by the rise of the input signal (or some external triggering signal) above some preset threshold value. If this threshold is set too low, random noise will trigger the system inadvertently; too high a threshold prevents recording of the initial rise of the desired signal. The digital recorder can be set to record "continuously" (new data coming into memory push out old data, once memory is full) until the trigger signal is received; then the recording process is stopped, thus freezing in the memory the data received *prior* to the trigger signal. An adjustable trigger delay allows operator control of the stop point, so that the trigger may occur near the beginning, middle, or end of the stored information.

While digital-memory waveform recorders are marketed without attached recording devices for their analog outputs, such packages are available as digital-memory oscillographs and oscilloscopes. The usual 40-Hz frequency response of pen/ink oscillographs (Sec. 12.4) can be extended to 20 kHz with a plug-in module[1] using waveform-digitizing principles. This makes the convenience and high recording quality of direct-writing instruments available for a very wide range of signals. Digital-storage oscilloscopes[2] are available in processing and nonprocessing forms. Processing types[3] include built-in computing power, which takes advantage of the fact that all data are already in digital form. Thus inclusion of interfacing and a microprocessor provides a complete system for information acquisition, analysis, and output. Processing capability ranges from simple functions (such as average, area, rms, etc.) to complete FFT spectrum analysis capability.

Nonprocessing digital scopes[4] are designed as replacements for analog instruments of both storage and nonstorage types. Their many desirable features lead some proponents to predict that ultimately they will displace analog scopes entirely (within the bandwidth range where digitization is feasible). The basic principle again rests on the scheme of Fig. 12.17. However, the scope operating controls are designed so that all confusing details are relegated to the background, and one appears to be using a conventional instrument (in fact, sometimes digital scope panels are simpler). (Also, most digital scopes provide switch-selectable analog operation as one operating mode.) One obvious advantage of digital operation is the "storage scope" capability provided by repetitively reading out the stored waveform, thus making transients repetitive and allowing their convenient display on the scope screen. (The CRT is an "ordinary," not a storage, type.) Furthermore, the voltage and time scales of the display are easily changed *after* the waveform has been recorded, which allows expansion (typically to 64 times) of selected portions for resolution of greater detail. A cross-hair cursor can be positioned at any desired point on the waveform and the voltage/time values displayed digitally on the screen and/or read out electrically. Some scopes use 12-bit converters, giving 0.025 percent resolution and 0.1 percent accuracy on voltage and time readings, which is much better than the 2 to 5 percent accuracy typical of analog scopes. "Split-screen" capability (simul-

[1] Gould Inc., Instruments Div., Cleveland, Ohio.

[2] Digital Oscilloscopes, *Electronics Test*, pp. 69–75, January 1979; P. C. Dale, The New Digital Scopes, *Prod. Eng.*, pp. 54–57, July 1980; M. Hurley, Oscilloscopes Go Digital, *Mach. Des.* pp. 96–102, May 22, 1980.

[3] Norland Instruments, Fort Atkinson, Wis.; T. G. Branden Corp., Portland, Ore.; Tektronix Inc., Beaverton, Ore.

[4] Tektronix Inc., Beaverton, Ore.; Nicolet Instrument Corp., Madison, Wis.; Gould Inc., Cleveland, Ohio.

taneously displaying "live" analog traces and replayed stored ones) enables easy comparison. The availability of pretrigger signal information (discussed under waveform recorders) is also a significant advantage. Display of stored data in both YT and XY format is possible. In addition to the fast memory readout used for CRT display, slow readout is obtainable to produce hard copy with external plotters. When more memory than the basic amount [typically 4,096 points (words)] is needed, a magnetic disk accessory allows expansion to 32,000 points. All digital-storage scopes are limited in bandwidth by the speed of their A/D converters; however, 20-MHz digitizing rates available on some scopes yield 5-MHz bandwidth, which is adequate for many applications.

12.11 DIGITAL PRINTERS/PLOTTERS

Wide utilization of computerized digital systems creates a large market for hard-copy devices such as printers.[1] Printers may be categorized as serial versus line, dot matrix versus whole-character image, and impact versus nonimpact. *Serial printers* produce a single character at a time, usually moving from left to right across a page (some recent printers move in *both* directions, increasing efficiency). *Line printers* appear to print an entire line simultaneously. *Page printers* print in a line-at-a-time mode, but can be stopped and restarted only on a page basis. Top speeds are about 200 characters per second for serial, 4,000 lines per minute for line, and 45,000 lines per minute for page printers.

   Dot-matrix printers form the desired characters by selective printing of dots in a row/column matrix (typically 5×7, 7×7, or 9×7). *Whole-character* printers form the entire symbol at a single stroke, as in a traditional typewriter. *Impact printers* bring the character element into forceful contact with the paper, sandwiching an inked ribbon between them. *Nonimpact printers* utilize electrophotographic ("laser/xerographic"), thermal,[2] ink-jet[3] electrostatic, or electrosensitive means to produce an image without contact or with minimal pressure.

   Whole-character impact printers are slow and noisy, but inexpensive, and they give good print quality and produce multiple copies (carbons). Dot-matrix impact printers give higher speeds and have graphics capability. Drum, chain, and band types of line printers offer high output rates. Thermal dot-matrix types are slow, cannot produce multiple copies (true of all nonimpact types), and

   [1] R. O. Huch, Printers: A Technology/Marketing Overview, *Digital Des.*, pp. 32–34, March 1981; Spotlight on Printers, *Digital Des.*, pp. 18–84, September 1978; I. L. Wieselman, Trends in Computer Printer Technology, *Comput. Des.*, pp. 107–115, January 1979.

   [2] W. Boles, Thermal Printhead Technology and Design, *Comput. Des.*, pp. 134–140, October 1979; T. R. Woodard, Thermal Printing with Semiconductor Printheads, *Digital Des.*, pp. 56–64, September 1980.

   [3] *IBM J. of R&D*, vol. 21, no. 1, 1977 (entire issue is devoted to ink-jet printer technology); J. D. Hill, Print Quality Tradeoffs in Ink Jet Technology, *Comput. Des.*, pp. 122–130, September 1979.

require special paper; but they are quiet and inexpensive and have graphics capability. Electrosensitive printers provide medium speed at low cost, but require special paper and produce low print quality. Electrostatic types have high speed, with versatile fonts and graphics capability, but need special paper and wet toner. Electrophotographic printers give high speed, high resolution, and quiet operation, but their high cost and maintenance necessitate high-volume applications. The several ink-jet types employ plain paper, are quiet, and range from 30 characters per second to 45,000 lines per minute; but they have reduced print quality and reliability at high speeds.

Digital XY plotters[1] are intended mainly for graphics, but do provide alphanumerics ("printing") as annotation for the graphics. Drum, flatbed, and hybrid types are available. Ink-pen and electrostatic writing methods presently dominate the market. Pen units provide high print quality and accuracy, ability to draw on both paper and plastic film, programmable multicolor capability, and low material cost; but they are slow and cannot easily color in large areas. Electrostatic plotters offer high speed, but produce lower quality and lack color capability.

12.12 MAGNETIC TAPE AND DISK RECORDERS/REPRODUCERS

The magnetic recorder/reproducer has a number of unique features derived mainly from its ability to record a voltage, store it for any time, and then reproduce it in electrical form essentially identical to its original occurrence. Recording methods include the direct, FM, pulse-duration modulation (PDM), and digital techniques.[2] We consider first the direct and FM modes of operation as applied to tape devices.

Figure 12.18 shows a functional diagram of a tape recorder/reproducer, and Fig. 12.19 shows a closeup of the record and reproduce heads. A current i proportional to the input voltage is passed through the winding on the record head, producing a magnetic flux $\phi = K_\phi i$ at the recording gap. The tape (thin plastic coated with iron oxide particles) passes under the gap, and the oxide particles retain a state of permanent magnetization proportional to the flux existing at the instant the particle leaves the gap. (Actually the applied flux and induced magnetization are not proportional because of the nonlinearity of the magnetic-hysteresis curve. Effectively, however, a close linearity is obtained by a

---

[1] B. Hirshon, Plotters: Charting a Course for the 80's, *Digital Des.*, pp. 30–31, May 1981; G. Lynch and R. Kaplan, Low-Inertia Plotter Has Microcomputer Control, *Mach. Des.*, pp. 128–129, January 8, 1981. See also L. G. Brunetti, A New Family of Intelligent Multi-Color X-Y Plotters, pp. 2–5; M. L. Patterson, Speed, Precision and Smoothness Characterize Four-Color Plotter Pen Drive System, pp. 13–19; and L. P. Balazer, Pen and Ink System Helps Assure Line Quality, pp. 20–25—all in *Hewlett-Packard J.*, Hewlett-Packard Corp., Palo Alto, Calif., September 1977.

[2] Magnetic Tape Recording Technical Fundamentals, Bell & Howell Datatape Div., Pasadena, Calif., 1979.

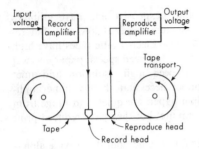

Input
voltage — Record amplifier — Reproduce amplifier — Output voltage

Tape transport

Tape — Reproduce head — Record head

Figure 12.18 Tape recorder/reproducer.

high-frequency bias technique.[1]) Thus, with a sinusoidal input signal

$$i = i_0 \sin 2\pi\, ft$$

and a tape speed of v inches per second, the intensity of magnetization along the tape varies sinusoidally with distance x according to

$$\text{Magnetization} \triangleq m = K_m K_\phi i_0 \sin\left(\frac{2\pi f}{v}\, x\right) \qquad (12.9)$$

where $m = K_m\phi$. The wavelength of the magnetization variation is then v/f inches. For example, a 60-Hz signal at a 60 in/s tape speed gives a wavelength of 1 in. If the tape with this signal on it is passed under the reproduce head, then a voltage proportional to the rate of change of flux bridging its gap will be generated in its coil. Note that since the output voltage depends on the rate of change of flux, if a dc current at the input had produced a constant tape magnetization, the reproduce head would have given *zero* output.

Thus the technique described above, the so-called direct recording process, can be used with varying input signals only, with about 50 Hz being the usual lower limit of frequency. Furthermore, since the reproducing head has a differentiating characteristic, the reproduce amplifier must have an integrating characteristic in order for the system output to be proportional to the input. An upper frequency limit also exists, because at sufficiently high frequencies, for a given reproduce gap and tape speed, one wavelength of magnetization will become equal to or less than the gap width. Then the average magnetization in the gap will be zero, and no output voltage will be generated. For example, at the fastest common tape speed, 240 in/s, and a gap width of 0.00008 in, this occurs at 3.0 MHz. Actually, the system is usable to only about half this frequency with reasonable accuracy. The frequency range of the direct recording process is approximately within the band 100 Hz to 2 MHz. The direct recording process does not give particularly high accuracy. Essentially this is limited by the signal-to-noise ratio, which is of the order of 25 dB (about 18 : 1). The rather high noise level is the result of minute defects in the tape surface coating to which the direct recording process is sensitive.

[1] Ibid.

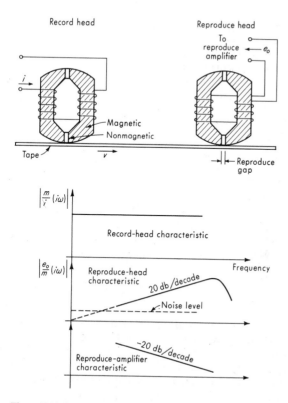

Figure 12.19 Record and reproduce heads.

When more accurate recording and response to dc voltages are required, generally the FM system is employed. Here the input signal is used to frequency-modulate a carrier, which is then recorded on the tape in the usual way. Now, however, only the *frequency* of the recorded trace is significant, and tape defects causing momentary amplitude errors are of little consequence. The frequency modulators employed here are similar in principle to those discussed under voltage-to-frequency converters and subcarrier oscillators in Chaps. 10 and 11. However, the frequency deviation for tape recorders is ± 40 percent about the carrier frequency. The reproduce head reads the tape in the usual way and sends a signal to the FM demodulator and low-pass filter, where the original input signal is reconstructed. The signal-to-noise ratio of an FM recorder may be of the order of 40 to 50 dB (100 : 1 to 330 : 1), indicating the possibility of inaccuracies smaller than 1 percent. By using sufficiently high carrier frequencies (432 kHz), the flat (± 1 dB) frequency response of FM recorders may go as high as 80,000 Hz at 120 in/s tape speed. To conserve tape when high-frequency response is not needed, generally a range of tape speeds is provided. When the tape speed is changed, the carrier frequency is altered in direct proportion. This makes the recorded wavelength of a given dc input signal the same, no matter what tape speed is being used, since ± 40 percent full-scale frequency deviation is utilized in all cases. Signals may be recorded at one tape speed and played back at any of the others without change in magnitude, but with a compression or an expansion of the time scale. A common set of specifications might be as follows:

Tape speed, in/s	Carrier frequency, kHz	Flat frequency response ± 0.5 dB, Hz	RMS signal-to-noise ratio
120	108	0–20,000	50
60	54	0–10,000	50
30	27	0–5,000	49
15	13.5	0–2,500	48
$7\frac{1}{2}$	6.75	0–1,250	47
$3\frac{3}{4}$	3.38	0–625	46
$1\frac{7}{8}$	1.68	0–312	45

Multichannel tape recorders are found with up to 14 channels on one 1-in-wide tape. Input to tape recorders is generally at about a 1-V level, and so most transducers require amplification before recording. The maximum time-base change of about $60:1$ $(120/1\frac{7}{8})$ shown in the table can be increased even further by rerecording the signal. For example, the original signal can be recorded at 120 in/s. Then it is played back at $1\frac{7}{8}$ with the output of the reproduce amplifier feeding the input of the record amplifier of another machine running at 120 in/s. If now this tape is played at $1\frac{7}{8}$, an overall slowdown of $4,096:1$ is achieved. An example application of tape slowdown is the recording of a 20,000-Hz signal on tape at 120 in/s and playback at $7\frac{1}{2}$ in/s into an optical oscillograph (frequency response to 2,000 Hz) for a permanent record. Also tape slowdown may be utilized when digital computations are to be performed on high-speed analog data. The digital equipment is very accurate, but may not handle rapidly varying inputs; thus one can analog-tape-record the data and then play them into the digital processing equipment at reduced speed. Digital waveform recorders perform these slowdown/speedup operations more conveniently than tape recorders, as long as the data record is not too long. The storage and playback feature of tape recording is employed widely in simulation. A particular environmental condition, such as vibration in an aircraft, is measured and tape-recorded in the actual environment. Then the tape is brought into the simulation laboratory where the environmental parameter (say, vibration) is repeated by playing the tape into a vibration shaker.

Wide utilization of digital data systems requires that digital data often be

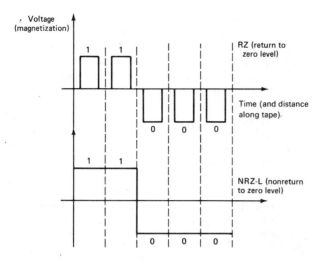

Figure 12.20 Digital recording formats.

directly recorded on magnetic-type or -disk systems.[1] Pulse-code modulation (PCM), using one of several coding schemes, is employed extensively to record the data, which are assumed to be in the binary form of 0s and 1s (a high voltage level and a low level). The actual magnetic-recording technique is similar to direct analog (rather than FM) in that the response does not extend to zero frequency. Codes of the return-to-zero-level (type RZ) (see Fig. 12.20) are wasteful of bandwidth since each bit requires *two* level transitions. Non-return-to-zero-level (NRZ-L) codes are more frugal of bandwith, but present recording problems since a long string of 1s or 0s is essentially "dc" which cannot be recorded properly. The frequency content of the signal depends, of course, on both the sequence of 1s and 0s and the tape speed. A rule of thumb says that the number of level changes per unit time should be about 1.5 times the low-frequency cutoff of the recorder. Thus for a 100-Hz cutoff, we would require a combination of tape speed and level changes such that there would be (on the average) 150 level changes per second. Most digital recording is done with the basic NRZ-L coding scheme or improved versions of it. One such version separates the NRZ-L datastream into 7-bit words; inverts bits 2, 3, 6, and 7; and adds a parity bit to make an 8-bit word. The parity bit is chosen so as to make the total number of 1s in the 8-bit word an odd count. Advantages of such a scheme include reduction of dc content and a guarantee that there will be at least one transition every 14 bits.

[1] Magnetic Tape Recording Technical Fundamentals, Bell & Howell Datatape Div., Pasadena, Calif., 1979; Parallel Mode High Density Digital Recording Technical Fundamentals, Bell & Howell Datatape Div., 1979.

Since digital recorder heads (just as analog recorders) can have multiple tracks, data can be recorded in serial, parallel, or serial-parallel format. In serial format, frequently used in instrumentation applications, data are recorded in a continuous stream on a single track. In parallel format, common in computer applications, data are recorded simultaneously on several tracks and the data have a relationship across the width of the tape, such as using an 8-track head to record 8-bit words. Serial-parallel mode is a process in which two or more serial datastreams, longitudinally recorded, also have a relationship across the width of the tape. For example, the number 10011101 could be recorded on two tracks by using the code that alternate bits employ alternate tracks; thus one track would contain 1010, and the second would hold 0111. This technique is utilized in instrumentation recording at very high data rates.

Specifications for digital tape recorders vary widely, but it may be useful to indicate present upper limits of performance. Both tape-packing density and speed of data transfer are of interest. An 84-track (2-in-wide tape) high-density recorder can transfer 600 million bits per second at 150 in/s tape speed and store 440 billion bits on a 9,200-ft reel of tape.

While magnetic-disk recording[1] encompasses a wide range of equipment, we mention briefly only the flexible ("floppy") disk class since it is quite common in instrumentation systems. Compared with tape cassettes or cartridges, a floppy disk ($5\frac{1}{4}$- or 8-in diameter) provides better random-access capability—160 ms (average) compared with 20 s—since cassettes typically move at $1\frac{1}{8}$ in/s while an 8-in-diameter floppy disk at 350 r/min reads at about 80 in/s. Track-to-track access time is as low as 3 ms, 8-bit words can be read at 250,000 words per second, and total storage ranges from 400 kbytes to 1 Mbyte.

12.13 SPEECH INPUT–OUTPUT DEVICES

At the human-machine interface, the vast majority of communication relies on the human sense of vision. Thus the meters, recorders, etc., of this chapter produce a visible output. In some applications *audible* input/output would be advantageous. Some manufacturing tasks require handling of parts and keyboard entry of computer data. If the data entry were by voice, the operator's hands could be kept free for parts manipulation. As one might suspect, voice output[2] is much

---

[1] Flexible Disk Drives, *Digital Des.*, pp. 46–60, April 1980; H. Shershow, The Flexible Future of Rigid Disk Drives, *Digital Des.*, pp. 52–60, May 1980; W. Floyd, Alternatives Abound When Assembling Flexible Disk Systems, *Digital Des.*, pp. 22–30, March 1979; B. Klevesahl and R. Stromsta, The Double-Sided Floppy Is Reborn, *Mini-Micro Syst.*, pp. 159–163, May 1980.

[2] J. Smith and D. Weinrich, Designer's Guide to Speech Synthesis, *Digital Des.*, pp. 53–57, February 1981; Voice Synthesis Productional Services, CL-600, Texas Instruments, Houston, Tex.; Semiconductors Speak, *Natl. Anthem*, no. 17, Natl. Semiconductor, Santa Clara, Calif.; P. Thordarson, Design Guidelines for a Computer Voice Response System, *Comput. Des.*, pp. 73–82, November 1977.

easier to achieve than voice input[1] (speech recognition); thus cost/capability has progressed much further in the speech output area. Presently available digital voice-output modules can duplicate accurately the natural inflection, intonation, and expression of the original voice. Thus one can supply the device manufacturer with an audio tape of the desired message, as spoken by the desired speaker, and they will produce a digital device which accurately reproduces the message on command. Also modules are available with a prerecorded and fixed "menu" of basic words (a typical one has numbers 0 to 12, letters A to Z, and 140 words). These basic items can be strung together arbitrarily by a program to create phrases and sentences. The sound of such devices is clearly intelligible, but will not be as natural as that of devices "custom-made" from an audio recording. Both classes of devices rely on clever data-compression techniques to drastically reduce the needed digital memory. "Brute force" digital recording (as practiced in high-fidelity music recording) would require about 96,000 Bd, while the data-compression techniques reduce this figure to about 1,200 Bd. This is what makes the whole scheme practical, since now messages of useful length (minutes) can be stored in memories of acceptable cost.

Digital voice-recognition modules can be obtained with vocabularies of 40 to 100 isolated words/phrases. For good accuracy, words must be clearly enunciated; however, some machines have "training" schemes that make them adaptive to a particular speaker's voice so that machine accuracy improves as a word is repeated. Some capability for understanding short sentences with a limited vocabulary exists, but continuous recognition of arbitrary speech seems to lie far in the future.

PROBLEMS

12.1 Calculate the ratios of peak value to rms and average value to rms value for (a) direct current, (b) a sine wave, (c) a square wave, (d) a half-wave-rectified sine wave, (e) a full-wave-rectified sine wave, and, (f) a train of rectangular pulses that are on 10 percent of the time and off 90 percent of the time.

12.2 Derive Eqs. (12.4) to (12.7).

12.3 Solve for $(i_g/e_s)(D)$ in the system of Fig. 12.10, neglecting inductance.

12.4 Taking inductance into account in the system of Fig. 12.10, find $(\theta_o/e_s)(D)$ and $(i_g/e_s)(D)$. Show the possible shapes of frequency-response curves for these transfer functions.

BIBLIOGRAPHY

1. Notes on the Julie Ratiometric Method of Measurement, Julie Research Laboratories, New York, 1964.
2. M. H. Aronson: "Handbook of Electrical Measurements," Instruments Publishing, Pittsburgh, Pa., 1961.

[1] B. Hirshon, Computer "Ears" Available to OEM Designers, *Digital Des.*, pp. 14, 17, May 1981; E. K. Yasaki, Voice Recognition Comes of Age, *Datamation*, pp. 65–68, August 1976.

3. L. W. Dean: Potentiometer Specifications, *Instrum. Contr. Syst.*, p. 73, January 1965.
4. S. A. Davis: Analog Voltmeters, *Electromech. Des.*, p. 48, November 1963; p. 44, December 1963.
5. R. Bergeson: Feedback Stiffens D'Arsonval Movement, *Contr. Eng.*, p. 121, September 1964.
6. J. W. Martin: Error Analysis in Measuring RMS Voltages, *Electro-Technol. (New York)*, p. 38, April 1965.
7. R. J. Erdman: DC Microvolt Measurements, *Instrum. Contr. Syst.*, p. 91, January 1964.
8. R. T. Hood: Measuring Current in High-Energy Arc Jets, *Instrum. Contr. Syst.*, p. 99, January 1964.
9. J. F. Keithley: Electrometer Measurements, *Instrum. Contr. Syst.*, p. 74, January 1962.
10. F. C. Martin: RMS Measurement of ac Voltages, *Instrum. Contr. Syst.*, p. 65, January 1962.
11. W. H. Schaeffer: The Six-Dial Thermofree Potentiometer, *Instrum. Contr. Syst.*, p. 283, February 1961.
12. A. L. Ispas: Interpretation of Magnetic Tape Recorder Specifications, *Instrum. Contr. Syst.*, p. 97, July 1964.
13. History of Magnetism, *Readout*, Ampex Corp., Redwood City, Calif., August–September 1961.
14. R. E. Morley: Time Compression Disk, *Instrum. Contr. Syst.*, p. 108, July 1964.
15. J. McElwain: Long-Term Magnetic Tape Recording, *Instrum. Contr. Syst.*, p. 111, July 1964.
16. E. D. Lucas: Miniature Tape Recorders, *Contr. Eng.*, p. 53, December 1964.
17. G. H. Schulze: Tape Recording Errors, *ISA J.*, p. 61, May 1964.
18. D. R. Davis and C. K. Michener: Graphic Recorder Writing Systems, *Hewlett-Packard J.*, October 1968.
19. W. R. McGrath and A. Miller: Fine-Line Thermal Recording on Z-fold Paper, *Hewlett-Packard J.*, February 1972.
20. C. A. Conaldson and C. A. Gustafason: Easier and Brighter Display of High-Frequency Signals, *Hewlett-Packard J.*, May 1968.
21. J. Johnson: Sampling Oscilloscope Techniques, *Electro-Technology*, September 1968.
22. W. Farnbach: A Scrutable Sampling Oscilloscope, *Hewlett-Packard J.*, November 1971.
23. A New World of Measurements for the Oscilloscope, *Tekscope*, Tektronix Inc., Beaverton, Ore., January 1971.
24. The Oscilloscope with Computing Power, *Tekscope*, Tektronix Inc., Beaverton, Ore., March/April 1973.
25. Fiber-Optic Cathode-Ray-Tube Visicorder Records DC to 1 MHz Responses, *Elec. Inst. Dig.*, January 1967.
26. P. Lowe: Graphic Data Recording: Fiber-Optic CRT, *Digital Des.*, October 1973.
27. R. L. Dudley and V. L. Laing: A Self-Contained, Hand-Held Digital Multimeter, *Hewlett-Packard J.*, November 1973.

THIRTEEN
ENGINEERED DATA ACQUISITION AND PROCESSING SYSTEMS

In the simpler measurement applications in which we monitor, control, or perform experimental engineering analysis with only a few channels of data and/or do only minor data processing, often a system "casually" assembled from available sensors, amplifiers, filters, recorders, etc., is adequate. However, when requirements become more comprehensive and stringent, data acquisition and processing systems of various types are commercially available and may be necessary. Such systems vary with respect to several factors:

1. Is data processing mainly analog or digital?
2. Is there one amplifier per channel, or is a single amplifier timeshared by multiplexing?
3. If the system is digital-computer-based, is it self-contained (includes the computer) or just a "front end" that can be interfaced with various computers?
4. Is the system a general-purpose type that can be ordered off the shelf, ready for immediate use, or will it be tailored to the specific application and thus require some system engineering before delivery and some debugging after delivery?
5. Are the signals changing relatively slowly, or is fast dynamic response required?

To illuminate these and other questions, we describe a few examples of commercially available systems of several representative types.

13.1 A VERSATILE, MODULAR SYSTEM EMPHASIZING ANALOG SIGNAL PROCESSING

Figures 13.1 through 13.5[1] show some selected features of a system which, while using digital control logic and digital readout (so it can be interfaced easily to a digital computer), relies heavily on analog methods for data processing. A system to meet the needs of a particular application is assembled from a mainframe (Fig. 13.1a), signal conditioning modules (Fig. 13.1b shows examples), signal processing modules (examples in Figs. 13.2 and 13.3), and readout modules (example Fig. 13.4). Mainframes can accommodate up to 60 plug-in modules and provide the necessary power supplies and interconnection facilites. Fifteen nondedicated connections are provided for any special features or alterations that might be needed. The 28 standard connections are divided into a *"housekeeping" bus* (power supplies, system reference voltage, synchronizing and timing functions), the *digital communications bus* (13 conductors used for intrasystem transmission of logic signals, such as scaling commands, channel addresses, etc.), and the *called signal bus* (transmits analog data signals "on request" from selected source modules to a common receiving element, usually a digital indicator or converter).

A wide range (12) of transducer *signal conditioning modules* (some of which are shown in Fig. 13.1b) are obtainable. Thermocouple conditioners accept low-level thermocouple signals and provide reference-junction compensation, amplification, and linearization to produce a standard ± 5-V dc full-scale output signal, linear with temperature. For transducers that require excitation and/or bridge completion (strain gage, LVDT, resistance thermometer, etc.), these facilities are provided in the module. All modules offer two ± 5-V dc outputs: a continuously available "prime output" and a "called output", which is transmitted onto the system *called signal bus* when and only when the module receives a logic call command. [Some signal conditioner modules include modest *processing* capability: Fig. 13.2a shows a two-channel LVDT unit that provides sum $(A + B)$ or difference $(A - B)$ functions useful in thickness and taper gaging. This approach is more economical than employing two separate LVDT conditioners and a general-purpose sum/difference module.] Note that a system of this type utilizes the "amplifier per channel" design philosophy. This is more expensive than a multiplexed system, but it provides continuous (rather than sampled) data, which preserves all the speed of response inherent in the transducer/conditioner.

A comprehensive family of 27 *signal processing modules* which perform analog operations and logical decision making also can be obtained. Figure 13.2b shows the use of three product/ratio modules in an electric-motor test stand. Functions $k(xy)$, $ky/|x|$, kx^2, and $k\sqrt{|x|}$ may be chosen on these modules. Five additional modules are shown in the spring-testing system of Fig. 13.3a. An op-amp type of integrating module computes absorbed energy in the shock-absorber test system of Fig. 13.3b. When integration over many minutes or even

---

[1] Catalog 9000-3, Daytronic Corp., Miamisburg, Ohio, 1981.

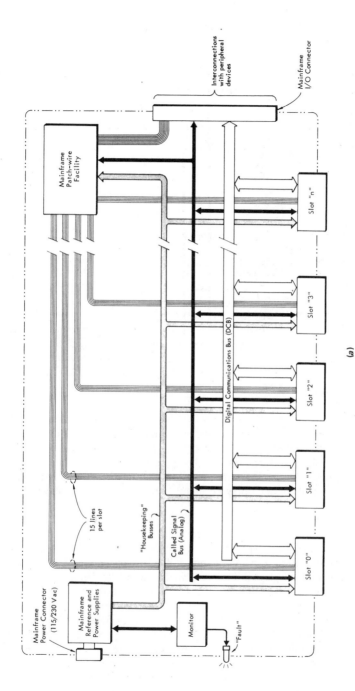

Figure 13.1 Data acquisition system architecture and signal conditioning.

843

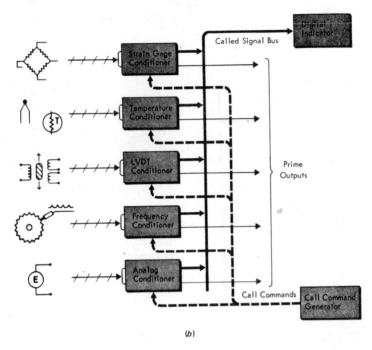

(b)

Figure 13.1 (*Continued*)

hours is needed, a drift-free integrator using voltage-to-frequency converter/counter techniques is available. A ripple/average module uses high-pass filtering (followed by rectification and low-pass filtering) to get a dc voltage indicative of ripple. A separate output uses low-pass filtering to find the average value. Either or both outputs may be useful, depending on the application (see Fig. 13.3c).

Many types of readout display/recording are possible. The dynamometer system of Fig. 13.4a employs simultaneous digital display and analog plotting. Digital indicators can be provided with binary-coded decimal outputs at little extra cost to interface with printers. Video display units (VDUs) provide highly readable CRT displays in flexible formats (Fig. 13.4b). Up to 100 channels of scanned measurement data can be received, stored, and updated continuously. The screen format is determined by internal programs stored in PROM chips; space is provided for 40 such chips. Users can design their own page formats on a convenient worksheet, which is then sent to the factory where the PROM is programmed. Any one of 40 possible formats can be called up at will by manual or program command. Since a standard code is used, the ASCII output card needed for the VDU interface is also usable for interfacing with teletypes, digital tape recorders, and digital computers (see Fig. 13.5). When channels are scanned and digital data are transmitted, the scan rate is 1,500 channels per second.

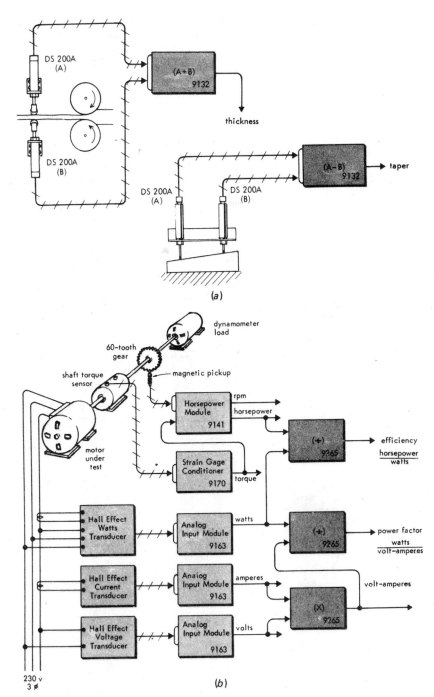

Figure 13.2 Applications of signal processing modules.

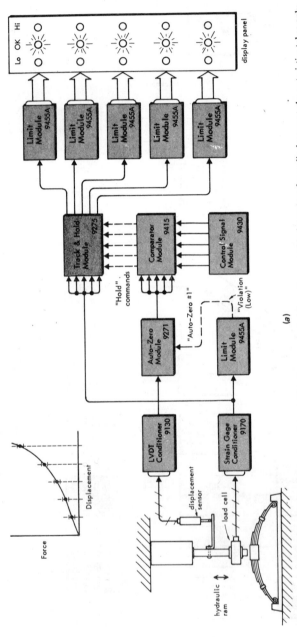

Figure 13.3 Applications of signal processing modules. (a) Testing an automobile spring assembly for proper force-displacement characteristics takes only 2 s in this production-line application of the model 9271. When the descending ram makes initial contact with the part, the displacement channel is instantly auto-zeroed by the model 9271. As the displacement signal subsequently reaches each of five test points—preset on model 9430 control signal module—the model 9415 comparator module will issue a logic "hold" command to model 9275. In this way, the system captures the five individual *force* values corresponding to the five *displacement* test points. These held values are then examined by five 9455a limit modules for conformance with appropriate high and low limits at each point. ULI logic outputs are transmitted to a display panel, where five green lights signify an acceptable part.

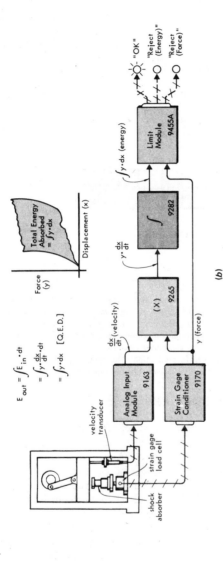

Figure 13.3 (*Continued*) (*b*) This system tests automobile bumper shock-absorbers for proper operation, measuring the total energy absorbed (area enclosed in graph) during an impact cycle delivered by a flywheel and crank arm. Substandard *viscous damping*—the result of low fluid or a leaky port—will be reflected by a *low* amount of energy absorbed, while a plugged port will cause an *excessive* force buildup during the stroke. Either condition will be detected by the model 9455A limit module, which will activate an appropriate "reject" alarm.

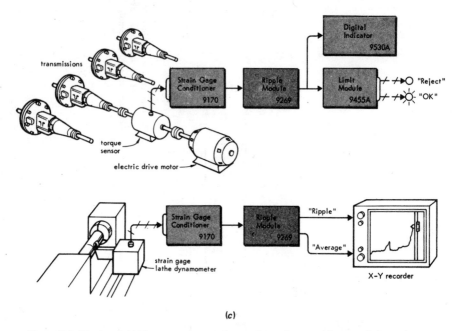

(c)

Figure 13.3 (*Continued*) (*c*) The top system: newly manufactured automobile transmissions are spun up by an electric motor and checked for roughness of torque input. Uneven torque loading resulting from improper fit, broken gear teeth, burrs, or foreign matter, is detected by the system and used to actuate a reject indicator. The lower system aids machining research by producing an *xy* plot of tool chatter versus cutting force under various conditions of cutting speed, tool dress, material hardness, etc. A similar technique can be adopted in production machining operations to monitor tool wear and to indicate the need for sharpening or replacement prior to spoilage of work.

13.2 COMPACT DATA LOGGERS

Although this class of system accepts a number of options and accessories, it can be properly considered an off-the-shelf item that can be put into service with a minimum of effort and delay. The term *compact* is descriptive: a typical unit provides 60 channels of data in a $20 \times 40 \times 60$ cm box weighing about 20 kg. Most manufacturers offer local or remote add-on scanners to expand to about 1,000 channels. Scan rates are modest usually (1 to 20 channels per second), and though versatile signal *conditioning* is provided, signal *processing* capability is limited to simple functions such as "$(mx + b)$ scaling", time averaging of single channels, group averaging of several channels, and alarm signaling when preset limits are exceeded. However, most units do allow interfacing to computers, where versatile processing is possible.

Data loggers of this class utilize a built-in microprocessor to control the internal operations and carry out calculations. The design philosophy is to multiplex all the data channels through a single amplifier–A/D converter, which is

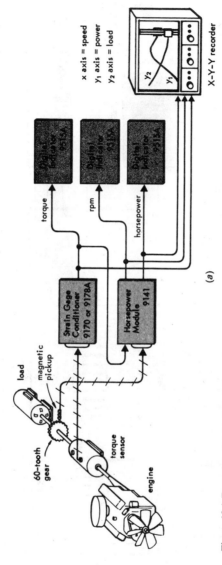

x axis = speed
y₁ axis = power
y₂ axis = load

X–Y–Y recorder

torque

rpm

horsepower

Strain Gage Conditioner 9170 or 9178A

Horsepower Module 9141

load

magnetic pickup

60-tooth gear

torque sensor

engine

(a)

Figure 13.4 Examples of output modules.

849

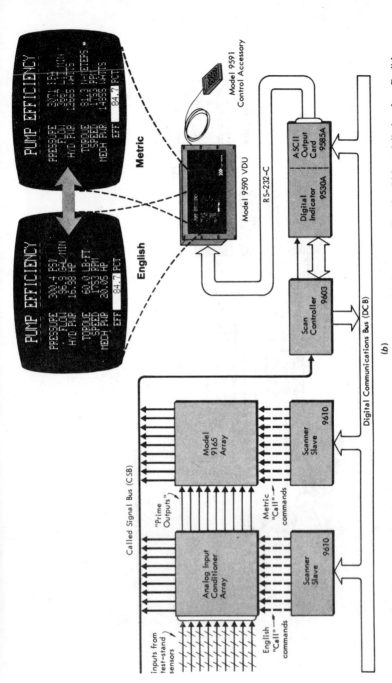

Figure 13.4 (*Continued*) (*b*) *Top right*: This system achieves, at quite reasonable cost, the instant rescaling of multichannel displays from English to metric units.

Lower half: Scaled in English units by the signal conditioner modules, the transducer input signals can be called for VDU display by conventional techniques. The "prime outputs" of the conditioners (and of any other analog processing modules as well) can then be used as input signals for an array of Model 9165 Five Channel Analog Input Modules, from which they are "callable" under a *second set of channel numbers*, with scaling and spanning adjustments appropriate for selected metric units. The VDU will contain two display pages. These pages are identical, except that the fixed text of one will indicate *English* units, while that of the other will indicate *metric* units. Channel-number assignments, of course, will be such that while both pages call the same measurement data, each will call this data under the set of channel numbers scaled for the units to be

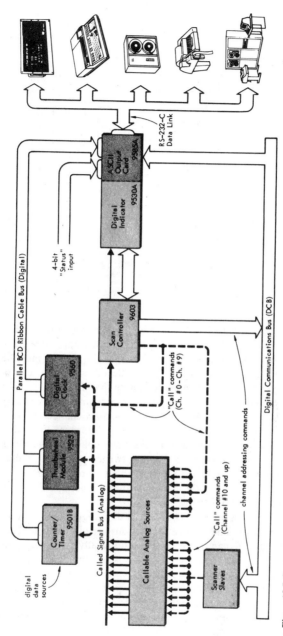

Figure 13.5 System output communications.

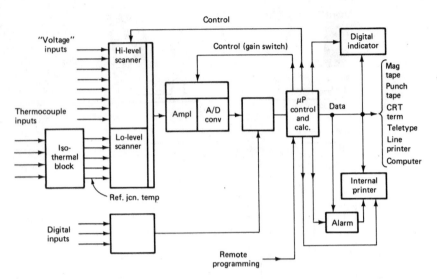

Figure 13.6 Compact data-logger configuration.

automatically ranged or gain-switched under program control to accommodate the signal level of each channel (see Fig. 13.6). This approach greatly reduces costs, but rules out applications in which fast-changing signals must be observed, since a (typical) 5 channel per second scan rate takes 12 s to scan 60 channels before returning to any given channel. Also note that if we plan to calculate a gas density from a pressure on channel 1 and a temperature on channel 60, the "time skew" of 12 s can cause density errors if the signals change too rapidly. Often multiplexers (scanners) are offered in both general-purpose (two-wire) and low-level (two signal wires plus shield) versions since millivolt-level signals, such as from thermocouples, generally use a shielded, twisted pair of conductors. A three-wire-scanner system can reduce errors from about 10 to 1 μV. Electromechanical reed switches are used frequently in such scanners since speed requirements are modest but low noise is important.

Since thermocouples are very common in data-logger applications, reference-junction compensation and linearization options are always available. Reference-junction compensation can be offered economically and accurately for any mixture of thermocouple types by use of an isothermal connection block. This thermocouple terminal block is designed to have a uniform (± 0.05°C) temperature (the reference-junction temperature) over its length. Block temperature is allowed to drift with ambient conditions, but is measured (often with a junction semiconductor sensor, since these work well near room temperature). This reference-junction temperature is sent to the microprocessor, where temperature/voltage data for each thermocouple being employed are stored and the necessary correction is calculated. The microprocessor also stores equations which curve-fit the thermocouple tables (over the desired range) for each of the thermocouple types, providing software linearization. For resistance thermometers the data logger provides constant-current excitation and software linearization.

The system amplifier and A/D converter are crucial elements for overall system accuracy. Since data-logger inputs vary widely in voltage range, while the A/D input typically is fixed at ± 10 V, the microprocessor sets the amplifier gain at a proper value as each channel is sampled (some loggers also provide automatic ranging). These range selections are entered from the front panel when the logger is programmed for the particular application. Programming of this and other functions is very simple; no computer languages need be learned. A typical set of ranges and resolutions would be ± 40 mV (1-μV resolution), ± 400 mV (10 μV), ± 4 V (100 μV), and ± 40 V (1 mV), with input impedance of 200 MΩ (except 10 MΩ on 40-V range). Zero drift is kept negligible by an automatic zero system. Frequently the A/D converters are of the dual-slope or voltage-to-frequency type, since conversion speed is modest and these integrating converters give good noise rejection. Fast or slow scanning rates (say, 15 versus 3 channels per second) may be chosen to allow a tradeoff between speed and accuracy since integrating A/D noise rejection improves as integration time is increased.

Readout by means of a built-in digital indicator and two-color printer (prints alarm values in red) of channel number, data, and time of day is standard. When the built-in printer is employed, printer speed (2 to 4 channels per second) limits overall system speed, even though scanning without printing may be possible at 15 to 20 channels per second. Readout format is selectable by front-panel programming. Some units provide 5-year nonvolatile program memory which preserves stored programs in case of power failure. Interface options for external magnetic tape, punched tape, CRT terminals, line printers and computers usually are obtainable.. Electrical programming (from a remote terminal, controller, or computer) which duplicates manual front-panel programming capabilities is possible also.

13.3 INSTRUMENT INTERCONNECTION SYSTEMS

Whereas the systems of Secs. 13.1 and 13.2 are intended mainly to coordinate data gathering from basic sensing devices, in this section we are concerned with the interconnected operation of instruments such as DVMs, counters, digital thermometers, signal generators, etc.—devices which may themselves include sensors but also are capable of higher-level system functions. Such an approach may be advantageous for a number of reasons. If only 5 or 10 channels of data are needed, a data logger with 60-channel capacity may be uneconomical. Perhaps many of or all the instruments were purchased for their individual capabilities and now we wish to be able to combine them in different ways at various times to meet specific needs as they arise. Such interconnected operation has been greatly facilitated by the almost-universal inclusion of microprocessors in electrical instruments for the purpose of "managing" instrument functions. And so instrument designers find it economical to build in comprehensive facilities for interconnection even if the instrument is marketed mainly as a stand-alone device.

One such system[1] is based on a "computing printer" which includes a scanner that connects with up to nine external instruments. The built-in microcomputer allows programming of many useful functions in addition to straightforward scanning and printing of the nine channels. One of the channels may be chosen to carry out one of four special functions: limits, units, math, or trace. *Limits* is a standard high/low alarm function. *Units* converts (and annotates the printout) to any desired engineering units. *Math* calculates either the difference or percentage difference between a selected reference value and the input data. *Trace* produces a graphic plot (rather than a printed table) against time for the selected channel, with programmable values for the left/right edges of the graph and alarm outputs when the graph exceeds either edge. Such graphic output is possible because the printer is a thermal dot-matrix type. Six other print modes may be selected:

Trace interval: graphic plot at programmed time intervals.
Interval: prints at programmed time intervals
Interval/limit: prints at programmed intervals; monitors for alarm when not
    printing
Single: prints one reading
Continuous: continuously prints up to three readings per second
Remote: prints continuously when signaled from rear panel

When one of the nine instruments is *itself* a multichannel device (say, a 10-channel digital thermometer), addition of an external scanner to the computing printer allows the system to scan and log the 10 thermometer channels before moving on to the next instrument.

Instrument interconnection capability is further enhanced when the individual instruments conform to IEEE-488, the standard digital interface for programmable instrumentation. This industry standard, officially introduced in 1975, has been incorporated into the products of many manufacturers to facilitate the development of interconnected instrument systems. A number of computer-based instrument controllers are available to take advantage of the IEEE-488 compatibility now present in many measurement and control devices. One[2] such unit, which combines minicomputer processing power with IEEE-488 interconnection, provides dual IEEE-488 busses and dual RS-232-C busses. (The earlier RS-232-C EIA (Electronic Industries Association) standard is still much used for bit-serial I/O devices such as modems and data terminals.) Provision of dual busses offers a number of advantages. The IEEE-488 standard allows 20 m of cable and 15 instruments per system; the dual bus doubles both these numbers, in addition to

[1] Computing Printer Model 203A, John Fluke Mfg. Co., Mountlake Terrace, Wash.; *Appl. Bull.* AB-50, John Fluke Mfg. Co.

[2] H. Draye, IEEE-488 Controller Promotes Building of In-House ATE, *Electronics*, pp. 147–152, June 1980; *Fluke Review*, John Fluke Mfg., Mountlake Terrace, Wash., May 1980; Fluke 1720A Instrument Controller, *Publ.* A0087D-10U8006/SE EN.

providing redundancy in case of partial system failures. It also allows partitioning the total system into "slow" and "fast" groupings to improve overall system throughput. For example, a digital multimeter might require the controller's attention only briefly to transmit a single reading, whereas a printer may hold that attention for long periods while multiple data points are transmitted. Allocating slow components to one bus and fast to another bus means that each can operate at an optimum rate.

To ease programming, the system's minicomputer utilizes a modified version of the Basic programming language, as do most such systems. (A special assembly language can be obtained when maximum machine speed is necessary.) The enhancements to standard Basic include a set of IEEE-488 commands for device control and special CRT sequence commands for controlling a video display. The main read/write memory is 60 kilobytes (24 kilobytes available to user) with a built-in floppy disk providing 175 kilobytes of mass storage. Optionally, an additional 128 or 256 kilobytes of electronic RAM (transfer rate of 130 kilobytes per second, 4 times as fast as a floppy) may be installed. Clearly, a system of this type provides a much higher level of computing power than the "computing printer" described earlier. Operator/programmer input and output are by means of a detachable keyboard and built-in CRT display. A unique feature of the display is a transparent, touch-sensitive switch matrix in the form of a thin plastic overlay on the CRT screen. Sixty rectangular areas, each acting as an independent, touch-sensitive switch, are provided. By writing any desired "key label" (and an enclosing "box") on the CRT behind one of these areas, one can create a "keyboard" whose labeling can be altered dynamically under program control, guiding the operator through the desired test sequence. With such a device, the keyboard may be utilized for program development by a skilled programmer, who builds into the program the CRT displays and prompts for operator response that are needed to guide the test operator through all the steps. When the debugged system is handed over to the test operator, the keyboard may be detached, if desired, preventing any inadvertent or intentional alteration of the program. The CRT display/switch now acts as a user-friendly input/output device, simplifying the operator's tasks, reducing errors, and increasing testing speed. Such a system meets the needs of both the skilled programmer and the less sophisticated test operator.

13.4 SENSOR-BASED, COMPUTERIZED DATA SYSTEMS

In this section we describe hardware/software which is commercially available at several levels of "completeness," ranging from single-board computer "front ends" to stand-alone systems complete with powerful minicomputers and high-level programming languages.

The simplest and least expensive devices will require considerable electronics/computer expertise on the part of the user, access to a microcomputer development system, and sufficient engineering time to integrate the interface and computer into a working overall system. Clearly, ready-made complete systems avoid these difficulties, but the user may pay the price of increased cost and performance that may not have been optimized for the specific application. These make/buy decisions are not always simple since the initial cost advantage of assembling a system from "cheap" components may be lost as development time extends beyond original estimates, because of unforeseen hardware/software difficulties.

Since microcomputer system design is a specialized area beyond the scope of this text, we give only a brief description of *microcomputer interface boards*, using Fig. 13.7[1] as an example. These are available from several manufacturers to interface with most of the popular microcomputers. To reduce program storage requirements and execution times (at the expense of memory address area), often memory-mapped I/O is employed. Note that analog inputs (up to 32 single-ended or 16 differential) are processed through a multiplexer, programmable-gain amplifier, sample/hold, and A/D converter in very similar fashion to those of the data logger of Fig. 13.6. However, since we wish to handle high-frequency signals, a successive-approximation (rather than dual-slope) A/D converter (maximum throughput rate 28 kHz) and an electronic (rather than reed switch) multiplexer are necessary. Two (optional) D/A converters are obtainable for driving analog recorders, generating analog control signals, etc. All connections to the microcomputer are made simply by plugging both units into the same card cage. While hardware problems are reduced to a minimum by use of such interface cards, recall that a major effort will be required to develop the software, since many microcomputers do not support convenient high-level languages. Also, the overall system throughput rate must be less (and often *much* less) than the 28-kHz figure given above, since software execution time must be added to the A/D conversion time.

Figure 13.8[2] shows a single-board, microcomputer-based data acquisition system designed to accept multichannel analog and digital inputs and provide digital output to a host computer (usually a mini or mainframe supporting high-level languages such as Basic or Fortran) through a standard serial communications port (RS-232C or 20-mA current loop). The on-board microcomputer unburdens the host computer by allowing supervisory control. It performs data-acquisition control, linearization, conversion to engineering units, limit checking, interface control, and data output formatting. The analog channels are scanned continuously (15 or 30 per second), and the resultant data are stored in microcomputer RAM. The data in RAM are refreshed on a continuous basis (the latest data are kept in memory), so that requests for data from the host are serviced immediately. Upon receipt of a transmit command, the microcomputer [via the UART (Universal Asynchronous Receiver Transmitter)] begins transmitting a string of data in ASCII format to the host. No programming of the micro-

[1] D. P. Burton and A. L. Dexter, Microprocessor Systems Handbook, p. 193, Analog Devices, Norwood, Mass., 1977.

[2] Model μMAC-4000, *Publ.* C603b-15-1/81, Analog Devices, Norwood, Mass.

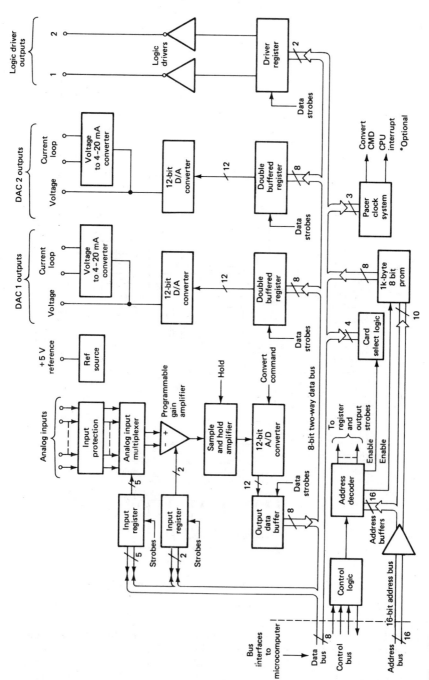

Figure 13.7 Microcomputer interface board.

857

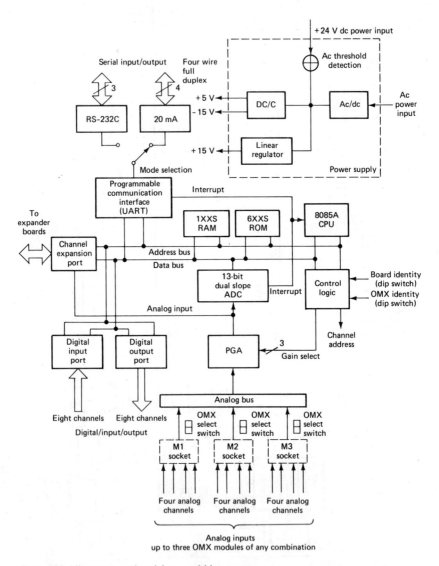

Figure 13.8 Microcomputer-based data acquisition system.

computer is necessary; it is preprogrammed by firmware to respond to host commands according to the simple command set of Fig. 13.9. The 12 channels of analog input are broken into three groups of four, and convenient four-channel plug-in modules for thermocouples, RTDs, strain-gage transducers, etc., are available. Up to three expander boards can be controlled by a master board, creating a "cluster" (see Fig. 13.10). Up to eight clusters can be operated from the same

A typical command ("C" protocol) is specified below:

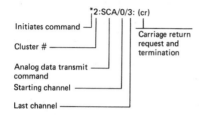

(This command instructs the μMAC-4000 to transmit the latest data from channels 0 through 3.)

A typical μMAC reply is specified below:

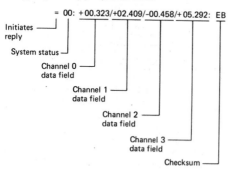

Analog data commands:	Function
CHANNEL n	Transmit channel n data
SCAN n, m	Transmit channel n through channel m data
FSCAN	Transmit data from channels set by ACTIVE command
TEMP n	Transmit temperature of cold junction associated with channel n

Digital I/O commands:

SET p, b	Set digital output bit b on port p
CLEAR p, b	Clear digital output bit b on port p
TEST p, b	Read state of input or output bit b on port p
DOUT p, d	Load the 8 output bits on port P to the value d, where d represents two hexadecimal digits
DIN p	Read state of the 8-bit data on port expressed in two hexadecimal digits

Limit commands:

LIMIT n, LL, HL	Set hi and low limits to channel n
RESET	Clear LIMIT EXCEEDED bit in the status word

Task supervision commands:

RUN	Start continuous channel scanning routine
HALT	Stop continuous channel scanning routine
SKIP n	Skip channel n
ACTIVE n, m	Continuously scan channels n through m inclusive
STATUS	Transmit status word

Protocol modifying commands:

FILL n	Add null characters in transmitted data to provide delay between carriage return and line feed
DEGREE F	Transmit temperature data in degrees Fahrenheit
DEGREE C	Transmit temperature data in degrees Celsius

Figure 13.9 Command set for data acquisition system.

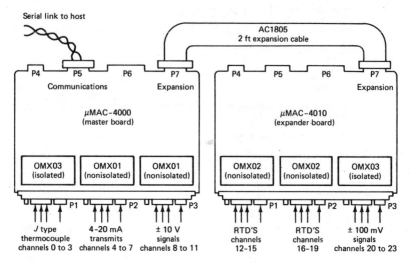

Figure 13.10 System expansion by clustering.

host, providing expansion to 384 channels. For distributed applications (Fig. 13.11) one can use either the party-line (multidrop) configuration (to save wiring costs) or the radial (point-to-point) connection for higher throughput rates. If it is assumed that the host computer supports a convenient high-level language, then assembly and programming of a complete system based on this approach should be quick and easy.

For those applications in which a sensor-based measurement and control system with comprehensive and easy-to-use computer processing is wanted with a minimum of user engineering effort, complete stand-alone systems such as that of Fig. 13.12[1] are available. Analog and digital input/output is through plug-in signal conditioning cards; space for 16 is provided (extension chassis allows expansion to 256 cards). A wide selection of card functions are available, allowing easy interfacing with all kinds of sensors and control devices. Single cards often are themselves multichannel devices with a multiplexer on the card; thus two levels of computer-controlled multiplexing are present, slot multiplexing (chooses the card slot desired) and card channel multiplexing (selects the channel wanted on the chosen card). For example, a single digital input card provides 16 channels (bits) while the analog input card has 32 channels (single-ended) or 16 channels (differential). This analog card has its own programmable-gain amplifier (PGA) (gain = 1, 16, 256) which, combined with the PGA of the central controller (gain = 1, 2, 4, 8), allows versatile selection of channel gain under program control. Thermocouple cards are four-channel units which share a common reference-junction compensation circuit and have fixed gain. Linearization is ac-

---

[1] MACSYM2 System Digest, Analog Devices, Norwood, Mass.

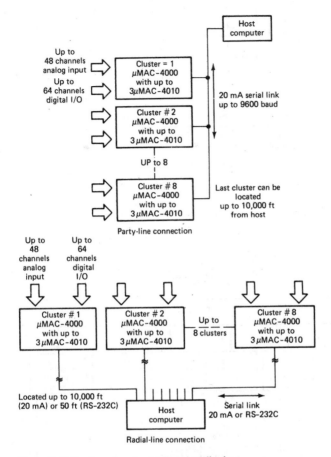

Figure 13.11 Configurations for distributed applications.

complished in software by a general-purpose polynomial subroutine. A fast (25-μs conversion time) successive-approximation A/D converter allows rapid scanning and storage of analog input (mixed channels at 2 kHz, single channel at 4 kHz).

A minicomputer specially designed for measurement-control applications has 32,000 words of 16-bit MOS random-access memory, augmented by 105 kilobytes of cartridge tape mass storage. Programming is in Macbasic, a version of Basic specially enhanced for easy system operation. For example, the statement

$$V = AIN(2,1) - 1.53$$

assigns to the variable V the value of the analog voltage on channel 1 of the analog input card in I/O slot 2, minus 1.53. Similarly,

$$AOT(1,3) = 3.42*1.3$$

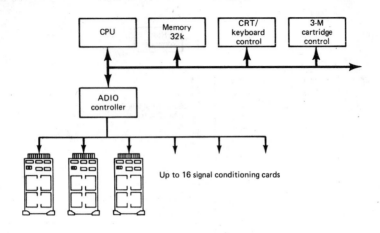

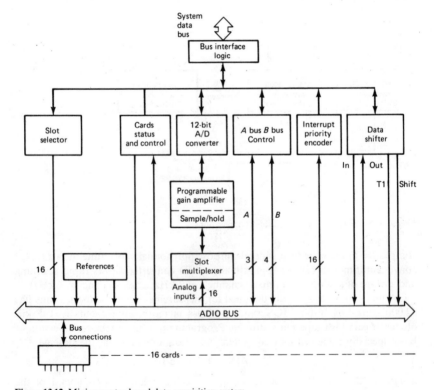

Figure 13.12 Minicomputer-based data acquisition system.

places a voltage of 4.44 V on channel 3 of the analog output card in I/O slot 1. For the digital input variable DIN,

$$I' = DIN(3,1)$$

takes the digital logic level from channel 1 of the digital input card in I/O slot 3 and places it in I', where I' is an integer variable. For the digital output variable DOT,

$$DOT(8,3) = X$$

turns on channel 3 of the digital output card in I/O slot 8 if X = 1 and turns it off if X = 0. System timing functions are eased by the availability of statements such as

$$WAIT\ 5.6$$

which causes the program to wait 5.6 s before proceeding to the next statement.

A common requirement of many applications is the ability to perform several operations or "tasks," independently of one another, in the same program. Examples include the monitoring of several analog signals in the laboratory or control of several process loops. To provide for this requirement, Macbasic is structured as a "multitasking" language, and it contains the necessary key words for implementation. Tasks are groups of Macbasic language statements that are defined as a task, and they are executed each time that task is activated unconditionally or by the satisfaction of a condition (such as an external event) or on a periodic basis. Up to 18 tasks may be defined at a given time, and if more than one task is active at a given moment, the active tasks will share resources and run simultaneously unless priority is assigned to a particular task. If a task completes its operation, it can "DISMISS" itself and return its resources to the system until the task is reactivated.

A simple Macbasic program might go as follows:

10 X = AIN(8,0) Acquire data on channel 0 of analog input card in I/Ō slot 8.

20 PRINT X   Print data.

30 AŌT(0,1) = .5*X   Scale data and output scaled value on channel 1 of analog output card in I/Ō slot 0.

40 IF X > 5 DŌT(1,5) = 1   Activate alarm connected to channel 5 of digital output card in I/Ō slot 1 if X > 5

50 WAIT .5   Wait 0.5 s.

60 GŌ TŌ 10   Acquire next value.

While this program is valid, certain capabilities are added if we convert it to the multitasking mode. These include the possibility of examining and modifying variables and/or listing the program *while it is executing*, which is a useful interactive feature, as noted:

10 TASK 1,40   Define task 1 which starts on line 40.

20 ACTIVATE 1   Activate task.

30 STŌP   Stop primary task.

40 K = 0.5   Set numerical value.

50 L = 5   Set numerical value.

60 X = AIN(8,0)   Input analog data of channel 0.

70 AŌT(0,1) = K*X   Output scaled data on channel 1.

80 IF X > L DŌT(1,5) = 1   If X > 5, sound alarm.

90 WAIT .5   Wait 0.5 s.

100 GŌ TŌ 60   Acquire next value.

All programs contain a primary task, such as lines 10 and 20 in the above example, which define and activate the original program task. Once the primary task is completed, control is returned to the keyboard (line 30), and now the user can interact with program variables and tasks and/or list the program. Variable L, for example, could be modified while the program is running simply by typing, say, L = 6. An example with three separate tasks provides further details on multitasking.

Task 1. Channels 0 through 9 of an analog input card in slot 0 are to be examined every 5 s.

Task 2. If any input is greater than 5.5 V, an alarm connected to channel 5 of the digital output card in slot 1 is turned on.

Task 3. The alarm remains on until an operator activates a switch connected to a process interrupt card, channel 2 in slot 3. (This is defined as an *event*.) When the switch is activated, the alarm turns off and the current value of each analog input channel is printed.

10 TASK 1,100   Define task 1, starting on line 100.

20 TASK 2,200   Define task 2, starting on line 200.

30 TASK 3,300   Define task 3, starting on line 300.

40 ACTIVATE 1 PERIOD 5   Activate task 1 every 5 s.

50 ACTIVATE 3 ON EVENT (3,2,1)   Activate task 3 when channel 2 of interrupt card in slot 3 goes to 1.

60 STŌP   Control is returned to keyboard for user interaction with program variables, tasks, etc.

100 K = 5.5   Set alarm limit value.

110 FOR I = 0 TŌ 9   Start standard Basic loop.

120 IF AIN(0,1) > K ACTIVATE 2   Check for alarm condition.

130 NEXT I   End standard Basic loop which scans channels 0 to 9.

140 DISMISS   Place task 1 on alert status, pending next activation in 5 s.

150 GŌ TŌ 110   Restart here—go to repeat the scan.

200 DŌT(1,5) = 1   Turn on alarm.

210 DISMISS   Task 2 complete.

220 GŌ TŌ 200   When reactivated, repeat.

300 DŌT(1,5) = 0   Turn off alarm.

310 FŌR I3 = 0 TŌ 9   Start standard Basic loop.

320 PRINT"CHANNEL";I3;" = ";AIN(0,I3)   Print each channel value.

330 NEXT I3   End standard Basic loop for printing channels 0 to 9 when alarm has sounded and been turned off.

340 DISMISS   Task 3 complete.

350 GŌ TŌ 300   Repeat when reactivated.

Of course, the above examples cannot explore all the programming possibilities of the Macbasic language, but they should make clear the ease with which the numerical and logical operations needed to implement a desired measurement/control system can be achieved with a high-level language and computer hardware specifically designed for such applications.

The standard system includes a keyboard and built-in CRT display, giving stand-alone programming capability. Optional features include IEEE-488 and RS-232C communications, floppy-disk hardware/software, interactive graphics, printers, and remote interactive CRT terminals.

BIBLIOGRAPHY

1. A. T. Snyder: Airborne Recorder and Computer Speed Flight-Test Data Processing, *ISA J.*, p. 44, July 1958.
2. E. J. Kompass: Information Systems in Control Engineering, *Contr. Eng.*, p. 103, January 1961.
3. J. P. Knight et al.: Low-Level Data Multiplexing, *Instrum. Contr. Syst.*, p. 86, August 1963.
4. L. W. Gardenhire: Selecting Sample Rates, *ISA J.*, p. 59, April 1964.
5. W. T. Botner: Digital Data Gathering System for Blowdown Wind Tunnel, Sandia Corp., Albuquerque, N. Mex., *Rept.* SCR-23, 1958.
6. J. K. Slap: Recording and Processing Test Data, *Contr. Eng.*, p. 177, September 1959.
7. E. Pacini: How Raytheon Cut Test Analysis Time from Days to Hours, *Instrumentation*, vol. 17, no. 1, Honeywell Inc., Philadelphia, Pa., 1964.

8. P. Westercamp: Computing Power Station Performance, *Contr. Eng.*, p. 72, December 1963.

9. Digital Data System Takes 15,625 Engine Samples a Second in Saturn Rocket Static Tests, *Contr. Eng.*, p. 19, October 1963.

10. W. C. Hixson: Instrumentation for the Pensacola Centrifuge Slow Rotation Room I Facility, *NASA, CR*-53341, 1964.

11. K. C. Sanderson: The X-15 Flight Test Instrumentation, *NASA, TM X*-56000, 1964.

12. J. D. Jones: High-Speed Low-Level Data Acquisition, *Instrum. Contr. Syst.*, p. 96, April 1965.

13. H. Gruen and B. Olevsky: Increasing Information Transfer, *Space/Aeron.*, p. 40, February 1965.

14. S. H. Boyd: Digital-to-Visible Character Generators, *Electro-Technol.* (*New York*), p. 77, January 1965.

15. W. Clifford: Digital Voltmeter Data Systems, *Instrum. Contr. Syst.*, p. 105, December 1964.

16. W. E. Schilke: The Analysis of Transmission and Vehicle Field Test Data Using a Digital Computer, *Gen. Motors Eng. J.*, p. 19, October-November-December 1964.

17. G. A. Korn: "Minicomputers for Engineers and Scientists," McGraw-Hill, New York, 1973.

18. A Computer for Instrumentation Systems, *Hewlett-Packard J.*, March 1967.

19. K. A. Fox et al.: A Human Interface for Automatic Measurement Systems, *Hewlett-Packard J.*, April 1972.

20. P. J. Torpey: Minicomputerizing Analog Data Collection, *Contr. Eng.*, June 1970.

21. J. V. Wait: Sampled Data Reconstruction Errors, *Inst. & Cont. Syst.*, June, 1970.

22. J. V. Dirocco: Signal Conditioning for Analog-to-Digital Conversion in Instrumentation Systems, *Elec. Inst. Dig.*, May, 1970.

23. D. P. Allen: How to Choose Data Acquisition Systems, *Contr. Eng.*, November 1969.

24. R. K. Kaminski: Computer Diagnosis of the VW, *Inst. Technol.*, September 1972.

25. J. Lum: Multiplexing Transducer Outputs: Low-Level Multiplexer or Amplifier per Channel, *Elect. Inst. Dig.*, pp. 5–8, June 1971.

26. L. Payne: Intrinsically Safe Data Acquisition, *Inst. Tech.*, pp. 55–59, December 1976.

27. D. Grant: Attaining Microprocessor Interface Compatibility with DAC and ADC Devices, *Comput. Des.*, pp. 158–161, December 1980.

28. R. B. Lake: Acquiring Data at High Speed with a Minicomputer System, *Digital Des.*, pp. 62–65, January 1981.

29. Mayo's Medical Mini-Micro, *Datamation*, pp. 207–216, March 1980.

30. V. L. Laing: Instrument System Provides Precision Measurement and Control Capabilities, pp. 3–8; J. S. Epstein and T. J. Heger: Versatile Instrument Makes High-Performance Transducer-Based Measurements, pp. 9–15; T. J. Heger et al.: Plug-in Assemblies for a Variety of Data Acquisition/Control Applications, pp. 16–22; V. C. Jones: Desktop Computer Redesigned for Instrument Automation, pp. 23–32—all in *Hewlett-Packard J.*, July 1981.

INDEX